Encyclopedia of American Business History and Biography

Iron and Steel in the Twentieth Century

Edited by

Bruce E. Seely
Michigan Technological University

A Bruccoli Clark Layman Book

☑ Facts On File

Encyclopedia of American Business History
and Biography:
Iron and Steel in the Twentieth Century
Copyright © 1994 by Bruccoli Clark Layman, Inc., and
Facts on File, Inc.

Library of Congress Cataloging-in-Publication Data

Iron and Steel in the Twentieth Century/edited by Bruce Seely
    p. cm — (Encyclopedia of American business history and
biography)
    Includes biographical references.
    ISBN 0-8160-2195-3

CIP 93-073377

British CIP information available on request

Designed by Quentin Fiore
Printed in the United States of America
10 9 8 7 6 5 4 3 2 1

To Nancy, Michael, and Karen

Encyclopedia of American Business History

Railroads in the Age of Regulation, 1900–1980, edited by Keith Bryant (1988)

Railroads in the Nineteenth Century, edited by Robert L. Frey (1988)

Iron and Steel in the Nineteenth Century, edited by Paul F. Paskoff (1989)

The Automobile Industry, 1920–1980, edited by George S. May (1989)

Banking and Finance to 1913, edited by Larry Schweikart (1990)

The Automobile Industry, 1896–1920, edited by George S. May (1990)

Banking and Finance, 1913–1989, edited by Larry Schweikart (1990)

The Airline Industry, edited by William M. Leary (1992)

Iron and Steel in the Twentieth Century, edited by Bruce E. Seely (1994)

Contents

Contents

Contents

Contents

Foreword

The Encyclopedia of American Business History and Biography chronicles America's material civilization through its business figures and businesses. It is a record of American aspirations — of success and of failure. It is a history of the impact of business on American life. The volumes have been planned to serve a cross section of users: students, teachers, scholars, researchers, and government and corporate officials. Individual volumes or groups of volumes cover a particular industry during a defined period; thus each *EABH&B* volume is freestanding, providing a history expressed through biographies and buttressed by a wide range of supporting entries. In many cases a single volume is sufficient to treat an industry, but certain industries require two or more volumes.

The editorial direction of *EABH&B* is provided by the general editor and the editorial board. The general editor appoints volume editors whose duties are to prepare, in consultation with the editorial board, the list of entries for each volume, to assign the entries to contributors, to vet the submitted entries, and to work in close cooperation with the Bruccoli Clark Layman editorial staff so as to maintain consistency of treatment. All entries are written by specialists in their fields, not by staff writers. Volume editors are experienced scholars.

The publishers and editors of *EABH&B* are convinced that timing is crucial to notable careers. Therefore, the biographical entries in each volume of the series place businesses and their leaders in the social, political, and economic contexts of their times. Supplementary background rubrics on companies, inventions, legal decisions, marketing innovations, and other topics are integrated with the biographical entries in alphabetical order.

The general editor and the volume editors determine the space to be allotted to biographies as major entries, standard entries, and short entries. Major entries, reserved for giants of business and industry (e.g., Henry Ford, J. P. Morgan, Andrew Carnegie, James J. Hill), require approximately 10,000 words. Standard biographical entries are in the range of 3,500–5,000 words. Short entries are reserved for lesser figures who require inclusion and for significant figures about whom little information is available. When appropriate, the biographical entries stress their subjects' roles in shaping the national experience, showing how their activities influenced the way Americans lived. Unattractive or damaging aspects of character and conduct are not suppressed. All biographical entries conform to a basic format.

A significant part of each volume is devoted to concise background entries supporting and elucidating the biographies. These nonbiographical entries provide basic information about the industry or field covered in the volume. Histories of companies are necessarily brief and limited to key events. To establish a context for all entries, each volume includes an overview of the industry treated. These historical introductions are normally written by the volume editors.

We have set for ourselves large tasks and important goals. We aspire to provide a body of work that will help reduce the imbalance in the writing of American history, the study of which too often slights business. Our hope is also to stimulate interest in business leaders, enterprises, and industries that have not been given the scholarly attention they deserve. By setting high standards for accuracy, balanced treatment, original research, and clear writing, we have tried to ensure that these works will commend themselves to those who seek a full account of the development of America.

— *William H. Becker*
General Editor

Acknowledgments

This book was produced by Bruccoli Clark Layman, Inc. Karen L. Rood is senior editor. James W. Hipp is series editor for the *Encyclopedia of American Business History and Biography Series*. Dennis Lynch was the in-house editor.

Photography editors are Edward Scott and Timothy C. Lundy. Layout and graphics supervisor is Penney L. Haughton. Copyediting supervisor is Bill Adams. Typesetting supervisor is Kathleen M. Flanagan. Darren Harris-Fain and Julie E. Frick are editorial associates. Systems manager is George F. Dodge. The production staff includes Rowena Betts, Steve Borsanyi, Barbara Brannon, Ann M. Cheschi, Patricia Coate, Rebecca Crawford, Margaret McGinty Cureton, Denise Edwards, Sarah A. Estes, Joyce Fowler, Robert Fowler, Laurel M. Gladden, Angeala Harwell, Jolyon M. Helterman, Tanya D. Locklair, Ellen McCracken, Kathy Lawler Merlette, John Morrison Myrick, Pamela D. Norton, Thomas J. Pickett, Patricia Salisbury, Maxine K. Smalls, William L. Thomas, Jr., Jennifer Carroll Jenkins Turley, and Wilma Weant.

Walter W. Ross and Brenda Gross did library research and were assisted by the following librarians at the Thomas Cooper Library of the University of South Carolina: Linda Holderfield and the interlibrary-loan staff; reference librarians Gwen Baxter, Daniel Boice, Faye Chadwell, Cathy Eckman, Gary Geer, Qun "Gerry" Jiao, Jean Rhyne, Carol Tobin, Carolyn Tyler, Virginia Weathers, Elizabeth Whiznant, and Connie Widney; circulation-department head Thomas Marcil; and acquisitions-searching supervisor David Haggard.

Introduction

The American iron and steel industry was the foundation of the U.S. economy for much of the twentieth century, the bellwether of American business, and a key to victory in two world wars. After finally surpassing the British in pig-iron output in 1897, the American industry had entered the twentieth century the largest and most efficient in the world, and even as late as 1946 *Fortune* magazine commented that "steel is still the nation's basic industry." Not surprisingly, steel always has mirrored the basic trends of the American economy, including the growing dominance of large firms, the changing patterns of business-government relations, the struggles between management and labor, and the uncertainties of technological change. Since 1970, steel's problems have typified those of American industry in general, including difficulties with foreign rivals and lagging productivity. Because of steel's economic prominence, its successes and failures often have assumed titanic dimensions; confrontations between steel companies and government or labor often have appeared absolute. As Inland Steel chairman Clarence Randall observed in 1949, "It seems to be the destiny of steel that we shall be a crisis industry."

The decline of American steel since 1970 does not negate its previous importance to a healthy domestic economy. In every mill town a smoky pall meant prosperity, for as a famous aphorism put it, as steel went, so went the nation. Mark Reutter appropriately opens his book *Sparrows Point: Making Steel* (1988) with John Gunther's words from *Inside U.S.A.* (1947): "What is steel? America!" Yet steel's importance reaches beyond economics; few industries have provided better examples of hard-charging entrepreneurs or better images of technological mastery for a nation fascinated with machinery. Behind a flare of red in the night sky were masses of red-hot coke pushed from an oven, streams of molten iron running into ladles, the pyrotechnic display of a Bessemer converter, a continuous coil of wire or sheet, huge plates, or bars emerging from a rolling mill. In open-pit iron mines in Minnesota, on lake freighters, and in the mills everything happened on a gigantic scale; the corporations that controlled the manufacturing and distribution of steel were among the nation's greatest. Steelmaking, in short, exemplified many facets of modern American society.

The entries in this volume survey the essential features of the industry's development in the twentieth century, beginning with the formation of the United States Steel Corporation in 1901 and tracing the most important officials and organizations in the industry into the mid 1980s. Through more than eight decades the most significant feature of the steel industry was change — in corporate organization, technology, labor relations, and international standing. But the biggest change was the ultimate fate of the industry. In 1901 American steelmakers still celebrated the heady thrill of overtaking Germany and England as the world's leading producer. No one would have guessed that by the 1980s their successors would be struggling to remain in business. Such reversals are not unknown in business history, but this one happened very quickly. As always, the American steel industry tended to the spectacular.

The Iron and Steel Industry in 1901

In 1901 the American iron and steel industry exhibited several deeply rooted patterns. First, steelmaking was a highly capital-intensive enterprise, with production centered in Pittsburgh, in the Mahoning River valley in Ohio, and on the south shores of Lakes Michigan and Erie. Mines and mills in Alabama and northern New York made smaller contributions. No matter where people decided to make steel, however, the high fixed costs involved in assembling raw materials and labor in mills with capacities large enough to ensure profits required huge investments of money and commitment. Second, patterns of production were becoming firmly fixed, so that most changes in technology after 1900 were incremental. Most of the iron ore used in steel production came from the Mesabi Range in Minnesota and the smaller ore ranges of northern Michigan; it was shipped south by boat on the Great Lakes and transferred to rail for movement to the mills. Coal from West Virginia and western Pennsylvania was turned into coke — almost pure carbon — to fuel huge blast furnaces that smelted the ore into pig iron. The pig iron was converted into steel in open-hearth furnaces, which in 1901 were overtaking

Bessemer converters as the primary method of making steel. The molten steel was poured into ingots, which were fabricated into rails, bars, wire, pipes, plates, sheets, and structural shapes in specialized rolling mills, often in separate establishments or giant forge shops.

The figures dominating the iron and steel industry in 1901 were Andrew Carnegie and his partner, Henry Clay Frick. In the late nineteenth century Carnegie had discovered the benefits of low-cost production based on continual technological improvement. Frick had recognized the value of vertical integration, combining his coal and coke holdings with Carnegie's mills and later with iron-ore properties. Carnegie Steel enjoyed economies of scale that permitted the firm to undersell competitors and profit under any market condition. Carnegie's fierce, cutthroat competitive style, reflected in his disdain for pools that were intended to stabilize the market, became the ethos of the industry. Other producers, such as J. P. Morgan's Federal Steel, copied Carnegie's policy of integration.

Still, in 1901 most steel companies were not so large as Carnegie, especially those in the Ohio and Alabama steel districts, and most were not completely integrated. Many operated only blast furnaces and sold their pig iron on the open market. Others combined iron-making and steelmaking capabilities and supplied fabricators that produced a single range of products — sheet, wire, pipe. Even Carnegie Steel produced mostly rails and structural shapes, leaving the market for finished-steel products to other companies, many of them small and local. But this pattern was changing as mergers pulled regional producers into national holding companies and the big steelmakers built fabrication facilities to make finished-steel goods.

Small blast-furnace operations and integrated giants alike shared several basic assumptions. First, all steel producers believed that the only role for government in the industry was to provide a high tariff to keep out foreign steel. Equally durable was the universal assumption that steel masters should run their mills without outside interference. This belief in managerial prerogative shaped a similarly monolithic view about labor. Steel managers after 1900 uniformly sought complete freedom to set wages and control their workforces. Before 1892 this power was limited somewhat by the Amalgamated Association of Iron, Steel, and Tin Workers, one of the strongest craft unions in the country. But a disastrous strike at Carnegie's Homestead Works in 1892 crippled that union, which began a long decline. Steel companies deliberately manipulated the ethnic and language differences of immigrant steelworkers to hinder union organizing. By 1900 organized labor maintained only a toehold in the industry.

One result was another fact of life in the iron and steel industry in 1900 — horrible working and living conditions and long hours at miserable wages. Labor historian David Brody's classic study, *Steelworkers in America: The Nonunion Era* (1960), demonstrates that the increasing use of technology did not necessarily provide a better life for workers. Backbreaking tasks were mechanized, yet hours of labor increased, with blast-furnace and open-hearth workers usually toiling 12 hours a day for 13 consecutive days before earning a 24-hour respite. The day off was achieved through the infamous "long turn," which required that men move from the day shift to the night shift by working for 24 hours straight. Bad housing in company towns completed the pattern of subservient existence for most steelworkers.

The Formation of U.S. Steel

The most important event in the history of the American iron and steel industry was the formation of the United States Steel Corporation in February 1901. It culminated certain industry patterns even as it initiated others. One of the former was the trend toward larger-scale business organization driven by the increasing scale of technology. A merger wave in the 1890s resulted in the formation of several large firms organized along product lines, including financier J. P. Morgan's American Bridge Company and his National Tube Company, the Moore Group of firms comprising the National Steel Company (a producer of crude steel) and three finished-steel producers, Illinois Steel, and American Steel and Wire. In 1899 the Republic Iron and Steel Company — the "rolling mill trust" — was formed by combining 34 small companies in Ohio, Indiana, Illinois, and Alabama. The Jones & Laughlin Steel Company (J&L) was formed in 1900 by bringing together two family firms organized 40 years earlier.

Driving these combinations was the desire for stability. Carnegie's rivals tended to fear the competition associated with the steel industry's volatile boom-and-bust cycle. Made cautious by their enormous capital investments, they disdained Carnegie's passion for cutting costs and running full. By the turn of the century Carnegie was perceived by steel producers and financiers alike as a danger to the industry. Morgan, the central figure in the formation of U.S. Steel, clearly sought stability not only for his own steel companies but also for the industry as a whole in conducting the negotiations that reined in Carnegie's competitive instincts. In return for about $480 million Carnegie retired from the steel business. U.S. Steel was created by combining Morgan's firms with Carnegie's, producing a holding company comprised of eight major and many smaller firms and some 200 plants. Capitalized at more than $1.4 billion, the new entity was the first billion-dollar concern in American corporate history.

U.S. Steel culminated the drive to huge iron and steel enterprises that began in the 1870s, but it was not just another trust. Its formation changed the entire industry. First, its size gave new meaning to the term "big business." Although not a monopoly, U.S. Steel at the time of its formation controlled 37 percent of the nation's pig-iron capacity, 74 percent of the Bessemer output, and 55 percent of the open-hearth capacity; it accounted for at least 50 percent of every major steel product. This situation forced competitors to develop similar scope by building or acquiring mills and mines in order to become integrated steelmakers. Bethlehem Steel Company, for example, which had been purchased by U.S. Steel president Charles M. Schwab in 1901, became the lead firm in an integrated holding company, the Bethlehem Steel Corporation, in 1904, a year after Schwab had left U.S. Steel. Schwab made Bethlehem Steel the second largest American steelmaker by 1920. Once-large companies such as Cambria Steel and Pennsylvania Steel succumbed to Bethlehem; smaller firms expanded. J&L, for example, embarked on a major capital-expansion program in 1905, building the enormous Aliquippa Works and diversifying its product line.

At the same time, the immensity of U.S. Steel posed problems. In December 1901 Schwab, a Carnegie protégé and the first president of the new firm, crowed that "the great steel industry of this country is now first of all . . . [and] is justly entitled to be throned and crowned." But this situation insured constant scrutiny of the new corporation. As *Iron Age* explained in an editorial on February 14, 1901, "It is certain that the new consolidation will lend much support to the anti-trust agitation, and will be its shining mark. Keen eyes will watch every move, and unscrupulous demagogues and an unbridled yellow press will distort even the most trivial incident." The merger was defended by *Iron Age,* which claimed that such a combination would prevent a disastrous struggle between Morgan and Carnegie for supremacy in the industry. Schwab's position, however, was that U.S. Steel was a combination — which he considered a natural outcome of growth in the industry — not a trust, which he considered doomed to failure.

But to many people the arguments of Schwab and *Iron Age* seemed problematic. As the Age of Reform dawned, muckraking journalists and reformers were set to battle corruption and the imbalance of power that they thought large organizations represented. Critics were aghast at the size of U.S. Steel; Morgan himself worried that the company would be dismembered by the government. Credit for its survival must be given to Judge Elbert H. Gary, a lawyer with a background in steel whom Morgan chose to run the corporation. Gary redefined the goals, behavior, and attitudes of the American steel industry.

Shaping the Industry: The Gary Years, 1901–1927

It is difficult for one person to move a bureaucracy, so an individual can rarely be credited with shaping the destiny of a large corporation. Rarer still is the person who dominates an industry. Yet Gary was such a person. Morgan appointed Gary to head U.S. Steel's finance committee in 1901 (he presided until 1927) and made him chairperson in 1903, but Gary soon spoke for the industry. He was a lawyer, not a steelman, a rarity in the industry. Indeed, steelmen had relatively little influence at U.S. Steel, as financiers dominated the board and set corporate policies. The crucial issue facing the company was clear: would the government let it survive? Ida M. Tarbell wrote that Gary believed that if the corporation were "allowed to continue in business until it has been proven that the intentions of its managers are good, that there is no disposition to exercise a monopoly or to restrain legitimate trade," then it could survive. He set out to forestall antitrust action with a new style of business conduct.

Fear of antitrust informed almost every decision at U.S. Steel in its early years. First, Gary established a new standard of ethical and moral behavior by board members. He published prices annually and stuck to them, hoping to achieve price stability. U.S. Steel rarely used its enormous power directly against competitors, as Carnegie's strategy of cutting costs and running full was abandoned. Gradually U.S. Steel lost market share in every line, so that by 1920 U.S. Steel possessed less than 50 percent of the market. When executives schooled in the Carnegie tradition — including Schwab and William Corey, the first two presidents of U.S. Steel — objected to the new rules, they were eased out of their positions. As he replaced Carnegie's men who had come up from the mills with a feel for machinery and a passion for technological improvement, Gary also replaced their enthusiasm for technological efficiency and innovation. Finally, Gary recognized that the treatment of workers could attract unwanted public attention, so he adopted several new programs, including sale of stock to employees in 1902 and a variety of welfare programs after 1911. Backed by Morgan, Gary overturned what he labeled "the old-fashioned ideas of doing business," and the era of welfare capitalism began.

Gary's quest for stability soon affected the entire industry, as he encouraged other companies to abandon Carnegie's style of management. He used several means to achieve this goal. In 1907, having overcome most resistance at U.S. Steel, he inaugurated the "Gary dinners," inviting leading steelmakers and reporters to discuss output, labor problems, and prices. All participants insisted that only common problems were to be discussed. But Gary had great leverage, although he cloaked his comments in moral terms as he encouraged cooperation among the firms in the steel industry. After complaints about the legality of the dinners, Gary in 1910 formed a new trade organization, the American Iron and Steel Institute (AISI), as a forum for what he came to call "friendly competition." Gary, president of the group, preached the same message he had delivered to U.S. Steel's managers — cooperation and stability were better than open competition. And the AISI became one of the most effective trade associations in the country.

Another element in Gary's search for stability was a pricing formula that eliminated the advantages of geography. Even in 1900 the steelmaking advantages of the southern shore of Lake Michigan (lower transportation costs for iron ore and finished products when compared to Pittsburgh and Youngstown) were apparent. U.S. Steel's mills were not on the lakes, but Gary championed basing point pricing to compensate. Under this system, freight costs were computed from a set location — Pittsburgh was the first basing point — regardless of where the steel was produced. Gary's goal was the elimination of competition based on price.

The AISI and basing point pricing were presented as serving the whole industry; everyone

gained if the boom-bust cycle could be tamed. At the same time, these developments also preserved U.S. Steel's leadership. Gary always appeared benevolent, but he used power when necessary. During the Panic of 1907, for example, U.S. Steel held to its published prices while many competitors slashed theirs. When the panic eased, however, U.S. Steel kept low longer than other companies, punishing those who earlier had broken the line. The Gary dinners began shortly thereafter. But in case other companies had missed the point, U.S. Steel in 1909 broke ground in Gary, Indiana, on the shore of Lake Michigan for what became the world's largest and most efficient steelmaking complex.

Gary also made an effort to limit technological innovation. Once Carnegie's hallmark, new technologies were dangerously destabilizing in Gary's world. Often expensive, new methods could only be explored by the largest firms, who then gained major price advantages over smaller firms. Unwilling to risk appearing a predator, Gary discouraged innovation at U.S. Steel. Thus the Grey mill, which produced structural beams, was rejected by U.S. Steel; Bethlehem adopted it and collected royalties from other producers, including U.S. Steel. By-product coke ovens pioneered in Germany in the late nineteenth century also were adopted slowly. Electric furnaces slowly replaced the crucible process for making tool and alloy steels, after the first was installed in 1906 at Halcomb Steel in Syracuse, New York. By 1913 the United States had ten furnaces, compared to 121 in the rest of the world.

Incremental changes in existing technology were more in keeping with Gary's ideas on innovation. Thus blast furnaces, open-hearth furnaces, Bessemer converters, and related equipment all evolved deliberately after 1900. Blast furnaces and open hearths grew in size, while material handling equipment grew more important. In rolling mills electric motors replaced steam engines over a long period of years, accompanied by much debate. As important, the AISI became an important mechanism for disseminating technical information, so that even small companies could learn about the newest developments. Managed in this way, technological change was controlled.

Another feature of Gary's strategy involved what historians Thomas McCraw and Forest Reinhardt called "Losing to Win." In other words, U.S. Steel deliberately sacrificed its market share to prove a lack of monopolistic intent. This strategy, unimaginable to Carnegie, permitted other steel companies to grow rapidly. One example was Inland Steel, a small firm organized in 1893 by the Block family. In 1901 the company moved to Indiana Harbor on Lake Michigan, where a fully integrated steel plant took shape. Armco also began in this period, incorporating in 1899 as the American Rolling Mill Company of Middletown, Ohio. George Verity carved a niche for Armco by developing special-purpose steels. Youngstown Sheet and Tube was organized in late 1900, with James A. Campbell the central figure who developed the vision of a large, integrated steel producer. Finally, Phillips Sheet and Tin Plate (the forerunner of National Steel) was organized in 1905 and grew into an integrated operation under the direction of E. T. Weir. By 1920 the company had been renamed Weirton Steel. U.S. Steel made no hostile moves as these companies became its competitors.

By 1905 the major steel corporations had taken shape, and the list of the largest steel companies began to change [See Table 1]. Several leading companies in 1900 were absorbed into newer firms — Bethlehem alone absorbed four of the top ten on 1904's list — and only J&L, Wheeling Steel (built around LaBelle Iron Works in 1920), and Colorado Fuel & Iron survived as large independent steelmakers. In every case, the new companies grew rapidly through a combination of internal expansion and acquisition under vigorous entrepreneurs. Every firm copied U.S. Steel and developed iron and coal resources and transportation facilities.

One should not, however, interpret Gary's decision to give ground to competitors as evidence that U.S. Steel intended to give up its domination of the industry. Rather, "friendly competition" worked because Gary wanted it to work and had the power to make it work. Gary always led by example, sticking to published prices and preaching cooperation. But as he promoted the new ways of thinking, he always held a trump card — the immense power of U.S. Steel. That Gary spoke for the industry was not an accident. He was completely sincere about cooperation, creating the AISI as the institutional mechanism. Other steel executives seemed equally genuine in their respect for Gary and never challenged his leadership. But Gary was firmly in charge.

The one threat to this structure was federal antitrust legislation. Here, too, Gary sought to achieve stability through cooperation, but he met with less success. His actions were based on the enduring assumption that government had no right to interfere. But many citizens and public of-

Table 1
Largest Steel Companies, Selected Years

RANK	1904	1916	1926	1938	1950
1	U.S. Steel	U.S. Steel	U.S. Steel	U.S. Steel	U.S. Steel
2	J&L	J&L	Bethlehem	Bethlehem	Bethlehem
3	Cambris	Pennsylvania Steel	Youngstown Sheet & Tube	Republic Steel	Republic Steel
4	Lackawanna	Bethlehem	J&L	J&L	J. & L.
5	Colorado Fuel & Iron	Cambria	Inland Steel	National Steel	National Steel
6	Pennsylvania Steel	Lackawanna	Wheeling Steel	Youngstown Sheet & Tube	Armco
7	Republic Iron & Steel	Republic Iron & Steel	Republic Steel	Inland Steel	Youngstown Sheet & Tube
8	Maryland Steel	Youngstown Sheet & Tube	Colorado Fuel & Iron	Armco	Inland Steel
9	Tennessee Coal Iron and RR Co.	Colorado Fuel & Iron	McKinney Steel	Wheeling Steel	Wheeling Steel
10	LaBelle Iron Works	Inland Steel	United Alloy Steel	Colorado Fuel & Iron	Colorado Fuel & Iron

SOURCE: Gertrude G. Schroeder, *The Growth of the Major Steel Companies, 1900–1950* (Baltimore: Johns Hopkins University Press, 1953), p. 199.

ficials doubted the intent and legality of U.S. Steel. Suspicion of monopoly was inevitable, and government officials remained dubious of Gary's efforts to portray U.S. Steel as a good trust. An adversarial relationship with the government existed from the time the corporation was formed, and historian Paul Tiffany has noted in *The Decline of American Steel: How Management, Labor, and Government Went Wrong* (1988) that suspicion has colored government-business relations in steel ever since.

Gary tried to ease the strain. He cultivated good relations with President Theodore Roosevelt, and an inquiry by the U.S. Bureau of Corporations proceeded according to the ground rules they developed. Moreover, the report, which suggested that the valuation of U.S. Steel was inflated, was not immediately released. Similarly, when U.S. Steel acquired the Tennessee Coal, Iron & Railroad Company (TCI) during the Panic of 1907, Gary and Frick obtained the approval of Roosevelt and his attorney general, Charles J. Bonaparte. That Gary failed to inform Roosevelt that the purchase lessened U.S. Steel's competition in the South might suggest the danger of accepting altruism at face value. But, as historian Melvin Urofsky observed in *Big Steel and the Wilson Administration: A Study in Business-Government Relations* (1969), Gary defined cooperation to mean compliant government officials.

Not everyone endorsed Gary's one-sided view of cooperation. Both Presidents William Howard Taft and Woodrow Wilson were less trusting than Roosevelt, and their suspicion lay behind an antitrust suit brought against U.S. Steel from 1911 through 1920. Attorney General George W. Wickersham warned a shaken Gary about the government action in October 1911; a congressional committee probed U.S. Steel's formation, valuation, and corporate behavior at the same time. Here was the real test of Gary's claim that U.S. Steel was a "good trust," and in the end Gary's effort to alter corporate behavior helped save the company.

The crucial factor in U.S. Steel's victory was its declining market share in every product line. As important, a parade of steel executives from competing firms testified that U.S. Steel was not predatory. Rather, they applauded Gary for bringing stability to the industry. Even Roosevelt testified on Gary's behalf. In the end the majority opinion, released in 1915, found no monopolistic intent in the company's formation and no effort to restrain trade: "Where no competitor complains, and much more so, where they unite in testifying that the business conduct of the Steel Corporation has been fair, we can rest assured that there has been neither monopoly nor restraint." The Gary dinners were judged illegal, but since they had stopped before the suit was brought, that point was moot. Thus Gary's strategy was vindicated.

The government's decision to appeal the case further poisoned the atmosphere between government and the steel industry, and the hostility continued through the start of World War I. The industry, for example, got excited about the rate revisions of the Underwood Bill in 1912, even though protection was hard to justify when American steel was so strong. Old habits died hard, so the industry fought. Yet in other areas the Wilson administration shifted from confrontation after 1915. The bill creating the Federal Trade Commission was modified to meet industry objections, while the Webb-Pomerene Act, which permitted companies to cooperate in overseas sales efforts, was designed to help American firms. Nonetheless, steel executives continued to claim that they were victims of unfair government regulations. Gary, for example, charged in his 1915 presidential address to the AISI that "there has been in substantial degree, an open hostility to business which has paralyzed many of its legitimate efforts." The legacy of Gary's suspicion remains.

It might seem that the history of the steel industry in the early twentieth century revolved solely around U.S. Steel. There was, in fact, much activity that took place outside the nation's largest corporation. Total steel production steadily increased, from 15 million tons in 1901 to almost 22.5 million in 1905. Output lagged after the recession in 1907 and recovered slowly until 1912. Despite the general slowness the newer steel firms showed considerable dynamism, with the vigor of their efforts meriting emphasis. Bethlehem Steel grew rapidly, acquiring Pennsylvania Steel and with it the new Sparrows Point plant near Baltimore. Bethlehem also pursued iron-ore projects in Cuba and further expansion that made it the second-largest steel company in the country. Other companies had similar records, for the caution of U.S. Steel was not initially copied by other firms.

World War I produced no immediate windfall for steelmakers. Production sagged in 1914 before jumping when war orders arrived the next year. In 1917 and 1918 the industry set a production record of more than 50 million tons, while profits as a percentage of sales reached levels never repeated. Figured as return on investment, the industry averaged 28.9 percent in 1917 and

20.1 percent in 1918. J&L and Midvale Steel & Ordnance posted returns exceeding 43 percent in 1917. U.S. Steel made more than $512 million and paid over $200 million to its stockholders during 1917 and 1918, including a 14 percent dividend on common stock in 1918. For the entire period from 1916 to 1920 U.S. Steel's operating profit was 25 percent, Republic's was 24.7 percent, and Youngstown's was 37.1 percent. The war even permitted two of Carnegie's protégés, Schwab of Bethlehem and Corey of Midvale Steel & Ordnance, to revive traditional managerial practices. Bethlehem evaded neutrality laws, taking war contracts from England, including one for submarines that were fabricated in a Canadian shipyard using Bethlehem Steel. Schwab actually became something of a celebrity when he took over the ailing federal shipbuilding program.

World War I also offered another test of Gary's structure for the steel industry, as the industry operated under rules that seemed to confirm Gary's lectures on cooperation. The government quickly learned that cooperation between business and the administration was necessary to fight an industrial war. Businessmen entered government service to direct agencies that coordinated production to meet wartime needs, and Bernard Baruch's War Industries Board played the largest role in guiding the economy during the war. Baruch asked Gary to develop the steel industry's role in this cooperative effort, and in reply the AISI housed numerous committees dealing with prices, production, and raw materials. J. Leonard Replogle, originally with Cambria Steel, directed the government's activities.

The industry's experience under Gary permitted it to adjust easily to this cooperative structure. Not everything went smoothly, as steelmakers failed to meet demand in Europe and America. In addition the industry heard complaints about prices, priorities, and the allocation of steel due to short supplies. The industry also utterly refused to honor any of government's promises to work with organized labor. On the whole, cooperation meant that business usually had the upper hand, and steel executives missed few chances to point out that steel had provided a model for the wartime economy.

Steel and Labor, 1900–1925

While the war brought production records, huge profits, and cooperation, it also exposed one activity — labor policy — where Gary's strategy delivered less than promised. Labor relations was one of the areas where Gary felt pressured to prove that U.S. Steel was a good trust. The 1892 strike had greatly weakened the Amalgamated Association, but critics of U.S. Steel watched for evidence of predatory behavior. Not surprisingly, then, Gary made better treatment of workers a highly visible activity, despite opposition from board members who had grown up with Carnegie Steel. A stock-purchase plan for employees was introduced in 1903, and at least one subsidiary introduced a pension plan at about the same time. A compensation fund for those injured or killed, started by Andrew Carnegie, was expanded. Company towns were not new, but U.S. Steel was especially proud of the playgrounds and other amenities it provided for workers in Vandergrift, Pennsylvania. In 1911 a company-sponsored welfare program brought improved sanitary conditions and hospitals to the mills, while industrial safety became a part of the program; U.S. Steel introduced the slogan "Safety First!" By 1920 U.S. Steel had spent $71 million on welfare work.

Other companies, including Bethlehem and Armco, developed less expensive welfare schemes, and the industry was proud of itself. An editorial in *Iron Age* in 1902 argued, "It can safely be claimed that employers have not aimed to deprive their employees of rights and privileges. . . . On the contrary, efforts are continually being made by employers to improve the condition of those employed by them, to make their lives enjoyable and in every way to make them contented with the conditions which surround them." Gary, who couched his arguments in moral terms, took particular pride in U.S. Steel's welfare efforts, boasting to the AISI in 1912, "I make the assertion, gentlemen, that in no line of industry, at any period in the history of the world in any country, was labor on the whole better treated in every respect than it is at the present time by the employers of labor in this great industrial activity."

It is difficult to accept such statements at face value. It was no coincidence that U.S. Steel promoted its welfare activities in 1911, the year the antitrust suit was filed. Moreover, historian Gerald Eggert has shown that Gary and the U.S. Steel board never kept their commitments to shorten hours and end the seven-day week. Improvements in working conditions were not allowed to interfere with profits. Nor would U.S. Steel allow workers to organize unions. The classic expression of this view came at an early meeting of the executive committee, when one of the members explained his rule for dealing with workers: "If a workman sticks his head up, hit it."

The Amalgamated Association of Iron, Steel, and Tin Workers found this out in 1901, when it struck U.S. Steel. The union hoped to act before the giant company pulled itself together, and walked out at several subsidiaries. Within a month it had been badly defeated, due in part to poor leadership, and it lost its right to represent men in several mills. Worse, the corporation consolidated its holdings, closing older, less efficient mills, many of them union strongholds. *Iron Age* asserted correctly, "It is doubtful if the association will ever be as strong as it was before the strike started."

The Amalgamated Association found itself in more trouble in 1909 after U.S. Steel announced it would no longer recognize any union. A 14-month strike against the American Sheet and Tin Plate subsidiary failed after the union overestimated the corporation's reliance on skilled workers. Other confrontations followed at intervals at other steel companies through 1916, but no companywide walkouts took place. No walkout was successful, and the strength of the union eroded further. Steel companies also scored a public-relations victory with their successful safety programs. By 1916 U.S. Steel reported that serious accidents had dropped 43 percent; Inland reported a drop of 55 percent, and J&L a decline of 71 percent. In addition the steel corporations avoided the brutality that marked earlier strikes.

Yet a show of force was possible when the need arose after World War I. The war stirred discontent among workers. First, the government promised workers they could organize, a move the industry fought totally. Bethlehem simply rejected it, as did most other steel companies, although labor shortages brought better pay. By early 1920 hourly wages for common labor had jumped from 20 cents in 1913 to 46 cents. Even the recruitment of southern blacks was not enough to keep wages down; moreover, some white workers resented the presence of blacks. Workers also were captivated by wartime democratic rhetoric, which they sought to apply in the mills. Finally, they objected to the continuation of the 12-hour shift after the war ended. It had been abandoned in some areas of the mills after 1910 but had been restored in 1917.

These grievances fueled a relatively successful organizing campaign by a new union calling itself the National Committee for Organizing the Iron and Steel Workers. William Z. Foster and John Fitzpatrick of the American Federation of Labor guided this attempt, bypassing the Amalgamated Association by organizing all steelworkers, not just the skilled crafts workers. Membership amounted to about 20 percent by the summer of 1919; 98 percent of the members favored a strike after Gary refused even to meet with the union. A few labor leaders feared staging a strike too soon, but on September 21, 275,000 workers left the mills.

The companies used every power at their disposal, including political control in many steel towns, state militias, and private detectives. They waged a propaganda battle, successfully linking Foster's radical past to growing public fears of immigrants and the Russian Revolution. Once branded as radical, the organizing campaign was doomed, but the companies made sure by using force and intimidation. That the industry was not completely shut down also weakened the strikers, as did discord between the Amalgamated Association and the National Committee. On January 8, 1920, the committee admitted defeat. The companies boasted about defeating dangerous "agitators sow[ing] vicious doctrines," according to the *Iron Trade Review,* but an investigation by the Interchurch World Movement found no evidence of bolshevism among the workers. Even a Senate investigation concluded that the workers "had a just complaint relative to the long hours of service on the part of some and had the right to have that complaint heard by the company." But organizing efforts were finished until the 1930s.

After 1920, steel companies set about restoring the relatively stable labor relations that had prevailed before the war. One of the companies' concerns was the end of open immigration, for quotas introduced in 1924 had cut off the traditional source of steel labor. Not surprisingly, welfare programs resumed in the 1920s; a few companies introduced employee-representation plans (ERPs) — actually company unions. The first, at Colorado Fuel & Iron, appeared after bloody confrontation at Ludlow in 1916, but Bethlehem Steel was the largest company to introduce such a scheme. Bethlehem's plan created a grievance structure but provided no mechanism for discussing wages; the company operated the organization.

Gary highlighted these activities as proof of the steel industry's moral behavior, arguing in his usual self-righteous manner that they lived up to the motto he had chosen for the AISI — "Right Makes Might." In 1924 he told the institute, "Let us so far as the steel industry is concerned always do the right thing. Let us be just to everyone, even though we are not treated fairly from Governmental authorities, from Congressional representatives, or even from some parts of labor. Let us maintain a record and a reputation for intending to do the right thing."

Such a claim was particularly ironic in 1924, however, given that a controversy had erupted in 1922 concerning working conditions in the industry. In spite of good intentions, about 20 percent of the industry's workers, especially those at blast furnaces and open hearths, still labored for 12 hours a day with the long turn coming every two weeks. U.S. Steel had announced the end of the 12-hour day with great fanfare in 1910, but furnace workers were never included, and others went back to 12 hours during the war. Reports of the strike publicized this practice, and a public outcry followed. President Warren G. Harding, prodded by Secretary of Commerce Herbert Hoover, personally asked Gary to try to change this situation. The AISI appointed a committee of leading executives, which reported several months later that a labor shortage prevented change. Gary believed the uproar was simply a ploy for a wage increase, but the industry could not stand the public clamor. Harding then asked Gary in June 1923 to remove the 12-hour day when conditions permitted. Gary finally relented, although he did not set a timetable. But by the end of the year the eight-hour day prevailed in the mills. Five years later *Iron Trade Review* reported that improvements in productivity as well as in machinery had covered any higher wage costs.

Steel in the 1920s

The grudging nature of the industry's decision to end the 12-hour day shows more than a contradiction between rhetoric and action about workers. Also visible is a complacent outlook in the steel industry that appeared in the 1920s. U.S. Steel especially had seen the threats that had concerned it deeply for two decades disappear. In 1920 the Supreme Court upheld the rulings of the lower courts against the government's antitrust suit. Although the Federal Trade Commission embarked on a major investigation of basing point pricing in 1921, the steel industry faced few problems in dealing with government during the decade. Worries about unions also faded after 1920. The result was increased adherence to Gary's calls for stability.

Predictably, U.S. Steel best exhibited this resistance to change. To be sure, the company devoted significant resources to facility improvements, spending nearly $550 million on

tube, sheet, and plate mills at Gary, Indiana; a general expansion at its TCI subsidiary; a wide-flange beam mill at Homestead; modernization of National Tube's plants; and an alloy bar mill at the South Works in Chicago. Even so, the corporation's production remained relatively flat, at shipments of 14 to 15.5 million tons of steel a year. Profits also held steady, jumping sharply only in 1929, even though there was a generally growing market for steel. Production surpassed wartime records every year after 1925, peaking at 63 million tons in 1929.

Some commentators have argued that Gary's policies were outdated by the 1920s and served the corporation poorly. *Fortune,* for example, analyzed Gary's legacy in a four-part series in 1936, presenting U.S. Steel as a very inefficient holding company, with much duplication of effort within its subsidiaries. Central direction of corporate activities was limited to Gary's finance committee, which controlled all expenditures and undermined the supposed autonomy of subsidiary presidents. *Fortune* argued that this situation drove away ambitious young managers. Above all, the company avoided any radical change, as was most evident in a capital-spending program that focused on accepted technology. Indeed, U.S. Steel did not even have a research laboratory until 1927.

But if U.S. Steel was not especially vigorous, was the problem widespread in the industry? Evidence of entrepreneurial vigor in steel during the 1920s does exist. At the beginning of the decade, Schwab and Grace were building a company to rival U.S. Steel, as Bethlehem grew by acquiring Lackawanna Steel, Cambria Steel, and Midvale Steel and Ordnance. The company also opened extensive ore deposits in Chile and expanded the Sparrows Point plant in an effort to rival the Gary Works. Armco also showed some aggressiveness, developing the technological breakthrough of the period — the continuous hot-strip mill. The company introduced the process in the mid 1920s and installed an improved version in 1927. When combined with a cold-reduction mill for producing galvanized sheet (introduced at Wheeling Steel in 1929), a revolutionary technological process was in place.

Two other developments suggest that much of the vitality in steel during the 1920s was found on the organizational front, not in the mills. In 1927 Cyrus Eaton, an outsider to the industry, became the leading stockholder in the Republic Iron and Steel Company, which proceeded to grow quickly by acquiring several other companies. In

1929 Eaton brought in Tom Girdler, president of J&L, to consolidate the renamed Republic Steel, which was the third-largest steel company in the country. The other organizational development was the formation of National Steel by the most aggressive steel man in the 1920s, Ernest T. Weir of Weirton Steel. Weir operated outside the industry's mold, for he was an aggressive salesman willing to cut prices. He also expanded Weirton's facilities, becoming the first to license Armco's continuous hot-strip mill in 1926. In 1929 Weir engineered a merger between Weirton, George Fink's Great Lakes Steel Corporation of Detroit, and the M. A. Hanna Company of Cleveland, an iron-ore supplier and shipper. National Steel Company, the fifth-largest steelmaker, was well positioned to serve the auto industry, for Fink's move into Detroit may have broken a gentlemen's agreement among the big companies not to build there.

The fact that Weir was exceptional indicates that the steel industry had generally adopted Gary's operating strategy by the 1920s. For example, efforts to export steel were never vigorous, and one joint program involving Youngstown and several other midwestern companies was abandoned. Equally indicative was a continued pattern of incremental technological innovation, with the lone exception being the hot-strip mill. Furnaces and materials-handling equipment increased in size; electric furnaces finally replaced the crucible steel process. But the important development, the hot-strip mills, attracted little attention from the big companies. Only Weirton and U.S. Steel licensed Armco's technology by 1927, and U.S. Steel envisioned the hot-strip mill that it installed at the Gary Works as a supplement to traditional hand-production methods. In 1928 U.S. Steel built new hand-fed sheet mills, but it was not alone in staying with proven technology, for Bethlehem had added 24 traditional mills in 1926. In that year Bethlehem finally created a research lab, which, as Mark Reutter relates in *Sparrows Point: Making Steel,* Grace ordered to stay away from "exotic experimentation." In fact, only Armco had a real research lab.

In keeping with this conservative pattern, the performance of Youngstown Sheet and Tube was typical of the 1920s. Having guided the company through dramatic growth before 1920, James Campbell now shifted attention from building a company to outmaneuvering larger competitors. His goal was to grow through merger. In 1921 Campbell sought to combine seven medium-sized

firms — Republic, Lackawanna, Cambria, Inland, Brier Hill, Steel and Tube Company of America, and Youngstown. This plan fell apart when Bethlehem purchased Lackawanna and then Cambria. More concerned than ever about its position, Youngstown unsuccessfully sought a merger with Inland in 1928. Finally, the company began talks with Bethlehem in 1930. But court action held up the merger until the Depression was so far advanced that Bethlehem could not afford the move. In the meantime Youngstown had deferred most capital expenditures. Generally, the company drifted for most of the decade.

Similar pictures could be painted at other steel companies. Wheeling Steel and Pittsburgh Steel, for example, showed little disposition to grow. J&L showed more activity after a reorganization ended its history as a family firm in 1923, but the company could not regain its position as the third-leading steelmaker. In general the smaller companies showed more vigor than the larger companies, but the industry as a whole reflected order and stability.

In August 1927 Gary died, and an era in the steel industry ended. His most important legacy was the transformation of the competitive pattern of steel, the replacement of Carnegie's cutthroat strategy with "friendly competition." Although to a large degree he continued Carnegie's legacy of hostile relations with government and labor, Gary introduced control of interindustry competition through pricing and institutional connections. He had reshaped the steel industry. *Iron Trade Review* evaluated Gary's career on August 18, 1927, and then asked the right question: "He faced problems peculiar to his day. He solved those problems by strength of vision, courage, and ability. He constantly had the welfare of the industry at heart. He performed his work well. . . . Who, in years to come, will rise from the ranks of industry to take his place as the natural leader of the great iron and steel industry?"

The Steel Industry in the Great Depression

Like other American industries, steel was devastated by the Great Depression. Markets shrank precipitously, except for light-rolled sheet products used in automobiles and home appliances. After 1932 sheet accounted for at least 40 percent of total steel production, up from 25 percent in 1929. This shift hit U.S. Steel and Bethlehem — primarily producers of heavy structural shapes, pipe, and rails — especially hard. But no one escaped, as production dropped from 63 mil-

Table 2
Net Income of Selected Steel Companies, 1932
(millions of dollars)

National Steel	1.663	Crucible Steel	- 3.614
Ludlum Steel	- 0.099	Colorado Fuel & Iron	- 4.253
Allegheny Steel	- 1.052	J. & L.	- 7.910
Armco	- 2.030	Republic Steel	- 11.261
Pittsburgh Steel	- 2.501	Youngstown Sheet & Tube	- 13,272
Otis Steel	- 2.830	Bethlehem Steel	- 19.404
Wheeling Steel	- 3.274	U.S. Steel	- 71.272
Inland Steel	- 3.320		

SOURCE: *Iron Age,* 131 (March 16, 1933): 8.

Table 3
Profits for Selected Steel Corporations, 1929–1940
(millions of dollars)

Year	U.S.S.	Bethlehem	Republic	National	J. & L.	Armco	YS&T
1929				12.6			
1930	104.4	23.8	-3.522	8.4	9.1	0.1	7.0
1931	13.0	0.1	-9.034	4.4	-2.3	-3.1	-7.0
1932	-71.2	-19.4	-11.261	1.6	-7.9	-2.0	-13.3
1933	-36.5	-8.7	4.049	2.8	-5.4	0.7	-8.3
1934	-21.7	-0.6	3.459	6.1	-3.7	1.0	-2.7
1935	1.1	4.3	4.456	11.1	-0.4	4.3	1.6
1936	50.5	13.9	9.587	12.5	4.1	6.4	10.6
1937	94.9	31.8	9.044	17.8	4.8	8.2	12.2
1938	-7.7	5.3	-7.998	6.7	-5.9	-1.3	-0.7
1939	41.1	24.6	10,671	12.6	3.2	7.6	5.0

*Actual operating deficit was $6.3 million. Profit resulted from special income due to sales of assets and tax adjustments.

SOURCE: William T. Hogan, S.J., *An Economic History of the Iron and Steel Industry in the United States* (Lexington, Mass.: Lexington Books, 1971), volume 3; *Fortune,* 5 (June 1932): 31.

lion tons in 1929 to 15.2 million in 1932. The industry ran at only 20 percent of ingot capacity in May 1932, 12 percent in July. Recovery began in 1933, yet the industry utilized only 25 to 50 percent of its capacity over the next three years. An announcement in *Iron Age* that production had reached 61 percent of capacity in March 1936 was a cause for celebration. In January 1937 the figure stood at 80 percent, but with the recession of 1937–1938 steel production sank to 27 percent of ingot capacity. The war in Europe finally raised capacity utilization to 89 percent in October 1939.

This production roller coaster was reflected in the industry's profitability. The worst year was 1932, when only one company reported a profit, and the top 20 firms lost a combined $149.5 million [see Tables 2 and 3]. U.S. Steel accounted for almost half that loss, but every company had problems. Two firms, however, performed better than average during the early 1930s. Inland operated at higher percent capacity than the industry average — 22.3 percent compared to 19.5 percent industrywide in 1932; and 72.7 percent compared to 48.7 percent in 1935. The company benefited from its geographic advantage, a construction program initiated in 1931, and continued sound management by the Block family. The surprise performer of the 1930s, however, was National Steel, the only steelmaker to post a profit every year. Weir provided excellent leadership, evidenced by the firm's response to the changing steel market of the 1930s. National's continuous hot-strip mills and cold-reduction facilities added at Weirton in 1927, and a second mill installed at its Great Lakes steel plant in Ecorse, Michigan, in 1930, gave it a marked advantage. Inland also had adopted the new technology, building a larger strip mill in 1931; both companies added additional hot-strip mills in 1936 and 1937. Other decisions reflecting attention to marketing considerations helped these two companies. The location of the Great Lakes plant offered one major advantage, while Inland merged with Joseph T. Ryerson & Son, the nation's largest steel-warehousing operation, in 1935.

Leadership in the Steel Industry

The performance of Inland and National notwithstanding, the industry was generally not well prepared for the Depression, and leadership during this period often seemed mired in traditional responses. Of course, few business leaders had answers to the economic disaster of the 1930s. Moreover, steel's problems were especially difficult, being a capital-intensive industry dependent on large volumes for profits. Even so, few steel executives rose to the occasion during the Depression. Schwab of Bethlehem replaced Gary as steel's spokesperson, but he was clearly past his prime. Grace ran Bethlehem with a group of technically sound assistants who shared his insular view of the world; they were determined to make quantities of steel without outside interference. In a 1941 article *Fortune* favorably compared the company to U.S. Steel, crediting an executive bonus system for high productivity. But that same bonus system produced a scandal in 1930 and 1931, when court proceedings revealed that Grace's compensation for 1929 was an astounding $1.6 million. With the Depression at its depths, Bethlehem's managers appeared in the worst possible light. But throughout the 1930s

they exhibited little sensitivity to public perception and even less imagination in dealing with the challenges facing the firm.

Executive initiatives at other firms also seemed limited. J&L had special problems. It produced less sheet steel than any major producer, having abandoned that market completely from 1926 through 1931. Now it paid the price for that decision, when sheet was the only solid market for steel. Youngstown also was poorly situated in the light flat-rolled market because it had delayed improvements during the 1920s. The company did not build hot-strip mills until 1935 and 1939, and it struggled through the decade. Armco struggled against geography, paying higher transportation costs for ore and coal because its main mill was not located on the lakes or the Ohio River. But Armco at least was strong in light-rolled steel; another asset was its research laboratory, which introduced a new galvanizing technique in 1937 that permitted shaping the steel after the tin coating had been applied. In the end, Armco outperformed the industry.

Republic Steel shared Armco's disadvantages but managed to do better under Girdler, who kept the firm afloat when Eaton's empire collapsed. The company showed profits in both 1933 and 1934 because of the company's light-steel capacity. Girdler rationalized the new acquisitions, lived off inventories, and recruited excellent financial officers. After 1935 Republic's management turned aggressive, acquiring several new firms, including Corrigan, McKinney, a semifinished steel producer based in Ohio; Gulf States Steel in Alabama; Witherbee, Sherman, an iron ore producer in northern New York; and Truscon of Ohio, the nation's largest maker of steel building products. Girdler won national recognition for this performance.

The difference between success and failure, then, was at least partly managerial capability. It was not surprising that Myron Taylor, U.S. Steel's chairperson after 1932, devoted significant attention to management reform. He made retirement mandatory at age sixty-five and hastened the departure of many longtime executives. He replaced them with younger men, including Edward Stettinius, Jr., who was thirty-three when he arrived from General Motors as vice-chairman in 1934. Benjamin Fairless, Taylor's choice to run Carnegie-Illinois Steel, the largest U.S. Steel subsidiary, came from Republic in 1935; he was forty-five. With the help of these and other young executives, many from outside the industry, Taylor reorganized U.S.

Steel completely, erasing its holding company structure. But the changes came too late to prevent a string of losses in the early 1930s.

Another indication of leadership problems was the industry's response to the challenges of the early 1930s. At first, steel executives echoed the soothing assurances of other business leaders about a quick recovery. Schwab's presidential address to the AISI in 1931 was typical. "Fear has been lessened," he announced. "There will be no collapse. . . . I have seen us pull through too many crises to be overwhelmed by the situation of the past few months." This message rang hollow by August 1932, when Taylor proclaimed that "the buying movement cannot long be delayed."

When words proved inadequate, other strategies were explored. Steel firms traditionally responded to a downturn in steel sales by laying off workers and cutting wages. After 1930, however, the largest steel companies tried something different. While many workers were laid off, Bethlehem and U.S. Steel especially tried to keep as many on the payroll as possible with share-the-work programs. Thus only 18,938 of the 159,032 employees at U.S. Steel in 1932 worked full-time, and the company spent $16 million on direct welfare. Taylor considered these decisions the corporation's finest, proof of big business's social responsibility. Even so, employment in steel dropped to 210,000 in 1932, compared to 420,000 in 1929.

The benefits of work sharing were limited by wage cuts, a more traditional corporate response to hard times. Common labor earned 44 cents per hour in 1930, but only 33 cents per hour in 1932. White-collar workers absorbed similar pay cuts, but short workweeks and wage cuts meant that workingmen could not feed their families. The cuts were opposed by a few steel executives — notably Weir of National Steel and James Farrell, president of U.S. Steel. In a sharp speech to the AISI in May 1931, Farrell commented that "it is a pretty cheap sort of a business when the largest industries in the country . . . [have] men who are working three days a week, and then cut that three days a week another 10 percent." Five months later U.S. Steel joined the companies criticized by Farrell and cut wages. The supreme irony, however, was that Schwab and Grace of Bethlehem, perhaps the highest-paid men in the industry, drew their full six-figure salaries.

Other responses to the Depression were proposed at the 1931 AISI meeting. Farrell and others demanded that companies end price-cutting, rebates, and other competitive measures in an effort to stabilize the market. Others blamed Congress

for failing to restore public confidence. An editorial in *Iron Age* on July 14, 1932, entitled "Fiddling While Rome Burns" captures this sentiment. Weir complained about government tinkering and meddling, delivering this message repeatedly after Roosevelt's election with the warning that interference retarded recovery. Another traditional response by steel executives were calls for protection, especially after 1932. Grace put it succinctly in November, calling foreign steel "a real menace," and Schwab echoed this view, even though imports accounted for only 1.4 percent of the American market.

Taken together, these ideas suggest that steel's leadership tried to respond but really had little new to offer during the Depression. Yet answers did begin to appear to the question posed by *Iron Trade Review* in 1927: Who would be the next leader of the industry? No one man would dominate as Gary had, but two individuals did move to center stage; both were aggressive and attracted attention when their firms outperformed larger rivals. Girdler of Republic was one who moved into the spotlight, serving as president of the AISI from 1937 to 1939. Weir, already recognized for what *Fortune* in June 1932 had called a "sensational" performance in turning a profit at National Steel, also rose to the forefront. In 1935 the *Nation* labeled him "the most successful and the most unpopular man in the industry," citing his eager embrace of new technology, the mill opening in Detroit, and his policy of aggressive price-cutting. He succeeded Girdler as AISI president in 1939. Both championed conservative, even reactionary, policies built on the industry's traditional assumptions. Weir, the only steel executive who had founded the company he ran, expressed most vocally the belief that government could not tell him how to run his company. Girdler had an equally sharp tongue when attacking government involvement in business affairs. Although progressive in setting policy at their own firms, both men were largely reactive in proposing prescriptions for the industry as a whole. To do otherwise, they believed, would result in cutthroat competition.

At U.S. Steel, Taylor's youth movement staffed in large part by managers from nonsteel companies offered an alternative to the industry's hard line. These men generally did not share the virulently antigovernment stance of other steel executives. Rather, they continued one aspect of Gary's policy — namely that U.S. Steel could not make decisions simply by considering profit and loss: they had to take a larger view. Taylor's share-the-work program reflected this thinking, and other initiatives followed. By showing social responsibility they hoped to end the confrontational nature of business-government relations in regard to steel. Thus two managerial responses to the challenges of the 1930s appeared in regard to the steel industry.

Regardless of their philosophy, all steel managers in the 1930s faced a dilemma, needing to make technological changes during a period of vanishing demand. There was little new construction during the 1930s, and only Inland and National added significant steelmaking capacity. But every steel company was forced to build continuous hot-strip mills and cold-reduction mills for producing sheet steel. Then in 1937 an electrolytic galvanizing process added another improvement. Between 1926 and 1937, 27 continuous strip mills were installed, and 51 cold-reduction facilities for producing tinplate were built from 1930 through 1939. It became easier to justify capital expenditures after 1936, especially as the tension in Europe increased. Even so, the capital demands were enormous for steel companies just emerging from years of red ink, as expenditures for new equipment and construction totaled about $840 million between 1935 and 1938.

Steel, Government, and Labor: Rewriting the Rules

The differences in managerial style in steel showed best in business-government relations during the 1930s. The pattern of distrust had been well established since 1901, and the New Deal only increased those suspicions. Few steel executives accepted government's emerging new role gracefully. As their most cherished assumptions regarding managerial prerogatives came under attack, they resisted almost every governmental initiative. Federal officials viewed this recalcitrance as politically motivated, especially after Weir and others became active in the Liberty League. As if the industry did not have enough problems, steel and and government spent the Depression fighting each other.

The New Deal started calmly enough. President Franklin D. Roosevelt's early programs were met with cautious acceptance by the industry — which was one indication of the industry's desperation. Even so, Charles Schwab's address to the AISI in May 1933 reflected the industry's wariness. He promised that steelmakers would applaud any cooperative response to the Depression. Especially welcome was the price stability promised by the National Industrial Recovery Act

(NIRA). But Schwab also stressed repeatedly that voluntary action was best and that legislative initiatives would be "most unfortunate."

Under the NIRA the steel industry set out to develop a code of conduct that prescribed prices, wages, and other matters of operation. The AISI again coordinated the various committees that oversaw the code; six heads of leading steel corporations made up the administrative committee: William A. Irvin of U.S. Steel (chair), Grace, Girdler, Weir, Leopold Block of Inland, Hugh Morrow of Sloss-Scheffield Steel and Iron, and William Jennings Filbert, U.S. Steel's longtime financial officer. Not surprisingly, men long accustomed to independence chafed at operating under the NIRA; prices proved very difficult to set. Yet cooperation eventually came rather easily to the industry, given Gary's legacy and the experience of World War I.

Other changes also proved easy to swallow. One example was a revision of the tax law in 1934 that required holding corporations to file a single tax form, not a form for each subsidiary. This requirement forced Bethlehem and U.S. Steel to reorganize. But in the area of labor reform steel executives refused to accept government initiatives, and a collision was inevitable. The industry totally rejected the right of workers to organize, as required in section 7(a) of the NIRA. It also opposed the Wagner Act, which guaranteed workers' rights to collective bargaining. Charles Hook of Armco warned in June 1933 that "the happy relationship which has existed between employer and employee in this country during the last ten years . . . should not be endangered by any wording in the bill."

Open conflict between government and steel on labor questions first began during preparation of the Steel Code in 1933. Industry leaders attempted to modify section 7(a) by including a provision calling for open shops. Only when confronted by Secretary of Labor Frances Perkins did they relent, all the while insisting they would not deal with unions. This point was emphasized on August 15, 1933, when Perkins gathered steel executives in her office to sign the steel code. When the six steel men saw William Green, the president of the American Federation of Labor (AFL) whom Perkins had also invited, they marched from the room and refused to return. They argued that joint signatures would be interpreted as dealing with a union. Perkins later wrote, "I felt as though I had entertained 11-year old boys at their first party rather than men to whom the most important industry in the U.S. had been committed." The code was signed only after President Roosevelt called Schwab and Taylor to his office for a talk. Even then, the industry countered 7(a) by expanding the then-forming employee-representation plans (ERPs) modeled on those at Bethlehem. By the end of 1934, 93 ERPs were in place, compared to seven in 1932.

Organized labor was buoyed by the NIRA, and the confrontation between labor and the steel industry was soon renewed. The first test of strength involved miners working in coal mines owned by the steel companies. John L. Lewis, head of the United Mine Workers (UMW), pushed hard to represent the so-called captive mines, most of them owned by U.S. Steel. After the cases went to court and after U.S. Steel eschewed an all-out fight at the last hour, 20 of the 29 mines voted for UMW representation in 1934.

A second confrontation grew from a rank-and-file movement in the largely moribund Amalgamated Association. Emboldened by the New Deal, men at many mills, often using ERPs as vehicles, sought more-direct recognition at the end of 1933. A walkout at National Steel and strike threats elsewhere demonstrated the restiveness of many workers. But the leaders of the Amalgamated Association failed to support the militant workers and even threw out of the union the Weirton chapter that had led the strike. Thomas Bell captures the flavor of these developments from the workers' point of view in his 1941 novel, *Out of This Furnace*. Weir and Girdler, on the other hand, openly defied the National Labor Relations Board and later the National Steel Labor Board. This struggle marked the beginning of Weir's bitter fight with the Roosevelt administration and made him the most visible figure in the industry.

The leadership of the Amalgamated Association and corporate resistance stalled labor's push in 1934. But the situation for workers was changing, for many were developing a collective spirit within company unions, especially after the government ruled that workers should choose their own officers. This spirit grew even as the companies refused to discuss wages and other questions with ERP leaders. Against a rising tide of labor militancy, steel companies in 1935 offered paid vacations and increased wages. But many workers recognized the defensive posture of the companies. Journalist John Fitch, who had covered the 1919 strike, captured the essence of the change in the steel towns in the title of an article in *Survey Graphic* in February 1936: "A Man Can Talk in Homestead."

In 1935 there was cause for hope among labor with the passage of the Wagner Act, which guaranteed workers the right to choose their own representation and formed a National Labor Relations Board (NLRB) to enforce its provisions. An organizing drive was inevitable, and a stronger effort to protect the interests of labor was undertaken in 1936 by John L. Lewis's new organization, the Committee for Industrial Organizations (CIO). The CIO was created to organize the mass-production industries — namely automobiles and steel. The CIO formed the Steel Workers Organizing Committee (SWOC) to conduct the organizing drive, placing UMWA vice-president Philip Murray in charge. He was backed by $500,000 from the mine workers, vice-presidents Clinton Golden and Van A. Bittner, treasurer David J. McDonald and several hundred CIO organizers.

The drive began at a mass meeting in Mc-Keesport, Pennsylvania, on June 21, 1936. The SWOC's initial target was Carnegie-Illinois Steel, U.S. Steel's largest subsidiary. The AISI issued a statement declaring, "The steel industry will oppose any attempt to compel its employees to join a union or to pay tribute for the right to work." During 1934 and 1935 *Iron Age* had claimed that ERPs were favored by 90 percent of the steelworkers. At various government hearings the companies enlisted workers, often officials in the ERPs, whose testimonies supported that claim. Yet by the end of the year SWOC had enlisted 125,000 steelworkers.

The breakthrough came in early 1937, when U.S. Steel's Taylor met secretly with Lewis. Taylor had decided not to fight for several reasons. First, a recovering economy was producing the first profits for U.S. Steel in years. Taylor also feared the company would lose foreign orders. At the same time, he was sure that contacts with the international steel cartel would enable the company to raise prices to recover the cost of any contract without fear of increased foreign imports. Pragmatically, Taylor saw little hope of winning a confrontation completely, and even success would come at a very high cost. Both his moral and business senses told him a fight was not seemly. By February, Murray and Fairless of Carnegie-Illinois Steel had worked out the basic features of an agreement. A contract was signed on March 2, 1937, setting off an explosion at other steel firms, which had been kept in the dark. Within two months 110 steel companies signed similar contracts covering more than 300,000 workers.

Not every company gave in, however. In early May the SWOC struck the J&L Aliquippa plant before the company agreed to an election supervised by the NLRB. The SWOC won. Crucible Steel signed after being threatened with a strike, and Pittsburgh Steel, Wheeling Steel, Allegheny Steel, and Sharon Steel also signed contracts. A core of resistance remained in Little Steel — Bethlehem, Republic, Inland, Armco, Youngstown Sheet and Tube, and National. All told, the six companies employed about 186,000 workers. They fought the SWOC relentlessly. The La Follette Civil Liberties Committee in the Senate later revealed that the firms purchased weapons, tear gas, and other equipment and paid networks of informers, private-detective agencies, and local government officials. Girdler of Republic spoke for Little Steel when he announced that he would not sign a contract with union officials even if an agreement was reached.

On May 26, 1937, the SWOC struck Republic, Youngstown, and Inland. The conflict was violent, especially at Republic's mills. The worst incident occurred in Chicago on Memorial Day, when police charged a crowd of strikers, killing ten. Bethlehem was drawn into the strike on June 10, when steelworkers struck in sympathy with railroad workers at its Johnstown plant. President Roosevelt convened a special Federal Steel Mediation Board, but the companies refused to sign a contract with the SWOC and rejected elections run by the NLRB. Perkins recorded later that she had never dealt with management like that of the steel industry's; during the strike in 1937 Grace and Girdler refused even to speak with her. Roosevelt asked Girdler for cooperation in setting up the mediation board, but when asked if he would accept Roosevelt as an arbitrator, he shouted back, "No!" In the short run this defiance led the SWOC to abandon the strike at the end of June. Even so, membership rolls showed 526,000 members in late 1937.

Government action slowly defeated Little Steel, as numerous NLRB rulings and court decisions backed the SWOC. Murray continued organizing efforts, which produced a final round of strikes and NLRB elections. A Supreme Court ruling in 1941 left the corporations with no legal footing, but the ensuing negotiations between management and labor were very difficult. A contract was signed in 1942, with government officials having been involved in the negotiations so as to ensure continued wartime steel production. Only Weir defeated the SWOC, when his employees chose the company union to represent them instead. National's workers were granted everything demanded by the SWOC and more.

For all of its resistance the industry had bought five years of freedom from the union, but at a cost of public respect and further animosity from the government. The SWOC, on the other hand, became the United Steelworkers of America (USWA) at its annual convention in May 1942. Murray, who remained president, had successfully directed one of the longest and most difficult organizing drives in the history of American labor. The legacy of that bitterness was long visible in labor relations in the steel industry.

Steel and the Government, 1938–1945

The bitterness of the unionization drive spilled over into business relations with government during the late 1930s. Again, two different approaches to dealing with the government were evident. On one side was U.S. Steel and its "industrial statesmen," men like Taylor and Stettinius who maintained close ties to Roosevelt aide Harry Hopkins and other government officials. In the style of bankers and gentlemen, they favored talking with government leaders rather than shouting at them. On the other side were Weir, Girdler, and Grace, who denounced Roosevelt, the New Deal, the NRA, and any labor legislation. Weir's rhetoric became even more shrill after the SWOC appeared. He obstructed every government agency dealing with labor and fought the union at every turn. He also widened the attack, blaming Roosevelt's policies for the recession of 1937–1938, blasting what he called the political domination of business, and accusing a "reactionary White House clique" of not wanting a recovery, since government spending gave it political power. After Roosevelt denounced "economic royalists," Weir proudly claimed, "I am an economic royalist." By 1933 Weir charged that Roosevelt wanted war. Girdler was not silent either, as shown by a speech he gave to the AISI in May 1939: "Even now in peace times while our government berates the dictatorship abroad, it persists in domestic policies leading straight down the road to dictatorship in the United States."

Weir's approach was typical of long-held beliefs in the steel industry. Another traditional assumption was that steel companies were best run by men with practical experience. Before 1900 this was, of course, the rule, but the rise of giant corporations gave bankers, lawyers, and eventually accountants control over policy. This situation was especially so at U.S. Steel. Gary was a lawyer, and the board was dominated by bankers, thanks to U.S. Steel's special relationship with the House of Morgan. That banker Taylor succeeded Gary and that Stettinius, a man with no steel background and whose father was a Morgan banker, was appointed president in 1938 confirmed this pattern. Some analysts blame the continued poor performance of U.S. Steel during the 1930s on overly cautious bankers ignorant of steelmaking.

The issue sounds very similar to recent discussions about the causes of problems in American manufacturing. David Halberstam, for example, argues in *The Reckoning* (1987) that the rising influence of "bean counters" and the demise of production experts since 1950 has accounted for many of the problems in the automobile industry. But the issue was alive in steel in the 1930s due to the differences between U.S. Steel and most other steel firms, which were run by men with operating experience. Don M. Watkins, editor of *Blast Furnace & Steel Plant,* a trade journal covering the technical side of steel production, consistently argued that steel men should run the industry. In May 1935 he wrote, "The trouble is that there are not enough men like Mr. Weir with real steel mill training at the heads of companies in position to fight back [against government criticism]." This attitude was evident in the March 1936 article in *Fortune* about U.S. Steel, which discussed the hiring of Fairless, a traditional steel man, from Republic in 1935. The business journal commented, "It looks as though the Depression has taught Mr. Taylor something that Gary never learned: that a steel company should be run by steel men as well as by lawyers and bankers." Grace told *Fortune* in 1941 that a basic principle at Bethlehem was that operating men, not bankers, ran the company. In explaining the company's long-term success, Grace added, "The management has always had its own way."

To the industry, the ideal leader of a steel company was a tough-talking, self-made man who had worked up through the mills. A college degree was not necessary — Grace and Girdler had diplomas, but Weir and Fairless did not. Long experience in mill operations, however, was a prerequisite, and every head of the AISI in the 1930s — Schwab, Grace, Girdler, and Weir — matched this profile. Each loudly opposed the changes forced on the industry and was willing to resist at almost any cost. Priding themselves for rigidity and an uncompromising stance, they were angered by the bankers and lawyers of U.S. Steel who had sought

nondestructive solutions to unionization. Given the tradition of toughness in other steel firms, U.S. Steel's solution was a sign of weakness and compromise, not a model for the industry.

Apart from labor issues, however, the lines between Taylor and Stettinius and steel men Weir, Girdler, and Grace were less clearly etched. Most important, U.S. Steel executives were no more eager for government intervention in their business than were their colleagues at National or Bethlehem. Exacerbating the unease that all felt about government's role in the economy were the hearings held by the Temporary National Economic Committee (TNEC) starting in 1938. The committee originated in the severe recession of 1937–1938, which some of Roosevelt's advisers blamed on monopolistic tendencies in the economy. If this could not be shown, at a minimum the committee hoped to demonstrate the dangers of concentrated economic power. Steel was targeted because of its violent political resistance to the New Deal as well as because of its organization and business practices. New Deal economists considered steel a classic case of administered pricing, where large firms set prices with little regard for market considerations. Another subject of inquiry was basing point pricing, which the Federal Trade Commission had opposed since the 1920s.

The industry was hostile to the hearings that began in late 1938. In February 1939 the committee requested data on distribution and pricing of ten steel products from 54 firms that controlled 90 percent of ingot capacity. Later the companies were asked to supply similar information for one month during 1937 and another during 1938. In late 1939 a group of steel executives testified before the committee; U.S. Steel's presentation attracted favorable comment from business periodicals. In the end the committee left behind voluminous records, and the Justice Department filed several antitrust suits. But steel was not affected, although a change in basing point pricing was made by the industry as the hearings began.

The TNEC confirmed the belief of some steel officials that the industry was under siege from the government. Moreover, another issue emerged during the hearings — steel capacity and the location of that capacity — that provoked further clashes between government and steel executives. As the Roosevelt administration turned attention to rearmament, the supply of steel became an important consideration. The government wanted to ensure an adequate supply for war, but the industry feared overbuilding, a phenomenon it had experienced after World War I and during the Depression. Most analysts predicted a war in Europe would be short, so the industry urged caution. The Roosevelt administration criticized the industry's hesitation and implied a lack of patriotism was behind the lack of cooperation. On this note of continuing controversy, the steel industry entered World War II.

The Steel Industry at War, 1940–1945

One of the more amazing chapters in American industrial history was the performance of American industry during World War II, as firms that had been moribund through the Depression operated at full capacity by 1942. Several major steel producers actually exceeded average annual capacity figures as the industry set new production records. U.S. Steel led all producers, shipping 110 million tons of finished steel from 1940 through war's end. Bethlehem produced more than 73 million tons of steel, running at 98.1 percent of capacity. In 1942 J&L pushed its facilities to 103 percent of capacity; Republic reached 100.4 percent of its capacity in 1943. As a whole, the American steel industry produced almost 90 million tons of finished steel during the peak year of 1944 and 427 million tons from 1941 through 1945.

This achievement pushed the industry's plants and its workers to their limits. Iron ore production jumped sharply after submarine warfare halted ore imports in 1942 and 1943, and domestic supplies made up the difference. The Lake Superior mines increased output from 73 million tons in 1940 to 105 million tons in 1942. U.S. Steel, which owned the largest mines on the Mesabi Range, alone produced more than 260 million tons of iron ore from 1940 through the summer of 1945, appreciably depleting its high-grade iron deposits. Another indicator of the production effort was the utilization of older mills and technologies. The number of active beehive coke ovens rose from 15,150 in 1940 to more than 18,500 in 1941; in 1942, beehive coke accounted for 8.3 percent of the nation's coke supply. Also, crucible steel production was revived a final time, accounting for 2,300 tons in 1941.

More important in raising output was a huge construction program. Little innovation took place during the war, although the introduction of National Emergency (NE) steels that conserved scarce alloying materials represented an impressive metallurgical achievement. The main order of the day was the brute-force effort to add new mills. Between 1940 and 1945, open-hearth ca-

pacity increased form 81.6 million tons to 95.5 million tons; 51 electric furnaces were built; and 22 new blast furnaces went on line.

The funds for expansion came from both the steel corporations and the government's Defense Plant Corporation (DPC). The cost was $2.6 billion, almost equally divided between the two sources. The government also permitted steel companies to depreciate new capacity on an accelerated five-year schedule. Even mills paid for with government funds were operated by existing companies. Republic Steel, which received some $160 million from the DPC, added a major new facility at South Chicago, Illinois, using $92 million in government funds. U.S. Steel spent $927 million on capital improvements during the war, drawing $470 million from the government. The DPC also funded two brand-new steel mills — one at Fontana, California, operated by Henry Kaiser, and the other in Geneva, Utah, built and operated by U.S. Steel — while supporting an ore-development project in northeast Texas that grew into Lone Star Steel. Government funds were behind Armco's expansion of its Sheffield Steel subsidiary in Houston, Texas. The location of these mills reflected the government's desire to move steel production westward.

Amid the emphasis on production, industry leaders had other concerns. First, the war involved government agencies much more directly in the operation of the industry and individual firms than had the New Deal. Federal officials not only provided construction funds, but they also allocated materials, established priorities, oversaw wages, and, above all, set prices. Genuine national planning of the economy developed after 1942, and extensive controls guided most steelmakers's decisions. Moreover, excess-profits taxes combined with price controls to limit profits sharply, and every major steel firm saw profits decline as volume rose. Governmental controls worked rather smoothly in steel, thanks again to the industry's experience with cooperation, and the AISI repeated its role as the forum for the industry. That former steel officials such as Edward Stettinius, Jr., had charge of allocation efforts and other programs also smoothed matters. But as the most important industry to the war effort of the nation, steel was also one of the most controlled.

The other wartime development concerned the position of labor. The grudging acceptance of the SWOC by "Little Steel" in 1941 and 1942 indicated that some firms had not fully accepted the union. Realizing this, labor leaders hoped to use the government councils running the economy to gain legitimacy. Walter P. Reuther of the United Auto Workers and Philip Murray both sought inclusion in government decision making and put forward sound proposals related to production. Murray's assistants Clinton Golden and Harold Ruttenberg authored plans for cooperation in the workplace that would today be labeled industrial democracy. The union also proposed more efficient ways of operating the mills. But in almost every instance labor was frozen out, for the idea of worker participation was anathema to steel's management from the plant level to the boardroom.

Labor journalist John P. Hoerr argues in *And The Wolf Finally Came: The Decline of the American Steel Industry* (1988) that the inability to develop a cooperative relationship between labor and management during the war haunted steel. The traditional adversarial relationship of labor and management, intensified by unionization in the 1930s, was set in concrete. The opening that might have developed after Taylor's recognition of the SWOC in 1937 was closed when Benjamin Fairless was promoted to U.S. Steel's presidency in 1940. Fairless, who proved a more forceful leader than chairman Irving S. Olds, favored a more traditional, confrontational relationship. He also wanted to close the rift with "Little Steel," and in 1945 he agreed with Grace to work together in postwar labor negotiations. Olds was an industrial statesman in Myron Taylor's mold and rebuffed the idea of breaking the union. Nonetheless, U.S. Steel adopted a harder line in labor relations during the war. The steel industry was again unified in its view of labor.

Steel in the Postwar Era

At war's end, steel executives were understandably proud of their contribution and claimed some responsibility for the Allied victory. Walter S. Tower, president of the AISI, told the institute in 1945:

> Through five fateful years the steel plants in this country have stood as a bulwark protecting human liberty and enlightenment from the threat of totalitarian darkness. It has been the task of those plants to provide material for building the greatest war machine that the world has ever seen.... [T]hat has been conspicuously well done.

Equally significant, however, was the opportunity that the American steel industry faced in the immediate postwar period. U.S. producers had no real rivals and enjoyed unprecedented superiority.

Almost two-thirds of the world's steel production in 1945 came from the United States. The first hints of economic recovery in 1947 dropped the American share to 57 percent, but rebuilding war-shattered nations clearly guaranteed a market for steel for several years.

This same postwar circumstance — American dominance in steel — also contained the seeds of later problems. Too many in the industry failed to appreciate how unique this situation was. Because they sold every ton of steel they made during this decade, they assumed they would always sell their steel. Quantity, not quality, became the driving concern. And because steelmakers confronted so few competitors in world markets, they assumed that that situation, too, was permanent. Finally, because they had the most efficient mills in the world, many steel men assumed that only Americans knew how to make good steel. When innovations did appear, they were dismissed if they originated outside the United States. In other words, American steelmakers in the 1950s were supremely confident of their position. Other American businessmen also assumed that the conditions of 1946–1959 represented a new normality, so the overconfident steel executives were not alone. Like other American industries, steel grew complacent.

The industry's self-assurance also was reflected in the continuation of traditional attitudes toward government and labor. In a word, toughness was again the preferred trait. Steel executives viewed government with unreserved hostility and chose a completely confrontational approach to labor. They were determined to ensure their managerial prerogatives. As a result, the admixture of big steel corporations, the United Steelworkers of America, and the federal government ensured that the steel industry would be the locus of repeated public controversies after 1945.

Yet if the roots of steel's later problems were visible — especially in hindsight — in the immediate postwar era, not all of the blame rests on steel's doorstep. Geographer Kenneth Warren has shown that American mills located in the Ohio and Mahoning river valleys paid a transportation penalty compared to mills on the Great Lakes. Pricing patterns masked this situation, however, and the industry remained centered in traditional locations. The fact that American firms resisted innovation only exacerbated the situation, so that over time American steel companies ceased to be the low-cost producers they once were.

Government policies after the war also were partly responsible for the industry's later problems. Steel's hostility to government was reciprocated by many government officials, and large firms were repeatedly investigated and called before congressional committees. Moreover, government policies on steel capacity and foreign aid were made with little consideration of long-term consequences. In the suspicious climate of the 1950s the concerns of steelmakers about such matters seemed selfish, given the preeminence of American firms. But these policies eventually proved disastrous to American producers. Historian Paul A. Tiffany has argued that mutual suspicion prevented finding common ground or discussions of mutual interests. The absence of cooperation eventually proved devastating. In short, the decade after the war was crucial to American steel.

In 1946 the steel industry's worst nightmare failed to appear: the country did not sink back into depression. Steel production declined only slightly at war's end and returned to near-record levels in 1947. In fact, as 15 years of pent-up demand and huge wartime savings launched consumers on a buying spree, government economists turned from emphasizing work-relief programs to controlling inflation. Steel executives had hoped to discard wartime controls quickly and were irritated at their retention.

Against this backdrop came the first postwar confrontation between the steel industry and government. The steelworkers' union had accepted government wage guidelines during the war, but in late 1945 it decided to push for wage increases when the end of overtime reduced the take-home pay of most workers. The union also wanted to reassert the authority it had lost during the war. Most industry executives viewed these negotiations as confrontations, although only a few favored breaking the union. *Fortune* aptly describes the atmosphere in 1946 as the "routine flexing of muscles in front of the bathroom mirror every morning." The article continues: "Collective bargaining in steel is still in the hollering stage . . . because both labor and management . . . seem to suffer from an obsession that to lose your toughness is something akin to losing your honor." The management position was strengthened by an agreement between Fairless and Grace that prevented the union from splitting the industry. Instead, pattern bargaining emerged to ensure that no company gained an advantage. U.S. Steel negotiated first, and other companies followed that settlement.

In early 1946 the first postwar strike occurred, after U.S. Steel argued that government price controls kept it from recouping the cost of a wage increase. Thus began a curious minuet of labor, industry, and government, including President Harry S Truman. U.S. Steel rejected a government compromise on wages and price hikes, and on January 17 about 550,000 union members closed down the entire industry. Henry J. Kaiser, who accepted the compromise, blasted steel executives for not being "smart enough" to reach an agreement. A settlement reached on February 15 was very close to the original offer, with wages rising 18.5 cents per hour and steel prices climbing $5 per ton.

The 1946 settlement set the pattern for steel's labor settlements until 1959. First, the public considered the industry's behavior selfish, greedy, and inflationary, potentially dangerous to the whole economy. Second, the antagonism between the industry and labor and government was publicly renewed and soon grew more bitter. The posturing about toughness by both steel and labor leaders marked all later negotiations; friendships across the line were discouraged because one was made to feel disloyal for having "contact with the enemy." Third, the steel industry settled on a policy of linking wage increases to higher steel prices. This strategy had several effects. It challenged government agencies concerned about inflation. It also removed any pressure on steel firms to increase productivity through new technologies. Why cut costs when prices could be raised, often more than required to pay for the contract settlement? As long as outside competitors and government officials did not challenge this arrangement, it provided steel executives with a new mechanism for guaranteeing stability.

The Truman administration, however, vigorously resisted price hikes to cover wage increases. Truman convinced many Americans that steel costs were a primary cause of inflation. As wage increases in 1947 and 1948 sent prices higher, the White House commented sharply; Truman once remarked that steel prices were always too high. The industry responded with invective asserting managerial prerogatives and critiquing every move in Washington, from tax policy to welfare programs. Most steel executives hoped for Truman's defeat in 1948.

The president, however, had a winning issue in his claim that business leaders did not understand that their private decisions had public importance, and steel was in the spotlight. Tiffany shows that steel executives with scant political sophistication exhibited little sensitivity to this argument. Truman went on the offensive. The Justice Department launched antitrust suits, including one against U.S. Steel for seeking to acquire a small West Coast steel fabricator. The Federal Trade Commission renewed its opposition to basing point pricing in 1947, and the industry gave way after the Supreme Court overturned the cement industry's pricing scheme in 1948. Hearings on steel prices by Congress's Joint Economic Committee in 1948 did not change the public's perception of steel as stubborn and greedy; even *Iron Age* criticized U.S. Steel's poor job of answering critics. Benjamin F. Fairless lamely complained that the charges were emotional and political, but Tiffany adds that the failure to grasp this was part of steel's problem.

The election of 1948 suggested that the American public agreed with President Truman, and steel leaders were put on the defensive as the confrontation intensified. The administration constantly scrutinized steel prices. The strike of 1949, in which pensions were a major issue, followed the industry's rejection of a compromise from a special Steel Industry Board appointed by Truman. Steel's argument that pensions were too expensive seemed spurious, since every executive had a pension plan. After a 42-day walkout the price hike that followed triggered new congressional probes. The Joint Economic Committee issued a majority report proposing to regulate steel as a public utility. The House Subcommittee on Monopoly Power (the Celler committee) examined the effect of big business on the economy. The industry responded forcefully, but the cold war and steel's connection to national security probably explained why nothing came of these hearings. To the public, steel remained arrogant.

Inflaming this situation were government complaints about steel capacity. Beginning in 1947 a few government officials warned of potential shortages; mills were running at more than 90 percent of capacity. The industry hesitated to overbuild, although it was expanding. From 1946 to 1949 U.S. Steel's capital expenditures totaled $976 million; Bethlehem spent $318 million, National $117 million, and Inland $96.1 million. The issue exploded in another confrontation in 1949, when President Truman threatened to build government steel mills. Supply shortages during the Korean War, despite record production of 96 million tons in 1950 and 105 million tons in 1951, seemed to prove the industry wrong. Moreover, government controls reimposed in 1950 further irritated steel executives.

In 1952 the three sides clashed. Labor sought higher wages and a union shop; the companies wanted higher prices and an open shop; and the government feared an inflationary settlement and production stoppages. No one compromised during negotiations that began in November 1951, and a strike date neared in April. On April 8 President Truman seized the steel industry in order to keep the mills running. The industry fought back, labeling Truman a socialist; speeches by Inland's Clarence Randall rebutting the president's position were especially effective. The Supreme Court ruled the seizure unconstitutional on June 2, precipitating a strike that lasted until June 26. The settlement gave the union wage hikes and a limited closed shop; the industry raised prices. President Truman was damaged politically. But in the long term, everyone lost, for this bitter struggle confirmed the basic pattern of hostility between steel executives and government officials.

Steel in the 1950s

The steel seizure was a factor in the public's rejection of the Democrats in November 1952. Certainly the industry anticipated better relations with President Dwight D. Eisenhower, which really meant it hoped to get its way. And in some respects those hopes were realized. During the mid 1950s the industry earned record profits and faced less resistance to steel-price hikes. Behind the scenes, however, problems that would confront the industry in the 1960s were strengthening. Only some blame can be traced to failure of steel executives to get the policies they wanted from the Republicans.

One result of government fears about steel capacity in the late 1940s was a massive steel-expansion program in the 1950s. U.S. Steel, for example, announced plans for a new mill in eastern Pennsylvania and broke ground on March 1, 1951, for the Fairless Works. With a capacity of 2 million tons, the mill had cost the corporation more than $500 million by the time it was opened in 1953. But this was only part of the corporation's building program, which totaled $3.47 billion from 1950 through 1959. Other companies also expanded. National Steel increased capacity from 3.9 million tons in 1945 to 7 million in 1960. J&L spent $683 million during the 1950s to upgrade its obsolete facilities. All told, American steelmakers increased capacity from 100 million ingot tons in 1950 to 148.5 million in 1960. Total investment from 1952 through 1959 was $8.714 billion.

Most of this money bought existing technology for established plants. At Sparrows Point, Maryland, for example, Bethlehem in 1956 began adding 2 million tons of capacity to the 6 million tons in place. The company built seven massive open-hearth furnaces, a tenth blast furnace, a battery of coke ovens, and slab, sheet, plate, and pipe mills at a cost of $200 million. According to Mark Reutter, the company boasted that the new facilities were "a near duplicate" to the existing plant. Thus the bigger open hearths involved few modifications, while the blast furnace did not include such improvements as top pressure, which increased fuel efficiency.

The industry as a whole continued this traditional pattern of incremental technological innovation. Blast furnaces grew slightly taller and output was increased by raising temperatures. Open-hearth furnaces remained the leading method of making steel, reaching their high point of 105 million tons in 1955, 91 percent of American steel output. Capacity increased from 87 million tons to 126.6 million tons, but the number of open-hearth furnaces declined as they increased in size. U.S. Steel, for example, had 342 furnaces in 1945 producing 30.3 million tons of steel; in 1960, 256 furnaces produced 39.9 million tons. A similar change occurred in rolling mills, especially hot-strip mills, as the power driving them rose from an average of 7,000 horsepower to 14,000 horsepower. The new mills rolled ingots weighing 50 tons, up from 15 tons, at higher speed.

To feed their mills American steelmakers sought new iron-ore supplies to replace the high-grade Mesabi deposits that were being depleted. The industry pursued two options. First, it developed methods of beneficiating, or enriching, low-grade ores by sintering or by producing taconite pellets. Second, American steel companies searched worldwide for new deposits of high-grade iron. Both steps were expensive. The pellet plants coming on stream in the late 1950s cost far more than untreated ore. But major iron discoveries in Venezuela, Brazil, and northern Canada also proved expensive to develop due to climate, distance, and geography. Some analysts, in fact, have blamed the cost of ore projects for the inability of steel companies to generate enough capital for improvements during the 1960s.

With their new mills, the nation's steel producers set a record of 117 million tons in 1955, with totals of 115 and 112 million tons the following two years. The United States accounted for 46.6 percent of the world's total steel production in 1950 and 39.4 percent in 1955. Even this relative decline, which reflected the economic recovery of Western Europe, Japan, and Russia, seemed

of minor consequence. Reutter observed that in 1953 Bethlehem alone produced almost a million tons more than all of the German mills combined. But lost in the congratulations that followed these achievements were two new technologies pioneered in the 1950s, the basic-oxygen process and the continuous caster. Both innovations offered major economies over existing processes, but their development was costly and slow. Typically, leading U.S. producers showed little interest in either process.

The basic-oxygen furnace (BOF) was developed in Austria, and two commercial plants were installed in 1953. The process injected oxygen from the top into a vessel filled with molten iron and scrap, producing steel much more quickly than an open-hearth furnace. The plant was less expensive and had lower operating costs. The process resembled the Bessemer process, and American steelmakers had experimented with oxygen in Bessemer converters since affordable bulk oxygen appeared in 1929. Yet the first American oxygen furnace was not installed until 1954, at McLouth Steel outside Detroit. In April 1955 an article in *Steel* announced, "Oxygen Steelmaking Arrives," but not until 1957 did a major company — J&L — build an oxygen furnace. By 1960 the process accounted for only 2 percent of the nation's steel output. It was instructive that a U.S. Steel engineer who visited the Austrian steel plants in the mid 1950s was reprimanded for doing so without authorization. Ironically, American steel executives, concerned that Japanese demand for scrap for open-hearth furnaces would increase scrap prices, encouraged the Japanese to look at the new technology even as they continued to build open-hearth furnaces.

Continuous casting developed much more slowly. Traditionally, steel ingots were transformed into shapes in several steps. First, molten steel was poured into large molds to form ingots. The ingots were rolled into slabs or blooms, which were in turn rolled into beams, bars, pipe, wire, plates, or sheets. A continuous caster produces slabs or blooms directly from molten steel, saving time and energy as well as reducing the steel lost in rolling and reheating. Republic Steel explored this process in the mid 1940s, and Allegheny Ludlum built a pilot plant in 1948. But Americans were very cautious, and an article in *Steel* in 1948 proclaimed, "No Steelmaking Revolution: Continuous Casting of Steel Will Not Revolutionize Methods." Thus this process also was developed in Europe. A study by Inland in the mid 1950s reinforced the prevailing skepticism about the technology, and an *Iron Age* report on the company's research in July 1956 declared, "Continuous Casting Too Big a Gamble for Large Mill to Try." The first commercial continuous-casting unit did not appear in the United States until 1962, and, as with the basic-oxygen process, a small producer — Roanoke Electric Steel — built it.

The American steel industry listed many reasons for moving slowly, including the unproved nature of the technologies, uncertainty over actual cost savings and future capacity needs, the small size of early models, and the existing commitment to relatively new open-hearth facilities. But other factors also played a role, including self-satisfaction in the industry. In the mid 1950s profits reached their highest level ever (more than $1 billion in 1955), as did production. Few competitors were visible on the horizon. Moreover, from 1952 through 1955 steel enjoyed calmer relations with the union and with the federal government. Price increases caused much less uproar. In short the industry seemed stable. Against this backdrop and the industry's long-term assumption that new technology threatened stability, steel men saw little reason to move precipitously. One need not claim, as some have, that monopoly caused the industry to ignore new technology. Rather, steel executives chose a course that was perfectly rational at the time: get the most out of existing investment.

That decision, however, had dire consequences, especially after 1955, when relations with both labor and government began to get sticky. The Eisenhower administration disappointed the industry on two counts. First, it favored a liberal trade policy and refused to erect barriers to imported steel. This stance was disconcerting as the European Coal and Steel Community — forerunner to the Common Market — and government-subsidized steel operations appeared abroad. These developments raised the specter of foreign competition. Imported steel rarely amounted to more than a million tons annually in the 1950s, but a rapid increase in world steel capacity caused American producers to keep an anxious eye overseas. Domestic producers had never exported to any great extent, but U.S. firms were losing a safety valve for excess capacity. By 1960 the American share of world output had dropped to 26 percent, and its share of world trade in steel stood at 7 percent, as compared to 15 percent in the mid 1950s.

The second disappointment was the use of American foreign aid to build mills overseas, a situation that was particularly galling to U.S. com-

panies. But during the cold war the goal of constructing a "bulwark against Communism" warranted the economic development of Europe and Asia. This meant building steel mills, and American aid for steel facilities amounted to $1.411 billion from 1947 through 1960. Japan received the most assistance: $204 million. Moreover, the U.S. government pushed to get Japan into the General Agreement on Tariffs and Trade (GATT) structure in 1955, a move that enhanced Japan's capability to export steel and other commodities. As Paul A. Tiffany has shown, neither the industry nor government found a way to discuss, much less develop common approaches to, these developments. But the result was that some foreign producers who began exporting steel to America had begun with U.S. help.

Labor relations also deteriorated after a respite in the mid 1950s. McDonald (Murray's successor) and Fairless had seemed to forge an amicable relationship that culminated in their triumphant tour of U.S. Steel plants after the 1952 strike. But the new harmony was not deeply rooted. John Hoerr asserts in *And the Wolf Finally Came* that relations in the plants remained confrontational. McDonald made little effort to alter this situation. Under Murray the union was highly autocratic, but McDonald seemed most concerned with staying in power. Political machinations, cronyism, and favoritism dominated an organization that stressed loyalty to McDonald. Moreover, McDonald set basic contract goals in competition with Walter P. Reuther and was determined to equal or surpass the gains achieved by the autoworkers.

Because the companies passed wage increases on to consumers, agreements were reached in 1954 and 1955 without a problem. A short strike occurred in 1956 when the industry attempted to limit the wage increase, but this attempt proved difficult in the face of record profits. Over three years, wages jumped 30 percent, and prices rose $21 per ton. This time, however, the administration asked the industry to moderate its price increases amid fears of inflation. In addition a new round of hearings was launched by Senator Estes Kefauver's subcommittee on antitrust and monopoly. A jump in imports occurred as consumers hedged against a strike, an ill omen.

The stage was set for the longest strike in the industry's history. McDonald, who nearly lost a union election in 1957 to an unknown local official, was determined to win a big settlement and protect workers against technological change. The industry wanted a contract that required no price hike. A new negotiator, R. Conrad Cooper of U.S. Steel, enforced this harder line. President Eisenhower vowed not to interfere. On July 15, 511,000 steelworkers walked out after inconclusive negotiations. They returned to work on November 7 after a Supreme Court ruling. Not until January 4 was an agreement signed, providing a 39-cents-per-hour increase over three years, more than the industry favored. Secretary of Labor John Mitchell and Vicepresident Richard M. Nixon played major roles in hammering out the accord.

The strike had severe ramifications. Steelmakers promised federal officials not to raise prices; moreover, the Kefauver committee watched events closely. Equally important, however, was that demand for steel had dropped from the highs of 1955–1957 to 85 million tons in 1958, 93 million in 1959, and about 100 million annually through 1962. At the same time, steel imports became a factor in the American market for the first time, edging up to 1.7 million tons in 1958 and jumping to 4.4 million in 1959. Thereafter imports stayed near 3 million tons — approximately 3 percent of the U.S. market. No longer could the industry automatically pass its cost increases to consumers, and steel companies entered the 1960s in a weakened position. All of the hidden problems were about to appear, and the decline of the American steel industry began.

American Steel, 1960–1990: Paying the Price

For American steelmakers the years since 1960 have been a disaster. From a position of world domination, they were reduced by the 1980s to a struggle for existence. Numerous factors and several culprits help account for the changes in the industry's status, including managerial arrogance and even incompetence; poor investment choices; the absence of a cooperative relationship with labor, with responsibility shared by labor, management, and government; structural changes in the U.S. and world economy; the technological backwardness of American industry; the advantages available to foreign producers who had help from their governments; the policy choices of the American government, which rested on a history of suspicion and hostility; and the absence of cooperation between the industry and government. Different commentators assign different weight to each, but taken together the evaluations explain the demise of an industry that once was the nation's most important.

Robert Crandall is one of the observers who argues that the American steel industry never recovered from the drop in demand in 1958 and the lengthy strike of 1959. Of course hindsight helps our view in these matters. In terms of steel production things did not look so gloomy once demand had recovered in 1963. Production increased from about 100 million tons annually during 1960–1962 to 127 million in 1963; it reached 141 million tons in 1969. The American record came in 1973, when the industry produced more than 150 million tons of steel.

But Crandall and others note that this apparent good health was an illusion. First, the upswing in demand led to huge capital investments for improved facilities and new technology. The sums involved were staggering. U.S. Steel devoted $4.225 billion for improvements between 1960 and 1969. During the same period Bethlehem spent almost $2.8 billion; National and Inland each spent more than $1 billion; and J&L more than $900 million. Yet the earnings of steelmakers failed to support these expenditures. The average annual rate of return in steel during the mid 1950s — 8 to 9 percent — was not equaled during the 1960s. U.S. Steel reported rates of return of 5 percent, Inland about 7 percent. Annual profits for J&L were about 6 percent from 1964 to 1966 but declined to 2.1 percent in 1969. Steel companies therefore borrowed heavily. Republic's long-term debt was $318 million in 1969; Bethlehem's was $419 million; and U.S. Steel's reached $1.25 billion in 1968.

This shaky financial situation was especially frightening because the industry was finally confronting changes in the basic technology of steelmaking. The new developments of the 1950s could not be ignored, and out of necessity the industry abandoned its go-slow approach to technological change. Almost no process was unaffected. An increasing percentage of the iron ore used in the industry was taconite pellets from huge deposits of low-grade ore in Minnesota. But this beneficiation process was expensive: one plant built by Pickands Mather in 1958 cost $300 million. By 1969, 20 pellet plants were in operation, producing 52 million tons. Larger coke ovens began to appear; blast furnaces showed amazing increases in size and temperature of operation that produced impressive gains in fuel economy; and rolling mills were computerized.

The replacement of open-hearth furnaces by basic-oxygen furnaces was the major technological challenge for steelmakers in the 1960s. The other important development, not as far advanced but recognized as equally important by the end of the decade, was continuous casting. The leading American steel firms had stood in the background while overseas competitors perfected these techniques in the 1950s; now they had to catch up. Their hesitation in the 1950s may have been fully rational given the state of those companies, especially the heavy investment in open hearths. As U.S. Steel's financial vice-president Robert Tyson put it in 1964, "Nobody who has efficient open hearth furnaces is going to throw them out to buy oxygen furnaces. We waited until we needed to replace old capacity." By waiting the industry also believed it could be sure oxygen furnaces were proven and then install larger units. In other words it wanted to avoid the risk of pioneering.

Yet Tyson's view of economic rationality, which justified conservative investment choices, put American firms behind in the innovation process, for they developed little experience with the new technology. Another view of rationality was required, for accepting technological innovation often means replacing perfectly acceptable, even profitable, established procedures with new and unproved facilities. Such thinking replaces lower short-term profitability with a strategy of long-term survival. This was the rationale of Carnegie, whose maxim was to seek the lowest cost of production. But such a strategy sacrifices stability, the basic goal of American steel executives after Gary. And even had the industry been disposed to adopt a strategy of lowest-cost production, that strategy became almost impossible to pursue in the 1960s. Tighter profits meant that many companies simply could not afford technological change. What made this all so unfortunate was the volume of capital that the industry had spent in the 1950s on older technology. The claim of some critics that the industry bought the wrong technology in the early 1950s overstates the case, since new techniques were unproven. But that investment certainly imposed a burden on American firms in the 1960s.

Smaller companies moved first to install both of these key innovations. Isolated instances aside, such as U.S. Steel's experimental work with continuous casters at Gary starting in 1962, Big Steel waited until the technologies were well developed. For oxygen furnaces this had happened by 1963; continuous casting was proven by the end of the decade. By delaying, however, firms were forced to move quickly later. Steel from oxygen furnaces accounted for only 3.4 percent of the nation's total in 1960, but that figure had jumped

to 48.2 percent by 1970. Open-hearth production, on the other hand, declined to only 36.9 percent of American steel in 1970. Continuous casters appeared more slowly, mainly because of the technical challenges. In 1972, 45 continuous casters were in operation, accounting for less than 5 percent of American steel production. Even by 1980 continuous casters produced only 20.3 percent of American steel. In Japan in 1980, however, 59.5 percent of the steel was continuously cast, and that year in West Germany the figure was 46 percent.

Even after American steel producers installed the new technologies, however, they failed to gain full cost savings, because most companies installed them in older mills. U.S. Steel, for example, explored the Q-BOP process, a German modification that permitted oxygen furnaces to be erected in older open-hearth shops. Limited capital was the cause. Only one greenfield mill was built in the 1960s — Bethlehem's Burns Harbor Plant east of Chicago on Lake Michigan. This integrated facility included one of the largest blast furnaces in the world, oxygen furnaces, and rolling and finishing mills; the company had spent more than $1 billion on it by 1972. U.S. Steel, on the other hand, repeatedly delayed plans to build a new mill at Conneaut, Ohio, blaming lack of capital and poor profits.

This combination of limited profits, intense pressure for new technology, and less-than-optimal installations was dangerous enough. But further complicating the picture for American steel executives were two other developments. First, foreign competitors, many with modern mills staffed with labor earning much less than American workers, began shipping significant quantities of steel to this country. Imports rose steadily, from 5.4 million tons in 1963 to 10.7 million in 1966 and 18 million in 1968. This flood of imported steel prevented the industry from generating more revenue through price increases. It also forced steel firms to alter market projections, curtailing plans for new plants even as it became more imperative to install the newest technologies.

Another competitor appeared on the domestic scene: the minimill. Initially, minimills were facilities with electric furnaces that melted scrap and produced up to 250,000 tons of steel a year. The first such operation was Northwestern Steel and Wire, which installed electric furnaces in the 1930s. But the technological developments of the 1960s opened other avenues for enterprising small firms. The key was adding continuous casters and small-bar mills to the electric furnaces. The savings in energy and steel produced by continuous casters enabled minimills to produce reinforcing rods, small bars, and angles far less expensively than could large firms using traditional technology. The mills could be located closer to markets and often were operated by nonunion crews. These circumstances gave minimill operators enormous cost advantages. By 1971 William T. Hogan had identified 38 minimills in operation in this country in such places as Florida and other southern states, the West Coast, and even the Midwest.

The appearance of minimills produced a bifurcated structure in the American steel industry. Older integrated steel firms seemed sluggish, technologically backward, and daunted by outside forces. Minimills appeared dynamic and technologically innovative. Most of the continuous casters installed by 1972, as well as the newest electric furnaces, were installed at the small mills, which were among the most efficient in the world. Management at firms such as Nucor resented the characterization of the industry as backward, claiming that not all steelmakers copied the integrated giants. A few large specialty steel firms such as Lukens also ran against the grain.

Traditional steelmakers faced even more problems with both labor and government, further compounding the already-problematic situation. The industry was shaken by the 1959 strike, and a desire to avoid another industrywide walkout colored labor relations for the next 20 years. Everyone had observed that hedge buying before the strike caused imports to surge; observers also noted that imports did not drop back afterward. One response was a joint congressional working committee, the Human Relations Committee, intended to develop better understanding between the two sides. It had only a minimal effect, although the 1962 agreement was reached early and without a strike.

It seemed at times, however, as though management was more willing to give in to labor's demands on salary issues. In 1963 steelworkers with high seniority gained an unprecedented "sabbatical" vacation of 13 weeks; cost-of-living clauses became standard in contracts. When Abel defeated McDonald in 1965 one reason was the rank and file's dissatisfaction with the relatively low increase in wages gained in 1962. Abel therefore sought and won a big contract in 1965. The Experimental Negotiating Agreement adopted in 1974 locked in this pattern, by guaranteeing wage increases as well as cost-of-living adjustments. In return the union promised not to strike; unresolved issues were arbitrated. The goal was to eliminate hedge buying, and the agreement covered the contracts of 1974, 1977, and 1980. The net result was the elimination of strikes until the mid 1980s, at a cost of steadily escalating wages. Yet the volume of imports did not slow.

Many commentators have argued that excessive wage hikes were a central factor in the loss of competitiveness by American steel. Certainly American wages in steel were the highest in the world even as productivity plummeted. But at least as important was the fact that the no-strike agreement did not remove the confrontational tone in labor-management dealings. Labor relations deteriorated in many mills, with workers as alienated from the union as from the companies that paid their wages. Especially damaging was a loss of concern for quality by both sides. In short, the ability of American steel companies to respond to its problems was not enhanced by developments related to labor. Both sides share the blame.

The industry's relations with government also showed little potential for providing assistance. In 1962 the continued suspicion between steel and government produced another major confrontation. President John F. Kennedy was as determined to avoid an inflationary settlement as the industry was to avoid a strike, and federal officials were intimately involved in the negotiations. The Kennedy administration believed that management had pledged not to raise prices, because the settlement was moderate. When U.S. Steel's Blough announced a price increase, which some in the industry followed, the administration was furious. Investigations were threatened and other actions considered before Blough backed down.

As in earlier confrontations the altercation with the Kennedy administration damaged the industry. First, open hostility between the steel industry and the federal government was renewed. Kennedy and other Democrats judged the industry selfish and duplicitous just as the need for cooperative action to deal with problems was increasing. The intensity of government pressure — dubbed jawboning — further reduced the industry's room to maneuver at a time when it was beginning to face greater imports, lower profits, and a capital squeeze. Scrutiny remained intense afterward, and steel prices were not raised for almost three years. Whether the industry could have raised prices in any event is not clear, given the appearance of imports. It was apparent, however, that the government showed little willingness to believe steel's assessment of its plight until late in the 1960s.

The emergence of other government programs that affected the steel industry only increased the siege mentality of some steel executives. For example, the civil-rights movement eventually led to federal observation of hiring practices by U.S. Steel and other companies as well as action against the seniority rules favored by the union. But the most bitter struggle concerned environmental pollution, a major issue for an industry whose manufacturing processes were inherently dirty. Just when the industry was strapped for capital, new demands were made to install expensive air- and water-treatment equipment. Between 1968 and 1980 the industry committed $4.4 billion to pollution controls; in new facilities up to 10 percent of start-up costs went to pay for environmental systems. Stricter air-pollution laws actually helped some firms decide to replace open hearths (next to coke ovens, the worst sources of air pollution) with oxygen furnaces.

Industry leaders also disliked the pattern of government enforcement of pollution laws, which was sometimes uneven. The rigidity of interpretation often depended on the economic health of individual firms. Furthermore, industry responses to demands for pollution control varied. Bethlehem cultivated the image of a responsible corporate citizen, acting before it was required and promoting its newer, cleaner mills. U.S. Steel, on the other hand, took a hard line and litigated almost every decision by the Environmental Protection Agency and local governments, installing pollution-control equipment only when legal options had been exhausted.

That the industry and the government managed to cooperate on a program to limit imported steel in the late 1960s was rather a miracle, given the history of hostility. One cannot help but notice the contradictory stances of many steel executives during these years: they decried government interference in contract negotiations and pollution control, while demanding protection from less expensive and in some cases higher-quality imported steel. That steel, the bastion of managerial prerogatives, sought protection was a signal of the industry's problems. But in 1968 the industry was desperate, with imports capturing 16.7 percent of the American market.

The first call for import restrictions came in early 1966 when Leslie B. Worthington, U.S. Steel's chairperson, asked for a temporary tariff to combat "unfair imports." Over the next two years steel executives repeated this suggestion before Congress, although quotas soon found broader support within the industry. A crucial development came in June 1968 when I. W. Abel and Thomas Patton of Republic sat together before a congressional committee and asked for quotas. Shortly thereafter the German and Japanese governments offered to place voluntary restrictions on exports to the United States; eventually the figure for imports was set at 14 million tons, to be divided between the European Coal and Steel Community and the Japanese. The figures for 1970 and 1971 were 14.7 and 15.4 million tons.

The voluntary restraints worked for two years, but several countries exceeded their quotas in 1971. In addition, steel from countries outside the agreement also increased, bringing imports to a record total of 18.3 million tons. Another agreement, with higher quotas, took effect in May 1972. After 1973, however, a surge in demand permitted everyone to ignore the system, and in 1977 imported steel totaled 21 million tons. The industry considered bringing suit to stop the dumping; others proposed countervailing duties. In the end, the government established trigger price mechanisms designed to eliminate dumping in American markets, but the industry felt government enforcement was lax. When steel demand fell off with the recession of the early 1980s, a new voluntary-quota system was introduced. The significant point, however, is that imports have continued to account for at least 20 percent of the American market. Ironically, an industry with a strong free-market tradition has resorted to government trade barriers.

Through the 1970s the industry struggled. In 1971 the Soviet Union produced more steel than American mills, eclipsing this country for the first time since the 1890s. When demand jumped in 1973, there was talk of steel shortages, and expansion plans were dusted off in some quarters. The industry investigated new technologies, such as direct reduction of iron and vacuum degassing of steel. Pollution restrictions forced steelmakers to explore new methods of coke production, including closed systems that prevented discharge into the atmosphere and pipeline charging. Water-treatment facilities were introduced using exotic methods such as biological systems with bacteria that eat many of the worst contaminants.

But demand for steel slowed in 1975, ending discussion of expansion as output fluctuated through 1981. The industry earned profits as it had every year since 1940, although it remained one of the least profitable sectors in American manufacturing. Profits in 1977 were slim, and Youngstown Sheet and Tube expired and merged with Republic — a harbinger of things to come. Double-digit inflation, prompted by higher energy costs, hit the industry hard, as did rapidly increasing cost-of-living adjustments pegged to the rate of inflation.

These problems paled, however, against the difficulties caused by the recession in 1982, the worst since the 1930s. Production fell below the figure for 1958, and the entire face of the American industry began to change. In 1975, 20 fully integrated steel companies operated 47 full-scale steel mills. By 1985 only 14 companies remained, with 23 integrated steel mills. U.S. Steel alone closed 5 plants, including such venerable sites as Homestead and Duquesne. In 1974 American steel producers had operated 163 blast furnaces; from 1982 to 1987 fewer than 50 were in blast. In 1974 steel capacity in the United States stood at 160 million tons. By 1987 it had dropped to 112 million tons on its way down to a predicted figure of 100 million tons. In 1970, 512,000 people were employed in the industry; even in 1980, 428,000 people still had jobs. By 1987 only 163,000 of those jobs remained. Bethlehem was typical, as employment dropped from 83,800 in 1981 to 32,900 in 1988. And these figures do not begin to convey the impact of economic dislocation on steelworkers and their communities.

During the 1980s restructuring was the talk of the industry. Companies scrambled to survive as losses totaled $3.3 billion in 1982 and a staggering $4.1 billion in 1986. Overall, the industry lost money every year from 1982 through 1986, and again in 1988. It was worse than the 1930s had been. The losses forced several mergers, including the combination of Republic and J&L into LTV Steel in 1983. National Steel — now National Intergroup — sold Weirton Steel through an employee stock-ownership plan (ESOP) in 1982. U.S. Steel changed its name in late 1986 to USX to demonstrate that it was not just a steel company. Several firms seeking access to new technologies or capital developed joint ventures with Japanese or Korean steel companies.

Another change has been the concessions accepted by the United Steelworkers. As old mills closed and jobs drained out of the industry, the union faced intense pressure to offer concessions to companies fighting for survival. Faced with threats of further plant closures, the USWA agreed to end pattern bargaining and permitted significant discrepancies in wages from firm to firm. Every steel company cut wages, changed work rules, and cut benefits. A few companies began to develop cooperative relations between management and labor, but old patterns died hard. In 1986, the union drew the line on concessions at U.S. Steel, which continued its confrontational style of labor relations. The union believed the company was out to break the union, and a 184-day strike followed, the industry's longest. Yet it attracted little national attention and had little impact on the economy, a situation that would have been unimaginable 15 years earlier. At the end of the strike, the union had its pride, the company its tough image. But steel was no longer the mainspring of the American economy.

By the late 1980s business journals and newspapers reported that signs of life remained in big steel in this country. With reduced payrolls, smaller workforces, and more-modern mills, several companies reported profits again. Moreover, occasional reports suggested the enormous strides that had been taken. Inland's mill at Indiana Harbor, Bethlehem's at Burns Harbor, and U.S. Steel's much slimmer facility at Gary were among the most efficient in the world. Wheeling-Pittsburgh seemed on the verge of returning to profitability after having sought bankruptcy protection, while Weirton Steel, operated by its employees, also struggled on. Employment rose to 169,853 in 1989. No one has predicted a full recovery, and most agree that the glory days of steel are gone. In 1947 Wilfred Sykes had told the AISI, "There is no substitute for steel." By 1990 that truism no longer applied to the American steel industry.

The Importance of Individuals

In steel's rise and fall, one might ask, how important were the people whose biographies fill this volume? One answer comes from the perspective of scholarship in American business history. For three decades Alfred Chandler's conception of the role of management and structure in big business has dominated thinking in the field. Yet an essay in the spring 1990 issue of *Business History Review* by Harold Livesay called attention to the popular business magazines, which frequently focus on the individual entrepreneur and dominant business manager. The two points of view are not mutually exclusive, although Chandler's emphasis on strategy, structure, and scope leads one to emphasize organizational structure more than the apparently faceless managers in determining a firm's fate.

Certainly many of the problems of the American steel industry can be accounted for by organizational shortcomings. Another perspective, which presents the recent competitive difficulties of steel as the inevitable fate of a mature industry, helps explain other aspects of steel's fate. Both reinforce the sense that individual managers were less important than the general development of the industry or the organizational structure of specific firms. And in fact very few individuals in this book stand out for shaping the industry. There is a basic sameness in many of the biographies, and compared to the colorful, hard-driving entrepreneurs of the nineteenth century, steel executives of the twentieth rarely showed flash or eccentricities. Many of them seem perfectly interchangeable, and even assessing the contribution of certain individuals can be difficult, given the size of the firms they managed.

In its December 12, 1935, issue, *Iron Age* released a survey of 176 top steel executives who represented firms with 95 percent of the nation's steel capacity. Three-quarters of them began in production, almost half of them as laborers; 25 started as engineers or chemists; another 26 began as clerks or stenographers. Only 13 entered as executives. More than half the field spent entire careers with one firm. Of the 5,664 man-years spent in steel companies, 4,481 were with the company they headed. And even those who shifted firms moved to other steel companies; only 46 had ever worked outside the steel industry.

There are, of course, obvious differences with recent steel executives. Steel managers now universally have college degrees, and have for half a century. Laborers can no longer climb to the top. But many traditional traits remain visible in 1990. One common thread is that steel executives tend to stay with one company. Recent problems in steel have made it easier for some companies to look outside for top executives, but the pattern has not changed much. There has, however, been a greater chance since the 1950s that the executive suites will be filled by lawyers or accountants. Even so, a widely held belief remains in the industry that operating people are best suited to run steel companies, a belief that accounts for the resistance and resentment toward accountants and others who have reached top management positions. There is one last pattern, not contained in

the survey but evident nonetheless: steel executives exhibit not only a conservative management and political style, they also share views about government and organized labor.

How important are these long-running patterns? For one thing, the basic continuity may help explain why change has come slowly in the steel industry. More important, the focus of this encyclopedia series puts the spotlight on individuals, showing how they make a difference for good or ill. There are not many instances of true greatness in the steel industry, but there were still giants on the earth in the days after Andrew Carnegie. A few men put an indelible stamp on steel, shaping their organizational and economic environment. Others worked creatively within the structure that they inherited. Too many showed little imagination or insight into their world and failed to realize when the familiar structure of the past needed to be changed. Since World War II the giant steel corporations have seemed to lack the dominant personality or "charismatic entrepreneur" that Livesay identified as the driving force behind highly successful and innovative firms. In part this is due to the problems and diminishment of the steel industry, in part because we lack the historical distance to see them. But even this situation indicates the importance of individuals in American business.

Editor's Note

There is no secret formula for identifying the important individuals in the history of any industry. A few are obvious even to the nonspecialist, but most required some thought. I began by compiling a list of more than 500 names from historical studies, such as William T. Hogan's multivolume economic history of the industry, and from trade journals such as *Iron Age* and *Steel*. Advice came from the editorial board of the series, and each author was asked to identify holes in the coverage of his or her area of expertise. In the end, several factors guided final selections. Limited availability of biographical material was one consideration, and several important figures who should have been represented are not. Others were omitted because of space constraints; others because no author could prepare the entry.

But one of the most important factors that affected the selection was the fact that the iron and steel industry has received relatively limited treatment by business and economic historians. This may sound surprising given the industry's obvious importance. There are accounts of various incidents in the steel industry's history — World Wars I and II; Truman's seizure of the steel industry in 1952; the confrontation with Kennedy over prices in 1963. The United Steelworkers Union also has received some attention. Many economists and journalists have explored the recent problems besetting the industry since the 1960s, although many lack historical perspective. Recent accounts by Hoerr, Tiffany, Donald Barnett, Crandall, Dennis Dickerson, Hogan, John Strohmeyer, and Mark Reutter offer fine insights to the industry's development.

Yet this recent attention cannot substitute for the absence of detailed case studies of individual corporations and biographies of the most important leaders. For the most pivotal figure of all — Gary — there is only Tarbell's uncritical treatment written in 1925 after extensive talks with Gary. A few — Tom Girdler, for example — wrote autobiographies. But in the end, there is much unexplored territory in the steel industry. To an unfortunate extent, these lacunae are the product of the reluctance of the big steel companies to open their archives to historians. A few internally prepared histories emerged at the fiftieth anniversaries of several companies or to celebrate other occasions. But professional historians usually find steel-company records closed. It should be noted, however, that Armco, Bethlehem, Inland, McLouth, U.S. Steel, and Wheeling-Pittsburgh assisted contributors to this volume. Only Bethlehem, however, has made its historical records accessible, although financial problems forced the company to close its splendid library. Its records are being housed in museums or libraries.

The holes in the historical record are reflected in the choices and the coverage of the industry and its people in this volume. Inevitably, most of the men profiled — and it is an exclusively male roster — were the top executives of the Big Steel companies. An effort was made to include operating, financial, sales, and research men who did not reach the boardrooms, but in most cases limited biographical information complicated this task. Even many top officials proved elusive. Thus the typical biographical entry in this volume is 1,000 words long, rather than 2,500 words as in other volumes in the series. More people are covered, but in less depth than many might prefer. It was surprising to find that we know more about many individuals who built the industry in the nineteenth century than about the faceless managers of giant corporations in the twentieth century. Even less information, however, could be found for the leaders of smaller companies. The companies could be treated in brief accounts, but very few chief executives could be chronicled.

Another caveat for users of this volume is a product of the size and scope of the steel industry. Integrated steelmakers were involved in mining as well as producing steel; they also operated transportation networks and fabricated the metal into finished products. But many nonintegrated firms also fabricated and produced finished and unfinished steel products. Unfortunately, both steel fabricators and companies engaged in mining and lake transportation, as well as steel producers attached to other industries (i.e., International Harvester's Wisconsin Steel and Ford's Rouge Steel), are not being covered in this volume because of space constraints. Also, the large engineering companies whose existence was tied to the steel industry (i.e., Mesta Machine, United Engineering and Foundry, Wean Engineering, and so on) could not be treated here. Some attention has been given to firms in the specialty-steel business, and several important minimills are represented. But the goal was to provide representative coverage of firms in these areas, not complete treatment. In the end, the emphasis of this volume rests on the large integrated steel companies and their executives. Responsibility for the final selections, and holes or gaps in those selections, rests with the editor.

I would especially like to thank all of the contributors for their efforts, as well as those who could not prepare entries but who offered guidance, suggestions, encouragement, and other assistance. Special thanks go to general editor William Becker and series editor James Hipp. Finally, I want to recognize Jane Koski and Barbara Wilder, interlibrary-loan librarians at Michigan Tech. This volume would not have appeared without their continuously cheerful assistance in the face of an almost endless stream of requests.

American Steel Production, 1900–1989
(in tons)

YEAR	OPEN-HEARTH FURNACE	BESSEMER CONVERTER	ELECTRIC FURNACE	CRUCIBLE	TOTAL
1900	3,805,911	7,836,942		112,629	11,410,928
1901	5,215,066	9,758,898		110,335	15,090,426
1902	6,370,256	10,234,967		126,305	16,740,920
1903	6,529,500	9,623,968		114,726	16,279,175
1904	6,617,146	8,802,237		93,398	15,523,074
1905	10,047,914	12,254,340	10,039	114,501	22,426,821
1906	12,298,063	13,748,930	16,105	142,815	26,205,913
1907	12,935,704	13,067,655	15,764	146,982	26,166,105
1908	8,777,136	6,850,766	6,868	71,267	15,706,037
1909	16,233,208	10,450,477	25,701	120,238	26,819,624
1910	18,485,050	10,542,305	61,975	136,979	29,226,309
1911	17,470,488	8,901,596	35,783	109,371	26,517,238
1912	23,274,410	11,567,249	23,701	136,099	35,001,459
1913	24,191,923	10,691,191	38,902	135,773	35,056,979
1914	19,235,646	6,967,348	30,947	100,653	26,334,594
1915	26,520,594	9,281,679	79,452	127,436	36,009,161
1916	35,185,278	12,386,124	189,865	145,255	47,906,522
1917	38,246,760	11,737,555	341,643	141,922	50,467,880
1918	38,594,518	10,501,384	573,096	128,925	49,797,923
1919	30,182,537	8,144,149	433,892	71,201	38,831,779
1920	36,592,522	9,949,057	566,370	80,937	47,188,886
1921	17,460,578	4,497,851	190,897	8,527	22,157,853
1922	32,826,061	6,629,614	387,563	32,039	39,875,277
1923	40,207,616	9,502,179	577,777	49,368	50,336,940
1924	35,366,632	6,607,541	484,429	25,170	42,483,772
1925	42,598,627	7,530,837	689,273	21,910	50,840,747
1926	45,575,016	7,766,716	729,930	17,352	54,089,014
1927	42,636,535	6,934,734	746,018	10,120	50,327,407
1928	49,407,631	7,414,618	898,531	8,701	57,729,481
1929	54,155,235	7,977,210	1,065,603	7,442	63,205,490
1930	39,255,073	5,639,714	686,111	2,523	45,583,421
1931	25,210,714	3,386,259	460,255	1,733	29,058,961
1932	13,336,210	1,715,925	270,044	722	15,322,901
1933	22,827,473	2,720,246	471,747	763	26,020,229
1934	26,534,838	2,421,840	404,651	595	29,181,705

YEAR	OPEN-HEARTH FURNACE	BESSEMER CONVERTER	ELECTRIC FURNACE	CRUCIBLE	TOTAL
1935	34,401,280	3,175,235	606,471	719	38,183,705
1936	48,760,463	3,873,472	865,150	914	53,499,999
1937	51,824,979	3,863,918	947,002	1,046	56,636,945
1938	29,080,016	2,106,340	565,627	7	31,571,990
1939	48,409,800	3,358,916	1,029,067	931	52,798,714
1940	61,573,083	3,708,573	1,700,006	1,024	66,982,686
1941	74,389,619	5,578,071	2,869,256	2,313	82,839,259
1942	76,501,957	5,553,424	3,974,540	2,010	86,031,931
1943	78,621,804	5,625,492	4,589,070	146	88,836,512
1944	80,363,953	5,039,923	4,237,699	25	89,641,600
1945	71,939,602	4,305,318	2,456,704	24	79,701,548

American Steel Production 1946–1989
(in thousands of tons)

YEAR	OPEN-HEARTH FURNACE	BESSEMER CONVERTER	ELECTRIC FURNACE	BASIC-OXYGEN FURNACE	TOTAL
1946	60,712	3,328	2,563		66,603
1947	76,874	4,233	3,788		84,900
1948	79,341	4,243	5,057		88,640
1949	70,249	3,947	3,783		77,978
1950	86,263	4,435	6,039		96,836
1951	93,167	4,891	7,142		105,200
1952	82,847	3,524	6,798		93,168
1953	100,474	3,856	7,280		111,610
1954	80,328	2,548	5,436		88,312
1955	105,359	3,320	8,050	307	117,036
1956	102,841	3,228	8,641	506	115,216
1957	101,658	2,475	7,971	612	112,715
1958	75,880	1,396	6,656	1,323	85,255
1959	81,669	1,380	8,533	1,864	93,446
1960	86,368	1,189	8,379	3,346	99,282
1961	84,502	881	8,664	3,967	98,014
1962	92,957	805	9,013	5,553	98,328
1963	88,834	963	10,920	8,544	109,261
1964	98,098	858	12,678	15,442	127,076
1965	94,193	526	13,804	22,879	131,462
1966	85,025	278	14,870	33,928	134,101
1967	70,690	NA	15,089	41,434	127,213
1968	65,836	*U.S. Bessemer production ceased in 1968.	16,814	48,812	131,462
1969	60,894		20,132	60,236	141,262
1970	48,202		19,931	63,330	131,514
1971	35,518		20,950	63,932	120,400
1972	34,898		23,710	74,592	133,200
1973	39,112		27,747	83,242	150,800
1974	35,405		28,703	81,592	145,700
1975	22,154		22,620	71,826	116,600
1976	23,434		24,576	80,000	128,000
1977	20,048		27,817	77,435	125,300
1978	21,372		32,195	83,433	137,000

YEAR	OPEN-HEARTH FURNACE	BESSEMER CONVERTER	ELECTRIC FURNACE	BASIC-OXYGEN FURNACE	TOTAL
1979	19,082		33,939	83,279	136,300
1980	13,054		31,166	67,615	111,835
1981	13,452		34,145	73,231	120,828
1982	6,110		23,158	45,309	74,577
1983	5,951		26,615	52,050	84,615
1984	8,336		31,369	52,822	92,528
1985	6,428		29,946	51,885	88,359
1986	3,330		29,390	47,885	81,606
1987	2,666		33,989	52,496	89,151
1988	5,118		36,846	57,960	99,924
1989	4,442		25,154	58,348	97,943

SOURCES: American Iron and Steel Institute, *Annual Statistical Report* (1940, 1950, 1960, 1970, 1980, 1990).

Iron and Steel Imports and Exports, 1900–1989
(in tons)

YEAR	ALL IRON AND STEEL EXPORTS	ALL IRON AND STEEL IMPORTS
1900	1,154,270	209,955
1901	700,852	221,297
1902	372,399	1,206,811
1903	326,679	1,778,797
1904	1,167,674	266,398
1905	1,009,243	416,454
1906	1,325,740	584,428
1907	1,301,981	1,662,358
1908	964,272	206,956
1909	1,239,709	363,984
1910	1,537,952	510,730
1911	2,187,725	262,457
1912	2,947,596	225,072
1913	2,745,635	317,260
1914	1,549, 554	289,775
1915	3,542,608	282,396
1916	6,105,881	319,871
1917	6,439,067	330,201
1918	5,375,281	169,110
1919	4,399,698	322,264
1920	4,935,137	410,857
1921	2,213,549	123,615
1922	2,095,270	725,855
1923	2,134,062	764,240
1924	1,924,824	577,240
1925	1,902,404	956,094
1926	2,186,574	1,083,081
1927	2,201,969	721,775
1928	2,889,549	774,212
1929	3,063,075	736,081

YEAR	ALL IRON AND STEEL EXPORTS	ALL IRON AND STEEL IMPORTS
1930	2,004,710	527,280
1931	848,690	397,331
1932	378,415	361,665
1933	582,950	337,175
1934	996,911	254,113
1935	982,010	377,950
1936	1,381,186	591,841
1937	3,914,923	509,064
1938	2,422,261	267,573
1939	2,798,881	319,947
1940	8,719,805	62,445
1941	7,166,055	41,598
1942	7,279,447	41,971
1943	7,291,074	39,416
1944	6,150,988	82,090
1945	5,109,229	
1946	5,030,656	70,310
1947	6,876,420	153,489
1948	4,697,405	472,804
1949	5,055,063	472,280
1950	3,093,474	2,029,516
1951	3,620,927	3,519,397
1952	4,456,885	1,696,537
1953	3,434,054	2,481,879
1954	3,157,961	1,253,797
1955	4,553,275	1,466,613
1956	5,216,427	1,999,875
1957	7,006,649	1,853,455
1958	3,463,345	2,136,195
1959	2,128,103	5,523,556
1960	3,470,466	4,087,583
1961	2,794,433	3,926,322
1962	2,518,586	4,993,207
1963	2,846,423	6,521,884

YEAR	ALL IRON AND STEEL EXPORTS	ALL IRON AND STEEL IMPORTS
1964	4,205,689	7,701,477
1965	3,089,326	11,963,742
1966	2,277,572	12,778,054
1967	2,167,773	12,813,017
1968	2,782,411	19,563,182
1969	5,939,379	15,443,558
1970	8,139,579	14,609,369
1971	3,546,918	19,629,645
1972	3,605,576	19,558,224
1973	5,059,570	17,008,297
1974	7,132,334	17,962,674
1975	4,146,350	13,919,245
1976	3,864,816	16,293,728
1977	3,218,000	21,658,123
1978	3,499,284	24,068,286
1979	3,762,318	20,326,554
1980	5,107,338	17,878,020
1981	3,774,725	22,639,380
1982	2,587,985	18,770,002
1983	1,764,738	19,301,384
1984	1,726,592	29,548,073
1985	1,569,222	27,632,542
1986	1,452,254	24,237,800
1987	1,707,717	23,836,367
1988	2,757,389	25,659,254
1989	5,374,432	22,056,070

SOURCES: American Iron and Steel Institute, *Annual Statistical Reports* (various years).

Iron and Steel in the Twentieth Century

Iorwith Wilbur Abel

(August 11, 1908 – August 10, 1987)

by Richard W. Kalwa

University of Wisconsin — Parkside

CAREER: Staff representative, Steel Workers Organizing Committee (1937–1942); director, District 27 (1942–1952), secretary-treasurer (1952–1965), president, United Steelworkers of America (1965–1977); vice-president (1965–1977), president, industrial union department, American Federation of Labor and Congress of Industrial Organizations (1968–1977).

Iorwith Wilbur Abel was president of the United Steelworkers of America (USWA) during the union's most powerful and prosperous years. He rose from the ranks to revitalize and democratize a union whose leaders had become too far removed from the concerns of USWA members. He negotiated steady increases in wages and benefits, increased union membership, and became a nationally recognized labor leader. His most innovative accomplishment, the elimination of strikes in the steel industry, was also his most controversial, since it caused dissatisfaction among steelworkers at the same time that it raised pay levels.

He was born on August 11, 1908, in Magnolia, Ohio, one of four children. His parents were John Franklin Abel, a blacksmith, and Mary Ann Jones Abel. He attended local schools, graduated from Magnolia High School, and attended Canton Actual Business College. In 1925 Abel went to work as a molder at the American Sheet and Tin Mill Works in Canton. Later he worked for the Canton Malleable Iron Company, the Colonial Foundry, and the Timken Roller Bearing Company in Canton.

The Timken foundry closed in 1930 at the beginning of the Depression, and Abel found work in a brickyard for 16 cents an hour, 12 hours a day. He would recall later, "[t]hat miserable job helped straighten out my social thinking and pointed me in the direction I was to travel the rest of my life." In 1936 he joined the Steel Workers Organizing Committee (SWOC) of the Congress of Industrial Organizations (CIO) and helped to organize Local 1123 at Timken. He held various positions in the local, eventually becoming its president.

Abel came to be known as "the biggest union hell-raiser in Canton," in one year leading 42 wildcat strikes. He participated in the so-called Little Steel strike in 1937. Although the Little Steel companies — those integrated firms smaller than the United States Steel Corporation — were successful in defeating the SWOC, Abel felt that the strike was ultimately a victory for organized labor. The violent struggle led to congressional investigations made by the La Follette Committee, which put an end to the use of espionage, arsenals, police, and the National Guard by the steel companies.

Abel was appointed as a staff representative of the SWOC by its first president, Philip Murray, in 1937. In February 1942 Murray named him director of USWA District 27, which covered the Canton area. Later that year the first convention of the USWA formally elected him to that post. He was active in the Ohio CIO council and the CIO Political Action Committee. During World War II, Abel served as a regional panel member on the War Labor Board, and also on the War Production Board and the War Manpower Commission.

Following the death of Murray in November 1952, the USWA executive board met to designate the top officers of the union and selected Abel as secretary-treasurer, the Union's second-highest position. By then, he had gained a reputation for being hardworking and modest, with few political enemies. He was officially elected as secretary-treasurer by a membership referendum in February 1953, running on a ticket led by USWA presidential candidate David McDonald.

As the chief financial officer of the union, Abel argued in favor of an unpopular dues increase at the divided 1956 convention, and criticized the opposition Dues Protest Committee for not confining their disagreement to proper channels. When his opponent in the 1957 referendum failed to sign his acceptance of nomination, how-

I. W. Abel (right) swearing in his successor, Lloyd McBride, as president of the United Steelworkers of America, June 1977 (photograph by James Klingensmith, Pittsburgh Post-Gazette)

ever, Abel wired the candidate for authorization in order to correct the error prior to the election deadline. This incident earned him the nickname "Honest Abel." In the 1961 union election no opposition candidate contested his position.

By 1964 Abel's dissatisfaction with McDonald's presidency induced him to declare his candidacy for head of the USWA. Abel took issue with the centralization of power that had occurred within the union under McDonald, particularly in the area of contract negotiations through the Human Relations Committee. Despite his high office Abel himself had been excluded from McDonald's decision making. A key feature of the campaign was the contrast in image between the vain, flamboyant McDonald, who had never worked in a steel mill, and the solid, down-to-earth Abel, regularly described as "rockhewn," "weathered," and "white-haired and bass-voiced." His running mates for vice-president and secretary-treasurer were Joseph Molony and Walter Burke, both district directors, and his campaign manager was Joseph Germano, director of the largest district in the union, Chicago-Gary.

Abel charged McDonald with practicing "tuxedo unionism" and declared his intention "to restore collective bargaining to its rightful place — the collective bargaining table — and take it out of the plush hotel suites, the country clubs, and the businessmen's clubs and place it back in the hands of elected representatives of the membership." He targeted for criticism the 1962 and 1963 settlements, which had brought no wage increases, neglected local issues, and excluded rank-and-file participation. Abel's fiery denunciations of McDonald's presidency caused leading newspapers and business groups to express concern that unrest within the union and an Abel victory would lead to instability and strikes. Abel's popularity among steelworkers prevailed, and he won by 308, 910 to 298,768, and was sworn into office in June 1965.

His first order of business was the negotiation of a collective agreement with the steel industry. This settlement, a three-year contract reached following months of negotiations and the intervention by federal mediators under President Lyndon B. Johnson, was reached in Washington on September 6, 1965. It provided the first wage increase since 1961. At the September 1966 convention of the USWA, a new structure of bargaining was approved. Power to ratify contracts in steel was given to the Basic Steel Industry Conference (BSIC), a body composed of the several hundred local union presidents in the industry. The conference procedure was extended to the aluminum, can, and metal fabricating industries as well, and a membership vote was required for approval of strikes. These changes were aimed at increasing the involvement of steelworkers in the bargaining process.

Civil rights became another important issue within the union. Abel had attended the first convention of the Negro American Labor Council in 1960, and had been endorsed for president by the Ad Hoc Committee, a group of black USWA leaders. He had been accused of racism during his campaign, however, and the NAACP directed criticism at the union for not vigorously fighting segregation and increasing job opportunities for blacks in the steel industry. At the August 1968 union convention, the Ad Hoc Committee, pointing out the scarcity of black officers and staff within the USWA, advocated the appointment of a black vice-president. The proposal was rejected as a reverse form of discrimination, however, and Abel declared, "I didn't hold office all these years as a Welshman but as a steelworker."

Under Abel's leadership, the USWA began to extend its involvement into new areas. Mergers with several unions, including the 75,000-member Mine, Mill and Smelter Workers, increased the size and diversity of the USWA. The absorption of these workers in 1967 made the steelworkers the major union in the copper industry, where that year it headed a coalition of unions representing 50,000 workers in a nine-month strike. The massive cost of strike support necessitated a special dues convention in March 1968, which introduced a temporary assessment and tied dues to base pay.

Abel had been elected as a vice-president and executive council member of the American Federation of Labor and Congress of Industrial Organizations in December 1965. He was appointed to the federation's Farm Workers Organizing Committee in 1966, as chair of its Economic Policy Committee in 1967, and replaced Walter Reuther as president of the AFL-CIO industrial union department in 1968. He was regularly suggested as a successor to George Meany as president of the AFL-CIO. Abel was the AFL-CIO representative on the Labor-Management Advisory Committee under the Johnson administration. Johnson also

appointed him as the labor representative on the National Advisory Commission on Civil Disorders in 1967, and as a member of the U.S. delegation to the United Nations in the same year.

The USWA allied itself with the steel industry in seeking import quotas, and in June 1968 Abel testified before the House Ways and Means Committee, arguing that unrestricted free trade interfered with collective bargaining and reduced labor to a "commodity." In the 1968 national election, Abel strongly supported Hubert Humphrey, a personal friend, for the presidency, and served as head of the AFL-CIO committee for Humphrey. He dissociated the USWA from a union official who ran as a delegate for Robert Kennedy in the Democratic primary, and resigned from the board of the Americans for Democratic Action to protest the organization's endorsement of Democratic liberal candidate Eugene McCarthy. Later, Abel refused to attend meetings of the McGovern-Fraser Commission on Democratic party delegate selection, on which he served as the labor representative. He attended the 1972 Democratic Convention as a delegate for Henry Jackson and did not support the party's nominee, George McGovern, in the presidential election.

In 1969 Abel was obligated to face a challenge for the presidency of the USWA. The opposition candidate, Emil Narick, was a relatively unknown staff lawyer who was able to draw on several sources of discontent within the union. The 1968 steel contract had provided for a six-percent wage increase, but the absence of cost-of-living protection caused membership dissatisfaction. Much of the rank and file, particularly younger steelworkers, felt that local issues were being neglected, and criticized what they perceived as Abel's excessive involvements outside the union. Abel's stands on civil rights caused a backlash of resentment among many white union members. Abel caused further irritation when he transferred Narick to an isolated research job. Abel defeated Narick in the February 1969 referendum by a margin of 257,651 to 181,122, but a majority of only 60 percent, a voter turnout of under 40 percent, and strong support for Narick in the large steel locals indicated a weakening of support in the union for Abel.

In negotiating the 1971 steel contract, however, Abel won a wage increase of 30 percent over three years and a cost-of-living adjustment, significantly raising his popularity among the rank and file. In the same year, he was appointed to serve as a labor member on the advisory Pay Board as part of President Richard M. Nixon's National Stabilization Program. But Abel resigned along with the other union representatives in 1972 in protest of the perceived antilabor bias of the government's anti-inflation policy. This move further endeared him to members, and in the February 1973 USWA referendum Abel was reelected president without opposition.

The layoffs caused by industry stockpiling prior to the 1971 negotiations, as well as his increasing political security, motivated Abel to examine the possibility of avoiding the use of strikes to gain concessions. The union executive board had rejected a similar proposal in 1967, but he began meetings in 1972 with R. Heath Larry of the United States Steel Corporation to discuss the concept. In March 1973 the Experimental Negotiating Agreement (ENA), a procedure to substitute binding arbitration for strikes in steel bargaining, was approved by the union executive board and a quickly scheduled Basic Steel Industry Conference meeting. In the meantime, a film depicting the threat of foreign steelworkers, *Where's Joe?*, was shown to workers, and Abel appeared in U.S. Steel ads urging greater plant-level cooperation.

The ENA was widely hailed by the industry as a milestone in mature collective bargaining, but it drew strong opposition within the union. Opponents claimed that an executive board statement in December 1967 had declared that strikes would not be eliminated without approval by a membership referendum, and challenged the ENA in federal court. The judge's decision, issued weeks before the April 1974 bargaining deadline, observed that the leadership probably failed to provide "the optimum widespread dissemination throughout the union membership," but found the procedure for adoption of the ENA to be legal. The contract reached under the ENA provided for a 35-percent increase over three years without recourse to arbitration. Abel later lashed out against his opponents at the USWA convention in September 1974, calling them "self-appointed, self-anointed saviours ... who are trying to invade us." As a member of the National Commission for Industrial Peace that year along with U.S. Steel's Larry, he expressed concern that excessive union democracy had interfered with responsible leadership and effective bargaining.

The September 1976 union convention was the last over which Abel presided, since a USWA mandatory retirement rule prevented him from seeking another term of office. He delivered a keynote speech on the goal of "lifetime security" for steelworkers, a comprehensive guarantee of either employment or income in the face of economic change. In part this proposal was intended to assist the candidacy of Lloyd McBride by diverting attention from the militant campaign led by Ed Sadlowski, a critic of Abel's. He had threatened to resign in the event of a Sadlowski victory, but this was made unnecessary by McBride's election in February.

Abel's last accomplishment as USWA president was the negotiation of a steel contract in April 1977. The main feature of the agreement, which provided a 30-percent increase in compensation over three years, was an Employment and Income Security Program. The plan coordinated and extended the protection of earnings and benefits, and provided for early retirement.

Upon his retirement, Abel moved to Sun City, Arizona, where he continued his activism. In 1979 he founded the Union Club. For former union members, this organization of over 1,400 lobbied for and concerned itself with the needs of the elderly. Responding to neighbors' initial misgivings about this venture, he wryly noted, "Old folks are not supposed to think, just to grow old." His wife, Bernice N. Joseph, whom he had married in 1930, died in 1982, and he was married a second time to Martha L. Turvey. In Saint Louis for the funeral of McBride in November 1983, Abel supported Frank McKee as interim president among executive board members. He backed Lynn Williams in the USWA special election in 1984, however. In an interview given in the last years of his life, Abel interpreted unfavorable public attitudes toward labor as a result of not remembering the early years of struggle and sacrifice, and of the younger generation's belief in the generosity of employers. He died of cancer on August 10, 1987, in Malvern, Ohio.

Publications:
"Man Steel Is Watching," *Business Week* (March 27, 1965): 48;
"Crisis Bargaining Admitted, but System Needed," *Steel,* 157 (December 6, 1965): 43–44;
"Contemplates Stock Options, Salaries, Security," *Steel,* 158 (May 23, 1966): 43;
Labor's Role in Building a Better Society (Austin, Tex.: Bureau of Business Research, 1972);
Our Future Is at Stake (United Steelworkers of America, 1973);

"Basic Steel's Experimental Negotiating Agreement," *Monthly Labor Review,* 96 (September 1973): 39–42;

Collective Bargaining — Labor Relations in Steel: Then and Now (New York: Columbia University Press, 1976);

"Comment," in *Forging a Union of Steel: Philip Murray, SWOC, and the United Steelworkers,* edited by Paul F. Clark, Peter Gottlieb, and Donald Kennedy (Ithaca, N.Y.: ILR Press, 1987), pp. 103–107.

References:

George Bogdanich, "Never a Kind Word for Abel," *Nation,* 216 (May 7, 1973): 591–594;

Bogdanich, "Steel: No-Strike and Other Deals," *Nation,* 219 (September 7, 1974); 171–174;

James A. Craft, "The ENA, Consent Decrees, and Cooperation in Steel Labor Relations: A Critical Appraisal," *Labor Law Journal,* 27 (October 1976): 633–640;

Joseph C. Goulden, *Meany* (New York: Atheneum, 1972);

John Herling, *Right to Challenge: People and Power in the Steelworkers Union* (New York: Harper and Row, 1972);

John P. Hoerr, "The Steel Experiment," *Atlantic Monthly,* 231 (December 1973): 18–34;

Hoerr, *And the Wolf Finally Came: The Decline of the American Steel Industry* (Pittsburgh: University of Pittsburgh Press, 1988);

David J. McDonald, *Union Man* (New York: Dutton, 1969);

George J. McManus, *The Inside Story of Steel Wages and Prices, 1959–1967* (Philadelphia: Chilton, 1967);

Jack Stieber, "Steel," in Gerald G. Somers, ed., *Collective Bargaining: Contemporary American Experience* (Madison, WI: Industrial Relations Research Association, 1980): 151–208;

Lloyd Ulman, *The Government of the Steel Workers' Union* (New York: Wiley, 1962).

Archives:

The records of USWA District 27, Canton, Ohio, from 1938 to 1972 are found in the Pennsylvania State University Libraries, University Park. The research notes and manuscripts for John Herling's book, *Right to Challenge,* are included in the John Herling collection at the Pennsylvania State University Library and are particularly useful to steel-labor studies. The papers of John Rooney, an assistant to Abel from 1958 to 1975, are in the Ohio Historical Society collections in Columbus, Ohio.

Avery C. Adams

(December 15, 1897 – December 11, 1963)

by John A. Heitmann

University of Dayton

CAREER: Laborer, assistant manager of sales, Trumbell Steel Company (1923–1928); sales manager, Republic Steel Company (1928); vice-president of sales, General Fire-Proofing Company (1928–1937); sales manager, Carnegie-Illinois Steel Company (1937–1938); vice-president, sales, Inland Steel Company (1939); vice-president, sales, United States Steel Corporation (1939); president, Pittsburgh Steel Company, Ltd. (1950–1956); president (1956–1958), chairman and chief executive officer, Jones & Laughlin Steel Company (1958–1963).

Avery C. Adams, who served as president, chairman, and chief executive officer of the Jones & Laughlin Steel Company (J&L) during the late 1950s and early 1960s, was born December 15, 1897, in Youngstown, Ohio. After graduating from Choate School, serving in the navy during World War I, and attending Yale for one year in 1920, Adams started his long career in the steel business in 1923 with the Trumbell Steel Company of Warren, Ohio. During the next 30 years Adams worked for a host of different employers, causing him to remark later in life that "Rolling stones gather no moss. I have concluded that I learned something different from every company I was with." Adams's successful career and his frequent job changes were, nevertheless, atypical in a tradition-bound industry that usually rewarded those with long service within a single organization.

Adams rose from laborer to assistant manager of sales at Trumbell, and in 1928 he briefly became sales manager at the Republic Steel Company before he moved to the General Fire-Proofing Company where he became vice-president of sales. During the late 1930s Adams held top sales positions with Inland Steel Company, United States Steel Corporation, and its subsidiary, Carnegie-Illinois Steel Company, but he made his big mark in the industry when he assumed the presidency of the Pittsburgh Steel Company, Ltd., in 1950. By 1956 the firm's capacity and sales had virtually doubled, and Adams seemed to be the logical choice to replace the retiring C. L. Austin at J&L.

During the late 1940s and early 1950s Austin and Ben Moreell had transformed J&L, replacing obsolete equipment with modern production facilities and diversifying its product line. Between 1947 and 1956 J&L added 1.3 million ingot tons to its capacity and was in the process of adding another 700,000 tons when Adams signed on. Furthermore, new products in hot-rolled, sheet and strip, and cold-finished steel along with wire and tubular goods were now being made. It was intended for Adams, widely known to friends and colleagues as Ave, to sell the goods that now could be turned out in the revamped J&L plants.

Yet J&L's performance during the Adams years (1956–1963) proved to be lackluster at best. From a net income high of more than $50 million in 1955, income slipped to a low of $23.2 million in 1958, with several subpar years thereafter. Several major capital improvements took place during the period, but the firm never fully exploited its early advantage in basic oxygen steelmaking. Indeed, rather than sustaining the technological momentum created by Ben Moreell during the late 1940s and 1950s, Adams emphasized increased sales, even though he was well aware of the inability of American steel to penetrate foreign markets and compete against the rising volume of cheap steel made overseas entering the United States.

Particularly after 1960 Adams became increasingly wary of federal government intervention in the steel business related to price and labor policies. During one crisis he coined the word "mythomaniacs" to describe Washington bureaucrats who exaggerated steel profits to justify arguments supporting government seizure of the industry. In 1961 Adams sharply criticized the Kennedy administration for its determination to hold the line on the steel industry. He wrote Kennedy that the freezing of steel prices was "a major step toward price-fixing by Government decree and constitutes [a] dangerous impairment of our free economy."

Avery C. Adams was a breed of super salesman rooted in the American steel industry of the pre–World War II period. *Iron Age* commented in 1950 that "One thing Ave knows is steel. He has made friends all over the place and he can properly be called an up-from-the-bottom man in an industry he loves and serves." But with the rise of new global challenges a different type of leader would be called upon.

Publications:

"Mills Use More Hot Metal in Steel Making," *Iron Age,* 180 (November 21, 1957): 88–89;

"Push-button Steel: Hot Strip Steel Rolled by Full Automatic Control," *ISA Journal,* 155 (February 1958): 66–67;

"Steelmaker Looks at Products and Marketing; Interview with A. C. Adams," *Iron Age,* 186 (November 24, 1960): 53.

References:

William T. Hogan, S.J., *An Economic History of the Iron and Steel Industry in the United States,* 5 volumes (Lexington, Mass.: Lexington Books, 1971);

"Iron Age Salutes Avery C. Adams," *Iron Age,* 165 (March 23, 1950): 42.

The Allegheny-Ludlum Steel Corporation

by Michael Santos
Lynchburg College

The Allegheny-Ludlum Steel Corporation was the result of the 1938 merger of two innovative specialty steel producers, the Allegheny Steel Company and the Ludlum Steel Company. Both firms had been leaders in the field of specialty steel for years at the time of the merger, largely because of innovative management and attention to research and development. These characteristics became the hallmarks of the newly created Allegheny-Ludlum Steel Corporation, and in no small measure account for the company's success in the decades after its formation.

The Ludlum Steel Company began in 1854 when James Horner acquired the Pompton Furnace and Foundry. Four years later James Ludlum formed a partnership with Horner, and in 1864 the name of the company was changed to Horner and Ludlum. The firm's only plant was located in Pompton, New Jersey, and from 1864 until 1906 manufactured high carbon steels for springs, files, and cutting tools.

In 1906 Edwin Corning became involved in the company through family investments, and under his leadership the firm began a period of rapid growth. The Pompton facility closed and its equipment was moved to Watervliet, New York. Becoming president of the company in 1910, Corning discontinued the use of crucibles to melt steel and installed electric furnaces that had been designed by company engineers. The company changed its name to Ludlum Steel Company in 1915, and in the subsequent ten years expanded its product lines to include stainless steels, silcrome valve steels for automobile and airplane engines, and nitralloy. In 1929 the company acquired the Atlas Steel Corporation, a producer of tool and alloyed steels with a plant at Dunkirk, New York. With production of quality steels being its main objective, the Ludlum Steel Company designed its own melting furnaces and established a system of control to assure accurate and uniform product composition.

Similarly, the Allegheny Steel Company began as a family enterprise and grew quickly under an aggressive management team that emphasized product quality. Like many iron and steel entrepreneurs of the late nineteenth and early twentieth centuries, the firm's founder, Henry E. Sheldon, epitomized the American Dream. Born in 1861, Sheldon was orphaned at the age of nine and grew up in the Episcopal Home for Boys in Pittsburgh. At fourteen he became a machine shop apprentice and at nineteen began work with Kirkpatrick and Company, a steelmaker in Leechburg, Pennsylvania. Within seven years of his employment at Kirkpatrick, Sheldon had risen to plant superintendent. At the age of twenty-eight he married the daughter of Alfred Hicks, who later became his business partner. In 1901 Hicks and Sheldon founded the Allegheny Steel and Iron Company, which was reorganized as the Allegheny Steel Company four years later.

Under Sheldon's leadership the firm grew dramatically. In 1901 it consisted of one open-hearth furnace, one bar mill, and four sheet mills, all capitalized at $300,000. Through a series of mergers and expansions Allegheny Steel's product line, facilities, and tonnage output grew. Between 1905 and 1935 the firm acquired the Interstate Steel Company, the Reliance Tube Company, the Western Tool and Forge Company, the Delaware Seamless Tube Company, the West Penn Steel Company, the Lamination Stamping Company, and the West Leechburg Steel Company. In the same period it pioneered in the development of specialty steels, invested heavily in research, and marketed its goods by emphasizing quality. Allegheny Steel was the first company in the United States to make stainless steel, and the firm broke new ground with the manufacture of alloyed steel, in particular chrome and silicon steels.

After the 1938 merger, which made the company the largest producer of stainless and specialty steels in the world, Allegheny-Ludlum continued to emphasize research. In 1939 and early

1940 the company announced the development of three new product lines that the industry hailed as significant breakthroughs in alloy metallurgy. The first, D.B.L. steel, was a new tool steel of the tungsten-molybdenum type; the second, Anodic Lustre, produced metal with a more brilliant finish than was possible using mechanical means, a fact particularly significant for parts that were impossible to finish by grinding and polishing. The third and most significant breakthrough was Pluramelt steel, a product that intermelted two or more metals of different composition so that they formed a continuous inseparable structure. These and other developments allowed Allegheny-Ludlum to take advantage of the increased demand for alloy steel during World War II. From the manufacturing of airplane engines to armor plating, Allegheny Metal met the production needs and became an important part of the U.S. war effort.

Continued innovation and expansion characterized Allegheny-Ludlum after the war. Concerned with the costs of labor disputes, Allegheny-Ludlum executives carefully tied company growth to unique programs in labor relations that sought to increase employee understanding of the workings of the corporation and "make them feel they belong." The first of these programs was begun in 1947. Disturbed by the results of a survey taken during the 1946 steel strike that showed two-fifths of the steelworkers and one-fifth of the public blaming management for the walkout, Allegheny-Ludlum officials launched a community relations program based on four principles: "(1) We have a genuine concern for the welfare of our employees and the communities in which they live; (2) Our wages are good; (3) Our profits are fair; (4) Continuous price increases which continuous wage increases might make necessary would endanger the security of the company, its employees, their communities, and the nation."

The concept behind the program was simple — improved communication among labor, management, and community would improve relations among the three and benefit each. The program was implemented in three phases: during the first phase, plant managers called meetings of their supervisory staff and gave them all the information they needed to take the message to the men under them. Programs were set up for the benefit of business leaders, union officials, clergy, and other community leaders at which Hiland Garfield Batcheller, the Allegheny-Ludlum president, and his entire staff presented the facts and figures

behind running the corporation. In the second phase, a special report for employees and friends was distributed in which "Al," a cartoon figure patterned after the company's star trademark, explained the firm's experience over the last year and its prospects for the immediate future. The third phase of the program consisted of a series of open-house visits for families and friends of Allegheny-Ludlum employees.

Despite such efforts, the firm experienced a bitter strike in 1952 that kept Allegheny-Ludlum shut down weeks after basic steel had reached an agreement with the United Steelworkers Union. The issue centered around workforce changes caused by facility improvements. Rather than abandoning the emphasis on communication begun in 1947, however, both sides held that the key to improved relations was better dialogue. Under the leadership of company president Edward J. Hanley and other Allegheny-Ludlum executives, an annual one-day general meeting to allow both sides to discuss concerns was begun. The program eventually evolved into three-day semiannual meetings between local union leaders and company and plant officials that gave both sides a chance to deal with local plant issues as well as industrywide problems and achievements. With the cooperation of union officials, company executives kept Allegheny-Ludlum competitive and dealt successfully with difficult issues, including what modernization meant for job security. The improved dialogue between labor and management made Allegheny-Ludlum a model of personnel relations by the 1960s.

Viewing personnel management as one part of an overall strategy of effective corporate management, Allegheny-Ludlum executives continued to invest heavily in research and development. Between 1945 and 1954 the company spent $96 million to expand and modernize its production facilities. In 1957 it began using a punch-card control system to insure uniform quality at its Brackenridge, Pennsylvania, rolling mill. In 1962 Allegheny-Ludlum expanded its research and development center at Brackenridge, turning it into a small steel mill as part of an effort to cut the time between experimentation and implementation of new technology. In 1963 the company went international with the establishment of a plant in Belgium. Investing over $80 million between 1964 and 1969, it modernized existing facilities and built a new melting complex at Natrona, Pennsylvania, near its Brackenridge works. Using computer technology to calculate raw-materials input to the converters, company engineers made the Natrona plant a model of production efficiency.

Following a trend among large corporations in the 1960s, Allegheny-Ludlum began a pattern of diversification that moved it into the production of items as diverse as golf clubs, fishing tackle, and power lawn mowers. With a goal of deriving half its revenues from operations outside the production of specialty steel, the company acquired Jacobsen Manufacturing in 1969. By the 1970s it had changed its name from Allegheny-Ludlum Steel Corporation to Allegheny-Ludlum Industries to reflect its new corporate image. The name was later changed to Allegheny International to suggest its multinational character.

In 1980 senior management in the firm's steel division and Tippins Machinery Company acquired the steel subsidiary and launched Allegheny-Ludlum Steel Corporation as a private business unit. The new firm operated with almost total freedom from its parent firm. On December 28, 1986, the company acquired all the common stock held by Tippins for cash and subordinated notes, thus becoming Allegheny-Ludlum Corporation.

Such changes in corporate structure did little to change the company's management style or philosophy, however. As Richard P. Simmons, president of Allegheny-Ludlum Steel, noted in 1981, "If we have a labor problem or productivity difficulties, it is the fundamental responsibility of management to solve them ... At Allegheny-Ludlum, the execution of this philosophy is people and attitudes."

References:
"Allegheny-Ludlum Talks to Its Union," *Steel,* 162 (February 12, 1968): 71–76;
H. G. Batcheller, "Let's Make Workers Feel They Belong," *Factory Management and Maintenance,* 106 (February 1948): 90–93;
E. J. Hanley, "Allegheny Takes on New Look," *Iron Age,* 174 (October 7, 1954): 71;
Charles Longenecker, "Allegheny Steel Company: Makers of Alloy Steels," *Blast Furnace and Steel Plant,* 26 (January 1938): 84–90;
Longenecker, "Allegheny-Ludlum Steel Corporation: Eastern Plants," *Blast Furnace and Steel Plant,* 27 (January 1939): 83–92;
Joseph L. Mazel, "Flying Solo: Allegheny–Ludlum Goes it Alone . . . and Successfully at That," *33 Metal Producing* (February 1982): 64–65;
"A Very Special Kind of Steel Producer," *The Magazine of Wall Street,* 124 (August 2, 1969): 25–27;
"What Labor Doesn't Know Can Hurt," *Iron Age,* 197 (June 2, 1966): 52–54.

Alloy and Stainless Steels

by Geoffrey Tweedale

Manchester, England

Strictly speaking, all steels are alloys, since even "ordinary" steel is a mixture of carbon and iron. The term *alloy steel,* which is synonymous with "specialty" steel, however, usually defines those steels that contain elements other than carbon in sufficient proportion to modify substantially some of their useful properties. For example, stainless steel, the most widely used alloy steel, contains about 12 percent chromium with less than 1 percent carbon.

Alloy steels were introduced into American steelmaking in the late nineteenth century, following their discovery in Europe. Julius Baur of the Chrome Steel Company of New York pioneered the use of alloy steels in the United States in the construction of the Eads Bridge between Saint Louis, Missouri, and East Saint Louis, Illinois, in 1870. Generally, however, before the 1880s only crucible steel, an expensive alloy that was produced in limited tonnages, was available if special properties were required. In England in 1868 tungsten had been introduced into the composition of crucible tool steel to improve its cutting properties, but, again, this was a specialist product not widely manufactured in America in the nineteenth century.

The age of alloy steels is generally regarded as having been launched by the English metallurgist and steelmaker Sir Robert A. Hadfield (1858–1940), who discovered manganese steel in 1882. The work-hardening characteristics of this steel (the result of adding 12 percent manganese), useful in railway track work and digging equipment, were exploited by the Taylor Wharton Iron and Steel Company of New Jersey. Using Hadfield's license, this firm cast the first alloy steel in the United States in 1892. In the 1880s Hadfield's discovery of silicon steel, which had excellent electrical properties that proved useful in the construction of transformers, also alerted engineers and steelmakers to the excellent results that could be achieved by alloying steels.

In both Europe and America the period from about 1890 to 1914 was one of intense activity in steel manufacturing; major advances were made in various engineering, tool, and stainless steel alloys. Also during this period leading alloy steel producers in the United States emerged — such as the Crucible Steel Company of America, Jessop Steel Company, and the Allegheny Steel Company. In the 1890s at the Bethlehem Steel Works F. W. Taylor (1865–1915) and Maunsel White (1856–1912) discovered ways of modifying and heat-treating traditional tungsten tool steel so that its cutting performance was greatly improved. Their development of chromium-tungsten cutting alloys led to the introduction of high-speed steel, which revolutionized engineering practices in the early twentieth century. At the Paris Exhibition in 1900 Taylor and White demonstrated the cutting abilities of their tool steel, representing perhaps America's first major contribution to alloy metallurgy and signaling the growing maturity of its specialty steel industry.

Meanwhile, Elwood Haynes (1857–1925), a metallurgist who had developed "Stellite," had investigated additions of chromium to low-carbon steel. By about 1915 he was ready to patent a corrosion-resistant steel. This product, soon known as stainless steel, was developed by Haynes with the help of English steelmakers in Sheffield, who had independently discovered it and were already commercially exploiting the alloy. Haynes and the Sheffielders licensed the product in the interwar period through a consortium, the American Stainless Steel Company, which was composed of the Bethlehem, Carpenter, Crucible, Firth Sterling, and Midvale steel companies. Another useful alloy was nickel steel, which was used in bicycle parts in the 1890s and would later be used by the automobile industry. In 1904 European advances in vanadium steel were brought to the attention of Henry Ford, who experimented with it for automobile parts. Production of these alloys was intimately bound up with the development of new steelmaking techniques, especially the electric furnace, which was introduced in America in 1906.

After World War I the production and use of these alloy steels greatly accelerated. Intensive research produced the so-called 18-4-1 (18 percent tungsten, 4 percent chromium, and 1 percent vanadium, with carbon below 1 percent) variety of high-speed steel, which further increased machining speeds. Mass-produced austenitic stainless steel, based on a chromium composition that included nickel, was available. The new material was exploited in the production of railway cars and in the facing for such prominent structures as the Chrysler Building in New York. By the 1920s alloy steels were used in oil well machinery, tractor parts, garden tools, aircraft engines, and buildings; and in the 1920s alloys were vital for America's high-growth industries, particularly automobile manufacturing. In 1923 the American automobile industry consumed more than 90 percent of alloy steel output, with the average passenger car using 700 pounds of alloy steel.

The major discoveries in alloy steels had mostly been made prior to World War I, but thereafter many new alloying elements were discovered and incorporated into standard products. Molybdenum, for example, became an important constituent of tool steel, particularly in the United States. Other discoveries have included the benefits of small additions of boron to increase the hardenability of steel of low alloy content. The pace of alloy development accelerated after 1945, alongside the rapidly growing demands of industry. Jet engine design, for example, required a continuous development of superalloys based on nickel. Research and development within the industry have also produced low cost ferritic stainless steel, precipitation-hardening alloys, low-expansion, high-temperature steels for gas turbines, duplex stainless steels, and superferritic alloys. New techniques have been introduced for melting and producing alloy steels; argon-oxygen-decarburization, vacuum-oxygen decarburization, ladle refining, and continuous casting have boosted bulk production of high-grade alloy steels.

In the 1970s and 1980s the market for composite materials became more competitive, but alloy steel demand and production have continued to grow. Stainless steel, the leading sector of the U.S. alloy steel industry, achieved a growth rate of .8 percent per annum in the 1980s. The alloy steel sector has maintained its unique position in the American steel industry; in 1980 specialty steels accounted for only 2 percent of steel tonnage, but 10 percent of the sales volume, in the United States.

Amalgamated Association of Iron, Steel, and Tin Workers

by James R. Barrett
University of Illinois at Urbana-Champaign

Founded in Pittsburgh on August 3, 1876, as an amalgamation of several craft unions representing skilled workers in the iron and steel industry, the Amalgamated Association (AA) grew to a peak membership of more than 24,000 by 1891, at which point the AA was one of the most prominent unions in the United States. The AA was also influential in the American Federation of Labor (AFL), which was founded in 1886. But because the AA accepted only skilled men, technological innovation threatened the union's status in the industry. As unskilled immigrants poured into the steel mills, management sought greater control over the production process and hiring policies.

The decisive confrontation between AA and management came in 1892 at the Carnegie Steel Company in Homestead, Pennsylvania. The strike turned violent, as first Pinkerton detectives and then troops were brought in to crush it. The AA's defeat ushered in a long period of decline for organized labor in the steel industry. By 1894 the union's membership had fallen to less than 10,000. Efforts to revive the union dimmed with its decisive defeat in a 1901 strike against the new giant in the industry, United States Steel Corporation. U.S. Steel wiped out the last vestiges of union organization in its plants when it declared an open shop in 1909, a position the firm upheld until 1937.

The AA's prospects flourished briefly during World War I when the union cooperated with other labor organizations interested in organizing the industry. The National Committee for Organizing Iron and Steel Workers coordinated the campaign and made good progress, particularly among the unskilled immigrant workers, until it too was defeated in a national steel strike in 1919.

While the AA's main obstacle may have been the intransigence of the steel employers, its own officials' conservatism and myopia also retarded efforts to unionize the industry. The advent of the National Recovery Administration and a friendlier political climate offered the AA prospects for success in 1934. However, the AA's president, Michael Tighe, hobbled AFL efforts in his refusal to waive his organization's exclusive jurisdictional rights over the industry's workers and in his demand for complete control over any organizing campaign. His was a narrow craft perspective which had outlived its use in an industry relying on mass production.

In the meantime the new Congress of Industrial Organizations (CIO) mapped out a plan to create one large union to represent all workers in the industry. A strong rank-and-file movement within the AA forced its leadership to join the CIO in spring 1936. The new federation chartered the Steel Workers Organizing Committee (SWOC), which led a vigorous and ultimately successful campaign to organize labor. In 1942 the AA surrendered its charter and was formally replaced by the United Steel Workers of America.

References:

Irving Bernstein, *The Turbulent Years: A History of the American Worker, 1933–1941* (Boston: Houghton Mifflin, 1971);

David Brody, *Steelworkers in America, the Nonunion Era* (Cambridge, Mass.: Harvard University Press, 1960);

Paul Krause, *The Battle for Homestead 1880–1892* (Pittsburgh: University of Pittsburgh Press, 1992);

Jesse S. Robinson, *The Amalgamated Association of Iron, Steel, and Tin Workers* (Baltimore: Johns Hopkins University Press, 1920).

American Iron and Steel Institute
by Donald F. Barnett

McLean, Virginia

The American Iron and Steel Institute (AISI) was established by United States Steel Corporation head Judge Elbert H. Gary in 1908. The institute incorporated earlier iron-making organizations — including the American Iron Association — dating back to 1855. The AISI was established to help promote steel interests, to collect and provide statistics on the industry (the Annual Statistical Report has been published since 1856), and to make possible the interchange of technical information among steelmen.

The AISI is a nonprofit organization representing firms engaged in ore mining and production, raw-steel production, and the finishing of basic steel products. Membership is open to all steel producers in the Western Hemisphere. AISI currently represents firms which account for more than 80 percent of raw-steel production in the United States.

AISI policies are set by a board of directors comprising the chief executive officers of member firms. The AISI president manages the institute, and AISI activities are carried out by committees whose members are experts in their respective areas.

The AISI has over the years collected and published statistics on the American steel industry; coordinated and funded research of new technologies and products; promoted steel use and developed steel standards; encouraged plant safety; and addressed issues dealing with the environment, energy, traffic, critical materials, employee relations, and health issues. The institute is also a lobbying force concerned with trade issues, taxes, government regulations, and other policy-making matters vital to the industry. As a public-relations organ, AISI has acted to improve the steel industry's image.

In its early days the AISI was also the focal point of industry efforts to coordinate steel pricing and to deal with the government on matters affecting pricing, labor relations, and so on. During World War I the AISI was the main coordinating agency for steel production, allocation, and pricing, as a committee chaired by Judge Gary worked with government officials to meet wartime demands. Similarly, the AISI occupied a central position in the efforts to establish and administer industrial codes during the New Deal. The trade association's officers frequently spoke for the industry, especially during periods of conflict with government, as in the late 1930s, the late 1940s, and the early 1950s. In the past AISI was often involved in hearings and legal action with the government that were related to matters of pricing, labor relations, or antitrust concerns. In recent years, however, AISI has avoided these sensitive issues and has concentrated much of its focus on trade and environmental issues.

AISI no longer represents the entire steel industry, with other trade associations representing minimills, specialty steel producers, and other special interests in the industry. AISI remains, however, the key American steel industry representative.

American Metal Market

by Bruce E. Seely

Michigan Technological University

American Metal Market is a daily publication that appears in tabloid format and is devoted to coverage of business aspects of the metalworking industries. It began publication in New York in 1882 as a weekly paper, launching daily publication (except Sunday and holidays) in 1899. As a daily, the paper offered the industry current reportage that monthly and weekly trade journals lacked. In 1901 the paper's name was changed to *American Metal Market and Daily Metal Market Report,* published Monday through Friday. The American Metal Market Company published the paper until 1925, billing it as "A complete daily report of the metal, iron, steel, ferro alloy, coal, and coke markets based on actual transactions." In 1926 Fairchild Publications acquired the paper and shortened the name to the original title, *American Metal Market,* and shifted publication to Tuesday through Saturday. This schedule remained in place until 1972,

when Fairchild absorbed *Metalworking News.* A Monday edition, entitled *American Metal Market Metalworking News Edition,* was launched to cover the metalworking industry, and in 1974 the Saturday edition was dropped. In 1984 the title of the Monday edition was changed to *AMM,* but in 1987 *Metalworking News* reappeared as an independent journal. Fairchild Publications continued to publish a regular Monday edition of *American Metal Market;* the paper has since appeared daily except on Saturday and Sunday. Through these many changes, this tabloid has provided some of the most thorough coverage of any trade publication in the iron and steel industry.

Reference:

David P. Forsyth, *The Business Press in America: 1750–1865* (Philadelphia & New York: Chilton, 1964).

American Steel and Wire Company

by Stephen H. Cutcliffe

Lehigh University

The American Steel and Wire Company, incorporated in New Jersey on January 13, 1899, was the third and largest consolidation of independent wire producers during the 1890s, an era in which there was a broad national trend toward mergers and the creation of large, vertically integrated trusts. Although enjoying only three years of independent existence prior to its acquisition by the United States Steel Corporation in 1901, the American Steel and Wire Company enjoyed a vir-

tual monopoly in the wire market and earned extensive profits for its shareholders.

With the development of barbed wire in the mid 1870s, the wire industry grew rapidly. Over 150 companies — jockeying for position and profit — slashed prices and undercut pooling arrangements, which resulted in an unstable business environment. In 1892 John W. Gates sought to enhance his competitive position and create a measure of stability in the industry by forming a

An American Steel & Wire ore unloader on the Cuyahoga River in Cleveland, Ohio (drawing by Jack Zajac, 1959)

$4 million holding company, the Consolidated Steel and Wire Company. Consisting of two wire mills and three barbed-wire manufacturers, the new firm was able to control approximately a quarter of the nation's barbed-wire output, as well as a wide range of other wire and nail products, all produced from its own steel rods.

Although the onset of depression in 1893 forced numerous small competitors from the field, prices continued to be depressed because of excess capacity, leading Gates to seek further consolidation. With the assistance of several partners and Elbert H. Gary as legal counsel, Gates approached the Morgan syndicate for enough financial backing to capitalize a $40 million trust of 14 major wire producers, many of which, like Consolidated, were already combinations of smaller companies. Although Morgan initially seemed responsive to the idea, his support wavered with the threat of war occasioned by the sinking of the USS *Maine* in February 1898. Morgan suggested putting the proposal on hold, but upon this suggestion several of the key participants refused to extend their options, and the deal collapsed.

Anxious to stabilize the industry, even if only to a limited degree, Gates and several of his midwestern producers agreed to form the American Steel and Wire Company of Illinois, with Gates as chairman of the board of directors. The new company was incorporated in April 1898 and capitalized at $24 million. The firm's 14 plants now produced approximately 75 percent of American wire products.

Still not satisfied with the extent of his control, and somewhat miffed at Morgan's reluctance to support his earlier plan, Gates turned to the banking firm of J. & S. Seligman for financial assistance in organizing a final combination capitalized at $90 million. American Steel and Wire Company of New Jersey absorbed the properties of its earlier namesake along with some 20 other important wire producing firms. When finally organized in January 1899, the new company controlled virtually the entire industry. Its vertical integration entailed Mesabi Range iron ore deposits, West Virginia coal deposits, steelmaking and wire producing mills, and manufacturing units for producing all types of wire fencing, steel rods, and drawn wire, as well as nails.

Gates's original vision was now complete. Although his closing of plants in 1900, putting 10,000 workers out of work, raised questions about his financial manipulations, the profitability of American Steel and Wire could not be disputed. During its approximately three years of independent existence it paid out over $9 million in dividends. This earning potential made it a more than desirable addition to U.S. Steel, into which it was consolidated in 1901.

References:
William T. Hogan, S.J., *An Economic History of the Iron and Steel Industry in the United States*, 5 volumes (Lexington, Mass.: Lexington Books, 1971);

Joseph McFadden, "Monopoly in Barbed Wire: The Formation of the American Steel and Wire Company," *Business History Review*, 52 (Winter 1978): 465–489;
Ida M. Tarbell, *The Life of Elbert H. Gary: A Story of Steel* (New York: D. Appleton, 1925).

Arnold Adams Archibald

(September 16, 1905 – June 24, 1980)

by John A. Heitmann

University of Dayton

CAREER: Metallurgical assistant, assistant works metallurgist, metallurgical engineer, salesman, assistant to vice-president (1935–1954), vice-president (1954–1965), administrative vice-president, Jones & Laughlin Steel Corporation (1965–1970).

The career of Arnold Adams Archibald is reflective of the twentieth-century American steel industry in transition. Once dominated by skillful businessmen with little or no technical training and by practical technical men who had worked their way up from the shop floor, the industry by the mid 1950s relied more on the leadership of university-trained experts like Archibald. This change was partly due to the growing realization within modern corporate circles that science and applied science could lead to increased profits.

Born in Nova Scotia, Canada, on September 16, 1905, to Lewis Edgar and Elizabeth May McCallum Archibald, Arnold Adams Archibald was educated at the Massachusetts Institute of Technology, where he received his B.S. degree in 1930. In 1935 he joined the Jones and Laughlin Steel Corporation (J&L) as a metallurgical assistant, and subsequently he was promoted to assistant works metallurgist, metallurgical engineer, and salesman. He served as assistant to the vice-president from 1935 to 1954.

Until the mid 1930s J&L was a technological follower rather than a leader within the American steel industry, but in 1936 the firm opened its first research facility, the Hazelwood Metallurgical Research Laboratory, where fundamental studies of the metallurgy of steel and process development were pursued. Indeed, J&L and many other American steel firms were slow to establish research and development laboratories, even though many electrical and chemical companies had recognized by the 1920s that research efforts could result in profit. The laboratory's first major breakthrough came in pioneering investigations related to the so-called Bessemer flame control process. The method used photoelectric cells and recording devices to monitor the reactions occurring in the Bessemer furnaces in order to scientifically determine when the steelmaking process was completed. This technique was licensed by J&L to other firms in 1941. As a consequence, no longer would the operator's judgment be absolutely crucial to the production of steel. Similar kinds of developments were concurrently taking place in other chemical and metallurgical process industries on the eve of World War II.

During the war years Archibald served as a dollar-a-year man with the War Production Board (WPB), and from 1946 to 1947 he was a member of the Civilian Production Administration. Upon returning to J&L, Archibald played an integral role in the firm's massive program of technological modernization. New equipment was installed, additional processes went on-stream, and modern plants outside of Pennsylvania were built, and as administrative vice-president of engineering, purchasing, and research Archibald directed the technical support necessary for this dramatic expansion. He oversaw the installation of new high-capacity furnaces to increase ingot production and the diversification of the product line to include cold-finished steel, stainless steel and wire, and tubular and tin-mill products. J&L's research staff was probing into new areas; for the highly competitive international industry in which J&L operated was maturing, and as a consequence opportunities to be taken advantage of were dwindling.

J&L's most noteworthy engineering achievement during Archibald's tenure as head of engineering and research was the successful installation of the basic oxygen process during the late 1950s. Offering rapid processing without sacrificing the quality of the product, the use of oxygen to eliminate silicon, phosphorous, manganese, and carbon from the steel was a technique first suggested by Bessemer in 1856 but not developed until the 1930s, after oxygen was made available in large quantities and at low cost. J&L was the first major American steel manufacturer to use this process, and, until the 1960s, its competitors did not follow suit.

References:
Douglas Alan Fisher, *The Epic of Steel* (New York: Harper & Row, 1963);
"Galvanizing Steel by a Continuous Process," *Business Week*, 1389 (April 14, 1956): 94–96;
William T. Hogan, S.J., *An Economic History of the Iron and Steel Industry in the United States*, 5 volumes (Lexington, Mass.: Lexington Books, 1971);
"Patent Fight Flaws Oxygen Steel," *Business Week*, 1476 (December 17, 1957): 186, 188.

Armco Incorporated

by Carl Becker

Wright State University

Incorporated in 1899, the American Rolling Mill Company (popularly known and later legally denominated as Armco) was an outgrowth of the Sagendorph Iron Roofing and Corrugating Company, a Cincinnati sheet metal plant on the verge of bankruptcy in the 1880s when George M. Verity became its manager. Verity revived the company and reorganized it in the 1890s as the American Steel Roofing Company and led it to a strong national position as a fabricator of iron and steel building products.

Nonetheless, Verity's company faced an uncertain future. Producers of flat-rolled and semifinished steel were merging into combines that put independent users of such steel — like Armco — at their mercy. Late in the 1890s Verity had taken the lead in an attempt to form a cooperatively owned association of independent fabricators to counter the rising power of the steel producers. Unable to organize the independent companies, Verity resolved to operate his own steel furnaces and rolling mills. Casting around for a location for his proposed plant, he received the offer of a site and a subvention of $75,000 from an industrial commission in Middletown, Ohio, a city of about 9,000 people 30 miles north of Cincinnati. He mustered the necessary capital of $500,000 and broke ground for construction of a mill at Middletown in 1900.

At the tapping of the first heat in 1901, the plant comprised a 30-ton open-hearth furnace, eight gas-fired heating furnaces, two 21-inch bar mills, four finishing mills, two cold mills, and a galvanizing department. It had the capacity for producing annually about 25,000 tons of bar and sheet products and employed around 200 men. Though unusual in that it integrated production from ingots to final steel products, Verity's enterprise seemed unpromising given conditions in the steel industry. Armco was a pygmy in a jungle in which giants held dominion. Even as operations began in Middletown, a gigantic merger of steel companies into the United States Steel Company was taking place; with its capital exceeding $1 billion and capacity of over three million tons annually, the new firm would stand as a "brooding omnipresence" on the landscape.

Smallness, though, had some inherent advantages for Armco. As a low-tonnage producer it could take up innovation that high-tonnage steel men might regard as niggling in its rewards. Precisely because it was small it had the agility and flexibility to move with celerity into new production — and engage in several distinct kinds of production. And it could use success in one market to move to both related and different markets, a strategy of output that would translate into company growth.

Indeed, from the outset Verity and his company achieved success and repute in the development of steels for a limited market. Armco's first such triumph came in the creation of electrical

This Armco Middletown Furnace was capable of producing 27,000 tons of molten steel per day.

steel sheets early in its history, in the opening decade of the twentieth century. The electrical industry, then in its infancy, had need for steel sheets possessing uniform metallurgical and magnetic properties for use with generators. Working closely with the Westinghouse Company, Armco metallurgists, notably Robert Carnahan, used trial-and-error methods to produce a low-sulphur, low-phosphorous steel sheet which would meet the requirements of the electrical men. With production entailing low tonnage and close quality control, Armco was well positioned to become a leader in the field. Later in the decade Armco metallurgists gave their company primacy in the production of corrosion-resistant steel for use as culverts. Through a series of laborious experiments, which caused much wear and tear on facilities, they succeeded in turning out Ingot Iron, which found markets throughout the world for culverts and wire fencing, and which became a metal for enameled sheets used in refrigerators and other household appliances in the United States. On the heels of that success, Armco men, in response to enhanced mechanization of production in the fledgling automobile industry, pioneered in the development of sheet steel with the finishing and drawing properties necessary for use by "bumping" machines that formed panels, fenders, and hoods for automobiles. Thus in little more than a

decade Armco established a niche in the industry through its speciality steels.

The company was succeeding in the marketplace too. By 1911 sales had reached $3.5 million, profits nearly $700,000. More than 1,700 employees produced nearly 100,000 tons of ingots per year. At about the same time in 1910 Armco was opening a new mill in Middletown — the East Works. Costing around $5 million, the new works, in contrast with the original mill, was able to employ "big mill" techniques in production, thus substantially reducing costs.

Contributing to success on the balance sheet was Armco's entrance into world trade. By 1912 the company was selling culvert steel in Argentina and Brazil. Agents in branch offices were soon selling Ingot Iron and other Armco products in, among other countries, Great Britain, France, Russia, Australia, and the Philippines. The company was also granting licenses to steel mills in Great Britain, Australia, and elsewhere to produce Armco's speciality steels. To direct sales abroad a special export department was formed in Middletown, which eventually became Armco International.

Though the coming of World War I gave Armco substantial profits, the war interrupted the company's course of production. First for the Allies and then for the United States, Armco turned

An Armco open-hearth plant

out 3.25 million shell forgings, as well as water jackets for Liberty motors, gun carriages, and other items. A forging shop became the centerpiece of activity, leaving the sheet mills idle. Sales, however, rose 400 percent from 1915 to 1918.

Well before the war ended Armco was looking forward to resumption of peacetime production. The company had long been concerned about the equilibrium of its production, especially access to pig iron, knowing that output and profits depended on a ready supply at stable prices. To reduce its vulnerability it acquired the blast furnaces, coal mines, and coke ovens of the Columbus Iron and Steel Company in 1917. It was one of a series of acquisitions aimed at keeping the various steps of production synchronous. Additionally, as postwar domestic orders for sheets used in a wide range of household appliances and orders from abroad for speciality steels accumulated, Armco installed new rolling mills at Middletown and at Zanesville, where it had purchased a small sheet mill in 1905.

Though necessarily attuned to market realities in acquiring and expanding facilities for production, Armco leaders — Verity and superintendents Charles Hook and John Butler Tytus — had a vision of a new mode of production, one unrelated to equilibration of furnaces and mills, that led the company to the purchase in 1921 of the Ashland Iron and Mining Company in Ashland, Kentucky. For a decade they had been discussing and infrequently experimenting with the development of the continuous hot-strip mill, which would render the laborious process of passing and repassing bars through manual mills obsolete. At Ashland they installed a mill, at a cost of about $10 million, that they resolved to make into a continuous mill. Directing the project, Tytus experimented one way after another and finally in 1924 was rolling out sheet steel on the first practical continuous mill in the history of steelmaking. It was a revolutionary innovation and gave Armco great prestige and profits — at least 26 companies installed continuous mills in the next 15 years under license from Armco — as well as enabled the steel industry to effect dramatic increases in productivity.

By the 1920s Armco leaders were also shaping innovative programs for their workers and home communities, notably the headquarters city of Middletown, where their company was the leading employer. In keeping with Verity and

Hook's belief that satisfied workers were more productive, Armco improved medical care in the mills, encouraged workers to organize committees to govern conditions affecting safety in the workplace, and voluntarily reduced the twelve hour shift to eight hours. The company sponsored workers' football, baseball, and basketball teams which played in inter- and intra-industrial leagues; aided the Armco Men's Association and the Armco Girls' Association, which organized various social, recreational, and cultural programs; and founded the Armco Band, an excellent industrial band under the direction of Frank Simon that played concerts throughout the Midwest. At Armco Park, a verdant expanse of rolling hills owned and maintained by the company, employees could gather for picnics, play softball and croquet, and pitch horseshoes. Children found there all kinds of playground equipment. With Verity in the vanguard, Armco executives were active in civic affairs, especially in ventures in Middletown intended to enlarge the public infrastructure — schools, libraries, and hospitals, for example. In no other community where Armco operated plants did the company exercise such influence over the common life. Not surprisingly, some citizens looked at Armco with jaundiced eyes, seeing it as a patriarchal force exerting excessive influence over the community.

As Armco reached a zenith in innovation, it entered a brief period of unparalleled growth. From 1925 to 1929 its sales rose from $34.3 million to $70.4 million, net earnings from $2.9 million to $6.1 million, and employment from 7,340 to 11,032. The company was also continuing its course of equilibrating and expanding its facilities. In 1927, needing more and better pig iron, it acquired several inoperative blast furnaces and coke ovens about 12 miles from Middletown and transformed them into a source for molten iron that was transported by rail to the East Works. Then in 1930 it purchased the Sheffield Steel Company in Kansas City, Missouri, thus gaining access to the market in the Southwest for bars, wire, and rods. That same year Verity relinquished the presidency that he had held for three decades, his tried and true friend Hook replacing him.

Though its record of growth and experienced leadership did not guarantee protection from the Great Depression, Armco weathered that crisis reasonably well. Production, payroll, and profits fell off appreciably early in the Depression, then rose to old levels by 1935, fell again during the re-

cession of 1938, and again moved upward late in the decade as a wartime economy took shape. Certainly Armco managers did not permit the fluctuating market to stay the course of innovation. Indeed, they could point to a record of achievement: the development of MULTI-PLATE, heavy corrugated sections bound together to serve as wide diameter culverts; production of the first stainless steel sheets by the continuous cold reduction method; the adaption of Tadeusz Sendzimir's revolutionary technique of applying zinc to steel sheets; the development of "Paintgrip," a phosphating process for making an effective paint-holding surface on galvanized sheets; and the creation of a special "non-aging" sheet steel with excellent ductility and drawing qualities for use on automobile bodies.

With the nation's entry into World War II in 1941, Armco again took up production of weaponry, notably light-gauge, face-hardened armor plate. Sales reached record levels, but earnings did not rise proportionately because of excess profits taxes, increased costs of raw materials and labor, and price controls. The war brought a kind of windfall to Armco in capital spending. With funds from the Defense Plant Corporation, Armco substantially expanded its Sheffield plant in Houston, Texas, which, of course, was engaged in defense production.

In the immediate postwar years Armco enjoyed continued growth. Production of ingots and castings rose from around 2.7 million tons in 1946 to about 3.9 million tons by 1950. From 1947 to 1950 net earnings jumped from around $25 million to $47 million, an 88 percent increase, while earnings in the entire industry grew 82 percent. The company was improving the quality of its speciality sheets and was enlarging its capacity at nearly all sites, particularly at the Sheffield plants in the West. It was giving increased attention to protection of its sources of raw materials by participating in cooperative ventures for access to ore deposits in the Mesabi Range and in Labrador — and engaging in various projects for the beneficiation of iron ore.

Through the 1950s Armco grew along rather traditional lines, with expansion of plants from Pennsylvania to Texas, diversification of fabricating products, and development of programs in access to and use of ore. It was also continuing to acquire other companies for purposes of diversification and equilibration of production. The most important acquisition came late in the decade in the purchase of the National Supply Company, a leading producer of drill rigs and other kinds of oil field equipment. By the end of the decade the company was able to report record-high sales of over $1 billion and net income of $77 million.

Not content to rest on its oars, Armco entered into new, ambitious programs in the 1960s. Early in the decade it was making a series of capital improvements, inter alia, at various plants — a blast furnace here, a plate mill there, and so forth. Midway through the decade the board approved a comprehensive plan, "Project 600," for capital spending, asserting that "the future will be served by those companies which anticipate it and take the steps to prepare themselves for it." Altogether, the project resulted in the expenditure of $800 million for improvements to or addition of foundries, electric furnaces, strip mills, and pollution equipment. Security for the future also came in a widening program of diversification. By various means — such as organization of new divisions, acquisitions of other companies, and cooperative ventures — Armco entered into production of super alloys and nonmetallic products, the leasing of industrial machinery and jet aircraft, and the sale of industrial insurance. It was a far cry from the production of 1901. At the same time, under the direction of a grandson of George Verity, the company was undertaking a vast program for the training of its managers.

As the 1960s came to an end Armco managers, shareholders, and employees could feel sanguine about the future as they viewed new programs and sales. In 1969, with sales of over $1.5 billion and profits exceeding $95 million, their company ranked third among all steel companies in the nation, surpassed only by U.S. Steel and Bethlehem Steel. Diversification had proved a powerful instrument, for in steel production it stood fifth in the nation.

Diversification and capital projects could not, however, protect Armco against the vagaries of the market or the insufficiencies of men. Like other integrated firms in the industry Armco had to work through a collapse in the market in the 1970s. As substantial increases in oil prices had a rippling effect, a downturn in automobile production ensued. Importation of steel was rising. Other debilitative forces were at play in the entire industry. Additionally at Armco, perhaps owing to slackness in management, the insurance business was falling at an alarming rate. Thus, sales and profits were disappointing early in the decade; but by 1977, as the economy generally improved — and buoyed by a wave of inflation — sales and

profits climbed to record levels of $3.5 billion and $120 million.

Despite the appearance of success and vitality Armco was heading toward dangerous and uncharted waters. After recording sales of nearly $7 billion and profits of around $300 million in 1981, it suffered a net loss of nearly $1.5 billion in 1982, 1983, and 1986; a recession, imported steel, and automation were all to blame. By resorting to restructuring and cost-cutting, the company was able by 1987 to recover a measure of what it had achieved in 1981, posting sales of $2.9 billion and profits of $117 million. In the process, it sold off 19 units or divisions, one for over $400 million. Employment from 1981 to 1988, affected in part by the sell-off, fell from 67,000 to 21,000. At Middletown, the symbolic hearth of the company, Armco was losing much of its economic and patriarchal sheet. Employment there declined from around 8,000 in 1981 to about 6,000 in 1988, with adverse consequences for retail establishments throughout the community; and the independent union, once a bastion of stability, was becoming more assertive, indeed bitter, as members saw their wages cut and the work force reduced — they now knew the meaning of the lyrics in a popular Bruce Springsteen song of the time: "foreman says these jobs are going, boys, and they ain't coming back, to your home town." In 1985 the company sent shock waves through the entire community when it moved the headquarters to Parsippany, New Jersey, purportedly to become more competitive with the "big boys." Three years later it gave the community another jolt, selling 40 percent of the Eastern Steel Division in Middletown to a Japanese firm, the Kawasaki Steel Corporation, which supposedly would infuse the division with the zeal and capacity for turning out higher-quality steel. All the while Armco was facing increasing criticism in the community for its alleged failure to control the emission of pollutants over Middletown. Clearly, the old steel company of 1901 no longer existed. A new steel company was on the move — but where it was going was not clear.

References:
Christy Borth, *True Steel: The Story of George Matthew Verity and His Associates* (Dayton, Ohio: Otterbein, 1941);
William T. Hogan, S.J., *An Economic History of the Iron and Steel Industry in the United States,* 5 volumes (Lexington, Mass.: Lexington Books, 1971);
Hogan, *The 1970s: Critical Years for Steel* (Lexington, Mass.: Lexington Books, 1972);

Hogan, *Steel in the United States: Restructuring to Compete* (Lexington, Mass.: Lexington Books, 1984);

Hogan, *World Steel in the 1980s: A Case of Survival* (Lexington, Mass.: Lexington Books, 1983);

R. C. Todd, Bennett Chapple, and W. D. Vorhis, comps., *Fifty Years at ARMCO: A Chronological History of the Armco Steel Corporation* (Middletown, Ohio, 1950).

Atlantic Steel Company

by Elizabeth C. Sholes

Industrial Research Associates

and

Thomas E. Leary

Industrial Research Associates

The Atlantic Steel Company is one of America's oldest and most successful minimills. Founded in 1901 by eight Atlanta businessmen, the company began as Atlanta Steel Hoop, manufacturing cotton ties and cooperage hoop from Pittsburgh steel. The company became Atlanta Steel in 1906, then incorporated in 1915 and adopted its current name.

During its first 50 years of operation, production was based on open-hearth ingots that fed the bar mill. In 1906 facilities included two 35-ton open hearths, a 25-inch blooming mill, a 14-inch billet mill, a 12-inch bar and rod mill, plus nail-making machinery. Despite modest beginnings, by 1935 the company was producing 70,200 tons of ingots per year and 100,000 tons by the end of the decade. The finished steel is still marketed under the distinctive brand "Dixisteel."

During the 1920s Atlantic Steel suffered a postwar downturn. Nevertheless, the company set out to expand production and cut costs. It electrified most processes, and subsequent savings in energy costs netted $1–1.25 per ton. Efficiency studies, inaugurated in 1915, continued to upgrade performance standards for equipment. These modifications, however, could not offset growing foreign competition in Atlantic's leading product lines, hoops and cotton ties. Even the company's forays into South American markets could not offset losses to foreign shipping. The Depression also hurt the company, but it recovered, in large part because the industry's Code of Fair Competition prevented ruinous price undercutting. Late in the 1930s Atlantic diversified, adding angles and purchased items to its inventory. The company was profitable, actually improving its market share and sales price, especially when the federal government removed Pittsburgh-base pricing after World War II. A further source of strength was the company's small debt.

By 1951 Atlantic had launched a $1.5-million program to install two Lectromelt furnaces and a combined bar and rod mill. By 1960 the new facilities completely supplanted the open hearths. In 1963 Atlantic added a $5-million Birdsboro blooming mill and soaking pits. Postwar growth was so steady that in the mid 1970s the company opened a second works in Cartersville, Georgia, with an electric furnace, continuous caster, and bar mill. The 1981 modernization of the original plant introduced a Morgan rod mill and a continuous caster. Atlantic Steel had by then all the facilities and characteristics of a minimill.

Atlantic's early success was in part due to a successful labor-relations program. Workers were given life insurance and disability coverage in 1925, and the comparatively small workforce made informal labor relations possible. In 1941, however, skilled workers went on strike, forcing Atlantic to sign with the steelworkers' union. Recent contracts, however, have included incentive plans with a negotiated base wage to keep overall labor costs down.

In 1979 Atlantic's profitability and solvency made it an attractive takeover target. It was acquired by Ivaco, a Montreal firm desiring a toe-hold in the U.S. steel market. Weathering the 1980s downturns, Atlantic Steel continues operation as a profitable electric steel plant.

References:
William T. Hogan, S. J., *A Comparison of Steelmaking in the United States* (Lexington, Mass.: Lexington Books, 1987);

Harry Richard Kuniansky, *A Business History of Atlantic Steel Company, 1901–1968* (New York: Arno, 1976);

M. P. Lawton, "A Southern Rolling Mill," *Blast Furnace and Steel Plant*, 12 (1924): 291–294, 305;

Robert S. Lynch, "What Steel Means to the South — and the South Means to Steel," *American Iron and Steel Institute Yearbook* (1956): 79–88.

James Bliss Austin

(July 16, 1904 –)

by Bruce E. Seely

Michigan Technological University

CAREER: Physical chemist (1928–1941), supervisory chemist (1941–1944), assistant laboratory director (1944–1946), director of research (1946–1956), vice-president of fundamental research (1956–1957), administrative vice-president, research and technology, United States Steel Corporation (1957–1968).

James B. Austin was a physical chemist who spent his entire career with United States Steel Corporation. Arriving when the corporation created its first research laboratory in 1928, he rose to the corporation's top research position in a career that spanned 40 years.

Austin was born in Washington, D.C., on July 16, 1904. He attended Lehigh University, and earned a degree in chemical engineering in 1925 before entering graduate school at Yale to study chemistry under the respected Dr. John Johnston. When Johnston went in 1928 to U.S. Steel to organize the corporation's research laboratory, he brought Austin, who had recently received his doctorate. Johnston gathered an impressive group of metallurgists in a corner of the headquarters of the Federal Shipbuilding and Dry Dock Company (a U.S. Steel subsidiary) in Kearny, New Jersey, where Austin would work alongside men such as E. C. Bain. Their primary task was to study the nature of steel, its structure, and its performance.

With its emphasis on fundamental research, the U.S. Steel laboratory afforded Austin an opportunity to pursue his interest in chemical thermodynamics. He also tackled questions relating to ferrous metallurgy, thermal expansion in metals

and refractories, the flow of heat in metals and refractories, and the absorption of gases on plane surfaces of metals. During his career he published 50 scholarly papers and was active in several professional organizations dedicated to the study of metals and ceramics.

In 1946 Johnston retired, and Austin became director of the laboratory. The promotion took Austin away from research and into administration and the U.S. Steel hierarchy. He became vice-president of fundamental research in 1956 and moved to corporate headquarters in Pittsburgh. In 1958 Austin replaced Bain as administrative vice-president for research and technology. Among his tasks was speaking for the corporation before scientific and technical audiences, an opportunity that Austin used to promote the importance of research to the steel industry.

But U.S. Steel was slow to adopt new steelmaking technologies — such as the basic oxygen process and continuous casting — that were about to transform the industry. Distrustful of innovation, U.S. Steel had never been a technological pacesetter, although it opened a new laboratory at Monroeville, Pennsylvania, in 1956. Under Austin this became perhaps the world's best fundamental steel research center. Yet U.S. Steel executives remained wedded to traditional steelmaking technology. Moreover, the enormous confidence generated by the success of U.S. Steel in the 1940s and 1950s fostered the belief that outside technologies were not worthwhile. This complacent and arrogant attitude complicated Austin's efforts in research. Even with three buildings dedicated to applied studies at Monroeville and a Chicago facility added later to study continuous casting, the company still lagged behind. Continuous casting was perfected at U.S. Steel in 1962, and the first oxygen furnace was installed at the Duquesne Works in 1964; but neither technique was enthusiastically embraced. Still, under Austin the research branch employed 1,800 people by 1969 and was responsible for some 1,300 projects. They had closed the gap in some areas, but U.S. Steel remained far behind the Japanese in the applied work that brought enormous improvements in productivity.

J. B. Austin retired in 1968, having earned wide recognition for his accomplishments from his peers. He had presented a special set of lectures at the American Society for Metals in 1941 and gave the 1946 Campbell Memorial Lecture of the American Society for Steel Treating and the 1952 Orton Lecture of the American Ceramic Society, of which he was an elected fellow. He was elected president of the American Society for Metals in 1954, fellow in 1969, and honorary member in 1975. He was also elected to the National Academy of Engineering, and Lehigh University awarded him an honorary doctorate in 1962. The British Iron and Steel Institute made him an honorary vice-president in 1961 and honorary member in 1964; he was also made an honorary member by the Metals Society of London and the Japanese Iron and Steel Institute. He received the Tawara Gold Medal from the Japanese organization in 1979. But Austin's most substantial professional affiliation was with the American Institute of Mining, Metallurgical, and Petroleum Engineers. He was voted a fellow in 1963, president in 1973, and honorary member in 1977. As president, he directed a study of the feasibility of merging with the American Society for Metals.

Away from U.S. Steel, James Austin exhibited the same inquiring mind and wide range of interests for which he was known in the laboratory. He collected Japanese prints, and the Carnegie Museum of Art in Pittsburgh made him an adviser on this subject. Austin was also a member of the Baker Street Irregulars, an organization of Sherlock Holmes enthusiasts. He published several spoofing papers in the pages of the *Baker Street Journal*, and his collection of Holmes material was featured in an exhibit on Sir Arthur Conan Doyle at the Lehigh University Library in May 1959.

Austin's wife, Janet Evans, earned a doctorate in chemistry at Yale in 1929. They married in 1930, and she pursued a career in chemistry as an abstractor for Chemical Abstracts and a researcher for Merck & Company. The couple also raised two children. Dr. Austin continues to reside in Pittsburgh.

Selected Publications:
"Equilibrium Pressure over Coexisting Salt Hydrates at Temperatures below 0°," *Journal of the American Chemical Society*, 50 (February 1928): 333–336;

and Ian Armstrong Black, "The Tesla-Luminescent Spectrum of Benzene," *Nature*, 125 (February 22, 1930): 274;

"Emission Spectrum of Benzene in the Region 2500–3000 A.," *Physical Review*, 35 (March 1, 1930): 152–160; also in *Journal of American Chemical Society*, 52 (November 1930); (December 1930);

"A Relation between the Molecular Weight and Melting Points of Organic Compounds," *Journal of the American Chemical Society*, 52 (March 1930): 1049–1053;

"Raman Effect in Liquefied Gases," *Nature*, 125 (March 22, 1930): 464;

"The Decomposition of Hydrocarbons in the Electrodeless Discharge," *Journal of the American Chemical Society*, 52 (July 1930): 3026–3027;

and Black, "The Chemical Behavior of Some Benzenoid Hydrocarbons in the Tesla Discharge," *Journal of the American Chemical Society*, 52 (November 1930): 4552–4557;

and Black, "The Emission Spectra of Some Simple Benzene Derivatives," *Journal of the American Chemical Society*, 52 (December 1930): 4755–4762;

"Temperature Distribution in Solid Bodies during Heating or Cooling," *Physics*, 1 (August 1931): 75–81;

"Thermal Expansion of Some Refractory Oxides," *Journal of American Ceramic Society*, 14 (1931): 795–810;

"Calculation of the Characteristic Frequency from the Coefficient of Compressibility," *Physical Review*, 38 (November 1, 1931): 1788–1790;

"The Heat Capacity of Some Hydrogen Halides at High Temperatures as Calculated from Raman Spectra," *Journal of the American Chemical Society*, 54 (August 1932): 3459–3460;

"Temperature Distribution in Solid Bodies during Heating or Cooling, A Correction," *Physics*, 3 (October 1932): 179–181;

"Heat Capacity of Iron, A Review," *Industrial and Engineering Chemistry*, 24 (November 1932): 1225–1235;

"Vacuum Apparatus for Measuring Thermal Expansion at Elevated Temperatures, with Measurements on Platinum, Gold, Magnesium, and Zinc," *Physics*, 3 (November 1932): 240–267;

"Entropy, Heat Content, and Free Energy of Iron," *Industrial and Engineering Chemistry*, 24 (December 1932): 1388–1391;

and R. H. H. Pierce, Jr., "Constitution and Thermal Expansion of Silica Coke-Oven Brick after Service," *Journal of the American Ceramic Society,* 16 (1933): 102–106;

and Pierce, "The Linear Thermal Expansion of a Single Crystal of Sodium Nitrate," *Journal of the American Chemical Society,* 55 (February 1933): 661–668;

"Thermodynamic Study of the Iron-Carbon Diagram: An Extended Abstract of a Paper by Korber and Odelson," *Metals & Alloys,* 4 (1933): 49–53;

and Pierce, "Thermal Expansion of Heat-Resisting Iron Alloys," *Industrial and Engineering Chemistry,* 25 (July 1933): 766–779;

and Pierce, "Linear Thermal Expansion and Alpha-Gamma Transformation Temperatures of Pure Iron," *Physics,* 4 (December 1933): 409–410;

and Pierce, "The Linear Thermal Expansion and Alpha-Gamma Transformation Temperatures (A3 Point) of Pure Iron," *Transactions of the American Society for Metals,* 22 (May 1934): 447–468;

and Pierce, "Linear Thermal Expansion and Transformation Phenomena of Some Low-Carbon Iron Chromium Alloys," *Transactions of the American Institute of Mining Engineers,* 116 (1935): 289–307;

"Dependence of Rate and Transformation of Austenite on Temperatures," *Transactions of the American Institute of Mining Engineers,* 116 (1935): 309–318;

and Pierce, "Determination of the Refractive Index of Vitreous Silica and the Calibration of Silica Refraction Thermometers Between 18° and -200° C.," *Physics,* 6 (January 1935): 43–46;

and R. B. Sosman, "An Apparatus for Measuring the Magnetic Susceptibility of Liquids and Solids at High Temperatures," *Journal of the Washington Academy of Sciences,* 25 (January 15, 1935): 15–32;

and Pierce, "Reliability of Measurements of Thermal Conductivity of Refractory Brick," *Journal of the American Ceramic Society,* 18 (1935): 48–54;

and Pierce, "Linear Thermal Expansion of Sodium Tungstate Between 20° and 600°," *Journal of Chemistry and Physics,* 3 (1935): 683–686;

"Physics and Chemistry of Firing Ceramic Ware," *Bulletin of the American Ceramic Society,* 14 (May 1935): 157–165;

"A Useful Integrated Form of the Equation for Calculating Change of Equilibrium with Temperature," *Journal of the American Chemical Society,* 57 (December 1935): 2428–2434;

"Use of Penetrating Radiations in the Measurement of the Porosity of Refractory Brick," *Journal of the American Ceramic Society,* 19 (1936): 29–36;

and Pierce, "Comparison of the Thermal Expansion of Used Silica Brick from an Insulated and an Uninsulated Open Hearth Furnace Roof," *Journal of the American Ceramic Society,* 19 (October 1936): 276–281;

and Pierce and Walter O. Lundberg, "Comparison of the Thermal Conductivity and Thermal Expansion of Coke-Oven Liners Made from Eastern and Western Quartzite," *Journal of the American Ceramic Society,* 20 (1937): 363–367;

"Linear-Thermal Expansion of 'Beta-Alumina,' " *Journal of the American Ceramic Society,* 21 (1938): 351–353;

Marion H. Armbruster, "The Absorption of Gases on Plane Surfaces of Mica," *Journal of the American Chemical Society,* 60 (February 1938): 467–475;

"Blast Furnace Operations," *Iron Age,* 141 (February 24, 1938): 30–31;

"Efficiency of the Blast Furnace," *Transactions of the American Institute of Mining Engineers,* 131 (1938): 74–99;

and R. L. Rickett, "Kinetics of Decomposition of Austenite at Constant Temperature," *Transactions of the American Institute of Mining Engineers,* 135 (1939): 396–416;

and Armbruster, "Absorption of Ethyl Iodide on a Plane Surface of Iron at 20°," *Journal of the American Chemical Society,* 61 (May 1939): 1117–1123;

"Methods of Representing Distribution of Particle Size," *Industrial and Engineering Chemistry, Analysis Edition,* 11 (June 15, 1939): 334–339;

"Insulation," *Metals Progress,* 36 (October 1939): 501–503;

"Factors Influencing Thermal Conductivity of Nonmetallic Materials," *A.S.T.M. Symposium on Thermal and Insulating Materials* (1939): 3–69;

and Pierce, H. Saini, and J. Weigle, "Direct Comparison on a Crystal of Calcite of the X-Ray and Optical Interferometer Methods of Determining Linear Thermal Expansion. Evidence of Differences among Calcite Crystals," *Physical Review,* 57 (May 15, 1940): 931–933;

and M. J. Day, "Heat Etching as a Means of Revealing Austenite Grain Size," *Transactions of the American Society for Metals,* 28 (June 1940): 354–371;

and D. S. Miller, "Magnetic Permeability of Some Austenitic Iron-Chromium-Nickel Alloys as Influenced by Heat Treatment and Cold Work, *Transactions of the American Society for Metals,* 28 (September 1940): 743–755;

and Day, "Chemical Equilibrium and the Control of Furnace Atmospheres," *Industrial and Engineering Chemistry,* 33 (January 1941): 23–31;

"An Equation Representing the Rate of Development of Rust on Galvanized Iron Sheets as Estimated by the A.S.T.M. Test," *Proceedings of the American Society for Testing Materials,* 41 (1941): 766–775;

and Day, "Chemical Equilibrium as Guide in Control of Furnace Atmospheres," *A.S.T.M. Symposium on Controlled Atmospheres* (1942): 20–56;

"Flow of Heat in Metals," Series of Five Educational Lectures Presented at the National Metals Congress in 1941 (Cleveland: American Society of Metals, 1942);

"Steelmaking Improvements Because of War Experiences," *Iron Age,* 155 (January 25, 1945); 118+;

"Introduction," Symposium on Determination of Hydrogen in Steel, *Transactions of the American Institute of Mining and Metallurgical Engineers,* 162 (1945): 353–354;

"Movies Record Blast Furnace Combustion Turbulence," *Steel,* 128 (February 12, 1951): 87;

"Significance of Equilibrium and Reaction Rate in the Blast Furnace Process," *Iron and Steel Institute Journal,* 167 (April 1951): 358–363;

"Research: Foundation of Steel's Progress Today," *Journal of the Franklin Institute,* 265 (May 1958): 385–394;

"The Role of Research: Better Processes and Better Products," in *Steel and Inflation: Fact vs. Fiction* (U.S. Steel, 1958), pp. 65–70;

"Three Centuries of Progress in the American Steel Industry," *Iron and Steel Institute Journal,* 189: (March 1962): 176–186;

"Steel Metallurgist, His Heritage and Future," *Journal of Metals,* 15 (August 1963): 560–562;

"Edgar Collins Bain: September 14, 1891 – November 27, 1971,"in *Biographical Memoirs,* volume 49 (Washington, D.C.: National Academy of Sciences, 1978).

References:

"A Biographical Appreciation," *Metals Progress,* 65 (February 1954): 82–84;

William T. Hogan, S.J., *An Economic History of the Iron and Steel Industry in the United States,* 5 volumes (Lexington, Mass.: Lexington Books, 1971), IV: 1669–1671;

"James Bliss Austin Heads U.S.S. Research Laboratory," *American Ceramic Society Bulletin,* 26 (January 15, 1947): 2–5;

Leonard Lynn, *How Japan Innovates: A Comparison with the U.S. in the Case of Oxygen Steelmaking* (Boulder, Colo.: Westview Press, 1982).

Edgar Collins Bain

(September 14, 1891 – November 28, 1971)

by Bruce E. Seely

Michigan Technological University

CAREER: Chemist, U.S. Bureau of Standards (1914); instructor of metallography, University of Wisconsin (1916–1917); chemical engineer, B. F. Goodrich Rubber Company (1917–1918); metallurgist, General Electric Company (1919–1923); research metallurgist, Atlas Steel Company (1923–1924); Union Carbide Research Laboratories (1924–1928); director, physical metallurgy research (1928–1935), assistant to vice-president of research and technology, United States Steel Corporation (1935–1943); vice-president in charge of research and technology, Carnegie-Illinois Steel Corporation (1943–1950); vice-president, research and technology (1950–1957), assistant executive vice-president of operations, U.S. Steel Corporation (1956–1957).

In the two decades prior to World War II, American metallurgists made impressive strides in furthering an understanding of iron and steel. A central figure during this creative era was Edgar Collins Bain, a pioneer in using X-ray diffraction techniques to advance the science of metallurgy.

Bain spent most of his career at the United States Steel Corporation.

E. C. Bain was born near Larue, Ohio, in 1891 and attended Wesleyan University and Ohio State University, where he received a degree in chemical engineering in 1912. In 1914 Bain took a position with the U.S. Geological Survey, switching quickly to the National Bureau of Standards, where discussions with staff scientists convinced him to return to graduate school; he earned his master's degree at Ohio State in 1916. He taught metallurgy at the University of Wisconsin for a year, but his real passion was research. A summer job at B. F. Goodrich turned into a position at the company's research lab, and Bain left the university in 1917.

During World War I Bain worked for the Chemical Warfare Services, where he continued his work started at Goodrich on gas masks. At war's end, however, Bain changed directions again, accepting a metallurgical research position in Cleveland at the National Lamp Works, a subsidiary of the General Electric Company. He worked under Zay Jeffries, the firm's consultant, taking high-speed photographs of tungsten filaments and studying the growth of grains in tool steels. In 1920 Jeffries suggested that Bain use X-ray diffraction apparatus to study metals, and Bain had found his field. He published his first papers in 1921 and pioneered the use of X rays to study crystal structures in metals.

Committed to metallurgy, Bain moved to the Atlas Steel Company in 1923, where he and Marcus Grossman studied tool steel, the function of chromium, and oil-hardened nondeforming tool steel. Financial troubles at Atlas in 1924 caused Bain to move to Union Carbide's Long Island laboratory, where he continued to study alloy steels. By 1928 he was recognized as one of the outstanding metallurgists in the country. In that year U.S. Steel was gathering the country's top metallurgists for its research laboratory, and Bain agreed to head a physical metallurgy unit at the lab in Kearny, New Jersey.

The steel corporation's research laboratory was organized along lines suggested by several leading industrial researchers. These consultants advocated conducting fundamental scientific investigations, while existing labs throughout the corporation dealt with more practical studies. In the long run this decentralization of steel research, as well as a suspicion of science among steel executives, isolated the new lab and limited its usefulness. But left in isolation Bain was free to follow his interests, and he concentrated his studies on how microstructures affect the quality of heat-treated steels. Heat-treating was a vital step in making very hard tool steels. Working with E. S. Davenport, Bain produced a classic paper on the subject, entitled "Transformation of Austenite at Constant Subcritical Temperatures." This research also demonstrated that grain size was a central factor in the hardenability of steel. Bain had opened the door for understanding alloy and heat-treated steels.

Professional recognition quickly followed. Bain received prizes from the American Institute of Mining and Metallurgical Engineers in 1929, the American Society for Treating Steel in 1931, and the American Iron and Steel Institute in 1935. In 1932 he had the honor of delivering both the Howe Memorial Lecture to the American Institute of Mining and Metallurgical Engineers and the Edward DeMille Campbell Memorial Lecture to the American Society for Steel Treating. He was elected president of the American Society of Metals in 1936, and in 1937 Ohio State presented Bain its Benjamin Lamme Gold Medal for eminence in engineering. Lehigh University awarded

him an honorary doctorate in engineering that same year.

U.S. Steel made Bain assistant to the vice-president for research and technology in 1935. Bain interviewed inventors who approached the corporation, advised the patent office, oversaw equipment requests and budgets, and served as technical adviser for a 37-minute film on steelmaking; ironically, the promotion ended Bain's career as a metallurgist, for he no longer spent time in the lab. In May 1938 the steel corporation reorganized its subsidiaries, and Bain moved to Pittsburgh, charged with coordinating research within the giant corporation.

World War II, however, absorbed more of Bain's time. In 1941 he headed an ad hoc committee in charge of developing a research program to improve materials used in military equipment. Soon he was on the National Research Council's War Metallurgy Committee, chaired by his old friend Zay Jeffries. He also served on the science advisory board of the Office of the Chief of Ordnance in the Army. Meanwhile, he directed U.S. Steel's efforts to find substitutes for scarce alloys — work that produced National Emergency (NE) steels.

In 1943 Bain became vice-president for technology and research at Carnegie-Illinois Steel Corporation, the largest operating branch of U.S. Steel. He was responsible for coordinating numerous research facilities as well as monitoring product quality. In 1950 U.S. Steel again reorganized its departments, placing all research programs under Bain's direction. Not surprisingly, in 1946 Bain chaired the first General Research Committee of the American Iron and Steel Institute. Despite U.S. Steel's hesitance to introduce new technologies, the company constructed a central research facility at Monroeville, Pennsylvania, in 1956, which included the Edgar C. Bain Laboratory for Fundamental Research.

In 1957 Bain retired from U.S. Steel. He had begun coordinating a research program that often suffered due to U.S. Steel's decentralization policies. Yet, during his tenure the company's research programs advanced the fundamental understanding of steel metallurgy. Bain's background in fundamental research, however, also may help account for the limited role U.S. Steel researchers played in developing two new technologies during the 1950s — the basic oxygen process and continuous casting. To be sure, Bain worked in a conservative corporate culture that disdained technological innovation. The central laboratory focused on questions at the microlevel while a number of small facilities throughout the corporation dealt with production questions. This structure offered limited resources and less guidance for dealing with the new postwar steel technologies. Bain at least built a central research facility for tackling both fundamental and practical questions.

After leaving U.S. Steel, Bain continued working on metallurgy questions. Despite a stroke, he consulted with H. W. Paxton and published a second revised edition of *Alloying Elements in Steel.* He also edited a fifth edition of his old friend Marcus Grossman's *Principles of Heat Treatment,* published a paper on Japanese swords based on his personal collection, and began his memoirs. He delivered the American Iron and Steel Institute's Schwab Memorial Lecture in 1952 and the Hatfield Memorial Lecture of the Iron and Steel Institute in London in 1955. He was elected to the National Academy of Science in 1954, two years after becoming the first American to receive the Grande Medaille of the Société Française de Métallurgie. He was also the first American to receive the Gold Medal of the Japan Institute of Metals, awarded him in 1964.

J. B. Austin, his longtime colleague at U.S. Steel, noted that Bain possessed boundless energy, an interest in almost any subject, and a love of puns, conversation, and music; he built his own stereo equipment, sang tenor in numerous choral performances, and played piano by ear. He derived deep satisfaction from working with his hands and had a special love of working with wood. He married Helen L. Cram in 1927, and they had a daughter and a son. Bain died on November 28, 1971, at his home in Edgeworth, Sewickley, Pennsylvania.

Selected Publications:

"Studies of Crystal Structure with X-rays," *Chemical and Metallurgical Engineering,* 25 (October 5, 1921): 657–664;

"What the X-ray Spectometer Tells Us about the Structure of Solid Solutions," *Chemical and Metallurgical Engineering,* 25 (October 5, 1921): 729;

and James R. Withrow, "The Relative Densities of Alkali-Metal Amalgams and Mercury," *Journal of Physical Chemistry,* 25 (October 1921): 535–544;

and Zay Jeffries, "Mixed Orientation Developed in Crystals of Ductile Metals by Plastic Deformation," *Chemical and Metallurgical Engineering,* 25 (October 26, 1921): 775–777;

"X-ray Data on Martensite Formed Spontaneously from Austenite," *Chemical and Metallurgical Engineering,* 26 (March 22, 1922): 543–545;

"Cored Crystals and Metallic Compounds," *Chemical and Metallurgical Engineering,* 28 (January 10, 1922): 65–69;

"The Nature of Solid Solutions," *Chemical and Metallurgical Engineering,* 28 (January 3, 1923): 21–24;

and M. A. Grossman, "On the Nature of High-Speed Steel," *Journal of the Iron and Steel Institute,* 110 (1924): 248–272;

"The Application of X-ray Crystal Analysis to Metallurgy," *Journal of Industrial and Engineering Chemistry,* 16 (July 1924): 692–698;

"The Nature of the Alloys of Iron and Chromium," *Transactions of the American Society for Steel Treating,* 9 (January 1926): 9–32;

and Grossman, "On the Nature of Some Low Tungstun Tool Steels," *Transactions of the American Society for Steel Treating,* 9 (February 1926): 259–270;

"Notes on the Atomic Behavior of Hardenable Copper Alloys," *Mining and Metallurgy,* 8 (February 1927): 151;

"X-rays and the Constituents of Stainless Steel," *Transactions of the American Society for Steel Treating,* 14 (July 1928): 27–50;

and Willis S. N. Waring, "Austenite Decomposition and Length Changes in Steel," *Transactions of the American Society for Steel Treating,* 15 (January 1929): 69–90;

and E. S. Davenport, "Transformation of Austenite at Constant Subcritical Temperatures," *Transactions of the American Institute of Mining and Metallurgical Engineers,* 90 (1930): 117–144;

"High Chromium Alloy Steels; Their Present Status," *Steel,* 87 (October 1930): 57–59;

and Grossman, *High Speed Steel* (New York: J. Wiley & Sons, 1931);

"Some Fundamental Characteristics of Stainless Steel," *Journal of the Society for Industrial Chemistry,* 51 (August 5, 1932): 662–667;

A Contribution from the United States Steel Corporation Research Laboratory on the Rates of Reactions in Solid Steel (New York: American Institute of Mechanical and Metallurgical Engineers, 1932);

"The Role of the Common Elements in Alloy Steels," *Iron Age,* 133 (April 5, 1934): 26, 82;

and Davenport and E. L. Roff, "Microscopic Cracks in Hardened Steel, Their Effects, and Elimination," *Transactions of the American Society of Metals,* 22 (April 4, 1934): 289–310;

and Davenport, "General Relations Between Grain-Size and Hardenability and the Normality of Steels," *Transactions of the American Society of Metals,* 22 (1934): 879–921;

"Some Characteristics Common to Carbon and Alloy Steels," *Steel,* 94 (May 28, 1934): 25–29, 36;

and Davenport, "The Aging of Steel," *Transactions of the American Society of Metals,* 23 (1935): 1047–1061;

and F. T. Llewellyn, "Low-Alloy Structural Steels," *Proceedings of the American Society of Civil Engineering,* 62 (October 1936): 1184–1200;

and J. B. Rutherford and R. H. Aborn, "The Relation of the Grain Areas on a Plane Section and the Grain Size of a Metal," *Metals and Alloys,* 8 (December 1937): 345–348;

and H. W. Paxton, *Alloying Elements in Steel* (Metals Park, Ohio: American Society of Metals, 1939);

Functions of the Alloying Elements in Steel (Cleveland: American Society of Metals, 1939);

"Iron and Steel: A Random Sampling of Ferrous Metallurgical Progress," *Mining and Metallurgy,* 22 (February 1941): 93–96;

"Conserving Alloying Elements Produces NE Emergency Steels," *Steel,* 112 (January 4, 1943): 265–266;

"Technical Trends in Steel Research," *Blast Furnace and Steel Plant,* 36 (October 1948): 1222–1225;

"Trends in Metallurgical Research in the United States," *Journal of the Iron and Steel Institute,* 181 (November 1955): 193–212;

"Nippon-to, an Introduction to Old Swords of Japan," *Iron and Steel Institute Journal,* 200 (April 1962): 265–282;

and Grossman, *Principles of Heat Treatment* (Metals Park, Ohio: American Society for Metals, 1964);

Phase Transformations and Related Phenomena in Steel (Metals Park, Ohio: American Society for Metals, 1973);

Pioneering in Steel Research: A Personal Record (Metals Park, Ohio: American Society for Metals, 1975).

References:

J. B. Austin, "Edgar Collins Bain," *Biographical Memoirs,* volume 49 (Washington, D.C.: National Academy of Sciences, 1978): 25–47;

"Distinguished Metallurgists," *Metal Progress,* 56 (September 1949): 338–339;

"Edgar Collins Bain," *Metal Progress,* 30 (December 1936): 46–49.

Basic-Oxygen Steelmaking Process
by Leonard H. Lynn

Case Western Reserve University

The Basic-Oxygen Steelmaking Process (BOSP) was first put into commercial operation by a small Austrian steelmaker in late 1952. Offering major construction and operating-cost advantages over the then-dominant open-hearth steelmaking process, the BOSP quickly became the technology of choice in new steelworks. By 1970 the BOSP was the most widely used steelmaking process in the world.

Steelmaking requires that carbon and other elements be removed from pig iron or scrap. The Bessemer and Thomas converter processes do this by blowing air from the bottom of a brick-lined steel shell through molten pig iron. The air reacts with elements in the pig iron to raise the temperature sufficiently to oxidize the carbon in the pig iron. These processes are cheap and efficient, but can only refine iron with sufficient silicon (in the case of the Bessemer process) or phosphorus (in the case of the Thomas process) for the necessary reactions to take place. Also, the steel they produce includes elements from the air that may not be desired. As a result the Thomas process was never widely used in the United States, and by 1950 the Bessemer process was responsible for

less than 5 percent of the total U.S. output of crude steel. The predominant steelmaking technology during the first half of the twentieth century was the open hearth, which accounted for some 90 percent of U.S. production by the early 1950s. Open hearths use heat from the burning of fuel and oxides from ore and scrap to refine steel. The process works slowly, so adjustments can be made in the metallurgy of the steel being produced. As a result the open hearth was the most flexible steelmaking process, able to refine a wide variety of irons and produce a wide variety of steels. Open hearths, however, were much more expensive to build than converters and also much more expensive to operate.

It had long been noted that if pure oxygen could be used in place of air in a steelmaking converter, the steelmaking process would be less demanding of the type of iron being refined and more able to produce steel with a tightly controlled metallurgy. Oxygen of sufficient quality was not available in large enough amounts until around 1930, and then it was not clear how the oxygen could be blown into the molten iron without damaging the containing vessels or the ports through which the oxygen was blown. There was also some uncertainty about how the oxygen would sufficiently mix with the iron to oxidize all of the carbon in it. Experiments were conducted in various parts of the world to develop an oxygen-blown steelmaking process, and in 1949 researchers in Switzerland finally succeeded. Like the Bessemer and Thomas processes, the BOSP makes use of a refractory brick-lined steel vessel filled with molten pig iron. Unlike the other two processes, however, the BOSP blows pure oxygen (instead of air) from above (rather than from below) the vessel and through the iron. Like the other two converter processes the BOSP equipment is relatively inexpensive to build and operate, for it does not require an apparatus to heat the iron, nor does it need fuel for heat. Unlike the Thomas and Bessemer processes it does not require iron ore that has a high silicon or phosphorous content to provide heat for a reaction. And since the BOSP uses pure oxygen it does not introduce nitrogen and other elements from air that may give the steel undesired properties.

In 1952 a then-small government-owned Austrian steelmaker, Voest, blew in the first commercial BOSP plant. Another government-owned Austrian steelmaker, Alpine, blew in another plant a year later. In 1954 a Canadian steelmaker, Dofasco, and a small U.S. steelmaker, McLouth, also put plants into operation. These early BOSP plants were small compared to many of the world's open-hearth steelworks, but in the late 1950s many firms put in larger-capacity plants, including Yawata and Nippon Kokan in Japan, and Jones & Laughlin Steel (J&L) and Kaiser in the United States. Later it was found that very large BOSP plants could be built. New, integrated iron- and steelworks with giant blast furnaces and huge basic-oxygen furnaces were constructed.

With the notable exception of J&L the major U.S. steelmakers did not install basic-oxygen furnaces until after many of their international competitors. In a widely cited article, Walter Adams and Joel B. Dirlam argued that the U.S. industry had wastefully invested large amounts of capital on obsolete open-hearth furnaces, when they could have installed more efficient BOSP capacity at lower cost. They concluded that this error was a result of an oligopoly which controlled the U.S. steel industry and dulled its competitive edge. Other scholars and industry representatives vigorously denounced this claim. Historians D. R. Dilley and D. L. McBride asserted that

A basic-oxygen furnace at the Colorado Fuel & Iron Company

the U.S. industry had moved with great speed to use the new technology. The percentage of U.S. steel produced by the BOSP was low by international standards, they argued, for several reasons that had nothing to do with managerial effectiveness. As the world's largest steel producer in the 1950s, the United States already had many modern open hearths in place. Since these hearths could not simply be abandoned, it was inevitable that the United States would for a while retain a relatively high percentage of its steel production in open hearths. Additionally, the early BOSP plants could only handle small tonnages of steel and, thus, were not clearly suitable for use in the large U.S. installations. It was natural that countries with smaller-scale facilities would be faster to introduce the BOSP. Finally, according to Dilley and McBride, attacks that claimed the U.S. industry was investing in the "wrong" technology because U.S. open-hearth production increased are misguided, because much of this increase could be attributed not to the construction of new open hearths, but to the increasingly efficient use of existing open hearths. In a major study of the U.S. steel industry and its major competitors in 1976 the U.S. Federal Trade Commission accepted this conclusion, though subsequent work cast doubt on it. Some of this later work pointed out, for example, that U.S. firms built far more open hearths than Japanese firms after the BOSP was clearly a viable alternative. Moreover, by 1970 the Japanese had scrapped the last of their open hearths, something that did not occur in the United States until the late 1980s.

While U.S. steelmakers may have been slower than their international rivals to adopt the BOSP, the new technology did spread with impressive speed through the integrated sector of the United States steel industry. Accounting for only about 5 percent of U.S. raw steel output in 1962, the BOSP reached 17.4 percent of output in 1965, nearly 50 percent by 1970, and over 60 percent in 1975. The open-hearth process, which accounted for nearly 90 percent of the raw steel produced in the United States in 1960, was eclipsed by the BOSP in 1970, and by 1980 the open hearth accounted for little more than one-tenth of U.S. output. With the rise of minimills — which rely on electric furnaces rather than blast furnaces — in the United States, electric furnace technology is beginning to challenge the preemi-

nence of the BOSP. In the late 1980s electric furnaces accounted for more than one-third of the steel produced in the United States, and their share was slowly increasing.

References:
Walter Adams and Joel B. Dirlam, "Big Steel, Invention and Innovation," *Quarterly Journal of Economics,* 80 (May 1966): 167–189;
D. R. Dilley and D. L. McBride, "Oxygen Steelmaking — Fact vs. Folklore," *Iron and Steel Engineer,* 44 (October 1967): 131–152;
Bela Gold, Gerhard Rosegger, and Myles G. Boylan, Jr., *Evaluating Technological Innovations* (Lexington, Mass.: Lexington Books, 1980);
William T. Hogan, S.J., *Minimills and Integrated Mills: A Comparison of Steelmaking in the United States* (Lexington, Mass.: Lexington Books, 1987);
Leonard H. Lynn, *How Japan Innovates: A Comparison with the U.S. in the Case of Oxygen Steelmaking* (Boulder, Colo.: Westview Press, 1982);
Paul Tiffany, *The Decline of American Steel: How Management, Labor, and Government Went Wrong* (New York: Oxford University Press, 1988).

Basing Point Pricing

by Donald F. Barnett

McLean, Virginia

Basing point pricing was a controversial mechanism once used in the American steel industries to determine transportation charges by means of a formula rather than by the actual cost of transportation. Normally the final price of a product, or delivered price, is the price at the factory — the f.o.b. (free on board) price — plus the cost of transportation to the consumer. Final prices from different producers to consumers vary, depending upon the location of the producer's factory. In industries that established a *single basing point,* however, all producers computed the transportation charges added to the f.o.b. price from one particular location, not from the factory of origin. Thus all producers collected the same transportation charges. Those producers located closer to

the consumer than the basing point collected phantom freight — transportation charges above the actual costs. But factories further from the consumer than the basing point absorbed a portion of the higher transportation charge. The structure became slightly more complicated when *multiple basing points* were established. Under this system, producers calculated transportation charges from the nearest, or *governing,* basing point.

In creating universal transportation charges, such pricing largely eliminated competition based on price and encouraged price stability, as long as all suppliers adhered to the formula and the base price was kept relatively rigid. Ideally all suppliers would quote identical delivered prices for their products, no matter where they or the consumer was located.

Basing point pricing appeared only in industries that were concentrated in one geographic region or in industries in which a single large firm established prices that prevailed at other firms. In steel both conditions applied: the industry was centered in western Pennsylvania and eastern Ohio, and the United States Steel Corporation established the price structure. Indeed, basing point pricing was championed by U.S. Steel head Elbert H. Gary in order to stabilize the industry. In 1903 a single basing point was established at Pittsburgh, and most large steel producers began to calculate transportation charges from Pittsburgh. The system was labeled Pittsburgh Plus. Chicago steel producers shipping to Pittsburgh, for example, absorbed all the freight charges but collected phantom freight on their sales in Chicago.

Chicago's emergence as a steel center to rival Pittsburgh led to the adoption of a Chicago Plus system for the Chicago area in about 1920. One reason influencing the steelmakers' move to multiple basing points was the Federal Trade Commission's (FTC) investigation of basing point pricing systems launched in 1920. At the conclusion of the study in 1924, the FTC ruled the Pittsburgh Plus system illegal. The industry responded by establishing several other basing points, but sellers still quoted delivered prices, not f.o.b. producer prices. The formulas were more complicated, but freight rates continued to be calculated from a base, not from actual freight rates from the sup-

plying mill. Transportation charges from the mill closest to the customer — the governing mill — determined the formula. Suppliers competing for business absorbed the cost of freight above that charged by the governing mill.

After multiple basing points were introduced in the steel industry in 1924, prices proved more volatile, but dramatic changes in the national economy brought on by the Great Depression also helped destabilize steel prices. Moreover, federal scrutiny of the entire concept of basing points continued. In 1938 Congress enacted legislation ratifying past FTC decisions and outlawed the practice. U.S. Steel appealed for a review of the 1924 FTC findings on Pittsburgh Plus. The review was delayed by pending court cases concerning basing point pricing in other industries, and U.S. involvement in World War II caused further delays in resolving the issue. In 1947 the FTC brought suit against the American steel industry, seeking to end all forms of formula freight pricing — including basing point pricing. A year later the Supreme Court upheld the FTC. Prices in the steel market have since been quoted f.o.b. from the supplying mill. Delivered prices have added actual freight charges from the mill to the customer.

Reference:

George Stocking, *Basing Point Pricing and Regional Development: A Case Study of the Iron and Steel Industry* (Chapel Hill: University of North Carolina Press, 1954).

Joseph Becker

(October 1, 1887 – February 26, 1966)

by Terry S. Reynolds

Michigan Technological University

CAREER: Office boy and clerk, law office, Essen, Germany (1901–1903); coke-plant operator and laboratory assistant, Victoria Mathias Collieries (1903–1906); coke-plant operator, chemist, Heinrich Koppers Company of Germany (1906–1910); chief chemist (1910–1912), superintendent of operations (1912–1917), superintendent of engineering and construction, consulting engineer, (1917–1925), vice-president and general manager (1925–1931), president, Koppers Construction Company of North America (1931–1936); vice-president and general manager, engineering and construction division (1936–1950), director (1950–1952), special adviser to management, The Koppers Company (1952–1966).

One of the key figures prior to World War II in researching new steel technologies, Joseph Becker helped introduce chemical-recovery coke ovens to America. As such, he was a leader in a new American industry that dealt in the recycling and selling of coke-oven gases.

Becker was born in Essen, Germany, on October 1, 1887, the son of Johann and Maria Duwentester Becker. His father was a member of the Essen police force. His formal schooling ended in 1901, and at age fourteen he began work as an office boy and clerk in an Essen law office.

Essen was a German center for iron and steel production, and Becker, having become disenchanted with his position in the law office, took a job in the Victoria Mathias Collieries in 1903. While studying chemistry in his spare time he learned to operate by-product coke ovens and soon became a laboratory assistant. In 1906 Becker was offered a position by Heinrich Koppers, founder of the Heinrich Koppers Company of Essen. Koppers had little use for college graduates, preferring those with practical experience to work in his plants and laboratories. After three years of working the coke ovens and assisting in the laboratory, Becker became a chemist at Koppers. In 1910 he was

Joseph Becker

named chief chemist, and he began to direct a study of the coking behavior of coals from around the world.

In 1907 Koppers had won its first American contract and installed a battery of by-product ovens at Joliet, Illinois, for the United States Steel Corporation. The success of these ovens led to further work in North America, and Koppers soon established offices in the United States to fill the contracts. Becker arrived in the United States in 1910 to take charge of the American Koppers laboratory in Joliet.

Becker quickly recognized that conditions in the American steel industry required coking units with a capacity larger than what was common in Europe, and he began to investigate differences in production yields between German and American plants. He was soon drawn out of the laboratory, however. From 1911 to 1913 nearly all of his time was spent solving problems at newly built Koppers coking plants in Illinois, Maryland, Indiana, Alabama, Ohio, and northwestern Ontario. He returned briefly to Germany in 1912 to marry Dorothea Molitor and was back in the United States that same year to become superintendent of all Koppers operations in North America. When Pittsburgh investors bought Koppers's American interests in 1914 and moved the company's headquarters from Chicago to Pittsburgh, Becker remained with the firm as its chief technical expert, becoming a U.S. citizen in 1919.

The onset of war in Europe in 1914 vastly accelerated the demand for by-product ovens in the United States. Escalating steel orders from warring European countries made the superior coking efficiency of by-product ovens more attractive than ever. Even more important, World War I created a profitable market for the chemicals of by-product coking ovens. The British blockade of Germany had cut off America's supply of German synthetic dyes and synthetic dye intermediates. Some chemicals, such as benzol, needed to produce synthetic dyes, however, were found in the by-product oven's chemical discharge, and a by-product chemical industry soon grew.

Becker and The Koppers Company played a critical role in the expansion of the American coking and by-product chemical industry during World War I. With Becker in charge of design and construction, Koppers built 3,378 coke ovens at 22 different plant locations during the war years, an average of one complete coking plant every 60 days. In addition, Becker supervised design and construction of by-product plants specifically designed to recover benzol and toluene to produce explosives. Some 90 percent of the by-product coke capacity installed in the United States during the 1914–1918 period came from Koppers and was in whole or part supervised by Becker.

Even before World War I Becker had begun making improvements in coke ovens. One of his first major innovations was the development of methods for using lean blast-furnace gas to heat coke ovens. This enabled by-product plant operators to divert the rich gas produced by the coke ovens to steel production or even for sale to city gas utilities. Becker also improved coke handling, simplified by-product plant design and operation, and patented a variety of methods that improved chemical by-product reclamation. In 1922 Becker invented the crossover flue system of heating coke ovens. This system enabled him to build a by-product oven that was larger, more efficient, and faster than previous designs. Called the Koppers-Becker oven, or the Becker oven, it won widespread acceptance in the United States. In the decades following its introduction on the market in 1923, more Koppers-Becker ovens were placed into operation in the United States than all other by-product ovens combined.

In subsequent years Becker made other contributions to the coking industry. In 1924, for example, Becker developed methods for coking the previously uncokable coals found in the American West, thus contributing to the birth of the West Coast iron and steel industry. In the same period he developed vastly improved methods for regulating the gas used to heat coke ovens and invented means for collecting and recycling the smoke which normally escaped when coke ovens were charged or pushed. Becker eventually took out 97 U.S. patents.

During World War II Becker was head of Koppers's large engineering and construction division, where he supervised the design and construction of coke-oven plants and a large styrene and butadiene plant at Kobuta, Pennsylvania, to produce synthetic rubber.

Becker was named to The Koppers Company board of directors in 1950. He retired in 1952 but was retained as a special consultant to management until his death in Pittsburgh on February 26, 1966. In addition to his career-long association with Koppers, Becker also served as a trustee of Eastern Gas & Fuel Association, as a director of NE Coal & Coke Company and the Connecticut Coke Company, and, briefly, as president of Koppers-Rheolaveur. In 1961 the Metallurgical Society of the American Institute of Mining, Metallurgical, and Petroleum Engineers established the Joseph Becker Award for distinguished achievement in coal carbonization.

Selected Publications:
and F. W. Sperr, Jr., *Recent Developments in By-Product Coke Oven Engineering* (Pittsburgh: Koppers Company, 1920);

"By-Product Coking with Particular Reference to New Combination (Koppers) Oven," *Blast Furnace and Steel Plant,* 10 (1922): 575–583;

"Modified By-Product Oven Design," *Iron Trade Review*, 71 (October 19, 1922): 1055–1062.

References:
The Becker Story (Pittsburgh: Koppers Company, 1952);

Horace C. Porter, *Coal Carbonization* (New York: Chemical Catalog Company, 1924);
Philip J. Wilson and Joseph H. Wells, *Coal, Coke, & Coal Chemicals* (New York: McGraw-Hill, 1950).

Charles Milton Beeghly

(October 6, 1908 –   )

by John A. Heitmann
University of Dayton

CAREER: Clerk, Metal Carbides Company (1931); clerk, sales trainee, Goff-Kirby Company, (1932–1933); sales trainee, Buffalo Slag Company (1934); salesman (1935–1939), sales manager (1939–1946), vice-president and director, Cold Metal Products Company (1946–1957); corporate director and president, strip-steel division (1957–1958), vice-president (1958–1960), president (1960–1963), chairman of the board and chief executive officer, Jones & Laughlin Steel Corporation (1963–1968).

As president and later chief executive officer and chairman of the board of the Jones & Laughlin Steel Corporation (J&L) during much of the 1960s, Charles M. Beeghly gained an enviable reputation within the American steel industry for his use of effective financial and cost-cutting measures. Beeghly's term as head of J&L coincided with the last profitable phase in the corporation's long history; for within a few years after his retirement the firm was confronted with enormous financial problems that would ultimately result in a dramatic scaling down of operations by the late 1970s.

Beeghly was born in Bloomville, Ohio, on October 6, 1908, to Leon A. and Mabel Snyder Beeghly. Beeghly was educated at Ohio Wesleyan University, where he received a bachelor's degree in 1930. Upon graduation Beeghly went to work as a clerk with the Metal Carbides Company of Newark, New Jersey. A year later he was hired by the Goff-Kirby Company in Cleveland, Ohio, as a clerk and sales trainee. In 1934 he became a sales trainee with the Buffalo Slag Company of Buffalo, New York. In 1935 he took a position with the Cold Metal Products Company, which his father

Charles Milton Beeghly

had founded in 1926. He steadily rose through the ranks at Cold Metal Products, and in 1939 he was promoted to sales manager. After a stint in the United States Army Air Corps from 1942 to 1945, Beeghly became a vice-president and director of the family-run firm in 1946.

In 1957 J&L purchased the company, and Beeghly was named president of J&L's strip-steel division and a corporate director. A year later he became J&L's executive vice-president, a position traditionally viewed as the stepping stone to the company presidency. In 1960 he replaced retiring Avery Adams as J&L president, and three years later Beeghly was named J&L chairman of the board and chief executive officer.

Teaming with salesman William J. Stephens, Beeghly directed a firm that, due to a modernization program implemented during the 1950s under the leadership of Adm. Ben Moreel, held a strong position in the steel industry. J&L's highly modern plants were equipped with basic-oxygen furnaces and could boast of extensive computer technology to control steelmaking processes and maintain inventories. Yet, despite the fact that J&L's market share increased from 4.9 to 5.8 percent between 1957 and 1963, Beeghly was concerned during the mid 1960s that increasing prices and labor costs might undermine J&L's position. The corporation's expansion into the Midwest further stretched costs, and by the late 1960s it had become clear that J&L, having just fallen on lean times, was ripe for a takeover.

Perhaps the most decisive turn in J&L's history took place in 1968 when Ling-Temco-Vought (LTV) purchased 80 percent of the firm's stock. Profits quickly plummeted, and J&L responded by initiating a series of cutbacks that prevented it from modernizing plant equipment and maintaining its share of an increasingly competitive international market.

After the LTV takeover Beeghly stepped down from his leadership position. He remains active in other civic and business matters, however, serving as director of the Dollar Savings and Trust Company of Youngstown, Ohio, the Mellon National Bank and Trust Company, and the Columbia Gas System and as trustee of Ohio Wesleyan University and the Carnegie Institute of Technology (now Carnegie-Mellon University).

Reference:

William T. Hogan, S.J., *An Economic History of the Iron and Steel Industry in the United States,* 5 volumes (Lexington, Mass.: Lexington Books, 1971).

Quincy Bent

(July 28, 1879 – May 5, 1955)

by Lance E. Metz

Canal Museum, Easton, Pennsylvania

CAREER: Assistant engineer (1901–1903), superintendent of furnaces, Lebanon, Pennsylvania, Pennsylvania Steel Company (1903–1909); assistant to president (1909–1910), general manager, Sparrows Point, Maryland Steel Company (1910–1918); vice-president, Bethlehem Steel Corporation (1918–1947).

Quincy Bent played a prominent part in the development of the Pennsylvania Steel Company and its subsidiary, the Maryland Steel Company. He also rose to prominence in the management of the Bethlehem Steel Corporation.

Bent was born on July 28, 1879, in Steelton, Pennsylvania. He was the son of Maj. Luther Stedman Bent and Mary Felton Bent. His maternal grandfather, Samuel Felton, was a former president of the Philadelphia, Baltimore and Wilmington Railroad and one of the incorporating partners in the Pennsylvania Steel Company. Felton also became the first president of Pennsylvania Steel, which had founded Steelton and built its plant there. Major Bent was an executive at the Steelton works, and he rose to become its general superintendent. Quincy Bent was destined from an early age to follow in his grandfather's and father's footsteps.

After attending local schools Bent entered Williams College, graduating in 1901. He immediately found employment with Pennsylvania Steel and was assigned a position at the Steelton works. In 1903 he was transferred to the company's furnaces near Lebanon, Pennsylvania, where he served as superintendent. He held this position until 1909, when he was promoted to the post of assistant to Frederick Wood, president of the Maryland Steel Company.

The Maryland Steel Company was a subsidiary of the Pennsylvania Steel Company. It had

Quincy Bent (photograph by Robert Yarnall Richie)

been founded in 1891 to manage the massive steel plant and company town that Pennsylvania Steel built at Sparrows Point, Maryland. Under the guidance of Wood, Sparrows Point rapidly became one of the largest and most productive steel plants in North America. At the age of thirty Bent was given complete responsibility for the management of the company town at Sparrows Point, a task which he fulfilled with efficiency and competence. His responsibilities were soon increased, and within three years he was also playing a large role in the management of the plant. In 1910 he was appointed general manager of the Sparrows Point plant, a position which he retained after Bethlehem Steel Corporation purchased the Pennsylvania and Maryland Steel Companies on February 16, 1916. During World War I Bent helped to convert the facilities of Sparrows Point to war production and supervised the doubling of Sparrows Point's productive capacity.

In 1918 Bent was transferred to Bethlehem's home office in Bethlehem, Pennsylvania, and was named vice-president of all of the company's steel plants. During the next decade his responsibilities were greatly increased as Bethlehem acquired additional plants, including the Cambria Steelworks at Johnstown, Pennsylvania, and the Lackawanna Steel Company of Lackawanna, New York, in 1922. During the next two decades Bent also as-

sumed responsibility for Bethlehem's many fabricating works and became head of its industrial research activities. He gained further recognition when he became a director of Bethlehem Steel. With the coming of the Great Depression, Bent was confronted with the difficult task of maintaining the company's facilities in a time of few orders and large layoffs. He successfully met this challenge.

During World War II Bent played a large role in the rapid mobilization of Bethlehem's resources for the war effort. He also served as a member of the Steel Advisory Committee of the Army-Navy Munitions Board and as the chairman of two committees of the War Production Board: the Metallurgical and Operations and Industry Advisory Committee and the Sub-Committee on Steel Industry and Steel Plant Expansion. During 1943 and 1944 he served as the chairman of the Joint United Kingdom, Canadian and United States Metallurgical Mission. For his efforts Bent was awarded the Gary Medal of the American Iron and Steel Institute in 1944. This award is the most prestigious honor given by the American iron and steel industry.

Bent retired from an active role in the management of the Bethlehem Steel Corporation in 1947. During his retirement he continued his sponsorship of the reconstruction of the seventeenth-century ironworks at Saugus, Massachusetts. Today the ironworks stands as a national historic site. He also served as a trustee of Williams College, which awarded him an honorary doctorate of laws in 1953. In 1946 he had received an honorary doctorate of engineering from Lehigh University. Bent was a member of both the American and British Iron and Steel Institutes. He also took an active part in the affairs of the American Institute of Mining Engineers, the American Academy of Political and Social Sciences, and the Army Ordnance Association.

Quincy Bent also played a prominent role in the civic life of the Bethlehem community. He was an active member of the Episcopal Church of the Nativity and served as a vestryman for 35 years. He also held a seat on the board of Bethlehem's Saint Luke's Hospital and served as its president from 1944 to 1949. He was a member of the Saucon Valley Country Club, and his estate, Wycliff Hills, near Bethlehem, was a noted showplace.

Quincy Bent married Deborah Norris Brock in 1910, and they had a son, Horace Brock Bent, who became a prominent attorney. Bent died in 1955.

Publications:

"The Use of Mayari Iron in Foundry Mixtures," *American Iron and Steel Institute Yearbook* (1912): 154–171;

"The Duplex Process of Steelmaking," *American Iron and Steel Institute Yearbook* (1916): 283–285 [Discussant];

"Effecting Economies in Use of Steel," *Steel,* 99 (October 12, 1936): 128;

"Modern Rolling Mill Practice in America," *Iron and Steel Institute Journal,* 138, no. 2 (1938): 412–421; abstract as "Rolling Mill Trends," *Steel,* 103 (December 26, 1938): 39–40;

Seventy-Five Years of Steel, Newcomen Address, 1939 (Princeton: Princeton University, 1939);

"Problems in the Manufacture and Use of Steel Products in the United States," *American Iron and Steel Institute Yearbook* (1939): 124–126 [Discussant];

"Manufacturing Problems Facing the Steel Industry in the Postwar Era," *American Iron and Steel Institute Yearbook* (1946): 63–67;

Early Days of Iron and Steel in North America (New York: Newcomen Society, 1950); abstract in *Blast Furnace and Steel Plant,* 39 (March 1951): 351–353.

Reference:

"Award of Gary Medal to Quincy Bent," *American Iron and Steel Institute Yearbook* (1944): 61–63.

Bethlehem Shipbuilding Corporation
by Lance E. Metz

Canal Museum, Easton, Pennsylvania

For much of the twentieth century the Bethlehem Shipbuilding Corporation and its successor, the Shipbuilding Division of the Bethlehem Steel Corporation, were world leaders in the construction and repair of vessels of all types. They also played a vital role in America's defense effort during World War II.

The Bethlehem Shipbuilding Corporation was an outgrowth of the United States Shipbuilding Company. U.S. Shipbuilding was to be a purchaser and operator of shipyards on the Atlantic, Gulf, and Pacific coasts; each of the shipyards would concentrate on the production of the types of vessels for which it was best suited. Unfortunately, many of these yards were poorly located, incompetently managed, and saddled with antiquated or badly maintained equipment. Despite these problems, U.S. Shipbuilding had begun operations by 1902. During that year Charles M. Schwab agreed to the merger of the Bethlehem Steel Company, which he had recently purchased, with U.S. Shipbuilding in return for $30 million in securities and a mortgage on U.S. Shipbuilding's properties.

U.S. Shipbuilding soon faced a severe financial crisis due to lack of orders, and Schwab refused to subsidize it by diverting profits from Bethlehem Steel. By October 1903 U.S. Shipbuilding had entered receivership, and it could not pay Schwab the interest due on his securities. To pro-

tect his interests Schwab began foreclosure proceedings to obtain U.S. Shipbuilding assets. Schwab's actions were protested by other U.S. Shipbuilding stockholders, who took legal action to protect their investments. A compromise was reached, and Schwab maintained control of Bethlehem Steel while Bethlehem acquired many of the U.S. Shipbuilding shipyards. These included such well-known facilities as the Fore River Yard in Quincy, Massachusetts, the Moore Yard in Elizabeth, New Jersey, the Harlan and Hollingsworth Yard in Wilmington, Delaware, and the Union Iron Works, which operated yards on San Francisco Bay, near downtown San Francisco, and at Hunter's Point at nearby Alameda. When Bethlehem Steel purchased the Pennsylvania Steel Company on February 16, 1916, it also gained control of its subsidiary, the Maryland Steel Company, with its large steel mill and shipyard at Sparrows Point, near Baltimore, Maryland. To consolidate all of its shipbuilding and repair facilities, Bethlehem Steel organized the Bethlehem Shipbuilding Corporation, Ltd., on October 30, 1917.

At the time of its organization Bethlehem Shipbuilding included 51 building slips and 75,000 employees. Besides the production of ocean-going vessels, the Bethlehem Shipbuilding Corporation also built gold dredges at the Union Iron Works and steel passenger railroad cars at Harlan and Hollingsworth. During World War I,

An ore boat (left) being repaired at the Bethlehem Shipbuilding yard in Baltimore, circa 1937

Bethlehem Shipbuilding began constructing destroyers for the U.S. Navy. The company built 35 vessels, more destroyers than the combined wartime output of all other U.S. government-owned and privately owned yards. Bethlehem Shipbuilding also built 16 submarines for the U.S. Navy and 4 additional ones for Great Britain. Between April 6 and November 11, 1917, Bethlehem Shipbuilding also completed nearly 625,000 deadweight tons of merchant ships, accounting for 22 percent of the total merchant-ship tonnage completed in the United States during World War I. Bethlehem Shipbuilding also operated for the U.S. Navy a ten-ship building yard at Squantum, Massachusetts, a similar building yard at Ridson in San Francisco, a boiler plant at Providence, Rhode Island, and a steam turbine plant at Buffalo, New York.

During the 1920s and 1930s Bethlehem Shipbuilding continued an active program of merchant and warship construction, including the completion of such noted vessels as the aircraft carriers USS *Lexington* and USS *Wasp*. When war broke out in Europe in 1939, the United States began a frantic program of naval rearmament. To better play its part in the national defense campaign, Bethlehem Shipbuilding reorganized and became

the Shipbuilding Division of the Bethlehem Steel Corporation, which was placed under the direction of Vice-president Arthur B. Homer. To meet the growing need for ship construction and repair facilities, Bethlehem greatly expanded its operations. When the United States entered World War II Bethlehem yards were allotted over 35 percent of the U.S. Navy's construction orders. The demand forced Bethlehem to build new yards, such as the Fairfield yard in Baltimore, which specialized in the rapid construction of cargo-carrying liberty ships; the Hingham yard near Boston, Massachusetts, which built destroyer escorts; and a yard in Staten Island, New York, which turned out destroyers. Other Bethlehem yards, such as the one in Alameda, fulfilled the U.S. Navy's need for combat-troop ships, while the Fore River yard continued its emphasis on large warships. Among the more notable of the fighting vessels that the Fore River Yard built were the battleship USS *Massachusetts,* the cruiser *Quincy,* and aircraft carriers such as the USS *Hancock, Wasp,* and *Lexington,* the later two carriers replacing earlier Bethlehem-built ships that had been lost in combat. Many tankers also were built at Sparrows Point.

The productivity of Bethlehem's yards during World War II was phenomenal. During World War I, it had taken at least 45 days to build a cargo ship; during World War II a liberty ship could be launched from a Bethlehem yard in ten days. Bethlehem also produced landing craft of all types. By 1943 Bethlehem had produced 1,050 ships for the United States and its allies, a record that was never surpassed by another shipbuilder.

After World War II Bethlehem shipbuilding facilities began a period of protracted decline, despite an initial burst of expansion which resulted in the creation of an overseas ship-repair facility in Singapore and the building of an oil-rig construction facility in Beaumont, Texas. By the 1960s Bethlehem was facing increased competition from foreign shipbuilding, and by the 1980s only the yards at Beaumont and Sparrows Point remained in operation. In 1989 the Sparrows Point shipyard completed what may have been its final ship, although it continues to solicit ship-repair work.

References:
Bethlehem Shipbuilding Corporation, LTD (South Bethlehem, Pa.: Privately printed by the Bethlehem Steel Corporation, 1922);
Robert Hessen, *Steel Titan: The Life of Charles M. Schwab* (New York: Oxford University Press, 1975);
John K. Mumford, *The Story of Bethlehem Steel, 1914–1918* (South Bethlehem, Pa.: Privately printed by the Bethlehem Steel Corporation, 1943);
Mark Reutter, *Sparrows Point: Making Steel, The Rise and Ruin of American Industrial Might* (New York: Summit, 1988).

Bethlehem Steel Corporation

by Thomas E. Leary
Industrial Research Associates
and
Elizabeth C. Sholes
Industrial Research Associates

Throughout most of the twentieth century Bethlehem Steel has ranked as the second-largest integrated steel company in the United States. The firm's principal lines of business have included the manufacture of iron, coke and its by-products, carbon and alloy steels in semifinished and finished forms, and the extraction, processing, and transportation of raw materials as well as finished products.

Bethlehem Steel was largely the creation of one man, Charles Schwab. Once the protégé of Andrew Carnegie, Schwab broke with the United States Steel Corporation in 1903 to develop a rival steel company on his own. In 1904 Schwab established Bethlehem Steel Corporation, named for its headquarters in Bethlehem, Pennsylvania. This corporation was a holding company that produced no goods but owned various operating units. Among these were a small specialty steel plant in the Lehigh Valley of eastern Pennsylvania and an inefficient iron ore mine in Santiago, Cuba. Both businesses were tottering on the verge of bankruptcy.

The nineteenth-century background of Bethlehem Steel's ancestors had reflected the problems encountered by the iron and steel industry east of the Allegheny Mountains. Guided by the eminent engineer and general superintendent John Fritz, the Lehigh Valley plant had begun by rolling iron and steel rails and later added other lines including heavy armor forgings and merchant plate. The company's directors, however, neglected the potential market for structural steel in nearby urban centers and also withdrew from plate and rail production. These questionable decisions of the 1890s were based on contemporary assumptions that eastern locations were less suited to steelmaking than were Pittsburgh and western districts. Schwab intended to overturn this prejudice.

During the formative years under Schwab, he and his closest ally, Eugene G. Grace, endeavored to move the company into the forefront of U.S.

Ore and coal boats arriving at Sparrows Point, Maryland, circa 1940 (drawing by Stow Wengenroth)

business. The first decade of the company's expansion strategy emphasized backward integration to secure raw material supplies; major product lines were added simultaneously. Schwab dispatched Grace to Cuba, where he reorganized production at the Juragua Mines. Other foreign ore sources figured in Bethlehem's plans after the 1913 acquisition of the El Tofo field in Chili. Shipments of ore from Chile to Baltimore and Philadelphia by means of the newly constructed Panama Canal created a long-distance transportation system that had an efficiency comparable to the Great Lakes network that supplied Pittsburgh's plants.

In 1908 Bethlehem introduced a product line that revolutionized commercial construction and moved the company into the structural market that had once tantalized John Fritz. An English immigrant, Henry Grey (1849–1913) had perfected a mill for rolling wide-flange I beams that dispensed with much of the need for fabrication of built-up structural elements. Schwab successfully scrambled to secure financing for this new venture during the economic panic of 1907. The Grey Mill became the centerpiece of the company's new

Saucon Division, also located in Bethlehem. Additional product lines positioned the steelmaker for significant penetration of markets for heavy capital goods, and profits from foreign defense contracts during World War I financed the next phase of growth.

Bethlehem undertook a buying spree of mergers and acquisitions over the next two decades. In 1916 the corporation purchased plants at Sparrows Point, Maryland, and Steelton, Pennsylvania, from the Pennsylvania Railroad. Between 1922 and 1923 Bethlehem bought Lackawanna Steel Company in New York and the Johnstown, Pennsylvania, works (formerly Cambria Steel) owned by Midvale Steel and Ordnance. Before these acquisitions Bethlehem already owned 11 subsidiary companies while Lackawanna had 13 and Midvale had 22; each group comprised a fully integrated operation. Through World War II, Bethlehem continued to add fabricating, ore processing, and finishing operations. In addition to its steelmaking and ship-building operations, Bethlehem possessed an impressive array of raw materials sources. The company held wholly owned and partially owned subsidiaries on the iron ranges around the upper Great Lakes. They continued exploiting their long-standing foreign ore sources in Cuba, Chile, Mexico, and Venezuela and held coal reserves in Pennsylvania and New Jersey. Transportation ventures included short line railroads and lake vessels for shipping both raw materials and finished products.

Bethlehem's ingot capacity rose from 1,129,000 tons in 1916 to 8,275,000 tons in 1930. Its major works each specialized in particular products and markets. The original Lehigh sites remained the premier sources for special forgings and continued to supply eastern centers with structural steel. Lackawanna was equipped with the famous Grey mills to serve construction sites around the Great Lakes. The New York plant also augmented the corporation's line of merchant bars used by makers of agricultural machinery and transportation equipment. Production of railway accessories was concentrated at Steelton, though both Lackawanna and Sparrows Point retained their rail mills. Of all of Bethlehem's plants, Sparrows Point grew most dramatically; its plate mills fed shipyards, and its tinplate furnished stock for canners. The historic Cambria works at Johnstown gave up rail manufacturing to concentrate on steel for its freight-car fabrication shops and lucrative customized product lines such as the Gautier specialty bar mills.

In the heavy steel markets, where the majority of Bethlehem's production was concentrated, demand sagged during the Depression. Smaller companies such as ARMCO, Weirton, and Inland had pioneered continuous strip mills capable of rolling lighter steels in long lengths for auto bodies, appliances, and containers. Bethlehem followed the trend during the mid 1930s and built strip mills at both Lackawanna and Sparrows Point. The production of these flat-rolled steels became the industry's most important stimulus for growth after World War II.

Even before the 1920s expansion, Bethlehem had carved out a central place in America's corporate hierarchy. Schwab's personal reputation as a steelman accounted for part of the company's eminence, but rapid growth and technological leadership also made the company an important economic force. Bethlehem was strategically situated within the financial community. From his days at U.S. Steel, Schwab had established cordial relations with the largest New York banks, including the J. P. Morgan banking pool, Chemical Bank, National City Bank (later Citicorp), and Chase Manhattan, as well as Girard Trust of Philadel-

phia. These institutions financed millions of corporate bonds that supplied Bethlehem with capital to underwrite their expansion programs. Until 1940 Bethlehem's "outside" directors included several bankers from Morgan Guaranty Trust, New York Trust, National City Bank, and various banks controlled by the Rockefeller family. After Schwab's death in 1939, Grace restricted the company's outside affiliations by gradually eliminating outside directors from the board who were not also company managers. Despite the company's lack of direct ties to the financial community, Bethlehem continued to raise impressive amounts of capital.

Through the 1950s Bethlehem's financial resources were sufficient to support the modernization of several plants simultaneously as well as the development of a large joint venture, Erie Mining, created to develop mining properties that would extract low-grade taconite ores for processing into higher-grade pellets. Expansion programs paid dividends; annual ingot capacity reached 20,500,000 by 1957. In the 1960s, however, those growth strategies changed. Bethlehem's half-century quest to rival U.S. Steel had failed to bring the company into the strategically important Chicago region despite two unsuccessful merger attempts with Youngstown Sheet & Tube, whose assets included a plant on Lake Michigan at Indiana Harbor. It was not until 1962 that Bethlehem finally secured its coveted Chicago-area location in Burns Harbor, Indiana. Here the company broke with its customary expansion strategy of acquiring and modernizing existing plants to build a brand new facility. By 1970 Burns Harbor had absorbed $1 billion of the firm's capital, far outstripping modernization costs at other plants and adding a hefty surcharge of $50 per ton of steelmaking capacity.

The saga of Burns Harbor exposed the dilemma of modernization during an era of increasing competition for capital and markets. The piecemeal improvements that Bethlehem had made periodically since the 1920s conserved capital, but the company had remained too conservative in this regard, failing to invest in increased efficiency with new processes and operations such as oxygen steelmaking and continuous casting which would have provided significant savings in labor and energy. As with the whole of the steel industry, Bethlehem found itself unable to fund expensive projects from its own retained earnings and found it equally difficult to attract outside bank capital, due to the modest returns on invest-

A backyard in the Bethlehem steelworkers' village at Sparrows Point, circa 1908–1910 (Hagley Museum and Library)

ment which are inherent in steel production and which were even less attractive where high capital costs eroded profits.

Schwab and Grace had approached expansion conservatively, raising and spending necessary capital but also retiring debt to protect profits. Directly after World War II Bethlehem's debt had shrunk to only $150 million, but two decades later escalated borrowing to construct Burns Harbor and to fund other projects had raised the debt considerably. By 1970 Bethlehem owed its creditors $1.7 billion. From 1970 to 1983 they borrowed an additional $1.7 billion from foreign banks as well as American credit sources. Debt retirement became a thing of the past.

Labor relations at Bethlehem reflected the conflicts over collective bargaining that reverberated through the first half of the twentieth century. Strikes in 1910 and 1918, for example, mirrored national concern over such issues as divi-

sions between supervisors and workers, labor's tactics for organizing large industries, and the legitimacy and limits of government intervention. The turmoil at Bethlehem was played out on a larger stage during the industrywide steel strike of 1919 and the Steel Workers Organizing Committee (SWOC) campaign of the 1930s. Bethlehem had introduced a limited form of collective bargaining, the Employee Representation Plan, in the aftermath of World War I, and Bethlehem's management steadfastly refused to negotiate with collective-bargaining agents that were independent of corporate control. Bethlehem ignored U.S. Steel's recognition of SWOC in 1937, and the ensuing "Little Steel" strike engulfed the Johnstown plant later that year. It took another series of strikes plus federal pressure on the company to maintain World War II defense production before Bethlehem capitulated in 1941 and recognized the steel-workers' union. Management and labor then settled down to haggle over the details of a bureaucratic industrial-relations system whose limitations became obvious during the postwar decades of prosperity.

After World War II there were significant changes in the markets for steel products. New materials such as plastics and aluminum began replacing steel in both capital and consumer goods. The trend escalated in the auto industry after the 1973 oil crisis sent car manufacturers on the quest for lighter body weights to secure higher mileage. Despite market shrinkage Bethlehem fared reasonably well financially until 1982, suffering only one bad year, in 1977. The company's growth strategy included product line diversification and overseas development, both of which aided profitability but never as extensively as in other steel firms such as U.S. Steel and National, which plunged into energy and financial services.

Bethlehem's dedication to steel manufacturing might have continued to serve the company well were it not for changes in the company's management perspective that were compounded by a national monetary crisis in the early 1980s. International currency speculation pushed the American dollar values very high relative to other world currencies, thereby making U.S. goods such as steel very expensive on the world market. Foreign-made commodities were cheaper and suddenly captured much of the domestic consumption. Basic U.S. manufacturing was badly hurt during the resulting recession of 1982–1983, and Bethlehem was no exception. From late 1982 through 1986 the company experienced its first unbroken string of losses since the Depression.

Unlike the 1930s, however, the 1980s generated new corporate retrenchment policies at Bethlehem. During the Depression Schwab and his corporate officers had scaled back production but had kept the plants working. Outmoded facilities were replaced with more technologically efficient operations. The company also had paid off existing debt to make the steel company more economically viable and capable of recruiting new capital with lower interest rates. In the early 1980s, however, the Bethlehem executive team headed by former Price Waterhouse accountant Donald Trautlein decided to close and demolish entire facilities, hoping to recover liquid capital from the federal government's generous settlements to companies that were going out of business. In one of the nation's largest shutdowns, the Lackawanna, New York plant was closed in 1983, save for three of its smallest operations. West Coast fabricating

and shipbuilding facilities were also phased out, and partial shutdowns occurred at the other active facilities. Consequently, when a general economic recovery got underway in 1984 and the overvalued dollar eventually declined, Bethlehem was left without the capacity to capitalize fully on improved domestic demand.

The company's restructuring was based on overall reductions in costs and capacity and a realignment of remaining plants and product lines. Raw steelmaking capacity was cut 30 percent, from 25 million tons as of 1977 to 17.5 million tons by the mid 1980s. The company temporarily dropped from second to third among the nation's steel producers. Total employment was slashed nearly twice as much as capacity, falling from 115,000 employees in 1975 to 48,500 by 1984. With the exception of Burns Harbor in Indiana, the surviving plants still contained a mixture of old and new equipment, the results of past modernization. Combining the most efficient operations often involved extensive interworks transfers of materials such as coke or semifinished steel, a far cry from the earlier goals of vertical integration and internal plant self-sufficiency.

Operations once controlled from the home office were decentralized into some two dozen semi-autonomous units which might include more than one plant. For example, the Bar, Rod and Wire Division (BRW) combined operations from Johnstown, Sparrows Point, and the remaining bar mill at Lackawanna. Walter F. Williams, Trautlein's successor as chief executive, pared down Bethlehem's bloated debt, boosted productivity, and preached a new gospel of improved product quality. However, productivity improvements were more attributable to payroll cutbacks than to overall plant modernization; employment as of 1991 spiraled down to 30,300.

Bethlehem emerged from its ordeal of drastic downsizing with renewed emphasis on its core steel business. Raw materials properties and other assets were pruned from the corporate balance sheet. Coated flat-rolled steel became the principal source of profits because these products were least susceptible to competition from minimills and imports. Since the mid 1980s Burns Harbor, which rolls automotive sheet, and Sparrows Point, which supplies the tinplate and construction markets, have benefited most from Bethlehem's relatively meager capital allocations for such equipment as continuous casters and galvanizing lines to apply corrosion-resistant coatings. Plants specializing in other products have been less fortunate; Bethlehem, however, did strive to provide a broader spectrum of rolled and forged shapes than some of its large U.S. competitors who concentrated almost exclusively on the flat-rolled sector.

Bethlehem has also recently taken the unusual step of forming joint ventures with foreign steelmakers (its nominal competitors) to acquire the scarce capital and improved technology required for plant modernization. Other U.S. firms had previously adopted this tactic, but for Bethlehem the results were mixed. The most ambitious proposal involved a project with British Steel PLC that would have upgraded facilities at Steelton for supplying steel to both the rail division and the structural division at Bethlehem, Pennsylvania. Failure to renegotiate a contract with the United Steelworkers Union, combined with a mounting recession, doomed this undertaking. After the British Steel deal collapsed in 1991, the fallout included the projected loss of 6,500 jobs, including the closing of iron and steelmaking operations at the Bethlehem, Pennsylvania, plant as well as the sale or shutdown of the entire Bar, Rod and Wire Division and the Steelton Trackwork Division. ISPAT, a multinational steelmaker based in Indonesia, subsequently attempted to purchase the BRW, but the offer fell through for much the same reason as the British Steel venture: inability to forge a labor agreement that would benefit both offshore investors and American workers.

References:
Robert Hessen, *Steel Titan: The Life of Charles M. Schwab* (New York: Oxford University Press, 1975);
William T. Hogan, S.J., *An Economic History of the Iron and Steel Industry*, 5 volumes (Lexington, Mass: Lexington Books, 1971);
Hogan, S.J., *Steel in the United States: Restructuring to Compete* (Lexington, Mass.: Lexington Books, 1984);
Thomas E. Leary and Elizabeth C. Sholes, *From Fire to Rust: Business, Technology and Work at the Lackawanna Steel Plant, 1899–1983* (Buffalo, N.Y.: Buffalo and Erie County Historical Society, 1987);
John Strohmeyer, *Crisis in Bethlehem: Big Steel's Battle to Survive* (Bethesda, Md.: Adler & Adler, 1986);
Paul A. Tiffany, *The Decline of American Steel: How Management, Labor and Government Went Wrong* (New York: Oxford University Press, 1988);
Melvin I. Urofsky, *Big Steel and the Wilson Administration: A Study in Business-Government Relations* (Columbus: Ohio State University Press, 1969);
Kenneth Warren, *The American Steel Industry, 1850–1970: A Geographical Interpretation* (Oxford: Clarendon Press, 1973).

Archives:
A Bethlehem Steel collection is found at the Hagley Museum and Library, Wilmington, Delaware, and at the Hugh Moore Historical Park and Museums, Incorporated, Easton, Pennsylvania.

Van A. Bittner

(March 20, 1885 – July 19, 1949)

by Richard W. Nagle

Massasoit Community College

CAREER: Union officer, United Mine Workers of America (1901–1942); organizer, Steel Workers Organizing Committee (1937–1942); assistant to the president, United Steelworkers of America (1942–1949).

Van Amberg Bittner, the son of Charles and Emma Ann Henck Bittner, was born on March 20, 1885, in Bridgeport, Pennsylvania. He began working in the coal mines when he was eleven years old and joined the United Mine Workers of America (UMWA). At sixteen years of age he was

elected president of his union local. While working in the mines, he continued his high-school education. In 1908, when he was twenty-three, Bittner became the vice-president of UMWA District 5, the Pittsburgh region. Three years later he was District 5 president. In the 1910s and 1920s he worked with John L. Lewis and Philip Murray as the UMWA's fortunes first rose with World War I and then fell in the 1920s. Bittner won a reputation as an effective organizer in the coalfields of Tennessee, Alabama, and West Virginia. By 1933 he was president of District 17 (West

Virginia) and a key subordinate of UMWA president Lewis.

When Lewis oversaw the creation of the Congress of Industrial Organizations (CIO) in 1935 and set out to organize the mass-production industries, Bittner was one of the UMWA officers most involved in the cause. Bittner went to work for the Steel Workers Organizing Committee (SWOC) in 1937; Phil Murray, SWOC's head, sent Bittner to oversee the Great Lakes Western region. Bittner was one of the signers of SWOC's contract with the United States Steel Corporation in March 1937. In the following months, though, Bittner and Murray suffered a bitter setback when the Little Steel strike failed. Bittner remained the head of the Western region as the SWOC used painstaking organization drives and elections overseen by the National Labor Relations Board to combat the open-shop policies of the Little Steel companies. In addition to his UMWA and SWOC posts, Bittner was the chairman of the CIO's Packinghouse Workers Organizing Committee and recruited among farm equipment workers in the late 1930s. The union's success was not realized until 1942.

The SWOC became the United Steelworkers of America (USWA) in May 1942, with Murray as its president. Bittner became one of the two assistants to the president, complementing Clinton S. Golden from the Eastern region.

When Lewis and Murray split in 1942, Bittner remained with Murray, the Steelworkers, and the CIO. During World War II he was a labor member of the National War Labor Board. In the postwar years Bittner continued to do several jobs at once. He kept his USWA post and helped direct the union's 1946 strike. He was national director of the CIO Organizing Committee and spent much of 1947 and 1948 leading a drive to recruit workers in the South; he concentrated on the textile and lumber industries. He was also vice-chairman of the CIO's Political Action Committee.

Bittner married Bertha Mae Walter in 1911. The couple had one daughter, Anna Mary, and lived in Crafton, Pennsylvania. Bittner died of heart disease in Pittsburgh on July 19, 1949.

Blast Furnace

by Robert Casey

Henry Ford Museum & Greenfield Village

The blast furnace produces pig iron — the intermediate product from which steel is made. In the furnace iron ore, limestone, and coke are subjected to a blast of hot air. The resulting chemical reactions separate the iron from its ore and produce two by-products — slag and blast furnace gas.

The most significant consequence of blast furnace development in the twentieth century was an enormous increase in the production of pig iron, as the table below illustrates:

Year	Number of Active Furnaces	Production (Millions of Tons)	Average Tons per Furnace
1900	232	13.8	59,402
1970	169	90.8	537,000
1988	50	55.7	1,114,000

Much of this increase was due to incremental improvements in blast furnace technology: higher hot blast temperatures; higher blast volumes; larger furnaces; improved cooling; blast furnace gas-powered blowing engines (circa 1903); steam turbine-driven turbo blowers (circa 1910); remote control clay guns for closing the tap hole (1914); and remote control drills for opening the tap hole (1950s).

Two important innovations were mechanical charging and the dry blast. M. A. Neeland's bucket system, installed at Duquesne, Pennsylvania, in 1895, became the first widely copied mechanical charging system. Arthur G. McKee's skip hoist system, patented in 1907, eventually became the standard of the industry. In 1905 James Gayley claimed that by using refrigeration to remove water from the air blast, the quantity of coke needed could be reduced, since heat would not be expending vaporizing moisture. Numerous dry blast operations were installed by 1920. It was later shown that control of the amount of moisture was more important than its absolute elimination, and many furnaces now add precisely regulated amounts of steam to the blast.

The rich iron ores of Minnesota's Mesabi Range, first mined in 1892, resulted in important changes in furnace design. The fine, crumbly texture of the Mesabi ores produced a fine dust that required major changes in the internal geometry of blast furnaces before the ores could be utilized to greatest advantage.

A growing trend throughout the century was the beneficiation, or upgrading, of raw materials. Early on Mesabi ore was washed to remove most of the silica. Beginning in the 1950s low-grade ores were processed into pellets with 60 percent iron content. Sinter — a mixture of flue dust, ore, coal, coke, and limestone particles — also became a major raw material component. By the 1980s, 95 percent of the ore used in blast furnaces was beneficiated.

Throughout most of the twentieth century, American blast furnaces were the most modern and productive in the world, but that leadership began to erode in the 1960s. By 1978 the largest furnace in the United States had stoves designed in Holland, a gas cleaner designed in Germany,

turbo blowers built in Switzerland, and a top designed in Luxembourg.

In recent years the cost of building blast furnaces has accelerated rapidly. In 1948 two modern furnaces could be installed for about $11 million. By 1969 a single furnace cost $50 million. Nine years later a state-of-the-art, 8,000-ton-per-day furnace cost $200 million. This enormous cost may convince major steel companies to become more interested in the various methods for direct reduction of iron ore.

References:

American Iron and Steel Institute, *Annual Statistical Report* (Washington, D.C.: AISI, 1988);

Bela Gold and others, *Technological Progress and Industrial Leadership: The Growth of the U.S. Steel Industry, 1900–1970* (Lexington, Mass. & Toronto: D. C. Henry, 1984);

William T. Hogan, S.J., *An Economic History of the Iron and Steel Industry in the United States,* 5 volumes (Lexington, Mass.: Lexington Books, 1971);

J. E. Johnson, Jr., *Blast Furnace Construction in America* (New York: McGraw-Hill, 1917);

United States Steel Corporation, *The Making, Shaping, and Treating of Steel,* seventh edition (Pittsburgh: U.S. Steel, 1957).

Blast Furnace and Steel Plant

by Bruce E. Seely

Michigan Technological University

Blast Furnace and Steel Plant was long a premier technical trade publication in the iron and steel industry. It provided superb coverage of technological change in the industry until 1971 when it ceased publication just as the industry faced severe technical challenges from abroad.

Blast Furnace and Steel Plant first appeared in March 1913, only to be absorbed into *Steel and Iron* in November 1914. *Steel and Iron* traced its origins to *Trade of the West,* begun in 1862. Its name changed at least five times over the next 50 years. It was *American Manufacturer and Iron World* in 1874–1905, *American Manufacturer* in 1905–1906, and *Industrial World* in 1906–1914. Despite the changes in name, this trade journal's scope of coverage remained the same — the iron trade and machinery building industry. From No-

vember 1924 to December 1915 the first issue of *Steel and Iron* each month was called *Blast Furnace and Steel Plant.* But in January 1916 *Blast Furnace and Steel Plant* reappeared as a separate publication. Andresen Company was the publisher. *Blast Furnace and Steel Plant* settled down to covering the technical changes under way in the industry. A monthly, it provided detailed accounts of new innovations, process improvements, and descriptions of individual mills, especially those with new facilities. With its offices in Pittsburgh the magazine was ideally located to cover the steel industry.

In about 1927, Steel Publications, Incorporated, took over the journal, along with several related trade publications. The masthead slogan of *Blast Furnace and Steel Plant* was "The Engineering Authority of the Steel Industry." It lived up to the claim especially after 1935. One strength of the magazine in these years was Charles Longenecker, who joined Steel Publications in 1926. He had an operating background with Cambria Steel and other companies and had contributed many technical articles on mill practice. For *Blast Furnace and Steel Plant* he produced monthly articles on every facet of the industry, including detailed descriptions of many leading steel plants. All of them contained brief historical summaries of the mill or company.

After 1935 Don N. Watkins was the editor, as well as president and treasurer of Steel Publications. He increased the size and coverage of the journal, but continued to focus on technical issues. He ignored, for example, the unionization drive of the Steel Workers Organizing Committee (SWOC). Watkins was on a first-name basis with the operating people in the leading steel companies, as was evident in the editor's monthly column, "My Turn." By 1940 companies that produced mill machinery — United Engineering and

Foundry, Mesta, Gathman Engineering, Freyn and Brassert — clearly saw this as their journal. Moreover, Watkins served for a time on early wartime materials committees, indicating the stature of the journal. He was succeeded as editor by Longenecker in 1943.

Blast Furnace and Steel Plant changed very little after World War II, but its best years were from 1935 to 1960. The suppliers of mill equipment — the primary advertising base of the journal — began to suffer by the late 1960s, as American steel companies cut back capital expenditures or turned to overseas machine builders. In an environment in which such dominant engineering firms as Mesta passed from the scene, the steel industry could not support overlapping trade publications, and *Blast Furnace and Steel Plant* published its last issue in March 1971. It was an omen of hard times to come for steel.

Reference:

David P. Forsyth, *The Business Press in America: 1750–1865* (Philadelphia: Chilton & New York, 1964), p. 225.

Joseph L. Block

(October 6, 1902 – November 17, 1992)

by Louis P. Cain

Loyola University of Chicago

CAREER: Sales department (1923–1927), assistant vice-president (1927–1930), vice-president (1930–1936), vice-president of sales (1936–1951), vice-chairman of board of directors (1949–1952), executive vice-president and chairman of finance committee (1952–1953), president (1953–1956), president and chief executive officer (1956–1959), chairman of the board, Inland Steel Company (1959–1967).

Joseph Leopold Block, the elder son of Cora Bloom and Leopold E. Block, was born in Chicago on October 6, 1902. His grandfather for whom he was named was one of the founders of Inland Steel Company. After graduating from the Harvard School for Boys in Chicago, Joseph attended Cornell University, where he was appointed an editor of the college newspaper while still an underclassman. By the end of his sophomore year, however,

he decided to marry. He left Cornell, joined the family firm, and, in time, served as its president and chairman of the board. His brother, Leigh, who dropped out of the University of Chicago in 1924 to join the firm, became its vice-president in charge of purchasing.

Block was assigned to the sales department of Inland Steel in 1923 and was named an assistant vice-president in 1927. In 1930 he was made a full vice-president and a director. In 1935 Inland merged with Joseph T. Ryerson and Son, Inc., the world's largest steel warehouse organization. Edward L. Ryerson became vice-chairman of Inland, while Joseph's father, Leopold, remained the chairman, and his uncle, Philip, remained the president. The following year, Joseph Block became Inland's vice-president in charge of sales. He held this position until 1951. During his term of office Inland's annual sales rose from $99 million to $519 million.

Joseph L. Block

In 1941 Block joined the steel division of the War Production Board in Washington as chairman of its Production Directive Committee. He became assistant director of its steel division in 1943 and deputy director a year later. While he was officially on leave from Inland, his annual salary from the government was "a dollar a year." The three annual paychecks for one dollar each were framed and displayed on his office wall.

In 1949, when Clarence Randall succeeded Wilfred Sykes as Inland's president, Block was promoted to vice-chairman of the board of directors. Three years later he was made executive vice-president and chairman of the finance committee, assuming the title held by his father at the time of his death.

Block became president of Inland Steel in 1953. For the two previous years Inland's outlay for expansion had been the lowest of the major U.S. firms. In 1955 Block announced plans for expanding Inland's Chicago operations. The plan called for the construction of a 19-story headquarters in Chicago's Loop; the erection of a large blooming mill at the Indiana Harbor works; and a landfill project to add 426 acres to the Indiana Harbor site, land on which three additional open-hearth furnaces and three new mills would be located. Old batteries of coke ovens were rebuilt, and new coking facilities were constructed. New soaking pits and a two-million-ton slabbing mill were put in place. New lines to produce galvanized-sheet steel and a variety of construction shapes were completed, including one for the production of wide-flange beams. The improvements to the Indiana Harbor plant were expected to increase Inland's ingot capacity from 5.2 million to 6 million tons within five years. By the end of 1959 Inland's measured ingot capacity was 6.5 million net tons.

Inland made additional acquisitions to flesh out and expand its finished steel fabrication facilities and product lines. In the mid 1950s Ryerson acquired five firms that increased Inland's warehousing and fabricating capacity and gave Ryerson access to additional regional markets. Inland Steel Products Division introduced product lines for the construction industry, and Inland Steel Container Company — which had been made a department of Inland in 1952 — expanded its facilities to produce a broad line of pails, drums, and other metal containers. The Cleveland Steel Barrel Company was acquired in 1955, and stainless steel containers were added to the product line.

When Randall retired in April 1956, Block became Inland's chief executive officer. He immediately confronted labor-relations issues, proclaiming in a 1956 address to the American Iron and Steel Institute that in his view the customary procedure in industry wage negotiations was for labor to "shoot for the moon" and for industry to offer labor nothing. In an interview given to *U.S. News and World Report* in June 1957, he expressed uncertainty as to where steel's wage-price spiral would end. During that period labor costs had risen substantially more than steel prices, and Block expressed the opinion that current prices were not adequate to cover the producers' costs. During the steel strike of 1959 Block became the spokesman for the major steel producers.

In January 1959 Block was elected chairman of the board of Inland Steel; the title had been vacant since Randall's retirement. This meant he was chairman of Joseph T. Ryerson and Sons, Inc., as well as Inland Steel Products Company. He continued as chief executive officer, but relinquished the presidency to John F. Smith, Jr. In the first quarter of 1960 Inland's earnings, sales, and shipments rose, approaching the high figures of the prestrike quarter of 1959. The expansion plans Block had announced in 1955 led to the construction of a 1,000-acre plant in Indiana Harbor. Completed in 1960, it was the fourth largest steel plant of its time, outside the Soviet Union. Inland sold 60 percent of its output in a 100-mile radius around the Indiana Harbor plant, encompassing the greater Chicago area — an area that used about one-third of the steel produced in the United States in 1960. Throughout the 1950s Inland was outperforming the rest of the industry.

In 1962 President John F. Kennedy's secretary of labor, Arthur J. Goldberg, persuaded the Steelworkers Union to accept a wage increase within the Kennedy administration's guidelines. Other steel firms decided to ignore the administration's wage-price guideposts and raised their prices by more than the approved 3.2 percent. Inland refused. Block stated that Inland's "decision was made in spite of, rather than because of, government intervention." Inland's reason for compliance was that because the demand for steel was weak at the time, Block did not believe the market would support the price increase. He also questioned the wisdom of a large industrywide price increase in the aftermath of a much smaller wage increase.

Inland, under Joseph Block's leadership, faced three major challenges in the 1960s. First, domestic competitors were increasing their production along the southern shore of Lake Michigan. Second, foreign competition in Inland's midwestern market was increasing as the Saint Lawrence Seaway opened Great Lakes ports to foreign-produced steel. Third, Inland was attempting to introduce new production methods, primarily the basic oxygen furnace (BOF).

In 1963 construction began on two basic oxygen injection steel furnaces. When completed, the furnaces produced a minimum of two million tons of BOF processed steel. This change in technology required Inland to use a higher volatility coking coal and pelletized iron ore in its blast furnaces. In 1960 pellets constituted only 1 percent of total ore used; by 1969, 74 percent of ore used was in pellet form. In addition, some ore was now being mined which was held in a taconite matrix, requiring a substantial investment in beneficiation and pelletizing facilities.

In 1967, at age sixty-five, Block retired from active participation in the day-to-day operations of the firm and turned the board chairmanship over to his cousin, Philip D. Block, Jr. In retire-ment, however, he remained involved with the affairs of the corporation.

Block has been a director of the American Iron and Steel Institute, the Commonwealth Edison Company, and the First National Bank of Chicago. He served as a trustee of the Committee for Economic Development and the Illinois Institute of Technology. He maintained a long involvement with the Chicago Association of Commerce and Industry, serving as its president from 1957 to 1959. He was a director of the Jewish Federation of Chicago from 1931 to 1952 and its president from 1947 to 1950. He is an ardent admirer of Abraham Lincoln and an avid collector of Lincolniana; the library of his home was rebuilt to house his collection.

Block frequently indulged his urge to write. He was a regular contributor to Inland publications. Letters containing his thoughts on Lincoln and on the steel industry frequently appeared on the editorial pages of Chicago's newspapers after his retirement.

Publications:
"The Growth of the Inland Steel Company," *Blast Furnace and Steel Plant,* 16 (September 1928): 1200;
"Joint Responsibility of Labor and Management," *American Iron and Steel Institute Yearbook* (1956): 91–101; also in *Commercial and Financial Chronicle,* 183 (May 31, 1956): 2605;
"Steel Industry's Greatest Challenge: Promotion of Our Industry's Freedom," *Commercial and Financial Chronicle,* 186 (November 14, 1957): 2117;
A Businessman's Viewpoint of the Problems of Chicago's Population Growth (Chicago, 1957);
"Outlook for Steel," *Commercial and Financial Chronicle,* 194 (July 27, 1961): 37–38;
"21st Century According to Steel's Crystal Ball," *Commercial and Financial Chronicle,* 205 (July 15, 1967): 2327.

References:
"Block Bows Out," *Newsweek,* 70 (November 6, 1967): 80–81;
William T. Hogan, S.J., *An Economic History of the Iron and Steel Industry in the United States,* 5 volumes (Lexington, Mass.: Lexington Books, 1971);
"Profitless Prosperity; J. L. Block," *Forbes,* 97 (May 15, 1966): 34.

Leopold E. Block

(January 13, 1869 – November 11, 1952)

by Louis P. Cain

Loyola University of Chicago

CAREER: Accountant-stenographer, Dreyfus, Block & Company (1894–1898); vice-president and director, Inland Steel Company (1898–1919); president, Buffalo Steel Company (1900–1909); treasurer (1909–1919), chairman of the board (1919–1940), chairman of the finance committee, Inland Steel Company (1941–1952).

Leopold Block was born on January 13, 1869, in Cincinnati, Ohio, the eldest son of Joseph L. and Rose Cahn Block. Joseph Block, one of the founders of Inland Steel, came to this country in about 1850 from the Alsace-Lorraine region in France, settled in Cincinnati, and helped make the Brown-Pollack Iron Company a success. After receiving his education in Cincinnati public schools, Leopold worked as an accountant-stenographer in the Pittsburgh partnership of Dreyfus, Block & Company, purveyors of scrap iron and steel. In 1898 he moved to Chicago to join his father and brother Philip at the Inland Steel Company. A wise salesman and manager, Leopold contributed significantly to the growth of Inland Steel.

In 1895, as a result of a difference of opinion over dividend policy, William M. Adams, one of Inland's founders, resigned from the firm and offered his stock for sale. Leopold purchased Adams's 150 shares, which meant the Block family controlled half the firm's 1300 shares. Two years later Inland purchased the failed East Chicago Iron & Forge Company for $50,000 and incorporated it separately as Inland Iron & Forge, a producer of railroad equipment. The following year Leopold Block moved to Chicago to manage this firm and was named vice-president and director of Inland Steel in 1898. In 1900 he added the position of president of Buffalo Steel Company of Buffalo, New York, to his list of duties, a title he held until 1909.

Leopold Block served Inland as vice-president until 1919 and as treasurer between 1909 and 1919 a job he took over from his brother Philip. In 1919, Leopold was named chairman of the board, replacing Alexis W. Thompson. Philip was simultaneously named president and his other brother, Emanuel, was named vice-president. The three brothers carefully guided the company through the interwar period, turning Inland Steel into one of the "Big Seven" steel producers in the United States.

Inland Steel owned Inland Iron & Forge for only three years, and during this time its capacity was increased from 25,000 tons to 70,000 tons of product per year. Inland further increased its assets when in 1901 Block agreed to a deal with the Lake Michigan Land Company. Lake Michigan Land had looked to offer 50 acres of duneland at Indiana Harbor to a company willing to spend at least $1 million in construction and development on the land. In 1901 Inland sold this plant for ten times what it had paid for it to help finance its new Indiana Harbor works. This is one of the first moves that earned Leopold his reputation as a shrewd financier.

As originally planned the Indiana Harbor works were to convert pig iron and scrap into ingot steel in two or three open-hearth furnaces, a blooming mill, and a bar mill. R. J. Beatty, who joined Inland in 1901 as a director and general manager, convinced Inland to substantially alter the number of products the mill would produce, changing the design and cost of the new works. Inland erected four 40-ton open-hearth furnaces, a blooming mill, a bar mill, and five sheet mills, at a cost of $2 million. The first ingot produced in the East Chicago industrial region was poured at the new Inland facility on July 21, 1902. By the end of 1902 the Indiana Harbor works were producing 19,343 tons of ingots yearly. On the eve of the Great Depression that figure was over 90 times greater, and the Indiana Harbor works had grown to cover 630 acres through purchase and the exercise of riparian rights.

At a time when steel companies were effecting larger mergers and industry commentators were predicting the demise of the independent steel company, Inland began to develop a unified, integrated organization. In 1906 Inland expanded its capitalization and authorized its first bond issue, $2.5 million of first mortgage bonds, to finance the purchase of the Laura Iron Ore Mine and construction of its first blast furnace, the Madeline #1 — named after Leopold's niece. The bonds were issued during the Panic of 1907, but despite the economic crash Leopold was able to secure the necessary funds for Inland. The Blocks brought Charles Hart and Alexis Thompson — both future Inland presidents — into the firm to assist with changes in Inland's corporate strategy.

In 1915 an expansion plan was adopted, doubling Inland's ingot capacity to 1.28 million tons by 1920. Once again, Leopold found a way to achieve growth through internal expansion rather than through merger. A new blast furnace with a capacity of 650 tons per day, the Madeline #3, was added along with a new battery of 40 coke ovens that made Inland self-sufficient in coke. A new open-hearth shop with ten furnaces and new mills — all of which were designed to run on electric power — were also part of Inland's new capital. Wilfred Sykes, a Westinghouse engineer and future Inland president, was brought in to assist with the electrification. By World War I, Inland had acquired its own fleet of ore boats and developed its harbor. After the war, the electrification of the mills was completed under a $70 million, five-year development plan.

In the early 1920s Inland upgraded its facilities in the expectation of expanding markets. Two mills were modified to expand the production of railroad iron and structural shapes. Beginning in 1923 Inland undertook another expansion, including the construction of a billet and slab mill, a merchant mill, and four new open-hearth furnaces, bringing the total number of open hearths to 26. In overseeing the financing of the expansion, Leopold authorized the sale of $10 million worth of 7 percent preferred stock and 168,450 shares of common stock. The authorized issue of common stock was increased to 1.5 million shares, and all existing common shares were changed from $100 par value to a no-par basis. The company sought to raise $17 million, using $13 million as new capital and the remainder to retire some of its small-bonded indebtedness. In 1928 all outstanding preferred stock was redeemed, and the debt was once again restructured.

Inland continued to insure its raw-materials position in the late 1920s. Iron-ore sources were increased through the execution of leases with the Langden and Snyder mines in Itasca County, Minnesota, and with the Greenwood mine in Marquette County, Michigan. Inland also acquired the Manistique Michigan Lime & Stone Company, consisting of three mines that became the Inland Lime & Stone Company. In 1930 Inland acquired the Wheelwright Coal Mine in the Elkhorn district of Kentucky, and leased over 14,000 acres of other Kentucky coal lands.

Increased outlets for the company's products were secured through the acquisition of Joseph T. Ryerson & Sons, Inc. of Chicago in 1935, the Milcor Steel Company of Milwaukee in 1936, and the Wilson and Bennett Manufacturing Company in 1939. Block, however, continued to move Inland toward vertical integration by strengthening the company's holdings in raw materials.

In 1940 Inland paid cash for the capital stock of the Nevada Corporation, which controlled four subsidiaries owning iron-ore mining properties in Iron County, Michigan. In 1937, Inland Tar Company was made a department within Inland when the remaining minority interest in the company was purchased. The Inland Steamship Company was dissolved in 1936, and Inland Steel purchased its three ore freighters at public sale. In 1939 Inland purchased from Republic Steel a lease on 80 acres of iron land in Iron County that became known as the Sherwood Mine. With the purchase of Hillside Fluorspar Mines in Rosiclare, Illinois, the integration of raw-material sources was completed. On the eve of World War II, Inland was an independent operation, capable of mining its own ore, coal, and limestone and marketing its final product to industrial consumers.

In 1940 Leopold Block resigned the board chairmanship and assumed the less taxing position of chairman of the executive committee, which he relinquished the following year to his brother, Philip. Leopold became chairman of the finance committee in 1941.

He was actively involved in the business and cultural life of Chicago, serving as a director of Commonwealth Edison, the First National Bank of Chicago, and Michael Reese Hospital. He was one of the leaders of the Chicago Museum of Natural History, now known as the Field Museum. Block married Cora Bloom on June 30, 1900. The couple had four children: Joseph, Leigh, Babette, and Elanor. Joseph and Leigh became an Inland president and vice-president, respectively. Block died on November 11, 1952.

Publications:
"Pittsburgh Plus Is Economic Law," *Iron Trade Review*, 72 (April 5, 1923): 1033;
"Code Demands Sacrifices for Common Good," *Iron Age*, 133 (May 31, 1934): 22.

References:
William T. Hogan, S.J., *An Economic History of the Iron and Steel Industry in the U.S.*, 5 volumes (Lexington, Mass: Lexington Books, 1971);
Selected Speeches, 1940–1967 (Chicago: Lakeside Press, 1967).

Philip D. Block

(February 16, 1871 – June 30, 1942)

by Louis P. Cain

Loyola University of Chicago

CAREER: Block-Pollock Iron Company (1888–1893); treasurer and plant manager (1893–1909), purchasing agent (1894–1901), vice-president (1901–1919), president (1919–1941), chairman of the executive committee, Inland Steel Company (1940–1941).

Philip Dee Block was born in Cincinnati, Ohio, on February 16, 1871, the second son of Joseph and Rose Cahn Block. He was educated in the public schools of Cincinnati, then joined his father at the Block-Pollock Iron Company, where he worked from 1888 to 1893. In that year he moved to Chicago with his father as part of the group that started the Inland Steel Company. Although only twenty-two years old, Philip was named treasurer and plant manager of Inland at an annual salary of $200. He was also named a member of the board of directors. A few months later, in January 1894, he was named purchasing agent.

On June 1, 1899, Philip married Celia F. Leopold. They had two children, Madeline, whose name was given to the company's first blast furnace, and Philip D., Jr., who became Inland's board chairman in 1967.

In spite of his youth Philip Block played an integral part of Inland's initial success. George H. Jones and Joseph Block went to solicit orders for the new plant, while Philip Block and John W. Thomas supervised the construction of the plant. The 40 carloads of second-hand machinery purchased from the Chicago Steel Works required a great deal of patching. The power for the plant came from an old high-speed steam engine Philip found in the Chicago City Hall.

By 1897 Inland's annual payroll was nearing $90,000, while its sales approached $350,000. Thus, when Philip heard that the East Chicago Iron and Forge Company was to be sold at a sheriff's auction, he quickly investigated adding its resources to Inland Steel. His report that the works could be put in shape and operated profitably convinced his brother Leopold to join the firm.

Philip became involved in all aspects of the company; he is reported to have been the second baseman on the company's first baseball team. He was promoted to vice-president in 1901, the rank given his older brother Leopold when he joined Inland in 1898. In 1909 he relinquished the treasurer's job to Leopold. On July 29, 1919, Philip was named president of Inland; Leopold became chairman of the board; and their youngest brother, Emanuel J. Block, became vice-president. Leopold served as chairman of the board until April 30, 1940. Philip Block served as president exactly one year longer. Then, at age seventy, he replaced Leopold as chairman of the executive committee, and Leopold was named chairman of the finance committee.

Like the other Blocks, Philip was active in the community, serving as a director of the First National Bank of Chicago and the Chicago Art Institute. He was the last surviving founder of Inland Steel Company when he died on June 30, 1942. The involvement of Philip and his brothers made Inland one of the most successful family firms in the United States.

Reference:

William T. Hogan, S.J., *An Economic History of the Iron and Steel Industry in the United States,* 5 volumes (Lexington, Mass.: Lexington Books, 1971).

Philip D. Block, Jr.

(June 25, 1906 – May 28, 1981)

by Louis P. Cain

Loyola University of Chicago

CAREER: Executive apprentice, blast furnace and coke plant departments, Indian Harbor works (1928–1931), executive apprentice, raw materials department (1931–1935), assistant vice-president (1935–1948), director (1942–1971), vice-president in charge of raw materials (1948–1956), senior vice-president (1956–1959), vice-chairman of the board (1959–1967), chairman of the finance committee (1963–1967), chairman of the board (1967–1971), chairman of the executive committee, Inland Steel Company (1967–1977).

The son of the Inland Steel Company's founders, Philip D. Block, Jr., was born in Chicago on June 25, 1906. He controlled the company during tumultuous years of increasing competition and continued the Inland tradition of expansion.

He attended the Harvard School for Boys in Chicago, Phillips Andover Academy, and the Sheffield Science School at Yale University, graduating in 1928. Following graduation he went immediately to work for Inland Steel in the blast furnace and coke plant departments. In 1931 he was moved to the raw materials department at the Chicago headquarters, becoming vice-president in charge of raw materials in 1948.

In 1935 Block had been named an assistant vice-president of the corporation. He was named a director in 1942 — the year of his father's death — and senior vice-president in 1956, becoming Inland's second ranking officer at that time. He was elected vice-chairman of the board three years later and, as a consequence of that appointment, became chairman of the board of the Inland Steel Container Company in Milwaukee, and vice-chairman of Joseph T. Ryerson

& Son, of Chicago, all subsidiaries of Inland Steel. In 1963 he was named chairman of the finance committee, a position previously occupied by his uncle Leopold Block, his father Philip Sr., and his cousin Joseph Leopold Block. In 1967 Joseph Leopold retired as Inland chairman of the board, and Block succeeded him. He held the position for four years until he reached age sixty-five, at which time he was named chairman of the executive committee until his retirement in 1977.

Increased competition in the 1960s had created an unstable price structure, causing profits to drop. Under Philip Jr.'s direction Inland adjusted its costs by increasing the efficiency of its operations and the output of those products in which it had a competitive edge. The company ended the decade with a substantial increase in both ingot capacity and the capacity to produce a wide variety of fabricated and basic products and shapes. Under Block's direction Inland began a program of diversification in 1970 that lasted throughout the decade. In order to plan this expansion, the Inland Corporate Development Group was created and reported directly to the chairman of the board.

From 1967 to 1971 Block was a member of the executive committee of the American Iron and Steel Institute and a director of the International Iron and Steel Institute. He was a director of the Continental Illinois National Bank between 1959 and 1972. At the University of Chicago he served as a life trustee, chairman of the visiting committee to the School of Social Service Administration, and chairman of several major fund-raising campaigns. He was a director of United Charities of Chicago, serving as president of that organization in 1951–1952. He served on the Health Education Commission of the State of Illinois, was a member of the executive committee of the Chicago Community Trust, and acted as a director of both the Jewish Children's Bureau of Chicago and of the Jewish Federation of Chicago. He also served as director of Inland Steel Urban Development Corporation and as an honorary director of the family firm.

He married Margaret L. Selz on November 28, 1934; they had two sons: Philip III (who

was vice-president in charge of purchases at his father's death) and Andrew. Block died in Chicago on May 28, 1981.

Publications:

"Present and Prospective Sources of Supply of Steelmaking Raw Materials," *American Iron and Steel Institute Yearbook* (1950): 271–272.

Roger M. Blough

(January 19, 1904 – October 8, 1985)

by George McManus

Pittsburgh, Pennsylvania

CAREER: Lawyer, White & Case (1931–1942); general solicitor (1942–1951), executive vice-president (1951–1952), vice-chairman (1952–1955), general counsel (1953–1955), chairman of the board and chief executive, United States Steel Corporation (1955–1969).

Roger Miles Blough, head of the United States Steel Corporation for 14 turbulent years, was born in the small western Pennsylvania town of Riverside on January 19, 1904. He grew up on a truck farm run by his parents, Christian Emanuel and Viola Nancy Hoffman Blough. His early education took place in Riverside's one-room school. He attended high school briefly in Johnstown, Pennsylvania, and finished his secondary education at Susquehanna Academy in Selinsgrove, Pennsylvania. He went to Susquehanna University in Selinsgrove, but twice he had to leave college to earn money for tuition: working in a Johnstown steel mill on one occasion, and teaching all eight grades in the one-room school on the other.

After graduating from Susquehanna in 1925, he spent the next three years teaching science and mathematics and coaching the basketball team at the high school in Hawley, Pennsylvania. He entered Yale Law School in 1928 and served as one of the editors of the *Yale Law Journal*. He received his law degree in 1931 and joined the Wall Street firm of White & Case, which had close ties with U.S. Steel. Irving S. Olds, who was a partner in White & Case during Blough's stay there, would later become chairman of U.S. Steel.

Roger M. Blough

In 1939 Olds assigned Blough to a team of 20 lawyers representing U.S. Steel in hearings before the Temporary National Economic Committee, which was conducting one of many government investigations into U.S. Steel's business practices. Due to its massive size and role as leader of the steel industry, U.S. Steel was frequently the target of antitrust allegations. As such, lawyers began to rise to positions of prominence at U.S. Steel during this period.

In February 1942 Blough joined what was then the United States Steel Corporation of Delaware as general solicitor. He is credited with playing a key role in the restructuring of the corporation — which was once a holding company with little control over its loosely knit subsidiaries — into a single operating company. The reorganization, which completed a process started by Chairman Myron Taylor in the 1930s, took effect in the early 1950s. Not surprisingly, Blough continued to climb steadily through U.S. Steel's hierarchy, becoming vice chairman in 1952, general counsel in 1953, and succeeding Benjamin F. Fairless as chairman and chief executive on May 3, 1955.

At this time the steel industry was still experiencing a postwar boom in demand. To meet the demand, steel companies began to expand. In 1958, however, just as the new capacity was being brought into operation, the steel market collapsed. Raw-steel output fell from 150 to 85 million tons, the lowest total since 1947. Despite a very weak market, with five steel companies losing money, steel wages were scheduled to go up 20 cents an hour.

That prospect brought speculation about a steel-price increase, a prospect that agitated officials in Washington, who disliked the postwar pattern of wage settlements and price hikes. Any price hike in those days brought charges of arbitrary pricing power and profiteering, although basic steel was usually the least profitable of the major industries. Citing a 170 percent rise in the price of steel between 1940 and 1958, the government did, indeed, accuse the industry of profiteering. Steel executives responded to the charges, arguing that prices had to go up because of the arbitrary, regular-wage increases placed by union contracts. Steel-labor costs rose 300 percent in the 20 years after 1940, and mill officials claimed that the wage escalation was partly due to government intervention in steel-labor negotiations. Blough summed up the steel industry's position when in 1958 he declared, "the matter of a price increase would not even come up if it were not for the very substantial employment cost increase we now face."

In late July a steel-price increase of $4.50 a ton was initiated by the Armco Steel Corporation and was followed by other steelmakers. The industry-wide increase provided more ammunition for those who charged the steel industry with anti-competitive tendencies.

In 1959 U.S. Steel and the industry took a stand against wage escalation. Although only 2 million tons of foreign steel had entered the domestic market in 1958, Blough correctly saw that continued wage-price inflation served as an invitation to foreign imports at a time when world steel-making capacity was rapidly rising. Accordingly, steel management proposed in 1959 that the labor contract be extended one year but with no change in pay and no cost-of-living provision. Most important, management sought deletion of the past-practice clause, which had limited management's right to change the sizes and responsibilities of work crews on the factory floor. The effort to relax this limitation was met with strong union resistance. Officials of United Steel Workers of America (USWA) told workers the companies were out to make wholesale job cuts.

The result was a strike that lasted 116 days and only ended when the administration of President Dwight D. Eisenhower intervened under the Taft-Hartley Act. U.S. Steel eventually signed a contract that raised wages 40 cents, but the long domestic strike had opened the door for steel imports. Shipments of foreign steel to the United States jumped from 2 million tons in 1958 to 5 million tons in 1959. Overseas mills gained a foothold in the United States that they would never relinquish. The steel strike was followed by a brief spurt in production to meet pent-up demand, but a lagging economy soon dropped mill operations to 50 percent capacity.

President John F. Kennedy moved into the White House, promising that his administration would get the economy moving: he would intervene in wage-price matters to prevent inflation and deal harshly with "administered" pricing — the simultaneous raising of prices by steel companies without overt collusion.

In 1961 the Kennedy team began making good on its pledge: tax reductions were to be used to stimulate the economy while price stability was to be achieved by securing from labor and management a promise of "responsible behavior" in negotiating contracts — an increase in wages should equal an increase in productivity. Price behavior was likewise tied to productivity. Highly productive industries could cut prices; less efficient industries would increase prices. The net effect would be stability.

Early in 1961 issues involving steel wages and prices were moving toward political center stage. Steel companies had absorbed two wage increases without price relief. There was speculation that a third pay hike, scheduled for October 1, 1961, would be followed by a steel-price increase. There was drumbeating on both sides of the steel-price issue. Thomas Patton, chairman of Republic Steel Corporation, claimed that "the basic factors which make a price increase necessary remain very strong." On August 23 Democratic senator Albert Gore, Sr., of Tennessee urged Kennedy "to bar a price rise by breaking up large companies to restore competition or to control prices if more persuasion failed." On September 6, 1961, Kennedy wrote the heads of steel companies, asking them to hold the price line and hinting that cooperation would bring government's help in restraining wage demands. The steel industry faced a new round of wage negotiations in 1962.

Roger Blough and other steel executives responded coolly to the Kennedy proposal. On September 21 Blough met with Kennedy at the White House. Blough later described his conversation with Kennedy in the January 29, 1963, issue of *Look* magazine, claiming that he had made no "commitment on price of any kind." Yet there was no steel-price increase in October, and the Kennedy administration was left with the impression that a deal had been made. On October 2 Secretary of Labor Arthur Goldberg urged steel labor and management to get started early on their contract negotiations. At the time, steel management doubted that the administration would really lean on labor. In management's view, previous cases of intervention suggested that the government would give labor what it wanted and then pronounce the settlement fair and noninflationary.

Bargaining began February 14, 1962, with general economic presentations in which the USWA emphasized large company profits, and the companies in turn pointed out the 12 percent increase in employment costs over the previous three years. Blough repeated his contention that steel companies were being squeezed by rising costs and that any further wage hikes would increase the pressure on management to raise prices.

The contract talks were broken off on March 2. Not until March 8 were the parties summoned back to the bargaining table by Kennedy. On March 31 a new steel-labor contract was signed. There was no pay increase, and the total cost of new fringe benefits was only 10 cents an hour, an increase of 2.5 percent. Speaking by

phone to the USWA wage-policy committee, President Kennedy hailed the agreement as "obviously noninflationary."

On Tuesday, April 10, 1962, the decision to raise steel prices was approved by U.S. Steel's 14-man executive committee, and at 5:15 P.M. that day Blough entered the White House to hand Kennedy a notice of a price increase. An across-the-board hike of $6 a ton — a 3.5 percent increase — was to take effect the next day. There were no fireworks during the White House meeting, but the Kennedy people struck back soon afterward. Administration officials contacted other steel companies, urging them not to follow U.S. Steel's lead and threatening them with loss of defense contracts. The administration also launched publicized investigations into steel-price fixing.

Kennedy felt betrayed by the steel men, for his economic program was strongly probusiness. Rather than stimulate the economy by increasing government spending, President Kennedy planned to get things moving by cutting taxes — including business taxes. The full fury of the administration came down on steel people on April 11 at a televised press conference in which Kennedy declared, "the American people will find it hard, as I do, to accept a situation in which a tiny handful of steel executives, whose pursuit of private power and profit exceeds their sense of public responsibility, can show such utter contempt for the interest of one hundred and eighty-five million Americans."

The next day Roger Blough had his own televised press conference. He cited U.S. Steel's low profits and noted that steel-labor costs had gone up 12 percent since 1958. He recalled his earlier statements on the need for price relief and reiterated that price increases were needed to modernize and to meet competition. But Blough's plodding style contrasted unfavorably with the dash and incisiveness of Kennedy. In terms of winning public support, it was no contest. Typical of the general reaction was the comment of the *St. Louis Post Dispatch*: "It looks very much as if the steel masters used the President and his Secretary of Labor for the purpose of beating down wage demands prior to a price decision they had in mind all along." Even Republicans were dismayed by the price action. "The increase at this time is wrong — wrong for Pennsylvania, wrong for America, wrong for the world," proclaimed Republican representative William Scranton of Pennsylvania.

On Friday, April 13, Inland Steel Company announced it would not go along with the higher prices. That effectively scuttled the increase, and at 5:00 P.M. that day U.S. Steel rescinded its new prices. Kennedy tried to restore harmony by claiming on April 18 that the "administration harbors no ill will" toward the steel companies. But the forces set loose by the confrontation would not go away overnight. For many years to come, any steel-pricing action would be subjected to special scrutiny and often would bring retaliatory action. Steel people were given a taste of things to come on June 25, when the phone records and expense accounts of U.S. Steel and nine other companies were subpoenaed by the Justice Department in connection to an April 27 price-fixing indictment.

Blough continued as U.S. Steel chairman until 1969. During his final five years he saw some of his earlier judgments vindicated: he was right in warning that the steel industry faced heavy capital needs. New technology had forced extensive replacement of refining and rolling facilities, and foreign competition was on the increase. Yet, despite identifying the cause of these problems at an early stage, Blough failed to come up with solutions for the steel industry. Steel imports rose from 3.6 million tons in 1960 to 7 million tons in 1970. Having alienated the government in 1962, however, steel companies had little success in securing government protection against foreign producers. Blough's effort to moderate inflationary forces was a disastrous failure. By 1970 U.S. Steel was signing a labor contract that raised employment costs 15 percent in the first year. That increase was followed by the announcement of an 8 percent increase in steel prices.

Many factors that contributed to U.S. Steel's downfall — such as Japan's industrial miracle and the rebuilding of Europe — were, however, completely outside Blough's control. It is fact, nevertheless, that in 1955 he took over the largest company in a strong steel industry, and during the 14 years of his leadership the company and the industry fell behind on technology, lost a portion of its market to imports, and saw wages increase beyond the control of management.

In assessing Blough's performance, one must confront a series of contradictions. Probably more than any business leader in modern times, Blough is identified with free-market, private-enterprise philosophy. Unlike many business people, however, he never displayed rabid antigovernment feelings. He asserted that prices should be set by the market, yet he led an industry that in those days was frequently insensitive to market forces. The most fundamental inconsistency in Blough's

position was with regard to foreign steel. During the time that he was the industry's leader, steel companies simply declined to compete with imports. The only remedy U.S. Steel proposed was government action in the form of quotas or enforcement of the trade laws. This thinking on the part of steel executives persisted as the import share of the domestic steel market rose to more than 20 percent. The policy of avoiding competition was unsuccessful, and it was inconsistent with free-market principles.

When Blough retired from U.S. Steel in January 1969, he rejoined White & Case. After retirement he was a founder of the Business Roundtable, a Washington organization concerned with the interests of industry. He died on October 8, 1985, in Hawley, Pennsylvania.

Publications:
"U.S. Steel Plans $3.25 Billion Capital Outlay," *Commercial and Financial Chronicle,* 183 (May 10, 1956): 2250;

"Inflation as a Way of Life; U.S. Steel's Efforts to Stem Inflationary Tide," *Commercial and Financial Chronicle,* 184 (November 22, 1956): 2175;

The Great Myth (U.S. Steel Corporation, 1957);

Great Expectations (New York, U.S. Steel Corporation, 1957);

"Steel Industry a Victim of Inflation; Administered Prices in Concentrated Industries," *Commercial and Financial Chronicle,* 186 (August 22, 1957): 800;

Steel In Perspective (New York: Newcomen Society, 1958);

"Tale of Two Towns [Cleveland and Dusseldorf]" *Commercial and Financial Chronicle,* 187 (May 22, 1958): 2290;

"Price and the Public Interest," *Commercial and Financial Chronicle,* 188 (September 18, 1958): 1113;

"Urges Halt to Cost-Push Inflation," *Steel,* 144 (March 3, 1959): 63–65;

"Steel Industry and a Billion Dollar Bundle," *Commercial and Financial Chronicle,* 189 (March 12, 1959): 1189;

Free Man and the Corporation (McGraw-Hill, 1959);

"Outlook for Steel and the Economy as a Whole," *Commercial and Financial Chronicle,* 192 (October 20, 1960): 1555;

"My Side of the Steel Price Story," *Look,* 27 (January 29, 1963): 19–23;

The Brighter Side (New York: U.S. Steel Corporation, 1964);

and David J. MacDonald, "Help Coming in Dumping Rates, Steelmen Told," *Steel,* 154 (March 23, 1964): 28;

and E. F. Martin, "Steel Leaders See Gains Overshadowing Problems," *Iron Age,* 197 (January 6, 1966): 36;

Business Can Satisfy the Young Intellectual (Harvard Business Review, Reprint Service, 1966);
"Business is a Public Affair," *Iron Age,* 198 (July 7, 1966): 44–47;
Government Wage-Price Guideposts in the American Economy (New York: New York University, School of Commerce, 1967);

"Look at Technology in the Steel Industry," *Automation,* 15 (January 1968): 23;
The Washington Embrace of Business (Pittsburgh: Carnegie Press, Carnegie-Mellon University, 1975).

Reference:
George McManus, *The Inside Story of Steel Wages and Prices, 1959–1967* (Philadelphia: Chilton, 1967).

Raynal Cawthorne Bolling

(September 1, 1877 – March 26, 1918)

by Gerald G. Eggert

Pennsylvania State University

CAREER: Lawyer in firm of Guthrie, Cravath & Henderson, New York City (1902–1903); legal staff (1903–1907), assistant general solicitor (1907–1913), general solicitor, United States Steel Corporation (1913–1917).

Raynal C. Bolling attained success early in life, joining the legal staff of the United States Steel Corporation only a year after admission to the bar and rising to the rank of assistant general solicitor by the age of twenty-nine and general solicitor of the firm by thirty-six. His promising career was cut short when he was killed in action during World War I.

Bolling's paternal ancestry was southern. The Bollings, from England, were planters in Virginia from 1660 until the Civil War. The Raynals, who were of French extraction, had fled Haiti to Charleston, South Carolina, during Toussaint-Louverture's uprising in the early 1800s. The son of Sanford Coley and Ada Leonora Hart Bolling, Raynal was born in Hot Springs, Arkansas. The family soon moved to Philadelphia. Bolling prepared for college at the William Penn Charter School and attended Harvard College in 1896. An exceptional scholar and athlete, he was elected to Phi Beta Kappa before graduating in 1900. In 1902 he earned his LL.B. degree from Harvard Law School and was admitted to the bar in New York.

After a year with the prestigious New York City law firm of Guthrie, Cravath & Henderson, Bolling joined the legal staff of United States Steel Corporation. By 1907 he had attained the office of assistant general solicitor. That same year he

Raynal Cawthorne Bolling

married Anna Tucker, the daughter of John C. Phillips of Boston. The couple had five children, Anna, Raynal C., Jr., Cecilia (who died in infancy), Diana, and Patricia. In 1909 they established themselves in one of the finest homes in then-fashionable Greenwich, Connecticut. There

Bolling organized a hunt club and was prominent in both social and civic activities.

While at U.S. Steel he became a member of, and sometimes chaired, the firm's safety committee. This group devised a voluntary accident-relief plan for corporation employees long before the states imposed similar mandatory programs on all firms. Bolling later assisted in framing New Jersey's compensation law, which served as a model for other states over the next decade. The safety committee went on to design many of U.S. Steel's other welfare benefits for employees.

In 1911 the United States brought suit to dissolve U.S. Steel under the Sherman Antitrust Act. Bolling became a key figure in the company's defense team from the initiation of the suit through the hearing before the U.S. Supreme Court. In 1913 he became general solicitor of U.S. Steel. He was also a director of the Tennessee Coal, Iron & Railroad Company and president of the Arkansas Farms Company.

Because of a fascination with aviation, Bolling in 1907 joined Squadron A of the New York National Guard and eventually organized the first National Guard airborne company. As World War I drew near he became an early champion of aerial preparedness. In 1916 his unit assisted Gen. John J. Pershing in the campaign against Pancho Villa in Mexico. Upon U.S. entry into the European war, Bolling entered the Reserve Officers Training Corps program at Plattsburgh, New York. He was commissioned a major in the Signal Corps and was sent on a mission overseas in June 1917 as a special representative of the Aircraft Production Board. Promoted to the rank of colonel in August 1917, Bolling was put in charge of all air service activities except those involving combat.

His death came in the spring of 1918. During an inspection by car at the front near Amiens, France, he fell under enemy fire. When he and his chauffeur took cover in two nearby shell holes, Bolling found his occupied by two German officers. He killed one but was in turn shot and killed by the other. He was at the time the highest ranking officer from the United States to be lost in combat.

In a tribute to Bolling, *Iron Age* declared that no young man in America had enjoyed brighter prospects.

Publications:
"Results of Voluntary Relief Plan of the United States Steel Corporation," *Annals of the American Academy of Political and Social Science*, 38 (July 1911): 35–44;

"Rendering Labor Safe in Mine and Mill," *American Iron and Steel Institute Yearbook* (1912): 106–113; also published as "Millions Spent for Improving Labor Conditions," *Iron Trade Review*, 50 (June 6, 1912): 1237–1239;

The United States Steel Corporation and Labor Conditions, Publication No. 685 (Philadelphia: American Academy of Political and Social Science, 1912);

"Opportunity in Our National Defense," *Collier's*, 56 (January 8, 1916): 74+.

References:
Henry Greenleaf Pearson, *A Business Man in Uniform, Raynal Cawthorne Bolling* (New York: Duffield, 1923);
Tribute to Bolling, *Iron Age* (April 25, 1918): 1090–1091.

Willis Boothe Boyer

(February 3, 1915 – January 31, 1974)

by Carol Poh Miller

Cleveland, Ohio

CAREER: Clerk, cold-strip department (1937), operating and accounting departments (1937–1942), assistant to the treasurer (1946–1951), assistant treasurer (1951–1953), treasurer (1953–1963), vice-president and treasurer (1960–1963), vice-president of finance (1963), vice-president of finance and administration, director (1963–1966), executive vice-president (1966–1968), president (1968–1973), chief executive officer (1971–1974); chairman of the board, Republic Steel Corporation (1973–1974); assistant manager (1942–1945), manager, Reconstruction Finance Corporation — Defense Plant Corporation Accounting (1945–1946).

Willis Boothe Boyer, president of Republic Steel Corporation from 1968 to 1973 and chief executive officer from 1971 until his death in 1974, was regarded as an unusually warm and friendly man for a top-ranked industrialist. He was "Bill" to all his associates, and he made a habit of quickly putting visitors on a first-name basis. Born in Pittsburgh, Pennsylvania, Boyer was the son of Pearce F. and Sarah Hester Boothe Boyer. He attended Mercersburg Academy and was graduated from Lafayette College in Easton, Pennsylvania, in 1937. Later, during his steady climb through the executive ranks at Republic, he attended the Case Institute of Technology, where from 1937 to 1942 he took courses in law and business and completed the advanced management program at the Harvard Business School in 1951.

Except for a brief stint with International Business Machines, Boyer spent his entire working life with Republic Steel. He joined Republic in 1937 as a clerk in the cold strip department of the Cleveland steel plant but quickly moved to the accounting department. He spent the war years as assistant manager, then manager, of the Reconstruction Finance Corporation — Defense Plant Corporation Accounting, which was formed as a partnership between the federal government and Republic to handle the accounting work of Republic's extensive Defense Plant projects. In 1946 he moved to Republic's treasury department, where his father served as vice-president comptroller. He was appointed assistant treasurer in 1951 and treasurer in 1953.

Boyer, as treasurer, and Republic chairman Thomas F. Patton made many changes within the company. They introduced computerized cost control and established a closer relationship between the sales and operating departments through regularly scheduled meetings. The introduction of basic-oxygen furnaces helped boost capacity and improve profits. In November 1963 the forty-eight-year-old Boyer was elected a director of the corporation and promoted to the newly created office of vice-president of finance and administration, giving him increased management responsibilities. Under the new alignment the offices of treasurer, secretary, managers of purchases and raw materials, plastics, and international projects divisions, and the general traffic manager reported directly to Boyer. Three years later Boyer was named executive vice-president, the number-two post, reviving a title held some 30 years ear-

lier by R. J. Wysor and giving him direct supervision over Republic's manufacturing division.

With a solid background in steel and finance and a personable and vigorous demeanor, Bill Boyer projected the image the industry was desperately seeking when he was named president of Republic Steel in May 1968 — a period of much turmoil in the steel industry. From his office in the Republic Building in downtown Cleveland, Boyer could see the lakefront docks, which served as a daily reminder of one of the industry's most pressing problems — the flood of imported steel. In 1968 more than 18 million tons of steel, representing about 17 percent of the domestic steel market, were imported. Relations with labor, fortunately, had normalized, but new contracts negotiated during the decade resulted in increased costs that the company, following the industry's historic showdown with President John F. Kennedy and the U.S. Congress, was unable to offset fully with increased prices.

In 1971 Boyer, then fifty-six years of age, was given the additional title of chief executive officer, succeeding Patton, who retired. "These are difficult times," Boyer told a *Cleveland Press* reporter in 1971, "and for the short run the steel industry faces serious problems due to imports of foreign steel, increasing costs of labor and a price level that does not provide adequate returns on the huge amount of capital now invested in facilities." Boyer added that Republic's facilities, following a decade of expansion and modernization, were "the best they have ever been" and predicted the company could look forward to improved operations and profits in coming years. Under Boyer's tenure Republic enjoyed sales in excess of $1 billion; its employees numbered 43,000.

Boyer was a director of the National City Bank of Cleveland and of the Procter & Gamble, Marathon Oil, and Sherwin-Williams companies. He was a member of the American Iron and Steel Institute and a vice-chairman of the Conference Board. Boyer was also active in civic affairs. He served as treasurer of the Welfare Federation and as vice-president and treasurer of the United Appeal (later renamed the United Way). He was a trustee of University Hospital in Cleveland and Case Western Reserve University. In 1970 Boyer headed a drive to raise money in the Cleveland area for the United Negro College Fund. He was an avid golfer and enjoyed playing tennis.

Boyer married Esther Greenwood in 1938. In the early days of their marriage, Mrs. Boyer scrawled messages on the hard-boiled eggs she packed in her husband's lunch. They had three sons: Willis Boothe, Jonathan, and Christopher. They resided in Shaker Heights, a suburb of Cleveland. Boyer died in Cleveland's Lakeside Hospital following a six-week illness, three days shy of his fifty-ninth birthday. His untimely death occurred during a period of strong sales and increasing profits at Republic. In 1974, the year of Boyer's death, Republic's earnings reached a record high of $170.7 million.

Publication:
"Problems of Financing New Steel Capacity," *Iron and Steel Engineer*, 34 (July 1957): 75–76.

References:
"Boyer Becomes President at Time of Steel Ferment," *Cleveland Press*, May 9, 1968;
"Boyer No. 2 Man in Top-Echelon Shift at Republic," *Cleveland Plain Dealer*, April 21, 1966;
"Republic: New President, New Image?" *Iron Age*, 201 (May 16, 1968): 11.

Thomas J. Bray

(May 1, 1867 – December 11, 1933)

by Carol Poh Miller
Cleveland, Ohio

CAREER: Pattern-maker, chief draftsman, Lewis Foundry & Machine Company, interim employment as pattern-maker, Riverside Iron Works (1883–1890); civil engineer, Ohio Steel Company (later Ohio Works, Carnegie Steel Company) (1894–1900); secretary, mechanical engineer, chief engineer, McGill & Company (later United Engineering & Foundry Company) (1900–1906); assistant to the president in charge of construction (1906–1907), vice-president, operations (1907–1911), president (1911–1928), Republic Iron & Steel Company.

During his 17 years as president of the Republic Iron & Steel Company of Youngstown, Ohio, Thomas J. Bray presided over the firm's expansion. Once a producer of rolled iron products Republic, under Bray, became on of the country's largest independent steel producers. Bray was born in Pittsburgh, Pennsylvania, the son of Thomas Joseph and Anna Jacova Collins Bray. His father, a native of Wales and a descendant of the Thomas Bray who was one of the founders of William and Mary College, emigrated to the United States in 1853. He worked as a mechanical engineer for the National Tube Company of Pittsburgh, where he developed the first welded steel pipe and designed and built that company's Riverside Iron Works at Benwood, West Virginia.

The younger Bray was educated in Pittsburgh's public schools, then served for three years (1883–1886) as a pattern-maker with the Lewis Foundry & Machine Company of Pittsburgh. He later served as a pattern-maker for the Riverside Iron Works and as chief draftsman for the Lewis Foundry, resigning from the latter position in 1890 to continue his education. Bray attended Lehigh University and earned a mechanical engineering degree in 1894. During the next six years he was employed as a civil engineer with the Ohio Steel Company of Youngstown, Ohio (later the Ohio Works of the Carnegie Steel Company) and as secretary and mechanical engineer with McGill & Company of Pittsburgh, an engineering firm specializing in pipe mills. He was chief engineer of that company in 1901 when it became part of the United Engineering & Foundry Company.

In 1906 Bray resigned from United Engineering to join the Republic Iron & Steel Company as assistant to the president in charge of construction. A year later, he was promoted to vice-president in charge of operations. In 1911 Bray was elected president of the company, a position he held until he resigned in 1928.

These were dynamic years for the steel industry and for Republic Iron & Steel. The company consisted of several wrought-iron plants that had merged in 1899. Under Bray's direction, Republic gradually shifted its focus to steel, building a large, centralized steel plant at Youngstown. During Bray's term as chief executive Republic built two blast furnaces, an open-hearth plant, by-product coke ovens, and a tube works; all contributed to Republic's rise as one of the largest and strongest independent steel producers in the country. In April 1928 Bray announced his resignation and retirement, which took effect on May 1 of that year. He was succeeded, briefly, by Elmer T. McCleary. Two years later the company merged with several other steelmakers and fabricators, forming the nucleus of the new Republic Steel Corporation.

Bray served as a director of the First National Bank and the Dollar Savings & Trust Company, both of Youngstown, and was a member of the executive committee of the Youngstown Sheet & Tube Company. He was a member of the American Iron and Steel Institute, the British Iron and Steel Institute, and the American Society of Mechanical Engineers.

Following retirement Bray worked actively on a project to build a canal from Beaver, Pennsylvania, which sits on the Ohio River, to Youngs-town, to reduce coal-delivery costs. In 1930 he was elected president of the Beaver-Mahoning-Shenango Improvement Association, organized to promote the canal project, which was never realized. Bray's leisure interests included golf and bridge. He was a Unitarian. On October 7, 1896, Bray married Isabel Matthews of Pittsburgh. They had three sons: Thomas Joseph (whose appointment as assistant general superintendent of the Republic Iron & Steel Company was announced concurrently with Bray's retirement as president), Theodore Matthews, and Charles William Bray. He died suddenly in his office in Youngstown on December 11, 1933.

Publications:
"The Importance of the Investment Factor in Sales Policy," *Year Book of the American Iron and Steel Institute* (1914): 114–120;
"The Investment Factor in Costs," *Iron Trade Review*, 54 (May 28, 1914): 967–68.

Reference:
Joseph G. Butler, Jr., *History of Youngstown and the Mahoning Valley*, 3 volumes (Chicago and New York: American Historical Society, 1921), II: 132.

Brier Hill Steel Company

by Bruce E. Seely

Michigan Technological University

The Brier Hill Steel Company, located in the Mahoning Valley near Youngstown, Ohio, grew out of a series of local iron-making companies formed in the late nineteenth century before it made the transition to a regional steelmaker in the early part of this century. Brier Hill was absorbed by Youngstown Sheet and Tube Company in 1923.

Brier Hill Steel traced its origins to the Akron Manufacturing Company, organized in 1838. Akron Manufacturing moved to the Youngstown vicinity in 1859 and was renamed the Brier Hill Iron Company. When the original charter ran out, the company was reformed as the Brier Hill Iron and Coal Company in 1868. The firm operated three blast furnaces, selling its merchant iron to Wheeling nail makers. It had a reputation as a progressive business, claiming to have installed the first Whitworth hot-blast stoves in the country as well as to have opened the first laboratory for chemical analysis. The company also attempted to work through the business cycle by stockpiling iron during slack times.

In 1878 Joseph Butler, one of the more colorful figures in the American iron and steel industry, joined Brier Hill as general manager. Butler had a background with other iron companies in the Mahoning Valley and eventually became a central figure in the Youngstown, and thus the national, steel community. Under his direction the company paid dividends for 25 successive years. The company also provided a training ground for such noted steel men as Julian Kennedy, a blast furnace and steel works designer and engineer; William B. Shiller, president of National Tube Company; and Carl Meissner, a chemist and engineer for United States Steel Corporation (U.S. Steel). In 1882 Brier Hill was reorganized again because its coal holdings had been virtually depleted. Renamed and recapitalized as the Brier Hill Iron Company, the firm continued to prosper into the early twentieth century. Butler became vice-president of Brier Hill in 1912.

In February 1914 W. A. Thomas finalized the creation of the Brier Hill Steel Company by consolidating the Brier Hill Iron Company with two sheet-steel rolling mills in Niles, Ohio, nine miles from Youngstown. The logic of the combination was clear — Brier Hill's two blast furnaces, built originally in 1889 and 1890 and rebuilt in 1896 and 1908, would gain an assured market while his own Thomas Steel Company and the Empire Iron and Steel Company's sheet mills gained an assured supply. The consolidation also brought together iron properties in the Lake Superior district and coal resources in Pennsylvania. The missing piece was a steelmaking facility to convert Brier Hill's pig iron, and in 1913 ground was broken for an open-hearth plant adjacent to the blast furnaces in Brier Hill. The seven 75-ton open hearths, a 40-inch blooming mill, and a six-stand billet and slab mill produced semifinished steel for the sheet mills when the open-hearth plant opened in 1914.

The new company prospered and acquired additional facilities. The oldest blast furnace was rebuilt, and five more open hearths soon were added. The Western Reserve Steel Company, a new operation with 28 sheet mills, was purchased, and in 1917 a third blast furnace was constructed. Also in 1917 a by-product coke plant was installed in Youngstown. By 1920 pig iron capacity stood at 600,000 tons and steel ingot capacity at 510,000 tons.

In the early 1920s Brier Hill was one of seven independent steel companies that figured in an unsuccessful merger attempt to create a large steel corporation able to compete with U.S. Steel and Bethlehem Steel. The plan failed after Bethlehem acquired two of the seven companies,

Lackawanna Steel and Midvale Steel & Ordnance. Youngstown Sheet and Tube Corporation then acted to improve its competitive position by purchasing two of the remaining firms, Steel & Tube Company of America and Brier Hill Steel. Youngstown paid $30 million in cash for Brier Hill, bringing most of the iron and steel operations in the Youngstown area under its control, although, almost immediately upon purchase, Youngstown Sheet and Tube sold the sheet mills in Niles because they were outdated.

Over the years the Brier Hill plant was a vital part of the Youngstown Sheet and Tube Company. The first electrically welded pipe production facility was installed at Brier Hill in 1930, and throughout the years improvements were made to the facilities. By the mid 1970s, however, the plant was increasingly vulnerable to imports; its location in the Mahoning Valley was uncompetitive compared to that of the low-cost mills on the Great Lakes. After 1975 operations at the plant were being closed by Youngstown Sheet and Tube, and when Jones & Laughlin Steel bought Youngstown Sheet and Tube in 1979 all operations at Brier Hill had ceased. Yet the plant was partially revived in the early 1980s by a minimill company. Hunt Steel acquired the property and installed a pair of 85-ton electric furnaces, a continuous caster, and modern pipe-making equipment — converting the old Brier Hill plant into the first minimill to make seamless pipe.

References:

Joseph G. Butler, Jr., *Recollection of Men and Events; An Autobiography* (Youngstown: by the author, 1925; New York: Putnam's, 1927);

H. Cole Estep, "Brier Hill Is Now Making Steel," *Iron Trade Review,* 54 (April 2, 1914): 627–628;

William T. Hogan, S.J., *An Economic History of the Iron and Steel Industry in the United States,* 5 volumes (Lexington, Mass.: Lexington Books, 1971), III: 983–987;

Hogan, *Minimills and Integrated Mills: A Comparison of Steelmaking in the United States* (Lexington, Mass.: Lexington Books, 1987), 31–34.

Eugene J. Buffington

(March 14, 1863 – December 9, 1937)

by Bruce E. Seely

Michigan Technological University

CAREER: Treasurer and director, American Wire and Screw Nail Company (1884–1898); treasurer, director, member of executive committee, American Steel & Wire Company (1898–99); president, Illinois Steel Company (1899–1932); president, Indiana Steel Company, (1906–1932); president, Gary Land Company, (1906–?).

Rising to the top from humble origins, Eugene J. Buffington was a typical United States Steel Corporation executive of the early twentieth century. He was born in Guyandotte, West Virginia, in 1863 and he attended Vanderbilt University in 1881–1883. He married Drucilla Nicholas in 1888 and had two children.

In 1884 Buffington helped found the American Wire and Screw Nail Company in Covington, Kentucky, and became its treasurer. By 1888 he was building a new mill in Anderson, Indiana, which the American Steel & Wire Company would acquire a decade later. Buffington became a director, treasurer, and a member of the executive committee. On January 1, 1899, Buffington was made president of Illinois Steel Company, a position he retained after the formation of U.S. Steel in 1901. He continued to head this arm of the new corporation until his retirement in 1932.

While heading Illinois Steel, Buffington built the largest steel mill in the world. U.S. Steel — deciding in 1905 that it had to replace older facilities and respond to a growing market in the Midwest — added capacity to Illinois Steel. Judge Elbert H. Gary, head of U.S. Steel, instructed Buffington to find a mill site. Waukegan and South Chicago were rejected as too expensive or too crowded, although the progressive tendencies of Illinois also may have affected that decision. Instead, Buffington chose a site in northern Indiana, several miles east of Hammond. Offering cheap land, a harbor for lake freighters, an inexhaustible water supply, connections to both trunk-line railroads and Illinois Steel's own line, and access to

Eugene J. Buffington

Chicago's huge labor pool, the site seemed ideal for industrial development.

Work began in 1906, the year the Indiana Steel Company was formed. Another subsidiary, the Gary Land Company, built the town. Buffington was president of both subsidiaries, and presiding over the transformation of the Indiana dunes into the city and mills called Gary marked the high point of his career. The modern plant was efficiently designed, and it included the world's largest rail mill. Several U.S. Steel subsidiaries located fabricating plants nearby, including Universal Portland Cement, which used blast furnace slag. By 1909, the year the first Gary steel was made, Gary was one of the largest industrial centers in the country.

The mill city itself was less successful. U.S. Steel, according to a 1909 article by Buffington, wanted to avoid creating any "suspicion of domination or control by the employer." The problems in Pullman, Illinois, some fifteen years before were not to be repeated. Even so, U.S. Steel guided Gary's growth, using as a model the American Sheet and Tin Plate Company, which successfully built a mill community in Vandergrift, Pennsylvania. The Gary Land Company laid out the town and built the downtown area and a section of nicer homes. The Gary Heat, Light, and Water Company, also headed by Buffington, won the utility franchises for the city. When plans to entice workers to build homes failed because of high land prices, the company reluctantly built houses. But the same steel men who laid out wonderfully efficient steel mills designed a city with few parks, no access to Lake Michigan, and a rigid grid system especially prone to congestion at shift changes. And in areas outside the company's control, development was haphazard. Historians Raymond Mohl and Neil Betten commented that "Gary's designers built only half a city and provided for only some of its citizens."

In the end, Buffington's city and mills reflected the thinking of Elbert Gary, who projected an attitude of social responsibility, but subordinated morality to profit. The town of Gary was not the only evidence of this attitude. For example, Buffington, in league with other subsidiary presidents, opposed the labor reform efforts of corporate vice-president William Dickson in 1910. Dickson campaigned to eliminate Sunday labor and the 12-hour day, but Buffington argued Illinois Steel would lose production and profits, a position Gary accepted.

Buffington's social conscience took a form common among leading industrial figures — community service. During World War I he served on the capital issues committee of the Chicago Federal Reserve Bank and was a director of the war camp community service. He had a long affiliation with the YMCA in Chicago, and supported his alma mater, Vanderbilt University, as a trustee and with a $100,000 donation in 1935.

Buffington also served on several corporate boards. He was a director of the U.S. Steel Corporation and its subsidiaries: H. C. Frick Coke Company, United States Coal and Coke, United States Fuel, Universal-Atlas Portland Cement, and the Indiana-Scully Steel Warehouse Company. He sat on the boards of the Continental-Illinois Bank & Trust Company, the South Chicago Savings Bank, and the Gary State Bank. He was a trustee of the Community Trust of Chicago, and the Chicago Sunday Evening Club.

Buffington was at heart a steelman, as U.S. Steel recognized by giving his name to both an ore carrier and the Indiana town where Universal-Atlas Portland Cement was located. A reflection of the esteem of his peers was his election as director of the American Iron and Steel Institute in 1928. As *Iron Trade Review* commented in 1926, "to few men has been given the opportunity to be associated with the construction of as much pig iron and finished steel capacity as Mr. Buffington." Myron Taylor, head of U.S. Steel during the Depression, asked him to retire in 1932 to open the way for younger men, and after forty eight years in the steel industry, Buffington said, "If I had my business life to live over, I would choose the steel industry . . . and association with the United States Steel Corporation. After so many years of service I naturally feel it is time for me to retire and take life more leisurely." Buffington died in Chicago after an operation on December 9, 1937.

Publications:
"Making Cities for Workmen," *Harper's Weekly,* 53 (May 8, 1909): 15–17;
"Contract Obligations," *American Iron and Steel Institute Yearbook* (1910): 101–102;
"Remarks," *American Iron and Steel Institute Yearbook* (1910): 122–123;
"Brief Remarks," *American Iron and Steel Institute Yearbook* (1928): 39.

References:
Gerald Eggert, *Steel Masters and Labor Reform, 1886–1923* (Pittsburgh: University of Pittsburgh Press, 1981);
"Industrial Man in the Day's News: Eugene J. Buffington," *Iron Trade Review,* 78 (January 14, 1926): 182.

Edwin O. Burgham

(August 5, 1889 – October 20, 1969)

by Glen V. Longacre III
Columbus, Ohio

CAREER: Foreman, Phillips Sheet and Tin Plate Company (1909–1923); metallurgist (1923–1925), assistant manager, Weirton Steel Company (1925–1930); plant manager (1930–1944), assistant general superintendent (1944–1948), general superintendent (1948–1953), vice-president for operations (1953–1955), president of operations (1955–1958), chairman of the board, National Steel Corporation, Weirton Steel Division (1958–1960); consultant to Weirton Steel Company (1960–1969).

Typical of many twentieth-century American steel executives, Edwin Oliver Burgham began his career working in the mills and rose slowly to the ranks of management. Burgham was born on August 5, 1889, in Leechburg, Pennsylvania, where his father, Oliver Higgins, had settled after emigrating from England in 1884. Burgham attended public schools in Parnassus, Pennsylvania, and in the Allegheny section of Pittsburgh, Pennsylvania. From 1905 to 1909 Burgham attended the School of Engineering at the Carnegie Institute of Technology in Pittsburgh. While at the Carnegie Institute he played shortstop for the varsity baseball team.

In 1909 Burgham took a position as foreman at the Phillips Sheet and Tin Plate Company in Clarksburg, West Virginia, where his father was employed as a plant manager. Phillips Sheet and Tin Plate had been incorporated in 1905 by James R. Phillips, Ernest T. Weir, E. W. Mudge, and other stockholders. The company's plant buildings, formerly owned by the Jackson Sheet and Tin Plate Company, housed eight stands of rolls — three for rolling of sheets and five for rolling tinplate.

From this start would eventually emerge the Weirton Steel Company in 1918. But the company Burgham joined was only just moving toward becoming a fully integrated steel producer. In 1909 limited space and insufficient water supply at the Clarksburg site had necessitated the expansion of

Edwin O. Burgham

facilities and operations to Holliday's Cove, West Virginia. The new location on the banks of the Ohio River was much larger, at first accommodating a tin-rolling mill, to which a blast furnace and blooming and billet mills were later added.

In 1925 Burgham was promoted to assistant manager of the tin mill in Weirton. That same year the company built the world's first 48-inch hot-strip mill, a step that continued the rapid growth of Weirton Steel. In 1929, however, Weirton Steel's corporate structure changed, and Burgham's career began to soar. Weirton Steel merged with the Great Lakes Steel Corporation and the M. A. Hanna Company of Cleveland, Ohio, to become the National Steel Corporation. The merger offered Burgham an opportunity to advance in the now much larger company hierarchy. From 1930 to 1944 he worked as plant manager in Weirton. Before his tenure the company had placed emphasis on handcraftsmanship in the production of tin sheets. Burgham, however, stressed the importance of new technologies and modernized the techniques of manufacturing tin plate, eliminating the older hand methods. Additional coke ovens were added in 1930. The company's first Bessemer converter was built in 1936, a second one in 1941. In 1944 he was promoted to assistant general superintendent, a position he held until 1948, when he was named general superintendent of the mill.

In 1953 Burgham became vice-president of operations for National Steel; two years later he moved into the president's office. In 1958 he advanced again to the chairman's office, following the death in 1957 of E. T. Weir, the company's driving force since 1905. As president and chairman, Burgham did not achieve the same level of visibility or importance that Weir had, but then few American steel men did; National had struggled to find a team of managers after Weir's death. Of Burgham, however, it can be said that as president he continued the company's position, established by Weir, as the most profitable firm in the industry, calculated as a percentage of total sales.

National's place in the industry owed much to the company's continuous capital expansion program launched in 1946. Burgham's background as a mill laborer gave him a practical knowledge of how to continue the improvement program and figured prominently in his promotion. National's two steelmaking facilities at Weirton and the Great Lakes plant near Detroit were completely rebuilt during these years. New blast furnaces doubled pig-iron capacity between 1951 and 1960, while the company radically increased the capacity of open-hearth furnaces from an average of 200 to 500 tons per heat. The company continued to focus on its light-rolled product line that grew from its pioneering role in the development of the hot-strip mill and expanded and improved its rolling mills as well. In 1959 the company authorized construction in the Chicago region of a $300 million facility named Midwest Steel Company. In 1960, at the end of Burgham's tenure as president, Weirton became one of the first American steelmakers to adopt the basic oxygen furnace.

Burgham was an ardent supporter of the Boy Scouts of America. He was a life member of the Region Four Committee of the Boy Scouts and an executive board member of the Fort Steuben Area Council. In addition to scouting, Burgham was involved in other civic projects, participating in the Boy's Clubs of America program, and, at different periods, as trustee for the Ohio Valley Hospital in Steubenville.

Burgham also served as president and director of the National Mines Corporation of Isabella, Pennsylvania; director of the National Steel Products Company of Houston, Texas; and director of the Midwest Steel Corporation, located in Portage, Indiana. All of these were National Steel affiliates. He was a member of the Iron and Steel Institute and the Association of Iron and Steel Engineers. He was also a member of Lions' International, the Shriners, the Rotary Club of Steubenville, and the Duquesne Club of Pittsburgh.

Burgham's first wife died in 1947. He was remarried on July 9, 1949, to Martha Edgington, former wife of Kenneth M. Hunt and daughter of Edwin E. and Martha McCauslen of Steubenville. He had an adopted son, Kenneth Morgan, by his second marriage. Burgham died in Steubenville on October 20, 1969.

References:
William T. Hogan, S.J., *An Economic History of the Iron and Steel Industry in the United States,* 5 volumes (Lexington, Mass.: Lexington Books, 1971);
Charles Longenecker, "Weirton Steel Company, Its 35 Years of Unusual Achievement," *Blast Furnace and Steel Plant,* 28 (August 1940): 773–775.

Joseph Green Butler, Jr.

(December 21, 1840 – December 19, 1927)

by Bruce E. Seely

Michigan Technological University

CAREER: Bookkeeper (1857–1858), financial manager (1858–1863), James Ward & Company, Niles, Ohio; business agent, Hale & Ayer, Chicago (1863–1866); manager, Girard Iron Company (1866–1878); general manager, Brier Hill Iron and Coal Company (1878–1912); vice-president, Brier Hill Steel Company (1912–1917); secretary, vice-president, Ohio Steel Company (1893–1906); chairman of the board, Bessemer Limestone & Cement Company.

Although he was a member of the fraternity of steelmen who brought the industry into the twentieth century, Joseph Butler reflected nineteenth-century attitudes. Trained by experience, he was devoted to a high tariff and freedom from government interference, and was totally opposed to organized labor. He maintained a devotion to history and writing which added to his colorful reputation among American steel executives of the early twentieth century.

Butler was born into the iron industry at Temperance Furnace, Pennsylvania, where his father and grandfather worked at nearby furnaces. To run a company store, his father moved in 1841 to Niles, Ohio, where Joseph attended school with William McKinley. At the age of thirteen he cut wood and made charcoal at the Tremont Furnace in Pennsylvania. A year later Butler was back in Niles as a clerk for James Ward & Company. He rose quickly in the company and in 1857 became the bookkeeper. His suggestions for improvements in Ward's rolling mill were rewarded with a promotion to mill manager.

In 1863 Butler accepted an offer from Hale & Ayer, a Chicago purchasing agency, and served as their agent at a Youngstown furnace for two years. But in 1866, he joined other local men in forming the Girard Iron Company. Their poorly constructed and designed furnace struggled, and the company barely survived the panic of 1873. A. M. Byers and Company of Pittsburgh provided an infusion of capital, and by 1878 only Butler remained of the original partners.

With the Girard Iron Company facing a continuing struggle, Butler decided to accept the post of general manager at the Brier Hill Iron and Coal Company at Youngstown, Ohio. Organized in 1859, the company produced iron for Wheeling's rail mills and posted steady returns, paying dividends for 25 years in one stretch. The company operated the first Whitworth hot-blast stoves in the United States, and was one of the first steel companies to build a chemical laboratory. The company also trained several leading steelmen, including Julian Kennedy, Carl A. Meissner, and William Shiller, later president of National Tube. The firm was recapitalized in 1882 as the Brier Hill Iron Company, and again in 1912 as the Brier Hill Steel Company; throughout the restructuring Butler remained as general manager and later became vice-president.

Butler also found time for other business activities. Beginning in 1890 he helped bring a Bessemer converter and rail mill to Youngstown to provide a market for Mahoning Valley iron. The project resulted in the creation of the Ohio Steel Company, which produced its first steel in 1895. In 1899 Ohio Steel became a part of W. H. Moore's National Steel Company, which was absorbed into United States Steel Corporation in 1901. Butler also backed the organization of Youngstown Sheet & Tube, and served as chairman of the board of the Bessemer Limestone & Cement Company. He even found time to win three terms on the Youngstown City Council, sit on the board of health, and serve repeatedly in the Youngstown Chamber of Commerce.

Butler was more than a leader of the Mahoning Valley iron industry. Nationally, he long led the fight to maintain a high tariff on imported steel and iron. He also was president of the Bessemer Pig Iron Association. But Butler most enjoyed his involvement in the American Iron and Steel Institute (AISI), organized by Elbert Gary in 1908.

Along with Charles Schwab, Butler became famous for his impromptu remarks at the semiannual AISI banquets. Adding to his reputation for wit were his writings and talks on the history of the iron industry. His seven books also included accounts of his life, William McKinley, and the Mahoning Valley. He was once president of the Mahoning Valley Historical Society.

In 1910 Butler unveiled plans for a National McKinley Birthplace Memorial in Niles, Ohio, in honor of his boyhood friend who had supported high tariffs. An act of Congress provided a charter and Butler raised subscriptions. He put up $100,000 himself for the McKinley library, and Henry Clay Frick donated $50,000. Construction of the monument started in 1915, and it was dedicated in 1917.

After retiring in 1917, Butler next provided Youngstown with an art gallery. He had started collecting paintings in the late 1870s and displayed them in his home. When a fire destroyed all but one of his pictures, Butler decided to build a museum. Finished in 1919, the Butler Art Institute was dedicated in 1920, with its benefactor furnishing an operating endowment for the $500,000 building, although he restricted the holdings to American artists.

Poor health, beginning with an auto accident in 1920, kept Butler from attending several AISI meetings, but he remained the grand old man of that organization through the mid 1920s, regaling the audience with his stories. A tie to the days before giant corporations made the industry so impersonal, Gary dubbed Butler "Uncle Joe." By 1920 few small companies survived, although the AISI banquets preserved the old camaraderie. Even Brier Hill was purchased by Youngstown Sheet & Tube in 1923.

Joseph Butler died on December 19, 1927. He had outlived his wife, Harriet Ingersoll, and two of his three children.

Publications:
A Catalogue of Indian Portraits in the Collection of Joseph G. Butler, Jr., Youngstown (Youngstown, Ohio: Vindicator Press, 1907);
Presidents I Have Seen and Known, Lincoln to Taft; and a Day in Washington's Country (Cleveland: Penton Press, 1910);
"Competition—Its Uses and Abuses," *American Iron and Steel Institute Yearbook* (1912): 39–43;

"Proposed McKinley Memorial at Niles, Ohio," *American Iron and Steel Institute Yearbook* (1912): 114–116;
"Remarks," *American Iron and Steel Institute Yearbook* (1913): 275–276;
"Remarks," *American Iron and Steel Institute Yearbook* (1913): 461–463;
"How to Hold the Home Market," *Iron Trade Review*, 54 (January 14, 1914): 7;
"Remarks," *American Iron and Steel Institute Yearbook* (1914): 508–512;
"Details as to What We Are Up Against," *Iron Trade Review*, 56 (January 7, 1915): 28;
"Remarks," *American Iron and Steel Institute Yearbook* (1915): 281–282;
"United States Must Make Choice," *Iron Trade Review*, 59 (November 2, 1916): 914;
"Fifty Years of Iron and Steel," *American Iron and Steel Institute Yearbook* (1917): 278–352;
A Journey Through France in War Time (Cleveland: Penton Press, 1917);
"Remarks," *American Iron and Steel Institute Yearbook* (1917): 409–412;
Fifty Years of Iron and Steel (Cleveland: Penton Press, 1918);
"American Steel in the World War," *American Iron and Steel Institute Yearbook* (1919): 274–292;
"Remarks," *American Iron and Steel Institute Yearbook* (1920): 224–231;
History of Youngstown and the Mahoning Valley, Ohio, 3 volumes (Chicago & New York: American Historical Society, 1921);
"Remarks," *American Iron and Steel Institute Yearbook* (1921): 235–238;
My First Trip Abroad (N.p.:The author, 1921; Cleveland: Penton Press, 1922);
Life of William McKinley and History of National McKinley Birthplace Memorial (Cleveland, 1924);
Recollection of Men and Events; An Autobiography (Youngstown, Ohio: Published by the author, 1925);
"Early History of Iron and Steel Making in the Mahoning Valley," *Iron Trade Review*, 77 (August 20, 1925): 425–428; (August 27, 1925): 481–484;
Autographed Portraits (Youngstown, Ohio: Butler Art Institute, 1927).

References:
"Butler's Brief," *Iron Trade Review*, 52 (January 16, 1913): 216;
"Joseph G. Butler, Jr., Dean and Historian of the Steel Industry," *Iron Age*, 120 (December 22, 1927): 1741;
H. Cole Estep, "Brier Hill Is Now Making Steel," *Iron Trade Review*, 54 (March 2, 1914): 627–637.

The A. M. Byers Company

by Michael Santos

Lynchburg College

Founded by Alexander McBurney Byers in 1864, the A. M. Byers Company grew to be the largest manufacturer of wrought iron pipe in the United States by the first decade of the twentieth century, and one of only two wrought iron producers in the world by the 1960s. A family-owned enterprise until 1956, the firm was shaped by the traditions and values of the Byers family.

Alexander McBurney Byers had entered the iron industry when company organization was personally controlled by iron masters who concentrated power in their own hands. Born in Mercer County, Pennsylvania, in 1837, Byers left his home at an early age to enter the iron business. By the age of sixteen he had become superintendent of Henry Clay Furnace and was reputed to be able to puddle iron better than some of his workmen.

Knowing the labor-intensive nature of wrought iron production and learning his craft at a time when the day-to-day operations of the iron plant were left to skilled craftsmen, Byers accepted a high degree of worker autonomy and control on the shop floor as a natural cost of doing business.

Unpredictable iron markets in the antebellum period taught Byers and other ironmasters that the only way to compete successfully was to control those elements of the industry that were controllable as closely as possible. Since workers held a monopoly of the skills necessary for production, labor costs were fixed. Technology and business structure were thus the major areas of innovation open to iron company executives.

Byers had integrated these lessons well by the time he incorporated his company with Joseph Graff in 1864. Having learned the importance of entrepreneurial control, Byers guided the company through several managerial changes over the next twelve years until February 1876, when he and his brother acquired sole ownership of the company. Firmly in control, Byers set about modernizing the company's business organization and technology. He supplied the firm's southside Pittsburgh plant with the latest equipment available, guaranteeing an annual output of 20,000 tons by 1879. By 1886 Byers had horizontally integrated his operations by acquiring a blast furnace at Girard, Ohio, to produce pig iron for his southside plant, and Lake Superior iron ore mines that provided both facilities with raw materials. Throughout his tenure as head of the firm, Byers enjoyed amicable relations with the Amalgamated Association of Iron and Steel Workers, the skilled iron workers' union. When steel began to challenge iron's supremacy in the 1880s, Byers, unlike other iron manufacturers, refused to convert to the cheaper metal. Vowing to produce "quality wrought iron pipe or bust," he carved out a stable, specialized market for his product. As an ironmaster who viewed the market as essentially erratic and uncontrollable, Byers believed that the types of capital outlays and restructuring necessitated by conversion to steel production were foolhardy.

When Alexander McBurney Byers died in 1900, his family continued his tough-minded conservative approach to business, maintaining control over all aspects of the firm's operations. As late as 1923 the Byers family held almost absolute financial control of the firm; until that year, stock was closely held and dividends were not made public. From 1900 through 1940 Byers's sons, Dallas Canon, Eben McBurney, and John Frederic, succeeded each other as head of the company, usually after an apprenticeship working on the shop floor. Byers's grandsons, J. Frick and Buckley, worked summer jobs in the company's mills and climbed their way to the top of the company hierarchy by the 1950s.

The company expanded and diversified operations to stay competitive throughout the twentieth century, often pioneering new technologies in wrought iron production in the process. Between 1903 and 1908 the company leased the property of the Clearfield Steel and Iron Company to supplement production of wrought iron being made for its Pittsburgh pipe works. In 1908 Byers opened the largest puddling mill in the United

States at Girard, Ohio, a facility that operated successfully until 1930. In that year the company opened a new plant at Ambridge, Pennsylvania, employing the first successful alternative to hand-puddling since its invention over 300 years earlier. The Aston, or Byers, process differed from earlier attempts to displace puddling by breaking the metallurgical process into its component parts, using different equipment at each stage of the process. To take advantage of demands for alloy steels generated by World War II, the company installed electric furnaces at the Ambridge facility in the 1940s.

Throughout its history the A. M. Byers Company experienced relatively good labor relations. Except for turmoil generated by rival unions in the early part of the century, the company negotiated with the Amalgamated Association of Iron and Steel Workers until 1930, when the Aston process made union puddlers obsolete. Even then the company was careful to find places for most of the men who had been put out of work by the new technology. When the Steel Workers Organizing Committee (the predecessor to the United Steelworkers Union) began its campaign in April 1937 to win recognition from the independent steel companies and smaller producers and fabricators, the A. M. Byers Company was one of the first firms to sign a union contract.

In 1956 General Tire and Rubber Company acquired control of the A. M. Byers Company in a friendly takeover. Apparently uncomfortable with Byers's traditional management style, General Tire officials began to dismantle the wrought iron capabilities of the company and use Byers as a holding company to acquire a wide range of smaller firms. Lay-offs and cutbacks accompanied a major corporate restructuring so that by the early 1960s all of the old Byers management team had either retired or left the company. In 1966 the firm closed its 102-year-old Pittsburgh plant. Three years later, when labor refused to accept a second "give-back" contract in three years, the company shut down the Ambridge mill, and with it, the firm's wrought iron division. General Tire's move away from wrought iron production made Alexander Byers's words seem prophetic: "quality wrought iron pipe or bust."

References:
James Aston and Edward B. Story, *Wrought Iron: Its Manufacture, Characteristics and Applications,* fourth edition (Pittsburgh: A. M. Byers, 1959);
Michael W. Santos, "Brother against Brother: The Amalgamated and Sons of Vulcan at the A. M. Byers Company, 1907–1913." *Pennsylvania Magazine of History and Biography* (April 1987): 195–212;
Santos, "Iron Workers in a Steel Age: The Case of the A. M. Byers Company, 1900–1969," Ph.D. dissertation, Carnegie-Mellon University, 1984;
Santos, "Laboring on the Periphery: Managers and Workers and the A. M. Byers Company, 1900–1956," *Business History Review,* 52 (Spring 1978): 113–133.

Archives:
The A. M. Byers Company Metallurgical Division Records Collection and the A. M. Byers Company Personnel Records Collection are located in the Archives of Industrial Society, University of Pittsburgh Libraries, Pittsburgh, Pennsylvania.

By-Product Coke Ovens

Terry S. Reynolds

Michigan Technological University

Having replaced the traditional beehive coke oven, the by-product coke oven was the primary technological advancement in the American coke industry in the twentieth century.

Beehive ovens, usually built of brick and shaped like a dome, formed the basis of the American coke industry during the nineteenth century. Coking plants consisted of long rows of beehive ovens, which were filled with raw coal through top openings. After ignition, the coal in the oven smoldered while the air flowing into the interior was closely regulated. The controlled heat purged the coal of volatile substances, typically one-third of the weight of coal, leaving, after 48 to 72 hours, a high carbon residuum, called coke. Usually enough heat was left from previous charges to

A by-product coke plant in Clairton, Pennsylvania, 1940s

start the process again. At the turn of the century more than 95 percent of the coke produced in the United States came from this process.

Developed in late-nineteenth-century Europe, the by-product coke oven did not replace the beehive oven in the United States until the early twentieth century. In an attempt to introduce the soda-ash industry to the United States, the Solvay Process Company first used by-product coke ovens at its chemical plant at Syracuse, New York, in 1893. The oven's first application to metallurgy, however, came four years later in 1897. Like beehive ovens, by-product ovens were arranged in rows, or batteries, and their operation was similar. The ovens heated raw coal, expelling its volatile contents. But, where beehive ovens had been heated internally, the box-shaped by-product ovens were heated by external flues, and pipes carried off the volatile gases to an adjacent by-product recovery plant instead of releasing them into the atmosphere.

Some of the gases from by-product ovens were saved and used as fuel; most of the gases were condensed and distilled to produce chemicals such as tar, ammonia, benzol, toluol, and naptha. But by-product reclamation was not the

only advantage of the ovens. They gave higher coke yields (70–75 percent instead of 65 percent), reduced coking time from 48 to 24 hours, and yielded a purer and more uniform coke. They could also coke some coals the beehive oven could not.

Despite these advantages by-product ovens were not immediately adopted. Beehive ovens could be built more quickly and much more inexpensively than by-product ovens — major advantages in regions where coal was abundant and capital scarce. And, because low quality coals which could not be treated by beehive ovens were initially used in by-product ovens, some steel manufacturers were prejudiced against the coke that by-product ovens produced. Moreover, prior to 1914 little demand for the by-product chemicals existed. In 1910 by-product ovens accounted for the production of a little over 7 million tons of coke, only 17 percent of the nation's output. And beehive ovens continued to be constructed at a rapid pace. In 1910 there were over 100,000 beehive ovens and only 4,000 by-product ovens.

By 1920, however, by-product ovens produced 60 percent of all coke in America and had clearly replaced the beehive oven. Several factors contributed to this shift in technology. First, the growing concentration and size of the American iron and steel industry made investment in the capital-intensive but more efficient by-product oven appear steadily more attractive. Second, during the 1910s markets emerged for many of the oven's chemicals. Demand for explosives in World War I made by-product chemicals such as benzol, toluol, phenol, and naptha commercially profitable. Benzol, a central component of aniline dyes, solvents, and varnishes, also found additional use when the decline of German chemical imports permitted the growth of an American organic chemical industry. At the same time, the rapid growth of the automobile industry created a widespread market not only for benzol as a component of automotive paints and varnishes but also pitch, used for roads.

The American iron and steel industry virtually completed its shift from beehive to by-product ovens in the 1920s. By-product coke ovens increased their share of coke production from 60 percent in 1920 to over 94 percent in 1930. The rapid employment of the by-product coke oven was in part due to the industry's desire to adapt European-developed technology to American conditions, with American emphasis on large-scale, rapid production. Many of the key developments along these lines were made by the Koppers Company. For example, the Koppers-Becker oven, introduced in 1922, reduced coking time from 24 hours to approximately 18 hours and increased capacity by 50 percent. During this period Joseph Becker, Koppers' chief design engineer, developed ovens that could use lean furnace gases while permitting the higher quality volatile gases from the coking process to be diverted to other uses, such as heating open hearth furnaces. Increased attention was also devoted to mechanization of charging and unloading the ovens and the chemical reclamation portions of the by-product plant, especially since the bulk of revenues from by-product plants came from chemicals rather than the coke.

After 1930 the main improvements made to the by-product oven were the result of attention to detail. Designers developed several systems, for example, to insure more uniform heating of the ovens and added prewashing to improve performance. The size of by-product ovens slowly increased. Although the by-product ovens constructed in the early 1970s remained about the same in width as their early-twentieth-century counterparts (15 to 18 inches), they were almost double the depth (approximately 50 feet) and triple the height (approximately 20 feet). The early Koppers ovens held 4.5 tons of coal; the average 1950s oven had a capacity of 16 tons. And between the 1930s and 1970s coking time was decreased further, from 18 to 15.5 hours.

The transition from beehive to by-product ovens brought several changes to the coking industry. The geographic location of coking shifted. Low-cost beehive ovens had traditionally been located close to the mine head, but the capital-intensive by-product plants were typically located near the blast furnaces. The integration of coke production into the steel plant also sharply reduced the role of the independent merchant coke producer. Beehive coke production had always been a haphazard art, but the increased attention to chemical components and energy flows required by the by-product oven transformed coke production from a largely practical, empirical technology to a technology with a strong scientific component.

References:

W. H. Blauvelt, "The By-Product Coke Oven and Its Products," American Institute of Mining and Metallurgical Engineers, *Transactions,* 61 (1919): 436–453;

Fred Denig, "Industrial Coal Carbonization," in *Chemistry of Coal Utilization,* edited by Homer H. Lowry (New York: Wiley, 1945), pp. 774-833;

C. S. Finney and John Mitchell, "History of the Coking Industry in the United States," *Journal of Metals,* 13 (April 1961): 285–291; (May 1961): 373–378; (June 1961): 425–430; (July 1961): 501–504; (August 1961): 559–561;

Bela Gold and others, *Technological Progress and Industrial Leadership: The Growth of the U. S. Steel Industry, 1900–1970* (Lexington, Mass.: Lexington Books, 1984);

William T. Hogan, S.J., *An Economic History of the Iron and Steel Industry in the United States* (Lexington, Mass.: Lexington Books, 1971);

William Gilbert Irwin, "By-Product System of Coke Making," *Cassier's,* 12 (1897): 581–592;

C. D. King, *Seventy-Five Years of Progress in Iron and Steel* (New York: American Institute of Mining and Metallurgical Engineers, 1948);

R. S. McBride and F. G. Tryon, *Coke and By-Products in 1919–1920* (Washington: Government Printing Office, 1922);

Harold E. McGannon, ed., *The Making, Shaping, and Treating of Steel,* ninth edition (Pittsburgh: United States Steel, 1971), pp. 106–177;

Eugene T. Sheridan and Joseph A. DeCarlo, Coal Carbonization in the United States, 1900–1962 (Washington, D.C.: U.S. Government Printing Office, 1965).

James Anson Campbell

(September 11, 1854 – September 20, 1933)

by Larry N. Sypolt
West Virginia University

CAREER: General manager and treasurer, Youngstown Ice Company (1881–1891); superintendent, Warren Mill, Trumbull Iron Company (1891–1894); superintendent, Pomeroy Mill of Union Iron and Steel Company (1894–1897); general superintendent, Mahoning Valley Iron Company (1897–1899); manager, Youngstown District, Republic Iron and Steel Company (1899–1900); vice-president and general manager (1900–1902), acting president (1902–1904), president (1904–1930), chairman of the board, Youngstown Sheet and Tube Company (1930–1933).

It is rare in the age of large corporations to be able to identify a single individual as the creator of a company. Yet clearly James Anson Campbell is the man who made Youngstown Sheet and Tube Company a success. Youngstown, Ohio, became the second largest steel-producing city in the United States, largely because of Campbell's effort.

Campbell was born in Ohlton, Ohio, the son of John and Caroline Jones Campbell, on September 11, 1854. He was of Scotch-Irish ancestry. He was educated in public schools in Niles and Girard, Ohio, and at Hiram College in Ohio. Campbell then taught school and prepared for the U.S. Military Academy at West Point, but financial necessities prevented him from attending.

Faced with this disappointment, Campbell worked in the hardware business in Youngstown for about five years. From 1881 to 1891 Campbell was general manager and treasurer of Youngstown Ice Company, which he may have helped found. He entered the iron industry in the Mahoning Valley in 1890, becoming superintendent of the Warren Mill of the Trumbull Iron Company in 1891. Three years later Campbell became superintendent of the Pomeroy Mill of Union Iron and Steel Company, serving in that position until 1897. He was general superintendent of the Mahoning Valley Iron Company from 1897 to 1899 and manager of the Youngstown District for

James Anson Campbell

the Republic Iron and Steel Company in 1899–1900.

In the fall of 1900 James Campbell and George D. Wick resigned from their executive positions with Republic Iron and Steel. Campbell had decided to go into business with William Wilkoff, who had just ordered construction of four sheet mills which he intended to erect at Niles, Ohio. Before these plans were complete, however, the two men were approached by several people, including George D. Wick, who suggested they form a stock company instead of a partnership and go into the iron business on a larger scale. Campbell and Wilkoff accepted Wick's proposal and expanded their plans to include a tube mill. The capital was fixed at $600,000, divided into 6,000 shares at $100 each.

Within a week of filing the articles of incorporation, all of the stock issued by the Youngstown Sheet and Tube Company had been purchased. Nearly all of the stockholders were residents of Youngstown or nearby communities. Several of the leading stockholders had made handsome profits on their investments in the Ohio Works, another steel plant organized in the 1890s; others had mercantile, banking, and real estate holdings in the Mahoning Valley and were greatly interested in other forms of investments. But the most prevalent reason for investing in Youngstown Sheet and Tube was an abiding belief in the new company's future success; the stockholders confidently placed their faith in both Campbell and Wick.

Campbell had acquired considerable knowledge of the iron industry during his decade as plant superintendent at various Mahoning Valley iron and steel plants, and his reputation for being a hard worker served him well. He was known to be a gentleman of inflexible integrity, great tenacity, and firmness, and his honesty and fairness endeared him to important people in the steel industry. Thus, when the first meeting of the stockholders of Youngstown Sheet and Tube was held on November 28, 1900, the board of directors elected Wick president and Campbell vice-president and general manager.

Four weeks later the board met again to increase the capital stock of the new venture from $600,000 to $1 million. The original amount could not have financed the construction of the plant that was needed. Campbell, in making these recommendations, was already demonstrating the vision and courage that were his chief assets. He saw clearly that the age of mass production had come, and that to be successful the new company would have to meet the stiff competition of a new era of giant corporations. During the 30 years he headed Youngstown Sheet and Tube, Campbell repeatedly initiated and carried through programs of expansion that almost took the breath from some of the more timid stockholders, who were used to making business decisions on a smaller scale at the local level.

In May 1902 President Wick submitted his resignation to the board of directors. The strenuous work involved in organizing the new company had left him in poor health. A new president was not chosen immediately: the directors instead formed an executive committee headed by Campbell. Two years later Campbell was chosen president and served in that position for the next 25 years.

The first four years of business were building years for Youngstown Sheet and Tube, but finally the company showed a $700,000 profit in 1904. During Campbell's presidency (1904–1930), the company made a profit every year with the single exception of 1914. On January 1, 1906, stockholders received their first dividend. Although slow in coming, these continued without interruption for the next 25 years, a record of which few companies can boast. Even the severe financial panic that began in the fall of 1907 had little effect on Youngstown Sheet and Tube. Their losses were basically "write-offs."

Under Campbell's direction Youngstown Sheet and Tube Company became an integrated steel producer, manufacturing a wide line of finished steel products; and the stockholders had the confidence in Campbell to vote capital increases for needed equipment and expansion. Moreover, they supported Campbell's decision to grow through acquisition during the 1920s. In 1923 Brier Hill Steel Company, another Mahoning Valley firm, was acquired by Youngstown Sheet and Tube, as was the Steel and Tube Company of America, with its plants at Indiana Harbor, South Chicago, and Evanston. Additional coal and iron ore properties were added in the 1920s. As a result, production jumped from 122,000 tons of sheet and pipe in 1901 to 3.12 million tons of ingots in 1935. The capitalization had similarly climbed from the initial $600,000 to $300 million.

Campbell, however, was disappointed at the failure of a proposed merger of Bethlehem Steel Company and Youngstown Sheet and Tube in 1930–1931. He had vigorously supported the merger — the third proposed combination involving Youngstown Sheet and Tube during the 1920s — in the belief that the company was too small to stand alone. In negotiations between the two companies Bethlehem promised to spend $350 million on improvements at Youngstown Sheet and Tube plants. The prospect of large additions being made at their plants left many at Youngstown Sheet and Tube feeling apprehensive over the possibility of a merger. When the merger failed Youngstown Sheet and Tube lost a valuable source of capital. But, just as potentially damaging, the company had made no effort to add the newest technology — a continuous hot strip mill — in order to help offset the loss of capital in a failed merger. In short, the company was poorly prepared for the Depression.

Youngstown Sheet and Tube, however, had little serious labor trouble during its formative years. Only two strikes had occurred during Campbell's presidency; both were short and free of violence. He established an industrial relations department, separate from the operating department, to prevent and redress wrongs affecting the company's work force. A representation plan for employees was agreed upon by management and representatives of the workers with the purpose "to provide effective communications and means of contact between the management and men on matters pertaining to industrial relations, to insure justice, maintain tranquility, and promote the common welfare." It was Campbell's expectation that through this department every wage earner would receive justice in all matters concerning his relations with the company.

It is only possible to estimate what the company and Campbell meant to the Mahoning Valley and Youngstown. During its first 30 years, Youngstown Sheet and Tube manufactured and sold almost $2 billion worth of iron and steel products and paid nearly $500 million in wages. It also paid many more millions in dividends and taxes and contributed to many public and private institutions. Campbell served as a trustee of Hiram College and the Youngstown Chamber of Commerce. He also played key roles in other Youngstown businesses. He was a director of several banks, the American Iron and Steel Company, Mahoning Ore and Steel Company, Crete Mining Company, Balkan Mining Company, Carbon Limestone Company, and other commercial ventures in the city. Clearly, Campbell was a central figure in the Youngstown business community.

On September 11, 1929, Campbell turned seventy-five years old. A testimonial dinner was held in his honor at the Hotel Ohio in Youngstown, where he was complimented by President Herbert Hoover. Other distinguished guests included Charles M. Schwab and Eugene G. Grace,

chairman and president, respectively, of Bethlehem Steel. The steel industry had shown its admiration by making Campbell an honorary vice-president of the American Iron and Steel Institute long before. Now they came to honor the man who had made his company one of the three largest steel producers in the United States.

On January 1, 1930, Campbell became chairman of the board at Youngstown Sheet and Tube. He was succeeded as president by Frank A. Purnell, who had acted as assistant president for several previous years. Campbell served as chairman until his death on September 20, 1933.

Publications:
"Remarks," *American Iron and Steel Institute Yearbook* (1914): 283; (1917): 236; (1921): 22–26, 27–29; (1927): 228–229;

Iron Trade Review, 69 (September 9, 1921); 543–545. [Testimony to Senate concerning tariff bill]

"Railroads Need More Freedom," *Iron Trade Review*, 69 (November 3, 1921): 1160–1162;

"Country Enters Period of Higher Wage Level," *Iron Trade Review*, 71 (September 28, 1922): 834;

"Business Outlook," *Iron Age*, 113 (June 26, 1924): 1881;

"Now Is the Time for Thrift," *Iron Trade Review*, 76 (February 5, 1925): 394;

"Meeting the New Competition," *Iron Trade Review*, 79 (July 8, 1926): 111–112;

"Production and Distribution Changes Set Pace for Steelmakers," *Iron Trade Review*, 80 (January 27, 1927): 292–293.

References:
50 Years in Steel—The Story of the Youngstown Sheet and Tube Company (Youngstown, Ohio: Youngstown Sheet and Tube Company, 1950);

Charles Longenecker, "Indiana Harbor Plant of the Youngstown Sheet and Tube Company," *Blast Furnace and Steel Plant*, 23 (January 1935): 46–56;

James T. McCleary, *Biographical Directory of the American Iron and Steel Institute* (New York: American Iron and Steel Institute, 1911), p. 13.

Dennis Joseph Carney

(March 19, 1921 –)

by Elizabeth M. Nolin

Paeonian Springs, Virginia

CAREER: Metallurgical observer, National Tube Company, United States Steel Corporation (1942–1943); U.S. Navy (1943–1946); development metallurgist, Duquesne Works; physicist, chief development metallurgist, superintendent of electric furnace department, superintendent of open-hearth shop, South Works (circa 1949–1956); division superintendent, steel production, Duquesne Works (1956–1959); assistant general superintendent (1959–1963), general superintendent, Homestead Works (1963–1965); vice-president in long-range planning (1965–1968), vice-president of applied research, U. S. Steel (1972–1974); vice-president of operations (1974–1975), executive vice-president and director (1975–1976), president (1976–1985), chief operating officer (1976–1977), chief executive officer (1977–1985), chairman of the board of directors, Wheeling-Pittsburgh Steel Corporation (1978–1985); president, Intra-Continental Construction Company (1985–).

Dennis Carney made a name for himself as the executive who saved Wheeling-Pittsburgh Steel Company in the late 1970s and early 1980s. He brought to the company a long background in operations at the United States Steel Corporation and a firm determination to modernize Wheeling-Pitt. He was very sure of himself and met with some success, but his style antagonized other executives and the union. In 1985 he stepped down from Wheeling-Pittsburgh Steel.

Dennis Carney was born in Charleroi, Pennsylvania, on March 19, 1921, the son of Walter Augustus and Ann Nandor Carney. He married Virginia M. Horvath in 1943, and they had five children. His wife died in 1984.

In 1942 Carney received a bachelor of science degree in metallurgy from Pennsylvania State University. After graduation he found employment at U.S. Steel in Pittsburgh, but in 1943 he entered the navy as a lieutenant. He served first in the Naval Research Laboratory and then in the Bureau of Naval Intelligence until 1946. In 1949 he earned a Ph.D. from Massachusetts Institute of Technology — an unusual background for someone who would work in the production end of the steel industry.

Carney was employed at U.S. Steel until 1974, climbing the corporate ladder through a series of research and production positions. After starting as a metallurgist, Carney filled a series of positions at the company's South Works in Chicago before returning to Pittsburgh in 1956 to serve as division superintendent of steel production at the Duquesne Works. In 1959 he became assistant general superintendent at the Homestead Works; he was promoted to general superintendent in 1963. This rapid rise through the ranks of U. S. Steel was a reflection of Carney's capabilities in both management and steelmaking. In 1965 Carney moved into the corporate headquarters as vice-president in long-range planning. In 1968 he resumed research activities, and became vice-president of applied research in 1972. He served in this position for the next two years.

Frustrated when it became apparent that he would not reach the top of U. S. Steel, Carney left in 1974 to become vice-president of operations at Wheeling-Pittsburgh Steel Corporation. Wheeling-Pitt offered both challenges and opportunities, as the company had rarely been successful since it was created in the merger of Wheeling Steel and Pittsburgh Steel in 1968. Carney brought an exuberant style and dynamic thinking to the company and, consequently, was promoted every year. In 1975 he became executive vice-president and a director of the company. In 1976 Carney was promoted to succeed Robert Lauterbach — who himself had begun to try and turn the company around in 1970 — as president, a position which he held until his retirement in 1985. His promotion was followed by a succession of new titles, becoming chief operating officer (1976–1977), chief executive officer in 1977, and chairman of the board of directors in 1978, again following in Lauterbach's footsteps. While rising rapidly, Carney set out to refine the image of Wheeling-Pitt — a company one analyst described in *Business Week* as "the used-car salesman of the steel industry."

Carney's strategy at Wheeling-Pitt was to modernize completely the company's steelmaking facilities. From 1974 to 1984 the company spent $806 million, of which $563 million went toward the installation of two continuous casters and other improvements in 1982–1983. The problem Carney faced was luring investment dollars even as the company was losing money. Carney proved effective at attracting new investors, including Nisshin Steel of Japan with which Wheeling-Pitt entered into a joint venture. He also won a federal loan guarantee to support construction of a rail mill at Monessen, Pennsylvania. Finally, he won wage concessions from the steelworkers, which materially enhanced the company's competitive position. By 1984 Wheeling-Pitt had the most modern basic steelmaking operations in the country and was vastly more competitive than it had been a decade earlier; the company was winning loyal customers. To many observers and company employees, Carney was the savior of the firm. Even *Business Week* noted that "Wheeling-Pitt, far more than most major U.S. companies, remains dependent on the efforts of just one man."

Yet, even as Carney was praised at a dinner in Monessen in the spring of 1984 for saving 4,000 jobs, his efforts were not met without controversy. Brashly self-confident, he gave little room to subordinates; he was intolerant of errors. Some considered him hard to work for, and by 1984 he had run through four different number-two executives. His ruthless managerial style and reputation as an industry maverick seemed to serve the company well, but *Business Week* argued that some of Carney's solutions "have often raised problems of their own." When the company asked for further wage concessions in 1985, the union resisted and successfully made Carney the issue in the resulting strike lockout. The contest between management and labor was very bitter, as the company declared bankruptcy and abandoned the new rail mill when it closed the Monessen plant completely. The union, however, felt it had won a victory. It had forced Carney's resignation, a union representative was soon appointed to the company's board, and wages were not cut as drastically as Carney had proposed. It was an unfortunate end for a brilliant and ambitious steelman, who clearly had enabled Wheeling-Pitt not only survive to but prosper.

One measure of Carney's accomplishments was the awards he received from professional societies to which he belonged. In 1957 he was named a member of the first technical exchange task force between the Soviet Union and the United States, and toured Russian steel plants. As a fellow of the American Society of Metals, he was given the Grossman award in 1959 by the Pittsburgh chapter; he had been a trustee of the chapter since 1972. The American Institute of Mining, Metallurgy, and Petroleum Engineers honored Carney with two different awards: the McKune award in 1951 and the Benjamin F. Fairless award in 1978. His other professional memberships include both the American and British International Iron and Steel Institutes and the American Iron and Steel Engineers. Dennis Carney also belongs to three honorary fraternities: Sigma Xi, a science fraternity, and Tau Beta Pi and Sigma Nu, both engineering fraternities. He is a member of the board of directors of Wheeling College. Carney co-authored a book, *Gases in Metals* (1953), and published many technical articles.

After his retirement from Wheeling-Pittsburgh Steel Corporation in 1985, Carney moved to Fort Lauderdale, Florida, where he has since served as president of Intra-Continental Construction Company.

Publications:
and others, "Sampling and Analysis of Liquid Steel for Hydrogen," *Journal of Metals,* 188 (February 1950): 404–413;

and A. D. Janulionis, "Examination of Quenching Constant," *American Society of Metals Transactions,* 43 (1951): 480–493;

and R. L. Stephenson, "Sampling and Testing of Sinter," *Mining Engineering,* 5 (March 1953): 309–311;

and D. P. Smith, L. W. Eastwood, C. E. Sims, *Gases in Metals* (Cleveland: The American Society of Metals, 1953);

"Pneumatic Steelmaking Processes," *Blast Furnace and Steel Plant,* 43 (June 1955): 635–640; (July 1955): 753–760; (September 1955): 1006–1010; (October 1955): 1039–1041;

"Some Factors Affecting Open Hearth Operations," *Journal of Metals,* 7 (January 1955): 39–50;

"Factors in Slag Reduction Detailed for Stainless Melting Practice," *Journal of Metals,* 7 (December 1955): 123–132;

"Nickel-Free and Low Nickel Austenitic Stainless Steel," *American Iron and Steel Institute Regional Technical Papers* (1955): 103–114; also *Blast Furnace and Steel Plant,* 43 (December 1955): 1377–1380;

and A. C. Ogan, "Large Electric-Arc Furnace Steelmaking Experience in U.S.A.," *Iron and Steel Institute Journal,* 189 (August 1958): 307–314.

References:

"Carney Named Wheeling President," *American Metal Market* (February 27, 1976): 1, 44;

John P. Hoerr, *And the Wolf Finally Came: The Decline of the American Steel Industry* (Pittsburgh: University of Pittsburgh Press, 1988), pp. 455–463;

William T. Hogan, S. J., *Minimills and Integrated Mills: A Comparison of Steelmaking in the United States* (Lexington, Mass.: Lexington Books, 1987); "The Maverick Who Could Save Wheeling-Pitt." *Business Week* (June 4, 1984): 84, 88.

Archives:

Wheeling-Pittsburgh Steel Corporation, *Annual Reports,* are located in the West Virginia Collection, West Virginia University.

Central Alloy Steel Corporation

by Carol Poh Miller

Cleveland, Ohio

When the Republic Steel Corporation was formed in 1930 it represented the merger of six companies, with the old Republic Iron & Steel Company as its nucleus. One of these six was the Central Alloy Steel Corporation, the largest producer of alloy and stainless steels in the United States. As part of the Republic organization, Central Alloy became known as the Central Alloy District, with extensive works at Canton and Massillon, Ohio.

Central Alloy Steel Corporation formed in 1926 as a result of the merger of the United Alloy Steel Corporation (formerly United Steel Company) of Canton, Ohio, and the Central Steel Company of Massillon, Ohio. Cleveland industrialists Cyrus Eaton and William G. Mather sponsored the consolidation; they subsequently held directorships in the new company, which was headquartered at Massillon. In 1929 Interstate Iron & Steel Company of Chicago was added to Central Alloy, an acquisition which would later give Republic an important foothold in Chicago at the time of the 1930 merger.

Both United Alloy Steel and Central Steel pioneered the American production of high-quality alloy and stainless steels. United Alloy Steel Corporation, located on the east side of Canton, had begun operations as United Steel Company in August 1904. The plant was initially intended to furnish sheet bars to two local steel-fabricating companies. But in response to increasing demand for stronger and lighter steel for automobile manufacture, the company began to manufacture alloy steel, producing the first U.S.-made vanadium steel in 1906 at the request of Henry Ford. "We

tried every steel maker in America," the automaker later recalled in his autobiography, *My Life and Work* (1922), "not one could make vanadium steel.... I found a small company in Canton, Ohio. I offered to guarantee them against loss if they would run a heat for us. They agreed. The first heat was a failure.... I had them try again, and the second time the steel came through."

In 1916, facing strong competition from the nearby Central Steel Company of Massillon, the United Steel plant was enlarged with additional furnaces and bar mills and equipped for cold drawing, turning, and heat-treating of steel. The same year Edward A. Langenbach and others organized the United Alloy Steel Corporation, buying out the old United Steel. In 1921 United Alloy Steel merged with the Berger Manufacturing Company, a steel fabricator, and the United Furnace Company, which supplied pig iron to United Alloy Steel. United Alloy Steel, meanwhile, began construction of a second plant that would increase the company's annual production of high-grade steel by 400,000 gross tons. Products included electric- and open-hearth steel, chrome-vanadium steel, open-hearth carbon steel, and Enduro stainless steel.

The Central Steel Company of Massillon, Ohio, was organized in 1914 by a group of Canton businessmen to supply steel to the Massillon Rolling Mill Company, which had been incorporated several years earlier and was soon absorbed by Central. Following an offer of land from the Massillon Board of Trade, the Central plant was erected in 1914–1915. Benjamin F. Fairless, who later would become president of the United States

Steel Corporation, worked on the construction of the Central plant. He was successively promoted, serving as president and general manager at the time of its merger with Republic. In 1916 Central Steel acquired the National Pressed Steel Company of Massillon, a producer of pressed structural steel.

Following the merger of United Alloy Steel and Central Steel in 1926, Central Alloy Steel received large orders for Enduro stainless from Henry Ford, General Motors, and for the construction of New York City's Chrysler, Radio City, and Empire State buildings.

For over half a century Republic's Central Alloy District remained a leading producer of car-bon, alloy, stainless, and high-strength specialty steels. In 1984 Central Alloy became part of the LTV Steel Company.

References:
Henry Ford and Samuel Crowther, *My Life and Work* (Garden City, N.Y.: Doubleday, 1922);
Tom M. Girdler and Boyden Sparkes, *Boot Straps: The Autobiography of Tom M. Girdler* (New York: Scribners, 1943);
John H. Lehman, *A Standard History of Stark County, Ohio,* 3 volumes (Chicago & New York: Lewis, 1916).

CF&I Steel Corporation

by H. Lee Scamehorn

Historic Learning and Research Systems

The CF&I Steel Corporation of Pueblo, Colorado, traces its origins to Gen. William Jackson Palmer's Denver and Rio Grande Railway, which, beginning in 1872, linked Denver and other cities with the Colorado territory's principal mining areas. An affiliate of the Denver and Rio Grande Railway, the Colorado Improvement Company was created in 1872 to extend the narrow-gauge line westward from Pueblo to coal mines in the vicinity of present-day Canon City. The Southern Colorado Coal and Town Company, a similar enterprise, financed the construction of a rail carrier to the coalfields at Walsenburg and Trinidad, in southern Colorado. The Colorado Improvement Company and Southern Colorado Coal and Town Company were combined in 1880 to form the Colorado Coal and Iron Company for the purpose of erecting an integrated iron and steel plant at Pueblo.

In part because of antiquated technology and also because transcontinental railroads opened the American West to eastern products, the Pueblo steelmaker had great difficulty competing in what it had hoped would be its own geographic market for rails and other cast, machined, and rolled iron and steel products. Consequently, the company derived its profits, if any, from the mining of coal and the manufacture of coke for smelters of precious metal ores. Competition in the fuel trade, particularly from John C. Osgood's Colorado Fuel Company, forced the Pueblo enterprise into a merger that created the Colorado Fuel and Iron Company in 1892. Osgood controlled that enterprise, the region's largest distributor of fuel as well as its only integrated producer of iron and steel.

In an attempt to expand and modernize the steelworks, Osgood experienced a cash-flow problem, which forced him to turn to George Jay Gould and John D. Rockefeller for funds in order to avoid bankruptcy. The two eastern financiers assumed control of the western steelmaker in 1903 and shared the direction of its affairs until 1908, when Gould sold most of his interest. The senior Rockefeller turned over the management of family investments to his son, John, Jr., in 1911. Until 1945 the younger Rockefeller, as the principal stockholder, shaped the policies of the western steel company.

The company gained notoriety in 1913–1914, when it led the operators in the southern Colorado field in opposing the coal miners' demand for, among other things, recognition of the United Mine Workers of America (UMWA). An often violent strike began in September 1913, and came to a head in April 1914 with the so-called Ludlow Massacre, in which two women and 11 children were killed when a miners' tent colony was burned to the ground during fighting between miners and the Colorado National Guard. Public

The Pueblo plant of the CF&I Steel Corporation, July 1975

reaction to the clash prompted John D. Rockefeller, Jr., to institute a new labor policy, the Employee Representation Plan, better known as the Rockefeller Plan. Under the plan a company union was created which served as a model for other U.S. industries during and after World War I and brought relative peace to the southern Colorado coalfields. CF&I miners were unenthusiastic about the scheme, and eventually took advantage of the National Recovery Act to win recognition for the UMWA in 1933. At the steelworks, however, the Rockefeller Plan remained in effect until the War Labor Board ordered its suspension in 1942, at which time the United Steelworkers of America became the bargaining agent for employees in the mills at Pueblo.

CF&I's operations encompassed at one time properties in four states, making it for many years the largest owner of company towns in the American West. Investments in communities were necessary to attract and retain competent employees at the firm's more than 60 coal and iron ore mines and limestone quarries. The corporation owned all of the land, dwellings, and other structures, including company stores. A subsidiary, the Colorado Supply Company, operated over a period of half a century some 50 retail stores and two wholesale outlets. That enterprise was dissolved in 1936; its mercantile units were made a CF&I operating division. The

last of the company-controlled coal camps, Valdez, Colorado, was closed in 1960. The last company store in Sunrise, Wyoming, was shut down in 1966, and Sunrise, the last company town, was dismantled in 1983.

Staggered by the Great Depression, the Colorado Fuel and Iron Company was forced into receivership, from which it emerged in 1936 as the Colorado Fuel and Iron Corporation. It enjoyed prosperity during World War II. That, together with the postwar outlook for steel, enabled the Rockefeller interests to sell the enterprise in January 1945 to Charles Allen and Associates.

Under new owners, CF&I absorbed Wickwire Spencer Steel Company in an effort to transform two regional firms into a steel producer of national status. When that goal proved unattainable, CF&I devoted its resources to upgrading technology as a way of strengthening its position in the western market. In recognition of the firm's focus on metallurgical production, its name was changed to CF&I Steel Corporation in 1966.

Allen sold the enterprise to the Crane Company, a New York conglomerate, in 1969. That holding company retained control of the Pueblo firm until 1985. The steel recession of the early 1980s forced the western enterprise to restructure, which afforded the parent enterprise an opportunity to spin-off the subsidiary to stockholders. The independent company under the leadership of Frank J. Yaklich, Jr., completed the process of restructuring, which began in 1983. Blast and basic oxygen furnaces were permanently shut down, along with by-product coke ovens. Coal mines were sold and iron mines and quarries closed. All improvement programs were canceled.

No longer a fully integrated operation, CF&I Steel became a minimill, producing steel in electric-arc furnaces which have about 600,000 to 700,000 tons of capacity. The company's rolling and finishing mills have continued to produce rails, seamless tubes, and wire products. CF&I's full recovery from the steel recession of the 1980s, however, depends on returning prosperity for rail-roads, the petroleum industry, farming, and construction.

References:
James B. Allen, *The Company Town in the American West* (Norman: University of Oklahoma Press, 1966);
Stuart D. Brandes, *American Welfare Capitalism, 1880–1940* (Chicago: University of Chicago Press, 1976);
William T. Hogan, S.J. *Global Steel in the 1990s: Growth or Decline* (Lexington, Mass. Lexington Books, 1991);
H. Lee Scamehorn, *Pioneer Steelmaker in the West: The Colorado Fuel and Iron Company, 1872–1903* (Boulder: Pruett Publishing Company, 1976).

Eugene Bradley Clark

(July 27, 1873 – July 29, 1942)

by Robert Casey

Henry Ford Museum & Greenfield Village

CAREER: Electrical engineer, Westinghouse (1894–1896); electrical engineer, assistant general superintendent, Illinois Steel Company (1896–1906); president, George C. Rich Manufacturing Company (1906–1907); president, Celfor Tool Company (1907–1916); president, Buchanan Electric Steel Company (1911–1916); president, Clark Equipment Company (1916–1942); president, American Sintering Company (1906–1942); president, American Ore Reclamation Company (1920–1942); president, Buffalo Sintering Corporation (1923–1942).

Eugene Bradley Clark was an engineer and manufacturer whose wide interests included pioneering work in electric steelmaking and recycling of blast-furnace flue dust.

Clark was born on July 27, 1873, in Washington, D.C. He attended public schools and entered the Sibley School of Engineering at Cornell University. During his senior year he edited the *Sibley Journal of Engineering*. He graduated with an M.E. in electrical engineering in 1894. He began his career as an electrical engineer with Westinghouse, moving on to a similar position with Illinois Steel Company in South Chicago in 1896. During his ten years with Illinois Steel, Clark rose to the position of assistant general su-

Eugene Bradley Clark

perintendent. Clark became acquainted with George Rich, a former Illinois Steel employee who had developed a twist drill that reduced the time needed to drill holes in railroad rails and had founded the George R. Rich Manufacturing Company to produce the drill. Rich's company, however, was experiencing financial difficulties. Clark took over the company's management and helped place the enterprise on sound financial footing. He was elected president of the company in 1906. The company changed its name to Celfor Tool Company in May 1907, and Clark continued as president.

Clark soon demonstrated his ability to anticipate industrial trends. He visited electric steelmaking operations in Europe, where the process was in its infancy; realized its potential; and formed Buchanan Electric Steel Company of Michigan to make steel castings. At this time there were fewer than 20 electric furnaces in the United States, producing less than 1 percent of America's steel. Clark's innovations included an improved method of fusing new furnace linings.

In December 1916 Clark merged Buchanan Electric Steel and Celfor Tool to form Clark Equipment Company, of which he remained president for the remainder of his life. Seizing upon the growing demand for motor vehicles, Clark Equipment specialized in the manufacture of trucking equipment and components, such as axles and transmissions. The company expanded to produce industrial forklifts and collaborated with the President's Conference Committee (PCC) — an alliance of transit company executives — in helping to develop a streamlined streetcar.

Clark retained his interest in the steel industry, searching for ways to recycle blast-furnace flue dust. In 1906 he assumed control of the American Sintering Company, which had a plant in Hubbard, Ohio. He organized the Buffalo Sintering Corporation in 1923 and beginning in 1920 was president of American Ore Reclamation Company of Chicago. Clark also served as a director of Liquid Carbonic Company, Upper Avenue National Bank of Chicago, and Associates Investment Company, South Bend, Indiana.

Clark's professional memberships included the National Association of Manufacturers, American Iron and Steel Institute, and the Society of Automotive Engineers. His first wife, Laura Wolfe, whom he married on October 28, 1899, died in 1917. They had four children. He married his second wife, Luella Mather Coon, on May 22, 1919. Eugene Bradley Clarke died in Chicago on July 29, 1942.

Publications:
"Electric Furnaces: Discussant," *Yearbook of the American Iron and Steel Institute* (1912): 75–79; also *Iron Trade Review*, 50 (May 23, 1912): 1173–1174;
"Treatment of Blast-Furnace Flue Dust," *Yearbook of the American Iron and Steel Institute* (1913): 314–326.

References:
William T. Hogan, S.J., *An Economic History of the Iron and Steel Industry in the United States*, 5 volumes (Lexington, Mass.: Lexington Books, 1971);
Frank T. Sisco, *The Manufacture of Electric Steel* (New York: McGraw-Hill, 1924).

Edmund Arthur Stanley Clarke

(January 21, 1862 – May 15, 1931)

by Thomas E. Leary

Industrial Research Associates

and

Elizabeth C. Sholes

Industrial Research Associates

CAREER: Assistant works superintendent, Union Steel Company (1890–1891); assistant general superintendent (1891–1895), general superintendent, South Works, Illinois Steel Company (1895–1899); general manager, Illinois Steel Company (1899); general manager, Deering Harvester Company (1900–1903); general manager of manufacturing, International Harvester Company (1903–1904); president, Lackawanna Steel Company (1904–1918); president, Consolidated Steel Corporation (1919–1922); secretary, American Iron and Steel Institute (1923–1931).

E. A. S. Clarke held a series of distinguished administrative positions with corporations and trade associations in the iron and steel industry. His 45-year career demonstrates the upward mobility and specialization of managerial functions that accompanied the expansion of private-sector bureaucracies.

Clarke's family background provided him with access to higher education as well as professional contacts. He was born in Ottawa, Canada, on January 21, 1862. His father, Thomas Curtis Clarke, was a noted engineer and builder who served as president of the American Society of Civil Engineers during the 1890s. After preliminary schooling in Philadelphia, young Clarke attended Harvard, where he received an A.B. degree in 1884. He subsequently took courses at Massachusetts Institute of Technology.

After graduation Clarke broadened his exposure to practical metallurgy in the laboratory of the Spang Steel and Iron Company (later renamed Spang-Chalfant) at Sharpsburg, Pennsylvania. When the laboratory's director, Robert Forsythe, joined the Union Steel Company of Chicago in 1885, Clarke accompanied him. In the course of being groomed for upper management, Clarke held positions in various departments at Union Steel before being promoted to assistant superintendent in 1890. His experiences in the many different departments proved to be an education in steelmaking that combined the theoretical aspects of formal learning with the practical experience of shop culture.

In 1889 Union Steel merged with two other rail producers in the Chicago area to create Illinois Steel Company, possessor of the largest production capacity in the United States at the time. It happened that the plant where Clarke worked was the new giant's weakest member. Fortunately for Clarke he was soon transferred to Illinois Steel's flagship, the South Chicago Works of the former North Chicago Rolling Mill Company (later renamed the South Works of United States Steel Corporation [U.S. Steel]). Situated at the mouth of the Calumet River, this facility expanded considerably during the Gay Nineties, and Clarke climbed steadily from assistant general superintendent (1891) to general superintendent (1895). In 1899 he was appointed general manager of Illinois Steel. The company had lost its independent position as a result of another merger

the previous year that created Federal Steel, presided over by Judge Elbert H. Gary and bankrolled by J. P. Morgan. Clarke resigned as general manager late in 1899 and embarked on a trip abroad.

Upon returning to Chicago in 1900 Clarke accepted the post of general manager with the Deering Harvester Company. His background and experience were certainly relevant to his new responsibilities. The agricultural machinery firm desired to stabilize its source of steel by constructing its own plant on the Calumet; the eminent consulting engineer Julian Kennedy designed the blast furnaces and Bessemer works. After International Harvester Company absorbed Deering, Clarke became general manager of the combine manufacturing department in 1903. He continued to oversee progress on the plant, which eventually became Wisconsin Steel, until he left the Chicago area to assume the presidency of Lackawanna Steel Company in 1904.

At his new employer's New York office Clarke filled a vacancy that had existed for ten months since the resignation of Walter Scranton. To staff its expanded operations Lackawanna had recruited heavily from the ranks of Illinois Steel; in addition to Clarke other alumni of Chicago's South Works included second vice-president and general manager, George L. Reis, as well as general superintendent, George Sheldon. As president Clarke also occupied a seat on Lackawanna's board of directors.

Clarke's primary mission involved piloting his company along the course set by industry giant U.S. Steel. Lackawanna's policies generally conformed with Gary's plans for stabilizing prices while minimizing potential government intervention against corporate collusion. Not surprisingly, Clarke was active in Gary's American Iron and Steel Institute (AISI), a forum for promoting industrial cooperation in place of competition. His 1912 paper, "Contract Obligations," spelled out the steps that steelmakers had taken to rationalize pricing procedures after the disruptive financial panic of 1907. During World War I the AISI enhanced its stature as an organization for mediating relations between businessmen and their counterparts in the federal government. Clarke served as secretary of the trade association committee that was acting as liaison with wartime regulatory agencies. He likewise spoke for the AISI on the necessity of formulating uniform personnel and relief policies that reflected a consensus of private and public priorities.

Circumstances more peculiar to Lackawanna Steel shaped other high points of Clarke's tenure as president. The company expanded its physical plant and rounded out the product line. A shipping subsidiary, Seneca Transportation, was formed to take advantage of Lackawanna's Great Lakes location by distributing more of its output via water. Backward vertical integration to control sources of raw materials included acquisition of the Ellsworth coal coking properties in Washington County, Pennsylvania. In 1916, however, Lackawanna disposed of its interest in the Cornwall ore banks, the largest iron mine in the East. During the same year a potential merger with Cambria Steel and Youngstown Sheet & Tube was aborted, due in part to the financial community's trepidation over Lackawanna's heavy-bonded debt.

Clarke resigned from Lackawanna Steel late in 1918 and subsequently became president of Consolidated Steel Corporation, a sales company formed to handle foreign trade for most of the larger producers except U.S. Steel. Under the recent Webb-Pomerene Act federal policy encouraged combined exporting companies similar to European cartels by offering exemptions from antitrust legislation. As president of Consolidated Steel, Clarke emphasized the critical importance of export sales in relation to the profits of American manufacturers. However, unfavorable foreign exchange rates and disagreements over shipments to Canada hampered Consolidated's effectiveness. The firm dissolved in 1922.

After leaving Consolidated Steel, Clarke entered the final phase of his long career. From 1919 to 1922 he had been treasurer of AISI. In 1923 he was elected that organization's secretary, a post vacant since 1920. He held the post for the remainder of his active life. In addition, he served on numerous AISI standing and special committees. Clarke also moved easily in New York social circles during his later years. He was a member of the Century, Metropolitan, Harvard, and India House Clubs. A varsity oarsman during collegiate days, his athletic interests included golf, tennis, and cricket.

E. A. S. Clarke died of pneumonia at his home in Rumson, New Jersey, on May 15, 1931. He was sixty-nine years old. He was survived by his wife, the former Louise Hall Ward of New York, whom he had married in 1890, and two daughters, Mrs. George M. Bodman and Mrs. Stanley Brown-Serman.

Publications:
"Discussion: Foreign Relations," *American Iron and Steel Institute Yearbook* (1910): 54–55;
"Contract Obligations," *American Iron and Steel Institute Yearbook* (1912): 23–30;
"Taking Care of Dependents of Absentee Employees Who Enlist for National Service," *American Iron and Steel Institute Yearbook* (1917): 112–118.

References:
"Changes in International Harvester Company's Officials," *Iron Age,* 74 (December 8, 1904): 32;
"Independent Steel Export Company," *Iron Age,* 102 (December 5, 1918): 1411;
Thomas E. Leary and Elizabeth C. Sholes, *From Fire to Rust: Business, Technology and Work at the Lackawanna Steel Plant, 1899–1983* (Buffalo: Buffalo and Erie County Historical Society, 1987);
"President Clarke of the Lackawanna Steel Company," *Iron Age,* 74 (November 3, 1904): 31;
"The Steel Plant of the Deering Harvester Company," *Iron Age,* 70 (September 18, 1902): 11;
Melvin I. Urofsky, *Big Steel and the Wilson Administration* (Columbus: Ohio State University Press, 1969).

Columbia Steel Corporation

by H. Lee Scamehorn

Historic Learning and Research Systems

The Columbia Steel Company was the first integrated producer of steel west of the Rocky Mountains. It had its origin in Pittsburg, California, 45 miles from San Francisco, where open-hearth operations commenced in November 1910. The enterprise was reorganized and recapitalized in 1922 in order to acquire the Utah Coal and Coke Company's fuel and iron ore properties. A blast furnace and 33 by-product coke ovens erected at Ironton, Utah, 50 miles south of Salt Lake City, turned out pig iron in 1924. The pig iron was shipped to the West Coast to be made into steel products for the regional market.

The United States Steel Corporation purchased control of Columbia Steel in 1929, giving U.S. Steel direct access for the first time to the expanding market of the West Coast. Renamed the Columbia Steel Corporation, the regional producer continued to make pig iron in Utah, most of which was shipped to plants for making steel at Torrance, California, and Pittsburg. The Columbia Iron Mining Company, a subsidiary of Columbia Steel, systematically explored and commenced the large-scale development of iron ore deposits near Cedar City, Utah.

The Ironton plant turned out about 200,000 tons of pig iron annually until World War II. A second furnace, relocated from Joliet, Illinois, more than doubled Columbia Steel's output. Four new batteries of coke ovens, 500 ovens in all, were built at Columbia, the firm's coal mine in Carbon County, 130 miles east of Ironton.

During World War II Columbia Steel was selected as an agent by the Defense Plant Corporation to erect a fully integrated plant in Utah, a location well beyond the reach of Japanese aircraft, to supply war industries on the Pacific Coast. The Geneva steel plant, located at Orem, Utah, eight miles from Provo on the eastern shore of Utah Lake, comprised three 1,100-ton blast and nine 225-ton open-hearth furnaces, 252 by-product ovens, blooming mills and plate mills. It had a capacity of 700,000 tons of ingots a year, and was operated under lease by U.S. Steel.

After World War II the Geneva Works was declared surplus and put up for sale. Henry Kaiser, who during the war had built an integrated plant at Fontana, east of Los Angeles, and the Colorado Fuel and Iron Corporation, operator of a similar but older plant at Pueblo, Colorado, sought to purchase the property, but after some equivocation U.S Steel, the wartime lessee, decided to buy the facility. The War Assets Administration accepted U.S. Steel's offer for the plant.

The corporation's two Utah properties were combined in the postwar years as the Columbia-Geneva Division. The plants at Ironton and Orem remained active producers of pig iron, some of which was made into steel locally, and the remainder continued to be sent to West Coast plants. Operations ended at Ironton in 1962. Six years later the property was given to Brigham Young University. After the blast furnaces, coking ovens, and sintering plant had been dismantled, the 386-acre tract was developed as an industrial park.

References:

G. Eldridge, "West Coast Steel Mill," *Steel,* 116 (April 2, 1945): 120, 123, 126; 117 (July 2, 1945): 108, 110, 158; (July 16, 1945): 134, 136, 174, 176, 178; (August 6, 1945): 130, 133, 159, 161–162, 164;

Gustive O. Larson, "Bulwark of the Kingdom: Utah's Iron and Steel Industry," *Utah Historical Quarterly,* 31 (Summer 1963): 248–261.

William H. Colvin

(May 20, 1897 – April 22, 1972)

by Geoffrey Tweedale

Manchester, England

CAREER: Clerk Harris Trust & Savings Bank, Chicago (1921–1925); manager, William H. Colvin & Company, Chicago (1925–1927); vice-president, Oklahoma Power & Water Company, Tulsa (1927–1930); partner, Pynchon & Company, Chicago (1930); secretary-treasurer (1931–1936), president, (1936–1945) Rotary Electric Steel Company, Detroit; president, (1945–1954) Crucible Steel Company of America.

William Henry Colvin, a leading figure in America's postwar specialty steel industry, was born in Chicago, Illinois, on May 20, 1897, the son of William Henry Colvin, a financier, and Bessie Small Colvin.

Colvin received his preparatory education at the University School, Chicago, and the Principia School in Saint Louis. During World War I he was commissioned a second lieutenant in the U.S.

Army infantry, before returning to Cornell University, where he graduated with an A.B. degree in 1920. His early business experience was in banking and finance. In 1921 he entered the bond department of the Harris Trust & Savings Bank in Chicago. In 1925 he was appointed manager of William H. Colvin & Company, the stockbrokerage firm founded in Chicago in 1898 by his father. Two years later he left to become the vice-president of the Oklahoma Power & Water Company, in Tulsa. Resigning in 1930, he returned to Chicago as a partner in the brokerage firm Pynchon & Company.

Colvin's introduction to steel manufacturing began in 1931, when he collaborated with his father and others in organizing and incorporating the Rotary Electric Steel Company, in Detroit, Michigan. He was initially secretary-treasurer of the firm, becoming its president in 1936. The firm was a specialty steelmaker; its early product was carbon steel bars, but by 1940 it was also manufacturing alloy steel bars. In this high-quality sector of the steel market — where tonnage was small, but the value was very high — there was room for the small, specialist producer who could rapidly adapt a product line to suit changes in demand. Colvin's company was therefore able to compete successfully with the giant specialty steelmakers at the other end of the market, where the Crucible Steel Company of America dominated the trade.

World War II was a profitable time for many steelmakers, but less so for Crucible Steel. The firm produced a wide range of specialty steels: tool, agricultural, wire, aircraft, automobile, and stainless steels, as well as more finished products for ordnance and shipping. But it had proved difficult to integrate the firm's diverse factories and huge product line, and consequently profitability was low. Crucible's net profits in 1946, for example, were only $527,000 on sales of $88 million.

In 1944 a reorganization of Crucible was begun by William P. Snyder, Jr., the chairman and a substantial stockholder. Recognizing that Crucible had become preoccupied with production at the expense of profits, Snyder succeeded in rousing the other directors to the need for a change in management. For that job he selected William H. Colvin, who became president of the Crucible Steel Company of America in 1945, succeeding Frederick B. Hufnagel.

Between 1946 and 1954 Snyder and Colvin created an entirely new executive group, borrowed $49 million, and used $25 million in profits to modernize the company and adopt more scien-

William Henry Colvin, Jr.

tific methods of management. Colvin reduced the number of Crucible's plants from ten to six, concentrated on ingot production at two plants instead of five, and progressively improved equipment. At the Midland Works outside Pittsburgh, where the bulk of Crucible's ingot production was concentrated, 113 new coke ovens were built, and 100 modernized; a hot-strip mill was installed alongside a train of cold-strip mills, placing Crucible in the flat-rolled stainless steel business. A modern blast furnace was added; two existing blast furnaces were rebuilt; eight open-hearth and five electric furnaces were rebuilt and enlarged to provide 250,000 extra tons of annual melting capacity; materials handling was improved by adding straddle trucks and lifts to the cranes and railroad cars; an agricultural products plant was installed to produce such items as sheared shapes and discs; and finally, Midland's rolling, annealing, and pickling equipment was modernized.

Meanwhile, the Park Works in Pittsburgh, where Crucible made many of its small batches of special alloys, was reorganized, as was the Syracuse plant, the company's primary tool-steel producer and its secondary producer of ingots. In 1945 a new plant was purchased in Pittsburgh for the railroad and industrial spring market. In 1948 a new plant for cold-rolled strip was added at Harrison, New Jersey, and the old ordnance plant there was converted for the making of magnets and other cast products. Trent Tube Company, a Wisconsin maker of stainless and high-alloy welded pipe and tubing, was acquired in 1948 through an exchange of stock. In 1950 Crucible joined with Remington Arms in ownership of Rem-Cru Titanium, a $6.6 million venture in the new field of melting and processing titanium. Colvin also expanded warehouse space, transferred the company headquarters from New York to Pittsburgh, and began pruning Crucible's more unprofitable items.

The result of these policies was apparent by the late 1950s. By 1957 Crucible had shown a dramatic increase over its 1946 sales and earning figures — tripling sales and increasing by 25 times the amount earned. By 1955 the company was paying dividends for the first time since World War II. In 1954, however, Colvin had decided that he was in need of a rest and retired, handing over the presidency to Joel Hunter, who continued his predecessor's program of modernization.

Colvin was also president and director of Rem-Cru Titanium, and a director of the Central & South West Corporation, the National Biscuit Company, and the Michigan Mutual Liability Company. He also served on the steel advisory committee of the War Production Board in 1942.

Colvin was married twice: on September 3, 1921, to Grace Ellett, who died in 1949, and on July 11, 1952, to Allis Ferguson. He had a daughter, Caroline, by the first marriage.

William Henry Colvin died in Sarasota, Florida, on April 22, 1972.

Publications:
Economic Power of Unions: Statements Presented to the Committee on Banking and Currency of the United States Senate, by William H. Colvin (New York: American Iron and Steel Institute, 1949);
Crucible Steel of America: 50 Years of Specialty Steelmaking in the USA (New York: The Newcomen Society in North America, 1950).

Reference:
Herrymon Maurer, "Crucible: Steel for Frontiers," *Fortune,* 55 (1957): 84–88.

Commission on Foreign Economic Policy
by Paul Tiffany
Sonoma State University

The Commission on Foreign Economic Policy (CFEP) was created in the summer of 1953 by President Dwight Eisenhower as a means of generating support for congressional extension of the 1953 Reciprocal Trade Agreements Act. The commission was to undertake "a broad-gauge study into the question of what [U.S.] foreign economic policy should be," but its real purpose was to support an extension of the free-trade laws that President Eisenhower felt were threatened by some of the more isolationist elements in Congress. Clarence Randall, president of the Inland Steel Company, was named to head the CFEP.

The commission solicited testimony from numerous groups and individuals. Among those were the American Iron and Steel Institute (AISI), which offered its views in December 1953. Representing essentially the vision of the larger integrated steel firms of the nation, AISI in effect recommended

adoption of a sweeping industrial policy in steel by the U.S. government. Noting that foreign steelmakers had recovered from the wartime ravages of the 1940s through close cooperation with their own governments, the AISI concluded that "the machinery does not exist in government today to inaugurate and carry out effectively a foreign economic policy which would best meet" the needs of the United States. As a consequence AISI recommended creation of the independent and permanent Foreign Trade Commission by Congress.

This proposed new commission would absorb the duties of the then-existing U.S. Tariff Commission, the Export-Import Bank, and the Committee for Reciprocity Information, thus becoming an institutionalized superagency to formulate, coordinate, and implement national economic policy regarding foreign trade. Through such an organization AISI believed that American steelmakers could better cope with the threat of rising offshore producers who themselves had obtained significant benefits from their own home governments and who appeared to be aiming their sights on American markets.

The Randall Commission, as the CFEP was generally known, took the views of the AISI into consideration along with the other 1,500 groups who testified. The resulting report to the president recommended an extension of the Reciprocal Trade Agreements Act, along with some new ideas generally to liberalize U.S. foreign-trade policy. Congress, however was not particularly sympathetic to such arguments in 1954, the year the recommendations were made, and all Eisenhower got was a one-year extension of the existing act. The American steel industry itself remained divided over the path that U.S. foreign economic policy should take in this period. Foreign imports were perceived as a threat — albeit a minor one — to many, so any departures from the long-held views of protectionism were approached only with great caution by many steel leaders. As such both AISI and the Eisenhower administration were unable sufficiently to focus public attention on a trade-policy debate. As a consequence no coherent and long-term foreign economic policy could be fashioned in the 1950s. The response to the growing strength of offshore steel producers remained ad hoc in nature. It would be another 20 years before steelmakers were able to better articulate a policy on steel imports into America — but by that time it was perhaps too late to stem the tide of such forces.

References:
Burton Kaufman, *Trade & Aid: Eisenhower's Foreign Economic Policy, 1953–1961* (Baltimore: Johns Hopkins University Press, 1982);
Paul Tiffany, *The Decline of American Steel: How Management, Labor, and Government Went Wrong* (New York: Oxford University Press, 1988).

Continuous Casting

by Thomas E. Leary

Industrial Research Associates

A post–World War II development in steelmaking, continuous casting was quickly adopted by foreign steel companies. The process bypasses many of the traditional stages of heating and cooling ingots. As such, the process has reduced operating costs while improving product quality. The large, integrated steel mills in the United States, however, were slow to adopt the new technology, lagging behind their foreign competitors into the 1980s, when much of the domestic market already had been lost to Japanese and European producers.

Successful continuous-casting techniques first were developed in the nonferrous branch of the metals industry. The German Siegfried Junghans devised a process that became common during the 1930s. His method featured a vertical, open-ended, water-cooled mold which oscillated to prevent the molten metal from adhering to its walls. Casting was continuous in the sense that the size of the ingot was not determined by the size of the mold; rather, given a steady supply of molten metal, the cast product would emerge like toothpaste from a tube. American manufacturers of brass and aluminum adopted the innovations, but application to ferrous products was inhibited by the properties of the metal. The melting point of steel was considerably higher and its thermal conductivity lower than nonferrous material.

To attain a high rate of production, continuous casting required rapid cooling to freeze the molten metal. In other words, casting speed was contingent on the cooling rate. The speed and uniformity of the casting rate also determined the surface quality and internal structure of the eventual product. As the steel passed through the mold, a solid surface formed, but the center of the billet remained molten until additional cooling upon exit, the most critical stage of the operation.

Continuous-casting facilities included cranes and ladle cars for handling molten steel, an intermediate reservoir or tundish to strain out slag as the metal entered the mold, the mold itself and water sprays

A continuous hot-strip mill producing steel sheets

that chilled the steel after it left the mold, rolls for drawing the steel from the mold, and a saw or torch for cutting the cast product into the requisite lengths. Early casters featured a single vertical mold, and the product was cut off while still in the vertical position.

Commercialization of continuous casting was handled by multinational consortia sharing patent rights and licensing agreements. In the United States the rights to the Junghans oscillating-mold patents were controlled until 1955 by Irving Rossi. His company, Concast, Inc., of New York, eventually licensed several suppliers for the U.S. and Canadian markets. Concast was, in turn, a branch of a Swiss corporation, Concast AG of Zurich, whose stockholders included Rossi and several European steelmakers and engineering companies. One of the American licensees was the Koppers Company, which equipped the first continuous-casting operation in North America at Atlas Steels in Welland, Ontario, in 1954.

Shortly after the Atlas operation proved successful, Inland Steel pondered an expansion program featuring both oxygen steelmaking and continuous casting. This would have been a revolutionary step for American mills at the time, but, after conducting trials with Atlas, Koppers, and Allegheny

Ludlum, Inland concluded in 1956 that enlarging the process to accommodate big, low-carbon slabs involved too many unknowns. The company went ahead with its original plan to install additional open hearths and a slabbing mill. Babcock & Wilcox undertook a pilot continuous-casting project with Republic Steel in 1948, and in 1962 they installed the first commercial billet caster in the United States at Roanoke Electric Steel, Roanoke, Virginia. Between 1964 and 1965 multiple-strand billet casters were erected at Roblin Steel in Dunkirk, New York, and at Armco's Sand Springs, Oklahoma, works.

In conjunction with electric-furnace steelmaking, continuous casting provided the technological basis that permitted minimills to challenge the large, integrated companies. Republic Steel's engineers who had assessed the potential significance of continuous casting after World War II did not expect it to be revolutionary. They wrongly guessed that the new technique would not lower production costs significantly nor promote decentralization of the steel industry into smaller units.

American steel companies, however, were slow to adopt the continuous-casting process in the production of wide slabs. Continuous casters for slab had to be paired with strip mills that were too expensive for smaller producers to install. Compared to billet production, the casting of big slabs also encountered more technical difficulties.

The first U.S. mill to install an experimental continuous slab caster was McLouth Steel of Trenton, Michigan, in 1963. Their principal customer was General Motors. Other installations at National Steel's tinplate plant in Weirton, West Virginia, and U.S. Steel's Gary Works followed in 1968. But the five-year lag between McLouth Steel's experimental program and the first signs of the industry's acceptance of the process typified the American steel industry reluctance to change with the times. By this time the attitudes of U.S. and Japanese steelmakers toward continuous casting, particularly for slabs, were not markedly dissimilar, however. Nippon Kokan installed Japan's first wide slab caster at its Tsurumi Works in 1967, and continuously cast steel accounted for only 4 percent of that country's output in 1968. But the Japanese were constructing several enormous new works, and Japan's percentage of continuously cast steel accelerated dramatically to 60 percent by 1980.

In contrast, casters in large U.S. mills were part of piecemeal modernization of existing works. Continuous casting represented 9 percent of U.S. output in 1975 and only 20 percent in 1980. Most of the state-of-the-art slab casters installed in U.S. mills during the 1980s, furthermore, were imported from Voest-Alpine of Austria, Mannesmann-Demag of West Germany, or Hitachi-Zosen of Japan.

References:

Keith W. Bennett and George A. Weimer, "World Continuous Casting Report," *Iron Age*, 227 (July 16, 1984): 49–58;

Hans Hofmeister, "The Continuous Steel Casting Process," *Blast Furnace and Steel Plant*, 49 (May 1961): 403–406;

C. L. Kobrin, "The Year of the Slab Caster," *Iron Age*, 201 (March 7, 1968): 101–108;

T. W. Lippert, "Continuous Casting," *Iron Age*, 145 (April 4, 1940): 31–39; (April 11, 1940): 44–51;

Lippert, "Continuous Casting," *Iron Age*, 153 (February 24, 1944): 48–65, 138;

Lippert, "Continuous Casting of Semifinished Steel," *Iron Age*, 162 (August 19, 1948): 72–80, 159–161;

Charles Longnecker, "Continuous Casting Process Now in Commercial Production," *Blast Furnace and Steel Plant*, 42 (November 1954): 1292–1294;

George McManus, "Continuous Casting: A New Generation Debuts in the U.S.," *Iron Age*, 229 (June 6, 1986): 19–29;

McManus, "The Direct Casting Controversy," *Iron Age*, 196 (August 5, 1965): 53–57;

McManus, "The Push Begins for Slab Casting," *Iron Age*, 204 (September 25, 1969): 105–112;

McManus, "Slab Casting: Caution Gives Way to Action," *Iron Age*, 199 (February 16, 1967): 93–100;

Kizo Takeda, "The Technical Development of the Iron and Steel Industry of Japan," *Blast Furnace and Steel Plant*, 54 (August 1966): 726–727.

Richard Conrad Cooper

(June 15, 1903 – October 2, 1982)

by Richard W. Kalwa

University of Wisconsin — Parkside

CAREER: Industrial engineer, American Associated Consultants (1929–1937); assistant vice-president, operations, Wheeling Steel Corporation (1937–1945); assistant vice-president, industrial relations, United States Steel Corporation of Delaware (1945–1948); vice-president, industrial engineering (1948–1955), vice-president, administrative planning (1955–1958), executive vice-president, personnel services, U.S. Steel Corporation (1958–1968).

Richard Conrad Cooper served as the chief labor negotiator for United States Steel Corporation at a time when that company exercised strong leadership over the entire steel industry in bargaining strategy. His engineering background and tough approach to negotiations made him a distinctive spokesman for U.S. Steel and other companies during a critical phase of the industry's evolution.

Cooper was born on June 15, 1903, the fifth of eight children, in Beaver Dam, Kentucky. His par-

Richard Conrad Cooper

ents were David Peter Cooper and Stella Taylor Cooper. When Cooper was four years old his father left his work as a coal miner, and the family moved to South Dakota, traveling part of the way by covered wagon. The family raised cattle near Pierre, living at one point in a sod house. Cooper rode broncos in rodeos for extra money as a young man.

He attended the University of Minnesota, where he studied engineering. He also played center for the football team and won a heavyweight boxing championship. In 1926 he received a B.S. degree and went to work as a field engineer for the Universal Portland Cement Company in Minneapolis.

In 1929 he married Irene V. Johnson, moved to New York, and began work with a consulting firm, eventually known as American Associated Consultants, which installed the Bedaux system of efficiency engineering for companies. This system was labeled a "speedup" device by organized labor. In 1937 Cooper was hired as assistant vice-president of operations by the Wheeling Steel Corporation. While at Wheeling, Cooper negotiated with John L. Lewis of the United Mine Workers of America (UMWA). Once, in response to a harangue by Lewis, he called the UMWA leader a "windbag." Challenged to a fight by Lewis, Cooper immediately accepted, but both men were dissuaded from the duel.

In 1943 Cooper became executive director of the Cooperative Wage Study, a group formed by 12 steel companies to study and reform the industrial wage structure. He was hired as the assistant vice-president of industrial relations for the U.S. Steel Corporation of Delaware in 1945, and was appointed to head the company group on the Joint Wage Rate Inequity Negotiating Committee. This company-union committee had been established in accordance with a War Labor Board directive that sought to eliminate wage inequities and standardize job classifications at each company. The committee reached a settlement in 1947. Cooper also served as an industry representative on the tripartite Steel Commission, mandated by the War Labor Board in 1945 to oversee inequity negotiations. Cooper's effort to establish a new incentive system based on what he called "a fair day's work" was less successful, and union opposition led to a compromise in 1952 which enabled the union to challenge incentive rates through grievances.

Cooper became vice-president of industrial engineering at U.S. Steel Company in 1948 and vice-president of administrative planning in 1955. Upon John Stephens's retirement in 1958 Cooper became vice-president of industrial relations for U.S. Steel. As chair of a four-man committee authorized to negotiate a contract for the 12 largest steel companies, Cooper assumed responsibility for the industry. Observers noted the contrast between the urbane, tolerant Stephens and his successor, characterizing Cooper as "taciturn, blunt, and strictly no-nonsense." His manner seemed to signify a new, tough approach in U.S. Steel's bargaining policy. In fact, Cooper and Chairman Roger Blough visited President Dwight D. Eisenhower prior to the 1959 negotiations to persuade him not to intervene.

Cooper has been associated with the industry's decision to raise the issue of work rules shortly before the labor contract expired, leading to a bitter, inconclusive four-month strike. Yet Cooper was inflexible in his determination to eliminate section 2B of the contract, which protected existing work practices. Cooper described the provision as "written and approved at four o'clock in the morning by six brain-weary men" and as having been excessively broadened by arbitration rulings.

He favored the cooperative approach taken by the Human Relations Committee in the wake of the 1959–1960 conflict, in which constructive problem-solving replaced the traditional adversarial approach. This period, however, came to an end with the battle for the presidency of the steelworkers union. As the chief industry negotiator in 1965, Cooper was disturbed by the union's internal turmoil. While local demands multiplied and hundreds of union officers descended on the national talks, Cooper complained publicly that the union was a "house divided against itself" that presented a "preposterous assembly of proposals." Following a settlement forced in large part by pressure from President Lyndon B. Johnson, he delivered a speech in which he called on Congress to "perform major surgery on the obese body of union power."

In 1967 Cooper proposed the idea of eliminating the use of strikes in the steel industry, but this was rejected by the leadership of the steelworkers. In an interview shortly before his retirement, he speculated that the uncontrolled use of union power would provoke the exercise of government control, "which could lead to the loss of the freedom to bargain collectively — even loss of the free enterprise system." His last service to U.S. Steel was to lead negotiations in 1968.

Cooper was a member of the American Iron and Steel Institute, the American Institute of Industrial Engineers, the Engineers Society of Western Pennsylvania, and the Pennsylvania So-

ciety for the Advancement of Management. He died on October 2, 1982, in Sewickley, Pennsylvania.

Publications:
"Wage Plan Involves Unit of Labor Management," *Steel*, 90 (April 11, 1932): 34–35;
"Basic Policy for Time and Method Studies: U.S. Steel Corporation," *Advanced Management,* 15 (April 1950): 2–4;
"Increased Productivity Through Incentives," *Iron and Steel Engineer,* 28 (August 1951): 138+;
"How Wage Inflation Hurts," *Steel* ,144 (February 16, 1959): 96;
"Steel Faces the Common Market," *Advanced Management,* 26 (November 1961): 5–8;
"Collective Bargaining at the Crossroads," *Iron and Steel Engineer,* 44 (December 1967): 113–115.

Published Speeches:
"Productivity in United States Steel," prepared for the Subcommittee on Antitrust and Monopoly of the Senate Committee on the Judiciary, Washington, D.C., August 1957, in *Steel and Inflation: Fact vs. Fiction* (N.p. U.S. Steel, 1958), pp. 55–63;
"Soft policies or firm?" (New York: United States Steel Corporation, 1961);
"Steel collective bargaining in the 60's," at the Labor and Industrial Relations Center, Michigan State University, East Lansing, Michigan, April 24, 1961 (New York: United States Steel Corporation, Public Relations Dept., 1961);
"What road ahead?" before the Printing Industry Association of Western Pennsylvania, University Club, 21 March 1968, Pittsburgh (New York: United States Steel Corporation, Public Relations Dept., 1968).

References:
James J. Healy, ed., *Creative Collective Bargaining: Meeting Today's Challenges to Labor-Management Relations* (Englewood Cliffs, N.J.: Prentice-Hall, 1965);
John Herling, *Right to Challenge: People and Power in the Steelworkers Union* (New York: Harper & Row, 1972);
John P. Hoerr, *And the Wolf Finally Came: The Decline of the American Steel Industry* (Pittsburgh: University of Pittsburgh Press, 1988);
Garth L. Mangum, "Interaction of Contract Administration and Contract Negotiation in the Basic Steel Industry," *Labor Law Journal,* 12 (September 1961): 846–860;
Grant McConnell, *Steel and the Presidency, 1962* (New York: Norton, 1963);
David J. McDonald, *Union Man* (New York: Dutton, 1969);

George J. McManus, *The Inside Story of Steel Wages and Prices, 1959–1967* (Philadelphia: Chilton, 1967);

Mark Reutter, *Sparrows Point: Making Steel: The Rise and Ruin of American Industrial Might* (New York: Summit, 1988);

Jack Stieber, *The Steel Industry Wage Structure* (Cambridge, Mass.: Harvard University Press, 1959);

John Strohmeyer, *Crisis in Bethlehem: Big Steel's Struggle to Survive* (Bethesda, Md.: Adler & Adler, 1986);

Paul A. Tiffany, *The Decline of American Steel: How Management, Labor, and Government Went Wrong* (New York: Oxford University Press, 1988).

Archives:

The speeches of R. Conrad Cooper published by U.S. Steel are housed in the Hagley Museum and Library, Wilmington, Delaware.

William Ellis Corey

(May 4, 1866 – May 11, 1934)

by Gerald G. Eggert

Pennsylvania State University

CAREER: Laborer, chemical laboratory assistant, puddler, roller (1882–1884), weigh-master (1884–1886), clerk in business office, J. Edgar Thomson Works of Carnegie Steel (1886); superintendent of plate mill and open-hearth slabbing mill (1887–1893), superintendent of armor plate department (1893–1897), general superintendent, Homestead Steel Works of Carnegie Steel (1897–1901); president, Carnegie Steel Company and National Steel Company (subsidiaries of the United States Steel Corporation) (1901–1903); president, United States Steel Corporation (1903–1911); chairman of the board, Midvale Steel & Ordnance Company (1915–1923).

William Ellis Corey, steelmaster and capitalist, began work in the mills as a boy. He was a junior partner of Andrew Carnegie by the age of thirty-one and president of the United States Steel Corporation at thirty-seven. He ended his business career as chairman of the board of Midvale Steel & Ordnance Company.

His *New York Times* obituary described Corey as having been "steel-born" in Braddock, Pennsylvania, the site of Carnegie's J. Edgar Thomson Works, and a town in which boys "heard the voice of the mills," just as New England boys heeded the call of the sea. His parents, Alfred A. and Adaline Fritz Corey, provided a comfortable life on the income from the father's coal business. William attended public school to the age of sixteen, then went to work, first as a grocer's boy and then at his father's coal tipple.

He attended night classes at Duff's Business School in Pittsburgh.

Corey entered the steel industry as a laboratory assistant in the chemical department of the J. Edgar Thomson Works in 1882. Determined to learn all aspects of steelmaking, he undertook stints at puddling, rolling, and working as a furnace man by day while studying chemistry and metallurgy by night. When he was seventeen years old he married Laura Cook, the daughter of a local miner; they had a son, Alan A. Corey.

Corey's determination paid off as he advanced from weigh-master in 1884 to a clerkship in the business office in 1886. He transferred a year later to the open-hearth department at Carnegie's Homestead plant and in 1887 became superintendent of the plate and open-hearth slabbing mills under general superintendent Charles M. Schwab. Impressed with Corey's ability to drive men to ever-higher levels of production, Schwab named him superintendent of the armor plate department in 1893. During this period Corey developed what would be called the Corey Reforging Process, which toughened armor plate and reduced its weight while increasing its resistance to projectiles. When Schwab moved to the presidency of Carnegie Steel in 1897, Corey succeeded him as general superintendent at Homestead. About the same time he became a Carnegie junior partner and superintendent of Carnegie's Carrie furnaces and Howard axle works.

The newly organized United States Steel Corporation absorbed Carnegie Steel in 1901.

Schwab became president of U.S. Steel, and Corey was elected president of Carnegie Steel and two other U.S. Steel subsidiaries — the National Steel Company and American Steel Hoop Company. Two years later Schwab resigned the headship of U.S. Steel, and on August 1, 1903, Corey took his place in what was then the highest salaried position in the United States. U.S. Steel enjoyed remarkable growth under Corey. Its labor force increased by half (from 150,000 to 225,000), while its annual payroll increased a third (from $130 million to $175 million). The firm built a new plant and a city for its workers at Gary, Indiana; acquired the valuable properties of the Tennessee Coal, Iron & Railroad Company; and leased the rich iron ore lands of railway magnate James J. Hill in Minnesota.

Corey, like Schwab before him, differed with U.S. Steel's chairman of the board, Elbert Gary, named to run the corporation by the firm's financier, J. P. Morgan. When the demand for steel fell sharply Corey advocated procedures taught by Carnegie: cut prices and wages and go on producing steel which will be disposed of when the market improves. By contrast, Gary urged stable prices and wages until conditions improved, meanwhile cutting production and laying off workers. As his differences with Gary grew, Corey's private life burst into the news. Once described as having "no fads except fondness for home," he fell in love with Mabelle Gilman, a musical-comedy star. Corey deserted his wife of 22 years and, following their divorce, married Gilman in 1907. The new Mrs. Corey preferred living in France, and after several unhappy years they divorced in 1910. Corey tendered his resignation as president of U.S. Steel in 1911.

Although the scandal had damaged the image of his colleague and occasional adversary, Gary told reporters that Corey was "too valuable a man to the steel business to be dispensed with on account of his private affairs." After four years of retirement Corey returned to business in 1915. The demand of the European nations for steel following the outbreak of World War I led Corey and other industrialists and bankers to form a new combine that would become second only to U.S. Steel in the industry. That firm, the Midvale Steel & Ordnance Company, brought together several Pennsylvania steel and armaments firms, including the existing Midvale Steel Company of Philadelphia, Worth Brothers of Coatesville, the Remington Arms Company of Eddystone, and, later, Cambria Steel of Johnstown. Corey brought into the management two friends who had risen through the ranks at Carnegie Steel and later gone to U.S. Steel with him, Alva C. Dinkey as president and William Brown Dickson as vice-president and treasurer.

Thanks largely to the war, Midvale cleared profits of nearly $100 million by the end of 1918. Despite strenuous efforts to overcome the firm's image as a speculative war venture, Midvale could not carve out a large enough place as a peacetime steel producer to survive. Profits gave way to losses during the economic downturn of 1921, and two years later the firm broke up. Bethlehem Steel took over Worth Brothers and Cambria, while Midvale resumed its identity as a separate firm in Philadelphia.

Physically Corey was powerfully built, round-faced, rosy-cheeked, and blond-haired. He had blue eyes and wore gold-rimmed glasses. A friend once described him as "one of the finest fellows, socially . . . but an icicle in business." Corey shunned publicity and, except for his publicized affair, lived quietly to the age of sixty-eight. His death on May 11, 1934, at his Fifth Avenue mansion in New York City was attributed to hardening of the arteries and pneumonia.

References:

Gerald G. Eggert, *Steelmasters and Labor Reform, 1886–1923* (Pittsburgh: University of Pittsburgh Press, 1981);
Obituary, *Iron Age,* 133: 30,
Obituary, *New York Times,* May 12, 1934.

Archives:

Considerable material on William Ellis Corey can be found in the diary and personal and business papers of William B. Dickson, his longtime friend and colleague at Carnegie Steel, U.S. Steel, and Midvale Steel & Ordnance. William B. Dickson materials are housed in Special Collections, Pattee Library, Pennsylvania State University, University Park, Pennsylvania.

Corrigan, McKinney Steel Company
by Carol Poh Miller

Cleveland, Ohio

The origin of the Corrigan, McKinney Steel Company reaches back to the 1880s, when Capt. James Corrigan (1848–1908) of Cleveland, Ohio, began building up a fleet of lake vessels to carry iron ore. Dalliba, Corrigan & Company (later Corrigan, Ives & Company) mined and sold iron ore to the growing industrial cities in Illinois, Ohio, and Pennsylvania. The firm prospered until the Panic of 1893, when it went into receivership.

Price McKinney (circa 1863–1926), a bookkeeper then serving as secretary of the firm, was appointed receiver. Under his direction the company developed into the largest independent seller of Lake Superior iron ore and branched into pig iron production. By 1900 the company operated blast furnaces in Cleveland, Ohio, Josephine and Scottdale, Pennsylvania, and Charlotte, New York. In 1908 Corrigan, McKinney bought 40 acres along the west bank of the Cuyahoga River in Cleveland; construction of the first of two stacks began the following year. The two furnaces, each with a daily capacity of 350 tons, were blown in during 1910 and 1912.

The difficulty of selling merchant pig iron to steelmakers, who increasingly operated their own blast furnaces, prompted Corrigan, McKinney to enter the steel business. Between 1913 and 1916 the company erected two additional blast furnaces, open-hearth furnaces, a coke plant, and rolling mills for the production of sheet bars, billets, and slabs. Hampered by the limited market for its semifinished steel, the company added 10- and 12-inch bar mills in 1927. The addition of these two merchant mills and the erection of a five-story office building at the plant in 1925 represented the company's last period of expansion prior to its acquisition by the Republic Steel Corporation in 1935.

At the time of its takeover by Republic, *Fortune* magazine called Corrigan, McKinney's relatively brief corporate history "as tragic and sensational as any in the annals of American business." The company was the creation of two men, James

Corrigan and Price McKinney. When Corrigan died in 1908 he left his son, James W. Corrigan, a 40 percent interest in the firm, to be held in trust by Price McKinney until the boy came of age. When McKinney changed the name of the firm to McKinney Steel Company in 1917 it infuriated young Corrigan, who challenged McKinney and won control of his inherited stock. While Corrigan, a notorious playboy, and his wife, Laura Mae, lived the life of socialites in London, McKinney ran the business, building one of the most efficient steel plants in the country.

In May 1925 James Corrigan returned to Cleveland and announced that he had acquired an additional 13 percent of the company's stock, giving him majority control. Corrigan took over as president, changed the firm's name back to Corrigan, McKinney, and set out to learn the steel business. Price McKinney, who had served as president since 1909, remained as a director but was rarely seen at the plant again; he committed suicide in 1926. Two years later James Corrigan died of heart failure, leaving majority control of the company in the hands of his widow. This is how matters stood in 1930 when Cleveland industrialist Cyrus Eaton, who was looking for additions to the new Republic Steel Corporation, engineered the purchase of Corrigan, McKinney by Cleveland-Cliffs Iron Company; five years later the firm was acquired by the Republic Steel Corporation.

"We had made a survey to discover the most logical place in the United States for a big steel plant," Republic Steel chairman Tom M. Girdler recalled in his autobiography, *Boot Straps* (1943). "Among the factors considered were water transportation for ore, short rail haulage for coal, and a strategic location for economical distribution of products. . . . As if some Delphic Oracle of economics had spoken, we knew that what we ought to get was the Corrigan, McKinney Steel Company." At the time of its merger with Republic, the Corrigan, McKinney plant included 4 blast furnaces, 204 by-product coke ovens, and 14 open-hearth furnaces capable of producing 1.1 million gross tons of steel ingots each year. Corrigan, McKinney improved Republic's access to the Great Lakes, gave it a better balance between its raw steel and finishing capacities, and boosted Republic's share of industry capacity from 7 to 9 percent. The merger was challenged unsuccessfully by the U.S. Justice Department.

References:

Tom M. Girdler and Boyden Sparkes, *Boot Straps: The Autobiography of Tom M. Girdler* (New York: Scribners, 1943), pp. 219–222;

William T. Hogan, S. J., *An Economic History of the Iron and Steel Industry in the United States,* 5 volumes (Lexington, Mass.: Lexington Books, 1971), III: 1228–1231;

Charles Longenecker, "Corrigan, McKinney Steel Company," *Blast Furnace and Steel Plant,* 24 (January 1936): 75–81;

"Pioneer Ore Firm Adds Steel Mills to Its Large Furnace and Iron Mine Holdings," *Iron Trade Review,* 61 (November 15, 1917): 1043–1054;

"Republic Steel," *Fortune,* 12 (December 1935): 76–83+.

Stewart Joseph Cort

(March 16, 1881 – September 23, 1958)

by Thomas E. Leary

Industrial Research Associates

and

Elizabeth C. Sholes

Industrial Research Associates

CAREER: Engineer, Duquesne Works, Carnegie Steel Company (1906–1909); superintendent, steel division, Cambria Plant, Cambria Steel Company (1909–1916) and Midvale Steel & Ordnance Company (1916); superintendent, open-hearth department (1917–1922), superintendent, Saucon Plant (1922–1928), general manager, Maryland Plant (1928–1947), vice-president, steel division, Bethlehem Steel Company (1947–1957).

Stewart J. Cort was an influential steel executive during an era in which the industry achieved its greatest prominence and prosperity. He spent nearly 50 years in plant and corporate management and was tied closely to the incremental expansion of the industry's leading production technique — open-hearth furnaces.

Cort was born in Ogle County, Illinois, on March 16, 1881, to Martha Henderson and Joseph Turney Cort. He attended public schools at Greensburg, Pennsylvania, in the bituminous-coal region. In 1906 he received a baccalaureate degree in metallurgical engineering from Lehigh University. His career in steel began that same year in the open-hearth department of the Duquesne Works of the Carnegie Steel Company — United States

Steel Corporation's largest subsidiary. In 1909 he climbed to the first rung on the executive ladder by becoming superintendent of the steel division at Cambria Steel Company's venerable works in Johnstown, Pennsylvania.

Cort consistently characterized himself as a "production man," his interests lying in the operational side of the industry. He was most interested in the basic open-hearth process, the method of producing carbon steel first introduced in the United States on a commercial scale at Carnegie Steel Company's Homestead Works in 1888. More steel was tapped from open-hearth furnaces within two decades of its introduction to America than from the Bessemer converters, whose pyrotechnics had first propelled the industry to prominence.

Cort's progress through the corporate hierarchy paralleled the expansion of open-hearth capacity at the plants where he was assigned. At Cambria annual ingot capacity doubled between 1908 and 1916 to 1,080,000 tons. In 1917 Cort became superintendent of the open-hearth department for Bethlehem Steel Company's Saucon Plant. Aggregate ingot capacity at Bethlehem's Lehigh and Saucon plants reached 1,500,000 tons by

Stewart Joseph Cort

1926, up by a third over the already substantial levels reached in 1916. When Cort took over in 1928 as general manager of the Sparrows Point Plant outside Baltimore, it was only the nineteenth-largest works in the country, with an ingot capacity of 1,020,000 tons in 1926; by 1948 the Maryland Plant had jumped up to second position with an annual output of 4,232,000 tons. As vice-president in charge of Bethlehem's steel operations after World War II, Cort oversaw the largest expansion in the corporation's history to date. More than 10,000,000 tons were added to annual ingot capacity as Bethlehem, along with other steelmakers, shook off fears of a postwar economic slump and projected its output on the basis of rising consumer affluence and cold-war tensions, which affected demand in the defense industry.

Cort maintained that the technology embodied in the open hearth could keep pace with growing demand and ever-more-stringent customer specifications. The wide variations in open-hearth practice and quality that he had noted in his technical papers of the 1920s had by the 1940s been partially corrected by advances in design, instrumentation, and testing. Toward the close of his career, Cort kept current with experiments using oxygen to improve combustion in the open hearth, but he did not anticipate any technological breakthroughs that would render the standard open-hearth methods obsolete.

Cort's professional standing gave weight to his opinions. As vice-president he also sat on Bethlehem's board. Cort was a member of the American Iron and Steel Institute (AISI), the Iron and Steel Institute of Great Britain, the Association of Iron and Steel Engineers, and the American Institute of Mining, Metallurgical, and Petroleum Engineers, which conferred on him the first Fairless Award in 1955, for outstanding leadership in manufacturing.

Cort was active in civic and charitable groups throughout the Baltimore area. His government service during World War II included stints with the National Manpower Commission and the Maryland State Defense Commission. His ties to his alma mater, Lehigh, through alumni affairs and trusteeships, were reinforced by his philanthropic activity, including the establishment of an endowment fund to provide scholarships. He was a member of the Saucon Valley Country Club, one of the fixtures in the lives of Bethlehem executives during the corporation's halcyon days. In 1910 Cort married Carolyn Myrtilla Schreiner, the daughter of a physician, from Mount Lebanon, Pennsylvania. The couple had two children: Carol and Stewart Shaw, who embarked on a career with Bethlehem's sales division that took him to the chairmanship of the company in the early 1970s.

Cort retired from Bethlehem on January 1, 1958, and died later that year on September 23 at the age of seventy-seven. Some months earlier Sparrows Point had tapped the first heat of steel from an expansion program that was designed to make his old plant the largest in the United States with an annual ingot capacity of 8,200,000 tons. The open hearths that came on line there during 1957–1958 were the last constructed at a major American steel plant. The open hearths eventually burdened Bethlehem with capital and operating costs higher than those associated with the basic oxygen process, an innovation then being adopted in Japan to compensate for limited supplies of scrap. Steelmaking in basic oxygen vessels did not become the basis for Bethlehem's modernization program until the mid 1960s. It later fell to Stewart S. Cort to dismantle many of the open hearths that his father had worked to perfect.

Publications:
"Discussion: The Basic Open-Hearth Process," *Year Book of the American Iron and Steel Institute* (1920): 367–372;

"Producer Coal Affects Steel Quality," *Iron Trade Review,* 67 (November 14, 1920): 1277–1278;

"Discussion: Open-Hearth Furnace Regenerators," *Year Book of the American Iron and Steel Institute* (1923): 410–413;

"Comparison of Open-Hearth Furnaces of Various Sizes," *Year Book of the American Iron and Steel Institute* (1926): 149–164;

"What Improvements in Quality and Production of Steel Have Been Made Since World War I?," *Blast Furnace and Steel Plant,* 29 (July 1941): 714–715;

"Always More Production," *Iron and Steel Engineer,* 26 (May 1949): 95–98;

"Progress in Production," *Iron and Steel Engineer,* 33 (May 1956): 142–144.

Reference:
Mark Reutter, *Sparrows Point: Making Steel: The Rise and Ruin of American Industrial Might* (New York: Summit, 1988).

Archives:
Archival materials are located in the Bethlehem Steel Collection, Hagley Museum and Library, Wilmington, Delaware; and the Bethlehem Steel Historical Collections, Hugh Moore Historical Park and Museums, Inc., Easton, Pennsylvania.

Stewart Shaw Cort

(May 9, 1911 – May 25, 1980)

by Lance E. Metz

Canal Museum, Easton, Pennsylvania

CAREER: Clerk, commercial research division (1937–1939), sales department, Pacific Coast division (1939), manager, commercial research and sheet and tin plate sales (1939?–1950), assistant general manager, sales, Pacific Coast division (1950–1952), general manager (1952–1954), vice-president in charge of Pacific Coast sales (1954–1960), assistant general manager of corporate services (1960–1961), vice-president of West Coast operations (1961–1963), director, (1961–1980), president, (1963–1973), chairman of the board, Bethlehem Steel Corporation (1970–1973).

Having spent most of his career with the Bethlehem Steel Corporation, Stewart S. Cort played a prominent role in the company's management. As Bethlehem's chairman during the 1970s, he presided over a period of prosperity, the last such period for the steel industry as a whole. Cort was born on May 9, 1911, to Carolyn Schreiner and Stuart J. Court at Duquesne, Pennsylvania, near Pittsburgh. At the time of his son's birth, Stuart J. Cort was employed as superintendent of an open-hearth shop with United States Steel Corporation. In 1917 he took a similar management position with Bethlehem Steel and moved his family to Bethlehem, Pennsylvania. Cort was educated at schools in Bethlehem and at the Blair Academy in Blairstown; in 1930 he entered Yale University, from which he graduated with a degree in economics in 1934. He then entered Harvard University's graduate school of business, from which he earned a master's degree in business administration in 1936.

After a brief stint as a salesman of prefabricated homes, Cort joined Bethlehem Steel in 1937 as a clerk in the company's commercial research division. By 1939 he had been transferred to Bethlehem Steel's Pacific Coast operations. He advanced quickly in a series of appointments to increasingly important positions, culminating in his appointment as vice-president in charge of Pacific Coast sales in 1954. In 1960 he was called back to Bethlehem, where he assumed the post of assistant general manager of corporate sales; in 1961 he was again promoted and returned to California as vice-president of Pacific Coast operations.

Cort was elected as the president of Bethlehem Steel in 1963. In 1970 he succeeded Edmund F. Martin as chairman of Bethlehem's board, a position he held for four years. During his tenure as both president and chief executive officer of Bethlehem Steel, Cort was witness to the company's peak years of prosperity and productivity. In 1973 Bethlehem produced a record 23,702,000 tons of steel and had a record income of $206,600,000. He continued his predecessor's expansion and modernization program by initiating major expansions of Bethlehem's production facilities in Burns Harbor, Indiana, and Lackawanna, New York.

To a much greater degree than his predecessors, Cort realized that Bethlehem and other steel corporations would have to become more environmentally sensitive. He committed Bethlehem to major investments in pollution-control measures. He was concerned with the increasing role that the policies of federal agencies would play in the future development of the steel industry. As a result, Cort made a strong and continuing effort to focus the attention of steel executives on the nation's capital in order to ensure that the industry's concerns would be heard. This belief in the importance of the interrelationship of government and business was a hallmark of both his business philosophy and his career. After his retirement as chief executive officer, he remained a Bethlehem director and adviser.

Cort was a member of the executive committee of the Yale Development Board and the executive council of the Harvard Business School Association, from which he received the Alumni

Achievement Award in 1974. He was also the chairman of the Radio Free Europe Fund in 1973–1974 and was an active supporter of the Boy Scouts. Cort died at his rural home near Bethlehem on May 25, 1980.

Publications:
"Alerts Industry to Dumping Threat," *Iron Age,* 192 (November 28, 1963): 58;
"Threat Is Now a Fact," *Automation,* 11 (August 1964): 135–136;
"Bargains or Booby Traps in Today's Steel Trade," *Commercial and Financial Chronicle,* 207 (June 20, 1968): 2440;
"Can the American Steel Industry Survive?" *Industry Week,* 169 (June 21, 1971): 44–50;
"Free Trade? Yes — But!" *Nation's Business,* 59 (July 1971): 54–57.

Reference:
"Cort of Bethlehem Steel," *Forbes,* 107 (May 15, 1971): 72.

Archive:
The Stewart S. Cort biographical file is in the Bethlehem Steel Historic Collection, Hugh Moore Historical Park and Museums, Inc., Easton, Pennsylvania.

Charles Raymond Cox

(April 27, 1891 – January 18, 1962)

by Bruce E. Seely

Michigan Technological University

CAREER: Accountant, Marwick, Mitchell & Peat Company (1914–1928); accountant, U.S. Shipping Board, Emergency Fleet Corporation (1918–1920); accountant, Crucible Steel Corporation (1920–1930); accountant, Babcock & Wilcox Tube Company (1930–1934); general superintendent, Ellwood City Works (1934–1936), vice-president, operations (1936–1941), executive vice-president, National Tube Company (1941–1943); president, Carnegie-Illinois Steel Corporation (1946–1950); president and director, Kennecott Copper Corporation (1950–1961).

Charles Raymond Cox followed an unusual path to an executive position in the steel industry. Before the 1960s it was production experience, not accounting expertise, that carried men to the

top of the steel industry. But Cox, an accountant, reached the top executive levels, and went on to head the largest U.S. copper producer. Central to his success were an affable personality and accommodating views on organized labor.

Cox was born in Schenectady, New York, in 1891. His father, Charles Mason Cox, was a business executive. He attended New York University, graduating in 1914 from the School of Commerce. His first accounting job was with the Marwick, Mitchell & Peat Company, but he served the Emergency Fleet Corporation from 1918 to 1920. After World War I he moved to the Crucible Steel Corporation in Pittsburgh. He stayed until 1930, then moved to the Babcock & Wilcox Tube Company's plant in Beaver Falls, Pennsylvania. In 1934 Cox joined U.S. Steel's National Tube Company. Apparently he had learned much about steel and tube production, for in a very short time Cox was general superintendent of National Tube's Ellwood City Works.

Cox moved quickly up the corporate ladder at this United States Steel Corporation subsidiary, becoming vice-president for operations in 1936 and executive vice-president in 1941. In 1943 he was named National Tube's president. National Tube was then a leading manufacturer of munitions and other war matériel. Cox gave the Christy Parks Works special attention, as it turned out 28 million shell forgings, finished shells, and bomb casings for the army and navy. President Harry S Truman later awarded Cox a Certificate of Merit for his contributions, citing increased steel, shell, and rocket production, and improvements in blast-furnace operations. For an accountant, Cox had become a good production man.

In 1946 Cox took over as president of Carnegie-Illinois Steel Corporation, bringing to the post an outlook on labor relations that differed from that of many longtime steel executives. When his promotion was announced, Cox commented on labor-management relations, saying, "We don't always agree on our problems but on the whole, I'm very strong for my men. If we ever get back to true collective bargaining, we'll get along fine." Other steelmen had made similar comments, yet Cox's sentiments were genuine. At another press conference held a month into his job, he strongly endorsed labor's right to strike, while predicting a steel settlement in 1947, largely because he had faith in labor organizer Phillip Murray, and because an agreement on job inequities had proved management and labor could work together.

His stand on organized labor was a reflection of Cox's affable yet forceful demeanor. In saluting him, *Iron Age* described Cox as a "tough hombre" who "doesn't like to tell someone twice what he wants. But yet he has done it and still does it in a way that draws people to him." But the writer added "there is no use trying to call Charley Mr. Cox. It just won't work." This type of personality was not common in the steel industry, and it marked Charles Cox as different.

This difference, combined with fine managerial abilities, attracted the attention of the copper industry. In September 1949 E. T. Stannard, president of Kennecott Copper Corporation, died in a plane crash. Cox replaced him in January 1950, moving, he said, for no good reason except for the challenge. He went with the best wishes of his U.S.

Charles Raymond Cox

Steel colleagues, whose send-off included a radio tribute to his bow ties and "fabulous cigars." In a typical gesture Cox and four stenographers worked all night to answer the flood of congratulatory letters and telegrams.

As president of Kennecott, Cox moved the company in several directions. He integrated operations by acquiring a refinery in New Jersey, building an electrolytic refinery at Baltimore, and acquiring the Okonite Company in Passaic, New Jersey. He expanded research efforts, especially in the areas of cost reduction, metal recovery from ores, and product development. As president he headed Kennecott subsidiaries, including the Quebec Iron and Titanium Corporation and the Orange Free State Gold Mining Companies, Ltd. Cox also inherited the position of leading representative of the copper industry, serving variously as director and president of the U.S. Copper Association and director of the Copper Institute; he was also active in the American Mining Congress. Throughout his career Cox retained his progressive outlook toward labor. In 1960 he warmly endorsed the call of Congress of Industrial Organizations counsel Arthur Goldberg for a National

Council of Labor-Management Advisors to discuss means of promoting full production and employment. The *New York Times* noted that Cox's support "contrasted with the coolness exhibited by the heads of other major companies to union moves for peace conferences divorced from contract negotiations."

From 1933 to 1938 Cox was an honorary member of the business advisory council of the Commerce Department. For many years he was a trustee of the Council for Economic Development and the U.S. Council of the International Chamber of Commerce. He was elected a director of the National Industrial Conference Board in 1950. He also served on several corporate boards, including Allied Titanium Corporation, Chase International Investment Corporation, Chase Brass & Copper Company, Pullman, Incorporated, and Knox Glass Company.

Cox was at different times a trustee of the National Safety Council, president of the Western Pennsylvania Safety Council, and a member of the business advisory council of the White House Conference on Highway Safety. He worked with the National Heart Fund campaign drives in 1959 and 1960. He supported New York University faithfully, chairing a development board to raise

$102 million in 1952 and heading a $75 million drive for the Washington Square campus in 1961. NYU awarded him an honorary doctorate of commercial sciences in 1950, citing Cox as a "superlative millman and industrial executive." Capping his list of honors was the Brotherhood Award of the National Conference of Christians and Jews in 1961.

Cox was married to Martha Knight in 1914. They had two children. He died in a fall from a train near his home in Darien, Connecticut, in January 1962, nine months after stepping down as president of Kennecott.

References:
"Ban on Public Utility Strikes is Urged by Steel Executive from Pittsburgh," *New York Times*, September 25, 1946, p. 23;
"Charles Raymond Cox," *Explosives Engineer*, 28 (May/June 1950): 68–69;
"Iron Age Salutes," *Iron Age*, 165 (February 9, 1950): 23;
"Labor Bid Backed by Industrialist," *New York Times*, December, 1960, p. 21;
School of Commerce, Accounts, and Finance, New York University, *Builders of Enterprise: (New York, 1950?);*
"Will Move Plant from Pittsburgh," *New York Times*, August 2, 1946, p.30.

George Gordon Crawford

(August 24, 1869 – March 20, 1936)

by Elizabeth C. Sholes

Industrial Research Associates

and

Thomas E. Leary

Industrial Research Associates

CAREER: Draftsman, Sloss Iron & Steel Company (1892–1894); chemist (1894–1895), assistant superintendent, blast furnace, Carnegie Steel Company (1895–1897); superintendent, blast furnace, National Tube Works (1897–1899); superintendent, blast furnace, Carnegie Steel Company (1899–1901); manager, National Department, National Tube Works (1901–1907); president, Tennessee Coal, Iron & Railroad Company (1907–1930); president, Jones & Laughlin Steel Corporation (1930–1935).

George Gordon Crawford directed the technological progress at the United States Steel Corporation (U.S. Steel) and the Jones & Laughlin Steel Company. Crawford was born in Madison, Georgia, on August 24, 1869. His father was an Atlanta physician who had been a surgeon with the Tenth Georgia Volunteers during the Civil War. Crawford graduated from the first class of the new Georgia School of Technology, where he earned a degree in engineering in 1890. Over the next two years he studied chemistry at Karl-Eberhard University in Tübingen, Germany.

With his graduate chemistry degree fresh in hand, Crawford accepted his first position — draftsman at Sloss Iron & Steel Company in Birmingham, Alabama. He soon moved to the Edgar Thomson works of the Carnegie Steel Company in Pittsburgh. After a brief stint as a chemist he advanced to engineering and became assistant superintendent at the blast furnace in 1895. His career moved him back and forth between the Edgar Thomson works and the National Tube mill in nearby McKeesport.

Crawford entered Edgar Thomson's blast-furnace department as the world-famous experiments conducted by works superintendent James Gayley were setting international performance and equipment standards. Gayley's work also established the Edgar Thomson works as a world-class blast-furnace operation. The furnaces were not without problems, however. In 1889 blast furnace C had exploded, killing several men, including founding superintendent Capt. W. R. Jones. While Crawford was in residence the furnace exploded twice more, and furnace H had a top blowout that killed eight and destroyed the cast shop.

These problems put Crawford in the forefront of inquiries into operating practices and the revision of standards. Under his direction the furnaces received extensive relining, and equipment was improved. As one of the industry's first technically trained managers, Crawford was in a particularly good position to utilize scientific and engineering advances emerging within the industry.

After two years at the Edgar Thomson blast-furnaces, Crawford was promoted to blast furnace and steelworks superintendent at National Tube Works, the company later consolidated by the J. P. Morgan interests.

His technical abilities garnered significant notice when he again returned to Edgar Thomson in 1899. While Crawford was superintendent, the blast-furnace department developed a safer metal skimmer patented by Carnegie Steel that prevented backflow and eliminated contact between hot metal and moisture. He continued to oversee other improvements until he returned for his second stint at National Tube in 1901. While there, Crawford planned and supervised a major overhaul of that works' blast furnaces, rolling mills, tube and pipe mills, and galvanizing operations. Simultaneously, he became an important advocate of the efficient use of by-product coke in blast-furnace operation.

During this time, National Tube was part of the Morgan takeover of Carnegie. Crawford's ten-

George Gordon Crawford

ure with U.S. Steel indicates that his early adaptations made him comfortable with change and able to deal with new organizational structures while maintaining control over daily operations. Crawford's technical and managerial expertise led to his selection as the first president of U.S. Steel's subsidiary, Tennessee Coal, Iron & Railroad Company (TCI&R), acquired in 1907. Crawford returned for the first time in 13 years to his native South, this time Alabama.

During his tenure in Birmingham he oversaw the expansion of different aspects of corporate operations: employee health and safety, social welfare, and housing programs. Largely due to the rural surroundings of TCI&R operations, U.S. Steel put emphasis on providing services to its employees, many of whom were black, all of whom were far from home. Significantly, this program also reflected U.S. Steel's strategy of avoiding antitrust action by behaving as a "good trust."

Crawford was a longtime member and director of the American Iron & Steel Institute (AISI). It was no coincidence that his chief affiliation within the institute was the welfare committee on which he served for many years after moving to Birmingham.

In 1930 Crawford resigned his presidency with TCI&R to pursue different challenges at the nation's third-largest steel company, Jones & Laughlin in Pittsburgh. During five years with this company Crawford oversaw several important improvements, including addition of a hydraulic 44-inch reversing blooming mill at the South Side Works. Widely touted as a major achievement in the trade papers, the mill was not completed until May 1936 — two months after Crawford's death.

At age forty-two, Crawford married Margaret Richardson. His professional life consumed much of his interest, and he was active in not only the AISI but also the American Institute of Mining and Metallurgical Engineers.

In 1935 Crawford retired from Jones & Laughlin and from active involvement in the industry. He and his wife returned to Birmingham, where Crawford enjoyed only a brief retirement.

He died a year later at his home on March 20, 1936.

Publications:
"Remarks," *American Iron and Steel Institute Yearbook* (1914): 515;
"Unusual Blast Furnace Boiler Plant," *Blast Furnace and Steel Plant,* 12 (January 1924): 2–4;
"South's Large Part in Exports," *Iron Age,* 117 (May 26, 1926): 1273–74 [Abstract of address to National Foreign Trade Convention].

References:
"J & L to Diversify Output with New Continuous Mill," *Steel,* 97 (December 23, 1935): 10–11;
"Samuel E. Hackett Succeeds George Gordon Crawford as president of Jones & Laughlin Steel Corporation," *Blast Furnace & Steel Plant,* 22 (June 1934): 330;
"Tennessee Coal, Iron and Railroad Company, Birmingham, Ala.," *Monthly Bulletin of the American Iron and Steel Institute,* 5 (March–April 1917): 62–69.

Crucible Steel Company of America
by Geoffrey Tweedale

Manchester, England

The Crucible Steel Company of America — now a division of Colt Industries — has been one of the dominant firms in the production of specialty steels.

After a decade of sluggish demand for high-grade steels, Crucible Steel was founded in 1901 through a merger of 13 steel producers. To complete its strategy of building a fully self-sufficient and integrated specialty steel producer, Crucible acquired blast-furnace facilities and purchased coal mines in 1911. In that year it organized the Pittsburgh Crucible Steel Company at Midland, Pennsylvania, to supply the primary material for the firm's products. In 1930 a 50 percent interest in Snyder Mining provided Crucible's ore supplies.

In 1939 *Fortune* described Crucible Steel as the United States Steel Corporation of the specialty steel industry. As was the case at U.S. Steel, managing Crucible's many factories and a huge product line, in which individual products were often tailored to specific customer's requests, proved difficult. Net profit varied greatly before 1915, and it

was only the demand caused by World War I that made Crucible's fortunes. Net profits soared from $1 million in 1914 to $13.2 million in 1916, a level which was roughly maintained until about 1920.

After the war, however, Crucible's reputation became tarnished by scandal. Herbert DuPuy, who received 5 percent of all manufacturing profits before deductions, was indicted (though later acquitted) of tax evasion in 1919. Horace S. Wilkinson, who ran Crucible with an autocratic hand until his death in 1937, manipulated the company's stock to flatter its assets sheet. Moreover, Wilkinson neglected the integration of the technical and selling activities of the company, with the result that in the interwar period the firm's earning record remained mediocre.

The vulnerability of Crucible in the interwar period, when competition from other specialty steel producers such as the Allegheny-Ludlum Steel Corporation became more intense, was eventually recognized by its management. Frederick B. Hufnagel, who became chairman and president in 1937, began reorganization, but this was interrupted by World War II, which once again boosted Crucible's profits. After the war the reform of Crucible could be delayed no longer, and it was carried through by William P. Snyder, Jr., and William H. Colvin.

By 1954, when Colvin retired, production had been concentrated, and a new executive group created. In 1957 Crucible ranked fourteenth among U.S. steel firms in ingot tonnage, producing 1.4 million tons, or about 1 percent of the industry's total capacity. But it was first in the production of tool steel, fourth in alloy steels, and fifth in stainless, and it was a leader in superalloys, precision castings, and stainless tubing.

In the 1960s and 1970s Crucible reflected the increasingly troubled nature of the American steel industry, as it faced foreign competition both abroad and in its own markets. It was a period of declining market share, boardroom intrigues, and mergers. In 1967 Joel Hunter became chairman, retiring from the posts of president and chief executive, which were assumed by John C. Lobb. The latter was recruited from the International Telephone & Telegraph Corporation and was brought to Crucible by Hunt Foods & Industries Inc., of Fullerton, California. Hunt Foods was the flagship company of Norton W. Smith, characterized in the trade press as a West Coast collector of companies and expensive paintings and who controlled about 23 percent of Crucible's stock. At this time the firm's annual meetings were suspended for two years while Smith battled with a group of dissident shareholders. Crucible ranked about fifteenth in tonnage in the U.S. steel industry, though its sales placed it in the top ten. By then it had lost its premier position to the Allegheny-Ludlum Steel Corporation, though it remained one of the largest of the small groups of specialty steelmakers in the U.S.

In 1968 the Crucible Steel Company was merged with Colt Industries Incorporated, becoming Crucible Alloy Division of Colt. In 1978 Charles Kurcina, who had joined the company in 1955 and worked his way up through the management ranks, became president of the division.

References:

William H. Colvin, *"Crucible Steel of America": 50 Years of Specialty Steelmaking in USA* (New York: Newcomen Society in North America, 1950);

"Crucible Steel," *Fortune,* 20 (November 1939): 74–79;

"Crucible Steel Company of America," *Iron Trade Review,* 34 (February 21, 1901): 13;

Crucible Steel Company of America, *50 Years of Fine Steelmaking* (New York, 1951);

Philip Dobbs, "Crucible Steel in Profitable Era," *Magazine of Wall Street,* 67 (October 19, 1940): 30–56;

Herrymon Maurer, "Crucible: Steel for Frontiers," *Fortune,* 55 (January 1957): 84–88;

Geoffrey Tweedale, *Sheffield Steel and America: A Century of Commercial and Technological Interdependence, 1830–1930* (Cambridge, U.K.: Cambridge University Press, 1987).

George Lewis Danforth, Jr.

(November 8, 1879 – ?)

by Robert Casey

Henry Ford Musuem & Greenfield Village

CAREER: Superintendent of steel production, Illinois Steel Company, United States Steel Corporation (1917–1918); assistant general superintendent (1918–1929), general superintendent, South Works, Illinois Steel Company, United States Steel Corporation (1929–1935); president, Open-Hearth Combustion Company (1935– ?).

George Danforth was a steel executive and equipment manufacturer who made major im-

provements to the open-hearth furnace. Danforth was born in Louisville, Kentucky, on November 8, 1879. After being educated in private and public schools, he went to Birmingham, Alabama, to take a position as a record clerk in the open-hearth department of the Birmingham Rolling Mill Company. He would spend the rest of his career associated with open-hearth furnaces. He soon moved on to become an open-hearth operator with Tennessee Coal and Iron Company, also in Birmingham. In 1903 Danforth left for Chicago to join the open-hearth department at the South Works of Illinois Steel Company. His rise was rapid. In 1905 he was made department superintendent. In 1917 he became superintendent of all steel production, and assistant general superintendent of South Works the following year. In 1929 Danforth was appointed general superintendent of South Works, a post he held until he left the company in 1935.

During his years at South Works, Danforth initiated or participated in the development of numerous improvements in open-hearth design and practice: the Naismith sloping backwall; the dolomite machine for repairing furnace bottoms; waste heat boilers, which salvaged heat from stack gasses to generate steam; venturi ports for improved combustion; and improved regenerators known as Danforth Checkers. In 1924 Danforth joined with the H. A. Brassert Company, a leading steel-mill engineering firm, to form the Open-Hearth Combustion Company and to control Danforth's patents. Danforth resigned from Illinois Steel in 1935 to become president of Open-Hearth Combustion. The company became a major supplier to the steel industry, at one time providing equipment and engineering designs used in over 90 percent of the open-hearth capacity in the country.

Danforth was a member of the American Iron and Steel Institute, and was on the editorial board of *Blast Furnace & Steel Plant* magazine from 1937 until his death. He married Jessie Norris on July 8, 1903. They had no children. The date of Danforth's death is not known.

Publications:
"Large Open Hearth Furnaces," *American Iron and Steel Institute Yearbook* (1929): 155–158 [Discussant];
"Rimmed Steel Quality," *Blast Furnace and Steel Plant*, 24 (September 1936): 781–82;
"Open Hearth Checkerwork Designs," *Blast Furnace and Steel Plant* 24 (October, November 1936): 899, 920–921, 981–983;
"Economics of Modernizing the Open Hearth Furnace," *Blast Furnace and Steel Plant*, 24 (December 1936): 1087, 1101;
"Open Hearth Design and Construction," *Blast Furnace and Steel Plant*, 26 (December 1938): 1183–1187; 27 (January 1939): 46–47;
"Modernization of Existing Open Hearth Furnaces and Auxillary Equipment," *Iron Age*, 144 (October 5, 1939): 47.

References:
William T. Hogan, S.J., *An Economic History of the Iron and Steel Industry in the United States*, 5 volumes (Lexington, Mass.: Lexington Books, 1971);
F. B. Quigley, "Open Hearth Furnace Regenerators," *Yearbook of the American Iron and Steel Institute* (1923).

Henry Fairchild DeBardeleben

(July 2, 1840 – December 6, 1910)

by Robert Casey

Henry Ford Museum & Greenfield Village

CAREER: Superintendent and general manager, Eureka Mining Company (1872–1873); president, Pratt Coal and Coke Company (1878–1882); president, DeBardeleben Coal and Iron Company (1886–1892); first vice-president, Tennessee Coal, Iron, and Railroad Company (1892–1894); vice-president, Alabama Fuel and Iron Company (1908–1911).

Henry Fairchild DeBardeleben was a key figure in the development of the Alabama iron industry. Using a fortune inherited from his father-in-

law, Daniel Pratt, DeBardeleben developed mines, constructed railroads and iron furnaces, and built company towns. His various holdings were the basis for the United States Steel Corporation's vast operations in the Birmingham District.

Henry DeBardeleben was born on July 2, 1840, the son of an Autauga County, Alabama, cotton planter. His father was the grandson of a Hessian captain who settled in South Carolina after the Revolution. DeBardeleben's father died in 1850, and a family friend, Daniel Pratt, became the boy's guardian. Pratt, a native of New Hampshire, had settled in Alabama in the 1830s, founded the town of Prattville, and made his fortune manufacturing cotton gins. DeBardeleben early in life exhibited the restless energy that was to make him a successful entrepreneur. He had little interest in education, preferring the woods to the schoolroom. Pratt employed him at various jobs, finally advancing him to superintendent of the gin works.

DeBardeleben's civil war service was brief. He enlisted as a private, fought at Shiloh, and was detailed out of the army to manage a bobbin factory at Prattville. In 1862 he married Daniel Pratt's sixteen-year-old daughter, Ellen. After the war DeBardeleben began to take a more active role in managing Pratt's other investments: railroads, mining, and iron making.

DeBardeleben's entry into the iron business was not auspicious. In 1872 Pratt and DeBardeleben acquired controlling interest in the Red Mountain Iron and Coal Company, changing the name to Eureka Mining Company. DeBardeleben was appointed superintendent and general manager, with responsibility for running the newly rebuilt blast furnaces at Oxmoor, Alabama, southwest of Birmingham. His inexperience and the Panic of 1873, however, combined in rendering the enterprise a failure.

The year 1873 also brought the death of Daniel Pratt, and DeBardeleben inherited part of the Pratt fortune. DeBardeleben gradually withdrew from active participation in the gin business and began prospecting for coal mines to develop. In 1878 he joined with mining engineer Truman H. Aldrich and railroad entrepreneur James Withers Sloss to form the Pratt Coal and Coke Company. Aldrich and Sloss owned lands west of Birmingham that held a coal deposit known as the Browne seam. DeBardeleben supplied the money to develop the mines, and the deposit was renamed the Pratt seam in honor of Daniel Pratt.

Over the next four years DeBardeleben threw himself into his task, building a railroad to transport the coal and developing markets for the product. In 1879 he persuaded Louisville investors to build the Birmingham Rolling Mills, producing bar, sheet, and plate iron. In association with T. T. Hillman he formed the Alice Furnace Company and built the first blast furnace in the Birmingham city limits. In 1881 his offer of Pratt coal at cost plus 10 percent induced James Sloss to build two blast furnaces on the east side of Birmingham.

Henry Fairchild DeBardeleben

By 1882 DeBardeleben was worn out from his labors and believed he had tuberculosis. When Memphis capitalist Enoch Ensley offered him $1 million for the Pratt Coal and Coke properties, DeBardeleben took the money and headed to his ranch in Texas to recuperate.

During the next four years DeBardeleben divided his time between Alabama and Texas, developing the Mary Pratt Furnace in Birmingham, running the Redding ore mine, and looking for further investment opportunities. In 1885 he convinced capitalists in Charleston, South Carolina, and London to back him in another venture, the DeBardeleben Coal and Iron Company, capitalized in 1886 at $2 million. DeBardeleben built the town of Bessemer, southwest of Birmingham, erected five blast furnaces, opened more coal mines, and sold building lots. Eager to get the town up and running, he transported and re-erected buildings from the 1884 New Orleans Exposition.

DeBardeleben soon had competition in the region, however. Tennessee Coal, Iron, and Railroad Company (TCI), which had purchased the former DeBardeleben properties from Enoch Ensley, was threatening to buy the Sloss operation and become the dominant force in the region. DeBardeleben, with his old ally Truman Aldrich, resolved to strike first. In 1891 they consummated a series of stock trades that combined their holdings with those of TCI. In 1892 a formal reorganization took place, with Nat Baxter as president, DeBardeleben as first vice-president, and Aldrich as second vice-president. This investment was to prove as disastrous for DeBardeleben as his Oxmoor experience of 1873.

Tennessee Coal and Iron was never financially strong and barely survived the Panic of 1893. In 1894 DeBardeleben, anticipating an upturn, attempted to gain control of a majority of TCI stock. He failed and lost his fortune in the process. Late in 1894 DeBardeleben resigned from TCI, which continued to struggle until it was acquired by the United States Steel Corporation in 1907.

The entrepreneurial spirit died hard, however. In 1905, with capital from Jessie M. Overton of Nashville, DeBardeleben and his sons organized the Alabama Fuel and Steel Company, which became Alabama Fuel and Iron in 1908. By the time DeBardeleben died on December 6, 1910, Alabama Fuel and Iron was the fourth largest coal producer in the state.

DeBardeleben was a flamboyant, free-spending gambler, whose strengths lay in promoting and organizing new enterprises. His penchant for risk taking, however, often left behind unwieldy organizations to be refined by more sober-minded managers.

References:
Ethel Armes, *The Story of Coal and Iron in Alabama* (Birmingham: Birmingham Chamber of Commerce, 1910);

Marjorie L. White, *The Birmingham District: An Industrial History and Guide* (Birmingham: Birmingham Historical Society, 1981).

William J. De Lancey

(June 2, 1916 –)

by Carol Poh Miller

Cleveland, Ohio

CAREER: Attorney, Cravath, de Gersdorff, Swaine & Wood, New York City (1940–1952); assistant counsel (1952–1954), assistant general counsel (1954–1959), general counsel (1959–1961), vice-president and general counsel (1961–1971), director (1968–1982), executive vice-president (1971–1973), president (1973–1979), chief executive officer (1974–1982), chairman of the board, Republic Steel Corporation (1979–1982).

William John De Lancey was born and raised in Chicago, the son of John Richmond and Louise Ella Hart De Lancey. In high school he worked as a newspaper reporter for the Elgin, Illinois, *Daily Courier-News.* "Otherwise," he recalled in later years, "I was in school studying until I became a lawyer." De Lancey received his bachelor and law degrees from the University of Michigan in 1938 and 1940, respectively, and was admitted to the New York Bar in 1941. Between 1940 and 1952 he was associated with the law firm of Cravath, de

Gersdorff, Swaine & Wood of New York City. During World War II he served as a naval air intelligence officer aboard an aircraft carrier.

De Lancey joined Republic Steel Corporation in April 1952 as assistant counsel and steadily worked his way to the top. He became assistant general counsel in 1954 and general counsel in 1959. He was elected as vice-president in 1961 and a director in 1968. Following the death of Chairman and Chief Executive Officer Willis B. Boyer in 1974, De Lancey, who had been named president a year earlier, was elected chief executive officer; the position of chairman had been temporarily abolished but was reinstated in 1979, when E. Bradley Jones was named president of Republic and De Lancey was named chairman.

William De Lancey took the helm at Republic Steel just as the industry entered a protracted downturn. A frequent speech maker, De Lancey publicly blamed the steel industry's woes on bad policies and mismanagement by the federal government. He criticized the government for creating inflation by overspending, imposing too many expensive pollution-control regulations, and allowing too much underpriced foreign steel into U.S. ports. In a speech to shareholders at the corporation's annual meeting in 1979, De Lancey even blamed the federal government for the industry's labor troubles, saying that expensive labor strikes "reflect the power which the federal government granted to trade unions in the 30s."

De Lancey retired from Republic in June 1982, following a first-quarter loss; the industry was in the midst of a protracted period of reduced steel demand. Like the rest of the domestic steel industry Republic was operating at less than 60 percent capacity. He pointed a finger at high interest rates and weak demand for steel, "aggravated by a flood of dumped and subsidized imports."

De Lancey served as a director of Ameritrust Corporation, Standard Oil Company (Ohio), Sherwin-Williams Company, Ohio Bell Telephone Company, Metropolitan Life Insurance Company, Reserve Mining Company, and Beatrice Pocohontas Company. His community activities included service as chairman of public television station WVIZ in Cleveland and as a trustee of the Musical Arts Association (Cleveland Orchestra), University Hospitals of Cleveland, the Cleveland Council on World Affairs, Case Western Reserve University, and the International Iron and Steel Institute.

In 1980 De Lancey was elected chairman of the American Iron and Steel Institute (AISI). In 1981 he received the Medal for the Advancement of Research from the American Society for Metals for his strong advocacy of steel-related research and development.

On July 10, 1940, De Lancey was married to Sally Ann Roe, a Boston native and graduate of Smith College. They had two children, Ann Louise and Mark Roe. De Lancey still resides in his longtime home in Shaker Heights and maintains an office in the former Republic Building in downtown Cleveland.

Publications:

Productivity Improvement — An Achievable Goal, remarks made to sales and marketing executives, Cleveland, Ohio, March 21, 1980 (Cleveland: Public Affairs Department, Republic Steel, 1980);

The Last Clear Chance, remarks made before the Association of Iron and Steel Engineers, Birmingham, Alabama, February 3, 1981 (Cleveland: Public Affairs Department, Republic Steel, 1981);

Japan and the U.S.A. — an American View, remarks made before the Tekko Kyokai, the Iron and Steel Institute of Japan, Tokyo, April 2, 1981 (Cleveland: Public Affairs Department, Republic Steel, 1981);

Steel — Protectionist or World Class Competitor?, speech delivered at a luncheon sponsored by the Houston Chamber of Commerce and American Iron and Steel Institute, Houston, Texas, October 20, 1981 (Cleveland: Public Affairs Department, Republic Steel, 1981).

References:

"De Lancey Blasts Government," *Cleveland Plain Dealer,* May 10, 1979, p. 5E;

"De Lancey Tags U.S. for Most of Steel Woes," *Cleveland Plain Dealer,* December 14, 1979, p. 6E.

William Brown Dickson

(November 6, 1865 – January 27, 1942)

by Gerald G. Eggert

Pennsylvania State University

CAREER: Manual laborer (1881–1886), payroll clerk (1886–1889) Homestead Works of Carnegie Steel; clerk, chief clerk, assistant general agent of ordering and shipping, Carnegie Steel Company (1889–1901); junior partner (1899–1901), member of board of directors and assistant to president (1900–1901) Carnegie Steel Company; assistant to president (1901–1902), second vice-president (1902–1909), first vice-president (1909–1911) United States Steel Corporation; vice-president and treasurer (1915–1923), Midvale Steel & Ordnance Company.

William Brown Dickson, steel executive, began in the industry as a manual laborer at the Homestead Works of the Carnegie Steel Company. He moved into clerical work, was sent to

William Brown Dickson

the Carnegie Steel headquarters in Pittsburgh, and by the age of thirty-three had risen to a junior partnership. Dickson held high offices at the United States Steel Corporation when it absorbed Carnegie Steel and later became a top executive at the Midvale Steel & Ordnance Company. He was a vigorous promoter of labor reform, fighting for, among other things, the eight-hour workday and the six-day workweek at U.S. Steel, and for an employee representation plan (ERP) at Midvale.

Born in Pittsburgh, Pennsylvania, Dickson's life began comfortably. He was one of 11 children of Irish-born John and Scottish-born Mary Ann McConnell Dickson. The Dicksons soon moved to Swissvale, south of Pittsburgh, where John owned a coal business inherited from his father. He also owned stock in a Pittsburgh bank. When that institution failed in the 1870s, he by law was liable for twice the face value of the stock he held. Although he put his older children to work, all ef-

forts to salvage the family fortune fell short as John's health failed.

From the age of nine William tended fans each summer in his father's mines. At eleven he quit school to take a job arranged by his parents as an errand boy for a Pittsburgh storekeeper. He worked at other jobs, including one at the Pittsburgh telephone exchange, before he entered the Homestead mills as a common laborer in April 1881, at the age of fifteen. There he worked the industry's infamous 12-hour shift, seven days a week, for five years. Transferred to clerical work at Homestead in 1886, he spent evenings at Duff's Business College in Pittsburgh and soon advanced to payroll clerk. In 1888 he married his childhood sweetheart, Mary Bruce "Mamie" Dickson, who despite the same family name was not related. They were to have six children: Susan, Emma, Thomas (who died in infancy), Eleanor, Charles, and Helen.

Dickson was summoned to Carnegie's main office in Pittsburgh in 1889 to help temporarily in the ordering and shipping department. The job proved permanent and rapid promotion followed: entry clerk, chief clerk, and assistant general agent which involved traveling to all parts of the far-flung operation and reporting directly to Carnegie. In 1899 Dickson received his junior partnership. A year later he sat on the board of directors and served as assistant to Charles Schwab, president of Carnegie Steel. When Schwab moved to the presidency of U.S. Steel in 1901, Dickson accompanied him, becoming second vice-president in 1902. Although Schwab resigned the next year, his successor was Dickson's friend, William E. Corey, who also had risen through the ranks at Carnegie Steel. Under Corey, Dickson became first vice-president in 1909.

From these executive positions, Dickson crusaded to improve working conditions. As a member of U.S. Steel's safety committee, he helped develop the "safety first" program, pensions for superannuated employees, and a voluntary employer-liability plan that automatically paid benefits to injured workmen or, if they were killed, to their surviving dependents. Meanwhile he served on the New Jersey commission that wrote and later enforced that state's pioneering liability law. In 1907 Dickson initiated a scheme to end the seven-day workweek at U.S. Steel by staggering shifts to give employees one day in seven off. He next campaigned to replace the standard two 12-hour shifts per day with three eight-hour shifts. The six-day reform, though ordered, was disregarded by top managers, and when Dickson forced a showdown, Elbert Gary, chairman of the board and de facto chief executive officer of the firm, sided with the managers.

Dickson's views on reform antagonized Gary. Gary favored "stable prices" regardless of changing market conditions and sought to achieve this through "harmony" — that is, by preaching and if necessary by coercing the acceptance of noncompetitive prices by companies throughout the industry. At the same time wage rates were to be maintained. Inventories would be reduced by laying off men and closing down plants as necessary. Dickson regarded "harmonizing" as unworkable and probably illegal and high wage rates accompanied by layoffs as unconscionable. Workers and their communities, he believed, could better survive wage cuts than unemployment in bad times.

Passed over for the presidency when Corey left U.S. Steel in 1911, Dickson resigned. During four years of unwanted retirement, he pressed for hours reform from outside the industry. All gains made, however, were offset by the enormous demand for steel that came with World War I. To take advantage of that demand, Corey, other steel men, armaments makers, and bankers formed a new combine, Midvale Steel & Ordnance Company in October 1915. Corey, the chairman of the board, invited Dickson to be vice-president and treasurer, and another onetime "Carnegie Boy," Alva C. Dinkey, to be president.

By this time Dickson believed that justice required giving laborers a voice in the workplace. Strongly opposed to regular unions, he favored employee representation plans (ERPs) or company unions. His efforts to sell Corey and Dinkey on the idea got nowhere until the International Association of Machinists (IAM) began organizing at Midvale. Corey and Dinkey feared that the federal government, which had granted workers in war industries the right to bargain collectively in return for not striking, would force Midvale to recognize the IAM. Accordingly they authorized Dickson to launch his ERP. Dickson hailed this as the beginning of "industrial democracy," knowing full well that his superiors had adopted it cynically and that Gary and most other steel executives opposed any concessions to unions, even company unions.

Midvale's ERP limped along, doomed as was the firm itself by the Armistice. As wartime profits plummeted, Midvale cut back on wages and jobs until 1923 when it sold out to Bethlehem Steel. Dickson, though only fifty-seven, this time found no escape from involuntary retirement. After two decades of frustration, he died at his home in Littleton, New Hampshire, at the age of seventy-six.

Publications:
Genealogy of the Dickson Family and Its Immediate Collateral Branches (Montclair, N.J., 1908);
"Betterment of Labor Conditions in the Steel Industry," *American Iron and Steel Institute Yearbook* (1910): 56–68;
"Can American Steel Plants Afford an Eight-Hour Turn?" *Survey*, 31 (January 3, 1914): 376;
"Eight-Hour Day and Six-Day Week in Continuous Industries," *American Labor Legislation Review*, 7 (1915): 155–167;
"Seven-Day Week Relic of Barbarism," *Iron Trade Review*, 60 (January 11, 1917): 143–147;
"War Problems of the Steel Trade," *Iron Age*, 102 (October 24, 1918): 1015–1016;
Democracy in Industry, third edition (Littleton, N.H., 1935);
"A Political Platform for 1936," (Littleton, N.H., 1935);
History of the Carnegie Veterans Association (Montclair, N. J.: Mountain Press, 1938);
Democracy at the Crossroads (Littleton, N.H.: 1939?).

Reference:
Gerald G. Eggert, *Steelmasters and Labor Reform, 1886–1923* (Pittsburgh: University of Pittsburgh Press, 1981).

Archives:
The diary and personal and business papers of William Brown Dickson are housed in the Special Collections, Pattee Library, Pennsylvania State University, University Park, Pennsylvania.

Alva C. Dinkey

(February 20, 1866 – August 11, 1931)

by Gerald G. Eggert
Pennsylvania State University

CAREER: Water boy, messenger boy, general handy boy, telegraph operator, Edgar Thomson Works of Carnegie Steel Company (1879–1885); machinist, Pittsburgh Locomotive Works (1885–1888); expert machinist, McTighe Electric Company (1889); secretary to general manager (1889–1893), electrician (1893–1898), superintendent of electric light and power plant (1898–1899), assistant to general superintendent, Homestead Works (1899–1901), general superintendent of Homestead Works (1901–1903), president, Carnegie Steel Company (1903–1915); president, Midvale Steel & Ordnance Company (1915–1923); president, Midvale Company (1923–1931).

Alva Clymer Dinkey, a steel industry executive, fit the classic "rags-to-riches" mold. He was born on a farm in Weatherly, Pennsylvania, the son of Reuben and Mary Elizabeth (Hamm) Dinkey. His father, who both farmed and worked on the railroad, was killed in an accident when Dinkey was only nine years old. When Dinkey was thirteen, his widowed mother moved with her six children to Braddock, in the steel district south of Pittsburgh, and opened a boardinghouse. She chose the area so that her sons might find jobs in the mills. One of her boarders, Charles M. Schwab, then at the beginning of his career in steel, befriended young Alva and later married the boy's sister Rana.

Dinkey attended the public school in Braddock only until he found work, while still only thirteen, as a water boy in Carnegie Steel Company's Edgar Thomson Works. He advanced to messenger boy and then to general handy boy for masons and bricklayers. In the meantime he studied telegraphy and in 1885, at the age of nineteen, became the mill's telegraph operator. Later that same year he found new employment as a machinist at the Pittsburgh Locomotive Works in Allegheny, and in 1888 he moved to the McTighe Electric Company of Pittsburgh where he worked as an expert machinist.

Alva C. Dinkey

Dinkey returned to Carnegie's employ late in 1889 as secretary to the general manager of the Homestead Works. It is unclear whether his brother-in-law, Schwab, who was general manager at Homestead from 1887 to 1889, hired him or recommended him. In any event, his future now more secure, Dinkey in 1890 married Margaret Stewart of Braddock; the couple would have four children: Margaret Elizabeth, Robert Eugene, Alva Charles, and Leonora Stewart. Not especially fond of office routine, Dinkey switched to the electrical department in 1893 and by 1898 was superintendent of the electric light and power plant. He purchased the town's electric plant, put it on a sound financial footing, and sold it at a

100-percent profit. He also organized a telephone system for the town of Homestead and nearby communities and later sold that too at a profit.

In 1899 William E. Corey, general superintendent at Homestead, named Dinkey his assistant. When Carnegie Steel became a subsidiary of the United States Steel Corporation in 1901, Corey moved to the new firm's headquarters and placed Dinkey in command at Homestead. Two years later U.S. Steel named Corey to its presidency and Dinkey to the presidency of Carnegie Steel, a post he held until 1915.

The outbreak of World War I in 1914 created an unprecedented demand for American-made steel, armaments, and munitions. To supply that market, bankers united with armaments makers and practical steelmen to form a new combine that ultimately would be second only to U.S. Steel in output. The new firm, Midvale Steel & Ordnance Company, brought together many Pennsylvania companies, including the Midvale Steel Company of Nicetown, Worth Brothers of Coatesville, Remington Arms Company of Eddystone, and Cambria Steel of Johnstown. Several of Carnegie's "Young Geniuses" who had risen through the ranks in his company and gone on into the management of U.S. Steel managed the new firm: Corey, former president of U.S. Steel, was chairman of the board; Dinkey, lured from Carnegie Steel, became president; and William Brown Dickson, onetime first vice-president of U.S. Steel, was named vice-president and treasurer.

Despite labor and materials shortages, high taxes, and disputes with workers, Midvale Steel & Ordnance prospered mightily during the war. After the Armistice, however, the firm quickly fell on hard times, and in 1923 it broke up — the old Worth Brothers plant and Cambria works going to Bethlehem Steel. The prewar Midvale Steel mill at Nicetown was taken over by Baldwin Locomotive, with Dinkey continuing on as president until his death in 1931. During the 1920s the firm annually sold about $6 million worth of steel products including locomotive parts, shaftings, high-pressure cylinders, and a variety of special heat- and corrosion-resisting alloys.

Dinkey's chief contributions to the industry were in the areas of inventiveness and production management. As an inventor he held patents on several devices: the Dinkey Reversing Controller which operated electric cranes and steel mill drives and soon became standard equipment throughout the industry, a rheostat, steel ties, a special process for making Bessemer steel, tongs for lifting materials, and an apparatus for charging and drawing furnaces. In its August 6, 1903, issue *Iron Age* credited Dinkey with revolutionizing rolling mill practice by introducing electrically driven feed tables that replaced hooks and tongs operated by hand, cutting the number of men required at the rolls from four to one. Elbert Gary, chairman of the board at U.S. Steel, regarded him as "one of the ablest all around steel men in the country."

With respect to labor relations, however, Dinkey shared the patronizing and conservative policies of most steel industry executives. Under federal pressure to bargain collectively with his employees at Midvale during the war, he reluctantly accepted an Employee Representation Plan. In addressing the representatives, he told them there was "almost no difference" between employees and managers; "Some of the best friends I have . . . [or] ever had have been workmen in the works," he declared. Every officer of the firm, he pointed out, had started in the shops and moved forward. Yet throughout his career he consistently sided with Schwab, Corey, and Gary in opposing such basic reforms as ending the seven-day workweek and the twelve-hour workday.

Dinkey died of heart disease in his residence at Wynnewood, near Philadelphia, at the age of sixty-five.

References:

Gerald G. Eggert, *Steelmasters and Labor Reform, 1886–1923* (Pittsburgh: University of Pittsburgh Press, 1981);

Robert Hessen, *Steel Titan: The Life of Charles M. Schwab* (New York: Oxford University Press, 1975), pp. 17–18, 231.

Archives:

The work and views of Alva C. Dinkey are frequently addressed in the diary and papers of his long-time colleague at Carnegie Steel, U.S. Steel, and Midvale Steel & Ordnance, William Brown Dickson; these are in the Special Collections of the Pattee Library, Pennsylvania State University, University Park, Pennsylvania.

Direct Reduction Ironmaking and Steelmaking
by Leonard H. Lynn

Case Western Reserve University

Integrated steelmaking — the processing and conversion of raw materials into a finished product — generally consists of four major processes: the production of metallurgical coke from coal; the use of the coke in blast furnaces to produce pig iron; the reduction of the carbon content of the pig iron in basic oxygen, open-hearth, or electric furnaces to produce steel; and the rolling and shaping of the steel into steel products. The direct reduction (DR) ironmaking processes replace the first two of these steps by making use of oil, coal (or a gas produced from these fuels), or natural gas in a shaft furnace, fluidized bed, or retort to produce pig iron from iron ore. The resultant pig iron can then be converted into steel in an electric furnace.

Direct reduction can offer substantial advantages over conventional ironmaking: the cost of facilities is lower, a plant's capacity is smaller, and the heavy pollution associated with the coking process is eliminated. In times when steel scrap is too expensive or difficult to obtain, direct reduction can provide an alternative processing method using electric furnaces. These potential advantages are highly contingent, however, on the prices of fuel and scrap. In the past the DR processes have been used mostly in countries where scrap is either scarce or too expensive, the costs of fuels such as natural gas are low, and steelmaking plants operate at relatively small capacity.

While direct reduced iron known as sponge iron has been produced since ancient times, as many as 100 "modern" DR processes have been developed since the 1920s. Most have not found widespread use. The most prevalent of the modern DR ironmaking technologies is the Midrex process, which was initially developed by the Midland-Ross Corporation of Ohio in the mid 1960s, and then by the Midrex Corporation, a subsidiary of the Korf group. In 1983 the Midrex Corporation was bought by Kobe Steel of Japan. At that time nearly half of the world's DR ironmaking capacity

was based on the Midrex process. Because the Midrex process relies on natural gas, however, it has not been widely used in the United States. Other popular DR technologies include the HYL, Armco, Purofer, Krupp, HIB, FIOR, and SL-RN processes. Differences among the processes center on the fuel and mechanical arrangements used to reduce the ore. Technologies based on coal rather than natural gas or oil have been viewed as most suitable for the United States. These include the KR (Kohle-Reduktion) and SL-RN processes.

Interest in direct reduction ironmaking increased in the United States during the 1980s as U.S. coke production capacity dwindled — capacity in 1990 was expected to be only a third of what it had been in 1965 — because of high costs and pollution-control regulations. In the late 1980s the U.S. steel industry showed considerable interest in developing DR technology to the point that it could be used to produce steel as well as iron. Such a development would provide "leapfrog technology" that would allow U.S. firms to jump to a new period of lower cost in which they would once again be internationally competitive.

References:

C. G. Davis, J. F. McFarlin, and H. R. Pratt, "Direct Reduction Technology and Economics," *Ironmaking and Steelmaking*, 9, no. 3 (1982): 93–129;

William T. Hogan, S.J., *Minimills and Integrated Mills: A Comparison of Steelmaking in the United States* (Lexington, Mass.: Lexington Books, 1987);

William T. Lankford, Jr., and others, eds., *The Making, Shaping and Treating of Steel*, tenth edition (Pittsburgh: Association of Iron and Steel Engineers, 1985);

Leonard H. Lynn, "Multinational Joint Ventures in the Steel Industry," in *International Collaborative Ventures in U.S. Manufacturing*, edited by David Mowery (Cambridge, Mass.: Ballinger, 1988);

George J. McManus, "Steel's Leapfrog Technologies Looking for a Prince," *Iron Age*, 229 (May 2, 1986): 32–44.

Electric Steelmaking

by Geoffrey Tweedale
Manchester, England

Prior to the twentieth century charcoal, fossil fuels, or an air blast were the favored methods for melting and refining steel. Developments in electrical engineering technology toward the end of the nineteenth century, however, led steelmakers to realize that electric melting was likely to be technically and commercially more efficient than the open-hearth and Bessemer processes.

In the nineteenth century Americans had made important contributions to electric metallurgy. Robert Hare, a Philadelphia chemist born in 1781, is credited with constructing the first electric furnace, which he used for converting charcoal into graphite. In 1886 Charles Hall had used electricity in making aluminum, and in building on his work two brothers, Alfred and Eugene Cowles, applied the technique to the production of aluminum bronze. In 1891 Edward G. Acheson used the electric furnace to produce a new grinding substance, carborundum. Edward Colby had also taken out basic patents on the electric furnace in the 1890s.

But the main developments in electric steelmaking took place in Europe. Particularly important was Sir William Siemens's construction in 1878 of the arc furnace, in which steel was melted between two conducting rods. By the end of the century French and Swedish engineers had succeeded in adapting this design for the production of good-quality steels. By 1903, when the Canadian government appointed a commission to report on electric steelmaking, European commercial production was well under way.

In 1906 the Halcomb Steel Company of Syracuse, New York, began commercial production of electric steel; and two years later so, too, did the Firth-Sterling Steel Company of Pittsburgh. By 1910 ten electric furnaces, usually of European design and less than 10-tons capacity, were in operation in the United States. By 1916 the number of electric furnaces in the United States had risen to 73. Yet, despite the availability of cheap hydroelectricity, America continued to lag behind Europe

An electric steelmaking furnace

in the development and implementation of the technology. The electric furnace did not win immediate acceptance, especially in the production of high-grade tool steel, and it was also plagued by technical problems in these early years. World War I, however, gave electric steelmaking considerable impetus. In the 1920s America emerged as the world's leading electric steel producer, and in the interwar period the process completely eliminated the older crucible method of melting. The electric furnace had been found ideal for melting scrap as well as alloy and tool steels.

Various technical improvements made after World War I, such as higher voltages and transformer capacities, and the demand for alloy and tool steels pushed electric steel production beyond the 1 million net tons mark in 1939. After 1945 electric furnaces changed from being small units mainly involved in speciality steel production to large units producing carbon steel. Due to increased power supply the late 1960s saw the introduction of 150-ton furnaces, and by 1969 annual electric steel output in the United States was over 20 million tons. In the 1970s the electric furnace became the number two steel producer in the United States, behind the basic oxygen converter. The electric furnace, integrated with small casting machines and bar mills, was responsible for one of the key developments of recent American steelmaking — the minimill.

References:

Report of the Commission Appointed to Investigate the Different Electro-Thermic Processes for the Smelting of Iron Ores in Operation in Europe (Ottawa: Department of the Interior, 1904);

William T. Hogan, S.J., *An Economic History of the Iron and Steel Industry of the United States*, 5 volumes (Lexington, Mass.: Lexington Books, 1971);

D. R. Lyon, R. M. Keeney, and J. F. Cullen, *The Electric Furnace in Metallurgical Work*, U.S. Department of the Interior, Bureau of Mines Bulletin No. 77 (Washington, D.C.: U.S. Government Printing Office, 1914);

Alfred Stansfield, *The Electric Furnace for Iron and Steel* (New York: McGraw-Hill, 1923);

Geoffrey Tweedale, *Sheffield Steel and America: A Century of Commercial and Technological Interdependence, 1830–1930* (Cambridge, U.K.: Cambridge University Press, 1987).

Experimental Negotiating Agreement

by Richard W. Kalwa

University of Wisconsin — Parkside

In 1967 R. Conrad Cooper of the United States Steel Corporation (U.S. Steel) proposed that, instead of striking, steelworkers should accept voluntary arbitration of labor-management disputes. The executive board of the United Steelworkers of America (USWA) rejected the proposal, noting that its passage would require a vote by the rank and file. Moreover, many USWA members believed that the right to strike was an inherent part of unionism.

The issue of voluntary arbitration reemerged in 1973. USWA president Iorwith W. Abel had survived an election challenge in 1969 and was reelected without opposition in 1973. Due to mandatory retirement he would not have to face another election, and it thus became politically viable for him to press for voluntary arbitration. He enjoyed popularity among the membership, for in 1971 he had negotiated a sizable settlement which restored the cost-of-living adjustment surrendered in 1962.

Abel believed that increasing foreign competition made it necessary for labor and management to enter into a new era of cooperation. The system of "crisis bargaining" had often caused companies to stockpile their steel to protect against the threat of a strike. Stockpiling, in turn, would open the domestic market to imported steel, and steel consumers would enter into long-term agreements with foreign suppliers. In 1971

this cycle had led to the layoff of 100,000 workers. The industry and the union further realized that regular government intervention had made "free" collective bargaining impossible.

In March 1973 the steel companies and the union leadership formulated an experimental negotiating agreement (ENA) which eliminated the use of the strike or lockout in negotiations. The agreement set a deadline of April 15, after which any bargaining impasse would be referred to a five-member arbitration panel for resolution in time for the August 1 contract-expiration date. In return for conceding the right to strike, the union received certain guarantees, which included a 3 percent annual wage increase and continued cost-of-living adjustments. The most important guarantee, however, recognized the right of the local unions to strike over local issues. The procedure would be "experimental" in the sense that it would be renewed from one round of negotiations to the next by mutual agreement of the parties.

The agreement was ratified by a majority of the Basic Steel Industry Conference, a body of about 600 local union presidents. Its application extended to over 300,000 unionized workers in the ten largest steel companies. In the 1974 negotiations the union took the guarantees of the ENA and won a three-year contract with a 35 percent increase in compensation, at a time when domestic steel shipments and profits were rising and the share of imports falling.

At the same time, however, opposition to the ENA was strong among many union members who felt that the leadership had given up its main bargaining weapon and had not submitted the proposal to a rank-and-file vote. Dissenting members challenged the no-strike agreement in federal court in 1974, and criticism of the ENA was a main theme of Ed Sadlowski's challenge to the administration in the union election of 1977. These attempts within the union to overturn the ENA, however, were unsuccessful.

But opposition to the ENA also increased among the companies. U.S. Steel's J. Bruce Johnston, chief negotiator for the industry, spoke out against the high cost of the agreement. The 1977 contract, which introduced an extensive income-security program, was followed by local strike votes and a four-month strike by 15,000 iron-ore miners. Although the companies agreed to use the no-strike procedure in the next round of negotiations, David Roderick's rise to the chairmanship of U.S. Steel consolidated industry opposition to

the ENA, and no decision was made on its renewal in the 1980 contract talks. In the concession bargaining of the following years the no-strike agreement was not used.

References:
Iorwith W. Abel, "Basic Steel's Experimental Negotiating Agreement," *Monthly Labor Review, 96* (September 1973): 39–42;

Abel, *Collective Bargaining — Labor Relations in Steel: Then and Now* (New York: Columbia University Press, 1976);

George Bogdanich, "Steel: No-Strike and Other Deals," *Nation, 219* (September 7, 1974): 171–174;

James A. Craft, "The ENA, Consent Decrees, and Co-operation in Steel Labor Relations: A Critical Appraisal," *Labor Law Journal, 27* (October 1976): 633–640;

John Herling, *Right to Challenge: People and Power in the Steelworkers Union* (New York: Harper & Row, 1972);

John P. Hoerr, *And the Wolf Finally Came: The Decline of the American Steel Industry* (Pittsburgh: University of Pittsburgh Press, 1988);

Hoerr, "The Steel Experiment," *Atlantic Monthly, 231* (December 1973): 18–34.

Fair Employment Practice Committee
by Dennis C. Dickerson
Williams College

As World War II approached, blacks had made little progress breaking the color barrier in major industry. Few were hired as skilled laborers, and even fewer were promoted off of the shop floor. A. Philip Randolph, president of the Brotherhood of Sleeping Car Porters, pressed President Franklin D. Roosevelt to end this injustice. With backing from Walter White of the National Association for the Advancement of Colored People (NAACP), Lester Granger of the National Urban League, and other black leaders, Randolph planned in 1941 to lead 100,000 blacks in a march on Washington, D.C. To avoid such a massive display of racial discontent, Roosevelt issued Executive Order 8802, banning discrimination in industries holding federal contracts and establishing the Fair Employment Practice Committee (FEPC) to enforce the policy.

Though successively placed in the Office of Production Management, the War Production Board, and later the War Manpower Commission, FEPC, strengthened by Executive Order 9346, found its final home in the executive office. With a chairman and seven part-time staff persons, the committee was empowered to investigate complaints of racial, ethnic, and religious discrimination in defense industries. With four divisions and 12 regional offices, and expenditures in the 1943–1944 fiscal year of $431,609, FEPC commenced its task of ensuring equal employment opportunity.

FEPC examiners held hearings across the nation to adjudicate disputes involving blacks, Hispanics, Orientals, Jews, Jehovah's Witnesses, Seventh-Day Adventists, and others. Between July 1, 1943, and December 31, 1944, the committee handled 5,803 cases with 3,188 coming from blacks. Some complaints resulted in wildcat strikes by black steelworkers in various Pittsburgh area plants where unfair treatment required immediate FEPC intervention. In 1944, for example, black laborers at the Clairton coke works in western Pennsylvania went on strike to protest discriminatory promotion practices. Because several steel plants depended upon these facilities, an FEPC investigator arrived to prevent a production shutdown. He initiated negotiations among black workers, the steel union, and plant officials which resulted in a satisfactory solution. FEPC staff also traveled to Philadelphia to aid black transit workers and to various West Coast communities to end racial discrimination in the shipyards. The committee had been so successful that blacks requested the opening of other regional offices.

With the end of World War II and the wartime employment emergency, Congress refused to renew funding for the FEPC. Even the backing of President Harry S Truman failed to win enough congressional support. In 1946 its work came to a halt. The establishment of various state and municipal FEPCs in the late 1940s and 1950s extended the agenda of the defunct federal agency.

References:

Dennis C. Dickerson, *Out of the Crucible: Black Steelworkers in Western Pennsylvania, 1875–1980* (Albany: State University of New York Press, 1986);

Final Report: Fair Employment Practice Committee June 28, 1946 (Washington, D.C.: United States Government Printing Office, 1947);

First Report: Fair Employment Practice Committee July 1943–December 1944 (Washington, D.C.: United States Government Printing Office, 1945);

William H. Harris, "Federal Intervention into Union Discrimination: FEPC and the West Coast Shipyards During World War II," *Labor History,* 22 (Summer 1981): 325–347;

Louis C. Kesselman, *The Social Politics of FEPC: a Study in Reform Pressure Movements* (Chapel Hill: University of North Carolina Press, 1948);

Paula F. Pfeffer, *A. Philip Randolph: Pioneer of the Civil Rights Movement* (Baton Rouge: Louisiana State University Press, 1990).

Archives:

The records of the Fair Employment Practice Committee, Regional Files, are at the National Archives, Washington, D.C.

Benjamin Franklin Fairless

(May 3, 1890 – January 1, 1962)

CAREER: Schoolteacher (1905–1907); surveyor, Wheeling and Lake Erie Railroad (1913); civil engineer, mill superintendent, general superintendent, and vice-president in charge of operations, Central Steel Company (1913–1926); vice-president and general manager (1926–1928), president and general manager, Central Alloy Steel Corporation (1928–1930); executive vice-president, Republic Steel Corporation (1930–1935); president, Carnegie-Illinois Steel Corporation (1935–1937); director, president (1938–1953), chairman of executive committee (1938–1952), chairman, chief executive officer (1952–1955), executive advisory committee, finance committee, director, United States Steel Corporation (1955–1961); president, American Iron and Steel Institute (1955–1960).

Benjamin Franklin Fairless, a steelmaker for almost half a century, was born in the small mining community of Pigeon Run, Ohio, on May 3, 1890, to Ruth and David Williams. His father, of Welsh descent, was a coal miner. Fairless was two years old when his mother was injured in an accident, and he was sent to live with his aunt and uncle, Sarah and Jacob Fairless, in Justus, Ohio. The couple legally adopted him, and he began his schooling in Justus. An avid businessman at the age of five, he bought newspapers at a half-cent a piece and sold them for a penny.

At fifteen, he earned $65 a year working as a janitor at the high school where he was a student. After graduating high school, he taught at nearby Rockville and at Navarre, Ohio, for two years to raise college tuition. He attended the University of

Wooster (now College of Wooster) in Ohio, playing semipro baseball between terms and on weekends. After two years at Wooster, he transferred to Ohio Northern University at Ada, Ohio.

After graduating with a degree in civil engineering in June 1913, Fairless went to work for the Wheeling and Lake Erie Railroad as a surveyor. The following spring, he got word that "Coxey's Army" was planning a march on Washington, and he took a trolley ride to Massillon, Ohio, to see the army assemble. From the train window, he observed construction activity for a new steel plant. He got off the train and applied for and received a job with the Central Steel Company of Massillon as an engineer.

Fairless advanced quickly in the company, becoming, in turn, mill superintendent, general superintendent, and vice-president of operations. In September 1926 Central Steel Company merged with the United Alloy Steel Corporation to form the Central Alloy Steel Corporation, and Fairless was appointed vice-president and general manager of the new company. Two years later, at the age of thirty-eight, he became the company's president and general manager.

In 1927 Henry Ford had abandoned his Model T, causing a slowdown in the auto industry awaiting the new 1928 models. At the same time, the steel industry was in a period of doldrums. With the announcement of the new Ford model, Fairless moved quickly and within two weeks had 15 of Central Alloy's 23 open-hearth furnaces operating, allowing Central Alloy to be first in line when the auto companies started looking for fast steel deliveries. In April 1930 Central Alloy merged with several other companies to form the Republic Steel Corporation, and Fairless was named executive vice-president of the new company.

Fairless's talents for organization and ability to step into delicate situations with well-timed suggestions brought him to the attention of Myron C. Taylor, United States Steel Corporation's chairman of the board. They worked together on arrangements for the National Recovery Act code under President Franklin D. Roosevelt in 1933. At about this time Taylor was preparing a reorganization program for U.S. Steel and was searching for an executive skilled in both operations and sales. In 1935 Taylor merged Carnegie Steel and Illinois Steel, the two largest U.S. Steel subsidiaries, and in September he chose Fairless to serve as president of the new Carnegie-Illinois Steel Corporation. There were differences of opin-

Benjamin Franklin Fairless

ion among business analysts on how successful Fairless would be at orchestrating the vast operations at Carnegie-Illinois. Some observers anticipated that he would be unable to keep peace among rival management organizations.

He proved successful at the U.S. Steel subsidiary, however, and in October 1937, at a quarterly meeting of U.S. Steel directors, he was elected director and president of the corporation, succeeding William A. Irvin on January 1, 1938. Shortly thereafter, in April 1938, he was named chief administrative officer of the corporation. He held that post until 1952, when he was made chairman of the board and chief executive officer; he continued as president until January of 1953, when Clifford F. Hood was elected to replace him. In 1955 Fairless was succeeded as chairman by Roger M. Blough and became president of the American Iron and Steel Institute.

From 1938 to 1955, under Fairless's guidance, U.S. Steel enjoyed one of its greatest periods of expansion, with capacity raised nearly 35 percent. Fairless's proudest achievement at U.S. Steel came in 1945, when he approved plans to explore and develop iron-ore deposits in Venezuela. In that year Fairless authorized $50,000 to cover the expense of aerial photography of the Venezuelan

Fairless breaking ground for the construction of the Fairless works

jungle on the recommendation of Mack Lake, a mining engineer and geologist. Upon completing his mission Lake recommended that further exploration and drill samples might confirm the location of the ore south of the Orinoco River. When Fairless was told that "maybe a half million" would be needed, his reply was "Go ahead."

This decision provided one source of iron ore for what was the nation's first and only postwar, fully integrated steel plant. Construction had begun on the Fairless Works in 1951 on a 4,000-acre site along the Delaware River at Morrisville, Pennsylvania, and was considered one of the most modern steel plants in the world.

The son of a coal miner, Fairless prided himself in having an intimate understanding of the problems and aspirations of the workingman. While Fairless was an executive of U.S. Steel, his brother, John Williams, was a member of the United Steelworkers. In commenting on labor relations Fairless often stressed harmony and chastised those who sought "to divide management and labor in warring and irreconcilable factions." He often disagreed with steelworkers leader Philip Murray yet planned and carried out with Murray's successor, David J. McDonald, a joint management-labor tour of U.S. Steel facilities.

In his autobiography, *It Could Happen Only in the U.S.* (1956; compiled from two October 1956 issues of *Life*), Fairless claimed to have warm relations with steel's labor leaders. Yet in 1941 he and John L. Lewis disagreed bitterly on the issue of the closed shop. During Fairless's tenure three strikes were called in the steel industry by Murray, and, in each, Fairless was one of the chief negotiators.

As a leader in American industry and in particular as a spokesman for steel, Fairless often testified at congressional hearings on big business. At times he was placed in the position of having to defend U.S. Steel against contradictory charges that U.S. Steel was too large yet did not expand quickly enough to meet the nation's demand. To such charges Fairless once answered, "No one has yet invented an accordian-pleated steel plant that will contract conveniently under the glowering eye of the Department of Justice, and then expand obligingly in times of national peril."

In June 1942 he was appointed a member of the advisory staff of Major General Levin H. Campbell, Jr., chairman of ordnance of the War Department. He also was a member of the Iron and Steel Industry advisory committee of the War Production Board. On June 20, 1946, he was awarded the Medal of Merit by the War Department for his services in breaking production bottlenecks during World War II. In September 1956 President Eisenhower named him chairman of a citizens committee to study the nation's foreign-aid programs. Along with a committee of prominent citizens he made a 52-day tour of 21 countries. He also served as a member of the President's Board of Consultants on Foreign Intelligence Activities.

Gregarious and sports-minded, Fairless enjoyed a wide circle of friends and retained his lifelong interest in baseball. His activities in many professional, private, and civic clubs were celebrated by a host of awards he received. He was a director of many organizations, including the American Ordnance Association, the Mellon National Bank and Trust Company, the Pittsburgh Consolidation Coal Company, the First Iron Works Association, and the Pittsburgh Pirates Baseball Club. Fairless also served as national cochairman of the National Conference of Christians and Jews in 1955 and as chairman of a national committee for the Eisenhower Memorial Library.

He often spoke publicly in defense of the free-enterprise system, gaining a reputation as an outspoken champion of large-scale industrial enterprise. He used his position as president of the American Iron and Steel Institute (1955–1960) for this purpose, and many of his speeches were widely distributed in mass-circulation publications.

In 1912 Fairless married Jane Blanche Trubey, with whom he had a son, Blaine. She died in 1942. He was married in October 1944 to the former Mrs. Hazel Hatfield Sproul of Philadelphia. Shortly before his death, Fairless and his second wife were divorced. Fairless died on January 1, 1962, after a long illness.

Publications:

"Savings with Alloys Universal in Scope," *Steel,* 99 (October 12, 1936): 123, 126;

"Scientific Approach to Our Problems," *American Society of Metals Transactions,* 26 (December 1938): 1154–1160;

"Standardization in the Steel Industry," *American Iron and Steel Institute Yearbook* (1939): 46–55;

"Steel Must Go Where Most Needed," *Iron Age,* 151 (December 17, 1942): 100;

"Steel Wage Increase May Force Rise in Prices," *Commercial and Financial Chronicle,* 158 (December 9, 1943): 2350;

"Steel Wage Controversy," Commercial and Financial Chronicle, 162 (September 20, 1945): 1341;

Steel Wages, Prices and Reconversion (New York, 1945);

Statement by Benjamin F. Fairless . . . on the Settlement of the Strike against the United States Steel Corporation (Pittsburgh, 1946);

The Steel Strike (New York: American Iron and Steel Institute, 1946);

"What Democracy Did for Me," *American Magazine,* 145 (February 1948): 21;

"Steel Capacity Is Adequate," *Commercial and Financial Chronicle,* 168 (December 30, 1949): 2759;

and E. M. Voorhees, "U.S. Steel Officials Defend Price Advances," *Commercial and Financial Chronicle,* 171 (January 26, 1950): 441;

"Steel, Iron Ore Supplies," *Chemical Engineering,* 57 (June 1950): 270;

"Detour Ahead," *Vital Speeches,* 16 (June 15, 1950): 529–533;

"Steel Industry Will Do the Job," *Commercial and Financial Chronicle,* 172 (September 21, 1950): 1083;

"Steel for the Future," *Atlantic,* 186 (December 1950): 30–33;

Guilty before Trial (New York: U.S. Steel Corporation, 1950);

Man's Search for Security, Address before the Illinois Manufacturers' Association, Chicago, December 13, 1949 (N.p., 1950);

A Report to the Boss (New York: U.S. Steel Corporation, 1950);

Business, Big and Small, Built America, Statement before the Subcommittee on the Study of Monopoly Power of the House Committee on the Judiciary, Washington, D.C., April 26–28, 1950 (New York: U.S. Steel Corporation, 1950);

Your Business and Mine (New York: U.S. Steel Corporation, 1950);

"Your Business and Mine: The Petroleum Industry Is Assured of Resources by Steel Industry," *Midwest Engineer,* 3 (January 1951): 6–7;

Target for Termites (New York: U.S. Steel Corporation, 1951);

"Does Labor Want Equality of Sacrifice?," *Iron Age,* 168 (November 22, 1951): 46;

How Not to Kill a Flea (New York: U.S. Steel Corporation, 1951);

"Your Stake in the Steel Crisis," *Vital Speeches,* 18 (May 1, 1952): 418–420;

The Great Mistake of Karl Marx (New York: U.S. Steel Corporation, 1952);

Ask and It Shall Be Given to You (New York: U.S. Steel Corporation, 1952);

The Boy Next Door (New York: U.S. Steel Corporation, 1952);

The Power to Destroy (New York: U.S. Steel Corporation, 1952);

Whose Sacrifice? (New York: U.S. Steel Corporation, 1952);

"Year-end Statement," *Blast Furnace and Steel Plant,* 41 (January 1953): 107;

"Taxation: The Power to Destroy America's Future," *Reader's Digest,* 62 (February 1953): 113–116;

"Clear Case of Financial Malnutrition," *Commercial and Financial Chronicle,* 177 (April 16, 1953): 1637;

"Predicting Ourselves into a Recession," *U.S. News and World Report,* 35 (October 2, 1953): 93–95;

"Christian Vocation in Steel," *Christian Century,* 70 (November 11, 1953): 1293–1294;

"For Sale: U.S. Steel Corporation," *Reader's Digest,* 63 (November 1953): 17–18;

"Employees Stake in Stability of Steel Business," *Commercial and Financial Chronicle,* 178 (December 3, 1953): 2158;

What Kind of America? A talk at the Economic Club of Detroit, September 21, 1953 (New York: U.S. Steel Corporation, 1953);

"Who Profits from Free Enterprise?," *American Mercury,* 80 (March 1955): 117–124;

"Further Expansion Will Be Imperative," *U.S. News and World Report,* 38 (May 15, 1955): 102–103;

"Perfect Price of Steel," *Commercial and Financial Chronicle,* 182 (October 27, 1955): 176;

"Steel Industry Planning Wide Expansion," *Commercial and Financial Chronicle,* 183 (January 19, 1956): 279;

"Steel's Depreciation Problem," *American Iron and Steel Institute Yearbook* (1956): 43–67;

"It Could Happen Only in the U.S.," *Life,* 41 (October 15, 1956): 163-164+;

"Attaining Future Growth for America and Steel," *Commercial and Financial Chronicle,* 185 (May 30, 1957): 2510;

"Who Says Steel Industry Is Immune to Competition?," *Commercial and Financial Chronicle,* 189 (June 4, 1959): 2513;

"New Horizons for Steel and How to Reach Them," *Commercial and Financial Chronicle,* 191 (June 2, 1960): 2378;

"President's Address; Steel Industry in the '60s," *Iron and Steel Engineer,* 37 (August 1960): 180–181.

References:

"Benjamin F. Fairless, President U.S. Steel Corporation," *Iron Age,* 140 (October 28, 1937): 86B–86C;

"Ben Fairless Turns a Corner," *Iron Age,* 175 (May 12, 1955): 60–61;

"Boys, You Take it from Here, Says Myron Taylor," *Fortune,* 21 (March 1940): 64–67+.

Archives:

Research materials are located in the American Iron and Steel Institute Collection; Library, Fairless Works (Morrisville), and Public Relations Department (Pittsburgh), U.S. Steel Corporation; Speeches by Steel Men, Hagley Museum and Library, Wilmington, Delaware.

James Augustine Farrell

(February 15, 1862 – March 28, 1943)

by Elizabeth C. Sholes

Industrial Research Associates

and

Thomas E. Leary

Industrial Research Associates

CAREER: Salesman (1889–1892), sales manager (1892–1893), general manager, Pittsburgh Wire (1893–1899); head, foreign sales, American Steel and Wire (1899–1901); manager, foreign sales, United States Steel Corporation (1901–1903); president, U.S. Steel Products Corporation (1903–1911); president (1911–1927), president and chief executive officer (1927–1932), director, U.S. Steel Corporation (1932–1943).

James Augustine Farrell's lifelong career as a steelman encompassed experiences that ranged from general mill hand and salesman to president of the United States Steel Corporation. With particular expertise in foreign trade, he directed the company's growth during its formative years and oversaw the expansion of corporate paternalism as well as product modernization.

Born on February 15, 1862, in New Haven, Connecticut, his father, John Guy Farrell, was an Irish Catholic seaman and merchant ship owner. He was lost at sea with his ship, the *Monte Cristo,* when James was only sixteen, forcing the family into reduced circumstances. James left school and began work as a common laborer at a New Haven wire mill. He quickly advanced to mechanic then wire drawer. After ten years Farrell left New Haven for Pittsburgh, where he was employed again as a laborer at Oliver Iron & Steel Company. In 1882 he moved to Pittsburgh Wire, where he advanced rapidly; between 1882 and 1886 he became a foreman in charge of 300 workers. A year after his arrival in Pennsylvania he was already well enough established to marry and begin a family.

Although Farrell lacked substantial formal schooling, he gained an education by studying on his own. His knowledge of the wire trade earned him a promotion to the sales department as a traveling representative. In 1892 the company again promoted him, this time to the New York City office as sales manager, where he soon earned a reputation among customers for absolute reliability. In 1893 he was promoted to general manager.

During the 1893 panic, Farrell successfully steered his company through rough economic waters by expanding its export base to 50 percent of output even as domestic sales were in decline. While other companies were failing, Pittsburgh Wire inaugurated construction of a new plant in Braddock, Pennsylvania, under Farrell's supervision.

In 1899 Pittsburgh Wire merged with American Steel and Wire, a New Jersey corporation begun by John Gates in 1897. Farrell headed foreign sales for the expanded corporation, enlarging the overseas sales base he had developed for Pittsburgh Wire. In 1901 American Steel and Wire was absorbed into the new United States Steel Corporation. Gates retired from the scene, but Farrell elected to remain with the new giant as manager of his subsidiary's continuing export business.

Initially U.S. Steel's constituent companies handled their own foreign trade. Farrell proposed an alternative to what he believed was wasteful duplication of effort, and in 1903 the corporation created a centralized Products Export Company with Farrell as president. This company handled all overseas business from the various subsidiaries, establishing 60 warehouses in 55 countries in Europe, Asia, South America, South Africa, and

James Augustine Farrell

Australia. The operation became a full subsidiary, U.S. Steel Products Company, with Farrell still at the helm. To service this expanded overseas operation, Farrell built a fleet of steamships, and during the next eight years he tripled the corporation's foreign business to over 1.2 million tons per year, or 10 percent of U.S. Steel's business. By 1913 the company had increased its representation to 268 overseas agencies handling a broad product line.

Farrell's organizational talents and contribution to the expansion of U.S. Steel's status as a global producer no doubt explained his next promotion. In 1911 J. P. Morgan and U.S. Steel chairman Elbert H. Gary selected Farrell to replace retiring president William E. Corey. But corporate politics also figured in Farrell's choice. Neither Corey nor Schwab, the first two presidents, had graciously accepted Gary's overriding authority in U.S. Steel affairs, especially in regard to pricing and competitive policies. Several raucous disagreements and lesser squabbles resulted from the efforts of Schwab and Corey to operate as Andrew

Carnegie had taught "his boys." In choosing Farrell, Gary bypassed those with Carnegie connections, including William Dickson. Clearly Gary sought an end to skirmishes with the president, and Farrell certainly was more pliable than his predecessors. He also exercised limited authority as president; Gary called the shots while Farrell implemented policies related to steelmaking operations. Not until 1927 did Farrell become the corporation's chief executive officer, and then only after Gary died.

During his tenure as president, Farrell continued to promote the expansion of foreign sales outlets but concentrated as well on improving U.S. Steel's facilities. During his 21-year career Farrell saw output soar 480 percent from 6 million tons to 29 million tons per year. The company diversified vertically and horizontally, capturing new raw materials sources in coal and coke lands (1.15 million acres), extending railroad lines to 2,964 miles, and expanding its shipping fleet to 689 lake freighters, motor vessels, tugs, barges, and scows.

At the heart of the corporation's growth was Farrell's promotion of expanded works operations. In Pittsburgh, for example, U.S. Steel upgraded its equipment to keep abreast of demand and market expansion. It added an electric furnace at Duquesne while simultaneously upgrading blast-furnace and open-hearth facilities. The Edgar Thomson Works became the focus of increased rail production, and in 1912 the company installed Edgar Thomson's first open hearth to produce basic steel for rails. In 1915 the company added a heavy rail mill and in 1929 again upgraded the size, number, and capacity of the supporting open hearths. The Homestead Work's rolling mills also were remodeled in 1912, but the most dramatic addition was the defense-oriented 3-high, 100-inch "Liberty Mill" built in six short months during 1918. After the war Homestead upgraded its blast furnace stocking equipment, added new 100-ton open hearths, and electrified its structural mills. In 1925 Farrell was behind construction of Homestead's massive 44-inch blooming mill, the world's largest, and the 1926 construction of the 52–52–52 wide-flange structural mill, which was so massive that it became the centerpiece of a tour by visiting steelmen attending that year's American Iron and Steel Institute meetings in Pittsburgh. Farrell also supervised the corporation's diversification into specialty steels such as ferromanganese and ferrosilicon, large-scale structurals, sheet, tube, light- and heavy-gauge wire, tinplate, armor plate, shapes, and coke by-products. Through expansion U.S. Steel assured its continued dominance of the American steel market.

U.S. Steel's size and market dominance, however, raised suspicions among a public already fearful of the increasing influence of big business. In 1912, less than a year after Farrell assumed the presidency, U.S. Steel became the object of congressional scrutiny. The Stanley Committee conducted an inquiry into the industry's operations and U.S. Steel's practices in particular. Farrell testified at length, confounding expectations by agreeing, at least in part, that federal supervision of the industry could be beneficial. He rejected government-mandated price controls, however, claiming that no monopoly existed in steel and that small companies, not large, established market prices. He argued the United States should follow Canada and Germany in allowing price agreements that preserved weaker competitors and rejected assertions that companies such as his own were selling products more cheaply in foreign markets than in the United States. He asserted that the price "is cheaper to the [American] consumer than in any other country." Despite his assurances doubts were not assuaged, and in 1913 he again had to answer, in weeklong testimony during U.S. Steel's trial for antitrust violations, charges of monopoly and unfair industry collusion.

An Irish Catholic among mostly Scots Presbyterians, Farrell in some respects stood out from his steel colleagues. His views on business and labor were moderate — and in some cases liberal — when compared to the often reactionary opinions of most steel executives. Although never a member of the corporate, liberal National Civic Federation, he espoused many of the reformist principles later adopted by the NCF. He argued that industry owed labor pensions, workers' compensation, and decent working conditions. Under his supervision in the 1910s and 1920s, U.S. Steel improved safety and started significant welfare and apprentice training programs. But Farrell was not enamored of unions; during the 1919 strike he absolutely refused to negotiate. He also claimed responsibility for effecting changes that actually had originated with labor's demands by asserting that the end of the 12-hour day and the increases in wages stemmed exclusively from U.S. Steel's initiative and wisdom.

Despite his antagonism toward the labor movement, there is no doubt that during the Great

Depression his was the sole voice pressing for wage stabilization and the maintenance of work opportunities. He alone asserted that steel employees were not overpaid. Farrell contradicted his colleagues, saying "I think it is a pretty cheap sort of business when the largest companies ... are trying to maintain wages, for men who are working three days a week, and then cut that three days a week another 10 percent." He blamed the Depression on domestic malfeasance and spoke vigorously against debt financing, speculation and the crippling effect large corporations have on small businesses. Economic improvement would come, he claimed, only by curing the economy of "demoralization" caused by ruthless competition and "lack of consideration for the rights of the rank and file in industry."

Despite these admonitions aimed at fellow businessmen, Farrell was well liked in the industry. His success at U.S. Steel and his leadership at AISI, where he was vice-president for nearly three decades (1914–1943), led to his winning the first Gary Memorial Medal awarded in 1929. Three years later Farrell withdrew from daily involvement with the industry and the corporation. His retirement was in keeping with new board chairman Myron Taylor's plan to bring younger talent to the top of U.S. Steel. But a year before his retirement Farrell saw company sales soar to over $730 million and employment increase to 53,619 full-time and 150,955 part-time workers — all as the effects of the Depression were being felt.

Farrell's retirement did not sever his ties with the company; he remained a director until his death and was active as an honorary member of the British Iron & Steel Institute. He was chairman of the foreign commerce committee of the U.S. Chamber of Commerce and a director of the American Manufacturers' Export Association. As a ranking Catholic businessman he was honored twice by the Vatican, as a Knight of Saint Gregory and recipient of the Grand Cross of the Order of Malta.

In 1889 Farrell married Catherine McDermott, and the couple had five children. He inherited a passion for sailing from his father, and from 1923 to 1928 he was owner of the *Tusitala,* one of the last full-rigged sailing ships still active in sea trade. He died at the age of eighty-one at his home on March 28, 1943.

Publications:

"Contract Obligations," *American Iron and Steel Institute Yearbook* (1910): 99–101;

"Foreign Relations," *American Iron and Steel Institute Yearbook* (1910): 43–51;

"Contract Obligations," *American Iron and Steel Institute Yearbook* (1912): 35–38;

"Central and South American Trade as Affected by the European War," *American Academy of Political and Social Science,* 60 (July 1915);

"Building Up Our Overseas Trade," *Iron Trade Review,* 58 (February 3, 1916): 291–293;

"Trumpet Call to All Business Men," *Iron Trade Review,* 60 (January 4, 1917): 41;

The Future of Foreign Trade, an Address by James T. Farrell Delivered at the Fourth National Foreign Trade Convention, Pittsburgh, Pa., January 26, 1917 (New York: National Foreign Trade Council, 1917);

Foreign Trade in Winning the War: Address by James T. Farrell Delivered at the Foreign Trade Session of the War Convention of American Business, Atlantic City, N.J., September 20, 1917 (New York: National Foreign Trade Council, 1917);

The Foreign Trade Outlook, Address to Reconstruction Convention, Chamber of Commerce of the United States of America, Atlantic City, December 5, 1918 (New York: National Foreign Trade Council, 1918?);

American Maritime Policy, Address at the 6th National Foreign Trade Convention, Chicago, April 26, 1919 (New York: National Foreign Trade Council, 1919);

The Shipping Situation of the World, Address to the 10th National Foreign Trade Convention, New Orleans, May 4, 1923 (New York, 1923);

"Making Headway in Foreign Trade," *Iron Age,* 117 (April 27, 1926): 1190–1191;

"Foreign Trade Outlook Favorable," *Iron Age,* 121 (May 3, 1928): 1257–1258;

"Business Sound, Some Lines Exceeding 1928," *Iron Trade Review,* 85 (October 31, 1929): 1103, 1113;

"Greater Prosperity through Foreign Trade," *North American Review,* 229 (January 1930): 1–6;

"Business Gets Back to Sound Business," *Iron Age,* 125 (May 15, 1930): 1447–1448, 1501;

"Shorter Work Week Would Reduce Unemployment," *Iron Trade Review,* 86 (May 15, 1930): 98, 102;

"Price Advances Essential to Recovery, Declares Foreign Trade Council," speech to National Foreign Trade Council, *Iron Age,* 127 (June 4, 1931): 1840–1842;

"Supplementary Steel," *Fortune,* 4 (July 1931): 114, 116, 119;

"James A. Farrell Doubts Value of Economic Planning for Steel," *Iron Age,* 128 (November 5, 1931): 1103, 1198;

"Foreign Trade in 1932," *Fact Finding Pamphlets in Foreign Trade No. 12* (New York: National Foreign Trade Convention, 1932);

Foreign Trade and Other Addresses (N.p., 1933);

Russo-American Trade Relations (New York: National Foreign Trade Council, 1934);

Fallacy of Economic Self-Containment, by James A. Farrell . . . *Delivered at the First General Session of the National Foreign Trade Convention, Houston, Texas, November 18, 1935* (New York: National Foreign Trade Council, 1935);

The World Trade Outlook, by James A. Farrell, Address to the First General Session, National Foreign Trade Convention, November 18, 1936 (New York: National Foreign Trade Council, 1936);

"Sell American," *Saturday Evening Post,* 211 (July 9, 1938): 230.

References:

Gerald G. Eggert, *Steelmasters and Labor Reform, 1886–1923* (Pittsburgh: University of Pittsburgh Press, 1981);

"J. A. Farrell Heads Steel Corporations," *Iron Age,* 120 (December 29, 1927): 1796–1797, 1824;

"James A. Farrell — Master of Detail and Hard Work," *Iron Trade Review,* 49 (July 20, 1911): 112–115;

"James A. Farrell to Retire in April; His Successor Not Yet Selected," *Iron Age,* 129 (January 21, 1932): 251;

"President Farrell before the Stanley Committee," *Iron Age,* 89 (January 25, 1912): 275–276;

"President Farrell's Further Testimony," *Iron Age,* 89 (February 1, 1912): 293–294;

"The Steel Corporation Dissolution Suit," *Iron Age,* 91 (May 15, 1913): 1185, 1206–1207;

James Weinstein, *The Corporate Ideal in the Liberal State, 1900–1918* (Boston: Beacon Press, 1968).

Federal Steel Company

by Stephen H. Cutcliffe
Lehigh University

As part of the flurry of late-nineteenth-century mergers, the Federal Steel Company was officially organized in New Jersey on September 8, 1898. Incorporated as a holding company and capitalized at $200 million, Federal consisted of the following subsidiaries: Illinois Steel Company, which held five large steel plants in Chicago, Milwaukee, and Joliet; Minnesota Iron Company, the second largest ore producer in the upper Midwest and owner of the Duluth and Iron Range Railroad and Minnesota Steamship Company; Lorain Steel Company in Lorain, Ohio; Johnson Company, with a plant in Johnstown, Pennsylvania; and two railroads — the Elgin, Joliet and Eastern and the Chicago, Lake Shore and Eastern — with rail lines linking several of the steel plants in the Chicago area.

The motivation for the formation of Federal Steel came from the economic difficulties encountered by the Illinois Steel Company during the mid 1890s in the face of the depression and stiff competitive prices set by the Carnegie Steel Company. Judge Elbert H. Gary, who had been appointed general counsel to Illinois Steel during this period of economic stress, recognized that financial success and perhaps even survival would come only if the firm could be fully integrated. Although it did own some ore and coal reserves, Illinois still depended extensively on the Minnesota Iron Company as an ore supplier. Gary, who recently had been involved with attempts to form a nationwide trust in the wire industry, sought to gain control of Minnesota Iron, as well as additional manufacturing facilities and a transportation network that would allow Illinois to set prices competitive with those of Carnegie Steel. Turning to J. P. Morgan & Company for financial support, Gary worked out the details of what the *New York Commercial* would headline as a "Gigantic Combination" that was sure to start "one of the greatest contests for supremacy . . . ever seen" in the steel industry. Prompted by Morgan, Gary left his legal practice and became the head of Federal Steel in October 1898.

Federal Steel was originally capitalized at $200 million, but only approximately half of the amount authorized was actually issued, divided equally between preferred and common stock. Yet, it was a huge corporation with a workforce of 21,000. At the time of its founding Federal Steel had an annual ingot capacity of approximately 2 million tons, or 15 percent of the country's total, comparable to Carnegie Steel's capacity. In 1899, its first full year of operation, Federal's net earnings were approximately $9.1

million — or about half that of Carnegie Steel — and the following year, $11.7 million.

Federal's emergence as a direct rival to his interests led Carnegie to invite Gary, Henry Frick, and H. H. Porter to a luncheon meeting at which he proposed a pooling arrangement for rail production. Despite being new to the steel industry, Gary was a polished negotiator and demanded that any arrangement be "on an equal basis." Somewhat surprised by Gary's steadfast position, Carnegie ultimately agreed, and the arrangement drawn up that day was subsequently presented to and approved by the respective companies.

Despite the agreement by the nation's two largest steel producers, increased competition and tension remained in the industry; the two sides failed to establish an effective rail association, and additional trusts emerged willing to undercut prices. Gary, Morgan, and Carnegie ultimately recognized that a more extensive, fully integrated combination was necessary if the Federal and Carnegie companies were to survive. In February 1901 Federal Steel, Carnegie Steel, and other major steel-producing firms combined to form the first billion-dollar corporation, the United States Steel Corporation.

References:

William T. Hogan, S.J., *An Economic History of the Iron and Steel Industry in the United States*, 5 volumes (Lexington, Mass.: Lexington Books, 1971);

Ida M. Tarbell, *The Life of Elbert H. Gary: A Story of Steel* (New York: Appleton, 1925).

Ferro-alloys

by Geoffrey Tweedale

Manchester, England

The prefix *ferro* means the presence of iron. In the making of tool steels, cutlery steels, and other alloy steels, manganese, chromium, or silicon are added, not as pure elements — which would be too expensive — but as alloys of iron. The major ferro-alloys are: ferrosilicon (15 to 90 percent silicon); ferromanganese (80 percent manganese, 5 percent carbon, 1 percent silicon), spiegeleisen (about 18 percent manganese, 5 percent carbon); silico-manganese (70 percent manganese, 20 percent silicon, 1.5 percent carbon); and ferrochromium (70 percent chromium, 5 percent carbon). Some ferro-alloys are made in blast furnaces, but higher reduction temperatures are achieved in electric-arc furnaces and are therefore used more extensively.

Ferro-alloys have underpinned many major advances in steelmaking. In the nineteenth century, for example, Robert Mushet's addition of spiegeleisen to Bessemer steel paved the way for the spectacular success of that product. The availability of ferromanganese gave another English metallurgist, Sir Robert A. Hadfield, the opportunity to discover manganese steel and launch the development of alloy steels. Similarly, the introduction of stainless steel in the 1920s depended on the production of ferrochromes with sufficiently low carbon content. More recently, small quantities of ferroboron have been added to steel to improve its hardenability — the degree to which it will harden after quenching.

Karl Leroy Fetters

(November 28, 1909 – October 3, 1990)

by Bruce E. Seely

Michigan Technological University

CAREER: Assistant metallurgist, National Tube Company (1933–1936); open-hearth metallurgist (1936–1938), general metallurgist (1940–1941), special metallurgical engineer (1943–1950), assistant to vice-president of operations (1950–1956), assistant vice-president of operations (1956–1959), vice-president of research (1959–1970), vice-president for planning and technology (1970–1971), vice-president for technical services, Youngstown Sheet & Tube Company (1971); doctoral student, Massachusetts Institute of Technology (1938–1940); assistant professor of metallurgy and staff member, Metals Research Lab, Carnegie Institute of Technology (1941–1943); with Office of Scientific Research and Development (1944); consultant after retiring in 1971.

Karl Fetters was a longtime metallurgist with Youngstown Sheet & Tube Company whose career demonstrated how American steel companies slowly adopted scientific research and college-trained metallurgists. The American steel industry hesitantly moved into the era of industrial research in the twentieth century; prior to 1925 only Armco had developed a research facility. Beginning in the mid 1920s, however, scientific understanding of both iron and steel metallurgy and metallurgists began moving from university and corporate laboratories into the steel industry. Fetters's career exemplified this development.

Karl Leroy Fetters was born on November 28, 1909, in Alliance, Ohio, where he attended public schools. He developed an early interest in science and held a job in the chemical laboratory with the Morgan Engineering Company when he was sixteen. After attending Carnegie Institute of Technology, he graduated with a degree in metallurgical engineering in 1931. Carnegie Tech was then developing an academic research program in metallurgy, and new studies were being launched in the school's Metals Research Lab.

In 1933 Fetters became metallurgical assistant with the National Tube Company in Lorain, Ohio. Academic connections may have helped him land the job, for the chief metallurgist at National Tube was Frank Speller, who chaired an outside advisory group for the Metals Research Lab at Carnegie Tech. Fetters remained in Lorain until 1936, when he began his association with Youngstown Sheet & Tube Company as an open-hearth metallurgist. He left Youngstown after two years to pursue his doctorate at MIT. He was a research assistant in 1938 and an Open Hearth Fellow in 1939–1940. In 1940 he received the doctor of science degree from Massachusetts Institute of Technology (MIT) and was chosen for the honor fraternities Sigma Xi and Tau Beta Pi; his dissertation topic was "Equilibrium Between Molten Iron and Lime-Silicon-Iron Oxide Slags."

Fetters returned to Youngstown as general metallurgist for a year, then went back to the Metals Research Lab as assistant professor of metallurgy to head a research project for the Office of Scientific Research and Development (OSRD) on seamless gun tubes. The OSRD hoped to manufacture gun barrels like seamless pipe, by forcing a round of steel through a piercing machine to form a hollow tube. Only 25 percent of the barrels made this way were acceptable, however. Fetters discovered that, by not using the bottom third of steel ingots and by paying careful attention to the pouring temperature of steel and transit time to soaking pits, good barrels could be made 98 percent of the time. A plant built for this technique produced 40,000 40-millimeter and 34,000 75-millimeter barrels for a fraction of the cost of forging gun barrels. In 1943 Fetters returned to Youngstown Sheet & Tube, this time for good, although he continued to work with the OSRD in 1944.

In his subsequent career Fetters occupied a series of technical positions and was the leading scientific figure at Youngstown Sheet & Tube. As special assistant to the operating vice-president he wrote several papers on various aspects of open-hearth steelmaking, focusing on slag-metal reactions in basic open-hearth furnaces. One of these papers, coauthored with J. L. Mauthe (later chairman of Youngstown Sheet & Tube), won the American Iron and Steel Institute's medal for the best paper published in the institute's *Journal* in 1948. Fetters became assistant to the vice-president of operations in 1950.

Fetters continued his rise in the company's hierarchy, advancing to assistant vice-president of operations in 1956. In 1959 Youngstown Sheet & Tube finally created a separate research and development department, and Fetters was promoted to vice-president as head of the laboratory. Under his direction the company built its Technical Center. In 1970 he capped his career as vice-president for planning and technology, retiring the next year. He continued to act as a scientific adviser to then-president Frank Nemec, who stated at the time of Fetters's retirement, "We'll sorely miss this man who has contributed so much to the technical competence of our company."

Throughout his career Fetters was active in many scientific and technical organizations. He joined the American Institute of Mining, Metallurgical and Petroleum Engineers (AIME) in 1937, served on various AIME committees, and was elected president of the Metallurgical Society in 1962 and AIME president in 1964. He was made an honorary member of AIME in 1976. He played equally important roles in the American Society for Metals (ASM) and the American Iron and Steel Institute. He also belonged to the British Iron and Steel Institute, the West of Scotland Iron and Steel Institute, the Verein Deutscher Eisenhüttenleute (Society of German Ironmen), and the National Association of Corrosion Engineers. He was elected a fellow of the ASM, AIME, and the American Institute of Chemists. In 1965 he became the first steel industry engineer and first Ohio resident to be elected to the National Academy of Engineering. Fetters died at his home on October 3, 1990.

Publications:

"A Rapid Method of Correlation Applicable to the Study of Steelmaking Reactions," *American Institute of Mining Engineers Transactions,* Iron & Steel Division, 140 (1940): 166–169;

and J. Chipman, "Slag-metal Relationships in the Basic Open Hearth Furnace," *American Institute of Mining Engineers Transactions,* Iron & Steel Division, 140 (1940): 170–198, 199–203;

and Chipman, "Volubility of Iron Oxide in Liquid Iron," *American Society of Metals,* reprint no. 54 (October 1941): 20–24;

and Chipman, "Equilibria of Liquid Iron and Slags of Systems CaCo-Mg0-Fe0-SI02," *American Institute*

of Mining Engineers Transactions, Iron & Steel Division, 145 (1941): 95–112;

"War Production in Acid Open Hearths," *Industrial Heating,* 9 (July 1942): 894, 896, 898, 900;

and Margaret Dienes, "Silver Chloride as a Medium for Study of Ingot Structures," *Metals Technology,* 10 (August 1943) 1–11;

"Basic Steel — Session on Melting and Oxidizing," *AIME Electric Furnace Steel Proceedings,* 1 (1943): 104–133;

"Summary of New York Meeting Papers on Deoxidation Practice," *AIME Open Hearth Proceedings,* 27 (1944): 232–297;

and J. W. Spretnak and E. L. Layland, "Ingot Factors in Production of Seamless Gun Tubes," *Industrial Heating,* 13 (December 1946): 1996–1998;

and E. W. Mahaney, "Basic Open Hearth Slags," *Iron Age,* 159 (March 6, 1947): 62–66; (March 13, 1947): 64–68;

and J. L. Mauthe, "The Mineralogy of Basic Open Hearth Slags," *Yearbook of the American Iron & Steel Institute* (1947): 264–298;

and E. G. Hill, H. C. Smith, and C. H. Herty, Jr., "The Effect of Raw Materials on Steelmaking," *American Iron & Steel Institute of Regional Technical Papers* (1947): 47–56;

"Basic Open Hearth Slags," *AIME Open Hearth Proceedings,* 31 (1948): 181–230;

and M. M. Helzel and Spretnak, "Distribution of Nonmetallic Inclusions in Some Killed Alloy Steel Ingots," *American Society for Metals,* reprint no. 9 (1948);

"Basic Open-Hearth Slags — Mineralogy and Control," *Proceedings of the National Open Hearth Committee, Iron and Steel Division, American Institute of Mining Engineers,* 31 (1948): 190–202, 208–209;

"Stresses Optimum Conditions for Economy and High Production," *Steel,* 122 (February 23, 1948): 115;

"The Chemistry of Metallurgical Slags," *Yearbook of the American Iron & Steel Institute* (1949): 486–489;

"The Distribution of Oxygen between Molten Iron and Iron-Oxide-Silica Slags," *Yearbook of the American Iron & Steel Institute* (1949): 389–391;

and H. H. Hottel, "Semifinished Steel," in *ABC of Iron & Steel,* sixth edition (Cleveland: Penton, 1950), pp. 164–177;

and Hottel, "Production of Semifinished Steel," *Steel,* 126 (April 10, 1950): 102, 105–106, 108, 111–112; (April 17, 1950): 86, 88, 90, 93; (April 24, 1950): 86, 88, 90, 93;

"The Physical Chemistry of Steel Making," *American Iron and Steel Institute Regional Technical Papers* (1950): 15–24;

"The Physical Chemistry of Steelmaking," *Industrial Heating,* 18 (January 1951): 74, 76, 78, 80; (February 1951): 265–266, 268, 274;

"Manufacture of Low-Carbon Rimmed and Killed Steels for Deep Drawing," *Proceedings of the National Open Hearth Committee, Iron & Steel Divi-*

sion, *American Institute of Mining Engineers,* 34 (1951): 250–252;

"Line Pipe," *Gas,* 28 (April 1952): 138+;

"Steelmaking Developments during the Past Decade," *American Iron and Steel Institute Regional Technical Papers* (1955): 25–36;

"Practical Steel Plant Research," *Blast Furnace & Steel Plant,* 49 (February 1961): 164 –170;

"The Nature of Fatigue," *Journal of Petroleum Technology,* 16 (August 1964): 869–872;

"The Competitive Challenge of Foreign Steelmaking Technology," *Allegheny Regional Advisory Board Proceedings* (September 28, 1967): 20–23;

and others, "Extractive Metallurgy," in *Report of the Panel on Extractive Metallurgy of the Committee* on *Mineral Science and Technology, Division of Engineering, National Research Council* (Washington, D.C.: National Academy of Sciences, 1969).

References:

"AIME Awards and Honors," *Mining Engineering,* 28 (February 1976): 92;

"Metallurgical Society President, 1962," *Journal of Metals,* 14 (February 1962): 118;

"Metallurgist Assists in Manufacture of Gun Barrels," *Blast Furnace and Steel Plant,* 34 (December 1946): 1542.

Archives:

A biographical file is at the National Academy of Engineering, Washington, D.C.

William Jennings Filbert

(November 4, 1865 – February 4, 1944)

by John N. Ingham

University of Toronto

CAREER: Clerk (1881–?), chief accountant, Chicago & North Western Railway Company (?–1898); assistant auditor, chief auditor, Federal Steel Company (1898–1901); assistant controller (1901–1902), controller (1902–1932), vice-chairman, finance committee (1932–1934), chairman, finance committee (1934–1936), general consultant to executives, United States Steel Corporation (1936–1944).

William J. Filbert was the epitome of the new professional manager who emerged in large corporations during the first half of the twentieth century. Virtually unknown to the general public and certainly lacking the charisma and daring of the early steel entrepreneurs such as Andrew Carnegie, Filbert was nonetheless recognized by insiders as a highly significant player in the formation of the modern American steel industry. Known in Wall Street accounting circles as "the wizard of the age," Filbert helped modernize steel's financial policies.

William Jennings Filbert was born on November 4, 1865, in Palatine, Illinois, near Chicago; his father was president of the local bank. William was educated in the public schools and began his business career in 1881, when he secured a job as a clerk with the Chicago & North Western Railway. At that time there were few schools teaching accounting in America; it was a

William Jennings Filbert

profession learned best on the job, and the railroads provided the best training ground. Beginning in the 1850s, the railroads had developed elaborate systems to control costs in their far-flung enterprises. During his first several years at the railroad, Filbert learned well the lessons of modern cost accounting and was promoted to the position of chief accountant.

In 1898 the new Federal Steel Company was organized by J. P. Morgan and Elbert H. Gary and hired Filbert to apply the modern concepts of auditing and cost accounting in order to compete with Andrew Carnegie's mighty Carnegie Steel Company. He soon advanced to the position of chief accountant. In 1901, when Morgan began the massive task of organizing the gigantic United States Steel Corporation, Filbert was hired as assistant controller, and in 1902 he was advanced to the position of controller of the entire enterprise, a position he held for the next 30 years.

Filbert made two major contributions to U.S. Steel. The first was most noted and praised by his contemporaries, while the second, nearly unnoticed by the general public at the time, has given Filbert his place in the history of accounting practice in America. The emergence of giant corporations through mergers attracted much criticism from politicians and the press, who feared that this powerful new presence in American business threatened the spirit of competition. In answering these criticisms, U.S. Steel developed a policy of full financial disclosure. In 1902, Filbert's first year as controller, U.S. Steel brought out a 35-page financial statement that set a precedent in the corporate world. The reaction of Filbert's contemporaries was generally ecstatic, with the *Commercial and Financial Chronical* proclaiming that *"the United States Steel Corporation has nothing to conceal."*

Current financial analysts, however, have been less impressed with Filbert's accomplishment on this score. Benjamin Graham, David L. Dodd, and Sidney Cottle have commented: "When United States Steel Corporation was organized in 1902 it showed no goodwill or other intangibles on its balance sheet. Thus all its capitalization was ostensibly covered by its property account and working capital. Many years later, it was revealed that on less than $769 million of the plant account was a written-up item known as 'water.' "

Filbert's other contribution was more lasting. Working with the eminent academic accountant Arthur Lowes Dickinson, Filbert developed the first rudimentary entity theory for asset measurement, replacing the older proprietary theory, which had little application to the separation of ownership and control that had developed at major corporations at the turn of the century. According to Michael Chatfield, the proprietary theory "envisioned few parties at interest, close contact by merchants with their affairs, and data summarized only for the use of owners or creditors, who were assumed to have specialized knowledge of the business." The massive new corporations such as U.S. Steel, however, were legally distinct from their owners and managers, and there was typically a constantly changing group of shareholders. In developing the entity theory, Filbert and Dickinson emphasized that the source of financing was immaterial to the corporation, and all returns to every supplier of capital were distributions of profits and not "expenses" to the corporation. These ideas were later given fuller expression by William A. Paton in his *Accounting Theory* (1922). The entity theory was ultimately adopted by other major corporations.

Filbert was also active in the establishment of the U.S. Steel and Carnegie Research funds to aid in the development of new technical advances in the industry. In recognition of his innovative and important work, Filbert was elected a director of U.S. Steel in 1920, serving in that capacity until his death in 1944. In May 1931, in recognition of his contributions to the steel industry, Filbert received the Gary Memorial Medal of the American Iron and Steel Institute.

Filbert's greatest executive responsibilities with U.S. Steel came in the 1930s. In 1932 he was appointed vice-chairman of the finance committee, and two years later he was elected its chairman. His advancement came at a critical time for the corporation; Elbert Gary had died in 1927, and the steel firm struggled to reorganize. The corporation was also hit hard by the Depression. When he assumed the position of chief executive chairman of U.S. Steel in 1932, Myron C. Taylor increased the importance of the finance committee, finding in Filbert an important ally. Filbert retired in January 1936 to serve as a general consultant to the executives of U.S. Steel.

As professional management began taking control of giant corporations, men such as Filbert assumed greater responsibility in the decision making. Having risen to U.S. Steel's top ranks from an accounting office rather than from the floor of a steel mill, Filbert was part of a significant movement in the American steel industry. He died on February 4, 1944.

References:

Michael Chatfield, *A History of Accounting Thought*, revised edition (Huntington, N.Y.: Krieger, 1977);

David F. Hawkins, "The Development of Modern Financial Reporting Practices among American Manufacturing Corporations," in *Contemporary Studies in the Evolution of Accounting Thought*, edited

by Chatfield (Belmont, Cal.: Dickenson, 1968), pp. 247–279;
Gary John Previts and Barbara Dubis Merino, *A History of Accounting in America* (New York: Wiley, 1979).

George R. Fink

(November 1, 1886 – July 29, 1962)

by Alec Kirby

George Washington University

CAREER: Door puller, salesman, West Penn Steel Corporation (?–1922); salesman, sales manager, Wheelock, Lovejoy & Company (?); president, treasurer, director, Michigan Steel Corporation (1922–1931); president, director, Great Lakes Steel Corporation (1929); president (1929–1954), director, National Steel Corporation (1929–1955).

George R. Fink, steel manufacturer, was born in Brackenridge, Pennsylvania, on November 1, 1886. The son of a glassworker, Fink was educated in public schools before taking work as a door puller with the West Penn Steel Corporation in Pittsburgh. He became a salesman for West Penn and shortly thereafter moved to Boston to work as a salesman for the steel firm of Wheelock, Lovejoy & Company. He rose to the position of sales manager, then returned to West Penn to take charge of steel-sheeting sales in the Midwest.

In 1922 Fink struck out on his own, founding the Michigan Steel Corporation in Ecorse, Michigan, some 20 miles from Detroit. The firm, in which Fink served as president, treasurer, and director, produced steel sheets for the Detroit automobile industry. In 1929 Fink founded and incorporated the Great Lakes Steel Corporation, also at Ecorse. Capitalized at $20 million, Great Lakes Steel was to produce primary, semifinished, and finished steel articles. Fink began construction of a major steelworks on the bank of the Detroit River. The undeveloped land had deterred other potential steel producers, who doubted that the swampy ground could support a major steelworks. Convinced that the plan was workable, Fink proceeded with the construction of a major facility with a capacity of 600,000 tons of steel ingots.

George R. Fink

As construction proceeded, Great Lakes merged with the Weirton Steel Company and components of the M. A. Hanna Company, creating the National Steel Corporation. The three firms found the merger to be mutually beneficial: the M. A. Hanna Company had vast ore resources in the Mesabi Range, and a fleet of freighters for transport; the Weirton Company provided technological know-how and product diversity; and Great

Lakes was strategically located near Detroit. After the merger George Humphrey became chairman of the executive committee, while E. T. Weir became chairman of the board of directors. Fink assumed the presidency. Two years later National acquired Fink's Michigan Steel Company.

Fink remained keenly active in the Great Lakes plant, which expanded from its original capacity. National Steel provided the plant with the necessary financial base for expansion, as the well-integrated corporation weathered the Great Depression without once running in the red — the only major steel producer to do so. As the 1930s drew to a close, Great Lakes and National profited mightily from defense contracts. By 1941, 90 percent of National's facilities were devoted to the war program, and continued expansion, notably for increasing pig iron capacity, took place. As president of National, Fink energetically adapted Great Lakes to the changing markets of the World War II era. During and immediately after the war, Fink guided the firm into a position as exclusive producer of the Quonset hut, a prefabricated building sold worldwide. Recognizing the potential for large profits in the postwar construction industry, Fink established the Stan-Steel Corporation of Terre Haute, Indiana, to manufacture other preengineered steel buildings.

Outside his continuing business success, Fink maintained a wide variety of personal and professional interests. In 1941 the University of Michigan granted him an honorary doctorate. He was active in the Masonic order, the Detroit Athletic Club, and various country clubs in the Detroit area. He also bred Hereford cattle on his farm near Brighton, Michigan.

In 1954 poor health forced Fink to resign as president of National, although he remained on the board of directors for another year. He died on July 29, 1962, at his home in Grosse Pointe, Michigan.

Reference:

William T. Hogan, S.J., *An Economic History of the Iron and Steel Industry in the United States*, 5 volumes (Lexington, Mass.: Lexington Books, 1971).

Ben Fischer

(November 26, 1913 –)

by John P. Hoerr

Teaneck, New Jersey

CAREER: Various posts with Socialist party organizations, ending as organizational and campaign director (1933–1941); research director, Aluminum Workers of America (1941–1944); associate director of research (1944), director, arbitration department (later becoming contract administration department, 1960), assistant to president, United Steelworkers of America (1962–1979); adjunct professor (1981–1984), director, Center for Labor Studies (1982–1992), Distinguished Public Professor of Labor Studies, Carnegie Mellon University (1984–).

Ben Fischer, known widely as a "labor intellectual," served as a highly influential staff member of the United Steelworkers of America (USWA) for more than three decades. Born in Hoboken, New Jersey, on November 26, 1913, Fischer had no formal education beyond high school and a two-year course at the Rand School of Social Science, operated by the Socialist party in New York City. He joined the Young People's League, an arm of the party, and occupied various offices from 1933 to 1935. Later he was appointed head of the Socialist party of Michigan and helped non-Communist leaders gain control of the United Auto Workers in 1939. Fischer directed the 1940 presidential campaign of Socialist candidate Maynard Krueger.

Fischer concluded that organized labor provided more opportunity to accomplish economic and social change than socialist politics. He joined the Aluminum Workers of America (AWA), affiliated with the Congress of Industrial Organizations (CIO), in late 1941 as research director. When the small AWA merged with the Steelworkers in 1944, Fischer was appointed associate research director of the steel union. In 1946 he became an international representative, serving under the direction of Philip Murray. After

Murray's death in 1952, Fischer worked for a year as housing director of the CIO in Washington and returned to the USWA staff in Pittsburgh in late 1953.

In the 1940s and 1950s the Steelworkers did not have a highly structured staff organization. Fischer and a few other key international representatives performed various duties assigned by the president. Among other things, he represented the union in wage stabilization cases during and after World War II, participated in aluminum- and can-manufacturing negotiations, and served as a troubleshooter in many other labor-management situations. A proponent of resolving issues through problem solving rather than confrontational tactics, Fischer often was assigned problems that required delicate political and economic compromises. He did not shy away from recommending unpopular solutions to difficult issues, even if this meant incurring the enmity of some unionists. A pipe-smoking man of medium height, he had the demeanor of a university professor rather than a union staffer. But he could be a powerful, extemporaneous orator when necessary and usually spiced his conversation and speeches with a dry, biting wit.

One of Fischer's most important contributions involved the arbitration of employee grievances, the bedrock of a successful union-management relationship. After the United States Steel Corporation and the Bethlehem Steel Company established boards of arbitration with full-time umpires in the 1940s, Fischer was the USWA's main architect of an arbitration system that became a model for many other unionized industries. He set up a national cadre of arbitrators who could be called upon to hear cases arising in different regions. Within the union Fischer developed strategies and tactics that established precedents for dealing with complex grievance issues, such as the so-called 2B clause in steel labor agreements.

Section 2B, first negotiated in the 1947 U.S. Steel agreement and later adopted in other company contracts, restricted management's ability to change or eliminate existing practices at the plant level. When steelmakers began reducing the size of mill crews after World War II, the USWA objected and cited section 2B. Company officials contended that the provision was limited to a narrow range of practices and had nothing to do with the size of work crews. Fischer, however, developed a "strict construction" strategy and argued in arbitration hearings that, without qualifying language, section 2B must be held to cover crew size. Arbitrators at Bethlehem and U.S. Steel agreed with this concept and in the 1950s issued a series of decisions prohibiting managers from reducing crews unless they installed new equipment or technology or otherwise changed the "underlying circumstances" of the job.

It was never proved that these rulings resulted in widescale featherbedding, as some company officials charged. Undoubtedly, however, 2B made it more difficult for companies to correct manning mistakes of the past, leading to excessively large crews in some mills. The industry's demand for elimination of section 2B led directly to the 116-day national steel strike of 1959. When it ended, 2B remained in the contracts.

Fischer's role in grievance arbitration was formalized when the union created an arbitration department (the name later was changed to contract administration department) in 1960 and made him director. In the late 1960s and early 1970s, when the cost and slowness of most companies' grievance systems caused widespread unrest in the plants, Fischer developed an expedited procedure for arbitrating individual worker complaints aside from discharge cases. The first such system in American industry, it was negotiated in the 1971 steel agreement.

USWA president I. W. Abel named Fischer a personal assistant, a post he first held under President David J. McDonald and later under Lloyd McBride. In these years Fischer also performed many other duties. Although he never sat on top-level bargaining committees in the steel industry, Fischer helped identify troubling mill-floor issues and develop collective-bargaining strategies to deal with them.

In the 1960s Fischer began to urge union leaders to address the steel industry's formidable civil rights problems resulting from decades of discrimination in hiring and promoting women and blacks and other minorities. Finally, in 1973 he managed to frame a way of handling the problem; it involved replacing departmental seniority within plants (a system that kept minorities out of many of the choice jobs), with plantwide seniority. In 1974 the USWA, major companies, and federal agencies adopted this approach in negotiating a consent decree to correct past discrimination and prevent future discrimination. Fischer was appointed the union chairman of a joint committee that monitored enforcement of the decree over the next several years.

Turning sixty-five, Fischer retired at the beginning of 1979 under a mandatory retirement rule covering USWA employees. He began teaching labor courses at Carnegie Mellon University and in 1982 established and became director of the university's Center for Labor Studies. In 1984 he was named Distinguished Public Service Professor of Labor Studies.

Reference:

John P. Hoerr, *And the Wolf Finally Came: The Decline of the American Steel Industry* (Pittsburgh: University of Pittsburgh Press, 1988).

Florida Steel

by Robert W. Crandall

The Brookings Institution

Florida Steel is one of the oldest minimill steel companies in the United States. Although its origins as a steel fabricator date as far back as 1933, the current company was incorporated as a privately held Florida firm in 1956. In 1966 Florida Steel offered its common shares to the public, but in 1988 the company once again was taken private in a leveraged buyout by a management-led investor group.

The company operates five medium-sized minimill steel plants in the Southeast, each using scrap-charged electric furnaces and continuous casters to feed bar- and rod-rolling mills. The first of these plants was opened in 1958 in Tampa, Florida. Four others were opened in Florida, North Carolina, and Tennessee between 1961 and 1980, but the Indiantown, Florida, plant was closed in the wake of the 1982 recession. In 1988 the company purchased a minimill in Knoxville, Tennessee, from the Blue Tee Corporation. In 1990 Florida Steel's total annual raw-steel capacity at these five plants was 1.5 million tons.

Unlike some of the larger minimill companies, Florida Steel has maintained a narrow product line and geographic focus. Its steel plants produce only merchant bars, rods, and small structural shapes, eschewing the larger structural shapes and the flat-rolled products such as sheet, strip, and plate. Moreover, it continues in its original business — the fabrication of reinforcing steel at approximately ten plants in the Southeast — through wholly owned subsidiaries.

Florida Steel has continually reinvested to modernize its five plants. All but the North Carolina plant are nonunion and have substantially lower labor costs than those of integrated steel; all of the plants enjoy very high labor productivity.

References:

William T. Hogan, S.J., *Minimills and Integrated Mills: A Comparison of Steelmaking in the United States* (Lexington, Mass.: Lexington Books, 1987);
Iron and Steel Works of the World, ninth edition (London: Metal Bulletin, 1987).

William Uhler Follansbee

(December 23, 1859 – December 19, 1939)

by Kevin M. Dwyer

George Washington University

CAREER: Clerk, Rosedale Foundry (c. 1876–1879); bookkeeper/salesman, Park-Scott & Company (1879–1883); partner, James B. Scott & Company (1883–1894); secretary-treasurer, (1894–1916), president (1916–1927), chairman of the board, Follansbee Steel Corporation (1927–1939).

The exponential growth of the corporate culture of oligopolistic capitalism did not entirely erase institutions of family capitalism from the steel industry in the 20th century. Smaller firms that manufactured finished products were able to establish and retain their market share — particularly in local markets — through specialization and product differentiation. The Follansbee Brothers Company of Follansbee, West Virginia, was one such operation. William Uhler Follansbee, its most renowned officer, was respected in the Pittsburgh district for his good business sense and fair labor policies. On the national level, his frequent testimony before Congress established him as an authority on the tariff — an issue vital to the fledgling U.S. tinplate industry.

William U. Follansbee was very much a family man, and his company's story is in many respects a family history. He was born to Gilbert and Maria Jackson (Haynes) Follansbee in Pittsburgh, Pennsylvania, on December 23, 1859. His father was a shoe manufacturer in Pittsburgh. William and his brother — and future business partner — Benjamin Gilbert (born May, 15, 1851) attended Pittsburgh public schools, with Benjamin completing his studies at Pittsburgh's Newel Institute and William graduating from Central High School.

In 1878 Benjamin, who had served nine years as a supply agent on the Pittsburgh and Pittsburgh & Lake Erie Railroads, took work as a bookkeeper at Park-Scott & Company, a well-established tin-importing and -jobbing firm. In 1879 he secured employment at the firm for William, who had been working as a clerk in

William Uhler Follansbee

Pittsburgh's Rosedale Foundry since his graduation from high school. Park-Scott was split in 1883 after the death of a partner, with the importing and jobbing work falling to the newly organized James B. Scott & Company. Both William and his brother became partners in the new firm, with William acting as bookkeeper and salesman. After Scott died in 1894, the firm became the Follansbee Brothers Company. Benjamin served as its president, William as secretary-treasurer.

Throughout their early years, the Follansbee Brothers adapted with good success to a changing environment. After 1890 tariff policy and consolidation transformed completely the U.S. tinplate industry. The McKinley tariff, for which William

Follansbee had actively lobbied, set a high duty on Welsh tinplate and virtually gave birth to an American tinplate industry overnight. While Follansbee's import business declined over the course of the decade, the opportunity for entry into the production of tinplate grew dramatically. Savage price wars following the Panic of 1893 had tin producers desperate for price stability until 1898, when William H. Moore offered them corporate sanctuary in the American Tin plate Company, which held 90 percent of the nation's tinplate capacity. Tin prices not only stabilized thereafter but for some grades rose by over a dollar per box. The integration of American tinplate into the giant United States Steel Corporation in 1901 dropped tin prices somewhat, but in the better grades, especially, they remained high. With demand for higher grades of tinplate increasing, the steel trust had set itself up for "a barrage of new competition," and in 1902 Follansbee Brothers joined the fray.

The company purchased a 250-acre farm in Brooke County, West Virginia, adjacent to the Ohio River. By September 1904 a sheet mill with three 40-ton open-hearth furnaces was up and running, and the "model steel town" of Follansbee, West Virginia, grew up around it. High-grade tinplate and terneplate (lead-coated tinplate used for roofing) became the company staples, and its plant gained a reputation among tinplate makers and engineers for its production of higher grades at consistent costs. In 1906 the company reorganized, and Follansbee Brothers became Follansbee Steel Corporation; the West Virginia mill became the Sheet Metal Specialty Company, a subsidiary of the new corporation.

Growing demand for higher-grade tinplate spurred the expansion of the business. Specialty built two more finishing mills in 1910 to accommodate the excess capacity of its open hearths and by that time had established a sizable clientele which consumed over 250,000 boxes of Follansbee plate per year. William replaced his brother as president of the firm in 1916, and Benjamin became chairman, overseeing the family's other business concerns, the Follansbee Water and Light Company and the Brooke County Improvement Company among them.

Under William's direction, as president until 1927 and as chairman thereafter until his death, the company prospered. In 1929 a second plant was established nine miles upriver from the Specialty Works in Toronto, Ohio. Built on a 20-acre plot surrounded by over 400 acres of coal-rich land, the new works consisted of four 40-ton open-hearth furnaces, ten sheet mills, a 30-inch sheet bar mill, a bloom press, and the annealing equipment their specialty metals required. The fully integrated works, which were substantially upgraded in 1941, were laid out so that raw materials could be delivered at one end while the finished products were made ready for easy distribution at the other. Situated on the riverbank and having convenient access to the railroads, the plant was easily able to ship the 84,000 tons of black sheet produced there annually.

A Republican and a Presbyterian, Follansbee took an active role in both politics and religion. Apprenticed in the importing business, his opinion on the tariff was highly sought, and he testified before Congress in support of the McKinley and the Paine-Aldrich tariffs and in opposition to the Underwood tariff. He was a founding member of the Brighton Road Northside Presbyterian Church in Pittsburgh and senior elder at Pittsburgh's Shadyside Presbyterian Church. Follansbee enjoyed good standing with workers as well, and his practice of talking over labor grievances personally earned him respect for being, as one employee put it, "the highest type of Christian man." He was an avid baseball fan.

William Follansbee married twice: to Jennie Childs in 1885 and, following her death, to Ruth Harper in 1910. He had eight children, several of whom followed him into the business, which is still in operation. William Follansbee died on December 19, 1939, seven months after the death of Benjamin.

References:

Blast Furnace and Steel Plant, 29 (March 1941): 313–315.

Iron Trade Review, 47 (October 13, 1910): 681–682; 50 (March 7, 1912): 557–588; 72 (January 11, 1923): 153–158;

Naomi R. Lamoreaux, *The Great Merger Movement in American Business, 1895–1904* (Cambridge, U.K.: Cambridge University Press, 1985).

William Z. Foster

(February 25, 1881 – September 1, 1961)

by James R. Barrett
University of Illinois at Urbana-Champaign

CAREER: Secretary, Stockyards Labor Council (1917–1918); secretary, National Committee for Organizing Iron and Steel Workers (1918–1920); secretary, Trade Union Educational League (1920–1929); presidential candidate, Communist party, USA (1924, 1928, 1932); secretary, Trade Union Unity League (1929–1935); national chairman, Communist party, USA (1932–1957).

William Z. Foster was at the center of each of the key radical movements in twentieth-century America — socialism, syndicalism, and communism. Steelmen best remember Foster as the architect of a national campaign to organize the steel industry at the end of World War I and as the leader of the great 1919 steel strike.

Born in Taunton, Massachusetts, to Irish Catholic immigrants and raised in Philadelphia slums, Foster left school after the third grade and drifted around the country, working at jobs ranging from seaman to locomotive fireman. His intimate knowledge of a wide variety of jobs and his sensitivity to the mentality of common workers made him a talented labor organizer.

Foster joined the Socialist Party in 1900 and joined the Industrial Workers of the World (IWW) in 1909 during the Spokane Free Speech Fight of that year. In 1910–1911 he traveled throughout Europe, studying the ideas and strategies of the French and German labor movements. Upon his return to the United States, Foster split from the IWW over the issue of dual unionism, which he vigorously opposed. From the French syndicalist movement he adapted the strategy of "boring from within" the mainstream labor organizations to win them over to a revolutionary program, and he established a series of organizations to accomplish this aim. Both the Syndicalist League of North America (1912–1914) and the International Trade Union Educational League (1915–1917) remained small "militant minorities," but many of their participants and organizational strategies played important parts in later radical movements, including the American Communist party. By the beginning of World War I, Foster had emerged as a leading theorist of American syndicalism.

During World War I, Foster led a successful organizing campaign among slaughtering and meatpacking workers. The drive was notable because it was probably the largest and most successful effort to date to organize unskilled workers in a mass-production industry and, as such, it called for innovative organizing and structuring of the union. Foster and his associates established the Stockyards Labor Council in 1917 to facilitate cooperation among the various American Federation of Labor (AFL) unions represented in the industry. Workers from a wide array of trades and ethnic backgrounds were brought together under this umbrella organization. By the spring of 1918 the packinghouse workers had won major concessions from their employers, and Foster turned his attention to the nation's giant steel industry.

In June 1918 Foster's resolution to organize the steel industry won the unanimous support of the AFL's national convention. The National Committee for Organizing Iron and Steel Workers, with AFL president Samuel Gompers as its nominal chair, was set up to coordinate the organizational drive. Foster filled the critical position of secretary, responsible for the committee's day-to-day activities.

Foster's efforts enjoyed great success during the late summer of 1918 in the steel towns along the southern shore of Lake Michigan, but the campaign met stiff resistance from employers and local officials in the Monongahela Valley area and elsewhere in Pennsylvania. Still, Foster orchestrated a brilliant organizing campaign which had swept nearly 100,000 workers into union ranks by the spring of 1919. When efforts at negotiation with the steel companies failed in the summer of 1919, Foster urged strike action. A national steel strike commenced on September 22, 1919. Foster ran the strike from a small office in Pittsburgh but

Delegates to the National Committee for Organizing Iron and Steel Workers, Youngstown, Ohio, August 20, 1919. William Z. Foster is seated fourth from right.

also traveled from one steel town to the next, co-ordinating relief efforts and speaking to workers. After considerable violence against the strikers, particularly in the Pittsburgh area, the strike was abandoned on January 8, 1920. Blacklisted for many months following the strike, Foster analyzed his experiences in *The Great Steel Strike and Its Lessons* (1920). Foster visited the Soviet Union during 1921 and joined the American Communist party later that year, providing the communists with many of their most important contacts in the trade-union movement.

In 1920 Foster was instrumental in building the Trade Union Educational League (TUEL), an important radical opposition group within the AFL. Throughout the early 1920s, Foster worked with prominent labor activists to create a national Farmer-Labor party, but his communist strategies alienated his most important noncommunist supporters and undercut his position in the mainstream labor movement; by the mid 1920s Foster found himself increasingly isolated as the AFL stepped up efforts to repress dissent within the federation. The TUEL provided leadership in several of the most important strikes of the decade, however, and kept alive the radical perspective.

Having opposed it as dual unionism in 1928, Foster eventually became the leading figure in the Trade Union Unity League, a separate federation of revolutionary unions. He campaigned as the American Communist presidential candidate in 1924, 1928, and 1932 and directed the party's trade-union work throughout the 1920s and the early years of the Great Depression. A physical collapse during his 1932 presidential campaign severely limited his active involvement in the labor movement, however. Though he advised and provided important contacts for communist organizers in the steel industry and elsewhere during the Congress of Industrial Organizations (CIO) organizing drive of the late 1930s, he turned much of his attention to writing and remained a leading figure in the American Communist party.

Foster was the party's chairperson between 1932 and 1957, from the height of its popularity during the Depression and war years through its isolation and decline during the political repression of the 1950s. He played a central role in the expulsion of Earl Browder, architect of the party's Popular Front policies, and led the organization's reversion to a more sectarian perspective in the postwar era. His own physical decline in the late 1950s mirrored the demise of the communist movement in the United States. William Z. Foster died in Moscow on September 1, 1961.

Publications:

Insurgency or the Economic Power of the Middle Class: A Discussion between Wm. Z. Foster, Member of the I. W. W., Now in Europe, Formerly Spokane Correspondent of "The Workingman's Paper," of Seattle, and the Editor Hermon F. Titus (Seattle: Trustee Publishing, 1910);

and Earl C. Ford, *Syndicalism* (Chicago: W. Z. Foster, 1913?);

Trade Unionism, the Road to Freedom (Chicago: International Trade Union Educational League, 1916);

The Great Steel Strike and Its Lessons (New York: B. W. Huebsch, 1920);

The Russian Revolution (Chicago: Trade Union Educational League, 1921);

The Railroaders' Next Step (Chicago: Trade Union Educational League, 1921);

The Revolutionary Crisis of 1918–1921 in Germany, England, Italy and France (Chicago: Trade Union Educational League, 1921);

The Railroaders' Next Step — Amalgamation (Chicago: Trade Union Educational League, 1922);

The Bankruptcy of the American Labor Movement (Chicago: Trade Union Educational League, 1922);

Russia in 1924 (Chicago: Trade Union Educational League, 1924?);

and J. P. Cannon and Earl Browder, *Trade Unions in America* (Chicago: Daily Worker Publishing, 1925);

Organize the Unorganized (Chicago: Trade Union Educational League, 1925);

Russian Workers and Workshops in 1926 (Chicago: Trade Union Educational League, 1926);

Strike Strategy (Chicago: Trade Union Educational League, 1926);

Wrecking the Labor Banks: The Collapse of the Labor Banks and Investment Companies of the Brotherhood of Locomotive Engineers (Chicago: Trade Union Educational League, 1927);

Misleaders of Labor (Chicago: Trade Union Educational League, 1927);

The Watson-Parker Law, the Latest Scheme to Hamstring Railroad Unionism (Chicago: Trade Union Educational League, 1927);

Acceptance Speeches by William Z. Foster, Candidate for President, and Benjamin Gitlow, Candidate for Vice-President (New York: National Election Campaign Committee, Workers Library, 1928);

Victorious Socialist Construction in the Soviet Union (New York: Trade Union Unity League, 1930);

Fight against Hunger, Statement Drafted by C.P.U.S.A. and Presented to Fish Committe (New York: Workers Library, 1930);

Little Brothers of the Big Labor Fakers (New York: Trade Union Unity League, 1931?);

Toward Soviet America (New York: Coward-McCann, 1932);

The Words and Deeds of Franklin D. Roosevelt (New York: Workers Library, 1932);

and Browder, *Technocracy and Marxism* (New York: Workers Library, 1933);

Unionizing Steel (New York: Workers Library, 1936);

Organizing Methods in the Steel Industry (New York: Workers Library, 1936);

Industrial Unionism (New York: Workers Library, 1936);

The Crisis in the Socialist Party (New York: Workers Library, 1936);

What Means a Strike in Steel (New York: Workers Library, 1937);

and others, *Party Building and Political Leadership* (New York: Workers Library, 1937);

Questions and Answers on the Piatakov-Radek Trial (New York: Workers Library, 1937);

Railroad Workers, Forward! (New York: Workers Library, 1937);

A Manual of Industrial Unionism, Organizational Structure and Policies (New York: Workers Library, 1937);

From Bryan to Stalin (New York: International Publishers, 1937);

Organizing the Mass Production Industries (New York: Workers Library, 1937?);

Stop Wage-Cuts and Layoffs on the Railroads: A Reply to President T. C. Cashen of the Switchmen's Union of North America (New York: Workers Library, 1938);

Halt the Railroad Wage Cut (New York: Workers Library, 1938);

Your Questions Answered on Politics, Peace, Economics, Anti-Semitism, Race Prejudice, Religion, Trade Unionism, Americanism, Democracy, Socialism, Communism (New York: Workers Library, 1939);

Pages from a Worker's Life (New York: International Publishers, 1939);

The United States and the Soviet Union (New York: Workers Library, 1940);

What's What about the War: Questions and Answers by William Z. Foster (New York: Workers Library, 1940);

The War Crisis: Questions and Answers (New York: Workers Library, 1940);

Roosevelt Heads for War (New York: Workers Library, 1940);

Capitalism, Socialism, and the War (New York: Workers Library, 1940);

World Capitalism and World Socialism (New York: Workers Library, 1941);

Socialism, the Road to Peace, Prosperity and Freedom (New York: Workers Library, 1941);

The Soviet Union: Friend and Ally of the American People (New York: Workers Library, 1941);

The Path of Browder and Foster (New York: Workers Library, 1941);

The Railroad Workers and the War (New York: Workers Library, 1941);

Defend America by Smashing Hitlerism (New York: Workers Library, 1941);

and Robert Minor, *The Fight against Hitlerism* (New York: Workers Library, 1941);

Communism versus Fascism (New York: Workers Library, 1941);

The Trade Unions and the War (New York: Workers Library, 1942);

The U.S.A. and the U.S.S.R. (New York: Workers Library, 1942);

Smash Hitler's Spring Offensive Now! (New York: Workers Library, 1942);

and Browder, Israel Amter, and Max Weiss, *Speed the Second Front* (New York: Workers Library, 1942);

Steel Workers and the War (New York: Workers Library, 1942);

Labor and the War (New York, 1942);

From Defense to Attack (New York: Workers Library, 1942);

American Democracy and the War (New York: Workers Library, 1942);

2nd Front Now (New York, 1943);

Soviet Democracy and the War (New York: Workers Library, 1943);

For Speedy Victory: The Second Front Now (New York, 1943);

The Soviet Trade Unions and Allied Labor Unity (New York: Workers Library, 1943);

The People and the Congress (New York, 1943);

Text of the Speeches Delivered at Madison Square Gardens, New York, September 18, 1945, on the Occasion of the Anniversary of the Communist Party of the United States of America (New York: New Century, 1945);

The Strike Situation and Organized Labor's Wage and Job Strategy (New York: Workers Library, 1945);

The Rankin Witch Hunt (New York: New Century, 1945);

The Menace of American Imperialism (New York: New Century, 1945);

Organized Labor Faces the New World (New York: New Century, 1945);

The Coal Miners, Their Problems in War and Peace (New York: New Century, 1945);

Our Country Needs a Strong Communist Party (New York: New Century, 1946);

Reaction Beats Its War Drums (New York: New Century, 1946);

Problems of Organized Labor Today (New York: New Century, 1946);

and others, *Marxism-Leninism vs. Revisionism* (New York: New Century, 1946);

The Menace of a New World War (New York: New Century Publishers, 1946);

Workers, Defend Your Unions! (New York: New Century, 1947);

The Technique of the Mass Campaign (San Francisco: State Educational Committee, Communist Party of California, 1947);

The New Europe (New York: International Publishers, 1947);

Quarantine the Warmongers (New York: New Century, 1947);

The Meaning of the 9-Party Communist Conference (New York: New Century, 1947);

Organized Labor and the Fascist Danger (New York: New Century, 1947);

American Trade Unionism: Principles and Organization, Strategy and Tactics: Selected Writings (New York: International Publishers, 1947);

and others, *The Communist Position on the Negro Question* (New York: New Century, 1947);

United States of America vs. William Z. Foster, et. al. (New York, 1948);

N.Y. Herald Tribune's 23 Questions about the Communist Party Answered by William Z. Foster (New York: New Century, 1948);

Danger Ahead for Organized Labor (New York: New Century, 1948);

On Improving the Party's Work among Women (New York: New Century, 1948);

Labor and the Marshall Plan (New York: New Century, 1948);

Beware of the War Danger! Stop, Look and Listen! (New York: New Century, 1948);

The Crime of El Fanguito: An Open Letter to President Truman on Puerto Rico (New York: New Century, 1948);

The Twilight of World Capitalism (New York: International Publishers, 1949);

In Defense of the Communist Party and the Indicted Leaders (New York: New Century, 1949);

Outline Political History of the Americas (New York: International Publishers, 1951);

The Steel Workers and the Fight for Labor's Rights (New York: New Century, 1952);

History of the Communist Party of the United States (New York: International Publishers, 1952);

A Letter to Congress: Defeat the Anti-Labor Smith Bill! (New York: New Century, 1952);

Danger Signals for Organized Labor (New York: New Century, 1953);

The Negro People in American History (New York: International Publishers, 1954);

History of the Three Internationals: The World Socialist and Communist Movements from 1848 to the Present (New York: International Publishers, 1955).

References:

Edward E. Johanningsmeier, "William Z. Foster: Labor Organizer and Communist," Dissertation, University of Pennsylvania, 1988.

Arthur Zipser, *Workingclass Giant: The Life of William Z. Foster* (New York: International Publishers, 1981).

Lewis W. Foy

(January 8, 1915 –   )

by Elizabeth C. Sholes

Industrial Research Associates

and

Thomas E. Leary

Industrial Research Associates

CAREER: Purchasing, Johnstown, Pennsylvania (1936–1950), buyer (1950–1952), assistant to purchasing agent (1952–1955), assistant purchasing agent (1955–1961), assistant to vice-president, purchasing (1961–1963), vice-president, purchasing (1963–1970), director (1963–1980), executive vice-president (1970), president (1970–1974), chairman, Bethlehem Steel Corporation (1974–1980).

In becoming the sixth chairman of Bethlehem Steel Corporation in 1974, Lewis Wilson Foy oversaw one of the corporation's most dramatic periods of transition. During his tenure he helped shape the character of the company's fortunes and the nature of the steel industry in general.

Foy was born January 8, 1915, in Somerset County immediately southwest of Johnstown, Pennsylvania. He attended Duke University from 1933 to 1934, then dropped out to work. After serving in World War II he attended George Washington University from 1943 to 1944 and Lehigh University from 1947 to 1949.

In 1936 Foy began his career with Bethlehem Steel as a laborer at the company's Johnstown plant. When Bethlehem's recovery from the Depression proved fragile, Foy, single and living at home, volunteered for layoff. During his time away from Bethlehem he and a friend began a small golf driving range in west Johnstown. The venture prospered so that when Foy received his callback notice he was torn between long-term security and good money. He opted to return.

Foy won a position in purchasing and worked in that area for thirty-four years. He became enmeshed in a broad range of Bethlehem's business operations and, according to a colleague, was dedicated to "trying how to do things either better, faster, or cheaper." Despite not having a background in production, Foy was considered a dark horse candidate to succeed Stewart S. Cort as chairman in 1974. In that year Foy was named chairman, in part due to the support of his predecessor. Cort had maintained that Foy was eminently well qualified to be chairman since he had had the pivotal role in organizing all construction, equipment, and labor contracts needed to build Bethlehem's last "greenfield" plant in Burns Harbor, Indiana, and was conversant with all aspects of Bethlehem's business activities.

Foy inherited Bethlehem at an opportune time. Bethlehem was running at peak capacity as the twenty-fourth largest industrial company in America and the forty-eighth largest in the world. Steel markets were booming, and steel exports were up nationally by 73 percent. In his first public interview as chairman, Foy declared that Bethlehem's greatest challenge was to "expand the industry as rapidly as possible to meet the growing demand" and avoid the shortages that encouraged import dependence.

Three years earlier, as company president, he had voiced the same positive message before the American Iron and Steel Executives annual conference. He stressed the need for national increases in productivity that could emerge from what he saw to be a beneficial climate of opportunity. He encouraged management to work constructively with the new union-management contract agreements on productivity and to work with employees in reorganizing systems of productivity. He was enthusiastic about the improvements in productivity at older facilities, such as Lackawanna's No. 6 skin pass mill, as well as improvements made at new mills, such as Burns Harbor.

Three years after assuming the chairmanship, however, Foy faced disaster. The company suffered its first loss since the Depression and had to cope with a blizzard at Lackawanna and a devastating flood in Johnstown. Foy was driven to cut losses and elected to close "less profitable operations" without replacements on hand — the first time in the history of the company such drastic cost-cutting measures had been adopted. The shutdowns were economically beneficial since they netted the corporation $750 million in federal tax refunds. By 1978 the company was again solvent, and the crisis abated. The shutdowns, however, cost the corporation substantial production capacity, including their only continuous caster and the rail mill at Lackawanna.

Foy had extensive ties to the national banking community, which was growing uneasy about the steel industry's future. Foy sided with the banks in their call for greater rates of return on investment, a decades-old argument, but he elected to follow financial rather than production criteria in the search for cost saving and began reducing "surplus capacity." Foy's optimism concerning the long-term health of the steel industry did not wane, although he became more openly critical of "government encroachment" on private business practices and of government's simultaneous refusal to aid industrial stability and growth through subsidies and tax breaks.

In 1978 Foy was elected chairman of the American Iron and Steel Institute (AISI). As an industry spokesman he continued to assert that the industry's recovery was imminent, but by 1980 he was also acknowledging the necessity for further drastic remedies to correct problems caused by overproduction, slow market demand, and import growth. He determined that Bethlehem needed an entirely new direction, so that upon his retirement in 1980 he bypassed tradition to choose newcomer Donald Trautlein, a former Price Waterhouse partner and architect of the 1977 shutdowns in the steel mills. Foy believed Trautlein would continue to apply the requisite financial criteria to production operations that Foy himself had begun.

In addition to his role in AISI and membership in the International Iron and Steel Institute, Foy served on the Conference Board and Business Council, took part in two national policy planning organizations, and was involved with the Newcomen Society. Foy was a member of the United Negro College Fund and was active in the United Way. A trustee of Moravian College in Bethlehem, he received an honorary degree from Moravian, as well as from Lehigh University and the University of Liberia. He married Marjorie Werry in 1942. They have two daughters.

Publications:
"Open Letter," *Bethlehem Review*, 162 (1974): 2, 10;
"Productivity, the Key to Steel's Future," *Iron and Steel Engineer*, 48 (November 1971): 50–52.

Reference:
John Strohmeyer, *Crisis in Bethlehem: Big Steel's Battle to Survive* (Bethesda, Md.: Adler & Adler, 1986).

Archives:
Papers relating to Foy's life and career are in the Bethlehem Steel Collection, Hagley Museum and Library, Wilmington, Delaware, and the Bethlehem Steel Historical Collections, Hugh Moore Historical Park and Museums, Incorporated, Easton, Pennsylvania.

Elbert H. Gary

(October 8, 1846 – August 15, 1927)

by Stephen H. Cutcliffe

Lehigh University

CAREER: Law clerk, Illinois Superior Court (1868–1871); partner, law firm, Van Armen & Vallette, (1871); partner, law firm, E. H. & N. E. Gary, later Gary, Cody & Gary (1871–1898); judge of DuPage County (1882–1890); president, Federal Steel Company (1898–1901); chairman, executive committee (1901–1903), chairman of the board, United States Steel Corporation (1903–1927); president, American Iron and Steel Institute (1910–1927).

Judge Elbert H. Gary was the central figure in organizing and running the United States Steel Corporation during the first three decades of the

Elbert H. Gary (Hagley Museum and Library)

twentieth century. Although coming to the steel industry relatively late in life after having worked as an Illinois lawyer and a county judge, Gary came to the forefront of the rapidly consolidating American steel industry at the turn of the century by advocating policies of cooperation and of stable prices and wages. In so doing, he helped regularize the industry and shape it into the form it largely possesses today.

The youngest child of Erastus and Susan (Vallette) Gary, Elbert Henry Gary was born on October 8, 1846. The Gary family on both sides was originally of New England Methodist stock. Gary's father had moved to Warrenville, Illinois, in the early 1830s and in 1849 to Wheaton, Illinois. His mother had come to Wheaton from Stockbridge, Massachusetts, with her father, Jeremiah, in 1839. Both families had moved to Illinois in pursuit of richer, more extensive farmlands. Although Erastus Gary became a prosperous community leader, he continued to farm his land and raised his children to know the rigors of farming life. Gary's parents imbued in their children a strict Methodist moral code, which later in Gary's life most commonly surfaced in the form of moral preachments directed at his colleagues and subordinates in the steel industry. Gary attended

the local district school, where he received a basic education, lessons that were monitored and reinforced by his parents at home. Gary also learned the basic elements of music that matured into a lifelong interest in the opera.

At the time the Civil War broke out, Gary, then fourteen, was enrolled at Illinois Institute, a Wesleyan Methodist college in Wheaton that Erastus Gary had helped to found. Ida Tarbell reported that Erastus Gary's ambition for his son to "'go through, [and] have a college education'" was ultimately thwarted when, despite his age, Gary was able to join a regiment in 1864. Shortly thereafter his older brother Noah was severely wounded at Resaca, Georgia, and Erastus was called to Georgia to bring him home. Elbert was released from duty by his regimental commander and returned home after a scant two-month military career to tend the family farm. On his return from the South, Erastus Gary rented the farm and moved into town, at which time Elbert took up teaching school for a term, a task to which he did not find himself particularly inclined.

In the spring of 1865 his maternal uncle, Colonel Henry Vallette, and Vallette's partner, Judge Hiram H. Cody, invited Gary to read for the law in their Naperville, Illinois, law office. Gary had been exposed to law while his father had served for two decades as a magistrate in DuPage County. After a year of self-study and assisting his uncle and Cody, Gary in the fall of 1866 entered Union College of Law in Chicago, graduating two years later in June 1868 at the head of his class.

Upon his graduation Gary was appointed a deputy clerk of the superior court at the salary of $12 a week on the recommendation of his dean, Judge Henry Booth, who was also a family acquaintance. Within seven months he rose to chief clerk at the then-handsome salary of $45 a week. Feeling financially secure, Gary married Julia E. Graves of Aurora, Illinois, on June 23, 1869. They settled in Wheaton, the DuPage County seat, and Gary commuted daily to Chicago for the next thirty years of his legal career. (Gary's wife, with whom he had two daughters, died in 1902; in 1905, he was married for a second time, to Emma Townsend of New York City.)

In the spring of 1871 Gary surrendered his clerkship and entered into partnership with his uncle in the firm of Van Armen & Vallette. It is indicative of his aggressiveness that Gary opted for a base salary of $100 per month and the opportunity to take on any business he could attract for himself rather than earn a percentage of the firm's earnings. His widespread contacts formed during his years as a clerk quickly brought him considerable business, which was only momentarily slowed by the great Chicago fire of October 1871. The fire completely destroyed Van Armen & Vallette. After Van Armen chose to retire, Gary and his uncle each decided to set up business for themselves, with Gary by his own account earning over $2,800 in his first year. Within two years, business had grown to the point that Gary sought additional help and took on as a partner his older brother Noah. The law firm of E. H. & N. E. Gary soon became Gary, Cody & Gary on the inclusion of Judge Cody, who had retired from the circuit court.

As the firm's business expanded, Gary increasingly devoted himself to corporate law, becoming associated with numerous railroad and industrial corporations, such as Northwestern Elevated Railroad Company and the western department of the Baltimore and Ohio Railroad, serving them both as counsel — and frequently as a board member. Gary's legal career and reputation grew during the last two decades of the century: he was elected as a DuPage County judge for two terms (1882–1890) — earning him the lifetime sobriquet "Judge" — and was elected president of the Chicago Bar Association, serving from 1893 to 1894. Gary remained active in local community affairs, holding offices as president of Wheaton's town council and as the town's first mayor and as president of the Wheaton County bank. Gary's reputation for fair play and legal understanding resulted in steady advance within the profession, bringing him recognition and modest wealth. During the 1890s, however, he gradually became involved in the iron and steel industry, an involvement that eventually launched him on a second career of central importance to the twentieth-century steel industry.

His first major introduction to the steel industry came in 1892, when John W. Gates approached Gary — who had successfully argued a case in which Gates was the opposing litigant — for legal assistance in combining several wire-producing companies into the Consolidated Steel and Wire Company. Gates had become involved in the volatile and competitive barbed-wire industry and sought to achieve a degree of vertical combination to provide a competitive edge and some measure of stability.

Gary's expertise in corporate law readily enabled him to consolidate five key companies, including two wire mills and three barbed-wire manufacturers. Consolidated Steel and Wire Company was capitalized at $4 million and included the Iowa Barb Wire Company (Allentown, Pennsylvania), the Braddock Wire Company (Rankin, Pennsylvania), the Lambert and Bishop Wire Fence Company (Joliet, Illinois), the Baker Wire Company (Lockport, Illinois), and the Saint Louis Wire Mill Company (Saint Louis, Missouri). Although there were over 130 other wire producers, Consolidated now controlled over 25 percent of the barbed-wire production in the country. However, barbed wire was only part of the total output, as the company now produced a wide range of nails and wire products and, most important, did it from its own steel rods. This vertical integration enhanced its position in the evolving steel industry. Gary's involvement with Consolidated did not end with the legal merger, for he soon found himself serving on the board of directors. Although many observers distrusted Gates's motives and feared the corporation's overcapitalization, Gary's purchase of stock and his willingness to serve as a director, combined with Consolidated's subsequent prosperity, suggest the stability of the venture. *Iron Age* called it "one of the most important events in the history of the wire trade."

Gates slowly expanded his organization by acquiring additional small wire-product producers. Meanwhile, the onset of the depression of 1893 further reduced his competitors. Nonetheless, depressed prices, combined with excess capacity and an inability on the part of the remaining manufacturers to reach satisfactory or long-lasting pooling arrangements, led Gates to seek further consolidation. For legal assistance he again turned to Gary.

In the fall of 1897 Gates and several associates — including John H. Parks, an industrial promoter — began taking options on wire mills across the country. With these options in hand Gates and Gary turned to the firm of J. P. Morgan for financial backing. With a planned capitalization of $40 million, no one individual, or even a group of manufacturers, could handle such a venture. Indeed, this had been one of the limitations on the ability of the industry to integrate in the past. Gary approached Morgan through Charles H. Coster, one of Morgan's ranking lieutenants. With Coster's help, Morgan agreed to form the necessary syndicate to underwrite the venture, to be called the American Steel and Wire Company, with 200,000 shares of preferred stock and an equal number of common shares, each valued at $100. By early 1898 all plans appeared to be on track despite the magnitude and complexity of combining 14 wire manufacturers and producers, several of which were combinations themselves. Unfortunately, the sinking of the USS *Maine* in mid February raised the specter of war. The economic uncertainty inevitably linked to the possibilities of war, combined with engineering and financial reports that raised questions about declining profits for some of the companies, led Morgan to withdraw his support. Although Morgan, through Coster, suggested delaying the proposal until a more stable and economically propitious period, several of the major manufacturers withdrew from the negotiations, leaving the promoters with insufficient capital. Gates's subsequent telegram to his home office tersely summarized his disappointment: "The jig is up. Black Monday morning."

Although the jig was surely up with respect to the original plan, several of the midwestern promoters, including Gates, were determined to set up some sort of wire and nail trust, even if smaller than initially envisioned. With Gary's help the American Steel and Wire Company of Illinois was incorporated later that spring with a capitalization of $24 million, half in preferred shares and half in common stock. Gates was elected chairman, and Gary was appointed a director, although he did not share financially in the corporation's formation. The new company's 14 plants now produced three-quarters of the country's wire products. Subsequently Gates would engineer an additional combination in the wire industry by forming the American Steel and Wire Company of New Jersey with a $90 million capitalization in January 1899. With this final combination Gates assured himself virtual monopolistic control over the nation's wire industry.

Although Gary was not involved with this last step in the consolidation of the wire industry, his interest in steel had been whetted. Experience with the formations of Consolidated and the American Steel and Wire Company of Illinois, combined with his exposure to the industry as general counsel to the Illinois Steel Company, convinced him that vertical integration in the steel industry made good sense. The Illinois Steel Company, organized in 1899, was the first large consolidation in the industry. It consisted of the Union Steel Company, the Joliet Steel Company, and the North Chicago Rolling Mill Company, an outgrowth of the old North Chicago works of Eber B. Ward, whose mills had rolled the first Bessemer steel rails in 1865. Although the firm was initially quite profitable, the 1893 panic created financial difficulties and provided the occasion for the firm's solicitation of Gary's legal advice as general counsel.

Gary was of the opinion that for Illinois Steel to become more profitable and competitive with the industry leader, the Carnegie Steel Company, it needed to integrate backward and control more completely its sources of raw materials. The most important of these was the chief Illinois supplier, the Minnesota Iron Company, the second largest northern-Midwest ore producer. Gary had been approached by one of the corporation's directors, Nathaniel Thayer, for advice about buying the Elgin, Joliet, and Eastern Railway, which owned 180 miles of trackage in the Chicago area, including a link to the company's Joliet works. Gary, who apparently had impressed Morgan with his role in the projected formation of American Steel and Wire earlier in the year, approached the banker with a proposal for a fully integrated firm that included extensive ore properties, furnaces, mills, and its own transportation system. Morgan, convinced by Gary's production and cost figures, appointed a committee — which included among others Governor Flower of New York, Senator Spooner of Wisconsin, Robert Bacon, a partner in the Morgan firm, and Gary himself — to plan the merger. With Gary taking the lead, the details of the consolidation were worked out during the summer of 1898, and on September 9 the $200 million firm was legally incorporated in New Jersey. The Federal Steel Company, as the new firm was known, became a holding company controlling the operations of six subsidiaries: the Illinois Steel Company; the Minnesota Iron Company; the Lorain Steel Company; the Johnson Company; the Elgin, Joliet and Eastern Railway; and the Chicago, Lake Shore and Eastern Railway Company. To raise necessary operating capital, J. P. Morgan and Company also organized a syndicate to put together $14 million in cash in October.

The magnitude of the new corporation clearly impressed the newspapers; the *New York Commercial* referred to it as a "Gigantic Combination of Ironmasters." However, they also suggested that it would be "the beginning of one of the greatest contests for supremacy that the world has ever seen." Ultimately, of course, that competition would be resolved with the formation of the Unites States Steel Corporation three years later. In the meantime, however, Gary undertook the immediate task of coordinating Federal's new holdings. Although John Gates had been appointed president of Illinois Steel in 1894, he was widely distrusted by the steel fraternity because of his speculative intentions. For this reason the Federal merger had been conducted largely without his knowledge, and he was refused an official place in the new company. Not unexpectedly, then, Gates, who was already upset with Morgan for not supporting the American Steel and Wire project, adopted Gary's original plan and launched the American Steel and Wire Company of New Jersey. Thus, instead of Gates, it was to Gary that Morgan turned for leadership talent.

Gary was initially reluctant to give up his $75,000-a-year legal practice, but Morgan, who was highly impressed by Gary, ultimately convinced him to accept the Federal presidency, giving Gary free reign to pick the company's officers and directors and name the members of the executive committee as well as set his own salary. In October 1898, almost 30 years after receiving his law degree, Elbert Gary at age fifty-two began a new career in the steel industry, a career that would last 30 years until his death in 1927.

Gary, who had little patience with traditional pooling arrangements because of their inherent instability, sought to bring stability to the steel industry through extensive vertical integration. His experiences with the formation of American Steel and Wire, Illinois Steel, and now Federal Steel, combined with an abortive attempt to create a formal "Rail Association" in 1898, further convinced him that if Federal was to be a major force in the steel industry as originally envisioned, it would need to expand its holdings even further. Gary later claimed to have suggested to Morgan the combining of the Federal and Carnegie interests as early as 1898, but his proposal apparently solicited little interest from the financier until 1900. By that year the economic boomlet of 1898–1899, which had provided an expanding market sufficient for all, had begun to wane. In order to cut costs both Federal and National Steel announced plans to expand their basic production facilities rather than to continue buying bars and billets from Carnegie to meet their fabricating needs. Similarly, National Tube Company's plans to build its own basic iron and steel producing facilities, combined with the formation of American Steel and Wire, created a potential threat to the profitability of Carnegie Steel. Even though Andrew Carnegie had been considering retirement, his immediate and formal corporate response was to announce plans to integrate forward by building fabricating plants for each of the major fin-

ished products. Additionally, he planned to build his own railway line from Harrisburg to Pittsburgh to compete with the Pennsylvania Railroad in response to the railroad's announced intent to abolish all transportation rebates.

By late 1900 Morgan was convinced that the battle lines had been drawn for what in Carnegie's words would be "a question of the survival of the fittest." Although hesitant to create another major consolidation, Morgan clearly recognized the need for some action. On December 12, 1900, J. Edward Simmons gave a dinner in New York City in honor of Charles Schwab, the new president of the recently reorganized Carnegie Steel, in order to introduce him to bankers and businessmen outside the steel industry. Schwab's after-dinner remarks were focused on his notion of a larger, more fully integrated, more profitable steel industry. Whether Schwab's vision was the final catalyst or not, his remarks fell on receptive ears, for Morgan apparently became convinced that buying out Carnegie Steel and combining it with other major producers — a position Gary had been advocating for some time — offered a realistic possibility, if not a necessity, for survival.

In early 1901 Morgan, with the assistance of Schwab, Gary, and George W. Perkins, formulated a plan for combining eight major steel producers and fabricators in what was to be the first billion-dollar corporation, United States Steel. Schwab, over a game of golf, induced Carnegie to agree to sell his interests for $400 million (Carnegie ultimately received slightly over $492 million for his holdings), while Gary went about convincing the Federal board of the desirability of the combination. In addition to the other "Morgan group" interests — the National Tube Company and the National Steel Company — the new corporation included the somewhat overcapitalized "Moore" organizations — the American Steel Sheet Company, the American Hoop Steel Company, and the American Tinplate Company — which were in no position to withstand a financial donnybrook. The eighth member was John W. Gates's American Steel and Wire Company, which agreed to join only when the new corporation threatened to build its own wire-producing facilities. When the charter was issued on February 25, the new corporation was capitalized at $1.154 billion, but the addition, at Gary's suggestion, of the Rockefeller ore interests — the Lake Superior Consolidated Iron Mines — and of the American Bridge Company brought the total to $1.404 billion when the company began actual operations in early April.

U.S. Steel was created to stabilize the industry by reducing cutthroat competition. The clear necessity for integration coupled with Carnegie's desire to sell and Morgan's willingness to organize the combination and the necessary underwriting syndicate resulted in a company with control over approximately 50 percent of the nation's production. The problem of how to run the corporation still remained, however. U.S. Steel was in reality an amalgamation of many different companies, each of which in turn was a combination of smaller companies. As U.S. Steel's chief financial architect, Morgan determined, at least initially, that the corporation should be run by an executive committee modeled after Standard Oil's successful approach. He charged Gary with serving as its chair. Robert Bacon, a Morgan partner and director of Federal Steel, was appointed head of the finance committee. Schwab was named president, with responsibility for carrying out the executive committee's directions.

As chair of the executive committee and later as chairman of the board, Gary was to face four major problems in running U.S. Steel: consolidation of control within the corporation, establishment of a consistent labor-relations policy, institutionalization of a noncompetitive strategy within the industry, and protection from federal antitrust suits. He sought to achieve protection from antitrust suits by creating an image of U.S. Steel as a "good" trust with progressive labor policies and profit-sharing plans.

In the beginning, real power at U.S. Steel lay with the executive and finance committees rather than in the hands of any one individual. Generally these committees were split on the question of corporate policy. On one side were the experienced steel producers, led informally by Schwab, from the older "competitive" school. On the other were primarily bankers and lawyers who sought long-term profit, good public relations, and avoidance of trouble with the federal government. The latter group gravitated toward Gary as its spokesperson.

The two men quickly came into conflict. Gary distrusted Schwab's ties to Carnegie and his policy of aggressive price-cutting and competitive expansion. Rather, Gary sought price stability and increased efficiency of operation among the subsidiary companies. Schwab, who thought he knew best how to manage the corporation, later complained that "Judge Gary, who had no real knowledge of the steel business, forever opposed me on

some of the methods and principles that I had seen worked out with Carnegie — methods that had made the Carnegie Company the most successful in the world." Schwab quietly sought operational independence as president, but the executive committee under Gary's chairmanship refused to let Schwab overstep his bounds. As historian Robert Hessen notes, the committee introduced a resolution on July 1, 1901, that henceforth the president should "furnish this committee with full reports of the operations of the company, and submit for their information and decision all matters requiring their supervisors as set forth in the by-laws." The resolution, accepted by the committee, had the intended effect of reining in the maverick. A dejected Schwab told George Perkins, Morgan's partner and then secretary to the finance committee, "Don't ask me to drive uptown this evening. My nerves are not in shape to enjoy the drive, I am simply heartbroken." Schwab did not immediately offer to resign, and in the meantime Gary began to shift primary authority from the executive committee led by Schwab to the finance committee.

While Gary consolidated his control of U.S. Steel, Schwab undercut his own position by becoming entangled in several awkward situations, which brought negative publicity not only to himself but also to the corporation. Initially the public press circulated stories about Schwab earning a million-dollar annual salary, stories seemingly confirmed by his construction of a several-million-dollar, French chateau-style mansion in New York City. The reality, however, was that his salary was $100,000. Gary, who feared the resulting negative publicity, publicly quashed the rumor, a move that did not endear him to Schwab, who enjoyed the limelight. Following this squabble, Schwab traveled to France for a vacation, and he promptly purchased a new automobile and drove to Monte Carlo, where he engaged in some nominal play at roulette. The American press blew the story completely out of proportion, alleging that Schwab had broken the bank, but the damage had been done: according to historian Gerald Eggert, Carnegie was outraged and personally affronted, "as if a son had disgraced the family." He suggested Morgan remove Schwab as president, but Morgan refused to accept the resignation when offered. Nonetheless, Morgan increasingly supported Gary's ideas for running the corporation.

Gary's frustration with Schwab, occasioned by the latter's resistance to his policies of openness and efficient centralization, was clearly aggravated by Schwab's public conduct, so much so that in mid 1902 Gary went to Morgan, who Gary believed was not showing enough support for his policies, in effect lending tacit support to Schwab. Fearing Gary's resignation, Morgan promised Gary his full support, assuring Gary's control over the corporation. All that remained were for the details to play themselves out in the course of the next year.

During the ensuing year Schwab increasingly became involved with several businesses not related to U.S. Steel, including the purchase of Bethlehem Steel in 1901. Later he insisted on using Bethlehem's properties as collateral to protect his personal interests when it was merged with the United States Shipbuilding Company, a major consolidation of shipyards in 1902. Schwab received $30 million worth of bonds and stock in the new corporation. But when United States Shipbuilding began to founder in 1903, Bethlehem's board of directors, dominated by Schwab, refused to use the steel company's earnings to bail out the parent company. Schwab was vilified in the press for speculating in stock, and the shipbuilding scandal led to his resignation as president of U.S. Steel. Schwab, who was already suffering some anxiety and tension from the Monte Carlo scandal and Carnegie's rebuff, stepped down on August 4, offering ill health as an excuse. William E. Corey, who had been serving as president of Carnegie Steel — the position from which Schwab had himself come — replaced Schwab as U.S. Steel's president. Gary was made chairman of the board of directors, a newly created position that in effect gave him control of the corporation.

From the beginning, Gary sought to stabilize U.S. Steel by weeding out inefficient plants before building new ones, but by 1905 the corporation had determined that a modern, fully integrated steel plant was necessary to preserve its position within the industry. Construction at a site 25 miles east of Chicago on the shore of Lake Michigan was selected and duly named Gary. Two subsidiaries, the Indiana Steel Company and the Gary Land Company, were organized to erect the steel plant and the accompanying city, respectively. Construction continued over several years, and in late December 1908 the first pig iron was produced, with steel ingots following early in 1909. By the end of 1911, when most of the work was completed, approximately $78 million of capital had been expended in constructing the plant and city. At the time, the plant was the largest in the world, with a capacity of more than one million tons of steel ingots annually. More important, the layout design for complete integration made it extremely efficient. Gary, Indiana, was not a company town in the usual sense, but it did receive financial support from the corporation, which invested heavily in the city's infrastructure and support services.

Almost from his first day in office, Gary sought to establish uniform labor relations as part of his overall approach to common policies. Gary, like most of his steel-industry colleagues, was a staunch advocate of the open shop, but he had inherited a motley set of labor practices. Gary initially tried to caution restraint in reaching a policy decision by taking time to study the labor question, but the Amalgamated Association of Iron and Steel Workers forced his hand by threatening to strike unless the new corporation agreed to recognize the union at all its plants. When the Amalgamated finally did go out on strike in July 1901, it was in the face of a corporate executive committee divided over policy choices. Tarbell recounts that in Gary's mind "it was the worst time of the worst year to have any trouble." Gary, who had advocated a policy of "temporizing," took something of a middle road, believing the corporation should recognize those contracts where they currently existed but not extend them to new mills. On the other hand, the Carnegie-trained steelmen led by Schwab held the attitude that "if a workman sticks up his head, hit it." It was this issue of an appropriate labor policy that led to the executive committee's July 1 resolution to support Gary by requiring the president (Schwab) to be directly responsible to the committee and to its chairman (Gary) between formal meetings.

The Amalgamated strike was generally unpopular — even among steelworkers — shortlived, and a failure, crippling the union. The end result was that U.S. Steel adopted an open-shop policy, recognizing the union only in those few mills where it was already established. Gary remained committed to this position until his death, a position that in part led to the 1919 steel strike for which he has often been criticized. It was Gary's policy not to "confer, negotiate with, or combat labor unions as such. We stand for the open shop." Gary's paternalism led him to believe that the corporation "know[s] what [its] duty is": "We feel obligated . . . in keeping their wages up and in bettering their conditions and keeping them in a position where they enjoy life. We are the ones to lead in this movement." There was "no necessity for labor unions." In contrast to his stance on unions, however, Gary was much more accommodating with respect to other labor issues and policies that fell under the category of welfare capitalism, in which Gary himself believed "the corporation and subsidiaries have endeavored to occupy a leading and advanced position among employers."

Gary's long-term desire that U.S. Steel be perceived as a "good" trust led him, along with George Perkins, chairman of the finance committee after November 1901, to establish an employee bonus and stock-option plan, which became operative in January 1903. Unlike Schwab's later Bethlehem Steel bonus plan that was based on an individual's improvements in efficiency and paid immediately on a biweekly or monthly basis, U.S. Steel's plan was based on the corporation's general progress. A certain percentage of the corporation's profits were set aside and invested in shares of preferred stock, but if the worker was discharged or left the employ of U.S. Steel, the shares were forfeited. This portion of the plan was only open to management, wage earners' incomes being too low to be generally affected. However, the stock option purchase part of the plan was open to all employees. Here employees were entitled to invest in 7 percent preferred stock on an installment basis. With three-quarters of the entire work force earning less than $800 per year and with shares selling at $82.50 each when the plan was announced, it is not surprising that participation never averaged much above 15 percent. And most of the participants were from the group of more highly skilled, better-paid steelworkers. It was precisely this group, however, that was the most susceptible to unionization and to whom the plan was in large part directed. Clearly the plan also served as a public-relations device, but, for those who did act on their option and stuck with it, the plan was profitable.

Other more significant reforms — the establishment of a safety program, an accident relief plan, and a pension plan — affected the rank-and-file workers more directly. In 1906 Gary, with the assistance and prodding of William B. Dickson, who had been Schwab's assistant at Carnegie Steel, began to investigate the question of worker safety and accident compensation. Steelmaking was inherently dangerous, but prior to this time steel companies paid little attention to how costly and debilitating worker deaths and accidents could be. In November 1907 an article by William Hard titled "Making Steel and Killing Men" appeared in *Everybody's* magazine, revealing that 46

men had been killed and some 2,000 others "merely burned, crushed, maimed or disabled" at the corporation's South Chicago plant in 1906 alone. The negative press forced Gary and Dickson to swing into action quickly. By April 1908 they had established a safety committee, and the ensuing safety campaign reduced serious accidents by 43 percent within four years. U.S. Steel's safety program subsequently became a model for other companies industry-wide.

The accident-relief plan evolved out of the safety program and was financed by the corporation without cost to its employees. Adopted in January 1910, the plan covered all workers as long as they stayed with U.S. Steel. Company physicians had to attest to a worker's injuries, and the employee had to agree to waive any legal rights to sue U.S. Steel, but otherwise the plan included free medical treatment, partial pay during recuperation or 6 to 18 months' pay for permanent injury, and benefits to a victim's survivors if a worker was killed on the job. In 1911 a pension plan in existence for the old Carnegie Steel employees was extended to all U.S. Steel employees, providing at least a modicum of retirement security. Rounding out the corporation's welfare reform were improvements in sanitary facilities, corporate housing (which also had the negative effect of tying workers more completely to the company), and expanded community contributions. By the second decade of the century U.S. Steel had thus made major advances in improving its labor relations in the areas of working and safety conditions, accident relief, and pensions. Dealing with the question of reducing the hours of labor, however, was much more difficult.

The twelve-hour, seven-day work week with its infamous "twenty-four hour turn" was typical throughout the steel industry when Gary became head of U.S. Steel, and largely remained so for most of the next decade. At Dickson's urging the finance committee in 1907 approved a reduction in Sunday hours for all but the most necessary blast-furnace workers, a reduction relatively easy for the subsidiaries to accept when conditions were depressed. With the revival of business conditions in 1909, however, Sunday hours for many workers returned. Subsidiaries that sought to increase productivity and workers themselves who, unless they were to receive additional compensation, wanted to maintain their earnings level, indirectly conspired against an easy solution. The 1909 Pittsburgh Survey called public attention to the matter, and Dickson, recently appointed first vice-president, began to press Gary more vigorously for a reduction in the work week. Gary tried to avoid pitting corporate profitability against more humane labor conditions by preaching that "if you are sure of your morals, good policy follows." The 1910 strike at Bethlehem Steel over long hours and a subsequent federal Bureau of Labor Statistics investigation finally forced Gary's hand. In March he prepared a telegram to the subsidiary presidents supporting the 1907 resolution against Sunday labor. Although he was frustrated by the turn of events and angered by Dickson's persistent "meddling" in the affair, Gary, ever the consummate image maker, called in the authors of the Pittsburgh Survey in order to get the most public mileage out of the decision. By year's end less than 5 percent of the corporation's workforce was employed seven days a week.

Reducing the seven-day work week had been hard to achieve, but reducing the twelve-hour day would prove even more difficult. Gary was of the opinion that the eight-hour day was unnecessary and that if U.S. Steel were to reduce hours unilaterally — a move he believed would be fruitless for the workers, who desired twelve hours' wages — then other companies would respond competitively by increasing the length of their work shift. In 1914 Gary claimed the company could not afford to increase wages to compensate for the reduced hours and further noted that since over 50,000 workers were U.S. Steel stockholders their interests as part owners of the corporation had to be consulted as well. Several employee stockholders were "planted" at the 1914 annual meeting to support Gary and the twelve-hour day. U.S. Steel's contemporaneous labor reforms on other fronts also softened public criticism of the twelve-hour day. Only in 1918 did governmental pressure force the steel industry to accept the "basic eight-hour day." Even then, however, wartime demands for steel meant more laborers than ever were working a twelve-hour shift, although they did receive overtime wages for any hours worked beyond the eight-hour day. Gary perceived overtime as merely a scheme to increase wages. At war's end the twelve-hour shift continued to remain in effect, but the overtime wages did not.

The uneasy wartime truce between management and labor quickly unraveled as workers set out to consolidate the gains of reduced hours and union recognition and to bolster wages against inflationary living costs, while management sought to reestablish the open shop. The steel-union organization expanded rapidly and by the spring of 1919 stood at over 100,000 members, but the industry, led by U.S. Steel and Gary, refused to recognize and bargain with the union. Massive national strikes, fueled by postwar disillusionment, engulfed the country that fall, with approximately 250,000 steelworkers — about half the total industry workforce — taking part. By January 1920 the strike was all but over, however. In the end the open shop, labor espionage, importation of strikebreakers, and economic realities crushed the strike, thereby forestalling industry-wide union recognition for another two decades until the passage of the Wagner Act and the coming of World War II.

Other than a maintenance of the traditional open-shop status quo, little was directly gained by either side. Indirectly, however, the strike did lead to increased public pressure for reduced hours, to which Gary was able to evade response until 1923, when, at the strong urging of President Warren G. Harding, the eight-hour day was widely adopted in the steel industry.

Ultimately, Gary's stance on labor issues was a throwback to the paternalistic attitudes of management in the nineteenth century, when industrialists believed they knew better than their workers what was good for them — a managerial prerogative he clearly did not mean to relinquish. In addition, he rationalized his position against reducing hours by claiming it might add as much as 15 percent to U.S. Steel's cost of production to shift to the eight-hour day without reducing the real wages of its workers — and finding additional workers to make up for the reduction would prove difficult. In reality, the cost was about a third of Gary's inflated estimate, and the difficulty of finding workers proved nonexistent. It should be noted, however, that such an increase in cost might well have cut into the price of foreign steel exports, a market Gary had recently struggled to establish and of which U.S. Steel controlled 80 percent as of 1911. In the words of historical analyst Gerald Eggert, "Welfare reform clearly cost far less than hours reform."

Although Gary responded inflexibly to workweek and hours-reduction proposals, he insisted on maintaining wage stability, even in time of economic depression. This seeming inconsistency is more readily understood when viewed as part of Gary's overall approach to establishing industry-wide noncompetitive policies, which included maintaining both stable prices and wages. During the economic downturn of 1907–1909 Gary, with Morgan's support, thus sought to maintain wage levels in the face of declining orders. It should be pointed out, however, that many U.S. Steel employees were not fully employed due to plants running below capacity or being shut down completely in some cases, thereby reducing total labor costs to the corporation.

Gary's long-term strategy of institutionalizing a noncompetitive environment for the steel industry was based on his belief in the need for economically stable conditions and in the importance of avoiding any public perception of U.S. Steel as a monopolistic or "bad" trust. The corporation's control of approximately 50 percent of the nation's iron and steel production enabled it to adopt such a strategy without significantly effecting major hardship on its stockholders. Thus from 1901 on, U.S. Steel established fixed prices for its products and openly announced them in the trade journals. Gary attempted to get other producers to follow suit. Although its competitors would periodically undercut U.S. Steel's prices secretly, thereby garnering business for themselves that might well have gone to the larger firm, the policy had the effect of stabilizing the industry as a whole. It is likely that Bethlehem Steel, for example, would have been unable to weather the economic downturn following the panic of 1907 if U.S. Steel had aggressively cut prices to maintain high levels of output.

Maintaining price and wage stability was relatively easy during periods of economic prosperity, but as soon as a downturn in sales occurred it became difficult for Gary and the finance committee to maintain enthusiasm for their policies even among subsidiary heads. On November 20, 1907, Gary invited the heads of the U.S. Steel subsidiaries and of the independent producers to meet at the first of a series of dinners that have become known as the "Gary dinners." A total of 51 executives representing 90 percent of the industry attended. The object of the meeting, in Gary's words, was "to prevent the demoralization of business, to maintain as far as practicable the stability of business and to prevent, if I could, not by agreement, but by exhortation, the wide and sudden fluctuations of prices which would be injurious to everyone interested in the business of the iron and steel manufacturers." Gary sought to establish a spirit of trust, confidence, and cooperation in stabilizing the relationship between heads of companies in the often volatile industry. Gary recognized that U.S. Steel had "no lawful right . . . to make any agreement express or implied, directly or indirectly, with our competitors in business to maintain prices." Gary's dinners and a series of regularized subcommittees organized by type of iron and steel product, however, did provide a relatively high level of coherence and centrality to the industry with regard to uniform pricing. Gary was very careful never to talk about formal price "fixing," but the "interchange of information" and "understandings" reached did do much to stabilize business, at least until conditions changed in a particular area or product line. When conditions changed the appropriate committee would meet again to reach a new understanding. If too many tended to stray from the path, Gary would call the industry's leaders together for another "exhortation" on the importance of maintaining stability.

Generally, U.S. Steel's competitors recognized that destructive competition would lead to a "war of the survival of the fittest," which they could not hope to win. At the same time, out of fear that the federal government would prosecute them under the Sherman Antitrust Act if they grew too large, U.S. Steel refrained from crushing its competitors. Nonetheless, general economic difficulties, price-cutting by the smaller independents, and pressure from his own subsidiaries finally led Gary to cut prices in February 1909 at the third annual dinner. Although U.S. Steel gained most of the business, there was no rush of new orders; the tide turned much more slowly, helping to reinforce the wisdom of Gary's policy. As conditions gradually improved, Gary refused to raise prices to take advantage of the expanding market. He wanted to teach the competitors a further lesson that they should not undercut U.S. Steel's prices, for only U.S. Steel could run at a profit at the new lower price levels. During the following summer Gary gradually let prices move upward, thereby keeping U.S. Steel's competitors from failing. This middle-ground policy allowed most rivals, including Bethlehem Steel, to weather the depression. By October 1909 prices had fully recovered, and on the fifteenth of the month the industry hosted a banquet in Gary's honor at which he was presented a formal loving cup, somewhat ironically by Schwab. Gary was triumphant. The keynote theme — "The Old Order Giveth Place to the New" — had indeed proved true. Although competition by no means disappeared and Gary continued to rail against "savage" competition and counsel for cooperation, he was recognized now not only as U.S. Steel's head but also as the spokesperson for the entire industry. Two subsequent events consolidated and confirmed this role in early 1910.

The first of these events led to Gary's appointment as chief executive officer of U.S. Steel in March. In the previous November, Gary had requested John C. Greenway, general superintendent of the corporation's Oliver Iron Mining Company, to come to New York to discuss the subsidiary's business without the knowledge of the latter's superior, President William J. Olcott. When Olcott learned of the nature of the meeting, he fired Greenway. Gary countered by insisting on his reinstatement. The subsequent confrontation and power struggle between Gary and the central corporation-oriented lawyer-bankers on the one side and the traditional steel producers represented by Olcott and U.S. Steel president Corey on the other was ultimately resolved in Gary's favor, and Greenway was reinstated. Following this decision Gary requested further clarification of his status. A resolution adopted by the board on March 1, 1910, explained that the chairman was to be "the chief executive officer of the corporation and, subject to the Board of Directors and the Finance Committee, shall be in general charge of the affairs of the corporation." Real power was now consolidated in Gary's hands, a fact that was evident when President Corey resigned on January 3, 1911. In a different time, Dickson, the last of the Carnegie men and first vice-president of U.S. Steel, might have succeeded Corey. Instead, Gary appointed his own man, James A. Farrell, president of the U.S. Steel Export Company.

At the same time that Gary was consolidating his position within U.S. Steel, he moved to expand his informal dinners by establishing the American Iron and Steel Institute (AISI). Based loosely on the British Iron and Steel Institute, the AISI began organizing and held its first formal meeting in May 1910. Gary was elected the first president, a position he held until his death 17 years later. Gary clearly intended the AISI to institutionalize cooperation, information exchange, and, if not price-fixing, certainly price stabilization. "Primarily the Institute was organized, and should be conducted, for the benefit of its members," Gary asserted. "It should result in decided pecuniary advantage to all." Conduct should be founded "on the belief that healthy competition is wise and better than destructive competition." With the formation of the AISI, Gary became the de facto chief spokesman for the entire steel industry. Its leaders turned to Gary for advice and followed his lead in matters relating to government

negotiations, the final major role Gary was to play in heading U.S. Steel.

Throughout his tenure as head of U.S. Steel, Gary was concerned with protecting the corporation from federal antitrust suits; since the corporation's inception Gary had expressed the fear that U.S. Steel might be viewed as monopolistic and in restraint of trade because of its control of approximately 50 percent of the nation's steel output. Gary, therefore, insisted that the corporation's raison d'être was efficient (and profitable) integration, not monopolistic control of the entire industry. He opened the corporation's financial reports and annual meetings to public scrutiny and instituted welfare programs for his employees, all in the name of public openness and responsible operation.

In December 1902 Gary traveled to Washington at the invitation of President Theodore Roosevelt. The two leaders hit it off well at what would prove to be the first of an ongoing series of meetings while Roosevelt was in office. Following the dissolution of the Northern Securities Company in 1904, Congress ordered the Bureau of Corporations to investigate U.S. Steel. Keeping a promise he had made earlier to Roosevelt, Gary threw open the corporation's books and records in the hope of avoiding subsequent action under the Sherman Act. His only concern apparently was that information which might adversely affect stockholders not be given out publicly. Little came of the bureau's investigation over the next several years; U.S. Steel was still in the government's eyes a reasonably clean trust. Although Gary supported Roosevelt in his institution of formal suits against the Tobacco Trust and Standard Oil in 1906, Roosevelt's actions made him particularly cautious the following year during the October Panic when Morgan approached Gary and U.S. Steel with a request to buy the Tennessee Coal and Iron Company (T.C. & I.) to help stabilize the faltering stock market.

The request originated out of the imminent financial demise of the New York brokerage firm of Moore and Schley, which had accepted a large block of T.C. & I. stock as collateral for a loan used to purchase the independent steel producer. Moore and Schley had in turn used the stock for collateral on some $25 million in loans now falling due. With the entire financial community in turmoil and little ready money to be had, Morgan thus turned to U.S. Steel to purchase T.C. & I. and exchange its bonds for the Tennessee stock.

Gary and the U.S. Steel finance committee were favorably disposed toward the acquisition in general, although some members, including former Carnegie steelmaker Henry Clay Frick, believed T.C. & I. would not be a profitable addition. In the end the committee decided to absorb the independent steel company, provided the merger would not provoke an antitrust suit. To settle that concern Gary and Frick embarked on a special train for Washington, where Roosevelt in concert with Secretary of State Elihu Root agreed not to oppose the merger. News of the president's "tacit acquiescence" was leaked, the stock market rallied, and T.C. & I. became part of U.S. Steel.

The purchase of T.C. & I. helped resolve the immediate fiscal crisis but did not end the subsequent depression. It also further strained U.S. Steel in terms of the new company's need for an immediate cash influx and thereby made expansion of the corporation's new, fully integrated plant at Gary, Indiana, somewhat more difficult. The biggest impact of the acquisition, however, was felt when the federal government decided in 1911 to institute formal proceedings against U.S. Steel.

As long as Roosevelt was president, U.S. Steel and the Bureau of Corporations had maintained, in Gary's words, an "understanding that we will not be unnecessarily injured and that use will not be wrongfully charged without an opportunity to show the facts." But President William Howard Taft had a different view, and, following the investigation of a congressional committee headed by Augustus O. Stanley of Kentucky, the government filed formal suit. It was the government's contention that more than a desire to stop the panic underlay the acquisition of T.C. & I. The suit further argued that U.S. Steel wished to absorb a "company that recently assumed a position of potential competition." Gary was furious that the government should consider U.S. Steel in restraint of trade, and, rather than offer a plan for divesting itself of some of its holdings, he devoted himself to defending the corporation.

Two years of testimony from more than 400 witnesses began in 1912 and filled 30 volumes. The U.S. District Court's decision, when finally handed down in June 1915, held that U.S. Steel was not in restraint of trade as defined by the Sherman Act, although the judges agreed that the Gary dinners had been. Since the dinners had been suspended and the period of formal cooperation was over, dissolution of the corporation was deemed too severe a penalty. Not satisfied, the government appealed the case to the Supreme Court in 1917, but no majority opinion was returned. Because of wartime emergencies the case was not reargued until 1919 during the steel strike. When the final decision was handed down in March 1920, U.S. Steel again was found not to be in restraint of trade. Judge Gary's "good trust" and his policies had been vindicated. Although the Federal Trade Commission would subsequently try to prosecute U.S. Steel for engaging in a pricing policy called Pittsburgh Plus — the price of steel plus freight from Pittsburgh, the cheapest place of production — little came of it, for Gary in his capacity as president of the AISI supported doing away with the system, which had, in fact, predated the formation of U.S. Steel.

During the 1920s U.S. Steel and the industry as a whole experienced rapid economic growth. This growth was in part due to expansion within the automobile industry, but other industries, including construction and consumer goods, also contributed to the demand for steel products. Gary, as chairman of both the board and the finance committee, personally continued to run the corporation until his death from cardiac failure on August 15, 1927. After his death, organizational responsibilities were spread more evenly among three men: J. P. Morgan, Jr., as chairman of the board; Myron C. Taylor, as chairman of the finance committee; and Farrell, who continued as president.

Whether one views Gary favorably as savior of the steel industry, as did his earlier, largely sympathetic biographers, or more critically as a monopolist who overcentralized the industry and repressed labor, as have later historians, the Judge clearly was the dominant figure in the industry for 30 years. Through the first quarter of the twentieth century, Gary used his role as head of U.S. Steel and president of AISI to put into place his vision of the large, vertically integrated form and of stable, industrywide prices and cooperation, establishing the framework within which the steel industry would evolve.

Gary's approach to managing U.S. Steel was based on a belief in centralized control as the most direct route to efficient operation, although the company never attained the well-defined managerial hierarchy of other large corporations such as Bethlehem Steel. It took him nearly a decade, but he gradually assumed primary responsibility for overall direction of the corporation. He showed little interest, however, in overseeing day-to-day operations, preferring to leave that to his subordinates. Although one critic's charge that "the first blast furnace Gary ever saw was in the hereafter" was clearly rhetorically overdrawn, it is true that he dealt principally with questions of general policy and long-term planning and finance. That this created tensions within the corporation was reflected in Gary's early struggles with more-traditional steelmen such as Schwab, Corey, and Dickson, but in the end Gary won out, and his policies were vindicated. Even though in terms of market share U.S. Steel lost its dominant position, declining from approximately two-thirds in 1901 to about 40 percent at the time of Gary's death to about one-third during the 1930s, the company continued to expand its capacity, although somewhat unevenly, and remained highly profitable. Although the corporation might have grown even more dramatically had it pursued the more traditional competitive policies of a Carnegie, it also might not have survived the federal government's antitrust actions intact. Gary seems to have been quite willing to sacrifice market share to keep U.S. Steel whole in the face of federal antitrust suits, and, indeed, the outcome of those legal actions seems to have hinged in large part on the company's significant decline in market share since its formation.

In pursuing his vision, Gary was a master of the value of public relations. The disclosure of quarterly financial reports, his open stockholder meetings, the institutionalization of labor reform programs, and the promotion of stable prices and wages were all designed to create an image of U.S. Steel as a "good trust," one that could (and did) stand the test of federal scrutiny. Without denying the value of much of what he accomplished, Gary did rule with something of an iron hand, even in his early years, but certainly after 1911. Perkins's humorous 1909 after-dinner speech on how Gary ran a board meeting has more than just a little ring of truth to it:

> The Directors meet around a big table, the Judge at the end of the table. Before him are heaped great piles of books. The Judge raps for silence and the Directors are still. "The Secretary will read the minutes of the previous meeting," and the Secretary reads. Then the Judge says: "Gentlemen, here are the minutes of the Finance Committee and the minutes of the meeting of the Presidents of the subsidiary companies, the reports of the Controller and the Auditor. They are all open before you for your examination. Has anyone any question to ask? If not, a motion to adjourn is in order. It is moved and seconded that

we adjourn. The yeas have it. The meeting stands adjourned."

Just as Gary brooked little dispute regarding his prerogative at board meetings, neither did he extend any real power to labor, supporting to the end the principle of the "open shop" and only grudgingly accepting the necessity of the eight-hour day. Unionization and collective bargaining would have been a threat to Gary's authority and would have introduced an element of instability into running the industry. Similarly, his use of appointed investigative committees and open stockholders meetings suggested an image of democratic control, but the reality was that Gary generally exercised proxies for at least 95 percent of the voting stock.

Nonetheless, and equally as important, Gary was an honest man in an age not generally known for its industrial morality. Gary's preachments to his executives that industrial practices "are not good because they pay, they pay because they are good" reflect his Methodist and legal backgrounds and did much to bring balance and stability to a previously turbulent nineteenth-century industry as it coalesced into its modern form in the twentieth.

Publications:
"Presidential Address," *American Iron and Steel Institute Yearbook* (1910): 33–42;

Address of Judge Gary to the Presidents of Subsidiary Companies. October 11, 1911 (New York, 1911);

"Presidential Address," *American Iron and Steel Institute Yearbook* (1912): 131–134;

"Presidential Address," *American Iron and Steel Institute Yearbook* (1912): 259–261;

"Presidential Address," *American Iron and Steel Institute Yearbook* (1913): 11–20;

An Address before the American Iron and Steel Institute at Chicago, Ill., Tuesday, October 21st, 1913 (Philadelphia: G. H. Paine, 1913);

President's Address at the Sixth General Meeting, American Iron and Steel Institute, New York, May 22, 1914 (New York, 1914);

"Presidential Address," *American Iron and Steel Institute Yearbook* (1914): 289–300;

Address by Elbert H. Gary, Chairman, United States Steel Corporation, at the Member's Council of the Merchant's Association of New York, January 20, 1915 (New York: Evening Post Job Printing Office, 1915);

Address of the President, Elbert H. Gary ... discussion by Willis L. King ... delivered at the Eighth General Meeting of the American Iron and Steel Institute, New York, May 28, 1915 (New York, 1915);

Address by Elbert H. Gary, President of the American Iron and Steel Institute at Semi-Annual Meeting, Cleveland, October 22, 1915 (New York, 1915);

"Herbert Spencer's Over-Legislation," *Forum*, 54 (November 1915): 596–604;

Co-operation in American Business; A Vital Principle for Procuring Prosperity in the United States by Means of Industrial and Social Efficiency (New York, 1915);

"Presidential Address," *American Iron and Steel Institute Yearbook* (1916): 11–24;

Japan as Viewed by Judge Elbert H. Gary . . . Addresses at New York and Tokyo (New York: J. C. and W. E. Powers Print, 1916);

Address by Elbert H. Gary, President of the American Iron and Steel Institute at Semi-Annual Meeting, St. Louis, October 27, 1916 (New York: Evening Post Job Printing Office, 1916);

Address by Elbert H. Gary, President of the American Iron and Steel Institute at Annual Meeting, New York, May 25, 1917 (New York: Evening Post Job Printing Office, 1917);

"Presidential Address," *American Iron and Steel Institute Yearbook* (1917): 225–235;

"Presidential Address," *American Iron and Steel Institute Yearbook* (1918): 13–24;

Remarks by Elbert H. Gary at a Meeting of Steel Manufacturers, Waldorf-Astoria Hotel, New York, December 9, 1918 (New York, 1918);

"Presidential Address," *American Iron and Steel Institute Yearbook* (1919): 11–22;

Address by Elbert H. Gary, President of the American Iron and Steel Institute at Semi-Annual Meeting, New York, October 24, 1919 (New York, 1919);

"What We Must Do," *Forum*, 62 (December 1919): 601–610;

Address by Elbert H. Gary, President of the American Iron and Steel Institute at Annual Meeting, New York, May 28, 1920 (New York: Evening Post Job Printing Office, 1920);

Address by Elbert H. Gary, President of the American Iron and Steel Institute at Semi-Annual Meeting, New York, October 22, 1920 (New York: Evening Post Job Printing Office, 1920);

Principles and Policies at U.S. Steel Corporation. Statement by Elbert H. Gary, Chairman at Annual Meeting of Stockholders (New York, 1921);

Address by Elbert H. Gary, President of the American Iron and Steel Institute at Annual Meeting, Hotel Commodore, New York, May 27, 1921 (New York, 1921);

"Better Way to Plan Ahead," *System*, 39 (May 1921): 649–653;

Address by Elbert H. Gary, at Banquet of American Iron and Steel Institute, Hotel Commodore, New York, November 18, 1921, Marshall Foch, Guest of Honor (New York, 1921);

Address by Elbert H. Gary, President of the American Iron and Steel Institute at Annual Meeting, New York, May 26, 1922 (New York, 1922);

Address by Elbert H. Gary, President, American Iron and Steel Institute, at Semi-Annual Meeting, New York, October 27, 1922 (New York, 1922);

"Looking Forward by Looking Backward Twenty Years," *System*, 43 (January 1923): 31–33;

Address by Elbert H. Gary, President, American Iron and Steel Institute, at Annual Meeting, Hotel Commodore, New York, May 25, 1923 (New York, 1923);

Twelve-Hour Day: Letters between the President of the United States and American Iron and Steel Institute; also Interview with Elbert H. Gary, Chairman, in Response to Inquiries by Representatives of Twenty-Six Different Newspapers, July 6th, 1923 (New York, 1923);

"Presidential Address," *American Iron and Steel Institute Yearbook* (1923);

"What's Ahead for Business in 1924," *System*, 45 (January 1924): 19–22;

Address by Elbert H. Gary, President, American Iron and Steel Institute, at Annual Meeting, New York, May 23, 1924 (New York: American Iron and Steel Institute, 1924);

"Workers' Participation in Industry," *World's Work*, 48 (June 1924): 197–203;

Address by Elbert H. Gary, President, American Iron and Steel Institute, New York, October 24, 1924: "Pittsburgh Plus" (New York, 1924);

Comments on the Presidential Election . . . at the Request of the Newspapers (N.p., 1924);

"Why I See Prosperity Ahead," *System*, 47 (January 1925): 25–29;

Address by Elbert H. Gary, President, American Iron and Steel Institute, at Annual Meeting, New York, May 22, 1925 (N.p., 1925);

Addresses by Elbert H. Gary, President, American Iron and Steel Institute, at Semi-Annual Meeting, October 23, 1925, Hotel Commodore, New York City (N.p., 1925);

"Conditions that Russia Must Satisfy," *Current History Monthly*, 23 (February 1926): 628–629;

Remarks by Elbert H. Gary, Chairman, United States Steel Corporation at Annual Meeting of Stockholders, April 19, 1926 (N.p., 1926);

Address by Albert H. Gary, President, American Iron and Steel Institute, at Annual Meeting, New York, May 21, 1926 (N.p., 1926);

Remarks by Elbert H. Gary, Chairman, United States Steel Corporation, in Response to Request from Representatives of the Press for a Birthday Interview October 8, 1926 (New York, 1926);

Address by Elbert H. Gary, President, American Iron and Steel Institute, at Semi-Annual Meeting, New York, October 22, 1926 (N.p., 1926);

Cycles versus Common Sense (Philadelphia: Curtis Publishing, 1926);

The Public Be Informed, a Notable Interview Given to Rose C. Fold by Elbert H. Gary (New York: Doubleday, Page, 1926);

"Presidential Address," *American Iron and Steel Institute Yearbook* (1927): 11–18;

"A Review and Some Remarks," *Century*, 114 (May 1927): 32–38;

Notable Paintings by Masters of the English XVIII Century, the Barbizon & Old Dutch Schools, Together with Several Important Examples by XIX Century Artists; Collection of the Estate of the Late Judge Elbert H. Gary sold by Direction of the New York Trust Company, Executor (New York: American Art Association, 1928);

Works of Art, Furniture, Fabrics, Rugs, Bronzes, Sculptures and Chinese Porcelains: Collection of the Estate of the Late Judge Elbert H. Gary, sold by Direction of the New York Trust Company, Executor (American Art Association, 1928).

References:

David Brody, *Labor in Crisis: The Steel Strike in 1919* (Philadelphia: Lippincott, 1965);

Brody, *Steelworkers in America: The Nonunion Era* (Cambridge, Mass.: Harvard University Press, 1960);

Vincent P. Carosso, *The Morgans: Private International Bankers, 1854–1913* (Cambridge, Mass.: Harvard University Press, 1987);

Arundal Cotter, *The Authentic History of the United States Steel Corporation* (New York: Moody Magazine and Book Company, 1916);

Gerald G. Eggert, *Steelmasters and Labor Reform, 1886–1923* (Pittsburgh: University of Pittsburgh Press, 1981);

John A. Garraty, *Right-Hand Man: The Life of George W. Perkins* (New York: Harper, 1960);

Robert Hessen, *Steel Titan: The Life of Charles M. Schwab* (New York: Oxford University Press, 1975);

William T. Hogan, S.J., *An Economic History of the Iron and Steel Industry in the United States*, 5 volumes (Lexington, Mass.: Lexington Books, 1971);

Joseph M. McFadden, "Monopoly in Barbed Wire: The Formation of the American Steel and Wire Company," *Business History Review*, 52 (Winter 1978): 465–489;

Thomas K. McGraw and Forest Reinhardt, "Losing to Win: U.S. Steel's Pricing, Investment Decisions, and Market Share, 1901–1938," *Journal of Economic History*, 49 (September 1989): 593–619;

Maurice Robinson, "The Gary Dinner System: An Experiment in Cooperative Price Stabilization," *Political and Social Science Quarterly*, 7 (September 1929): 137–161;

Ida M. Tarbell, *The Life of Elbert A. Gary: A Story of Steel* (New York: D. Appleton, 1925).

Archives:

Addresses and Statements by E. H. Gary, 8 volumes, was compiled by the Business History Society and is located in the Baker Library, Harvard School of Business Administration.

James Gayley

(October 11, 1855 – February 25, 1920)

by Robert Casey

Henry Ford Museum & Greenfield Village

CAREER: Chemist, Crane Iron Company (1877–1880); superintendent, Missouri Furnace Company (1880–1882); blast-furnace manager, E. & G. Brooke Iron Company (1882–1885); blast-furnace superintendent (1885–1890s); manager, Edgar Thomson Steel Works, Carnegie Brothers & Company, Ltd. (early 1890s–1897); managing director, Carnegie Steel Company (1897–1901); first vice-president, United States Steel Corporation (1901–1908); president, American Ore Reclamation Company (1908–1920); president, Sheffield Iron Corporation (1917–1919).

James Gayley has been called the "Founder of modern American blast-furnace practice." During a career spent largely with Carnegie Steel and the United States Steel Corporation, he was responsible for significant improvements in furnace design, furnace operation, and raw materials handling.

Gayley was born at Lock Haven, Pennsylvania, on October 11, 1885. His father, Samuel, immigrated to Pennsylvania from northern Ireland. Shortly after Gayley's birth, Samuel, a Presbyterian minister, became pastor of a church in West Nottingham, Maryland, where Gayley was reared. He was educated at West Nottingham Academy and Lafayette College, earning a degree in mining engineering from Lafayette in 1876.

Gayley's rise to prominence was rapid. His first job was as a chemist at the Crane Iron Com-

James Gayley

pany in Catasauqua, Pennsylvania. In 1880 he became superintendent of the Missouri Furnace Company in Saint Louis and two years later moved on to the E. & G. Brooke Iron Company in Birdsboro, Pennsylvania, as blast-furnace manager. In 1885, at the age of thirty, he became superintendent of blast furnaces at Carnegie's Edgar Thomson works in Braddock, Pennsylvania. During the next fifteen years Gayley made his reputation as the preeminent blast-furnace operator in America.

When Gayley arrived at Edgar Thomson, Pittsburgh furnace operators had little concern for fuel economy. The proximity of coal mines made coal and coke cheap, and operators concentrated on driving their furnaces hard to get maximum production. Gayley bent his efforts toward reducing coke consumption, increasing iron output per cubic foot of stack volume, and increasing the life of furnace linings.

Gayley introduced numerous innovations while at Edgar Thomson: bronze cooling plates for blast furnace walls; an improved casting apparatus for use with Bessemer converters; charging bins for blast-furnace raw materials; compound condensing engines for blowing blast-furnaces; and the dry blast system for removing moisture from the furnace blast.

In the early 1890s Gayley became manager of the entire Edgar Thomson Steel Works and in 1897 was appointed managing director of Carnegie Steel Company. After Carnegie Steel was absorbed in the 1901 merger that created the United States Steel Corporation, Gayley was promoted to first vice-president in charge of raw materials and transportation at U.S. Steel. During that period he was responsible for the introduction of Hulett ore unloaders and contributed to the design of ore vessels adapted to the use of the Huletts.

By far the most important of Gayley's developments was the dry blast. Ambient moisture in the air blown into blast furnaces had to be heated and vaporized by the coke fueling the furnace. It had long been recognized that eliminating this moisture would allow the furnace to make better use of the coke. Gayley developed a refrigeration system that condensed the moisture, thus drying the air. He began his experiments at Carnegie's Isabella furnaces in 1894, but it was not until 1904 that he was ready to announce his results. In a paper titled "The Application of Dry-Air Blast to the Manufacture of Iron" delivered to the American meeting of the Iron and Steel Institute, Gayley enumerated the advantages of his process. During the 12 months preceding the introduction of the dry blast, production at Isabella averaged 341 tons of iron per day, with a coke consumption of 2,306 pounds per ton of iron. In the following 12 months production rose to 408 tons of iron per day, while coke consumption fell to 1,915 pounds per ton of iron.

In the ensuing years numerous dry blast installations were constructed, both in the United States and abroad, but many firms balked at the large capital investment necessary. Nevertheless, Gayley's efforts served to stimulate other improvements. As *Iron Age* asserted, "If the leading steel interests did not extensively invest in refrigerating plants, they steadily worked to secure the same results without making the considerable investment required for drying air."

In 1908 Gayley resigned from U.S. Steel and soon became president of American Ore Reclamation Company, a firm specializing in the construction of plants for sintering ore fines and flue dust. From 1917 to 1919 Gayley served as president of the Sheffield Iron Corporation, which briefly operated a furnace in Alabama. James Gayley died in New York City on February 25, 1920, and was buried in Colora, Maryland. He was survived by three daughters and his former wife, Julia Thurston Gardiner Gayley, whom he married in 1884 and from whom he was divorced in 1910.

Gayley's awards include honorary doctor of science degrees from the University of Pennsylvania and Lehigh University; the 1908 Elliott Cresson Medal of the Franklin Institute; and the 1913 Perkin Medal from the American Society of Mechanical Engineers. He was a member of the American and British Iron and Steel Institutes and the American Institute of Mining Engineers and a trustee of Lafayette College. In 1902 he donated the Gayley Hall of Chemistry and Metallurgy to Lafayette College.

References:
William T. Hogan, S. J., *An Economic History of the Iron and Steel Industry in the United States,* 5 volumes (Lexington, Mass.: Lexington Books, 1971);
"Death of James Gayley," *Iron Age,* 105 (March 4, 1920): 684 –685;
J. E. Johnson, Jr., *Blast Furnace Construction in America* (New York: McGraw-Hill, 1917);
L. E. Riddle, "Reminiscences of the First Application of Dry Blast," *Blast Furnace & Steel Plant,* 28 (May 5, 1940): 464 –470;
Woodward Iron Company, *Alabama Blast Furnaces* (Woodward, Alabama: Woodward Iron Company, 1940).

Tom M. Girdler

(May 19, 1877 – February 4, 1965)

by Carol Poh Miller

Cleveland, Ohio

CAREER: Sales engineer, Buffalo Forge Company (1901–1902); foreman, superintendent, Oliver Iron & Steel Company (1902–1905); assistant superintendent, Colorado Fuel & Iron Company (1905–1907); superintendent, general superintendent, Atlantic Steel Company (1907–1914); assistant to the general superintendent, assistant general superintendent, general superintendent, Aliquippa Works, Jones & Laughlin Steel Corporation (1914 –1924); general manager of operations (1924 –1926), vice-president of operations and director (1926–1928), president, Jones & Laughlin Steel Corporation (1928–1929); chairman of the board and director (1930–1956), president, Republic Steel Corporation (1930–1937); chairman of the board, Consolidated Vultee Aircraft Corporation (1942–1945).

In his rise to the position of chairman of the Republic Steel Corporation, the nation's third largest steelmaker, Tom Mercer Girdler observed firsthand the evolution of the steel industry from hand to continuous-strip mills, from Bessemer to open-hearth and electric furnaces, from nonunion to union labor, from the 12-hour to the 8-hour day. Organized labor despised him for his tooth-and-nail opposition to unions and for his role in

the bloody "Little Steel" strike of 1937; business saw him as one of the last rugged individualists in the industry, an image that Girdler liked to cultivate.

The third of five children, Tom Mercer Girdler was born on a farm near Sellersburg in rural Silver Creek Township, Clark County, Indiana, the son of Lewis and Elizabeth Mercer Girdler. He was named after his maternal grandfather, Dr. Tom Mercer, a country doctor in southern Indiana who delivered each of the five Girdler children at home. His father, a Civil War veteran and a stern taskmaster, was superintendent of the Falls City cement mill, one of several mills of the Union Cement & Lime Company owned by Girdler's uncle, Dexter Belknap. Girdler's father also raised and trained trotting horses, an avocation Girdler inherited.

As a boy Girdler did chores on the family farm, which provided fodder for the horses and mules used in mill operations. He attended school in a small country schoolhouse in Sellersburg alongside the children of the men who worked in the cement mill or its quarries. In the fall of 1891, when Girdler was fourteen, his family moved to Jeffersonville, Indiana, eight miles away, where they bought a house overlooking the Ohio River

Tom M. Girdler

and Kentucky. There Girdler entered the eighth grade at the Lloyd & Chestnut Public School. He attended Jeffersonville High School for one year, then enrolled in the du Pont Manual Training High School, across the river in Louisville, Kentucky, from which he was graduated in 1896. Each summer the family returned to the farm in Sellersburg, and Girdler worked at his father's cement mill, where he absorbed enduring lessons about management-labor relations.

"Father didn't have unions to contend with, nor did the men," Girdler recalled in his autobiography, *Boot Straps* (1943). Problems on the job were settled "man to man." "As nearly as I can define it, father's policy in dealing with labor was to pay the going rate of wages and give the men treatment at least a little better than they could expect elsewhere." If a man fell sick, he guaranteed their credit at the store, waived the rent for the company-owned house, and saw to it that the doctor came. "There was something feudal in the relationship," Girdler admitted, "something perhaps not to be defended on the Fourth of July.

Nevertheless, I was brought up by a father who believed that the troubles of the men who worked for him were his troubles."

Following his graduation from high school, Tom Girdler taught physics and woodworking at the school in exchange for postgraduate courses. He also found part-time work training stationary steam engineers in small industrial plants in Louisville. His wealthy aunt, Jenny Belknap, offered to finance his college education and in 1897 sent him to Lehigh University in Bethlehem, Pennsylvania. At Lehigh, Girdler was an outstanding student. In studying mechanical engineering, he demonstrated unusual technical aptitude and graduated second in his class. He sang in the glee club and the university choir, acted in plays put on by the Mustard and Cheese Dramatic Association, and was an editor of the semiweekly *Brown and White*. Eugene G. Grace, who would become chairman of the board of Bethlehem Steel Corporation, was a classmate of Girdler's at Lehigh, but it is not known whether the two men were acquaintances at the college.

Girdler graduated from Lehigh in March 1901 with a degree in mechanical engineering. He was expected to return to Indiana and take charge of his uncle's business. A fraternity brother, D'Arcy Wentworth Roper, arranged employment for Girdler with the Buffalo Forge Company, however. Roper was going abroad to open a London office for the company and needed an assistant. Girdler trained for two months, bid goodbye to his family, and left for London, where he shared a room in a boardinghouse with Roper and earned $12.50 a week selling heaters, blowers, fans, and forges in Great Britain, France, and Belgium. After almost a year in London, having grown homesick, Girdler returned home in March 1902 to accept an offer from another college friend, Cadwallader Evans, Jr., to go to work for Oliver Iron & Steel Company, owned by Evans's uncle and located in Pittsburgh.

At Oliver, Girdler was made foreman of the bolt shop at $83.33 a month. Almost a year later, he was appointed superintendent of the nut factory, where he oversaw a largely immigrant work force of more than 200. Girdler spent three years with Oliver. During this period, on November 4, 1903, he was married to Mary Elizabeth ("Bessie") Hayes of Louisville, a childhood sweetheart distantly related to his mother.

The college-educated steelman, something of a novelty at a time when most top executives prided themselves on having started with a wheelbarrow and a shovel, advanced rapidly. "I was a rare specimen in the steel business," Girdler wrote in *Boot Straps,* "for the simple reason that I had a technical education." In 1905 he accepted an offer to run the nut, bolt, and spike mills of the Colorado Fuel & Iron Company (CF&I) in Pueblo, Colorado, for which he earned $175 a month. He remained in Pueblo for two years.

As assistant superintendent of the rail-rolling mill, he had to switch from the day to the night turn (shift) each week and, in his words, "learned that the twelve-hour day of the steel industry was industrial barbarism." His health suffered, his weight dropped from 148 pounds to 125, and he observed the dangerous conditions that caused a dozen or more men to lose their lives each year at the Pueblo plant. At the same time, he learned much about making steel and developed a reverence for it that would stay with him for the rest of his life. In *Boot Straps,* Girdler described tapping an iron furnace in terms befitting a mystical experience:

> This dramatic recovery from raw earth of wonderful riches is, I think, the most splendid frontier of steel. If you ever watch the men make a cast you will understand the almost religious feeling with which any good steel man is bound into our craft. Whoever drills part way into that brick-hard plug of fire clay and then, with a long rod, strikes through to the prison of the terrific liquid feels not only hazard but immense power. Inside the furnace hundreds of tons of stock are pressing down but this is unseen so that the consequent bursting forth with a rush and roar of flame seems to be an escape of a live and angry sun.

Girdler was assistant superintendent of the merchant department when in June 1907 he followed the assistant general manager of CF&I to the Atlanta (later Atlantic) Steel Company in Atlanta, Georgia. There he accepted a job as superintendent of the rolling mills. A year later Girdler at age thirty was promoted to general superintendent of the small but growing steel plant, which employed 400. But Girdler was anxious to return to the East to the big steel centers.

Girdler heard about a vacancy at the Aliquippa Works of the Jones & Laughlin Steel Corporation, located 26 miles below Pittsburgh on the Ohio River. William Larimer Jones, nephew of company founder Benjamin Franklin Jones, hired Girdler as assistant to Fred Hufnagel, general superintendent of the plant. Built on the site of the old Woodlawn Amusement Park, the Aliquippa

Works was seven years old and still building when Girdler arrived in the spring of 1914. His assignment was to run the steel town of Woodlawn (later renamed Aliquippa), whose company-built houses spread over the steep hills overlooking the steelworks. Girdler was soon made assistant general superintendent of the Aliquippa Works. In that position he was the highest-ranking company official living in Woodlawn. Girdler compared his role in the Woodlawn community to that of a "political boss": "There was in Aliquippa, if you please, a benevolent dictatorship" with Girdler, the benevolent dictator, working to make Woodlawn "the best town in the world in which a steelmaker could rear a family." However, there were some who disagreed with Girdler's characterization of Woodlawn as the "perfect company town" and instead referred to the town as "the Siberia of the industry."

The Aliquippa Works consisted of a narrow, crowded strip of blast furnaces, open-hearth furnaces, a Bessemer converter, coke ovens, skelp and tube departments, and a blooming mill; other mills were producing tinplate, rods, wire, and nails. But as Aliquippa boomed, owing to war production, Woodlawn suffered from war shortages, especially fuel, and fell victim to Spanish influenza. Bessie Girdler became ill and died in the summer of 1917; Tom Girdler suffered from chills and what was eventually diagnosed as chronic appendicitis.

At the end of 1920 Fred Hufnagel left Jones & Laughlin, and Girdler succeeded him as general superintendent of the Aliquippa Works. Four years later he was named general manager of operations for Jones & Laughlin and left Aliquippa for a desk job in Pittsburgh; the Aliquippa Works was now a fraction of his responsibilities. In 1926 Girdler became vice-president in charge of operations. When William Larimer Jones died late that year, Girdler was passed over for the presidency, which instead went to Charles A. Fisher. "A lot of people influential in the company thought I ought to be made president," Girdler recalled in *Boot Straps.* The slight gave rise to Girdler's feeling that "there was a ceiling over my head in Jones & Laughlin." Following Fisher's resignation due to poor health, Girdler was named president of Jones & Laughlin in January 1928, but he held the position only until October 1929, when he left. By Girdler's account he resigned after having met with "sharp disagreement" over the prospect of a merger. He had urged a merger with the National Steel and Weirton Steel companies, and Cleveland quired. One or the other of the proposed mergers, Girdler believed, was vital to Jones & Laughlin's future if the company was not to be restricted in its future operations. Jones & Laughlin, however, maintained that Girdler was fired for dealing with Eaton.

At age fifty-two Girdler joined forces with Eaton — who had attempted, unsuccessfully, to recruit Girdler two years earlier — and another Cleveland industrialist, William G. Mather, who were putting together a new Midwest steel company. Monday, October 28, 1929, was Girdler's first day at work in Cleveland; the next day the stock market suffered its worst decline in history. From an office in Cleveland's Union Trust Building, Girdler worked quietly as a member of the advisory board of Continental Shares, Eaton's powerful investment trust, until the agreements forming the basis for the new Republic Steel Corporation were announced on April 8, 1930. The new steel company — the nation's third largest, behind U.S. Steel and Bethlehem — represented the merger of the Republic Iron & Steel Company of Youngstown; Bourne-Fuller Company, a bolt-and-nut concern in Cleveland; Central Alloy Steel Corporation, with plants in Canton, Massillon, and Chicago; and Donner Steel Company of Buffalo. Girdler was named chairman of the board. Elmer T. McCleary, president of the old Republic Iron & Steel Company, became president but died two weeks later, and Girdler assumed the additional responsibilities of president.

With the merger completed during the advent of economic depression, Girdler later admitted that he "could not have picked a worse time to begin organizing a lot of ill-assorted properties, mines, mills, and factories into one big steel company." Many of Republic's subsidiary companies were recognized as liabilities by the banks, which were reluctant to extend credit to Republic. For its first five years the company lost money and existed in what Girdler described as "an industrial hell" — often barely able to meet the payroll. As president of Jones & Laughlin, Girdler's annual salary and bonus in 1929 had totaled $350,000; at Republic, Girdler earned just $129,000 in 1935.

Girdler had persuaded a former Jones & Laughlin colleague, Rufus J. (Jack) Wysor, to help him organize Republic. He also recruited Charles M. White — who had held Girdler's old job as general superintendent of the Aliquippa Works — to join him. Wysor was appointed vice-president of operations and a director and was made presi-

dent in 1937; White was named assistant vice-president of operations, made vice-president of operations in 1935, and succeeded Wysor as president in 1945. Both were young, aggressive steelmen in the Girdler mold. They joined a management team *Fortune* described in 1935 as being "as smart a group of hard-hitting, high-powered executives as the industry can show." Despite its wealth of talent Republic was in the grip of a nationwide economic malaise: the firm's operating rate dropped to 18 percent, and annual sales plummeted from $100 million in 1930 to $48 million in 1932. Republic survived by shutting down operations, laying off employees, cutting costs, and relying on its inventories. Eaton, meanwhile, lost most of his wealth with the onset of the Great Depression and dropped out shortly after Republic's formation.

In 1935, to improve Republic's competitive position, Girdler arranged a merger with the Corrigan, McKinney Steel Company of Cleveland, a modern steel plant with massive ore reserves, and Trucson Steel Company of Youngstown, a leading steel fabricator. With the merger Republic shifted its corporate headquarters from Youngstown to the new Republic Building, part of Cleveland's massive new Terminal Tower development. Economic conditions improved, and by 1936 the corporation enjoyed a net profit of $9.5 million. "We had weathered the depression," Girdler later wrote, "but the great force that seemed bent on our destruction thereafter was the Government of the United States."

First, the Justice Department challenged Republic's acquisition of Corrigan, McKinney as a violation of the Sherman Antitrust Act. After a long and costly battle Republic won this fight. Next, labor problems loomed as John L. Lewis and the Congress of Industrial Organizations (CIO), which had signed a contract with the United States Steel Corporation, demanded that Republic sign also. Girdler adamantly refused to recognize the Steelworkers Organizing Committee (SWOC), pointing to the employee representation plan Republic had instituted a short time before. On May 11, 1937, the CIO called a strike against several independent steel producers, including Republic.

The so-called Little Steel strike of 1937 was bitter and violent. The worst violence occurred at Republic's Chicago plant, where, on May 30, 1937, a confrontation occurred between strikers and police protecting the plant. Police fired on pickets and sympathizers; ten were killed; one hundred others were injured. In a riot at Youngstown the following month, two were killed and thirty-eight injured. Girdler blamed "radical agitation" by leaders of the SWOC for the "Memorial Day massacre" in Chicago and insisted that police had only done their duty in resisting the onrush of picketers. Girdler often claimed that Republic was not opposed to its workers joining labor unions; however, he insisted that the unions had been infiltrated by communists who were bent on overthrowing the state and who coerced workers into joining organized labor movements. "Under no circumstances," Girdler declared, "will we give in and sign a C.I.O. agreement." In the midst of the strike and just two days prior to the bloodshed in Chicago, Girdler was elected president of the American Iron and Steel Institute (AISI), but apparently not without contest. The *Cleveland Plain Dealer* on May 28, 1937, speculated on the reasons for the three-hour, closed-door meeting of institute directors prior to the announcement of Girdler's election, reporting that "His reputation as a fighter was said to have caused concern in some quarters of the industry." Girdler served as AISI president for a single two-year term.

There should be no doubt that Girdler lived up to his reputation. At the end of June 1937 the Senate Committee on Post Offices and Post Roads was investigating charges that strike pickets had interfered with the delivery of U.S. mail to a Republic plant. At one stormy session Girdler called SWOC chairman Philip Murray "a liar"; asserted that Democratic senator Joseph F. Guffey of Pennsylvania "doesn't know what he is talking about"; and branded the CIO "a lawless, communistic and irresponsible organization." In a statement to the Senate Civil Liberties Committee on August 11, 1938, Girdler gave three reasons for his refusal to sign a contract with the CIO:

> First, we were convinced that an overwhelming majority of our employees did not want us to sign. . . . Second, we were convinced that a contract was but the first step toward a closed shop and the check-off, the ultimate goal of the C.I.O. . . . Third, we were convinced that the C.I.O. was not under responsible leadership, and that communistic influences were dominating its activities.

Girdler submitted to the committee, chaired by Senator Robert M. La Follette, Jr., what he purported to be "the records of 550 cases of violence and intimidation by the C.I.O." and urged the committee to "investigate these violations of the civil rights and liberties of American citizens." Later he railed against the La Follette committee, the National Labor Relations Board, and the Democratic National Committee, charging that they "supplied opportunities for the writing of untold columns of left-wing 'copy' to add force to the effort to manufacture a legend that Republic Steel Corporation was an exploiter of labor and that I, Tom Girdler, was a fanatical enemy of laboring men. This was a cold-blooded plot."

The Little Steel strike eventually collapsed. But the CIO continued to organize Republic workers, and Tom Girdler continued to fight the union. By 1942 a majority of Republic Steel workers had signed up with the CIO. Late in 1941 the War Labor Board certified that the CIO represented a majority of Republic Steel workers, and Republic began negotiations with the United Steelworkers of America. On August 11, 1942, five years after the costly strike, Republic signed a contract with the CIO at the order of the War Labor Board. Characteristically, Girdler — who once had said, "I have a little farm with a few apple trees, and before spending the rest of my life dealing with John Lewis I am going to raise apples and potatoes" — left the signing to other Republic executives. Two years later, in a speech to the Cleveland Chamber of Commerce, he continued to decry "collective bargaining by government edict at point of bayonet."

Long after the triumph of organized labor, Girdler held firm in his belief that the vast majority of American workers did not want to belong to unions and that they should not be forced to work in a "closed shop." He never retreated from the position that the government had forced the union on him. Almost no American industrialist of his time had worse press. Perhaps to win the public approval that eluded him, the "tough guy of industry" — as one journalist dubbed him — suspended his feud with the New Deal and organized labor during World War II to accept one of the top production assignments of the war. In 1941, eight days after Pearl Harbor, Girdler received permission from Republic's board of directors to take on the additional job of chairman of the board of Consolidated Aircraft Corporation and Vultee Aircraft, Incorporated. (Vultee owned control of Consolidated, and the two firms soon merged as the Consolidated Vultee Aircraft Corporation.) Girdler, then sixty-four, remained as chairman of Republic, dividing his time between Cleveland and Convair headquarters in San Diego, California.

Girdler had entered aircraft manufacture through his friendship with financier Victor Emanuel, who invited Girdler to join the board of Cord Corporation (later Aviation Manufacturing Corporation) in 1937. At Consolidated Vultee, Girdler introduced mass-production techniques to aircraft manufacture, and under his leadership the firm produced 25 percent of the nation's heavy-bomber output by the end of 1944, the most famous being the B-24 Liberator. Riding his personalized motor scooter, Girdler became a familiar sight on the grounds of the sprawling Consolidated plant. He liked to say that the country's safety depended on good performance by industrialists as much as by admirals and generals.

Girdler resigned as director and chairman of Consolidated Vultee on April 23, 1945, after almost three and a half years of service. "When I accepted this position," he wrote in his letter of resignation, "I made it clear that I would relinquish it when I felt my portion of the job was finished. I believe that time has now come. The defeat of Germany is assured." Girdler served Consolidated "without any increase in his personal income," according to a statement issued by Consolidated's board in accepting Girdler's resignation. Girdler's annual salary with Republic was then $275,000. The sum provoked a mild rebuke from Eleanor Roosevelt in her syndicated column "My Day": "I think [such high-paid executives] must accept the fact that the men who work in their organizations are going to expect to share in the same kind of benefits that the executive receives."

By 1942 Republic's steel output was devoted almost exclusively to war production. Since 1939 the steelmaker had expanded its electric-furnace capacity twelvefold, its output of ship plate and flat-rolled steel fivefold. Part of Republic's increased capacity was financed from the firm's profits, but most of the new facilities were Defense Plant Corporation installations financed by the U.S. government. "If the Rugged Individualist, the uncompromising Industrial Separatist in Girdler is spiritually compromised by the presence of about $150 million worth of government-owned facilities on Republic land," *Fortune* wryly commented in 1942, "he permits no outward signs."

Girdler presided over Republic's postwar expansion and modernization. By 1955 Republic had become the world's largest producer of alloy and light steels. It employed nearly 70,000 workers and was well positioned to tap the postwar boom in automobile production and consumer goods. In June 1955, at age seventy-eight, Girdler

relinquished his duties as Republic's chief executive officer, continuing to serve as chairman. He was succeeded by Charles M. White, president of Republic since 1945. Girdler retired as chairman in August 1956, the most profitable year in company history, Republic having shown a net income of $90.4 million.

Girdler liked to attribute his success to his organizing ability and his formula: "Find the right man for the job and let him run it." By his own definition Girdler was an executive rather than a production man; he formulated general policy and delegated its execution to others. Thomas F. Patton, longtime chief counsel for Republic who later became the company's president, recalled Girdler as a "talented nonconformist," a mentally and physically vigorous man who loved a good argument and hated "stuffed shirts." Girdler's standing within the executive ranks, however, was counterbalanced by the views of the rank and file, which was said to hate and fear him. Many recalled how Girdler whipped Republic into shape in 1930 by ruthlessly firing veteran foremen and mill hands.

The stocky, balding executive, voted "best story teller" in college, was a frequent and popular speaker, often addressing industry groups and educational institutions. A favorite theme of his was the conflict between "free enterprise" and a "collectivist" state in which industry was "regimented." "The country cannot travel on both roads in different directions at once," he told the Association of Iron and Steel Engineers, meeting in Cleveland in September 1938. Girdler criticized the government for deficit spending and flailed against "left-wing planners," government intrusion into the economy, and the principles of the New Deal in general. "Quite clearly," he told a reunion of alumni of Cleveland's Case School of Applied Science in June 1938, "the attempt is being made, whether by impractical humanitarians or destructive schemers, to abolish the system of private enterprise, private property, and private business."

Girdler, an Episcopalian, was married four times. With his first wife, Elizabeth Hayes, he had four children: Jane Belknap (Mrs. Alfred C. Dick); Joseph Hayes, who became general superintendent of the Atlantic Steel Company, the same position Girdler had once held; Mary Elizabeth (Mrs. Harry Davis); and Tom M., Jr., a manager with Republic's Union Drawn Steel Division in Massillon, Ohio. Tom Girdler's second and third marriages, to Clara Astley and Lillian C. Snowden, ended in divorce. In December 1942, a week after the termination of his third marriage, he married Helen R. Brennan, his secretary. Girdler was then sixty-five, his bride thirty-six. The couple resided at Greystone Farm in the Cleveland suburb of Mentor and spent frequent weekends at their horse-breeding farm in Kentucky. Following Girdler's retirement from Republic, they moved to Harleigh, a 13-acre estate near Easton, Maryland. Helen Girdler died in August 1964.

Girdler served as a director of the Goodyear Tire & Rubber Company and the Cleveland & Pittsburgh Railroad and as a trustee of the Case Institute of Technology. In 1955 Lehigh University awarded him an honorary doctor of engineering degree. Among his hobbies were golf, horseback riding, and fox hunting. He belonged to the Chagrin Valley Hunt Club, the Summit Hunt Club, the Kirtland Country Club and Union Club (Cleveland), the Duquesne Club (Pittsburgh), and the Iroquois and Idle Hour country clubs (Lexington).

Girdler died of a heart attack at his country estate in Easton on February 4, 1965 at the age of eighty-seven. He had been ill since a previous attack in August 1964. He was buried in Louisville, Kentucky.

Selected Publications:

"Remarks," *American Iron and Steel Institute Yearbook* (1930): 269–270;

"Remarks," *American Iron and Steel Institute Yearbook* (1934): 33–37;

"Record Steel Demand Approaches," *Steel*, 95 (October 8, 1934): 10–12;

"The Outlook for Steel," *American Iron and Steel Institute Yearbook* (1935): 50–55;

"How to Prevent Unemployment," *Nation's Business* 24 (May 1936): 23–24;

"Problems Confronting the Industry," *American Iron and Steel Institute Yearbook* (1936): 55–64; reprinted in *Iron Age*, 136 (June 4, 1936): 29–30;

"Blazing the Trail to Restored Profits," *Steel*, 99 (October 12, 1936): 122–123;

"Steel and the American Farmer," *Blast Furnace and Steel Plant*, 24 (November 1936): 991;

"South as a Producer and Consumer of Steel," *Manufacturers' Record*, 106 (June 1937): 32+;

"Engineers, Not Reformers, Make Jobs, Girdler Tells Case Men," *Iron Age*, 141 (June 19, 1938): 78;

"The Way to Industrial Peace," *Vital Speeches*, 4 (December 15, 1938): 264–267;

and R. J. Wysor, "Republic Steel Corporation Operates at Record Breaking Capacity," *Blast Furnace and Steel Plant*, 31 (January 1943): 131;

and Boyden Sparkes, *Boot Straps; The Autobiography of Tom M. Girdler* (New York: Scribners, 1943);

"A Half Century in Steel," *American Iron and Steel Institute Yearbook* (1953): 71–79;

"We Can Expect Improvement in Steel Industry," *Commercial and Financial Chronicle*, 179 (May 20, 1954): 2202.

References:

"Gary Medal Awarded to Tom Girdler," *American Iron and Steel Institute Yearbook* (1955): 144–145;

"Typical Steel Man," *Newsweek*, 4 (July 21, 1934): 30.

Alfred Stroup Glossbrenner

(June 6, 1901 –)

by Larry N. Sypolt

West Virginia University

CAREER: Laborer, American Sheet and Tin Plate Company (1930–1931); foreman, South Chicago Works, Illinois Steel Company (1932–1935); assistant superintendent (1935–1936), superintendent, Campbell Works (1936–1942); superintendent, Brier Hill Works (1942–1943); general superintendent, Youngstown District (1943–1947); assistant vice-president (1947–1950), vice-president of operations (1950–1956), president (1956–1965), chief executive officer (1963–1965), chairman of the board, Youngstown Sheet and Tube Company (1965–1967).

Alfred Stroup Glossbrenner followed the typical path of executives who reached the top at the Youngstown Sheet and Tube Company. He spent almost his entire career with the company, where he had an operations and production background. Once in the executive suite, Glossbrenner

Alfred Stroup Glossbrenner

faced difficult challenges related to a failed merger with Bethlehem Steel Corporation and to a changing competitive environment. He sought to restore the company to the position it had held before World War II and during the profitable years of the 1960s labored hard toward this goal.

Alfred Glossbrenner was born in Indianapolis, Indiana, on June 6, 1901, the son of Alfred M. Glossbrenner and Minnie Stroup. He attended the University of Wisconsin and several trade schools. On September 19, 1923, he married Romona Bertram, and the couple had two sons, Alfred Bertram and David Withers.

Glossbrenner's first job was in his father's printing plant in Indianapolis. He then went to work for United States Steel Corporation's American Sheet and Tin Company during 1930–1931 at the company's plant in Gary, Indiana. In 1932 Glossbrenner transferred to the corporation's South Chicago plant as a foreman.

In 1935 Glossbrenner moved to Youngstown Sheet and Tube Company as assistant superintendent of the Campbell Works hot-strip mill. This was the company's newest installation, and assignment to the plant was a sign that Glossbrenner was highly regarded by his new employers. The new mill quickly paid off, helping the company through the Depression. Glossbrenner advanced quickly, becoming superintendent in 1936. He

served in that capacity until 1942, when he was named superintendent of the Brier Hill Works. Glossbrenner continued to climb the corporate ladder at Youngstown; he became the general superintendent of the Youngstown District in 1943 and served as assistant vice-president of the company from 1947 to 1950. From 1950 to 1956 he served as vice-president in charge of operations. Glossbrenner's immediate supervisor, J. L. Mauthe, had also followed this track up through the corporate hierarchy, and when Glossbrenner was elected to the corporation's board in 1953, it was confirmed that he was heir apparent to Mauthe.

Mauthe had become president of Youngstown Sheet and Tube in 1953, and in 1956 he became chairman; Glossbrenner assumed the presidency, holding that post until 1965. During those nine years Glossbrenner continued the efforts of Mauthe to modernize facilities, and capital expenditures for that period totaled more than $608 million. Much of this spending came after 1959, however, when Youngstown Sheet and Tube resumed major improvements that had been delayed in 1954, when a merger with Bethlehem Steel was announced. Afraid to waste money on what might become redundant facilities, the company had put several projects on hold. The Justice Department, however, opposed the merger, resulting in litigation in which a final decision was not reached until December 1958, when the courts upheld the government's objections to the combination and plans for merging were dropped. Glossbrenner then had to make up for lost time. He told *Forbes* in 1965, "I don't know how much the merger attempt hurt us, but it did disrupt our organization and planning." Once the uncertainty was removed, "we got off our seats and started to work like hell."

It was well he did, for *Forbes* reported that in 1960 the company was "in sad shape." Glossbrenner continued the policy of his predecessors and devoted special attention to the Indiana Harbor Works, the company's lowest cost facility. He also installed an array of new finishing mills to broaden the firm's product line, including a six-stand cold reduction mill at Indiana Harbor — the world's first such installation — and a second continuous annealing line and a new galvanizing line at the Campbell Works in Youngstown. In 1960 alone the company spent $102.3 million, a record for a single year. In keeping with this push toward broadening markets, Glossbrenner brought R. E. Williams to Youngstown in 1961 as vice-president for sales. Williams had built a marvelous reputation in this area during a career with the United States Steel Corporation.

The building program continued through the 1960s, and in 1964 construction was started at Indiana Harbor on an 80-inch cold reduction mill, while plans for an 84-inch continuous hot-strip mill were announced. These improvements solidified Youngstown Sheet and Tube's position as a major producer of wide flat-rolled products in the Chicago market. Net income also rose substantially during this period to almost $350 million. During 1964 and 1965 the company recorded its largest profits — more than $50 million each year. *Forbes* hailed Glossbrenner for directing the "comeback" at Sheet and Tube.

Glossbrenner also made a substantial commitment to research, hoping to increase Youngstown's share of the total steel market. Research personnel increased from 20 to 140 between 1960 to 1965; a new research and development center with 80,000 square feet was built in 1964. If there was one problem with Gloss-

brenner's modernization program, however, it was its piecemeal nature. The plants in Youngstown still bore a heavy disadvantage in transportation costs compared to Indiana Harbor, although half of the steel production was still centered in the Mahoning Valley. The improvements did little to alter this basic fact. Furthermore, as *Forbes* noted in 1965, Youngstown had neglected making other improvements, for it was the only major company that had not adopted basic-oxygen furnaces. Glossbrenner bristled at this criticism, declaring, "we'll put them in, but right now there are other things to do." But in the long run the improvements not made at old plants would cost Youngstown dearly.

When Mauthe retired as chairman in 1963, Glossbrenner's title changed to chief executive officer. He held this post until late 1965, when R. E. Williams, executive vice-president since 1963, became president. Glossbrenner moved up to become chairman of the board. Williams's promotion represented a break with Youngstown Sheet and Tube's traditional policy of favoring executives with production experience.

Glossbrenner's responsibilities as chairman also included service on the executive committee of Sheet and Tube; he was a director of the Youngstown Steel Door Company, Dollar Savings and Trust Company, National City Bank of Cleveland, Bell Telephone, and the YMCA. A director of the American Red Cross and Youngstown Community Chest, Glossbrenner also held a seat on the board of the Lykes-Youngstown Corporation, the formal name of Youngstown Sheet and Tube after Lykes acquired the steel company in 1969.

His professional memberships included the American Iron and Steel Institute (AISI) and the Association of Iron and Steel Engineers (AISE). He served as honorary vice-president of the AISI and in 1950 was named president of the AISE.

References:
"Facing a Fight," *Forbes*, 95 (February 1, 1965): 28–29;
William T. Hogan, S.J., *An Economic History of the Iron and Steel Industry in the United States*, 5 volumes (Lexington, Mass.: Lexington Books, 1971).

Edwin Hays Gott

(February 22, 1908 – August 27, 1986)

by Dean Herrin

National Park Service

CAREER: Industrial engineer, Koppers Company, Philadelphia (1929–1936); industrial engineer, Ohio Works and Clairton Works (1937–1941), industrial engineer, assistant division superintendent of maintenance, assistant division superintendent of Central Mills, assistant general superintendent of services, Gary Works (1941–1949), assistant general superintendent, South Works, Chicago (1949), general superintendent, general manager of operations, Youngstown District Works (1951–1956), vice-president of operations (1956–1958), vice-president of production in Steel Producing Divisions (1958–1959), administrative vice-president of central operations (1959), executive vice-president of production (1959–1967), president and chief administrative officer (1967–1968), chairman and chief executive officer (1969–1973), director, United States Steel Corporation (1973–1980).

Edwin H. Gott, a native of Pittsburgh, Pennsylvania, served as president and subsequently as chairman and chief executive officer of the United States Steel Corporation from 1967 to 1973. He was born February 22, 1908, to Leonard and Isabel Dalzell Gott. After graduation in 1929 from Lehigh University with a degree in industrial engineering, Gott found employment in Philadelphia, Pennsylvania, as an industrial engineer with the Koppers Company. He began his long association with U.S. Steel in 1937 as an industrial engineer at the company's Ohio Works outside Youngstown. Gott spent the next 20 years shuttling among U.S. Steel plants in Clairton, Gary, Chicago, and Youngstown, always rising in the company's managerial ranks. Returning to his hometown of Pittsburgh in 1956, Gott successively served as vice-president of the company for steel operations, administrative vice-president for steel and coal operations, and executive vice-president for production. In 1967, a year after being named to the board of directors, Gott was elected to replace Leslie Worthington, who had retired, as president

Edwin Hays Gott

of the company. When U.S. Steel chairman and chief executive officer Roger Blough retired in 1969, Gott was elevated to replace Blough. He retired in 1973, when he reached U.S. Steel's mandatory retirement age of sixty-five.

When Gott assumed the presidency of U.S. Steel in 1967, the company was facing serious competition from both domestic and foreign steel producers. Even though U.S. Steel remained the giant in the industry, its percentage of domestic steel shipments had dropped from 29.3 percent in 1957 to 24 percent in 1966, and the company's sales and earnings were growing more slowly than almost all the other major steel companies'. The company was suffering from decisions made in the 1950s to expand capacity with older technological equipment. U.S. Steel particularly lagged in the production of light, flat-rolled steel for automobiles, appliances, and cans, which was becoming the industry's most profitable product line in the 1960s. From 1964 to 1966, 60 percent of U.S. Steel's steelmaking capacity was devoted to products that accounted for only 40 percent of the industry's shipments. Even though some of U.S. Steel's economic problems were caused by a rise in steel imports and by the expansion of formerly nonintegrated steel companies — which reduced U.S. Steel's market in semifinished steel — more than one industry analyst agreed with *Business Week,* which called the company "a bureaucratic dinosaur" in 1967.

Gott approached these problems on three fronts. Since taking over as vice-president of production in 1958, he had been heavily involved in plans for capital expansion, including the $1.8 billion spending program that was already in full swing when he took over as president of the company. As president and CEO Gott oversaw the adoption of newer technology, such as basic-oxygen furnaces and the newest electric furnaces, the increase of company outlay for improvements, and the creation of new raw material supply networks in Central and South America. Gott also increased U.S. Steel's capacity in light, flat-rolled steel with the erection of a new 84-inch hot-strip mill and a continuous casting unit at the Gary Works near Chicago. Always a production man at heart, Gott told *Forbes* in 1973, "It doesn't do you any good to sell steel if the production boys don't give you enough to fill your orders!"

As Gott increased production, he decreased bureaucracy. As executive vice-president he participated in reorganizing management at U.S. Steel in 1963, consolidating the operating and sales functions of seven divisions into one central operations division. As president, and later as chairman, Gott made early retirements more common and chose not to fill several vacant vice-presidencies. Total employment at U.S. Steel dropped from 205,000 in 1969 to 181,000 by 1972. In interviews Gott refused to discuss the balance of power among the traditional U.S. Steel triumvirate — president, chairman, and CEO, and finance committee head — but he clearly was comfortable with consolidating as much power at the top of the company as possible.

A third approach adopted by Gott to increase profits, and perhaps most important to U.S. Steel in the long term, was to intensify the diversification of U.S. Steel's products. From its beginning the company had been involved in a variety of nonsteel endeavors, but the company's financial commitment to areas such as chemicals, real estate, cement, and titanium increased during Gott's leadership. Symbolic of this incremental move away from steel production was the cover of U.S. Steel's annual report for 1968, which described the company as "A Diversified Producer of Materials and Services." One of Gott's lieutenants told *Business Week* in 1967, "The day is past when any industry is loyal to one material."

Despite Gott's efforts to reverse U.S. Steel's financial decline in the 1960s, the situation worsened in the early 1970s. Despite high production levels, profit margins declined from 4.5 percent in 1969 to 3.1 percent in 1971, and the company's share of the domestic steel market dropped 7 percent between 1969 and 1972. Increased steel imports, intense industry competition, and increased costs associated with plant modernization and the installation of pollution-control equipment all took their toll on U.S. Steel's profit outlook. In the long run plant modernization and corporate streamlining proved less successful than diversification in improving U.S. Steel's financial future.

Gott was described by colleagues as a hard-driving businessman with a passion for obtaining facts and information. He liked to keep a hand in all phases of the company's operation, but he also believed in corporate team play. He was actively involved with local community organizations, such as the Children's Hospital of Pittsburgh, and with the National Boy Scouts of America. He was a member of the American Iron and Steel Institute (AISI) and the Association of Iron and Steel Engineers, as well as many other organizations and social clubs. He served as chairman of the AISI from 1971 to 1973. He died on August 27, 1986, in Pittsburgh.

Publications:
"Eloquence for Change," *Iron and Steel Engineer,* 42 (December 1965): 143–147;

"Technological Software," *Iron and Steel Engineer,* 44 (December 1967): 94–96;

"Don't Sell Steel Short, Gott Urges," *Iron Age,* 207 (May 20, 1971): 53–55;

"Dis-loyal Opposition," *Mining Congress Journal,* 58 (January 1972): 90ff;

References:
"Big Man at Big Steel," *Forbes*, 99 (June 15, 1967): 73;
John Hoerr, *And the Wolf Finally Came: The Decline
of the American Steel Industry* (Pittsburgh: University of Pittsburgh Press, 1988);

William T. Hogan, S.J., *An Economic History of the
Iron and Steel Industry in the United States*, 5 volumes (Lexington, Mass: Lexington Books, 1971);
"New Boss at the Corporation," *Business Week* (July
8, 1967): 140–142.

Eugene G. Grace

(August 27, 1876 – July 25, 1960)

by Elizabeth C. Sholes

Industrial Research Associates

and

Thomas E. Leary

Industrial Research Associates

CAREER: Lifelong career with Bethlehem Steel. Crane operator (1899–1902), superintendent of yards and transportation (1902–1906), general superintendent, Juraga Iron Company subsidiary (1906), general superintendent, Bethlehem Steel Company (1906–1908); general manager and director (1908–1911), vice-president and director, Bethlehem Steel Corporation (1911–1913); president, Bethlehem Steel Company (1913–1916); president (1916–1945), chairman (1945–1957); honorary chairman, Bethlehem Steel Corporation (1957–1960).

Next to Charles Schwab, Eugene Gifford Grace was the most important figure in the growth and development of the Bethlehem Steel Corporation, one of America's premier iron and steel producers. During his career at Bethlehem Steel, which spanned 58 years, Grace managed the company's daily operations and saw Bethlehem grow into the nation's second-largest iron and steel manufacturer.

Grace was born in Goshen, New Jersey, on August 27, 1876, the son of John Wesley and Rebecca Morris Grace. His father was first a ship's captain, then a country storekeeper. The family's origins in America predated the Revolutionary War, with the first paternal ancestor, John Grace, settling in Philadelphia, Pennsylvania, circa 1700.

Grace showed an early interest in mechanical and electrical experiments. He received his preparatory education at Pennington Seminary in New Jersey and entered Lehigh University in Bethlehem, Pennsylvania, in 1895. He majored in elec-

Eugene G. Grace (Hagley Museum and Library)

trical engineering, and legend has it that he wandered through the Bethlehem steel mills seeking and occasionally receiving permission to work there in order to perform experiments for his engineering courses.

Grace was an outstanding student and an excellent athlete. He was captain of the varsity baseball team and was considered good enough to turn professional, an offer he turned down. He graduated from Lehigh in 1899 as class valedictorian with a degree in electrical engineering. Thirty years later Lehigh would confer upon him an honorary Doctor of Engineering as well.

On June 29, 1899, Grace began his career at Bethlehem Steel as an electric crane operator. The man who would later be known as "Mr. Steel" thus began his professional life in a laborer's job that paid less that $1.80 per day. Grace, however, made the most of his position, observing the flow of raw goods and finished products throughout the plant and suggesting means by which to improve materials handling in the yards. After six months he was promoted to the open-hearth department then promoted again in 1902 to superintendent of yards and transportation. In this capacity he met Charles Schwab.

Schwab had begun his own steel career under Andrew Carnegie, rising rapidly through the ranks to become the first president of the United States Steel Corporation. In 1903 a series of disagreements with Elbert Gary, chairman of U.S. Steel, culminated in a major upheaval that cost Schwab both power and money, and he elected to strike out on his own. He bought Bethlehem Steel, a small concern with decrepit works, manufacturing ordnance, rails, and some miscellaneous items. Schwab, however, was determined to build a steel company to rival his old companies, Carnegie Steel and U.S. Steel.

In 1904 Schwab took his first official tour of Bethlehem and encountered young Superintendent Grace, whose responsibilities included, among others, handling the storage of Schwab's personal railroad car. Schwab recognized Grace's talent, quick mind, and ambition as qualities that matched his own and elected to promote Grace's career with the same vigor that Andrew Carnegie had shown Schwab during his days at Carnegie Steel.

By September 1905 Schwab had chosen Grace to become the general superintendent of the Juraga Iron Company, an iron ore mining operation in Santiago, Cuba. Grace quickly assessed the situation, then mechanized operations as thoroughly as possible to save Bethlehem substantial money on every ton of ore. He returned to company headquarters in February 1906 and was appointed general superintendent of Bethlehem Steel Company, charged with constructing the company's new Saucon plant in South Bethlehem. The centerpiece of this new facility was the innovative

mill created to roll wide-flange beams for use in the construction of skyscrapers and other monumental structures. This new structural product was first known as the "grey beam" for its inventor but later was better known as the "Bethlehem beam." By 1910 the beam helped make Bethlehem the nation's leading supplier of commercial structural steel. Schwab energetically promoted the use of the beam in buildings such as the Chicago Opera House and Field Museum, while Grace oversaw the escalating production demand in Bethlehem.

The company's rapid growth led to a major reorganization. Grace was tapped once again for advancement, but he initially rejected Schwab's offer to become company president. Grace instead accepted the position of general manager, supervising all manufacturing, purchasing, and sales. Under his direction, Bethlehem began to expand its holdings, acquiring shipbuilding facilities in California and Massachusetts, a forge in Pennsylvania, and a machine company. The two largest acquisitions were steel plants at Sparrows Point, Maryland, and Steelton, Pennsylvania. By the end of 1916 Bethlehem was on its way to becoming a formidable opponent to U.S. Steel. As the company grew, Grace continued to rise through the company's ranks, becoming vice-president and director in 1911, president of the company in 1913, and president of the corporation in 1916. Grace's influence over the direction and character of the company's growth and operations became increasingly evident despite differences in temperament between him and Schwab. Schwab was the quintessential entrepreneur, the man of vision and financial acumen, with little interest in daily operations. Grace, on the other hand, was more remote and austere, the embodiment of the professional manager. It was Grace who oversaw the daily production, the man who hired, fired, and solved problems.

In 1910 both Bethlehem and Grace faced their first labor-relations test. The company's workforce asked for but did not receive time-and-a-half for overtime or the abolition of overtime outright. Schwab ignored the initial demand, which came from a minority of the workforce in only two machine shops. The strike built momentum, however, and it fell to Grace to cope with the growing unrest. He immediately sent a letter to the city police and officials threatening to hold them liable if the strikers caused any damage, despite the absence of actual violence. The resulting escalation of police presence created exactly the violence Grace feared. The upheaval led Grace and Schwab to muster local businessmen on behalf of the embattled company and against those whom Grace dubbed "outside agitators." The strike was eventually broken through threats and physical coercion, and Grace began a second career as a leading national spokesman against unions and collective bargaining. He also criticized the attempt by the government to establish parameters of worker-management interaction, despite his reliance on governmental control against threats to his company's interests.

When Schwab resigned as president of the corporation in February 1916, Grace succeeded him at the age of thirty-nine. At the time the industry was caught up in a production boom caused by World War I, and Bethlehem's early concentration in armor plate placed the company in the forefront of war manufacture. Other than the large quantities of plate, it also produced gun forgings and structural steel. Bethlehem's newly acquired shipyards in Massachusetts and Maryland became the nation's production leaders during the duration of the conflict.

While Bethlehem was running full tilt to meet wartime demands, Grace was determined to add still more capacity by acquiring still other shipyards, steel plants, and fabricating operations. Between 1915 and 1919, Grace virtually tripled the production of ingots and castings and more than tripled net tonnage. Grace and 15 other executives believed so strongly in Bethlehem's future that they personally endorsed a $15.5 million note to finance this expansion. As direct participation of the United States in World War I became inevitable, the deployment of critical resources became a national priority. Financier Bernard Baruch and U.S. Steel head Judge Elbert H. Gary formed a committee of steelmen to shape production and consumption goals. This committee included both Schwab and Grace.

In 1913 Grace openly admitted that Bethlehem, Carnegie Steel, and Midvale Ordnance — the three firms that produced armor plate — had colluded on defense-oriented pricing for the navy since the contracts were always split among them. Initially, the arrangement satisfied both the navy and the manufacturers. Grace, who was the architect of the plan, argued that competition meant that "I would then know I could not run my plant some years" if another firm got the bid. Instead he preferred the maintenance of the industry-government informal agreements in which the par-

ticipants set a "fair price" and shared in the production of armor.

Despite industry-government cooperation in pricing, Grace was unenthusiastic about government regulation. War labor legislation mandated that companies working exclusively on government contracts were to adhere to the 8-hour day. Bethlehem, however, announced a return to the 10-hour day with 8-hour rates and with no overtime for the difference. A month later in May 1918 a representative from the War Labor Board (WLB) arrived at Bethlehem to quell the disturbance caused by Grace's edict. Grace agreed to pay overtime but steadfastly refused to negotiate with his workers on any but an individual basis. When the WLB directed Grace to negotiate with labor representatives, he flatly refused to comply and rescinded his agreement to pay back overtime. Bethlehem eventually capitulated on the monetary issue, but Grace continued to oppose the right of government to intervene in what he saw as private labor-management relations. After the 1918 armistice he escalated his opposition to the WLB and undermined the wartime agreements, particularly those respecting the establishment of bargaining units. For all of his career Grace would remain an outspoken opponent to government regulation, even when he warmly embraced government assistance in financing the increase of productive capacity and in establishing programs to improve the business climate.

In his vehement articulation of laissez-faire principles, Grace sought to preserve Bethlehem's autonomy over its choices regarding labor relations, yet both government and organized labor applied pressures to which Grace had to respond. Bethlehem had long had a system of incentives and rewards related to individual performance for top managers; Grace benefited handsomely. Despite the unwieldy factors of workforce size and geographic diffusion, Grace vowed that each worker's personal merit and individual accomplishments could be measured and rewarded. He therefore created a program to placate his employees and forestall both union organization and governmental regulation — the Employee Representation Plan (ERP). A paternal system that allowed bargaining and the doling out of benefits, the ERP was considered by industrialists to be very progressive. Unions saw it as nothing more than a sweetheart agreement, or at best a company union. The ERP did nothing to protect Bethlehem from the upheavals of the 1919 strikes, but the plan endured until the Steel Workers Organizing

Committee's (SWOC) labor organizing efforts in the late 1930s. The ERP unquestionably provided many benefits to workers; even though it did not permit adversarial collective bargaining, it established grievance procedures, criteria for the election of employee representatives, a pension plan, paid vacation days, disability benefits, survivor benefits, and even an employee stock purchase plan.

Grace fought outside unionization in other ways. In 1920 he and U.S. Steel officials were accused of refusing to sell steel to any contractor employing union construction workers. They insisted that contractors have all work done by a new group, Iron League Erectors Association (ILEA), a totally open shop created by the steel companies themselves. Mandated use of ILEA erectors added 25 to 35 percent to building costs and led to a lawsuit against the steel producers. Grace stoutly defended Bethlehem's position of upholding the open shop and vowed publicly that he would adhere to this stance even if all construction dried up or if 95 percent of Bethlehem's own employees joined a union. Not even exposure before a governmental subcommittee could make him recant or withdraw the company's mandate to contractors.

With the end of the war Grace again sought mergers and acquisitions with the encouragement of Schwab. Wartime expansion whetted both men's appetites for new directions in company growth, and postwar plans involved capital widening and deepening. Bethlehem put renewed emphasis on backward and forward integration by acquiring coal and iron ore reserves and expanding shipping and steel capacity. Grace told the American Iron and Steel Institute (AISI) — the industry trade organization — that during the war the company had invested $1.65 billion in new plants and properties and in upgrading existing facilities. Plans for the future were stifled, however, by the low return on earnings — 5.5 percent per year — throughout the industry. Grace argued that the industry needed, and was entitled to earn, at least 10 percent, or there would be no incentive for companies to continue to try to raise huge sums of capital. He was not prophetic, however, since the industry never sustained double-digit profits yet continued to grow and expand for another 50 years.

Ever determined to grow, it was in these years that Bethlehem made its first overtures to Youngstown Sheet & Tube, a merger both companies desired but never consummated. Federal regulators considered it monopolistic and blocked the amalgamation, further deepening Grace's antiregulatory sentiments. In 1922, however, Bethlehem acquired Lackawanna Steel, south of Buffalo, then took over Midvale Steel & Ordnance, its old armor-plate ally, in 1923. It also acquired Midvale's Cambria Steel Company of Johnstown, Pennsylvania, which became an important Bethlehem operation. By 1929 the company had nine major steel works, five major shipbuilding and repair sites, and extensive holdings in iron, coal, limestone, and shipping.

Throughout this busy period both Schwab and Grace were active in the AISI. In its early years the organization met semiannually at lavish dinners hosted by Gary. The dinners, held at some of New York City's most elegant hotels, ostensibly were meant to serve as nonpolitical forums for technical and business condition information. Charges were raised and denied that the dinners were actually covers for industry-wide price fixing and market collusion; many of the presentations, however, were political in character and designed to hew concensus and discipline among members with respect to public actions on critical issues.

Grace was one of the most active members of AISI well into the 1950s. He made numerous presentations to the dinner meetings, was a member of the board of directors from 1934 to 1937, and served as AISI president in 1935–1936. In 1934 he was awarded the Gary Medal, named for the institute's founder and longtime president, an honor Grace shared with Schwab.

By the beginning of the Depression, Grace had brought Bethlehem to the status of world competitor. The Bethlehem-Lackawanna-Midvale merger alone created a complex of 46 companies, each a completely integrated operation. By arranging this merger and synchronizing production nationwide, Grace gave Bethlehem a $1 to $3 per ton price advantage in the market. In 1929, despite the onset of the Depression, Bethlehem set an all-time income record; the fourth quarter, which fell after the crash, showed the largest gain.

To reward Grace for a job so extraordinarily well done, Schwab gave his favored employee the usual salary-plus-bonus based on merit. Grace received $1.6 million for his work in 1929. In large part the rich remuneration was a manifestation of bountiful earnings, but stockholders challenged the reward as excessive. Other stockholders brought suit against the new proposed merger with Youngstown on the grounds that Grace's pay and the other bonuses to executives eroded Bethlehem's working capital. Midvale stockholders filed yet another suit against Bethlehem's directors for excessive compensation to Grace and the general front office staff, while still other stockholders were outraged because Bethlehem had not paid a single dividend from 1925 to 1928 but paid Grace and the others an average of $750,000 each year capped by the 1929 largesse. During the lawsuits Grace declared that his bonus was based on 1.5 percent, but of what sum Grace could not reply. Throughout this crisis Bethlehem's public relations director, Ivy Lee, later a Rockefeller intimate, urged full public discussion of the issue, but his advice was rejected. Grace as usual repudiated outside interference in corporate matters, even from his own stockholders. The net result, however, was an end to Schwab's control over bonuses and the creation of a cap on the percentage of earnings anyone — even Grace — could receive.

In the midst of the furor over upper-management compensation, Grace handled the daily operations now considerably affected by the deepening Depression. Despite stockholders' fears, Grace found sufficient capital to continue the corporation's expansion plans. He acquired McClintic-Marshall and Pacific Coast Steel and pursued, still unsuccessfully, the ever-elusive Youngstown merger. New facilities moved Bethlehem away from over-reliance on capital goods production, which was experiencing trouble in the 1930s. Now Bethlehem had excellent sheet-and strip-mill capacity for lighter steel production. Grace also began the construction of another strip mill at Lackawanna despite the poor state of the economy.

Neither Grace nor Schwab believed that the Depression would last. They were convinced that basic industry would both create an economic upswing and benefit from it, if only they could weather the storm. In 1930 Grace spearheaded two major employee benefit programs to help sustain their people and the mills. Grace and Schwab declared that there would be no layoffs and they opened an employee stock-purchase plan for the company's 7 percent preferred stock. Despite his antipathy to unions, Grace was committed to his people. In 1926 he had asserted that the company could not increase its profits by reducing the wages of its working men. That philosophy held firm in the company throughout the Depression years. In 1931, however, they found they could no longer adhere to an absolute no-layoff policy. In June, Grace began a new system in which all work was to be distributed among the employees. The

plan was especially welcome in Lackawanna, where the plant at times ran at only 2 percent capacity. The company promised a minimum living wage to 7,000 permanent employees and guaranteed some employment to every existing employee if only for a few hours a week. In Lackawanna the company also agreed to contribute to general city relief even in aid of those who did not work at the plant. Bethlehem created a new "unemployment department" to supervise the programs and distribute both work and direct aid. The plan carried Bethlehem through three devastating years before recovery began and kept all the mills operating, thereby avoiding the costs of shutdown and start-up. By late 1935 the company's orderbook began improving rapidly, and Bethlehem was ready to meet the increasing demand. Grace's farsighted strategies kept Bethlehem alive and able to capitalize on the rapid fluctuations he predicted would follow the downturn.

The enacted policies of President Franklin D. Roosevelt were met with ambivalence by Grace. On the one hand he vocally denounced the 1933 Securities Act as the cause of private capital shortages he deemed essential for recovery. On the other hand he and Schwab both welcomed the National Recovery Administration (NRA) because it suspended antitrust provisos and allowed companies to collude openly on prices and market shares. It was Grace's ideal business environment come true. In his 1935 presidential address to the AISI, Grace forcefully advocated industry cooperation with the NRA; he and Schwab always supported any government policy that increased industry's autonomy and self-direction. Grace was particularly pleased with another aspect of the National Industrial Recovery Act (NIRA): government support of the creation of a rationing program similar to that operating in wartime. With the AISI organized as the authority on regulating codes governing the steel industry under the NIRA guidelines, Grace became a prominent member of the administrative committee that set the pricing, marketing, and labor policies that covered 95 percent of the steel industry.

Grace's role as president of AISI gave him greater freedom to create steel industry policies than did his position at Bethlehem. In June 1936, for example, he participated in the opening efforts to create a truly international steel cartel. The Earl of Dudley, head of the British Iron and Steel Institute, came to the United States on the *Queen Mary* to gain the support of American companies for participation in the cartel. U.S. steelmakers expressed great "surprise" at his announcement. But an impetus for U.S. participation in the cartel came from the threat posed by European companies dumping cheap steel on American markets. The greatest offenders were German steel companies, one of which, Kloeckner Steel, obtained Public Works Administration (PWA) contracts to supply steel for the Triborough Bridge in New York City. The cartel, however, was to include German as well as French and British firms which would agree to fix prices and market shares to protect themselves from the impact of competition. A further incentive for U.S. involvement, which threatened to run headlong into American regulatory laws, came from changes in the balance of trade caused by the Trade Agreements Act (TAA) of 1934, which lowered tariff barriers to encourage free trade. The law threatened to undermine some domestic industries' entrenched market positions. On the other hand, exports, particularly of steel, comprised only a small portion of U.S. production. Once the TAA went into effect, imports rose 245 percent in the first quarter of 1936 alone.

To encourage the Americans to join the cartel the Earl hosted an elegant dinner aboard the *Queen Mary* that was attended by select steelmen, including Grace. The next day Grace reciprocated. Ultimately only Bethlehem, U.S. Steel, and Republic were invited to join, a fact that could have seriously jeopardized domestic steel operations. The three companies tried to allay the domestic steel industry's fears of unfair advantage by promising to cooperate with the cartel only in the export trade, thereby keeping within the confines of the Webb-Pomerene Act regulating collusion. In addition, the three promised that the cartel's invitation would be submitted to the directors of AISI, where Grace was currently president, before the three corporations actually joined.

Grace's stance toward governmental policies became less cooperative and increasingly resistant as he came to believe that Roosevelt's programs were hindering Bethlehem's private goals. The federal action Grace despised most was the 1935 National Labor Relations Act (NLRA), known more familiarly as the Wagner Act. The NLRA directed companies to accede to unionization if their employees voted to join. This was a red flag to Grace. In his 1936 AISI presidential address he vigorously attacked "excessive federal power" and outlined in 20 points his assertion of the supremacy of private direction and planning over governmental interference and labor unionism.

He also blamed excessive taxation for slowing the nation's recovery from the Depression and claimed that all existing reforms in hours, wages, and employment conditions had come not from government but from "a voluntary recognition of enlightened management." A year earlier he had summed up his entire attitude toward the NLRA in one word — "vicious" — and he stood firm in rejecting the 30-hour bill, social security, and other social reform measures. He claimed that the 1936 Council of Industrial Organizations (CIO) drive at Bethlehem had failed due to the success of Bethlehem's ERP, which, rather than establishing cost-of-living wage increases, revised wages when "costs get out of hand."

It is true that much of Bethlehem's success in recovering from the Depression by late 1935 stemmed from its own strengths and from the careful policies of plant use maintained by Grace and Schwab. Farsighted policies of capital expansion and careful investment coupled with useful ERP benefits and the unemployment assistance programs kept the company alive. The direction of Depression-era investment put Bethlehem into new product lines in sheet, tinplate, and lightweight finished steel aimed at a recovering consumer-goods market. The company's heavy capital goods did well, too. As Detroit recovered, not only did auto production pick up but so did construction. Bethlehem won the construction bid for Chrysler Corporation's new corporate headquarters in Detroit.

The 15-year investment plan Grace spearheaded covered virtually every phase of steel production. The new strip mills at Lackawanna and Sparrows Point quickly proved profitable. Decentralized auto production and a growing market in the South were particularly helpful to Sparrows Point expansion. By 1938 income was at an 8-year high, as were wages, which were 31 percent ahead of the 1929 boom levels. Employees, claimed Grace, had more money in their pay envelopes after 37 hours per week than they had had in 1929 after 48 hours. For these reasons Grace and his fellow steelmen disdained government injunctions and welfare programs. On the surface at least, Bethlehem had forged its own recovery and raised the living standards for all its employees. In the same breath, however, Grace asserted the impossibility of rolling back steel prices despite the return of prosperity. Bethlehem, he said, had made only 6 percent return on investment, so a price reduction of any size was clearly out of the question.

Grace assumed ever more responsibility at Bethlehem as Schwab's health deteriorated. Grace also became Schwab's defender as stockholder challenges to Schwab's competence — and to the size of his salary — were repeatedly raised. In 1938 the board of directors had to muster its proxy control to defeat a stockholder motion to force Schwab to retire. In 1939, however, Schwab died at the age of seventy-seven, leaving the corporation without his vitality and without a chairman. Nevertheless, the company continued apace with Grace at the helm.

Thus it was Grace alone who shepherded Bethlehem through the transition to the union era and oversaw the massive mobilization for war production. He continued to build upon the recovery after 1936 so that by 1939 the company was logging record peacetime orders. Ever the fiscal conservative, Grace proposed, as Schwab had before him, to use surplus income to retire debt. Bethlehem's credit and securities ratings were excellent in part because of this modest approach to indebtedness and to dividend disbursement. As had Schwab, Grace sought to shield the company from the influence of government and organized labor, and he was generally successful.

Another key factor in Bethlehem's financial security came from the network of interlocking relations the company had maintained on its board of directors. In Schwab's day Bethlehem was linked to several important financial circles, including the J. P. Morgan banks and the Rockefeller interests. In this one respect Grace rapidly and completely departed from precedent. By 1940 Bethlehem had only one outside director and soon had none. Grace had himself been a director of the Morgan Guaranty Bank but dropped that position in 1940, choosing to operate Bethlehem as a totally closed corporation, nevertheless seeing it through one of its most profitable periods ever.

The war years of course made enormous demands on Bethlehem's resources and plant capacity. The company's armor-plate and gun forgings were once again in demand; it supplied nearly one-third of the nation's needs. Bethlehem shipyards turned out record numbers of both merchant and combat vessels, a total of 1127 between 1939 and 1945. Those same ship-repair facilities turned around over 35,000 vessels for overhaul or reconversion. The company was ready to meet these demands, and by 1943 it had tripled the employment levels of 1939, occasionally running beyond 100 percent of capacity. Grace's careful amalgamation of resources over the years meant that demands could be met with few bottlenecks and relatively few problems overall.

The production necessities of World War II placed the same pressures on government and industry to cooperate that had existed during World War I and through the Depression. Despite the biting criticism Grace had heaped on the Roosevelt administration in the 1930s, Grace and Roosevelt found themselves allied on at least two occasions. Just prior to the war Bethlehem had been forced to shut down a blast furnace at Lackawanna, New York. The demand for pig iron was intense, so the federal government stepped in and completely financed construction of a new, larger-capacity furnace, which became Bethlehem's property after the war. Roosevelt also tapped Grace to be head of the Controlled Materials Plan that allocated scarce commodities for both military and civilian use. The group was similar to the committee formed by AISI during the 1930s, but this time Grace was the key player in charge of framing and running the national program. For steel, Grace devised a central dispatching center through which materials would be ordered and distributed. The plan worked extremely well during the war and kept commodities moving smoothly for both Bethlehem's production needs and those of other steel producers.

During the war Grace remained president of the corporation; the company functioned without a true chairman to replace Schwab. As president, Grace continued to exercise direct control over operations and, in visiting the company's many sites, was a visible presence. As a symbolic gesture of leadership and wartime sacrifice, Grace reduced his salary by more than $300,000 and expected others to follow suit. He nevertheless remained the country's second-highest salaried executive in 1940, and as the owner of substantial quantities of company stock he reaped the benefits of renewed production and profit that likely offset the salary losses.

Bethlehem and Eugene Grace also had to contend with an increasingly strong, unified, and government-backed union movement. Steel executives viewed the program advocated by SWOC as socialist at best, "bolshevik" at worst. Ironically, in 1923 Schwab had advocated unifying all the nation's steel companies into one giant concern, presumably under his tutelage, but in 1941 when CIO-SWOC head Philip Murray proposed exactly the same thing to Roosevelt, the hackles of steelmen rose. Bethlehem resisted SWOC pressure to unionize even more fervently, but the conditions of wartime production gave unions the extra legitimacy of government sanction. Industry-labor-government alliances on the War Labor Board (WLB) and War Production Board also gave labor a new and public voice.

These conditions were barely tolerated by Grace and his colleagues. In 1942 Bethlehem issued a vehement diatribe against the Marshall Field plan, a proposal to settle the strike between SWOC and the "Little Steel" companies. The strike was mediated by the WLB; a Bethlehem press release denounced the Field plan as a "cleverly camouflaged closed shop; and the public should not be and, as we believe, will not be fooled by it." In placing part of the blame for the Field plan at the door of the White House, the press release castigated President Roosevelt for having not lived up to his 1941 promise to leave closed shops unprotected. The article couched Bethlehem's position in terms of patriotism: "The real question before this country today is national security, not union security. Every man should have the right to work without being under pressure to participate in any particular scheme for any kind of membership security."

Bethlehem further tried to forestall unionization at its facilities by increasing workers' earnings — although that was done through extending hours — and liberalizing vacation plans. The 1941 resolution of the Little Steel strike, however, included an NLRB order to Bethlehem to discontinue its ERP, which the NLRB described as a company union. Bethlehem employees voted to have the steelworkers union represent them — only four years after 96 percent had supported the ERP in favor of the union. The SWOC victory led to the installation of grievance procedures and a shop steward network that did not depend on company benevolence.

Despite having lost a battle to organized labor, Grace enjoyed the increase of company profits brought on by World War II. Before the war began he had predicted Bethlehem was "in a position to be War Baby No. 1 as in the last War, but I can tell you that [we] don't want that kind of business." In gross sales Bethlehem crossed the $1 billion threshold in 1943, an achievement made possible by Grace's expansion and modernization policies as well as by sheer demand. By 1945 Bethlehem's annual capacity was rated at 13 million tons per year.

During this period Grace received recognition for his contributions to the industry. In 1942 he was awarded the Bessemer Gold Medal of the British Iron and Steel Institute in recognition of his role in promoting cooperation between the British and American steel industries and in aiding the British war effort through increased steel production. At home one of Bethlehem's new ore carriers was named for him, and in December 1945 he was elected chairman of the board — the culmination of a 46-year career.

After the war Bethlehem was ready to swing into production to meet the demands of rebuilding for a peacetime economy, and returning troops and a newly mobile society gave a boost to Bethlehem's production. Grace continued to exercise a great deal of control over daily decision making. His insistence on increasing capacity once again prepared the company to meet wartime demands brought on by the Korean War in 1950 without suffering from overcapacity during times of peace.

Grace constantly excoriated President Harry S Truman's labor policies and his threat to nationalize the entire steel industry. Grace feared the loss of control over his enterprise, especially in the pre-Taft-Hartley days when supervisors and managers might have been tempted to unionize. In a climate of labor victories over management, Grace sought to bind his salaried staff to the company by extending the paternalistic carrot once again, this time in the form of a management-only club, the Saucon Country Club, which provided loyal white-collar employees with an outstanding recreational facility. The bait worked.

Grace began to warn of foreign competition and the erosion of American markets due to the "prohibitive" costs the American steel industry confronted in the post–World War II era. He further warned of the collapse of the shipbuilding industry and of the decline of the entire steel industry if nothing were done to curtail high wages, cost-of-living increases, governmental interference — anything that Grace believed could hamper production decisions and erode profits.

But by 1950 various interests saw things differently and called for greater external regulation of the steel industry. One demand was to break up the large firms to encourage competition. A justification for this move was the resurrection of concern over the steel cartel. U.S. Steel, Republic, and Bethlehem were known to have joined in 1938, and it was claimed that they reformulated the alliance in the postwar years. By the early 1950s, public sentiment had grown not only against alliances with former enemy nations but also against monopolies in general. There were accusations that Bethlehem and the other two steel companies had participated in off-the-record meetings in Brussels with German companies. Furthermore, despite promises in the 1930s that cartel members would confine their cooperation to exports, the three companies had set standards for industry marketing and pricing policies that affected 90 percent of U.S. steel production. In a vital postwar economy with apparently limitless opportunity, smaller steel companies and the public looked askance on the notion of restricted market opportunities and fixed prices. The most damning evidence of U.S. participation in the cartel came from documents secured at cartel headquarters in Luxembourg in which the British representative praised U.S. Steel's Benjamin Fairless, Republic's Tom Girdler, and Grace for the "methods of secretly violating the antimonopoly laws of the United States."

Grace was, in addition, a strong voice on behalf of German reindustrialization. He argued that this process could be controlled and directed in nonmilitary ways, and exports to the United States could be limited. The prospect of a reindustrialized Germany did not sit well with many Americans, however, and German imports soon proved to be a critical source of competition to U.S. steel markets.

At the same time the government and other observers were worrying about Bethlehem's ties to foreign steel companies, the House Judiciary Committee was investigating the nature of ties among U.S. corporations that promoted collusion and threatened competition. Direct ties between competing companies had been outlawed for decades, but indirect links among those same competitors could be legally sustained as executives came together on the board of directors of third-party companies. The House investigation revealed that 16 of the top steel firms had ties to Bethlehem's old banking alliance, J. P. Morgan, and thus to each other. There was no illegality in these arrangements, but a corollary investigation of the industry by the Federal Trade Commission (FTC) was more serious. The FTC raised charges of price fixing on steel and railroad freight rates and of agreements on voluntary restraints of trade and facilities expansion. The charges carried serious penalties and caused the industry considerable worry.

Despite all of these headaches, Grace's Bethlehem continued to grow and prosper. The company enjoyed good financial health, sound market opportunities, good credit opportunities, and a warm welcome on Wall Street. Grace pressed to continue expansion, claiming that the nation's capacity would never catch up with demand. He met virtually every expansion target he chose, save the acquisition of or merger with the ever-elusive Youngstown Sheet & Tube.

Bethlehem maintained an important place within the steel oligopoly, and Grace's ability to influence market share, product line, prices, and uniform labor agreements underwrote the company's success. Consequently, Grace elected to downplay the technological innovations entering the postwar production arena and curtailed modernization of existing facilities. Instead he proposed a more exciting "greenfield" development — construction of a brand new steel mill — that extended the company's capital costs. Nevertheless, Grace left his company on solid ground. At the time of his retirement in 1957 most of his and Schwab's dreams were fulfilled and the turmoils that now afflict the steel industry were nowhere on the horizon.

Grace supported various educational institutions and programs and was an important fundraiser for his alma mater, Lehigh University. In 1940 he endowed Lehigh's Grace Hall as a center for student athletic, musical, and social functions. He was a longtime participant in the Lehigh Alumni Association and served as the president of the University's Board of Trustees from 1924 to 1956. Grace also supported Moravian College in Bethlehem. His interest in education led him to create Bethelehem's Loop program in which promising college graduates could enter a rapid management-training course to learn all facets of the company's operations before moving into management.

He was a member of AISI, the American Institute of Mining and Metallurgical Engineers, the Iron and Steel Institute of Great Britain, and the Society of Arts and Sciences. He was also a member of the Conference Board, a national policy planning association. Active within his community, he was involved with the operations of St. Luke's Hospital in Bethlehem.

Although Grace relinquished his interest in baseball at an early age, he retained a lifelong passion for golf. He gained national recognition as an amateur golfer and also for his golf course designs. He was a member of the Golf Architects Association, helping to design holes at various courses around Bethlehem and other locations. He was instrumental in bringing the U.S. Golf Association's Amateur Tournament to the Saucon Country Club in Bethlehem in 1951.

Grace married Marion Brown, daughter of a Bethlehem lumber merchant, in 1902, and they had three children, a daughter and two sons. The couple maintained a winter home in Aiken, South Carolina, where Grace golfed at the Palmetto Club. He also maintained memberships in some of the nation's foremost social clubs, the Maryland, Union, and Links clubs. He built a substantial collection of oil paintings, including works by Gainsborough and Reynolds. His interest in the arts extended to books, and he liked to cultivate an image as a man of letters. In 1947 he paid tribute to his old friend and mentor Charles Schwab in the inaugural AISI "Schwab Memorial Lecture" and wrote the first "official" biography of Schwab. It was later reprinted as a small, separate volume. Grace suffered a series of strokes in 1958 and died in Bethlehem two years later on July 25, 1960.

Publications:

"Manufacture of Ordnance at South Bethlehem," *American Iron & Steel Institute Yearbook* (1912): 172–181;

"Ordnance Manufacturing at the Bethlehem Works," *Iron Trade Review*, 51 (October 31, 1912): 829–831;

"We Are Better Friends Now," *Collier's*, 76 (August 1, 1925): 25;

"Distributed Prosperity," *Saturday Evening Post*, 199 (September 4, 1926): 3–4;

"After-dinner Address," *American Iron & Steel Institute Yearbook* (1926): 241–244;

"After-dinner Address," *American Iron & Steel Institute Yearbook* (1927): 263–265;

"Steel Workers and Consumers Benefit More than Investors," *Iron Trade Review*, 80 (April 21, 1927): 1022–1023, 1030;

"Industry and the Recovery Act," *Scribner's Magazine*, 95 (February 1934): 96–100;

"We Shall Maintain the Open Shop," *Steel*, 94 (May 28, 1934): 12;

"Presidential Address," *American Iron & Steel Institute Yearbook* (1935): 30–44;

"Political Experimentation Retards Recovery," *Iron Age*, 135 (May 30, 1935): 14 –19;

"Continued Experimentation Hinders Industrial Progress," *Steel*, 98 (June 1, 1936): 14+;

"Industry and the Public," *Vital Speeches*, 2 (August 1, 1936): 678–682;

"Industry and the Public," *American Iron & Steel Institute Yearbook* (1936): 30–45;

"Presidential Address," *American Iron & Steel Institute Yearbook* (1937): 29–37;

"Progress in Steel," *Iron Age*, 139 (January 7, 1937): 60–63;

"Workers Can Have the Representation They Want," *Steel*, 100 (May 31, 1937): 15–16+;

"Eugene G. Grace Pictures Steel Industry as on Most Solid Foundation in Years," *Iron Age*, 139 (June 3, 1937): 33–36;

"New Orders Exceeding Company's Capacity," *Iron Age*, 144 (November 2, 1939): 83;

"Reply," *American Iron & Steel Institute Yearbook* (1944): 56–57;

Charles M. Schwab (Bethlehem, Pa., 1947);

"Schwab Memorial Lecture," *American Iron & Steel Institute Yearbook* (1947): 33–66;

A Career, and How to Make Good in It (Bethlehem, Pa.: Bethlehem Steel Co., 1948);

"Award of Gary Medal," *American Iron & Steel Institute Yearbook* (1952): 59–63;

The Doorway to Stabilized Progress; Address to 63rd General Meeting of the American Iron and Steel Institute, May 26, 1955 (Bethlehem, Pa.: Bethlehem Steel Co., 1955).

References:

Robert Hessen, *Steel Titan: The Life of Charles M. Schwab* (New York: Oxford University Press, 1975);

Thomas E. Leary and Elizabeth C. Sholes, *From Fire to Rust: Business, Technology and Work at the Lackawanna Steel Plant, 1899–1983* (Buffalo, N.Y.: Buffalo and Erie County Historical Society, 1987);

John Strohmeyer, *Crisis in Bethlehem: Big Steel's Battle to Survive* (Bethesda, Md.: Adler & Adler, 1986);

Archives:

Papers relating to Grace's life and career are located in the Bethlehem Steel Collection, Hagley Museum and Library, Wilmington, Delaware, and the Bethlehem Steel Historical Collections, Hugh Moore Historical Park and Museums, Inc., Easton, Pennsylvania.

Thomas Carlisle Graham

(January 26, 1927–)

by John A. Heitmann

University of Dayton

CAREER: Various positions (1947–1974); president and chief operating officer (1974–1975), president and chief executive officer, Jones & Laughlin Steel Corporation (1975–1983); vice-chairman and chief operating officer, United States Steel Corporation (later USX Corporation 1983–1986); president, United States Steel division, USX Corporation (1986–1991); chairman and chief executive officer, Washington Steel Corporation (1991–1992); president and chief executive officer, Armco Steel Corporation (1992–   ).

Thomas C. Graham emerged during the mid 1970s as one of the American steel industry's best managers, responsive to the formidable challenges posed by imports and a sluggish national economy. An early advocate of labor-management participant teams (LMPTs), Graham spent much of his career at the Jones and Laughlin Steel Company (J&L) before moving to the ailing United States Steel Corporation — which subsequently became a part of USX — in 1983.

Graham was born in Greensburg, Pennsylvania, on January 26, 1927, and was trained in civil engineering at the University of Louisville, graduating in 1947. From 1947 to 1974 Graham rose through the ranks at J&L after starting as a draftsman in the raw materials division. Along the way he gained valuable experience related to steelmaking technology and production methods, serving at the Star Lake iron mining operation, the Pittsburgh general office, and the Hennepin works.

In 1974 he was elected president and chief operating officer at J&L and aggressively began to raise productivity levels by reducing the number of man-hours required to make a ton of steel. In part this was accomplished by slashing many white- and blue-collar jobs, but most significantly Graham also changed the management style at J&L; once described as "Prussian" in its policies regarding its workforce, J&L under Graham began to stress worker participation in its deci-

Thomas Carlisle Graham holding a press conference, December 28, 1983 (photograph by Joyce Mendelsohn for the Pittsburgh Post-Gazette*)*

sion-making process. Since steelmaking is a batch process employing large furnaces and massive fabrication apparatuses often separated by considerable distances, coordination among foremen and workers is key to successful operations. However, the antagonistic relations between management and labor that were so characteristic of the post–World War II steel industry made the plant a divided workplace, with rigid rules, featherbedding, waste, and inherent inefficiencies. Graham reversed this situation by forcing both mid-level managers and union workers to become more flexible. He brought these two camps together in labor-management participant teams (LMPTs) and cut the unnecessarily large workforce, trading pensions for smaller work crews. By 1983 J&L had 80 LMPTs throughout the corporation. Furthermore, J&L introduced new technology with the ultimate goal of cutting the cost per ton of making steel. Perhaps Graham's single greatest success during the 1970s was in combining two ailing companies, J&L and Youngstown Sheet and Tube in 1978. This union of the seventh- and eighth-largest steel producers in the United States enabled the marginally productive Hennepin, Illinois, J&L processing plant to receive raw steel from Youngstown's Indiana Harbor mill located near Chicago, thus enhancing the efficiencies of two economically troubled facilities.

An impatient man eager to move forward, Graham held views concerning labor which were undoubtedly shaped by his association with John H. Kirkwood, J&L's vice-president of industrial relations. Kirkwood maintained that participation was not merely all "giving on the part of workers" but that the blue-collar force possessed valuable insights and knowledge crucial to the firm's success and ultimate survival. In an industry that for generations was characterized by a wide gulf separating company men from the blue-collar workplace, this participatory management style proved to be revolutionary.

As Graham's reputation grew, he became a valuable commodity within the American steel in-dustry, and in 1983 he was lured away from J&L by U.S. Steel's David Roderick. Graham rapidly shook up U.S. Steel's complacent management structure and its traditional methods of doing business. He aimed to change U.S. Steel's time-honored goal of being an industry price leader, advocating a marketing rather than a production strategy whereby U.S. Steel would no longer attempt to supply every consumer with every possible product. U.S. Steel's top-heavy management was markedly reduced. Under Graham departments were staffed at minimal levels, with contracted help added on a temporary basis only if necessary. With these changes in place, U.S. Steel's break-even point fell from 75 to 60 percent of capacity.

It is difficult to say whether Graham's reforms will have a lasting impact on an industry in transition. Nevertheless, Thomas Graham's methods and philosophy are reflective of the emergence of a new American steel industry, rooted historically in the post–World War II era but seeking new directions to ensure its viability in a competitive global setting.

Publication:
"The Sales Outlook for Steel in 1976," *Iron and Steel Engineer,* 52 (December 1975): 61–62.

Reference:
John P. Hoerr, *And the Wolf Finally Came: The Decline of the American Steel Industry* (Pittsburgh: University of Pittsburgh Press, 1988).

Great Lakes Ore Transportation

by Terry S. Reynolds

Michigan Technological University

By 1900 the basic patterns of ore transportation on the Great Lakes were established. Railroads linked iron-mining regions to ports, mostly on Lake Superior. At these ports railcars dumped their ore into pockets on ore docks, from which the ore was loaded onto boats. The ore then passed through the Sault Sainte Marie Canal, and most was ultimately received by ports on the southern shore of Lake Erie, where mechanical unloaders emptied the boats and loaded the ore onto railroads that carried the ore to steel mills.

Although the flow patterns of Great Lakes ore transportation were established by 1900, the fleet that formed the backbone of the system was still in the throes of change. Between 1900 and 1910 more than 250 new boats were added, and in the same decade steam-powered, steel vessels all but completely displaced wooden sailing vessels. During that period, marine engineers improved the steel ore boat by providing wider hatches and unobstructed holds to speed up loading and unloading. They also steadily increased capacity. The first 475-foot-long boats, with capacities of 5000 to 6000 tons, had appeared in the late 1890s. The ore carriers launched between 1908 and 1910 were approximately 600 feet long and had capacities of 10,000 to 15,000 tons.

After 1910 the pace of change slowed significantly. From 1910 to 1930 only around four new vessels per year entered service. In 1908 the average ore cargo was 8300 tons; the figure for 1929 was only slightly higher — 9500 tons — and the 600-foot vessel had become standard. Even this slow pace of growth ended in 1929. The Depression of the 1930s brought a virtual halt to new construction; only four new bulk carriers were launched between 1930 and 1941. Even the renewed demand for ore during World War II increased construction only slightly, as 20 new vessels were added to the lake fleet between 1941 and 1945.

Neither did the immediate postwar period bring much growth to the fleet. Fearing a postwar recession, most shippers continued to use aging fleets with vessels averaging thirty-two years in age. Between 1945 and 1950 the U.S. bulk fleet on the Great Lakes declined from 335 vessels with a capacity of over 3.1 million tons to 261 units with a capacity of 2.3 million tons. Only the onset of the Korean War caused the Great Lakes fleet to grow again. Between 1950 and 1954 more than 20 new vessels were added, some new and some converted World War II oceangoing vessels. But expansion ended with the truce in Korea, and in 1955 the lake fleet resumed its decline. This decline accelerated after 1959 with the opening of the Saint Lawrence Seaway, which permitted offshore shippers to enter the Great Lakes. No new boats entered the U.S. lake fleet between 1961 and 1972, and the number of units in the fleet continued to drop to around 170 units in 1964 and 140 by the early 1970s. Moreover, by the 1970s the average age of vessels was around forty years.

The size of the U.S. bulk fleet declined, but some of the losses were offset by new technology. The average capacity of ore boats in service in 1950 was only around 11,000 tons; the vessels built in the 1950s were in the 20,000- to 25,000-ton range, the upper limit of docking facilities on the lakes and the locks at Sault Sainte Marie. Shipyards simultaneously made significant improvements in the operating performance of bulk carriers. One major postwar change was in the

boat-engine design. Practically all early-twentieth-century boats used hand-stoked, reciprocating steam engines. Vessels built in the late 1930s and 1940s had new, more efficient uniflow reciprocating engines. Oil-fired steam turbines became the predominant power source in boats constructed in the 1950s and were replaced by diesels after 1960.

In the 1950s the engines of many older lake vessels were improved. Many pre-1930 bulk carriers were refitted in the 1940s and 1950s with mechanical coal stokers, water-tube boilers, oil-fired boilers, steam turbines, or some combination of these. The average speed of lake vessels increased from around 11 to around 16 knots.

The completion of the Poe Lock at the Sault Sainte Marie Canals in 1969 permitted expansion in the size of lake freighters. In 1972 the first of the 1000-foot-long, 105-foot-wide ore boats entered service. They had carrying capacities of over 50,000 tons, nearly twice the previous maximum weight. By 1981, 13 of the 1000-foot boats, all diesel-powered, had entered service. The other major technological change in the post-1970 era was in the self-unloader. Self-unloaders had been used for decades, but not in the ore trade, since direct-shipping ores were difficult to unload by conveyor. The beneficiated, pelletized ores that replaced direct-shipping ores in the 1960s were much stronger and could be unloaded by conveyor, and by 1980 many lake boats had been retrofitted with self-unloaders.

The post–World War II era also brought a significant shift in traditional lake transportation patterns. Declining American reserves of high-grade iron and the opening of the Saint Lawrence Seaway did not deprive Lake Superior ports of their primacy as ore shippers because beneficiation permitted them to shift to low-grade ores. But these developments did eat into this primacy. In 1950 the international trade in ore was miniscule. Vessels coming up the Saint Lawrence contributed less than 1 percent of all ore shipped on the lakes. But by 1970 this figure had increased to 15 percent, while the share of ore shipments from Lake Superior ports had declined from 93 percent to 73 percent.

References:

James P. Berry, *Ships of the Great Lakes, 300 Years of Navigation* (Berkeley, Cal.: Howell-North Books, 1974);

Johan A. Buire, "Barrels to Barrows, Buckets to Belts: 120 Years of Iron Ore Handling on the Great Lakes," *Inland Seas*, 31 (1975): 266–277;

Gary S. Dewar, "Canadian Bulk Construction, 1945–1959," *Inland Seas*, 43 (1987): 24–42;

Dewar, "Canadian Bulk Construction, 1960–1970," *Inland Seas,* 43 (1987): 102–123;

Dewar, "Changes in the Existing Bulk Fleet, 1945–1970," *Inland Seas,* 45 (1989): 95–116;

Dewar, "Conversions of World War II Ships for Great Lakes Service," *Inland Seas,* 41 (1985): 28–45;

Dewar, "Rebuilding the Post-War Bulk Fleet," *Inland Seas,* 42 (1986): 84–101;

Donald A. Gandre, "Recent Changes in the Flow Pattern of Iron Ore on the Great Lakes," *Inland Seas,* 22 (1971): 247–259;

Bela Gold and others, *Technological Progress and Industrial Leadership: The Growth of the U.S. Steel Industry, 1900–1970* (Lexington, Mass.: Lexington Books, 1984);

William T. Hogan, S.J., *An Economic History of the Iron and Steel Industry in the United States,* 5 volumes (Lexington, Mass.: Lexington Books, 1971);

Karl Kuttruff, R. E. Lee, and D. T. Glick, *Ships of the Great Lakes: A Pictorial History* (Detroit, Mich.: Wayne State University Press, 1976);

Gerald Manners, *The Changing World Market for Iron Ore, 1950–1980* (Baltimore: Johns Hopkins University Press for Resources for the Future, 1971);

G. G. Skelley, "Recent Developments in Bulk Ore Carriers," *Skillings' Mining Review,* 40 (November 24, 1951), 1–2, 6;

Fred D. Vines, "Vessel Transportation of Iron Ore," *Skillings' Mining Review,* 50 (March 25, 1961): 4–7.

The Grey Mill

by Lance E. Metz

Canal Museum, Easton, Pennsylvania

The success of the Grey Mill was largely responsible for the growth of the Bethlehem Steel Corporation during the early decades of the twentieth century. Designed to produce continuously rolled wide-flange structural steel beams and columns, the Grey Mill helped to revolutionize the American construction industry by making possible the creation of bridges of longer span and buildings of greater height.

The Grey Mill was the product of a creative British engineer, Henry Grey, who was born in 1849 and immigrated to the United States in 1870. By 1880 he was the superintendent of the Cleveland Rolling Mills of Newburgh, Ohio, where he began experimental efforts to roll structural beams directly from ingots. The existing method of fabricating beams was expensive, for they had to be riveted together from half sections. In 1896, Henry Grey took a position with the Ironton Structural Steel Company in Duluth, Minnesota. At Ironton, Grey and his associate George W. Burrell continued the experiments, basing their efforts on earlier work conducted at Duluth by James and Levi York.

Within a year, Grey had developed a mill that could roll wide-flange beams and columns. To better market his mill, Grey moved to New York City and published an article describing his achievement in the June 17, 1897, issue of the *Iron Age.* Grey stressed that his wide-flange beam possessed greater strength and rigidity with less weight and production costs than did conventional fabricated beams. But not until 1902 did a steel company make use of Grey's process. In that year, Differdingen Steel Works in Luxembourg placed a Grey Mill in operation. Earlier Grey had offered his process to the recently formed United States Steel Corporation, but that firm's conservative finance committee rejected his proposal. However, Charles M. Schwab was more receptive and had earlier instigated Grey's presentation to U.S. Steel during his brief tenure as U.S. Steel's president.

When Schwab organized the Bethlehem Steel Corporation in 1904 by combining the Bethlehem Iron Company with the remnants of the U.S. Shipbuilding Company, he realized that the new concern would need a diversified domestic product line to complement its successful military sales. In 1905 he inspected the Differdingen installation and became convinced that Grey's process was a success. Early in 1906 he committed Bethlehem to build a Grey Mill in the United States. Henry Grey took an active role in the design of the Bethlehem Grey Mill. He received a fee of $83,000 for his services. In addition, he received royalties on the mill's product.

Eugene Grace supervised construction of the Bethlehem Grey Mill. Grace was confronted with a monumental task since the Grey Mill was the centerpiece of an entirely new division of the Bethlehem Plant that ultimately cost nearly $5 million. Despite a lack of working capital, Grace successfully completed his task and Bethlehem's Grey Mill in January 1908. To finance its completion and early operation Charles M. Schwab created a novel financial arrangement in which Bethlehem's supplier and principal customers largely subsidized the new mill. Additional capital to maintain its operation was provided by bank loans secured by Andrew Carnegie.

The first buildings to incorporate continuously rolled wide-flange beams and columns from the Bethlehem Grey Mill were the New York State Education Building at Albany and a sugar refinery in Boston, Massachusetts. However, it was the inclusion of 12,000 tons of Bethlehem Grey beams in the construction of New York's Gimbel's Department Store that first made this product nationally known. Within the next few years, Bethlehem Grey beams and columns became the standard in the construction industry.

The profits generated by the products of the Bethlehem Grey Mill helped to propel Bethlehem to a position as America's second-largest steel producer. It had no competitors as a producer of continuously rolled wide-flange beams and columns until 1927, when U.S. Steel built a Grey-type mill at its plant in Homestead, Pennsylvania. This installation, however, was in violation of Grey's patent rights which Bethlehem controlled, and U.S. Steel was forced to pay Bethlehem royalties for each beam that its mill produced.

Henry Grey died in 1913, and the original Grey Mill, although it had long since been joined by others, remained in operation at the Bethlehem Plant of the Bethlehem Steel Corporation until April 1988.

References:

Henry Grey, "A New Form of Structural Steel," *Iron Age* (June 17, 1897): 14;

Henry Grey, "A New Process for Rolling Structural Steel Shapes," *Engineering News,* 46 (November 21, 1901): 387.

Robert Hessen, *Steel Titan: The Life of Charles M. Schwab* (New York: Oxford University Press, 1975);

"Obituary of Henry Grey," *Iron Age* (May 1913): 1147.

Archives:

The Grey Mill File is located at the Bethlehem Steel Collection, Hugh Moore Historical Park and Museums Incorporated, Easton, Pennsylvania.

Samuel Everett Hackett

(July 21, 1877 –?)

by John A. Heitmann

University of Dayton

CAREER: Clerk, American Tin Plate Company (1898); clerk, Republic Iron & Steel Company (1899); purchasing agent, Joseph F. Ryerson & Son (1899–1916); office and warehouse manager (1916–1919), general manager of sales (1919–1923), vice-president (1923–1934), president, Jones & Laughlin Steel Corporation (1934–1938).

Samuel Everett Hackett was president of Jones and Laughlin Steel Corporation (J&L) from 1934 to 1938, one of the most turbulent periods in the firm's long history. Hackett was born on July 21, 1877, in Coralville, Iowa, and after completing high school and attending business college he began his career in 1898 as a clerk for the American Tin Plate Company in Chicago. In 1899, after a brief stint with Republic Iron & Steel Company, he gained employment with Joseph F. Ryerson & Son, where he rose to become manager of the order department and a purchasing agent. In 1916, Hackett left Ryerson & Son for J&L, where he held positions of increasing responsibility, including manager of the Chicago branch office and warehouse from 1916 to 1919 and general manager of sales in Pittsburgh from 1919 to 1923. In 1923 Hackett was chosen to be a J&L vice-president, and with the resignation of George Gordon Crawford in 1934 Hackett became president of the company.

The years immediately preceding the Great Depression were characterized by prosperity and expansion for J&L, and a significant number of improvements were made at facilities located in Pittsburgh and Aliquippa, Pennsylvania. Net income rose from $15.1 million in 1925 to $20.8 million in 1929, while the shipment of rolled steel products increased more than 300,000 tons during the same period. With the 1929 stock market crash, however, income and sales plummeted, and J&L's financial position was exacerbated by its lack of capacity in light flat-rolled materials, the steel commodity most in demand during the bleak years of the early 1930s. To rectify this situation, J&L, under Hackett's leadership, installed in its Pittsburgh works a 96-inch continuous hot-strip mill along with two high-speed tandem mills capable of cold-rolling sheet steel. This capital expenditure of nearly $24 million represented the largest single investment in J&L's history to that time and thus was a momentous decision for a firm that had reduced work hours and wages. With the exception of the disastrous year of 1938, the late 1930s were a time of steady financial improvement at J&L. But labor unrest shook the company to its roots and ultimately ruined Samuel Everett Hackett.

For some time labor had been trying to organize J&L, indeed the entire steel industry, and the company challenged this intrusion of its autonomy by using spies, firing suspected organizers, and generally making life uncomfortable for union sympathizers, especially in the company town of Aliquippa. The dismissal of ten union organizers, members of the Amalgamated Association of Iron, Steel and Tin Workers of America, led to a landmark court case in 1937 that tested the constitutionality of the National Labor Relations Act of 1935. The decision reinstated the workers and validated the union's organizing efforts but ultimately divided J&L into the two hostile camps of management and labor.

With labor relations in disarray and the company facing a loss of more than $5 million in 1938, Hackett resigned. His vacancy was filled by Wales-born Horace Edgar Lewis, who like Hackett had scaled the corporate ladder, but from the mills rather than the office. Furthermore, Lewis faced a much different set of challenges toward the end of the 1930s than did his predecessor. With a war in Europe seemingly imminent, J&L orders began to pick up, and net income rose significantly in 1939–1940. With the troubled Hackett years now behind it, J&L prospered during World War II while running at full capacity and under Lewis's leadership aggressively pursued the big three in the American steel industry into the postwar era.

Publication:
"Bright Future Assured for Steel Market," *Iron Trade Review,* 64 (May 22, 1919): 1,370, 1,372.

References:
Richard C. Cortner, *The Jones and Laughlin Case* (New York: Knopf, 1970);
William T. Hogan, S.J., *An Economic History of the Iron and Steel Industry in the United States,* 5 volumes (Lexington, Mass.: Lexington Books, 1971);
Rose M. Stein, "Aliquippa Celebrates," *Nation,* 144 (April 24, 1937): 466.

Edward J. Hanley

(February 27, 1903 – March 13, 1982)

by Michael Santos

Lynchburg College

CAREER: General Electric Company (1927–1936); secretary, Allegheny Steel Company, (1936–1941); treasurer (1941–1946), financial vice-president (1946–1949), director (1947–1982), executive vice-president (1949–1951), president (1951–1967), chairman of the board of directors (1962–1972), chairman, finance committee, Allegheny-Ludlum Steel Corporation (1972–1982).

As president of the Allegheny-Ludlum Steel Corporation, Edward James Hanley not only helped that company continue its role as leader in the production of specialty steels after World War II, but made the firm a model for effective personnel relations as well. Indeed, Hanley's managerial style and personal philosophy fit well at Allegheny-Ludlum, a company long known for its innovation and emphasis on quality of product.

There was little in Hanley's early background that suggested the course his career would take. Born on February 27, 1903, in Whitman, Massachusetts, the son of Francis J. and Mary Ellen Hanley, he grew up in a fairly well-to-do middle-class home. His father was a doctor who numbered among his patients the family of the future cardinal Francis Spellman. Hanley was afforded many opportunities as a boy, including access to a quality education. Graduating from Phillips Academy in Andover, Massachusetts, in 1920, Hanley went on to the Massachusetts Institute of Technology, where he received a degree in mechanical engineering in 1924. In 1927 he graduated from Harvard Business School with a master's degree and in that same year took a job with the General Electric Company at its Schenectady, New York, facility. His education completed and his career launched, Hanley married Dorothy Ward in 1930.

In 1936 Hanley became interested in a position with the Allegheny Steel Company. Visiting Pittsburgh on March 13, 1936, for an interview, Hanley was met by the company's treasurer and driven to Allegheny's principal mill at Brack-

Edward J. Hanley

enridge. It was, as Hanley later recalled, "a dismal day, the water [from the rivers] was rising rapidly." Having decided to accept the job offer, Hanley and his wife moved to Brackenridge, where his office was located. If the Saint Patrick's Day flood gave Hanley a negative first impression of western Pennsylvania, springtime did little to change it. It was, as Hanley noted, "a dismal looking country . . . We looked at a couple of places that looked like they hadn't been painted in I don't know how long. Of course, the topography is such that you have a house here, and another one will be up higher. In one of the houses we looked at, the smoke from its chimneys would be just about level with the first floor of the one above it. Paint deteriorated so fast with the smoke that was around at the time."

From this rather inauspicious beginning, Hanley's career with Allegheny Steel advanced quickly. As secretary of the company, he helped negotiate the merger between Allegheny Steel and Ludlum Steel Company in 1938, making the new company the largest producer of stainless and specialty steel in the world. Over the next thirteen years Hanley moved up the corporate ladder steadily, becoming treasurer in 1941, financial vice-president in 1946, and executive vice-president in 1949. In 1947 he became a member of Allegheny-Ludlum's board of directors. During these years Hanley learned the innovative management style that set Allegheny-Ludlum apart from other steel producers. In particular, Allegheny-Ludlum's attention to research and development and personnel relations gave Hanley the background he needed to assume the presidency of the firm in 1951.

Indeed, in 1947 Hanley was a part of the management team that introduced a unique concept in labor-management relations. Concerned by the results of a survey taken during the 1946 steel strike that showed both steelworkers and the general public had an overall negative opinion of the steel industry and its management, Allegheny-Ludlum's top executives launched a public-relations campaign designed to communicate four basic principles to the company's employees, their friends and families, and community leaders: "(1) We have a genuine concern for the welfare of our employees and the communities in which they live; (2) Our wages are good; (3) Our profits are fair; and (4) Continuous price increases which continuous wage increases might make necessary would endanger the security of the company, its employees, their communities, and the nation." What made this program significant was the involvement of the firm's entire management, from the president down to the foremen, in a concerted effort to improve management-labor relations and make employees feel that they "belonged."

As Hiland G. Batcheller, president of Allegheny-Ludlum, wrote, "There is no doubt that mass production techniques have revolutionized American industry. . . . We are proud of the products, but we are ashamed of the by-products — such as the transformation of labor from craftsmanship to rote, from pride of achievement to listless nonentity. . . . this starvation of the worker ego is a prime factor in labor's dissatisfaction, dis-

sension, and willingness to be led into slowdowns and strikes. . . . There was a time in America when the boss owned the works and ran them. He knew his workers personally, and their problems. They knew him, and they knew his business. They had real pride in their jobs and their company, and got deep satisfaction from their craftsmanship. The crux, then, is to make it possible once again for the worker to feel he is a part — a necessary part — of his company and to obtain the satisfaction of accomplishment from his labor."

In adopting Batcheller's philosophy, Hanley expanded and modified the company's labor policy to meet new challenges after he became president. Experiencing a bitter strike in 1952 over workforce changes caused by facility improvements, Allegheny-Ludlum was strikebound weeks after the steel companies had reached an agreement with the United Steelworkers of America. Hanley and other top Allegheny-Ludlum officials were convinced that the problem was caused by inadequate communications. Workers were striking, as Batcheller had noted earlier, because they had not been made to feel a part of the decisions that were redefining their work and their lives. To remedy the problem, management proposed a one-day general meeting at a central location to allow both sides to air problems and discuss solutions. The meetings quickly degenerated into grievance sessions and seldom dealt with specifics of local plant issues. Realizing this, both sides worked to modify the format, resulting in a three-day semiannual meeting at sites as diverse as Milwaukee, Pittsburgh, Fort Lauderdale, and Las Vegas. During the meeting guest speakers addressed issues of concern to the industry, plant tours were held, separate plant-level meetings between union and management officials were scheduled, and on the third day a combined meeting of both sides discussed issues not resolved in the plant-level meetings or issues with company-wide implications.

For Hanley, the program was a source of personal and company pride. Commenting to an *Iron Age* reporter in 1954, Hanley noted that the company's investment of $96 million in expansion and modernization of production facilities over the past nine years was only part of the reason for Allegheny-Ludlum's postwar success. He emphasized that the company's attention to the "so-called intangibles" of running a steel company had a lot to do with Allegheny-Ludlum's position in the industry. Hanley accounted for improved productivity by pointing to better labor relations, quarterly management sessions where supervisors at all levels got a chance to talk things out with the boss on any subject they wanted to raise, better community relations, and attention to accident prevention. At the time the article was written, the Brackenridge plant had logged 3 million man-hours without a lost-time accident, while other Allegheny-Ludlum plants had exceeded 2 million hours without a lost-time mishap.

The emphasis on communication and what one observer called "personal, elbow rubbing contact," worked to promote genuine respect and friendly relations between company and union officials. According to William J. Hart, district 19 director of the USWA, "Ed Hanley's a gentleman. He's a man with a feeling for people." Hanley took the need for communication and openness with his employees seriously. In 1963, for example, the company's Dunkirk and Watervliet, New York, plants were operating at a loss because of problems with crew sizes and incentive programs. Hanley took the problem to the union. As Hart recalled, "You know what companies do in situations like that. They close down. Ed Hanley didn't do that. He called us in. He explained the problem. He asked for our help." The result was that union and management cooperated on a successful rehabilitation effort to make both mills profitable.

Under Hanley's administration Allegheny-Ludlum spent $130 million on expansion and developed several new types and forms of stainless steel. In 1957 the firm introduced a punch-card control system to insure uniform product quality at its Brackenridge plant. In 1962 the research and development center at Brackenridge was expanded to cut the time between experimentation and implementation of new technology. In the ten years since Hanley had assumed the presidency, the company increased its spending on research by six times. Allegheny-Ludlum began using computers and programmed controls to cut costs and increase product consistency and in 1964 announced plans to invest $80 million over the next five years to further modernize existing facilities and build a new melting complex at Natrona, Pennsylvania.

By this time Hanley had been chairman of the board of directors for two years, and Allegheny-Ludlum had become an international corporation, dedicating a plant at Genk, Belgium, on November 7, 1963. Before retiring as president in

1967, Hanley helped found the Titanium Metals Corporation as an affiliate of Allegheny-Ludlum and the National Lead Company. Titanium Metals Corporation became the largest titanium producer in the world.

For Hanley, Allegheny-Ludlum's growth was intimately tied to the community's development. Reflecting the firm's emphasis on the importance of community relations, Hanley became a member of the Allegheny Conference and Pennsylvania Economy League, joining with other business, civic, and political leaders to encourage an urban renaissance in Pittsburgh. Hanley retired as chairman of the board in 1972 but continued to serve as a director of Allegheny-Ludlum, as well as several other corporations, until his death on March 13, 1982, at Presbyterian University Hospital in Pittsburgh.

Publications:
"Allegheny Takes on a New Look," *Iron Age,* 174 (October 7, 1954): 71;
A Steelman Speaks: Addresses by Edward J. Hanley, President, Allegheny Ludlum Steel Corporation (Pittsburgh: Allegheny-Ludlum Steel Corporation, 1961);
"Review of Specialty Steel Industry," *Iron and Steel Engineer,* 40 (November 1963): 81–84.

References:
"Allegheny-Ludlum Talks to Its Union," *Steel,* 162 (February 12, 1968): 71–76;
H. G. Batcheller, "Let's Make Workers Feel They Belong," *Factory Management and Maintenance,* 106 (February 1948): 90–93;
"Human Relations Program Gets Glad Hand from Union," *Iron Age,* 161 (April 29, 1948): 123, 125;
"Selling the Company to the Community," *Factory Management and Maintenance,* 105 (April 1947): 78–80;
"A Steel Man Looks Ahead — 'You Can't Afford Not to Move': The *Iron Age* Interviews E. J. Hanley, President, Allegheny-Ludlum Steel Corporation," *Iron Age,* 186 (September 15, 1960): 113–114;
"What Labor Doesn't Know Can Hurt," *Iron Age,* 197 (June 2, 1966); 52–54.

Archives:
An interview with Edward J. Hanley, by Nancy Mason, for the Pittsburgh Renaissance Project, October 26, 1972, is in the Pittsburgh Renaissance Project: The Stanton Belfour Oral History Collection, Archives of Industrial Society, University of Pittsburgh Libraries, Pittsburgh, Pennsylvania.

Charles Hart

(January 5, 1869 – May 23, 1950)

by Louis P. Cain
Loyola University of Chicago

CAREER: Assistant blast-furnace manager, Republic Iron & Steel Company (1899–1906); vice-president, Republic Steel Corporation (1906); president, Inland Steel Company (1906–1908); president, Delaware River Steel Company (1909–1940); president, Wrought Iron Company of America (1927–1930); director, Chester National Bank (1922–1940); president, Victoria Gypsum Mining and Manufacturing (1922–1940).

Charles Hart was born to Samuel and Ellen Yardley (Eastburn) Hart in Doylestown, Pennsylvania, on January 5, 1869. He attended the Friends School in Doylestown, then graduated with a bachelor of science degree from Swarthmore College in 1892, where he played four years of varsity football. He later pursued postgraduate work in chemistry at Lafayette College. On October 12, 1907, Charles Hart was married to Florence McCurdy, of Youngstown, Ohio; they had four daughters.

In 1893 Hart was hired as a chemist in the research laboratories of the Carnegie Steel Company in Braddock, Pennsylvania, where he gained expertise in blast furnace operations. After six years of work for Carnegie and other steel companies, he transferred to the Republic Iron and Steel Company of Youngstown, Ohio, in 1899 as an assistant blast-furnace manager. While at Republic, his primary responsibility was to complete the physical organization of the company following the merger of several small plants, the first step toward the formation of Republic Steel Corporation. As this work neared a successful completion in 1906, Hart was named a vice-president of Republic and attracted the attention of Inland Steel, which needed an individual with his talents.

Inland had determined to become an integrated producer, a step that required the construction of a blast furnace as well as the acquisition of coal and iron ore supplies. Charles Hart and Alexis W. Thompson were brought into the firm because of their expertise in such integrated oper-

Charles Hart

ations. Hart became the third president of Inland Steel in 1906. Thompson was named Inland's first chairman of the board the year following Hart's appointment as president.

In 1906 Inland expanded its capitalization and authorized its first bond issue — $2.5 million of first mortgage bonds — to finance the construction of the blast furnace as well as to acquire the Laura Ore Mine near Hibbing, Minnesota. The authorized capital stock was again increased in 1907; however, serious trouble accompanied the bond issue, so much so that Leopold E. Block, reputedly the most optimistic of Inland's officers, commented that he felt the company was unlikely to survive the panic of that year.

Inland altered its financial plan in order to complete construction of the blast furnace. On Thompson's recommendation, Inland mortgaged its property and issued $2.5 million in gold bonds. Sidney Love, a prominent Chicago broker and friend of Thompson's, undertook the sale. Following the crash of 1907 bond sales ceased and Love's company failed. Love issued Inland a promissory note in excess of $500,000 for the bonds he had sold. In anticipation of those funds Inland steadily increased its construction loans with the Continental and First National Banks in Chicago. By November 1907, the small, young company had borrowed in excess of $1 million, a very heavy loan. In the wake of the panic, Inland was unable to market the balance of its bonds at prices it could afford to accept. It proved necessary to halt construction of some of the smaller parts of Inland's program. By 1908, when the panic had completely subsided, the construction program, including Madeline #1, Inland's first blast furnace, was completed. The Madeline #1 furnace was blown in on August 31, 1907, when Inland was still unsure whether it could pay for it.

Hart served as Inland's president for less than two years, between September 17, 1906, and April 9, 1908, before returning to the East Coast. During that time, he lent the company his blast furnace expertise and saw it through two of the most difficult years in its history. The integrated operations he inaugurated at Indiana Harbor are still central to the company's strategy.

In 1909 he joined with his brother-in-law, Robert McCurdy, and Myron A. Wick in organizing the Delaware River Steel Company of Chester, Pennsylvania. He served as president of that firm until his retirement in 1940. The firm dissolved at the same time. During its life, Delaware River Steel utilized imported low-phosphorous ores to supply most of the foundries in the Chester area. On average, the firm produced a little more than 120,000 tons per year of quality pig iron.

Hart also was actively involved in the reorganization of the Wrought Iron Company of America in 1927. He served as president of this Lebanon, Pennsylvania, firm until 1930. As a director of the Chester National Bank he became the receiver of the Penn Seaboard Steel Company and the Keystone Plaster Company, both of Chester. He also served as president of the Victoria Gypsum Mining and Manufacturing Company from 1922 until his retirement. In 1934 he opened Charles Hart and Associates to "examine and report upon all phases of business propositions."

Hart was a charter member of the American Iron and Steel Institute, as well as many other professional organizations associated with his vocation. He was also a member of several professional organizations associated with his avocation, farming. Hart died in Bryn Mawr, Pennsylvania, on May 23, 1950.

Publications:
"World Iron Ore Supply Apparently is Without Limit," *Iron Trade Review,* 86 (April 10, 1930): 63–64;

Known Iron Ore Reserves of the World and Their Significance (Chester, Penn., 1939);

Refractory Requirements for the Electric Smelting of Iron Ores," *Bulletin of the American Ceramic Society,* 20 (February 1941): 53–56;

"Electrometallurgical Treatment of Ores," *Blast Furnace and Steel Plant,* 32 (June 1944): 673–674, 691–694;

"*The Manufacture of Electric Furnace Pig Iron,*" *Steel,* 115 (August 7, 1944): 111, 114, 140;

Memories of a Forty-niner (Philadelphia: Dunlap, 1946).

Reference:
William T. Hogan, S.J., *An Economic History of the Iron and Steel Industry in the United States,* 5 volumes (Lexington, Mass.: Lexington Books, 1971).

Robert B. Heppenstall

(January 12, 1904 – February 21, 1966)

by Kevin M. Dwyer
George Washington University

CAREER: Metallurgist, Heppenstall Forge and Knife Company (1925–1927); production manager (1927–1929), vice-president (1929–1939), president; (1939–1957, 1962–1966), chairman of the board, (1957–1966), the Heppenstall Company; president, Pennsylvania Chamber of Commerce (1945–1957); national board of directors, Boys Clubs of America (1946–1966).

Robert Bole Heppenstall was a second-generation American and a seventh-generation steelmaker who presided over his family's business and upheld its tradition of growth through innovation. As had the founding fathers of the American steel industry, Heppenstall combined a specialization in the techniques of production with keen management instinct and judgment.

Heppenstall was born on January 12, 1904, to Charles William and Rachel Eleanor (Bole) Heppenstall in Pittsburgh. After having attended Pittsburgh public schools, Heppenstall went on to Williams College, graduating with a bachelor of arts degree in 1925. He joined the family firm, the Heppenstall Forge and Knife Company, by then headed by his father, and in 1926 married Katherine Monroe of Pittsburgh, with whom he would have three children. He throve in the family business, rising by 1927 to the rank of production manager. At that position, his good technical sense served the company well, and in 1928 he introduced to the United States a high-frequency induction process for melting tool-steel ingots. In 1929 he became vice-president, a position he held for ten years at the firm, which, in 1930, had changed its name to the Heppenstall Company. In 1939, at the age of thirty-five, Robert succeeded his father to the presidency of the two-plant concern, which had a net worth of about $3.7 million, annual sales of close to $4 million, and 850 employees.

The Heppenstall Company had its beginnings in 1889 when Heppenstall's grandfather Sam Heppenstall became a partner in a new ven-

Robert B. Heppenstall

ture, the Trethewey Manufacturing Company of Pittsburgh. Sam Heppenstall had crossed the Atlantic in 1867 from Sheffield, England, leaving behind his family and his job as a sheet roller at the John Brown Company — where his father worked as a foreman — to become a rolling-mill superintendent at Henry Disston and Sons Company in Philadelphia. A decade later he moved to Pittsburgh where he supervised in the mills at Howe, Browne and Company before becoming involved in the company founded by Samuel Trethewey. Trethewey Manufacturing specialized at first in steel mill parts, rolling mill shears, and single- and double-frame steam hammers in particular. By the mid 1890s, however, it had acquired specialization in forging and knife manufacture, and its name was changed to Pittsburgh Forge and Knife Company in 1896. Sam Heppenstall, who became president of the firm in 1898, sold off the hammer production equipment to the Erie Foundry Company in 1902. He secured full control of the company by 1904, and in that year he changed its name to the Heppenstall Forge and Knife Company.

Sam Heppenstall established a sound formula for success in business that his son Charles and his grandson Robert faithfully followed and successfully executed: upgrade production techniques to meet the changing demands of a specialized market. Owing to its successful development of America's first steam-drop hammer, increased demand for its rolling-mill shears, steel knives, and rotary shears, and to several innovations in hydraulic technologies brought to these shores, Heppenstall Forge and Knife secured its niche and earned a reputation for quality in the specialized-parts market. The firm's specialty rolling mill forgings, turbine wheels, and shafts won a steady clientele by 1910, and, at the dawning of the American automotive revolution, its specialty die block held potential for considerable growth.

The family's business enjoyed great success and almost continual growth during Robert Heppenstall's time at the helm. In 1955 the firm acquired the physical assets of Midvale Steel Company to form Midvale-Heppenstall Company of Philadelphia, with Heppenstall serving as president. In expansion, the firm's traditional emphasis on adapting production to meet changing, more highly specialized demand assured the firm's continued success in the market. Heppenstall continuously updated the production technology, adopting a consumable-electrode vacuum arc steel-melting process in 1960; the vacuum arc furnace operated at the company's Midvale plant was the largest melting unit in the world. Its large, highly stressed, steel-alloy forgings met the traditional parts demand for steam and gas turbines but were also used in nuclear reactors, missiles, and rockets. The specialty steels produced at the Midvale plant met growing demands for use in die blocks, tools, and other industrial finished-steel products. Among the many trade brands of Heppenstall steels were Hardtem, Pyrotem, Pyroreal, Thermotem, Thermoreal, Moldtem, and "C" Annealed.

Robert Heppenstall vacated the presidency in 1957 when he became chairman of the board. But in 1962 he assumed the dual role of president and chairman, when the Heppenstall Company of Detroit and the Heppenstall Company of Bridgeport, Connecticut, were established as wholly owned subsidiaries. At that time the net worth of Heppenstall's four plants stood at about $12.4 million, with annual sales of $38 million and 2,580 employees.

During World War II, as the company prospered, Heppenstall had assumed a leading role in civic affairs. He worked as president of the Boys Clubs of Pittsburgh from 1943 to 1949, leaving the organization in good financial shape with expanded facilities and broad popular support. He was invited to sit on the Boys Clubs of America national board of directors in 1945, and he held that seat until his death; the national organization awarded Heppenstall its Bronze Medallion in 1957. Heppenstall was also a leading member of the Pennsylvania Chamber of Commerce, serving as its president from 1945 to 1957, and he retained close and permanent ties with several other charitable and religious affiliations throughout his life.

Arthur B. Homer

(April 14, 1896 – June 18, 1972)

by Lance E. Metz

Canal Museum, Easton, Pennsylvania

CAREER: Assistant to the general superintendent, Fore River Yard (1919–1921), manager of diesel engineering and sales, Bethlehem office (1921–1926), manager of sales, New York City office, Bethlehem Shipbuilding Company (1926–1931); assistant to the vice-president in charge of shipbuilding and repair (1931–1934), assistant vice-president of Shipbuilding Division (1934–1940), director, vice-president of Shipbuilding Division (1940–1945), president (1945–1957), chief executive officer and chairman, Bethlehem Steel Corporation (1957–1964).

With the exceptions of Charles Schwab and Eugene Grace, Arthur B. Homer probably did more than any other individual to shape the development of the Bethlehem Steel Corporation. Under his direction Bethlehem Steel not only successfully undertook the largest single corporate shipbuilding program in American history, but it also completed what remains America's most modern integrated steel plant. It also developed one of the world's most advanced steel research facilities.

Homer was born to Eleazer B. and Elizabeth Hough Homer in Belmont, Massachusetts, on April 14, 1896. At an early age he moved with his parents to Providence, Rhode Island. Frequent visits to the shore communities of Narragansett Bay during his childhood nurtured his love of the sea and sailing, an interest which would become a focal point of his adult life. After attending the Providence Technical School, Homer entered Brown University, where he majored in economics and engineering administration. He graduated with academic distinction from Brown in 1917. When the United States entered World War I, Homer volunteered for naval service. Commissioned as a lieutenant, he was assigned to the U.S. Navy's Submarine Service. As a junior officer on a submarine, he served several months of sea duty. In 1918 he was reassigned to serve as the engineering officer of the navy's first large fleet submarine, then

Arthur B. Homer in Sangus, Massachusetts, September 17, 1954 (Hagley Museum and Library)

being built by the Bethlehem Shipbuilding Company, a subsidiary of the Bethlehem Steel Corporation, at its yard at Quincy, Massachusetts.

When World War I ended, Homer remained at Quincy, and in 1919 he resigned his naval commission to accept a position with Bethlehem Shipbuilding as assistant to the general superintendent of the Fore River Yard at Quincy. His work at Quincy attracted the attention of Bethlehem's top management, and he was transferred to corporate headquarters at Bethlehem, Pennsylvania, to become manager of diesel engineering and sales for the shipbuilding division. Homer's abilities and diligence made him a protégé of Bethlehem's president, Eugene Grace. Homer's rise in the corporate hierarchy continued throughout the 1930s. In 1940 he was appointed to the position of vice-president of Bethlehem's Shipbuilding Division and simultaneously became a director of the corporation.

During World War II Homer was directly responsible for implementing the rapid expansion of Bethlehem's shipbuilding activities through both the enlargement of existing facilities and the construction of new shipyards at such locations as Hingman, Massachusetts, and Staten Island, New York. Bethlehem constructed more than 1,100 naval and merchant ships of all types during the war years, including the aircraft carriers USS *Wasp* and USS *Hancock* and the battleship USS *Massachusetts*. This massive shipbuilding program was the largest ever undertaken by a private corporation. Homer's key role in America's war effort was recognized with a Presidential Certificate of Merit and the Vice Admiral Jerry Land Medal of the Society of Naval Architects and Marine Engineers.

Homer's wartime achievements were instrumental in securing his appointment as the president of the Bethlehem Steel Corporation in 1945. During his tenure as president he guided Bethlehem Steel's adjustment to a peacetime economy and its rapid expansion during the 1950s. In 1957 Homer succeeded Eugene Grace as the Bethlehem Steel Corporation's chief executive officer, and in 1960 he became chairman of the board.

Between 1957 and his retirement in 1964 Homer was responsible for a major expansion of both the production and research facilities of Bethlehem Steel. Under his direction, Bethlehem designed and constructed a large integrated steel-making plant at Burns Harbor, Indiana. This plant, which specializes in the production of steel sheet products, enabled Bethlehem to establish a greater foothold in the midwestern market.

Homer also directed the consolidation of Bethlehem Steel's previously scattered research facilities at a new complex that was constructed on South Mountain overlooking the home office at Bethlehem, Pennsylvania. Resembling a college campus, the research complex was named in honor of Homer. In 1987 it was sold to Lehigh University, although Bethlehem Steel still maintains an active presence in a portion of these facilities.

Prior to his retirement, Homer served successively as vice-president and trustee of the American Iron and Steel Institute. He was also for many years an active member of its executive committee. In 1964 he received the institute's Gary Medal for outstanding achievements in the industry. Homer served as honorary vice-president of the Society for Naval Architects and Marine Engineers. An active member of both the New York Yacht Club and the Cruising Club of America, he was a noted yachtsman.

Homer married Sara Yocum on September 14, 1922, and the couple had three children. Homer retired to an oceanside home at Saybrook, Connecticut, in 1964. He died at Hartford, Connecticut, on June 18, 1972.

Publications:
"The Doorway to Stabilized Progress," *American Iron and Steel Institute Yearbook* (1935): 812–893;

"Bethlehem Embarks on Major Modernization Program at Johnstown Plant," *Iron and Steel Engineer,* 27 (July 1950): 132;

"Incentives for Production," *American Iron and Steel Institute Regional Technical Papers* (1951): 59–70;

"Steel's Future; Producers' Point of View: Address to Eastern Regional Conference of Financial Analysts Societies," *Commercial and Financial Chronicle,* 176 (December 4, 1952): 2121;

"The Doorway to Stabilized Progress," *American Iron and Steel Institute Yearbook* (1955): 81–93;

"Steel Industry Prospects," *Commercial and Financial Chronicle,* 181 (June 2, 1955): 2529;

"Meeting the Nation's Steel Needs," *Commercial and Financial Chronicle,* 184 (December 13, 1956): 2523.

Archives:
The Arthur B. Homer Clipping File is in Bethlehem Steel Corporation Collection, Hugh Moore Historical Park and Museums, Incorporated, Easton, Pennsylvania.

Clifford Firoved Hood

(February 8, 1894 – November 9, 1978)

by Bruce E. Seely

Michigan Technological University

CAREER: Technical apprentice (1915), sales manager, Packard Electric Company (1915–1917); U.S. Army (1917–1919); clerk, Worcester Works (1917), foreman (1919–1925), assistant superintendent (1925–1927), superintendent (1928–1932), assistant operations manager and operations manager, Worcester Division (1932–1935), vice-president, operations (1935–1936), executive vice-president (1936–1938), president, American Steel & Wire Company (1938–1949); president, Carnegie-Illinois Steel Company (1950); executive vice-president (1951–1952), president (1953–1959), director, member executive committee, United States Steel Corporation (1959–1967).

Clifford Hood was a quintessential steelman, rising from the ranks to become president of the largest steel company in the world. An electrical engineer by training, he never left the United States Steel Corporation after he arrived. He guided the expansion program of the company during the late 1940s and 1950s.

Hood was born February 8, 1894, the only child of Edward Everett and Ida Florence Firoved Hood, and was reared on farms in northwestern Illinois at the turn of the century. His curiosity about industry, he would later recall, was piqued by an electrical engineer who rented a room from his parents, leading him to study electrical engineering at the University of Illinois. He managed the school's annual electrical show of manufacturers' displays and graduated in 1915 with a bachelor of science degree in electrical engineering. His first positions were as a sales engineer and then assistant cable-sales manager at the Packard Electric Company in Warren, Ohio.

From selling electrical cable Hood moved into the manufacturing of steel wire in 1917, accepting a position as operating clerk with American Steel & Wire Company. This subsidiary of U.S. Steel manufactured all types of finished products, as well as producing much of its own steel. Six weeks after having accepted the job, however,

Clifford Firoved Hood (Fortune Magazine *photograph*)

he joined the army and served in the coast artillery as a private. In 1919 he returned from overseas duty in France a first lieutenant and went back to work with American Steel & Wire as a foreman in the electrical-cable department at the Worcester plant. Gaining experience in both steelmaking and fabricating operations, Hood rose steadily within the Worcester plant, becoming assistant superintendent in 1925 and superintendent in 1928. Hood became assistant operations manager for the Worcester district in 1932 and manager of the district the next year. His next move was to vice-president of operations in 1935, a promotion that took him to the company's Cleveland headquarters. In 1936 he was elected executive vice-president, and two years later became president of American Steel & Wire.

Due to U.S. Steel chairman of the board Myron Taylor's management reforms instituted during the 1930s, Hood for the next 12 years controlled the day-to-day operations at American Steel & Wire, which had a product line of some 70,000 items — from piano wire and Erector-set girders to watch springs and cyclone fence. The challenges of the job were increased by plant expansions during the war, especially at Worcester; Pittsburg, California; and Gary, Indiana. The company's adjustment to peacetime involved just as many challenges, and, in keeping with the changes in the postwar working environment, Hood developed a management-incentive plan that was adopted for the entire corporation in 1953.

In 1950 Hood became president of the largest U.S. steel subsidiary, Carnegie-Illinois Steel Company, the largest steelmaking operation in the world. One of his tasks was to help in a restructuring process that would combine all steelmaking operations into the U.S. Steel Corporation. For the first time U.S. Steel was not just a holding company. In 1951–1952 Hood was executive vice-president for operations of the new corporation, with his main concern being production. He oversaw the company's capital expansion program, especially the brand-new Fairless Works near Morrisville, Pennsylvania. This $450 million project was the largest steel mill ever constructed and served as U.S. Steel's first integrated operation in the eastern United States.

In December 1953 Benjamin Fairless stepped up to the chairman's office, and Hood replaced him as president of U.S. Steel, with responsibility over sales and production. Based in Pittsburgh, Pennsylvania, Hood oversaw a far-flung industrial empire. He told the *New York Times* that he wanted to do things better: "If we don't continue to improve, there's something wrong with us." In 1955 he launched the Search for a Better Way program; designed to improve human relations in the company, the program was billed a success by the corporate hierarchy. Hood also chaired the corporation's policy committees. An executive policy committee met weekly to find ways to lower costs and increase efficiency; the administrative committee, with a more general agenda, met monthly. In both cases, Hood prided himself on installing a cooperative system of decision making.

Unfortunately for Hood, however, in the 1950s U.S. Steel returned to a pattern — that had prevailed under Elbert Gary — of weak presidents and strong chairmen. Under both Fairless and Roger Blough, Hood enjoyed only limited authority. Fairless built an advisory group in New York that made basic policy decisions; Hood was not included. When Blough replaced Fairless in 1955, Hood found himself further removed from the power center. Typically, Blough announced a $2.5 billion modernization program in 1956 while Hood stood by silently. Instead, Hood presided over ceremonial occasions — the opening of contract negotiations, the tapping of the billionth ton of steel, and the dedication of a memorial to honor United Steelworkers' president Phillip Murray. When Hood's retirement was announced at the annual meeting in May 1959, a stockholder actually asked whether he had been in virtual retirement since February.

Hood remained on the board until 1967 but devoted his retirement to humanitarian and civic projects. A member of the National Council on Alcoholism, he launched and chaired an industrial advisory committee in 1960 that helped create hundreds of corporate programs. A Baptist, Hood staunchly advocated a free-enterprise system that adhered to Christian principles. He supported various organizations along this line, including the Freedoms Foundation, the American Cause and American Enterprise Institute, the Foundation for Economic Education, and the American Security Council Education Foundation. In 1976 he founded the Christian Businessmen's Fellowship.

Like many top U.S. Steel executives, Hood received many awards, including five honorary doctorates, a Horatio Alger Award in 1954, and the Illini Achievement Award in 1957. In 1969 Hood retired to Palm Beach, Florida, with his second wife, Mary Ellen Tolerton, whom he had married in 1943 after the death of his first wife, Emilie R. Tener, whom he had married in 1917. He had two children by his first marriage. Clifford Hood died in Palm Beach on November 9, 1978.

Publications:

The Meaning of Fifty Years in Business (Cleveland, 1942?);

"Steel Industry and the War Effort," *Blast Furnace and Steel Plant,* 30 (November 1942): 1271–1273;

"Steel Prices and Profits are Low, but Costs are Higher," *Iron Age,* 161 (February 5, 1948): 140–141;

Profit Engineering (New York: U.S. Steel Corporation, 1951);

From a Trickle of Iron (New York, 1952);

Reaching the Purple Twilight, address to the Case Institute of Technology, Diamond Jubilee, April 10, 1953 (New York: U.S. Steel Corporation, 1953);

also published in *Commercial and Financial Chronicle*, 177 (April 23, 1953): 1740;
More Than Bread Alone (New York: U.S. Steel Corporation, 1953);
"Shape of Things to Come," *Commercial and Financial Chronicle*, 179 (March 4, 1954): 986;
"The Next Heat of Steel," *American Iron and Steel Institute Yearbook* (1954): 67–77;
"Closing the New Inflationary Gap," in *Steel and Inflation: Fact vs. Fiction* (New York: U.S. Steel Corporation, 1958), pp. 23–26.

References:
"Blough, Hood Guide U.S. Steel," *Steel*, 136 (May 9, 1955): 48;
"Hood of U.S. Steel," *Fortune*, 47 (February 1953): 146–147;
"In Ben's Shoes," *Time*, 60 (December 8, 1952): 87;
"They're Reshuffling Big Steel," *Business Week*, (December 6, 1952): 27–28;
"United States Steel," *Fortune*, 53 (January 1956): 88–95, 198–200, 202–204.

Charles Ruffin Hook

(July 12, 1880 – November 14, 1963)

by Carl Becker

Wright State University

CAREER: Night superintendent (1902–1904), assistant general superintendent (1904–1910), superintendent (1910-1930), president (1930–1948), chairman (1948–1959), director, Armco Incorporated (1948–1960).

Unquestionably, George M. Verity, founder of the American Rolling Mill Company (ARMCO), was the heart and soul of that steel firm, leading it over three decades from a tenuous existence to a prominent position in the steel industry. Charles Hook, however, stood beside Verity in championing innovation in steelmaking and modernizing the relationship between management and labor.

Hook, born on July 12, 1888, in Cincinnati, Ohio, the last of four children, could trace his lineage to English ancestors who came to Massachusetts in 1633. His father, Henry Hook, served in the Union army during the Civil War; and his mother, Katherine Klusman Hook, was the daughter of a British army officer. Henry Hook was a carriage maker whose trade provided the family only a meager existence. The boy attended Walnut Hills High School in Cincinnati and worked at odd jobs until at the age of eighteen he went to work as an office boy at William Simpson's Riverside rolling mill. Seeing the work in the mill, he aspired to learn roll turning, or sheet rolling, a craft zealously guarded by masters to the point of secrecy. Somehow he persuaded the superintendent

Charles Ruffin Hook

of the mill to teach him the art of rolling on Sundays, when the mill was closed.

The Tin Plate Trust, organized late in the 1890s, acquired the mill, and Hook found himself transferred to Chicago and then to New York City as an apprentice accountant. He wanted to make steel, however, and asked to be assigned to a mill where he could learn more about steel. Surprised by his request but willing to grant it, his superiors sent him to the Morewood mill in Gas City, Indiana, as the clerk to the superintendent, who happened to be the same man who had tutored him in sheet-rolling at Riverside. Learning quickly, Hook soon became a foreman in the mill.

He renewed his acquaintance, too, with William Simpson, who introduced him to George Verity. Verity, who was just beginning operation of his rolling mill in Middletown, found in Hook a kindred soul, an avid listener to his talk of building a steel company whose managers and workers might work in cooperation with each other — the managers would apprise workers of operating policies and provide them with incentives, and the men would respond with honest labor — and give Armco a competitive edge over larger mills. Hook readily accepted Verity's offer in 1902 to become night superintendent of the Middletown mill.

Only twenty-two years old, slight and boyish, Hook met ridicule from older steelworkers in the mill as he perpetually queried them about production problems. He persisted, though, and learned, and gradually the men came to respect him, calling him "Charlie" instead of "the kid." Early in his career at Armco, Hook won recognition for his innovative methods of dealing with workers. He began a pioneering venture in employee representation — industrial democracy — when he asked to create voluntarily a committee to discipline men who damaged machinery, heretofore a task left to foremen and superintendents. Later, as assistant general superintendent, and believing that a gulf between management and labor was widening, he went to the local chapter of the Amalgamated Association of Iron, Steel, and Tin Workers of North America with a startling request to address a meeting of the union. Union officers, though disconcerted, invited him to attend. Taking the floor, he asked the union to elect an advisory committee to meet with him weekly to discuss how the several departments might operate more efficiently. Ultimately, each department selected an advisory committee to consult with Hook. Many workers coming to the mill from other plants expressed disbelief at the arrangement, unable to believe that owners and managers would deign to consult with them. But Hook persisted in his approach and indeed widened his concern. By 1919 he was becoming active in organiza-

tions that would later evolve into the American Management Association and chaired committees concerned with the question of employee representation. Only the Colorado Fuel & Iron Company exhibited anything like Armco's commitment to better labor relations.

Hook also proved to be an important figure in Armco's great innovation in steelmaking — the development of the continuous hot-strip mill. One day in 1904 John Butler Tytus, a recent graduate of Yale who had worked in his father's paper mill in Middletown, came to Verity seeking employment in the steel mill. Verity sent him to Hook, who recited a litany of the hard labor of the mill, looking to dissuade a man of Tytus's youth and literary education from entering a career in steelmaking. Tytus would not listen, and Hook, perhaps believing that he might teach him a lesson, took him on as a "spare hand"; as such, Tytus would have to assume the tasks of any member of a work crew absent from the mill.

Tytus proved himself an adept worker. Hook, following his progress, made him his first assistant and soon discovered that he was a perpetual experimenter. Tytus was especially interested in finding a means of reducing the laborious process of turning out sheet steel and saw an emulative model in the Fourdrinier paper machine used in his father's mill. His conversations with Hook often revolved around the continuous mill. But the two men had to share the idea from a distance beginning in 1906, when at Hook's recommendation Tytus was appointed superintendent of the Armco mill at Zanesville.

Hook continued in his daily work, spending much of his time dealing with problems attendant to the development of various speciality steels. He was particularly concerned about the rate of accidents in the Armco mills and encouraged workers to organize special committees to initiate safety programs. Awarding cash prizes for the reduction of accidents, Armco saw an improvement in its safety record and by 1917 had become the first company in the iron and steel industry to qualify for accident insurance under group rates. Hook was also taking the lead in reducing the infamous twelve-hour shift to eight hours.

Tytus had returned to Middletown in 1910 to install manual sheet mills at the company's East Works, and he and Hook, at Verity's encouragement, resumed their discussions about the continuous mill. Although now general superintendent of all Armco plants, Hook could not go much beyond talk in supporting Tytus because of the con-

The Armco management in 1959: Charles Ruffin Hook, chairman; Logan Truax Johnston, executive vice-president; and Ralph Larrabee Gray, president. A portrait of Armco founder George Verity is on the wall behind them (Fortune Magazine photograph).

ditions of production in the few years prior to World War I. The automobile industry provided substantial demand for sheet steel, and Armco could not spare resources for experimentation. During the war, however, with sheet mills lying dormant as the plant turned to the production of shell forgings, Tytus had the opportunity to use the idle mills as a laboratory in which he could test his concepts of running a continuous mill.

After the war, Verity and Hook decided to install an experimental continuous mill at Middletown, seeing in it the means of absorbing the excess capacity created by the enlargement of open-hearth furnaces during the war. Tytus was ready for the good work, but then the market intervened. Orders for light sheets for consumer goods of all sorts — washing machines, refrigerators, and so on — and for speciality steels permitted nothing but production for the day. In 1921 Verity took a decisive step, purchasing the Ashland

Iron and Mining Company in Ashland, Kentucky. At Ashland were blast furnaces, open-hearth furnaces, a blooming mill — but no sheet mills; here Verity envisioned the installation of an experimental mill. With Hook offering the conclusive resolution, the board at Armco authorized the expenditure of $10 million for the development of the experimental mill. Tytus directed planning for construction of the mill, presided over its installation at Ashland, and early in 1924 was successfully rolling sheets through it. It was a revolutionary innovation that gave Armco profits and prestige. All the principal steel companies in the nation installed similar mills in their plants under license from Armco in the ensuing years, with the result that production rose rapidly and costs declined appreciably in the industry. Tytus could rightfully lay claim to primacy in the development of the continuous mill, but Hook had been an important figure in setting the stage for success.

All through the fits and starts leading to production at the new mill, Hook also had to attend to the everyday business of steelmaking, of dealing with the men who worked at furnaces and mills. All the while, too, he remained on close terms with George Verity. In fact, he had wed Verity's daughter, Leah, in 1913. Together, Hook and Verity saw Armco enjoy continued growth through the 1920s, its production and profits reaching new heights in the decade prior to the Great Depression. By 1930 Verity, deciding to taste some of the fruits of his long labor, was ready to give up his presidency and turned it over to Hook, whose intellect and experience, Verity felt, would continue to direct the company along established lines.

In 1931, under Hook's direction, Armco developed Multi-Plate, heavy curved corrugated sections bolted together to serve as wide-diameter culverts. A year later the company produced the first stainless steel sheets to be processed by the continuous cold-reduction method. Midway through the decade, he brought together the Polish engineer Tadeusz Sendzimir and Armco in the building of a pilot plant in Butler, Pennsylvania, for the production of Sendzimir's revolutionary method of applying zinc to steel sheets, a process permitting the elimination of costly dipping of uncoated sheets in molten zinc in the making of galvanized sheets. He also had a hand in developing "Paintgrip," a phosphating process for creating an effective paint-holding surface on galvanized sheets. And during his presidency, he could point to Armco's development of a special sheet for automobile bodies, a "non-aging" steel sheet with ductility and drawing qualities that would not deteriorate with time.

As the chief administrator, Hook had the final voice in a series of decisions in the 1930s that enhanced Armco's domestic and foreign production. At Middletown he had Tytus employ unconventional engineering methods in revamping and modernizing the hot-strip mill. In Argentina, France, England, Mexico, and Venezuela he directed Armco to expansion of its operations and the creation of subsidiary companies. Giving his company more secure access to raw materials, he completed negotiation in 1939 for a long-term lease of magnetite ore deposits in the Mesabi Range.

As Hook championed innovation and expansion, he had to pilot Armco through the shoals and riffles of the Great Depression. Sales, profits, and employment had fallen, of course, early in the Depression; but by 1935, owing to its strong position in the sheet market and inflationary policies initiated by the government, Armco had experienced a substantial recovery. Falling prices during the recession of 1938 resulted in a net loss, but the advent of war in Europe and domestic defense spending helped redress the balance as the company entered the 1940s. Particularly important in the defense program was the expansion of a Sheffield plant in Houston, Texas. Using its own resources and funds furnished by the Defense Plant Corporation, Armco spent over $50 million to build coke ovens, open-hearth furnaces, and finishing facilities there.

Through the war, Hook saw to the fulfillment of defense contracts and met the opportunities offered by a bounding market. He could report an increase in output of ingots from about 2 million tons in 1940 to nearly 3 million by 1945 and a near-doubling of profits from around $7 million to $13 million in the same period — though governmental ceilings on steel prices acted as a constraint on net returns. He was especially at ease in working in an environment requiring cooperation between private and public sectors of the economy. Early in the 1930s he had become increasingly conversant with the ways of Washington, serving as a member of a special advisory council for the Department of Commerce. As the nation entered the war, he voluntarily offered the national government use of Armco facilities for the training of skilled craftsmen needed at an aircraft engine factory near Cincinnati.

Remaining adept in pursuing the interests of his company, Hook's crowning success came in 1945 when he brought about a merger with the Rustless Iron and Steel Corporation in Baltimore, a company in which Armco had been acquiring stock for several years. The companies had been working together to improve the quality of stainless steel; Rustless continued that work as a division of Armco. Nearing the end of his career in steelmaking, Hook had effected a protective diversification for his company. In 1948 he resigned the presidency of Armco, becoming chairman of the board and remaining as a director until his death in 1963. He had served George Verity as an able lieutenant, leaving his mark as a man who played an important role in giving the company and the industry a remarkable means of production. As president of Armco, he had directed it to continuing success in innovative ventures and in the market. When he assumed the presidency, the company ranked seventh and eighth nationally among steel firms in the various indices of sales and production; when he left his post, Armco had reached record levels in earnings and output and ranked fifth and sixth in the several indices measuring performance in the industry.

After his retirement, Hook continued to lead a busy life. He was an active member of various private and public commissions dealing with government and business. His most notable service came as chairman of the Second Hoover Commission in 1953. In 1955 he received the Gary Medal of the American Iron and Steel Institute for outstanding achievement in human and industrial relations. He was also an energetic advocate of programs for young people, especially the Boy Scouts.

He and his wife donated a large tract of land in Warren County, Ohio, to the Boy Scouts for a camp named Camp Hook.

Hook spent the last few years of his life with a daughter in Baltimore, Maryland. He died of cancer in 1963 and was buried in the Verity plot in Woodside Cemetery at Middletown.

References:

Christy Borth, *True Steel: The Story of George Matthew Verity and His Associates* (Dayton: Otterbein, 1941);
John Tebbel, *The Human Touch in Business: The Story of Charles R. Hook* (Dayton: Otterbein, 1963).

Hot Strip Mill

by Robert Casey

Henry Ford Museum & Greenfield Village

The most significant development in rolling mills in the twentieth century was the hot strip mill. Such mills are continuous: a hot slab passes through successive sets of rolls and is reduced to a thin strip that is rolled into a large coil.

Prior to 1927 steel sheets were produced by passing red hot bars back and forth through rolling mills by hand. As the sheets became thinner, they were stacked and rolled together in packs to increase the output per mill. Hand-rolled sheets suffered from an irregular surface finish and variations in thickness, and their size was limited by what men could handle.

Several unsuccessful attempts at continuous rolling were made in the late nineteenth and early twentieth centuries. The first successful step toward a continuous mill came at the Ashland, Kentucky, plant of American Rolling Mill Company, under the direction of John B. Tytus. Tytus developed his mill between 1921 and 1924. It consisted of a reversing blooming mill, a holding furnace, seven two-high rolling stands, a second holding furnace, four two-high stands, and three three-high stands. The mill produced sheets piled flat, rather than coils. The key to the success of Tytus's mill was the discovery that the surface of the rolls should be slightly convex and that each successive set of rolls should be slightly less convex.

In Butler, Pennsylvania, Harry M. Naugle and Arthur J. Townsend of the Forged Steel Wheel Company were at work on their own hot strip mill. Their mill produced a coil of thin strip and pioneered the use of large roller bearings and four-high rolling stands. The four-high design allowed small-diameter work rolls to be supported by much larger-diameter back-up rolls. This kept the work rolls from deflecting, thus producing a more uniform sheet. The small work rolls could also make greater reductions in strip thickness on each pass through the rolls. American Rolling Mill purchased Forged Steel Wheel in 1927 and thus acquired most of the patents necessary for building successful hot strip mills.

Despite its advantages the hot strip mill was not immediately adopted by the American steel industry. The United States Steel Corporation combined the new technology with traditional methods of sheet rolling by installing a continuous mill whose long strips were cut into sheets and finished on hand mills. Many steel men regarded the hand mills as more flexible, allowing for a wider variety of widths and gauges during the rolling. The advantages of the hot strip mill were too great to ignore, however, and, between 1930 and 1940, 20 large mills, ranging from 36 to 96 inches wide, were built.

A hot strip mill

In the postwar period many new mills were added, and older mills were rebuilt, but between 1960 and 1970 ten mills were constructed that were so superior to their predecessors that they are referred to as Generation II mills. These new mills rolled coils twice as large as first-generation mills, at twice the speed. With features such as electronic controls and automated roll changing, they substantially improved productivity and product quality.

The advantages of Generation II mills did not come cheaply: they cost between $80 and $135 million. Rising costs have limited the number of Generation II mills that have been built, and have led steelmakers to search for less expensive alternatives — one being the Steckel mill, or reversing hot strip mill. These mills place coilers on either side of a single-roll stand. The strip is passed back and forth through the rolls, being alternately wound on each coiler. For years Steckel mills could not match the quality of continuous mills, but that is changing. Since 1985 a $75 million mill in Tuscaloosa, Alabama, has used modern process controls to produce strip equivalent in quality to that produced on the best continuous mills.

In 1989 another alternative went into operation in Crawfordsville, Indiana, where Nucor Corporation is using a German-designed continuous caster to produce 1.5-to-2-inch slabs that pass through a 200-foot-long roller hearth furnace into a four-stand finishing mill.

References:

Bela Gold, William Peirce, Gerhard Resegger, Mark Perlman, *Technological Progress and Industrial Leadership: The Growth of the U.S. Steel Industry, 1900–1970* (Lexington, Mass.: D. C. Henry, 1984);

William T. Hogan, S.J., *An Economic History of the Iron and Steel Industry in the United States,* 5 volumes (Lexington, Mass.: Lexington Books, 1971);

The Making, Shaping, and Treating of Steel, seventh edition (Pittsburgh: U.S. Steel, 1957);

George J. McManus, "Taking the Wraps off Nucor's Sheet Mill," *Iron Age* (June) 1989;

William L. Roberts, *Hot Rolling of Steel* (New York: Marcel Dekker, 1983);

Norman L. Samways, "Tuscaloosa Steel Corporation — A Unique Market Mill for Hot Rolled Flat Products," *Iron and Steel Engineer,* 66 (March 1989); 19-27.

Max Don Howell

(October 18, 1887 – January 20, 1967)

by Donald F. Barnett

McLean, Virginia

CAREER: Traveling auditor and accountant, American Telephone and Telegraph Company (1912–1921); chief accountant, Ohio Bell Telephone Company (1921–1924); senior consultant, Lybrand, Ross Brothers & Montgomery (1924–1926); financial consultant, Beekman, Bogue & Clark (1926–1928); controller, Postal Telegraph and Cable Company (1928–1930); vice-president, Chemical Bank & Trust Company (1930–1937); vice-president, assistant to the president, secretary and treasurer, United States Steel Corporation, Delaware and Pittsburgh offices (1937–1950); vice-president and treasurer, United States Steel Corporation, New York offices (1950–1952); treasurer (1951–1952), executive director (1952–1953), executive vice-president (1953–1962), president, American Iron and Steel Institute (1963); administrative officer, Steel Management Association (1963).

Max Don Howell came relatively late in life to the steel industry, bringing with him experience in financial positions in utilities, accounting firms, and in banking. His arrival coincided with important challenges to the industry from the late 1930s through the early 1960s. First as treasurer of the United States Steel Corporation from 1937 through 1952, and then with the American Iron and Steel Institute (AISI) until 1963, Howell was at the center of some of the most exciting — and most disturbing — events in the industry's history.

Howell was born in Lockhaven, Pennsylvania, and attended Seneca Hills High School. In 1912 he earned his bachelor's degree in business administration from the University of Michigan. His initial inclination had been to study forestry because of his love of the outdoors, but Howell soon developed a special interest in corporate and business law, an interest that would serve him often during his career. His first job was with the American Telephone & Telegraph Company as a traveling auditor and accountant, a position he held until 1921, when he became chief accountant

at Ohio Bell Telephone Company. In this job he helped arrange the consolidation of the firm's two largest phone companies as well as organized the entire firm's accounting and statistical activities.

In 1924 Howell changed jobs, becoming senior consultant at the accounting firm of Lybrand, Ross Brothers & Montgomery in New York City; he was concerned with corporate reorganizations, refinancing, and accounting systems. He remained only until 1926, when he moved to the New York law firm of Beekman, Bogue & Clark, where his work continued to revolve around corporate finance. In 1928 he became controller of the Postal Telegraph and Cable Company, also located in New York. Two years later Howell entered the world of banking as vice-president of Chemical Bank & Trust Company in New York. Because he served as director of several companies in which Chemical Bank was involved, he remained in touch with corporate activities.

Although typical of the path accountants and financial executives followed, Howell's career track was not the usual route taken by aspiring young steel executives. Howell's career, however, closely resembled that of U.S. Steel chairman Myron Taylor. Both were interested in the financial side of corporate management; both had practical experience in running companies; and both entered banking but remained in touch with companies in different industries. Howell moved to U.S. Steel in 1937 as vice-president of the Carnegie-Illinois Steel Corporation — U.S. Steel's largest subsidiary. Taylor was continuing to reorganize the corporation; he had already shaken the company by bringing in a number of outsiders and even non-steel men, and Howell's appointment was in keeping with Taylor's program.

Howell found himself working with another man from outside U.S. Steel — Benjamin Fairless, hired in 1936 from Republic Steel to run Carnegie-Illinois. Over the next 25 years Howell's steel career closely paralleled Fairless's, and the two became good friends. In 1938 Howell became vice-president and treasurer for U.S. Steel Corporation of Delaware, the operating company, as well as the holding company until 1941, although he continued to serve Carnegie-Illinois. In 1944 he became an assistant to then-president Fairless, while remaining vice-president and treasurer, titles he would maintain through the corporate reorganization of 1950.

These were interesting and complicated times for U.S. Steel's management, due to the demands of wartime production and government regulation, pressures from the United Steelworkers, and the challenges of transition to a peacetime economy. Howell played an important role in all of these areas as the leading financial official in the company before retiring in 1952.

Even then, however, Howell's career remained linked with Benjamin Fairless when Fairless joined the AISI, the trade association formed by Elbert H. Gary in 1908. In 1951 Howell had become the AISI's treasurer; the next year Fairless retired from U.S. Steel to become the president of the institute. Howell then became executive director, with his title changing to executive vice-president in 1953.

Howell replaced Walter Tower, who had served the institute since 1933, often acting as spokesperson for the industry. Fairless now assumed that role, while Howell oversaw the many activities of the AISI, including its record gathering and statistical efforts, public relations work, and research program. The AISI, for example, had 24 technical committees by 1958. Howell's efforts ensured that the institute remained one of the best trade associations in the country.

The AISI also played visible roles in the continuing confrontations between the steel industry and the federal government, echoing the complaints of the large corporations about federal policies. Howell was involved in all of these events, including the confrontation with President John F. Kennedy over prices in 1962. After that struggle, Howell became the AISI's president but only held the post for a year before retiring in 1963.

Howell spent his retirement at Sea Island, Georgia, with his second wife, Isoline Campbell MacKenna, whom he had married on May 17, 1947.

Publications:

"Address of the Executive Vice President," *American Iron and Steel Institute Yearbook* (1953): 15–21;

"Production of Steel in 1953 Between 112 and 113 Million Tons," *Blast Furnace and Steel Plant*, 42 (January 1954): 107;

"Steel and Iron Production in 1953 Establishes Record," *Blast Furnace and Steel Plant*, 42 (January 1954): 41–42;

"Opening Address," *American Iron and Steel Institute Yearbook* (1954): 13–14;

"Address of the Executive Vice President," *American Iron and Steel Institute Yearbook* (1954): 55–56;

"The Steel Industry and the Institute," *American Iron and Steel Institute Regional Technical Papers* (1954): 99–108, 273–283;

"Prospects for 1955 Good; Production Likely to Increase," *Blast Furnace and Steel Plant*, 43 (January 1955): 41–42;

"Address of the Executive Vice-President," *American Iron and Steel Institute Yearbook* (1955): 1–6;

"Opening Address," *American Iron and Steel Institute Yearbook* (1955): 21–22;

"Steel's Dynamic Progress," *American Iron and Steel Institute Yearbook* (1955): 67–77;

"Review of Steel Industry's Progress," *Commercial and Financial Chronicle*, 181 (June 2, 1955): 2546–2547;

"Progress Report on the Steel Industry and the Institute," *American Iron and Steel Institute Regional Technical Papers* (1955): 1–6;

"Steel's Dynamic Progress," *American Iron and Steel Institute Yearbook* (1955): 67–77;

"Steel Industry Provides Record Tonnage of Steel in 1955 and Steps Up Planning for Expansion," *Blast Furnace and Steel Plant*, 44 (January 1956): 41;

"Continued Keeping Pace with the Steel Industry," *American Iron and Steel Institute Yearbook* (1956): 69–76;

"Steel Production May Set a Record in 1957," *Blast Furnace and Steel Plant*, 45 (January 1957): 45;

"Report on a Year of Progress," *Yearbook of the American Iron and Steel Institute* (1957): 59–65;

"The Institute's First Fifty Years — And a Look Ahead," *Yearbook of the American Iron and Steel Institute* (1958): 59–68;

"Annual Report of Institute Activities," *Yearbook of the American Iron and Steel Institute* (1959): 153–160;

"Continued Improvement Expected in Steel Production," *Commercial and Financial Chronicle*, 189 (January 15, 1959): 223;

"Recent Institute Activities," *Yearbook of the American Iron and Steel Institute* (1960): 171–179;

"The Institute Is You," *Yearbook of the American Iron and Steel Institute* (1961): 245–252;

"American Iron and Steel Institute — Servant of Steel Industry," *Yearbook of the American Iron and Steel Institute* (1962): 243–250;

"Ten Years of Institute Growth," *Yearbook of the American Iron and Steel Institute* (1963): 189–195.

References:
"Obituary," *New York Times*, January 21, 1967, p. 31;

"Personals," *Iron Age*, 140 (July 8, 1937): 75, 78.

Frederick B. Hufnagel

(October 31, 1878 – May 16, 1954)

by Geoffrey Tweedale
Manchester, England

CAREER: Draftsman, mill foreman, American Steel & Wire Company, Newburgh, Ohio, Works (1900–1901); draftsman, mill superintendent, rolling-mills superintendent, assistant-general superintendent, general superintendent, South Side Works (1901–1913), general superintendent, Alquippa Works, Jones & Laughlin Steel Corporation (1913–1920); president, Pittsburgh Crucible Steel Company (1920–1944); president (1926–1938, 1941–1944), chairman (1937–1944), Crucible Steel Company of America.

Frederick B. Hufnagel, a specialty-steel executive, was born in Mount Vernon, New York, on October 31, 1878, the son of Mary Imhof and Conrad Bernhard Hufnagel, a German watchmaker who had arrived in the United States from Germany in 1860 and had settled in Mount Vernon.

In 1900 Hufnagel graduated from Cornell University with a degree in mechanical engineering and took a job as a draftsman with the American Steel & Wire Company at its Newburgh, Ohio, Works. During his eight months at Newburgh he also worked as a mill foreman. In 1901 he moved to the South Side Works of the Jones & Laughlin Steel Corporation in Pittsburgh. In 1911, after having worked at various positions as a draftsman and mill superintendent, he was promoted to general superintendent of the South Side Works. He was transferred in 1913 to oversee Jones & Laughlin's Alquippa (Pennsylvania) Works. He was elected president of the Pittsburgh Crucible Steel Company in 1920.

The Pittsburgh Crucible Steel Company was a wholly owned subsidiary of the Crucible Steel Company of America, the largest of the country's specialty-steel producers at that time. Hufnagel was hired to expand the Pittsburgh plant and increase its earnings. Although the parent company was devoted to the specialty-steels market, the Pittsburgh company was a completely integrated tonnage-steel producer with its own officers and

Frederick B. Hufnagel

independent sales force. The Crucible Steel Company took very little of the Pittsburgh plant's production, allowing the latter to go its own way in catering to the demand for low-alloy steel from its open-hearth furnaces. For many years during the interwar period, in fact, the Pittsburgh Crucible Steel Company was more profitable than its parent.

In 1926 Hufnagel was elected president of the Crucible Steel Company of America, which was under the chairmanship of Syracuse financier and real estate dealer Horace S. Wilkinson. Crucible enjoyed a commanding position in the U.S. specialty-steel business in the 1920s: it supplied most of the country's requirements in tool steel and was the major producer of alloy steels for such growth industries as automobile manufacture. The company, however, was plagued with organizational and public-relations problems. Although later acquitted, Crucible's wartime chairman, Herbert DuPuy, had been indicted for tax evasion and was removed at the stockholder's insistence. Moreover, Crucible had difficulty coordinating its remarkably diverse product line, huge selling force, and many operating subsidiaries, which included nine steel mills scattered over four states, two coal mines, a water company, and a 50 percent interest in a Mesabi ore mine. In the interwar period Crucible's sales were high, and its profit margins were slim (or sometimes nonexistent), and the competition from other companies was becoming more severe. Under Wilkinson's autocratic hand Crucible was unable to move very far in either rationalizing production or formulating more-advanced sales strategies.

In 1937 Wilkinson died, and Hufnagel became chairman of Crucible. The problems he faced were immediately apparent: in 1937, a good year for the company, Crucible made $4 million profit on net sales of nearly $60 million; but in 1938 it lost $2.2 million on sales of $29 million. Crucible's average percentage of profits on sales during the 1933–1938 period was only 3.1 percent, while its increasingly successful competitors, such as the Carpenter Steel Company of Reading, Pennsylvania, were recording profits of over 10 percent.

In 1939 *Fortune* had described Hufnagel as "Stern but not without a twinkle, quiet spoken and aloof, [with] a reputation among the hard-drinking steel gentry of never more than toying with a cocktail and of always thinking twice over an investment." His background had been in tonnage steel; nevertheless, he was the first practical steel executive the company had had since its earliest days. He recognized the weaknesses in the way Crucible did business — finding it necessary, upon becoming president, to keep long, personal

sales tabulations to follow Crucible's myriad orders. He was also aware of the need to delegate more authority than had his predecessor. In 1938 he appointed as president Raoul E. Desvernine, a New York lawyer and author of an anti–New Deal tract, *Democratic Despotism* (1936), to help reorganize Crucible's marketing and labor relations.

Together they began tackling Crucible's sales and management problems. Hufnagel issued the company's first fairly complete annual report in an attempt to remove the firm's notorious reputation for secrecy. He jettisoned Wilkinson's more-dubious business practices, such as buying the company's own stock and carrying it on the asset side of the balance sheet. The Pittsburgh Crucible Steel Company was also merged as a property with the parent to simplify the corporate structure and reduce taxes. Other consolidations — closures of inefficient plants and removal of unprofitable items — were also planned.

America's involvement in World War II, however, once more plunged the company into a seller's market as the demand for its specialty steels, vital for the engineering and armament industries, soared. By 1943 sales had reached $209 million, resulting in a net income of about $5 million. In the following year Hufnagel retired with the basic structure of the company unchanged; its problems were further exacerbated by the war. It was left to Hufnagel's successors, William P. Snyder, Jr., and William H. Colvin, to find solutions.

Hufnagel was married in Pittsburgh on June 18, 1910, to Ceora Wilson; they had two children. He died in Philadelphia on May 16, 1954.

References:
William Colvin, *Crucible Steel of America: 50 Years of Specialty Steelmaking in USA* (New York: Newcomen Society, 1950);
"Crucible Steel," *Fortune*, 20 (November 1939): 74;
Crucible Steel Company of America, *50 Years of Fine Steelmaking* (1951);
"Crucible Steel in Profitable Era," *Magazine of Wall Street*, 19 (1940): 30;
Herrymon Maurer, "Crucible: Steel for Frontiers," *Fortune*, 55 (January 1957): 84–88.

Human Relations Committee

by Richard W. Kalwa

University of Wisconsin — Parkside

In the wake of the long, bitter strike of 1959, the steel companies and the United Steelworkers concluded that new methods were necessary to resolve disputes over complex and controversial bargaining issues. As a result the 1960 agreement provided for a joint committee to examine guidelines for wages and benefits, job classifications, incentive payments, seniority and layoffs, medical benefits, and any other problems identified by the parties. The companies and the union each named four members to this body, called the Human Relations Committee; it was cochaired by R. Conrad Cooper of the United States Steel Corporation and United Steelworkers president David McDonald.

The industry had opposed the use of outside experts, or "neutrals," preferring to remove discussions from the public and government attention of the past. Specific areas of concern were assigned to subcommittees, which utilized union and company staff personnel and drew representatives from various parts of the industry. The role of these subcommittees was to collect information, attempt to resolve differences, and make recommendations to the parent committee.

The parties agreed on certain ground rules to govern committee interaction. Discussions were to be confidential, nonbinding, and separate from the processes of contract negotiation and grievance handling. The committee sought to emphasize problem solving rather than confrontation, to develop information necessary for resolving issues, and to promote a cooperative attitude through ongoing meetings between the companies and the union.

The work of the committee provided the background for the negotiation of the 1962 steel contract, which was completed three months prior to the deadline and provided for only a 2.5 percent increase in labor costs in the first year. The agreement removed cost-of-living adjustments, stressed job security, and improved benefits for unemployment and retirement but did not raise wages. The report of the subcommittee on senior-

ity also provided joint recommendations for contract modifications in this area. This peaceful, noninflationary settlement evoked widespread approval of the methods adopted by the Human Relations Committee.

However, as the committee came to assume more responsibility for contract negotiations, union district directors and local officers were excluded from the process. The 1963 agreement, again with a small increase in benefits but no change in wages, was reached by top-level negotiators without formal notice. The committee did not set a specific termination date for the contract, and resolved that any jointly approved changes would be implemented immediately.

This centralization of bargaining authority, often at the expense of plant-level issues, combined with the economic recovery of the steel industry and the resulting worker militancy made the Human Relations Committee a central issue in the United Steelworkers' election in 1965. McDonald was defeated by I. W. Abel, who campaigned on the promise to increase membership participation in bargaining. Shortly afterward contract negotiations once again became adversarial, with a settlement reached only after the parties were summoned to the White House by President Lyndon B. Johnson. The new settlement made no provision to continue the Human Relations Committee, although an unofficial body, whose functions were limited to research, was retained.

References:

James J. Healy, *Creative Collective Bargaining: Meeting Today's Challenges to Labor-Management Relations* (Englewood Cliffs, N.J.: Prentice-Hall, 1965);

John Herling, *Right to Challenge: People and Power in the Steelworkers Union* (New York: Harper & Row, 1972);

Charles C. Killingsworth, "Cooperative Approaches to Problems of Technological Change," in *Adjusting to Technological Change*, edited by Gerald G.

Somers and others (New York: Harper & Row, 1963), pp. 61–94;

Grant McConnell, *Steel and the Presidency* (New York: Norton, 1963);

David J. McDonald, *Union Man* (New York: Dutton, 1969);

George J. McManus, *The Inside Story of Steel Wages and Prices 1959–1967* (Philadelphia: Chilton, 1967);

John A. Orr, "The Rise and Fall of Steel's Human Relations Committee," *Labor History,* 14 (Winter 1973): 69–82;

Jack Stieber, "Steel," in *Collective Bargaining: Contemporary American Experience,* edited by Somers (Madison, Wis.: Industrial Relations Research Association, 1980), pp. 151–208.

George Magoffin Humphrey

(March 8, 1890 – January 29, 1970)

by Alec Kirby

George Washington University

CAREER: Lawyer, Humphrey, Grant & Humphrey (1912–1918); assistant general counsel, later general counsel and junior partner (1918–1924), executive vice-president (1924–1929), president (1929–1952), chairman of the board, M. A. Hanna Company (1952–1953); United States secretary of the Treasury (1953–1957); chairman of the board (1957–1961), chairman of the executive committee (1961–1963), chairman of the finance committee (1963–1967), director, National Steel Corporation (1967–1970).

George Magoffin Humphrey, lawyer, industrialist, and secretary of the Treasury under President Dwight D. Eisenhower, was born on March 8, 1890, in Cheboygan, Michigan, the son of a lawyer. Described as tall and rugged, Humphrey earned the trust of Eisenhower despite having roots in the Republican party's conservative wing led by Robert Taft.

Humphrey earned his law degree from the University of Michigan in 1912. Following graduation, he worked for his father's law firm in Saginaw. He was impatient with mundane legal work, however, and wished to make his mark in industry. In 1918 he moved to Cleveland, Ohio, where he was appointed first as assistant general counsel and then as general counsel for the M. A. Hanna Company, a firm that was founded in 1892 as a Great Lakes shipping concern and, by the beginning of the twentieth century, had branched out into mining, iron and steel production, and shipbuilding and had acquired railroad interests. In 1924, two years after the firm was incorporated, he was elected executive vice-presi-

dent, and in 1929 he succeeded Howard M. Hanna as president. His promotion came at a critical time in the history of the Hanna company; the firm was steadily losing money. Humphrey's first task was to dispose of the company's unprofitable interests and invest in areas with greater profit potential. Humphrey's strategy was to guide the firm more firmly into the iron-ore, freighter, and pig-iron industries. He also launched initiatives into such widely dispersed fields as plastics and banking. In 1929 he assisted in merging the firm's interests in Lake Superior iron ore, freighters, and steel production with the Weirton Steel Company and the Great Lakes Steel Corporation. The resulting umbrella firm was the National Steel Corporation, of which Hanna acquired a 28 percent interest. Yet even after the National merger Humphrey pushed for additional acquisitions. Hanna soon became interested in coke and iron and coal industries throughout Pittsburgh and the Great Lakes. As a result of his aggressive strategies of restructuring, Humphrey became one of the most influential industrialists in the Midwest by 1953. He continued as president of Hanna and also served either as president or on the board of directors of a wide assortment of steel-related corporations, including the Pittsburgh Consolidated Coal Company, the world's largest bituminous coal producer.

Humphrey's experience as an industrial leader made him a natural candidate for an advisory role in public affairs. Beginning in the 1940s, he took an increasing interest in associations and organizations seeking to promote corporate concerns. His first important federal post came in

George Magoffin Humphrey

1946, when he became chairman of the Business Advisory Council of the U.S. Department of Commerce. Two years later he served as head of the Reparations Survey Committee to advise the Allied powers on the issue of dismantling German industrial plants. He became an increasingly prominent advocate of federal retrenchment and of a sound dollar. In 1948 he actively supported the presidential nomination of conservative Republican senator Robert Taft of Ohio. After Taft barely survived a challenge in the Ohio presidential primary by Harold Stassen, Taft's chances for the Republican presidential nod grew dim. For the next four years Humphrey was not active in presidential politics. However, after Eisenhower won the presidency in 1952, Paul Hoffman, who as head of the Economic Cooperation Administration chose Humphrey to lead the Reparations Survey Committee, joined Gen. Lucius Clay in persuading Eisenhower to nominate Humphrey as secretary of the Treasury. The president-elect announced his choice in November 1952.

The selection of Humphrey was a source of reassurance to the conservative wing of the Re-

publican party, which viewed high federal spending and taxes as a prescription for inflation. Leading the charge against the federal budget, Humphrey took aim at not only domestic spending but also expenditures for foreign aid and the military. His efforts had immediate effect, as the incoming administration reduced the $9.9 billion deficit projected by President Harry S Truman for fiscal 1954 to $3.1 billion. Although the ending of the Korean War accounted for much of the savings of that year, the Eisenhower administration managed to achieve a budgetary surplus by 1956. As a staunch believer in unfettered free enterprise, Humphrey also took the lead in arguing that wage and price controls instituted during the Korean War be abolished. Although this policy seemed to contradict his anti-inflation policy, Humphrey argued that a "tight money" policy would counteract inflationary pressures, giving the economy time to increase production to the point at which the increased supply of goods would soak up inflationary pressures. Eisenhower agreed, and the controls were abolished in early 1953. Humphrey saw in the modest inflation that ensued a vindication of his arguments.

Inflation nevertheless remained his highest concern, and he continued to believe that strict control of the money supply could forestall rapid price increases. His move to increase interest rates with 30-year government bonds at a relatively high rate of interest was denounced as tightfisted, and, indeed, the economy soon entered a recession. Humphrey later admitted that he had erred in advocating high interest rates. His fear of inflation even led him to temper his enthusiasm for tax cutting. Early in his stay at the Department of the Treasury he opposed both a reduction of income taxes and a move to eliminate the wartime excess-profits tax, arguing that this would increase the deficit and lead to higher prices. With Eisenhower's support the tax cut was deferred. Several months later he came out in support of the cut, believing that federal spending was on a downward trend — which in fact it was: in 1954 government expenditures declined for the first time since 1948, and the budget deficit had declined by one-third.

The most dramatic moment of Humphrey's tenure as secretary of the Treasury came in 1957. The administration's proposed budget for fiscal 1958 embodied what Eisenhower, one week after his 1956 reelection, had termed "modern Republicanism." The budget, the biggest ever submitted to Congress, contained provisions for federal school construction, the development of water resources, and the expansion of social security. Eisenhower optimistically believed that continuing prosperity and the resulting increase in tax receipts would enable the government to close the fiscal year with a budgetary surplus. On January 16, as the budget went to Capitol Hill, Humphrey told reporters that the money requested by the administration was "a terrific amount" and complained about the "terrific tax take we are taking out of the country." He also predicted that if spending were not cut over a long period, "you will have a depression that will curl your hair." Pundits were quick to point out what they perceived to be insubordination on Humphrey's part, yet the charge was incorrect. At that point Eisenhower was concerned that Congress would increase spending above the administration's requests, and Humphrey was merely following administration strategy in trying to prevent such increases. Humphrey's comments, however, fueled a move to cut Eisenhower's requests, and the final appropriations were $4 billion less than the president had requested.

Humphrey continued to defend not only the fiscal 1958 budget proposal but the overall financial management of the Eisenhower administration. Nevertheless, in June of 1957 he resigned as secretary of the Treasury and became honorary chairman of M. A. Hanna. His public troubles were not over, however. In 1962 Sen. Stuart Symington, chairman of the Senate Armed Services Subcommittee on National Stockpile and Naval Petroleum Resources, launched an investigation into government contracts with suppliers of government stockpiles. An immediate issue was a 1953 contract, signed four days before Eisenhower's inauguration, between the federal government and the M. A. Hanna Company, which resulted in tremendous profits for the firm and its ore-mining and -smelting subsidiaries. At the time of the contract, Treasury secretary–designate Humphrey still served as president of Hanna. Moreover, a series of subsequent contracts through the 1950s brought Hanna further lucrative government revenues. From retirement Eisenhower defended Hanna. Still, in 1963 the Symington subcommittee published a report charging that there had been waste and mismanagement of the stockpile program and hinted — although it did not directly charge — that Eisenhower administration officials had acted improperly in aiding stockpile suppliers. Humphrey in turn charged that the report was politically motivated.

Continuing as honorary chairman of Hanna, Humphrey also served as chairman of the board of National Steel. He also was active in a wide variety of public and professional organizations. He died on January 9, 1970, in Cleveland.

Publications:
and Shepherd Krech, eds., *The Georgia-Florida Field Trial Club, 1916–1948* (New York: Scribners, 1948); "National's Steel Capacity Grows," *Steel*, 141 (November 11, 1957): 14–15.

References:
Charles Alexander, *Holding the Line: The Eisenhower Era 1952–1961* (Bloomington: Indiana University Press, 1975); Charles Horsley, "A Power in the Capital," *New York Times,* January 21, 1970, p. 47; Nelson Lichtenstein, ed., *Political Profiles: The Kennedy Years* (New York: Facts on File, 1976); Eleanora Schoenebaum, ed., *Political Profiles: The Eisenhower Years* (New York: Facts on File, 1977).

Archives:
Papers of George M. Humphrey as secretary of the Treasury, 1953–1957, are at Western Reserve Historical Society.

Charles Lukens Huston

(July 8, 1856 – March 14, 1951)

by Bruce E. Seely

Michigan Technological University

CAREER: Clerk and bookkeeper (1875–1879), puddle-mill supervisor (1879–1890), partner, (1881–1890), plate-mill supervisor (1882–1891), supervisor, open-hearth department, Charles Huston & Sons [initially Huston & Penrose, which was succeeded by Huston, Penrose & Company] (1890); second vice-president, director, works manager, Lukens Iron & Steel Company (1890–1897); vice-president (1897–1908), director, Belmont Iron Works of Philadelphia (1897–1944); first vice-president, Lukens Iron & Steel Company (1897–1917); vice-president, works manager, (1917–1925), vice-president, Lukens Steel Company (1925–1951).

Charles Lukens Huston's career in the steel industry spanned more than 75 years, all with Lukens Steel. He devoted his attention to technical operations, striving constantly to maintain the company's reputation as one of the nation's leading plate manufacturers. He held patents on various improvements in the manufacture of steel. Active into his nineties, he was involved with Lukens Steel up to the time of his death.

Charles Lukens Huston was born July 8, 1856, in Coatesville, Pennsylvania, near the iron works that the family had already built into a leading supplier of iron plates for steam boilers and structural uses. His great-grandfather Isaac Pennock had formed the Brandywine Iron Works and Nail Factory in 1810; his grandfather Dr. Charles Lukens had leased the mill in 1817. His grandmother Rebecca Pennock Lukens ran the works after her husband died in 1825, and her son-in-law — and Huston's father — Charles Huston, became a partner in the company in 1849. Huston's father also was an innovator and recognized expert on plate production; his pioneering studies in the testing of the strength of plates permitted him to begin marking plates with their tensile strength.

Huston graduated from Haverford College with a bachelor of arts degree in 1875, and, after

Charles Lukens Huston

taking a supplementary course at a Philadelphia business college, entered the family firm. Since 1870 he had worked part-time during the summers shoveling coal and handling other chores at the mill. In 1875 he was made a clerk and bookkeeper at $8 per week. In a pattern not often found in family firms, the fourth generation of the family proved as energetic as the first. In 1879 Charles was given charge of the puddling works and made a partner in Charles Huston & Sons in 1881, joining his father and brother Abram.

213

Over the next decade his father was not active in the management of the company, and Charles's responsibilities increased. The brothers were always guided, however, by their father's words, "In conducting the business, never get behind the times. It is better in the iron and steel business to be ahead of the times rather than one whit behind." With that thought, the company began making steel in 1881. In 1885 Charles Huston & Sons installed a machine to spin dish-shaped steel heads for boilers from hot steel, and Charles steadily improved this technology, making spun heads a company specialty. In 1882 Charles took charge of the plate mills, overseeing construction in 1890 of the largest plate-rolling installation in the country, a three-high mill able to roll plates up to 120 inches wide. The rolls were 34 inches in diameter and driven by a Corliss steam engine. In 1890 the company was incorporated as Lukens Iron & Steel Company and Abram was elected president while Charles became second vice-president and works manager. In that capacity he built the first open hearths for Lukens in 1891–1892. Charles became vice-president in 1897, retaining the title of works manager, and continued making improvements in the company's technology and the layout of the mills. Eventually he held 23 patents, including several related to the production of spun heads. His emphasis on quality and technical expertise paid off. Lukens, for example, enjoyed a major advantage when selling plate for the fabrication of locomotive fireboxes, because its plates were large enough to make the firebox from a single plate, eliminating seams that might cause a boiler to fail.

Maintaining this type of advantage was Huston's main concern. Lukens boasted the world's largest rolling mills in the 1870s, 1890, and 1903, but other firms eventually surpassed these in size. But in 1915 Charles designed a plate mill that held the size record for 40 years. The technical obstacles were enormous. Huston initially planned a three-high mill 180 inches wide, thereby surpassing in size a 178-inch mill in Austria. But no one could cast the chilled-iron rolls 50 inches in diameter. Huston then designed a two-high reversing mill with four rolls. In this design, the two chilled-iron working rolls, each 34 inches in diameter and weighing 30 tons, were backed by larger supporting rolls of cast steel, 50 inches in diameter and weighing 60 tons each. This arrangement compensated for the smaller size of the working rolls and kept the working rolls from springing, a problem that pro-

duced plates of uneven thickness. As a bonus, the design permitted Huston to build a wider rolling mill — 204 inches wide. The United Engineering and Foundry built the 40-foot tall mill in accordance with Huston's designs.

In 1925 Huston stepped down as works manager, although he remained vice-president for another 25 years. His only other business interest was the Belmont Iron Works of Philadelphia, of which he was vice-president from 1897 to 1908, and director until 1944. He was active in engineering circles, belonging to the American Society of Mechanical Engineers, the American Institute of Mining and Metallurgical Engineers, and the American Society for Testing Materials. He was elected an associate member of both the American and British Iron and Steel Institutes.

Huston also was involved in civic and humanitarian endeavors, reflecting another side of the Lukens tradition — a sense of responsibility to the company's workers and to the community. From 1907 to 1940 he served as director of the poor for Chester County; he also was president of the board of trustees for Coatesville Hospital. He supported the China Inland Mission, and a leper hospital in the Belgian Congo was built largely through his support. He helped organize the YMCA in Coatesville in 1891, and was local president until 1916. After the building was destroyed by fire in 1913, he contributed half the cost for a new structure; he also served on the state executive committee for the YMCA and was state president of the organization in 1902. A religious man, he was a commissioner to the general assembly of the Presbyterian Church from 1910 to 1924. He chaired the Philadelphia district of the World's Christian Fundamentals Association in 1931 and was a vice-president and director of the Montrose Bible Conference Association. Huston married Annie Stewart of Savannah, Georgia, in 1895; they had four children. The two boys, Stewart and Charles, followed their father into Lukens Steel and were executives in the company when he died in 1951.

References:
"Builds World's Largest Plate Mill," *Iron Trade Review,* 59 (November 23, 1916): 1041–1044, 1055–1056;

Eugene DiOrio, *Lukens: Remarkable Past — Promising Future* (Coatesville, Pa.: Lukens, 1985);

"One Hundred and Thirty Years of Iron and Steel Making," *Iron Trade Review,* 46 (June 30, 1910): 1263–1269;

Julian C. Skaggs, "Lukens, 1850–1870: A Case Study in the Mid Nineteenth-Century American Iron Industry," Dissertation, University of Delaware, 1975;

Skaggs, "Rebecca Lukens, " in *Encyclopedia of American Business History and Biography: The Iron and Steel Industry in the Nineteenth Century*, edited by Paul Paskoff (New York: Facts on File, 1989), pp. 243–247;

Skaggs and Richard L. Ehrlich, "Profits, Paternalism, and Rebellion: A Case Study in Industrial Strife," *Business History Review*, 54 (Summer 1980): 155–174.

Archives:
   The Hagley Library in Wilmington, Delaware, is the repository for the records of the Lukens Steel Company.

Charles Lukens Huston, Jr.

(May 19, 1906 –October 2, 1982)

by Bruce E. Seely

Michigan Technological University

CAREER: Various positions in metallurgical and operating department (1929–1934), personnel relations staff, ARMCO Steel (1934–1939); director, personnel relations (1939–1942), assistant to the president, Lukens Steel Company (1942–1943); president, Lukenweld (1943–1946); executive assistant to the president (1946–1949), vice-president (1948–1949), president (1949–1970), chairman of the board, (1970–1974); director (1974–1981), director emeritus (1981–1982), Lukens Steel Company.

Charles Lukens Huston, Jr.

Charles Lukens Huston, Jr., was the fifth generation of his family to head the Lukens Steel Company. With an operating background in metallurgy and personnel relations, he maintained the firm's traditional record of good labor relations. But he also kept the specialty steel firm positioned as a technological and quality leader. Decisions made during his two decades as president helped account for the continued success of Lukens Steel at a time when many American steel companies fell prey to changing economic conditions and steel imports.

Born May 19, 1906, in Coatesville, Pennsylvania, Huston came from a steel family; his father, uncle, a long line of predecessors, as well as his brother, Stewart, held high management positions at Lukens Steel, headquartered in Coatesville. He attended the Hill School in nearby Pottstown and went on to Princeton University, graduating with a bachelor of science degree in 1928. He spent a year at Massachusetts Institute of Technology studying metallurgy and related subjects and later took night courses at the University of Cincinnati in 1931–1932.

Huston's postgraduate education reflected an understandable interest in the science of steel; he had been made a director of Lukens Steel in 1928. But in 1929 he accepted a position as a metallurgical assistant at ARMCO Steel. In 1931, the year he was promoted to standard practice adviser at ARMCO, he resigned his directorship at Lukens. He moved into an administrative post as assistant staff supervisor of employment and personnel relations in 1934. In 1939 he returned to Coatesville and was hired as director of personnel relations at Lukens Steel.

In running the company, the Huston family for over a century had stressed management's responsibilities toward the workers and the community and in marked contrast to many steel executives, Huston harbored no antiunion sentiment. Typically, he asked a heretical question in a 1943 talk to the American Iron and Steel Institute (AISI): "Have we lost our men's loyalty to the unions?" Clearly reflecting his roots in a small steel company, Huston argued that the steel companies had the "responsibility" to satisfy the "fundamental desires of the workers," so they would say, " 'This is a good place to work; I like it here.' "

In 1943 Huston was elected president of Lukenweld, a Lukens subsidiary that specialized in fabricating plate using welding technology. In 1946 he became executive assistant to the president, Robert Wolcott. He filled a vice-president's post at Lukens in 1948, and Huston became president of Luken Steel in 1949, a post he held until 1970.

Lukens expanded significantly after World War II, as did many U.S. steelmaking firms in what seemed a golden age for American industry. Between 1945 and 1965, Lukens spent $105 million, adding facilities to make sodium hydride and a physical testing laboratory, expanding the cladding and alloy departments, and making other improvements to the production and finishing facilities. In 1957 Huston announced the largest single expansion effort in the firm's history — $33 million that provided an electric furnace, a 140-inch slabbing mill, more soaking pits, and more processing facilities. After the company's 150th anniversary in 1960 a second 100-ton electric furnace, mold preparation facilities, and scrap handling equipment were added. A third electric furnace — capacity: 150 tons — was completed in 1965, as the company became the first in the nation to install computer controls for the melt shop and vacuum-degassing equipment. The leading position of Lukens in electric steelmaking was confirmed in 1973, as the company retired its last open hearths and added a fourth electric furnace.

Clearly, Lukens made significant advancements under Huston. The company was a leader in the specialty steel market, a leading supplier of plate to the navy, and the manufacturer of the largest spun heads in the country. A final measure of the company's success under Huston was a rise in annual sales from $55 million in 1949 to $192 million in 1973. In 1974 he stepped down as chairman but remained on the board. His successors inherited a sound company, with a firm niche in the specialty plate market and a reputation for technological innovativeness. Thanks to this strong base, Lukens was able to survive the upheaval in the steel industry that began in the mid 1970s.

Huston served on several other corporate boards, including Marine Midland Bank, Newport News Shipbuilding and Drydock Company, and Penn Plastics Corporation. He was active in the AISI and the American Management Association. He sat on the executive council in 1946–1947 and on the management council from 1950 to 1956. He was a representative at large on Princeton University's Alumni Council from 1958 until 1963, a trustee of Drexel University from 1950 onward, and chairman of the board from 1967–1971. At various times, he served on the Pennsylvania State Advisory Council for Mental Health and on the board of Coatesville Hospital; he was the first president of Family Service of Chester County. From 1959 to 1965, he was a member of the Canadian-American Committee, whose purpose was to improve business relations between the two countries. For this type of service he received an honorary law degree from Drexel University in 1972.

Huston married Nancy Gardner of Middletown, Ohio, in 1933, and they had three children. For recreation he enjoyed boating, golf, and tennis, and maintained memberships in the Nantucket Yacht Club, the Sankaty Head Golf Club of Nantucket, the Upper Merion Cricket Club, and several golf clubs near Coatesville. He maintained residences in both his home town and Nantucket, Massachusetts. He died on October 2, 1982.

Publications:
"Design for the Future of Industrial Relations in the Steel Industry," *American Iron and Steel Institute Yearbook* (1943): 295–304;

"What Management Expects from Advertising: Lukens Steel Co.," *Industrial Marketing,* 35 (July 1950): 49;

"Increased Productivity Through Human Relations," *American Iron and Steel Institute Yearbook* (1951): 101–108;

"Progress, Past and Future," *Iron and Steel Engineer*, 35 (August 1958): 69–72.

References:
"Lukens Steel Company — A Specialty Plate Steel Producer," *Iron and Steel Engineer*, 51 (April 1974): 51–57;
"Lukens Steel — The Heart and Lifeblood of a Town," *Iron Age*, 165 (March 30, 1950): 119, 225–228;

"Start-up of 150-ton Electric Arc Furnace at Lukens," *Iron and Steel Engineer*, 52 (December 1975): 36–44.

Archives:
The Hagley Library in Wilmington, Delaware, is the repository for the records of the Lukens Steel Company.

Illinois Steel Company

by Stephen H. Cutcliffe
Lehigh University

The Illinois Steel Company was the first major consolidation in the United States steel industry; its formation typified an early trend in the industry toward mergers and trusts. Illinois Steel was incorporated with a capital stock of $25 million in May 1889 and consisted of three leading midwestern steel producers — the North Chicago Rolling Mill Company, the Union Steel Company, and the Joliet Steel Company. The North Chicago Rolling Mill, which in 1872 became the eighth U.S. company to install a Bessemer converter, had its origins in Eber Ward's North Chicago Works. The North Chicago Works in 1865 had rolled the first Bessemer-produced steel rails in America, while the Union and Joliet Steel Companies were the sixth and ninth Bessemer plants at the time of their respective construction in 1871 and 1873. The American Iron and Steel Association *Bulletin* called the merger "the most important event that has ever taken place in the history of the American iron trade."

The combined operations included 14 blast furnaces, an iron-rolling mill, four Bessemer steel plants, and mills for a range of finished products including rails, beams, merchant iron and steel, and rods and nails. The addition of its own coke ovens and some ore and coal holdings made it reasonably well integrated. With a labor force of 10,000 and a productivity capacity of almost one million tons annually of steel ingots, it was, at least briefly, the largest steel producer in the world.

Although the company faced stiff competition from the Carnegie Steel Company and would suffer as a result of the Panic of 1893, for the first few years of its existence Illinois Steel was very successful. In 1890, its first full year of operation, Illinois Steel earned a profit of $2,261,025, while the return on capital stock between 1890 and 1892 averaged 8.9 percent. In 1893, however, business fell. Plants were idle or run at partial capacity, and the company suffered a deficit of more than $885,000. Although business picked up slightly in subsequent years, the total average return on capital stock for its decade of existence came to only 3.8 percent.

During this difficult period Judge Elbert H. Gary, a midwestern specialist in corporate law, who was later to play a central role in the formation of the United States Steel Corporation, was appointed the company's general counsel at the suggestion of H. H. Porter, an Illinois Steel board director. As Gary's understanding of and familiarity with the steel industry deepened, he became convinced that to compete successfully with the Carnegie interests, whose profits had continued to grow despite the Depression, the Illinois Steel Company would have to expand both its raw materials reserves and its manufacturing facilities. In short, it had to become more fully integrated to produce finished and semifinished goods at prices competitive with those of Carnegie.

To accomplish this end, Illinois Steel with the help of Gary sought to acquire the Minnesota Iron Company, its largest supplier of ore. Gary turned to J. P. Morgan & Company for financial backing in mid 1898, and with Morgan's support and blessing, the Federal Steel Company was incorporated in New Jersey in September, absorbing Illinois Steel and several other midwestern producers and steel-related concerns. Although its average earnings during this period of its independent existence had not been great, its size, diversity of product, and sources of raw materials made Illinois Steel a central component of the

Federal Steel consolidation, which in turn was absorbed into U.S. Steel in 1901.

References:
William T. Hogan, S. J., *An Economic History of the Iron and Steel Industry in the United States,* 5 volumes (Lexington, Mass.: Lexington Books, 1971);
Ida M. Tarbell, *The Life of Elbert H. Gary: A Story of Steel* (New York: D. Appleton, 1925).

Inland Steel Corporation

by Louis P. Cain
Loyola University of Chicago
and
Robert M. Aduddell
Loyola University of Chicago

Inland Steel Corporation was founded in Illinois on October 30, 1893. It was one of the ten largest firms in the U.S. steel industry by World War I and grew to be one of the "Big Seven" by the 1950s. Inland's reputation in the industry is defined by a strong commitment to family management — which has lasted much longer than the other major producers — an adherence to a conservative financial policy that promotes internal growth through retained earnings rather than through acquisitions financed by external borrowing, and the maintenance of a single production site from which it marketed the majority of its output over a relatively small area in the Midwest.

Inland began production in Chicago Heights, Illinois, in January 1894. The Chicago Heights Land Association agreed to provide six acres of land and $20,000 for buildings if the firm located there. The company's founders arranged the purchase of the entire complement of machinery from the bankrupt Chicago Steel Company for $8,000. Inland initially fabricated harrow teeth, plow beams, cultivator attachments, and other farm implements by purchasing used steel rails, splitting them, then rerolling the steel. By 1895 it was producing bar stock and rolled bed rails as well. Despite the fact that Inland was launched during a relatively severe business downturn, it earned a profit of $17,747 during its first full year of operation.

Inland's initial common stock issue was 1,300 shares at $50 per share. There were eight original stockholders: Joseph Block and his son, Philip D. Block, held 500 shares; George H. Jones, the principal organizer, and his father-in-law, Elias Colbert, held 300 shares; while William M. Adams, Joseph E. Porter, John W. Thomas, and Frank Wells held the balance. The original board of directors was made up of the stockholders, with the exception of Thomas. The company's initial officers were: Porter, president; Jones, vice-president and general sales agent; Adams, secretary; Philip Block, treasurer and plant manager; Joseph Block, purchasing agent; and Thomas, superintendent. Jones replaced Porter as president in 1898.

The speed with which the new venture was launched can be understood in terms of the skills the founders brought to the task. Joseph Block had been senior partner and general manager of the Block-Pollack Iron Company of Cincinnati, an iron merchandising firm that had supplied the Chicago Steel Works with rails. George H. Jones had been sales manager for S. P. Kimbark and Company, wholesale hardware jobbers. Joseph E. Porter had been engaged in manufacturing and distributing farm implements in Ottawa, Illinois. John W. Thomas had been the foreman of the Chicago Steel Works and brought an intimate knowledge of the actual machinery Inland used.

Iron ore being unloaded from the L. E. Block, *a ship in the Inland Steel Great Lakes fleet, 1938*

During 1895 there was a minor dispute among board members concerning dividend policy. The two Blocks together with Jones and Colbert sought a policy of low dividends and a high rate of profit reinvestment. Adams, Porter, and Wells sought larger dividends. The dispute was resolved by Adams's resignation from the board and from his position as secretary. Thomas took Adams's position on the board; Porter became secretary; and Leopold E. Block, Joseph Block's eldest son, bought Adams's shares, bringing the family's ownership to 650 shares.

In 1897 the East Chicago Iron and Forge Company failed, and Inland purchased it for $50,000, incorporating it separately as Inland Iron and Forge. A year later Leopold Block left a position with a family interest in Pittsburgh to manage the new concern. Over the three years Inland owned this company, Inland Iron and Forge increased annual capacity from 25,000 tons to 70,000 tons.

The modern era in the history of Inland Steel began in 1901, when Joseph Block accepted the Michigan Land Company's offer of 50 acres of land on the shores of Lake Michigan at Indiana Harbor, Indiana, to any firm willing to spend $1 million to develop the site for the production of open-hearth steel. The site, more than 15 miles from the Chicago Heights plant, was well served by rail and water transportation. The $1 million needed to develop the land, however, far exceeded the company's total assets, which at the time were $435,000; its gross fixed assets were $195,000.

The funds for expansion were acquired from several sources. Inland Iron and Forge was sold to Republic Iron and Steel for $500,000. Inland's authorized common stock was increased from 2,600 (to which it had been raised in 1899) to 15,000 shares. The largest new stockholder was R. J. Beatty, owner of the Midland Steel Company in Muncie, Indiana. Beatty sold his firm and joined Inland as a director and general manager of the new facility. He brought with him a group of experienced superintendents, foremen, and craftsmen.

During this period Inland decided to become an integrated producer, a step requiring the acquisition of coal and iron ore supplies and the construction of a blast furnace. The decision was mostly due to the increasing difficulty of purchasing scrap and pig iron of the precise grade and quality necessary to supply the firm's expanded open-hearth operations. Integration was accomplished in a series of orderly steps between 1904 and 1915. A coal-mining and coal-washing operation was acquired near DeSoto, Illinois; the Laura iron ore mine at Hibbing, Minnesota, was purchased, along with 28 acres of land contiguous to the original Indiana Harbor lot for the purpose of constructing a blast furnace. This furnace, named Madeline #1 in honor of Philip Block's daughter, had a rated capacity of 400 to 500 tons per year; it produced its first iron on August 31, 1907.

Charles Hart of Republic Steel and Alexis W. Thompson were brought into Inland because of their expertise in such integrated operations, Hart becoming the third president of Inland Steel in September 1906 and Thompson named Inland's first chairman of the board in October. Hart served as Inland's president for less than two years, returning to the East Coast in April 1908; Thompson was named to replace him.

During the next six years, six additional open-hearth furnaces, three sheet mills, a plate mill, a billet mill, a new battery of coke ovens, and a by-products unit were constructed, and a second blast furnace, Madeline #2, was built in 1912. As a result, each 24-inch mill rolled only one product rather than two or more, and this resulted in a substantial savings in set-up time and costs. Two additional ore mining operations, the Grace on the Mesabi Range and the Thompson on the Cuyuna Range, were purchased in 1911. The Inland Steamship Company was formed that year, and two carriers were purchased — the *Joseph Block* and the *N. F. Leopold*.

Inland's fixed assets grew at an average rate of 37 percent between 1901 and 1914, and this led to 18.9 percent growth in rated ingot capacity at the Indiana Harbor works between 1904 and 1916. During this period, Inland's annual profits averaged 13.4 percent of gross fixed assets. Inland relied on financing such growth through new issues of common stock and retained earnings to a far greater extent than the rest of the industry. None of this growth was achieved through merger, the more common approach in the industry.

Joseph Block died on December 6, 1914. In the twenty-one years he spent at Inland, the firm had become a force in the national as well as the midwestern steel industry. His three sons were all involved in the firm. Leopold and Philip were vice-presidents at the time of their father's death. In July 1919, when Alexis Thompson left the pres-

idency, Leopold was named chairman of the board, Philip became president, and the youngest son, Emanuel, was named vice-president. They led the firm until World War II. In time, several of Joseph's grandchildren would also be named to the highest levels of Inland's management.

Inland emerged from the war-oriented economy of 1914–1918 in a relatively strong position. It reaffirmed its commitment to production of basic steel at a single location and confined its activities to vertical integration within the steel industry. In 1917 Inland moved its incorporation from Illinois to Delaware, where more liberal requirements might, they believed, increase the firm's flexibility. By 1920 Inland was producing 2 percent of the nation's steel and had become the seventh largest steel firm with the second highest average rate of return on fixed assets — 21.3 percent between 1901 and 1920. The location at Indiana Harbor was ideal given its proximity to raw materials sites, the surrounding increase in demand, and the existing transportation network. Pittsburgh-plus pricing — an industry-wide pricing system adopted in order to decrease competition — operated through Inland's early years, and further contributed to the company's growth and increased profits.

From 1911 to 1920, the uncertainties introduced by the antitrust case against the United States Steel Corporation, including the possibility that Pittsburgh-plus pricing might be banned, led Inland to discuss a merger with six other firms — Youngstown Sheet & Tube, Republic Iron & Steel, and Lackawanna Steel Company among them — but the talks ended when Lackawanna was acquired by Bethlehem Steel in 1922.

Inland had continued to improve and expand the Indiana Harbor plant, and when the merger became unlikely, Inland made two vertical acquisitions in 1924. Substantial plant capacity was acquired from the Milwaukee Rolling Mill Company to produce rolled black sheets, and galvanized and painted roof materials. This plant, which used steel produced in Indiana Harbor, was renamed Inland Steel Company of Wisconsin. Inland also acquired the Red Top Steel Post Company of Chicago Heights. During 1925 Inland added another blast furnace with an attendant battery of coking ovens. This, together with expanded by-product operations and continued electrification at Indiana Harbor, was financed in part by the placement of $12.5 million in gold bonds due in 1945. As in years past, however, Inland continued to rely primarily upon retained earnings and depreciation accruals to finance expansion.

Due to postwar adjustments, Inland's sales began to falter after 1918. Sales for 1919 were just over half those of 1918; a strong rebound occurred in 1920, but 1921 proved to be a weaker year than 1919. The sales levels of 1920 were not to be reached again until 1924. Nevertheless, by the end of 1925 Inland had grown to become the nation's fifth largest steel firm as measured by basic steel ingots supplied.

The late 1920s were prosperous years for Inland, with profits reaching $11.7 million in 1929. Renewed discussions for a merger between Inland and Youngstown were given serious consideration in 1928, but the two firms failed to reach agreement on some critical matters.

The Great Depression proved as difficult for Inland as it was for the rest of the steel industry. Profits during the period of 1931–1935 were less than what they had been in 1929 alone. In 1933 the company unwittingly missed paying a dividend for the only time in its history.

The Depression was also a period in which Inland's product mix changed drastically. Inland's management correctly judged that lighter rolled products for consumer goods applications would prove more lucrative and offer greater opportunities for growth than its existing emphasis on heavy products for the capital goods industries. Such a shift in output, while maintaining the requisite balance between ingot production and fabrication facilities, required a substantial capital outlay. It was also likely to require early write-offs of existing mills. This is not a problem when a firm experiences a healthy cash flow. When cash flow is diminished, however, reorientation can be a costly and chancy strategy. The strategy proved successful when in 1930 Inland secured the rights, under license from the American Rolling Mill Company, to use the continuous process for rolling plates, wide strips, and sheets. Construction of a new rolling mill began in 1931, and the following year the products of this mill were headed for the automobile and consumer appliance fields. By 1935 the revenues generated by the new lighter products surpassed the heavier products.

In 1935 Inland merged with Joseph T. Ryerson and Son Incorporated, the nation's largest steel warehousing firm, under the stipulation that Ryerson not lose its identity. The size of Inland's board was increased by the addition of four Ryerson executives, including Joseph T. Ryerson and

Edward L. Ryerson, who was named to the newly created position of vice-chairman of the board.

In the late 1930s Inland entered a period of growth by acquisition, a departure from its usual strategy of growth by construction. Two important fabricating firms acquired during this period were the Milcor Steel Company in 1936, a producer of fireproof building products, and the Wilson and Bennett Manufacturing Company in 1939, a producer of steel drums, barrels, and pails. Like Ryerson, Milcor retained its company identity even though it became a wholly owned subsidiary of Inland. Wilson and Bennett had been an Inland customer of long standing, and this acquisition assured Inland a captive market for its new lighter rolled products.

The late 1930s were difficult years for labor relations. Inland did not come to agreement with the Steelworkers Organizing Committee (SWOC) in 1937 and was thus a target of the "Little Steel" strike of that year. It was not until the start of World War II that Inland signed an agreement with the union.

The three Block brothers severely curtailed their activities as World War II approached. Emanuel retired as vice-president in charge of purchases in 1938, a year before his death. Leopold retired as chairman of the board in 1940 and was replaced by Edward L. Ryerson. Philip retired as president in 1941 and was replaced by Wilfred Sykes. Sykes, who served as assistant to the president beginning in 1930, joined Inland as a specialist in electrification.

Inland, under Sykes, was faced with the need to expand capacity for war production. This required a difficult transition back toward heavy products. Inland operated at full capacity during the 1941–1946 period; its operating rate during the period was 104.8 percent. While profits were restrained by price controls and excess profits taxes, Inland reduced its long-term debt and paid dividends throughout the war. Its wartime income measured as a percent of long-term investment averaged 12.8 percent over the war years, the highest return earned by any of the top twelve steel firms.

Inland emerged from the war, as did most of the nation's industrial firms, with a good deal of deferred maintenance and some new capacity related to the war effort. The immediate postwar period, therefore, included years of reconstruction and of transition back toward lighter products. This required a large infusion of capital and manpower. Blast furnace #6 went into operation in

1942 — replacing #3 — and provided a slight net increase in the output of pig iron. Clarence Randall replaced Sykes as president in 1949, and Joseph L. Block, son of Leopold and grandson of Joseph, replaced Randall when he was named chairman of the board succeeding Edward Ryerson, who retired in 1953.

The 1950s were difficult years for the steel industry. The first three years of the decade were influenced by the Korean War, recessions followed in 1954 and 1958, then came the 116-day national steel strike in 1959. Inland continued to produce all of its basic steel at the Indiana Harbor works. It remained the company's only integrated basic steel production facility, but it experienced substantial expansion. From an ingot capacity of 3.575 million tons in 1951, Inland increased the mill's output to 6.5 million tons by 1959. Inland made additional acquisitions to flesh out and expand its finished steel fabrication facilities and product lines, building rather than purchasing basic steel capacity. In keeping with its previously established policy, Inland maintained self-sufficiency in raw materials and transportation facilities insofar as possible. At the other end of the production flow, Inland's fabrication and distribution subsidiaries expanded their operations. Throughout the 1950s Inland outperformed the industry in terms of earnings and the percentage of full capacity operating rates achieved.

Clarence Randall left Inland in 1956, and the job of chairman was left vacant until 1959, when Joseph Block filled the position; John Smith was named to replace him as president. Smith served until 1966 when he was replaced by Frederick G. Jaicks. Joseph Block retired in 1967 and was replaced by his cousin Philip D. Block, Jr. When Philip retired four years later, Jaicks was named chairman, and Michael Tenenbaum was named president.

Problems arising from the increase in competition during the 1960s were not unique to Inland, but Inland was the only major steel producer that did not challenge the wage and price guidelines set by the administration of President John F. Kennedy in April 1962. The guidelines limited wage and price increases to the rate of growth of labor productivity, generally estimated at 3.2 percent; the steel industry, however, attempted to raise prices by approximately twice that amount. The Kennedy administration persuaded the United Steelworkers of America (USWA) to accept a wage increase within its guidelines. Believing that the demand for steel was weak and would not sup-

port a price increase, Inland held the line on its prices. Joseph Block also questioned the wisdom of a large industrywide price increase in the aftermath of a much smaller wage increase.

Inland faced three major challenges in the 1960s. First, domestic competitors were increasing their production along the southern shore of Lake Michigan. Bethlehem built a new integrated mill at Burns Harbor, Indiana, that specialized in flat rolled products, plates, and tinplate products — the mainstays of Inland's product line. National Steel built a sheet and plate finishing operation adjacent to Burns Harbor. Lykes-Youngstown expanded its Indiana Harbor operation; Jones and Laughlin built a plant at Hennepin, Illinois, south and east of Chicago on the Illinois River; and U.S. Steel increased its capacity to produce hot strip at its nearby South Works.

Second, foreign competition in Inland's midwestern market was expanding, with a sixfold increase in landings of foreign-produced steel at Great Lakes ports between 1963 and 1968. Given that Inland had historically sold over half its products within 150 miles of its home base at Indiana Harbor, increased shipments of foreign steel through lake ports presented a particularly serious threat.

Third, Inland was adopting new production methods, primarily the basic oxygen furnace (BOF) process. In 1963 construction began on two basic oxygen injection steel furnaces. When completed, the furnaces produced a minimum of two million tons of BOF process steel. This change in technology required Inland to use a higher volatility coking coal and pelletized iron ore in its blast furnaces. In 1960 pellets constituted only 1 percent of total ore used; by 1969, 74 percent of ore used was in pellet form. In addition some ore was now being processed into a taconite matrix, a step that required a substantial investment in beneficiation and pelletizing facilities.

Increased competition in the 1960s created an unstable price structure with consequent downward pressure on profit performance. Inland adjusted its costs by increasing the efficiency of its operations and increased the output of those products in which it had a competitive edge. Inland ended the 1960s with a substantial increase in both ingot capacity and the capacity to produce a wide variety of fabricated and basic products and shapes.

In the 1970s the demand for steel did not rise as rapidly as the gross national product. Inland's capital stock continued to rise as did its profits, but the percent of capital investment funded internally fell. The company continued to diversify, typically through its acquisition of firms, such as the A. O. Smith-Inland Corporation, created in 1969, which produced powdered metal for sintering and glass reinforced plastics. In order to plan diversification, the Inland Corporate Development Group was created and reported directly to the chairman of the board.

As the 1970s ended, Inland remained a viable and growing firm. It was self-sufficient in raw materials. It had pioneered and was still operating a large integrated steel mill at Indiana Harbor, now capable of producing approximately ten million tons of raw steel per year in a mix of BOF electric and oxygen-enriched furnaces. The acquisition of Ryerson had enabled Inland to produce and market customized steel to meet the varied needs of its customers. Frank Luerssen was named president in 1978 and also became chairman when Jaicks retired in 1983. R. J. Darnell became president in 1984.

Inland experienced net losses from 1981 to 1985 — the first consecutive years of losses in its history — causing the firm to make a series of difficult decisions. It restructured its management while remaining an integrated steel firm. This involved both downsizing the operations while narrowing its product line toward the higher value-added sheet and strip steel. In 1988 Inland became involved in two joint ventures with Nippon Steel that opened new markets as well as broadened their existing markets. Entering the 1990s, a leaner Inland stood out as an exception to the average American steel company that had lost the ability to compete in world steel markets.

References:
Henry W. Broude, *Steel Decisions and the National Economy* (New Haven: Yale University Press, 1963);
Wayde Grinstead, *50 Years of Inland Steel, 1893–1943* (Chicago: Inland Steel Company, 1943);
William T. Hogan, S.J., *An Economic History of the Iron and Steel Industry in the United States,* 5 volumes (Lexington, Mass.: Lexington Books, 1971);
"Inland Steel: 1902–1956," *Blast Furnace and Steel Plant* (August 1956);
C. E. Longnecker, "Inland: A Midwestern Steel Company," *Blast Furnace and Steel Plant* (January 1930);
Gertrude G. Schroeder, *The Growth of Major Steel Companies, 1900–1950* (Baltimore: Johns Hopkins University Press, 1953);
Herbert Solow, "Inland Does It Again," *Fortune* (July 1958);
H. L. Travis, "Outstanding Steels: National and Inland," *Magazine of Wall Street* (October 7, 1939);
Kenneth Warren, *The American Steel Industry, 1850–1970: A Geographical Interpretation* (Pittsburgh: University of Pittsburgh Press, 1973).

Interlake Steel Corporation

by Michael Santos
Lynchburg College

The Interlake Steel Corporation was the result of the 1964 merger of the Acme Steel Company and Interlake Iron Corporation. The agreement that created the new corporation brought together two financially struggling companies and produced what at the time was the eleventh largest integrated steel firm in the United States. Since 1964 Interlake's growth and diversification have been nothing short of phenomenal. In 1970 the firm changed its name from Interlake Steel to Interlake, Incorporated, to reflect its broader product lines and corporate interests. Between 1967 and 1983 sales increased from $300 million to $1 billion. By 1986 the iron and steel divisions of Interlake had become superfluous to the overall focus of the corporation, and a new firm, the Acme Steel Company, was created as a producer of hot and cold rolled sheet and strip steel.

The history of the multinational Interlake Corporation began in 1880 with the founding of an obscure steel staple manufacturer that eventually evolved into the original Acme Steel Company. The firm came into being almost by happenstance, the result of three men working independently on the problem of product damage in transit.

M. E. McMaster owned a furniture store in Shelbyville, Missouri. Making a regular 20-mile round-trip to the railroad line over rough country roads to pick up furniture for his customers, McMaster was increasingly frustrated that the side rails of the bedsteads he ordered frequently did not withstand the journey from the shipper to his shop. Puzzling over the problem, he came upon the idea of making a steel staple with barbs in each end which, when driven into the ends of the two rails, prevented rubbing and damage in shipment and storage. Patenting his idea and the machine to produce it, McMaster moved to Quincy, Illinois, where he could run his machine by water power. McMaster modified his invention during the next few years by using different materials and expanding their functions, including developing brass staples for use as ornamental floor plates. The demand for his product grew such that in 1889 he incorporated his business as the Quincy Floor Plate and Staple Manufacturing Company.

No sooner had the firm incorporated than James E. MacMurray bought controlling interest in the company and became its president. MacMurray had earlier invented a machine that turned steel coil into nailed-on strapping to reinforce shipping boxes. Expanding the company's production line to include nails, staples for wire fencing, corn poppers, and cistern equipment, MacMurray renamed the firm the Quincy Hardware Manufacturing Company.

Meanwhile, in Chicago, William O. Swett had devised an elongated wire staple that could be used to hold the lid on a wooden bucket. Patenting his idea in 1884, he founded Acme Flexible Clasp Company the same year.

Having expanded his company and the demand for its products through aggressive salesmanship — recruiting orders from shippers as far away as London — MacMurray was ready to increase his firm's output capacity. In 1899 he persuaded Swett, then in his late seventies, to merge Quincy Manufacturing and Acme Flexible Clasp Company under the latter's name. Swett retired from the business shortly thereafter, leaving MacMurray president of the firm.

From 1900 to 1925 the company's size and production grew. Expanding its product line to include box bands and shingle bands, Acme found its suppliers of steel unable to produce a high quality steel to meet its needs. In 1905 the board of directors voted to acquire several small cold rolling mills that would allow the company to flatten round wire into split box strapping. Changing its name to Acme Steel Goods Company

in 1907, the firm was cold rolling enough steel by 1914 to meet all its needs. Increased demand for Acme products generated by World War I fostered further expansion, and in 1917 the board of directors decided to open a rolling mill in Riverdale, Illinois, with the idea of making all the steel the company needed and marketing the rest. By 1925 the firm had moved so completely into the manufacture of heavier lines of strip steel that its name was changed to Acme Steel Company.

Having carved out a specialized yet diverse market for its products, Acme Steel survived the Great Depression in remarkably good shape, paying dividends — although modest at times — on its stock throughout the 1930s. World War II increased demand for Acme goods as it did throughout the steel industry, and the company entered the postwar period in what appeared to outsiders to be good fiscal shape. Between 1946 and 1956 Acme expanded its facilities and holdings, acquiring the Morrison Stitcher line, the Howell Company and its subsidiaries, Geneva Modern Kitchens, Elgin Kitchens, the Stanley Corporation, and the Newport Steel Corporation. It opened the Acme Company of Canada in Scarborough, Ontario.

The appearance of growth, however, was deceptive. Acme's acquisition program represented an attempt by the company to stay solvent in an increasingly hostile market. As an unintegrated firm, Acme had to pay a higher price for the semifinished steel it bought than it could charge for the products it produced. Forced to modernize to stay competitive, the company acquired a $36 million long-term debt that threatened to bankrupt it by the early 1960s.

The Interlake Iron Company, meanwhile, was facing similar problems. Interlake had its origins in 1905 when the By-Products Coke Company was formed. Between 1905 and 1930 By-Products Coke had built several coke ovens and constructed plants for several steel firms in Indiana and Ohio. In 1930 By-Products Coke acquired the Zenith Furnace Company, the Toledo Furnace Company, and the Perry Furnace Company and changed the company's name to Interlake Iron Corporation. Primarily a pig iron producer, Interlake, like Acme Steel, was an unintegrated firm and by the early 1960s faced a declining market for its products.

With both companies confronting economic hard times, they began to negotiate a merger with an eye to creating an integrated steel company that could compete successfully in the iron and steel market. The logic for the move was simple: Acme Steel needed the pig iron and iron ore that Interlake produced, and Interlake could take advantage of the greater demand for steel and Acme's greater production capacity. In December 1964 the two firms combined to form Interlake Steel Corporation.

In 1967 the company hired Reynold C. MacDonald as its president and chief operating officer. MacDonald had made a reputation for himself as the vice-president of operations at Lone Star Steel Company, helping to make it one of the more successful small steel firms in the country. When MacDonald took over Interlake, the company was at a crossroads. It could invest in expanding its iron and steel facilities, but, as MacDonald noted, this would be costly. He proposed, and the board of directors concurred, that Interlake's future, like that of many other iron and steel firms at the time, lay in diversification.

Moving into the area of material handling, Interlake acquired Hoeganaes Corporation and Lodi Fab Industries, Incorporated, and in 1970 changed its name to Interlake, Incorporated, to reflect its new diversified look. Throughout the 1970s and 1980s the firm continued to grow, moving into such diverse areas as metal powders, plastics, and aerospace components and acquiring firms and facilities in England, France, Germany, Australia, and Canada.

In 1967, when MacDonald assumed the presidency of Interlake, 53 percent of the company's sales came from iron and steel. By his retirement in 1983, 75 percent of Interlake's revenue came from nonsteel manufacturing operations. The creation in 1986 of a new Acme Steel Company from Interlake's iron and steel divisions is perhaps the ultimate commentary on how far Interlake had come from the days when McMaster, MacMurray, and Swett experimented with steel staples and strapping.

Reference:
William T. Hogan, S.J., *An Economic History of the Iron and Steel Industry in the United States,* 5 volumes (Lexington, Mass.: Lexington Books, 1971).

Iron Age

by Bruce E. Seely

Michigan Technological University

Iron Age was one of the first industrial trade publications in the United States and one of the longest continuously published magazines of any kind. John Williams began publishing *The Hardwareman's Newspaper and American Manufacturers' Circular* in the summer of 1855 but changed the name to *Iron Age* in April 1859. Assisted by his son, David, he continued to provide equal treatment of the iron, hardware, and machine building industries. Williams's most important innovation was the publication of commodity prices gathered by correspondents in several cities, a service that aroused complaints from some hardware salesmen and manufacturers. By 1865, boasting a circulation larger than any paper of its kind, the journal had grown to blanket folio size in order to provide space for detailed coverage of personal, technical, and business developments in industry. A metallurgical column that appeared in 1869 was the first anywhere in English.

This coverage cemented connections between *Iron Age* and business leaders. Judge Elbert H. Gary of United States Steel Corporation, for example, saw trade journals furthering his goal of a stable steel industry. Their editorial positions often siding with the industry, the editors of *Iron Age* and its leading rival, *Iron Trade Review,* were the only journalists invited to the Gary dinners, which evolved into the annual meetings of the American Iron and Steel Institute (AISI). Articles covered technical and business developments in the trade; and regular columns provided production statistics, prices, ore shipments, and other statistical information. In attempting to emphasize the importance of the role the trade journal played in the industry, a commentator at a 1918 session of the AISI asserted that, "without its trade papers," the steel business would be reduced to a "condition of chaos."

Iron Age was always the leading paper, although *Iron Trade Review* emerged as a serious competitor after 1900. David Williams continued to publish his father's journal for 54 years, selling the paper to a group of business paper publishers in 1909. The new owners turned *Hardware Age* into a separate publication, but *Iron Age* was not affected. It continued to grow, adding an annual review of the state of the iron and steel industry. The magazine even won mention in a 1920s Broadway play in which someone needed to be hit with a heavy object. The line read, "Hit him with a copy of *Iron Age!*"

When the Chilton Company bought the Iron Age Publishing Company in December 1934, it acquired a journal that had outdistanced its competitors. *Iron Age* ran more advertising pages than any other magazine in the country, more than all other metal trades journals combined. It lived up to Chilton's claim that *Iron Age* was "The First Paper in the Metal Working Field." Circulation was 16,000 in 1940, with 1,100 of those abroad, and only about an eighth of the subscriptions went to the steel industry.

After World War II *Iron Age*, like many trade journals, reduced the depth of coverage in many feature columns. The personals section, for example, ran one-line notices of new positions rather than biographical sketches. But its news coverage of the industry remained journalistically solid, as was

widely recognized at the centennial of *Iron Age* in 1955. But when the iron and steel industry encountered rough times in the 1970s, *Iron Age* also declined. In 1977 it was renamed *Chilton's Iron Age*; in 1984 it began appearing as two separate, semimonthly publications: *Chilton's Iron Age — Manufacturing Management* and *Chilton's Iron Age — Metals Producer*. After 229 volumes the trade journal was renumbered starting in March 1986. In December 1987 Fairchild Publications purchased it, renaming it *Iron Age: The Management Magazine for Metal Producers*. *Iron Age* now appears monthly.

References:
A. O. Backert, "Relation of the Trade Papers to the Iron and Steel Industry: Discussion," *American Iron and Steel Institute Yearbook* (1918): 161;
David P. Forsyth, *The Business Press in America, 1750–1865* (Philadelphia: Chilton, 1964);
"The *Iron Age* Celebrates Its Centennial," *Printers' Ink*, 251 (June 24, 1955): 48–49;
"The *Iron Age, 1855–1955*," *100 Years of Metalworking*, special issue of *Iron Age* (June 1955): 2–16;
"The *Iron Age* Gears Up for a Second Century," *Sales Management*, 74 (May 20, 1955): 56, 58, 60, 61;
Roy Johnson, "Business and the Business Press," *Printers' Ink*, 118 (January 26, 1922): 41–42, 44;
"Making America Strong," *Sales Management*, 47 (December 15, 1940): 88.

Iron and Steel Engineer

by Bruce E. Seely

Michigan Technological University

Iron and Steel Engineer is the monthly publication of the Association of Iron and Steel Engineers. Almost from its inception, it has been one of the leading technical publications in the steel industry.

At the turn of the century electricity began to challenge steam engines as the preferred motive power of the steel industry. The opposition electricity encountered, however, led its advocates to organize as a professional group — the Association of Iron and Steel Electrical Engineers — in 1907. Membership was open to engineers from any company who could compare notes and learn how others were coping with the problems they faced. The association published proceedings be-

ginning in 1907, and a bulletin was started in 1919. In 1924 a monthly journal, *Iron and Steel Engineer*, replaced the bulletin, combining announcements of association business with technical papers and discussion.

Electrical subjects dominated the journal for many years, but *Iron and Steel Engineer* also began to treat other topics as the association expanded its interests to broader technical questions facing the industry. Reflecting this shift, the association changed its name to the Association of Iron and Steel Engineers in 1937. *Iron and Steel Engineer*, however, remained a technical publication. After 1925 it carried the association's proceedings. It reported on conferences and conventions of interest to engineers in the industry, printed technical papers, and devoted regular columns to furnaces, rolling, steelworks masonry, and maintenance. In addition, writers such as T. J. Ess, an engineer with 16 years of experience at Republic and other steel companies, prepared articles that described in detail the equipment and layout of individual plants. Ess became a fixture at the journal after joining the association in 1938 and remained so into the early 1960s. Another highlight has been the annual review issue, which provides a thorough treatment of changes in the technology of steelmaking. Initially appearing each January, after 1976 the review issue was moved to February.

As the product of a professional organization, *Iron and Steel Engineer* has always been less commercial than *Blast Furnace and Steel Plant*, the trade journal closest to *Iron and Steel Engineer* in content. Perhaps for similar reasons, a longtime feature was "The Open Hearth," which reported on career moves in the industry, especially among operating personnel. Only after 1971 did the personal columns become simple notices of career moves. Significantly, by the late 1960s an increasing number of papers came from European and Japanese engineers.

Iron Ore Beneficiation

by Terry S. Reynolds
Michigan Technological University

Beneficiation is the use of mechanical processes to increase the retrievable iron content and reactive properties of natural iron ores. Before 1900 almost the only method used to improve the quality of ores was crude hand sorting prior to shipment. Occasionally ores were crushed and screened to provide a more nearly uniform structure, but the amount of iron extracted was not altered. Most iron ores used in American blast furnaces were natural, or direct-shipping, ores.

Beneficiation slowly grew in importance, encouraged by the disappearance of high-grade natural ores. On the Mesabi Range in northern Minnesota, for example, the iron content of ore shipments dropped from 55 percent in the mid 1890s to about 51.5 percent by 1911. As iron content dropped, mine operators sought beneficiation processes to raise the yield of their ores. All of the processes introduced in the twentieth century involved crushing and grinding ores to fine particles so that impurities (usually silica) could be more easily separated from iron-bearing particles.

Thomas Edison and others had experimented with magnetic separation to improve the iron content of magnetic ores in the late nineteenth century with little success. Between 1900 and 1920, however, several mines in the eastern United States succeeded in applying magnetic separation commercially. But most American iron ores were nonmagnetic and required other forms of beneficiation. These methods relied on the differing densities of iron and silica to promote separation. The earliest techniques promoted separation by washing. When ore was agitated in water, iron and silica particles were partially separated due to their differing densities. Attempts to apply this technique failed before 1900, including a large plant erected in 1881–1882 near Negaunee, Michigan, that concentrated some 12,000 tons before it closed. The first commercially successful washing plants were placed in operation in Alabama around 1908, and the Oliver Iron Mining Company erected the first ore-washing plant on the Minnesota ranges at Coleraine in 1910 in response to the declining iron content of Mesabi shipments. By 1923, 18 washing plants and one magnetic concentration plant were operating in Minnesota. By 1940 washing had become a widely used and important beneficiation technique.

Some nonmagnetic ores, however, could not be beneficiated by washing. In the 1920s metallurgists developed jigging — the use of pulsing streams of water to float away impurities and secure more complete separation. The first commercially successful Minnesota jigging plant was built at Biwabik, Minnesota, in 1924, although it was predated by a jigging plant erected on the Gogebic Range in Michigan around 1900.

In the 1930s and the 1940s American metallurgists introduced more-complex high-density techniques for beneficiation, often adding these to older washing plants. In these processes crushed, screened, and washed ores were subjected to additional treatment, often involving chemicals which promoted separation between materials of different densities. The first successful high-density medium-separation plants were erected in the late 1930s on the Mesabi.

New nonchemical separation techniques emerged in the 1940s. Shortly after World War II, for example, Minnesota mining companies introduced the Humphrey spiral, which consisted of a helical trough with ports. Because of their differing densities, waste silica and iron concentrate could be drawn from different ports as a slurry containing both descended the trough.

By 1950 iron mining companies could combine beneficiation techniques and raise the iron yield of intermediate ores from around 40 to 45 percent to around 60 percent. Washing or jigging alone typically raised the content of ores from around 40 to 45 percent to around 50 to 55 percent. High-density methods could then increase iron yield further to around 60 percent, comparable to high-grade natural ores.

The proportion of ores beneficiated by one means or another grew slowly in the early twentieth century. In the 1910s probably less than 10 percent of all ores was beneficiated. This proportion had risen to only around 20 percent by 1940. The slow expansion was largely due to the continued availability of high-grade ores, an availability encouraged by tax laws. Minnesota was by far the largest iron producing state, and after 1914 Minnesota taxed mining companies based on the assessed value of their ore reserves. As a result, mining companies on the Minnesota ranges abandoned attempts to develop low- and intermediate-grade ores in favor of depleting their high-grade ores at as rapid a rate as possible, since the longer they stayed in the ground the longer the company paid high taxes.

World War II and the postwar economic boom, however, accelerated the exhaustion of high-grade ore reserves and caused the American iron and steel industry serious concern over future sources of ore supply. This pushed American iron mining companies in two directions. First they began to seek off-shore sources of high-grade iron ores, and imported high-grade ores from these sources began to penetrate the American ore market seriously in the 1950s. At the same time, American companies investigated the possibility of beneficiating the large reserves of domestic low-grade iron ores.

Most early beneficiation work had focused on improving high-grade or intermediate-grade ores. Low-grade iron ores, such as Minnesota taconite and Michigan jasperite, both with 25–35 percent iron content, had been largely ignored. There were two exceptions. In 1922 the Mesabi Iron Ore Company had attempted to beneficiate low-grade ores commercially. The process involved crushing and grinding very hard taconite, followed by a series of concentrating processes. The company produced around 150,000 tons of beneficiated ore, but declining ore prices forced it out of business in 1924.

This left only Minnesota's Mines Experiment Station, under the direction of E. W. Davis, in the field. Davis had begun to work on beneficiating taconite as early as 1913. In subsequent years he and his associates continued to try to develop the technologies necessary to make it commercially exploitable. Because of the hardness of Minnesota taconite and its low iron content, Davis and his associates had to grind taconite almost to the consistency of flour for existing beneficiation processes to be successful. This meant that an agglomeration process was necessary before concentrated taconite could be transported and fed into blast furnaces. The most widely used agglomeration process — sintering — would not work. Thus Davis and his associates set out to find an agglomeration process that would convert taconite concentrate into a blast-furnace feed superior to natural ores. They took the finely ground concentrate and mixed it with a binding agent, bentonite. As the mixture was slowly rotated in a heated drum, the bentonite and the taconite formed marble-sized pellets by the snowball effect. These pellets were baked in a kiln to a hardness suitable for transportation and blast-furnace feed.

In 1955 the Reserve Mining Company opened a small commercial plant to concentrate taconite and produce pellets using Davis's processes. Another small plant was erected in 1956, but taconite production expanded only at a moderate rate until the 1960s. Two events prompted more-rapid expansion. In 1960 blast-furnace tests conclusively demonstrated the great superiority of pellets to natural ore as a blast-furnace feed. And in 1963–1964 Minnesota modified its tax laws to provide incentives for the development of its low-grade ore reserves, with Michigan following suit. In 1955 American ore suppliers produced only slightly more than 1.1 million tons of pellets. This figure increased to 13.7 million tons by 1960 and grew rapidly to 58.5 million tons by 1969 and to more than 80 million tons by the mid 1970s. The introduction of taconite pellets enabled the Lake Superior district to maintain its position as the leading iron-producing district in the United States and halted, in the 1960s, the further growth of American dependence on imported iron ores.

In the quarter century after 1950, beneficiation drastically changed the nature of the American ore trade. In the early 1950s more than three-quarters of the ore fed to American blast furnaces had been raw ore. By 1964 the situation had reversed — more than three quarters of all ores shipped were beneficiated in some manner before being sent to market. The durability of pelletized iron ore in transportation and its excellence as a blast-furnace feed pushed this process even further in ensuing decades. By the 1980s only 3 percent of the material consumed in American blast furnaces was natural, direct-shipping ore, about 75 percent was pellets, and another 20 percent was sinter.

References:

Nathaniel Arbiter, "Mineral Beneficiation: A Perspective," *Centennial Volume: American Institute of Mining, Metallurgical, and Petroleum Engineers, 1871–1970* (New York: AIME, 1971), pp. 88–118;

T. B. Counselman, "Recent Developments in Iron Ore Beneficiation in the United States," *Blast Furnace and Steel Plant*, 33 (January 1945): 90–95, 156, 160–161;

E. W. Davis, *Pioneering with Taconite* (Saint Paul: Minnesota Historical Society, 1964);

W. H. Dennis, *Foundations of Iron and Steel Metallurgy* (Amsterdam: Elsevier, 1967);

M. M. Fine, "The Beneficiation of Iron Ores," *Scientific American*, 218 (January 1968): 28–35;

Bela Gold and others, *Technological Progress and Industrial Leadership: The Growth of the U.S. Steel Industry, 1900–1970* (Lexington, Mass.: Lexington Books, 1984);

W. A. Haven, "Iron Ore Beneficiation Assumes a Major Role in the Manufacture of Pig Iron," *Blast Furnace & Steel Plant*, 33 (January 1945): 81–83, 89;

William T. Hogan, S.J., *An Economic History of the Iron and Steel Industry in the United States*, 5 volumes (Lexington, Mass.: Lexington Books, 1971);

Charles L. Horn, Jr., *The Iron Ore Industry of Minnesota and the Problem of Depleted Reserves* (Minneapolis, Minn.: Charles L. Horn, Jr., 1956);

Lawrence A. Roe, *Iron Ore Beneficiation* (Lake Bluff, Ill.: Minerals Publishing, 1957);

E. H. Rose, "Iron Ore: The Big Picture," *Journal of Metals*, 13 (November 1961): 814 –819;

Peter Warhol, A. E. Matson, and Louis J. Erck, "Short History of Progress Used to Date on Intermediate Ores," *Skillings Mining Review*, 35 (February 1, 1947): 1–2, 6, 15.

Iron Ore Sintering

Terry S. Reynolds

Michigan Technological University

Sintering is an iron-ore beneficiation process that by controlled burning fuses small, iron-bearing particles and coke dust (fines) produced by iron mining, ore handling, ore beneficiation, and steel-mill operations. The rough, porous lumps resulting from the process are used to feed blast furnaces.

Sintering was first used extensively in nineteenth-century Europe to improve the chemical and physical characteristics of sulfide ores by heating them in cast iron "pots" to drive out the sulfur. Initially the process was applied only to nonferrous ores because of their greater per-pound value.

Around 1890 James W. Neill, an American mining engineer, attempted to apply sintering to pyrite iron ores on a commercial scale. The method he developed was similar to that used on nonferrous sulfide ores — heating batches of ore in large iron pots or boxes — but he abandoned it around 1902. Other metallurgists developed similar batch processes early in the twentieth century for sintering flue and blast-furnace dust. In these processes the coke content of the dust served the same function as the sulfur in sulfide ores — it provided the internal fuel necessary for the operation. None of these processes, however, enjoyed long-term commercial success. As batch processes, they required too much labor to charge and discharge the pots and to break the sinter into lumps of suitable size.

In 1906 A. S. Dwight and R. L. Lloyd, working at a lead smelter in Mexico, developed a continuous-process sintering machine, which replaced the cast iron pots, or "boxes," with relatively shallow iron trays, called pallets, mounted on an endless conveyor. A hopper distributed sulfide ores in the boxes. As the ore-filled boxes slowly moved through a combustion zone, the sulfur in the ore was ignited, and the heat roasted and fused the ores. The older batch processes had used an upward flow of heat which was not evenly distributed and which disturbed the upper layers of the ore fines. Dwight and Lloyd corrected these problems by drawing air down through the relatively thin layers of ore in the pallets. After passing out of the ignition region, the pallets dumped the fused fines — which were thin enough to be easily broken to appropriate size — at the far end of the machine and returned to be reloaded at the hopper; the cycle was then repeated.

The Dwight-Lloyd machine required some modification before it could be applied to iron ores. Because of lower per-pound value of iron ores, machines used to treat iron fines had to have a larger capacity than those used for lead and copper ores. And most iron ores, unlike nonferrous sulfide ores, did not have an internal fuel.

Metallurgists worked out these problems within a decade. They enlarged the capacity of Dwight-Lloyd machines from 30 tons a day to 50–100 tons per day (modern machines have capacities approaching 10,000 tons per day) and made mechanical operation more reliable. Solving the fuel problem was a little more difficult, however. Thus the first application of the sinter machine in the iron and steel industry was not to iron ores but to the fines produced by blast furnaces. These contained both reusable iron ore and an internal fuel in the form of coke dust. The first Dwight-Lloyd machine for sintering blast furnace fines was put into operation in 1911. The machine transformed dusts — which could not be recycled directly because they would have clogged blast furnaces — into a material with a suitable physical consistency. Iron and steel producers quickly sensed the advantages of sintering. Between 1920 and 1930, for example, they erected 32 sintering plants and 40 more in the 1930s.

Although American metallurgists experimented with sintering iron-ore fines as early as the 1910s, most of the early American sintering plants used only flue dust. Beginning in the late 1920s, however, the ore industry began to use sintering to reclaim ore fines, providing the fuel needed for the process by mixing ore fines with flue dust. The resulting sinter was then mixed with crude ores in a 40-to-60 ratio for charging blast furnaces.

Initially, sintering was introduced to reclaim materials otherwise lost. But studies by C. E. Agnew, J. T. Whiting, and others in the late 1930s and early 1940s indicated that sintering had other advantages: as a feed material it was superior both physically and chemically to natural ores. This discovery increased the use of sintering as a prefurnace treatment for iron ores.

The rapid expansion of sintering came in the 1950s and early 1960s as a result of declining American supplies of high-grade direct-shipping ores. To compete with imported high-grade ores, the American ore industry was compelled after World War II to boost the iron yield of its large reserves of low-grade ores. Initially, the beneficiation processes involved crushing, grinding, washing, screening, and otherwise treating low-grade ores. These operations produced rich iron-bearing particles which had to be agglomerated before being physically suitable as a blast-furnace feed. Sinter production thus increased from less than 8 million tons in 1950 to over 25 million tons in 1955 and to over 50 million tons in 1964. American blast-furnace burdens averaged 40 to 50 percent sinter in 1965 compared with only 10 to 20 percent sinter a decade earlier.

In the late 1960s the volume of blast-furnace feed provided by sintering began to decline, largely due to the rise of the pelletizing process of iron-ore beneficiation. Pelletized iron ore was physically and chemically a better blast-furnace feed and more easily transported. These characteristics also sharply reduced the volume of ore fines and blast-furnace dust, two of the principle raw materials used for sintering. In 1988 only 17 million tons of sinter were consumed by the American iron and steel industry, little more than one-third of the peak in the 1960s. Nonetheless, sintering remains an important auxiliary process in ore beneficiation, providing between 10 and 20 percent of American ore consumption.

References:

The Competitive Status of the U.S. Steel Industry (Washington, D.C.: National Academy Press, 1985);

W. H. Dennis, *Foundations of Iron and Steel Metallurgy* (Amsterdam: Elsevier Publishing, 1967);

Dennis, *A Hundred Years of Metallurgy* (Chicago: Aldine, 1964);

William A. Haven, "Iron Ore Beneficiation Assumes a Major Role in the Manufacture of Pig Iron," *Blast Furnace and Steel Plant*, 33 (January 1945): 81–83, 89;

William T. Hogan, S.J., *An Economic History of the Iron and Steel Industry in the United States*, 5 volumes (Lexington, Mass.: Lexington Books, 1971);

Michael O. Holowaty, "History of Iron Ore Sintering Recalls Variety of Experimentation," *Journal of Metals*, 7 (1955): 19–23;

Lawrence A. Roe, *Iron Ore Beneficiation* (Lake Bluff, Ill.: Minerals Publishing, 1957);

H. E. Rowen, "Development of the Dwight-Lloyd Sintering Process," in *History of Iron and Steelmaking in the United States* (New York: Metallurgical Society, 1961), pp. 57–60;

E. W. Shallock, "Thirty Years of Iron Sintering," *Blast Furnace and Steel Plant*, 28 (January 1940): 71–75; (February 1940): 169–170.

Iron Ore Supplies

by Terry S. Reynolds
Michigan Technological University

The story of American iron ore supplies in the twentieth century can be broken into two broad periods: pre–World War II and post–World War II. In the pre–World War II era practically all American iron ores were direct-shipping ores, produced domestically. In the postwar period beneficiated ores displaced direct-shipping ores, and imported ores became an important source of American iron.

Scattered, low-grade iron deposits are common in the United States. But by 1900 few blast furnaces were using these supplies. Iron ore mining was already geographically consolidated into a few major producing areas. By far the most important of these was the Lake Superior District. The six ore ranges in this district — the Marquette, Menominee, and Gogebic ranges in upper Michigan and Wisconsin and the Mesabi, Vermillion, and Cuyuna ranges in Minnesota — supplied 75 percent of all U.S. iron ore by 1900 and 86 percent by 1920; this pattern persisted throughout the twentieth century. The Mesabi Range, located around 65 miles north of Duluth, Minnesota, was the most important of the six Lake Superior District iron ore ranges. Blessed with rich ores located near the surface and accessible by open-pit mining methods, the Mesabi Range for many years produced more than half of all American iron ore.

Through most of the twentieth century the Southern District near Birmingham, Alabama, was the second leading supplier of domestic iron ores. In 1900 this region produced 16 percent of all American ores; however, demand soon shifted to the richer ores of the Lake Superior District and the highly mechanized, open-pit methods which could be used there. Despite increased production, the Southern District's contribution to American ore supplies slipped to 10 percent by 1920, and even this figure was misleadingly high because the iron content of the Alabama ores averaged 35 percent compared to 50–55 percent for Lake Superior District ores.

The Northeastern District, consisting of mines in the Adirondacks of New York and a few mines in New Jersey and Pennsylvania, typically provided 2 to 4 percent of American iron ores between 1900 and the 1940s. Mines in the western United States, during the same period, furnished less than 2 percent. Before World War II most ores shipped from all of these districts were direct-shipping ores that required little beneficiation other than simple washing.

Imported ores played a minor role in the American iron and steel industry before 1950. The few, small-scale imports that occurred before 1900 served a few furnaces and mills located on the eastern seaboard — such as Sparrows Point, Maryland — and came principally from Cuba, Spain, Algeria, and Canada. Cuban mines developed with American capital were the chief suppliers of imported ores in 1900, providing about half of the 897,000 tons of ore imported. That same year American ore production was 27.3 million tons, so imported ores provided only about 3 percent of domestic consumption. Imported iron continued to play a small role in American ore supplies through World War II, largely because of the high cost of transportation and the richness and abundance of domestic ores. Imported ores still accounted for only 2 to 3 percent of U.S. consumption in 1940, with Chile, Cuba, and Sweden the principal suppliers, in that order. Problems with transporting the foreign ores in part accounted for the low percentages; the only efficient bulk transport system for international iron ore prior to 1940 was developed by Bethlehem Steel to move Chilean ores to the eastern U.S. seaboard.

Until 1945, then, the American iron and steel industry was essentially self-sufficient in its supply of iron ore. Easy access to high grade ores was, in fact, one of its main international competitive advantages. This situation changed radically after World War II. The war effort consumed a half-billion tons of rich Mesabi ore and large volumes from other domestic sources as well. Postwar economic expansion continued the accelerated consumption of domestic ores so that by 1947 the Mesabi Range, for example, had an estimated reserve of only 25 years. This forced American ore producers and consumers to take steps to assure themselves of future ore supplies. In the 15 years following World War II, American iron and steel companies developed technology to beneficiate the large domestic reserves of taconite, which had 30 to 35 percent iron content; supported extensive domestic exploration for additional ore; and invested in foreign exploration for natural, high-grade iron ores.

By the late 1950s American ore producers had the ability to convert low-grade ores with 25–35 percent iron content into pellets containing more than 60 percent iron. This process, first applied commercially to Minnesota taconites in 1956, was rapidly adopted, and by 1970 there were 20 pelletizing plants operating in the United States and 12 in Canada. This enabled the Lake Superior District to maintain its position as the most important supplier of domestic iron ores. The district has continued to produce 75 percent or more of all domestic iron ore. In fact, the pelletization process had another effect as well. Because pellets worked better in blast furnaces than did natural ore, direct-shipping ores have largely disappeared in the domestic ore trade.

Attempts to discover new high-grade domestic iron ore supplies were not as successful as attempts to beneficiate low-grade ores. During World War II and into the 1950s, the Bureau of Mines investigated hundreds of iron deposits throughout the United States. These efforts yielded some additional reserves but no major new discoveries.

Similar exploration by U.S. companies abroad was much more successful. In the late 1940s and the 1950s American companies were in the forefront of ore exploration and discovery worldwide. Bethlehem and the United States Steel Corporation, for example, located large deposits of high-grade iron ore in Venezuela in the late 1940s. American companies had imported Canadian ore before 1945, but additional explorations in Canada by several companies yielded large new deposits of ore, especially along the Labrador-Quebec border. Republic Steel in 1949 acquired majority interest in the Liberian Mining Company and made several major discoveries in that country. American companies, somewhat later, also identified high-grade deposits in Brazil and Australia.

After World War II imports rose from under 3 million tons (less than 4 percent of domestic consumption) in 1946 to nearly 35 million tons (less than 4 percent of domestic consumption) in 1960. Initially, Canada and Venezuela were the leading sources, but ore came as well from Peru, Sweden, Chile, and Brazil. After Venezuela nationalized its iron ore production in 1975, that country declined in importance and was replaced by Brazil as the leading South American supplier. In 1988 Canada supplied approximately 45 percent of U.S. iron ore imports, Brazil about 24 percent, and Venezuela 18 percent, with only small volumes coming from other countries.

Until 1959 imported ores steadily increased their share of the American ore market, reaching 40 percent of the total. After 1960, however, the pelletizing process turned back the tide, and in recent years imported ores have supplied from 20 to 25 percent of American ore consumption. But if the Lake Superior District has held its own against foreign ore imports, other American ore districts were not so fortunate. The Southern District went into decline in the late 1950s because its brown ores were low in iron content and not readily beneficiated; high quality foreign ore could be brought to Alabama furnaces by river barge. The district's share of domestic ore production dropped from 10–15 percent to under 5 percent in the early 1960s. By 1970 all Alabama iron mines had closed. The Eastern District of New York and Pennsylvania barely held its own, producing a steady 3 to 5 million tons and 4–6 percent of U.S. ore production in the decades following World War II. Only the Western District grew in relative importance in the postwar period, due to the wartime decision to establish a steel industry near the West Coast. In the decade preceding World War II the western states had produced less than 1 percent of American iron ore. In the 1950s, however, the district's share grew to between 4.5 and 9 percent.

References:

Donald F. Barnett, *The Iron Ore Industry: Problems and Prospects* (Ottawa: Canadian Department of Energy, Mines, and Resources, 1982);

Barnett and Louis Schorsch, *Steel: Upheaval in a Basic Industry* (Cambridge, Mass.: Ballinger, 1983);

Donald A. Brobst and Walden P. Pratt, eds., *United States Mineral Resources*, Geological Survey Professional Paper 820 (Washington, D.C.: U.S. Government Printing Office, 1973);

Bela Gold and others, *Technological Progress and Industrial Leadership: The Growth of the U.S. Steel Industry, 1900–1970* (Lexington, Mass.: Lexington Books, 1984);

William T. Hogan, S.J., *An Economic History of the Iron and Steel Industry in the United States*, 5 volumes (Lexington, Mass.: Lexington Books, 1971);

Hogan, *World Steel in the 1980s: A Case of Survival* (Lexington, Mass.: Lexington Books, 1983);

Gerald Manners, *The Changing World Market for Iron Ores, 1950–1980: An Economic Geography* (Baltimore: Johns Hopkins University Press, 1971);

U.S. Tarrif Commission, *Iron Ore* (Washington, D.C.: U.S. Government Printing Office, 1959);

Kenneth Warren, *World Steel: An Economic Geography* (Newton Abbot, U.K.: David & Charles, 1975);

Sheldon Wimpfen, "Iron and Steel Imports," *Skillings' Mining Review*, 60 (March 6, 1971): 1, 8–10.

Iron Trade Review

by Bruce E. Seely

Michigan Technological University

Iron Trade Review and its successors were leading trade journals in the metals industry, second only to *Iron Age*. In its heyday, from 1900 through 1930, *Iron Trade Review* covered many events in the steel industry with even more detail than its rival.

The origin of this journal may be traced to *Iron Trade Review and Western Machinist*, first published in 1867. The publishers followed closely the model established by *Iron Age* a decade earlier. On April 26, 1888 the first number titled *Iron Trade Review* appeared. The new editors hoped to continue an identification with the western steel industry, publishing in Cleveland rather than New York. In March 1901 the Iron Trade Review Company was purchased by the Iron & Steel Press Company in which John Penton was vice-president. He became president in 1867. Over the years Penton Publishing Company published a series of trade journals in the metal trades.

With correspondents in the major iron and steel centers, *Iron Trade Review* reported every week on prices and events in the industry with enormous detail. By 1900 it was second in reputation to *Iron Age* and for a time even equaled its rival. It absorbed *The Iron and Machinery World* in 1906, and in 1908 Penton titled the first January issue the "Yearbook of Industry," thereby introducing an annual review of the iron and steel trade. By 1917 *Iron Trade Review* had 30 correspondents in the United States, and the company opened an office in London in 1920.

With the first issue of 1930, the journal took the name *Steel*. It still provided excellent coverage of the trade and is a fine source for tracing historical developments, but it never quite equaled *Iron Age* as the voice of the industry. After World War II both the quantity and quality of attention devoted to plant expansions and people declined, a shift common to all trade journals. The journal also grew smaller in the postwar years. Yet between 1930 and 1970 the magazine won more editorial awards than any other business publication. With the last issue of 1969 Penton Publishing renamed *Steel* in an effort to target a broader audience. The publishers argued that conglomerates and rampant diversification meant that few firms operated in a single industry and trade papers should follow suit. Thus *Steel* became *Industry Week*, a general business paper for industrial managers. In

1976 it went from a weekly to a semimonthly publication.

References:

"Charles J. Stark Appointed Editor," *Iron Trade Review*, 60 (February 15, 1917): 430–431;

David P. Forsyth, *The Business Press in America, 1750–1865* (Philadelphia: Chilton, 1964);

John A. Penton, "Relation of the Trade Papers to the Iron and Steel Industry: Discussion," *American Iron and Steel Institute Yearbook* (1918): 167–169;

William M. Rooney, "Change — The Only Sure Thing," *Steel,* 165 (December 29, 1969): 222–227;

"*Steel* Wins 'E' Award," *Steel,* 150 (April 2, 1962): 71.

William Adolf Irvin

(December 7, 1873 – January 1, 1952)

by Bruce E. Seely

Michigan Technological University

CAREER: Western Union messenger, part-time (1886–1888); telegraph operator, assistant freight and passenger agent, Pennsylvania Railroad Company (1888–1894); shipping clerk, plant superintendent, P. H. Laufman & Company (1894–1901); cost clerk, head of cost department, American Sheet Steel Company (1901–1904); chief of the sheet division, cost department (1904–1905), chief clerk, operating department (1905–1909), assistant to the vice-president, operations (1909–1924), vice-president, operations, American Sheet & Tin Plate Company (1924–1931); vice-president (1931), president (1932–1938), vice-chairman of the board (1938–1939), member of finance committee and board of directors, United States Steel Corporation (1939–1952).

William Adolf Irvin, like many top executives in the steel industry before 1930, was a self-made man. He climbed from the bottom of the sheet-steel industry and built a reputation as a production specialist that carried him to the top executive ranks at the United States Steel Corporation in 1931. After working under U.S. Steel chairman Myron Taylor through the tumultuous 1930s, he spent the 1940s leading the National Safety Council.

Irvin was born to Sophia Bergan and David Sylvester Irvin in Indiana, Pennsylvania, on December 7, 1873. After his father's death, Irvin at the age of eleven began working part-time jobs. As a result, his formal education came from night courses taken at the Indiana State Normal School. Irvin was a messenger for Western Union; then, as had Andrew Carnegie, he became a telegraph op-

William A. Irvin

erator for the Pennsylvania Railroad in 1888 and later was promoted to assistant freight and passenger agent.

In 1894 Irvin moved into the steel industry as the shipping clerk for P. H. Laufman Company, a manufacturer of sheet steel and tinplate in Apollo, Pennsylvania. By the time the American Sheet Steel Company acquired Laufman in 1900, Irvin was plant superintendent. After that combination was absorbed by U.S. Steel in 1901, Irvin was transferred to the cost department in the general offices of the sheet-steel company in New York. He became chief of the sheet division in the cost department when the American Steel & Tin Plate Company was formed in 1904. A year later he moved back to Pittsburgh as chief clerk of the operating department, a post he held until 1909, when he was promoted to assistant to the vice-president for operations.

American Sheet & Tin Plate was a finishing company, turning ingots from other U.S. Steel subsidiaries into steel sheets using a labor-intensive process that had changed little since 1800. U.S. Steel closed many old sheet mills after 1901, but Irvin still watched over mills in western Pennsylvania, eastern Ohio, and West Virginia. But just as he was promoted to assistant to the vice-president, a brand-new sheet mill opened at Gary, Indiana. From 1909–1931, as assistant to the vice-president for operations and then as vice-president for operations himself, Irvin oversaw operations there and at all other mills of American Sheet & Tin Plate.

During the 1920s this subsidiary was one of the most profitable in the corporation, as demand grew rapidly due to expanding automobile and appliance industries. But the corporation made a blunder in the 1920s, when it ignored the continuous strip mill, a new mechanized steel-production technique that radically lowered labor costs. Instead, the corporation expanded capacity by upgrading older hand mills. Even when U.S. Steel's first strip mill was installed at Gary in 1926, it was merely to replace a preliminary step in production, with hand mills providing final finishing. This decision reflected U.S. Steel head Elbert H. Gary's dominance over decision making and conservative policies regarding innovation. The company paid a high price for this conservatism in the 1930s, when it was all but excluded from the leading market for sheet — the auto industry.

Irvin was more successful in implementing a wage policy for American Sheet & Tin Plate that stabilized the income of its workers. *Iron Age* credited this plan with labor harmony during the 1920s and claimed it also produced a higher wage scale.

In 1931 Irvin was promoted to vice-president of operations for the entire corporation, a move that took him back to New York. Six months later, U.S. Steel chairman Myron Taylor tapped him to succeed James Farrell as president of the corporation. Irvin

was a surprise selection; *Fortune* even got his middle name wrong. In introducing himself to the press in 1932, Irvin set the tone for his low-key executive style: "I want you to understand that giving out statements for the company will not be part of my job." Taylor then fielded questions for Irvin. Irvin testified before the Senate Interstate Commerce Committee on basing points in 1936 and issued a few public pronouncements regarding the company union and wage increases. Otherwise, his name rarely surfaced.

His influence was widely felt, however, within U.S. Steel, as *Fortune* explained in 1932: "He may not make Steel's speeches or Steel's decisions. But a big enough job for anybody is making Steel's steel." Especially important was Irvin's knowledge of sheet steel. Because U.S. Steel adopted continuous strip mills late, its share of the sheet market was only 6.5 percent in 1932 and 10 percent in 1933, and an expansion program in 1935 targeted sheet production. Modern equipment was installed at Gary, the South Works in Chicago, and the McDonald plant in Youngstown; a brand-new mill at Dravosburg, Pennsylvania, was dedicated in December 1937 as the Irvin Works. This highly efficient mill had an anual capacity of 100,000 tons of tinplate and 500,000 tons of sheet-steel.

The dedication of the mill was Irvin's last act as president, becoming vice-chairman in 1938. He served until April 1939, although he remained quite active on the board of directors and finance committee through 1951. Irvin also embarked on a career in the safety field. Injured in a steel-mill accident in his youth, Irvin reorganized the National Safety Council in 1942 and chaired its trustees until his death. During the war he served on the War Production Fund to Conserve Manpower, and after 1945 he turned his attention to traffic fatalities. His only other official activity in retirement was as a member of the executive committee of the Willys Overland Motor Company.

A typical steel man, Irvin's recreational pursuits were golf, hunting, and fishing; for many years, he made moose-hunting trips to Canada. He also kept a kennel of setters at his home outside of Pittsburgh. He was married twice, having five children with Luella May Cunningham, whom he married in 1893. After she died in 1908, he married Gertrude Gifford in 1910.

Publications:
"Steel Institute to Broaden Scope as Leaders Assert Faith in Future," *Iron Age,* 129 (May 26, 1932): 1176, 1198;

"Wagner Bill Would Foment Discord and Hinder Recovery," *Iron Age,* 133 (May 31, 1934): 18–19;

and I. L. Hughes, "Company's Position Put Before Carnegie Workers," *Iron Age,* 133 (June 7, 1934): 46, 63;

"A Message for the Metal Congress," *Iron Age,* 134 (September 27, 1934): 11;

"A Review of the Year," *American Iron and Steel Institute Yearbook* (1935): 45–49;

"Swelling Imports Threaten Domestic Price and Wage Structure," *Iron Age,* 135 (May 30, 1935): 22–24;

"Steel and the Steel Industry," *American Society of Metals Transactions,* 23 (December 1935): 1142–1146;

"Steel and Oil in New Era," *Steel,* 98 (May 18, 1936): 20–21;

"Competition from Imports of Foreign Steel Products," *American Iron and Steel Institute Yearbook* (1936): 46-54;

"Changing Trends in the Steel Industry," *American Iron and Steel Institute Yearbook* (1938): 39–45;

and T. M. Girdler and E. T. Weir, *Where Does Steel Stand? Addresses at American Iron and Steel Institute, 47th General Meeting, New York, May 26, 1938* (New York: American Iron and Steel Institute, 1938);

"Killed, Not in Action: 102,000," *New York Times Magazine,* June 28, 1942, pp. 12–13;

"The Steel Industry's Place in Safety," *American Iron and Steel Institute Yearbook* (1944): 96–102.

References:

"An Important Job," *Fortune,* 5 (June 1932): 38–39;

William T. Hogan, S.J., *An Economic History of the Iron and Steel Industry in the United States,* 5 volumes (Lexington, Mass.: Lexington Books, 1971);

"William A. Irvin Breaks Ground for New Mill That Bears His Name," *Iron Age,* 139 (May 27, 1937): 70;

"William A. Irvin to Become President of Steel Corporation," *Iron Age,* 129 (March 17, 1932): 687, 694.

Francis Kenneth Iverson

(September 18, 1925 –)

by John N. Ingham

University of Toronto

CAREER: Research physicist, International Harvester (1947–1952); technical director, Illium Corporation (1952–1954); director of marketing, Cannon-Muskegon Corporation (1954 –1961); executive vice-president, Coast Metals (1961–1962); vice-president/general manager (1962–1965), president, chief executive officer, director (1965–1985), chairman, chief executive officer, Nucor Corporation [formerly Nuclear Corporation of America] (1985–).

As a student in metallurgy at Purdue University, Iverson and his class visited one of the huge, integrated steel producers in the Chicago-Gary area, later recalling that "we actually had to step over workers who were sleeping there. I decided then and there that I didn't want to work for a big steel company." Instead, he became the "father of the minimill," and, in the words of historians Donald Barnett and Lewis Schorsch, in so doing "reinvented the steel industry."

Iverson was born on September 18, 1925, in rural Downers Grove, Illinois. He attended Northwestern University in 1943–1944 and transferred to Cornell, where he earned his bachelor's degree in engineering in 1946. A year later he received a master's degree in metallurgy from Purdue. Iverson then took various jobs in the metals industry, working as a research physicist for International Harvester in Chicago from 1947 to 1952 and as technical director of the Illium Corporation in Freeport, Illinois, from 1952 to 1954. In 1954 he became director of marketing for Cannon-Muskegon Corporation in Michigan, and from 1961 to 1962 he served as executive vice-president at Coast Metals in Little Ferry, New Jersey.

In 1962 Iverson joined Nuclear Corporation of America as general manager of its Vulcraft division. Located in Florence, South Carolina, Vulcraft made steel joists and girders. His was a successful, money-making division, within a parent company that suffered continual losses in the production of diodes, leasing equipment, and radiation detectors. With sales of $22 million in 1965, Nuclear lost $2 million and was moving toward a Chapter 11 bankruptcy. Later saying that he had "got the job by default," Iverson was appointed Nuclear's chief executive officer in 1965.

Upon assuming control, Iverson conducted salvage operations to keep the firm afloat. He sold off or shut down more than half the divisions and reduced the corporate staff from twelve to two — himself and the vice-president of finance — and moved the headquarters from Phoenix, Arizona, to Charlotte, North Carolina. His surgery worked wonders: within a year the company showed pretax profits of $1.3 million, and it remained in the black for the next two decades. Under Iverson's leadership the company reached the top of the joist-manufacturing field two years later but was spending 56 percent of every sales dollar on steel costs. It was purchasing 80 percent of its steel from European steelmakers, who were using scrap metal to manufacture steel in minimills with electric-arc furnaces and other new technological advances. Iverson was convinced that if he built a minimill he could make his own steel and make it more cheaply.

Virtually all steel made in America at that time was produced in large, integrated mills, following time-honored American patterns of achieving economies of scale in mass-production industry. Minimills, on the other hand, typically produced only billets, the smallest semifinished shapes, which were then suitable for rolling into bars, small structural shapes, and rods. At the time when Iverson was looking to ensure an inexpensive steel supply, the minimill could not produce slabs and sheets, which were the mainstays of the integrated mills. The minimills simply charged scrap metal into an electric furnace to produce molten steel, and, unlike the integrated mills of the 1960s, they used continuous casting for all of their output. The continuous caster not only bypassed the reheating before the primary rolling of ingots, but also produced steel with more consistent metallurgical properties than the ingot process. Many integrated mills in Europe, and especially Japan, were using continuous casting in the 1960s, but America's large mills were notoriously slow to adopt the process. Nucor, serving as its own general contractor, broke ground for its new minimill in Darlington, South Carolina, in September 1968. Nine months later, it was ready to start pouring steel.

With no experienced steelworkers in the area, Iverson recruited whatever local talent was available — sharecroppers, grocery clerks, schoolteachers, and carpenters. None had ever been in a steel plant before, which for Iverson was a mixed

Francis Kenneth Iverson

blessing. On the one hand, they were nonunion and willing to work for lower wages than their counterparts in the integrated mills. On the other hand, when the furnace was first tapped and the molten steel came pouring out, they were so frightened that they ran from the building. After a time, however, these workers developed skills, and by the early 1980s Nucor had grown into the most profitable carbon-steel manufacturing operation in the world. In 1984 it produced 1.5 million tons of steel, with sales of $660 million and profits of $44.5 million.

Currently in the American steel industry, it is in the minimills that the most important technological advances for the industry are forged. One of the major technological advances is in the area of electric-arc furnaces. Using water-cooled furnaces, ultra-high power, oxygen-enrichment, ladle metallurgy, and a host of other improvements, the minimills have significantly improved electric-furnace technology. In addition, through the use of continuous-billet casting they have also made significant gains, and other refinements have reduced costs. Minimills, and especially Nucor, have pioneered the direct-rolling of finished shapes, a goal long desired by integrated steel producers. The motto of minimills has been to "build tight, build quick, and build cheap," in order to get the best technology on line as quickly and as cheaply as possible. This has also led to the development of smaller plants with narrower product lines and fewer redundancies, enhancing the profitability of their operations.

Another area of critical importance for minimills is labor. In 1981 it cost an American integrated steel plant $393 per ton to ship wire rods; it cost the average minimill $284. Of that $109 difference, $71 was accounted for in labor costs. The key difference was not wage rates, because Nucor workers on average earned about 20 percent more than their counterparts in unionized, integrated mills. The greatest variation was actually in the number of man-hours it took to produce each ton of steel. Nucor, one of the most productive firms in the industry, was producing almost twice as much steel per man-hour as the large plants. Most analysts have pointed to the fact that Nucor, and many other minimills, are not unionized as the reason for this difference. Iverson does not agree: "I've heard people say that Nucor is proof that unions per se have a negative impact on worker productivity. That's nonsense! . . . The real impediment to producing higher-quality product more efficiently isn't the workers, union, or non-union; it's management." Iverson introduced a system of incentives to boost productivity which, in one form or another, includes all employees, management and workers alike.

Iverson also established a system of job security similar to that found in Japanese industry. Once hired, if a person works hard and well at Nucor, he is guaranteed a lifetime job. Iverson has attempted to avert layoffs, to assure the worker stability. Programs like these have built company loyalty among Nucor's workforce, and even Japanese visitors have been impressed. As one Japanese steel executive commented, "I'm not used to seeing that in an American steel mill."

As Nucor came to dominate markets in its limited product areas, it began to expand into other products. In 1977 it began manufacturing the steel deck used in the floors and roofs that its joists supported. In 1979 Nucor became the first, and for a few years the only, minimill to manufacture cold-finished bars, a specialty product used to make shafts and machined precision parts.

Iverson's biggest breakthrough, however, came in the development of a new technology for making thin, two-inch slabs at his mills. Nucor invested over $235 million in a new mill at its Indiana plant which allowed the company to move beyond being strictly a minimill producer of low-end products to becoming a supplier of flat-rolled steel, like that used by the auto companies. Such technology and product development put Nucor in the "big leagues" of steel producers for the first time.

The mill was opened in 1989, with a capacity to produce about 800,000 tons annually — not enough, as one of the Nucor executives said, "to get auto companies too excited right away." Nonetheless, it kept Nucor on the cutting edge of the newest technology in the industry. As Iverson commented, "We're not seeing the demise of the steel industry, we're seeing its transformation into a high-tech business." Nucor's new technology in this area gave it tremendous advantages over integrated mills. Using a casting machine developed by a West German company, the new technology allows the steel to be taken directly from the mill to where it is rolled into products, thereby saving energy, time, and capital. Furthermore, the slab does not have to be cooled and reheated, as is the case at large, integrated plants. Therefore, according to Iverson, Nucor will save about $50 to $75 a ton.

And Iverson is, as one person commented, the "paladin" of this trend. Tom Sigler, head of Continental Steel, a Nucor competitor, has said: "Right now Ken Iverson is 'The Force' in steel. He's shown that the small companies with the new technology are paving the way — not the Inlands or the U.S. Steels. He's shown that despite all you hear, there's still a future in steel."

Publication:
"Bright Future Seen for Vacuum Investment Castings," *Iron Age,* 178 (July 19, 1956): 120–122.

References:
Walter Adams, *The Structure of American Industry: Some Case Studies* (New York: Macmillan, 1950);
Donald F. Barnett & Louis Schorsch, *Steel: Upheaval in a Basic Industry* (Cambridge, Mass.: Ballinger, 1983);
Barnett & Robert W. Crandall, *Up From the Ashes: The Rise of the Steel Minimill in the U.S.* (Washington, D.C.: Brookings Institute, 1986).

Frederick G. Jaicks

(July 26, 1918 –)

by Robert M. Aduddell
Loyola University of Chicago

CAREER: General manager, Indiana Harbor works (1959–1962), vice-president of manufacturing and research (1961–1966), president and chief operating officer (1966–1971), chairman of the board, chief executive officer, Inland Steel Corporation (1971–1983).

Frederick Gillies Jaicks, the tenth president of Inland Steel Corporation, was born in Chicago on July 26, 1918, the son of Andrew and Marjorie Gillies Jaicks. After graduating in 1940 from Cornell University with a bachelor of science in mechanical engineering, he took a job with Inland, an Illinois corporation, as a production worker in the steelmaking department. With the exception of the U.S. Naval Reserve, in which he served during World War II, Inland was to be Jaicks's only employer.

Between 1940 and 1959 Jaicks held a wide range of increasingly important positions within the steelmaking department. In 1959 he was named the general manager of the Indiana Harbor works. In 1962 he became vice-president of manufacturing and research. The following year he was elected to the board of directors.

Jaicks was elected president of Inland Steel in 1966, replacing John F. Smith, Jr., who retired because of ill health. He served as president and chief operating officer until July 1971, when he was named chairman of the board and chief executive officer following the retirement of Philip D. Block, Jr., in June. Michael Tenenbaum was named president, succeeding Jaicks.

In 1966, when Jaicks became president, Inland produced 6.9 million tons of ingot steel. By contrast, when Jaicks joined the company in 1940 production was only 2.9 million tons. In 1966 Inland reached $1 billion in net sales for the first time, as compared to approximately $142 million in 1940. Sales slipped below $1 billion in 1967, but never again. Inland's fixed assets had also grown, but its managerial philosophy remained constant. All basic steel was produced at a single

plant, the Indiana Harbor works. The majority of this steel was sold within 150 miles of that plant. Growth at this site was almost entirely financed by retained earnings. Inland made acquisitions only to secure raw material supplies or to gain steel-consuming or -selling subsidiaries.

During the period Jaicks was president, Inland and other companies in the steel industry began to face increasing foreign competition. Between 1965 and 1971, 13.6 million tons of steel were imported annually, 40 percent of which landed at the various Great Lakes ports that served as Inland's primary marketing area. The quantity of foreign steel entering Inland's market was equivalent to the output of a second Inland-sized steel facility. With the foreign producers undercutting the prices sought by domestic producers, Inland began to explore the prospects for diversification. The recommendation for profitable diversification targets was made directly to the chairman of the board by a newly-appointed corporate development staff.

The first steps toward diversification were finalized in 1969, when Inland purchased a 30 percent interest in On-Line Decisions, an Oakland, California, firm that specialized in designing and marketing decision-making software for managerial purposes. The same year Inland formed the Inland Steel Urban Land Development Corporation. This entity held both Scholz Homes Incorporated of Dayton, Ohio, and Jewel Builders of Columbus, Ohio, which Inland had acquired. Scholz had a wide range of interests in building prefabricated homes, while Jewel was involved in building single- and multiple-family homes and developing urban real estate.

Even as it diversified, Inland decided to remain primarily a steel-producing firm, but under more stringent competitive conditions. The company upgraded its research and development facilities in 1967, moving from an older plant at Indiana Harbor to a new one at East Chicago. New steelmaking eqipment included a basic oxygen furnace with a 2-million-ton-per-year capacity. It supplied one-third of Inland's steel needs. At the same time, an old open-hearth furnace that had been in operation since 1902 was closed. A new billet mill with ancillary facilities — including computer-controlled processing, a new plate-finishing operation connected to the existing 76-inch hot strip mill, and an 80-inch hot strip mill — were also placed in service. These and additional new facilities required Inland to issue $50 million in 25-year first mortgage bonds. This expansion continued in 1968 as plans progressed for Inland's first electric-furnace steelmaking plant and continuous casting machine for billets.

In 1969 Inland put a 12-inch bar mill on-line that could be expanded to serve as a processing unit for the proposed electric furnace. As a result, Inland's original Chicago Heights plant, the one in which the company began its operations in 1893, was closed and sold. Its bar output was no longer required.

During Jaicks's presidency Inland maintained its customary stance of independence with respect to raw materials supplies. By 1967 the Wabash ore mining operation was at full capacity, with Inland, four other domestic firms, and three foreign firms sharing the output of the Labrador mine and the Pointe Noire, Quebec, pellet plant. The Black River, Wisconsin, taconite mine and pelletizing operations became fully operational in 1969.

For his efforts in the steel industry, Jaicks was awarded the Benjamin F. Fairless Award from the American Institute of Mining Engineers in 1971, the Gary Medal from the American Iron and Steel Institute in 1977, and a medal by the International Iron and Steel Institute in 1984.

Frederick Jaicks retired from Inland in August 1983, shortly after his sixty-fifth birthday. Frank Luerssen succeeded him as chairman of the board. Jaicks continued to serve on the company's board of directors. He also served as a director of Amoco, Champion International, R. R. Donnelley, and Zenith. He served on the board of trustees of several important Chicago institutions: Illinois Institute of Technology, the Museum of Science and Industry, and Rush-Presbyterian-Saint Luke's Hospital. Illinois Institute of Technology awarded him an honorary degree, as did Saint Joseph College, De Paul University, and Valparaiso University.

Publications:
and others, "Continuous Casting of Three Types of Low Carbon Steel," *Journal of Metals Transactions,* 9 (August 1957): 1057–1072;
"Productivity: The Name of the Game," *Journal of Industrial Engineering,* 19 (July 1968): 11–13;
"Money: The One Raw Material Steel Lacks," *Iron Age,* 214 (September 2, 1974): 34–35.

References:
"Facing the Future in a Troubled Industry," *Fortune,* 84 (July 1971): 13;
William T. Hogan, S.J., *An Economic History of the Iron and Steel Industry in the United States,* 5 volumes (Lexington, Mass.: Lexington Books, 1971);
"Jaicks Says Steel Won't Delay Expansion," *Purchasing,* 79 (July 8, 1975): 15;
"What Steel Critics Choose to Ignore," *Iron Age,* 217 (May 17, 1976): 23–30.

Jessop Steel Company

by Geoffrey Tweedale
Manchester, England

The Jessop Steel Company has been throughout the twentieth century a small but important independent producer of high-grade alloy steels. The firm was founded in Sheffield, England, in 1793. By the nineteenth century it was one of the foremost crucible steel makers of its day and plundered the American market for business in tool and saw steel. By the end of the century, however, the emergence of U.S. crucible steel makers and a high tariff on English imports had severely impaired Jessop's dominant position in the U.S. market. In 1901 Jessop had begun to construct a crucible steel works on a 400-acre site at Washington, Pennsylvania, close to America's major steelmaking district in Pittsburgh. Fueled by natural gas, the furnaces were designed to specialize in sheet and bandsaw steel and had an annual capacity of about 5,000 tons. The move to establish a Washington plant was largely made to circumvent the tariff, but it may also have been triggered by the merger of America's major specialty steelmakers into the Crucible Steel Company of America. The Sheffield directors also saw the American subsidiary as a way to escape from the English steel industry's restrictive labor practices.

Jessop soon established itself in the American market, though its connection to Sheffield proved short-lived. In about 1920 the firm was sold to American interests, perhaps because profits had never been high enough for the English parent company, which had itself been absorbed by another concern. In the interwar period Jessop upheld its reputation as a premier producer of fine steels: electric furnaces were installed, and the firm expanded its product line to include tool and stainless steels and various kinds of die and surgical steels.

After World War II stainless steel became an increasingly important part of Jessop's product line, so that by the 1960s half its sales came from this source, with a third from tool and alloy steels and the rest from carbon and other steels. By then the company was producing its own alloyed bar from its subsidiary, the Green River Steel Corporation in Owensboro, Kentucky, which had been acquired in 1957. Jessop's sales were $65.2 million in 1967 with the firm producing about 135,000 tons from its two plants. Jessop's output was sold extensively to the aircraft and automobile industries for routine sheets and forgings. It also served more exotic markets: for example, the atomic power plant of the USS *Will Rogers* was built by Westinghouse using Jessop steel.

In 1968 Jessop was purchased by Athlone Industries, a New York–based maker of industrial products and apparel. Thereafter Jessop became even more strongly identified with the production of stainless-steel plate, where there was a strong demand in pollution control, food processing, and petrochemical applications. In 1978 sales from Jessop's two steel mills had reached nearly $84 million. In the 1980s the company also increased its capacity for producing tool steel, though stainless steel remained its major product. By 1987 Jessop, with sales of $170 million, accounted for approximately 55 percent of Athlone's sales and 78 percent of its operating income. It had become the second largest producer of stainless-steel plate in the United States, surpassed only by the Eastern Stainless Corporation.

References:

"Jessop Emphasizes Tool Steel Line," *American Metal Market*, 87 (13 March 1979): 1, 20;

"Jessop Steel Company Puts a New 18-In. Mill on Production," *Blast Furnace and Steel Plant*, 32 (March 1944): 331–342;

"Jessop Steel Promotes Metallurgical Research by Erecting a New Laboratory," *Blast Furnace and Steel Plant*, 32 (August 1944): 943–948;

Geoffrey Tweedale, *Sheffield Steel and America: A Century of Commercial and Technological Interdependence, 1830–1930.* (Cambridge, U.K. & New York: Cambridge University Press, 1987).

Archibald Johnston

(May 30, 1865 – 1947)

by Lance E. Metz

Canal Museum, Easton, Pennsylvania

CAREER: Laboratory assistant, physical testing department (1889–1890), assistant to general superintendent, forge department (1890–1898), assistant general superintendent, Bethlehem Iron Company (1898–1901); general superintendent (1902–1904), president, Bethlehem Steel Company (1904–1908); first vice-president, Bethlehem Steel Corporation (1908–1916).

Archibald Johnston played a significant role in the early development of the Bethlehem Steel Corporation. He also became a prominent civic leader, helping to create the consolidated City of Bethlehem, Pennsylvania.

Archibald Johnston was born to Martha (Stroman) and Joseph Johnston on May 30, 1865, at Phoenixville, Pennsylvania. The family moved to Bethlehem, Pennsylvania, before he turned three. His father rose from master mechanic to serve two terms (1885–1887 and 1890–1893) as a democratic member of Bethlehem's borough council. Johnston's family were members of the

Moravian church, which had founded Bethlehem in 1741, and he received his elementary and secondary instruction in Moravian schools.

Johnston received an engineering degree from Lehigh University in 1889. While at Lehigh, Johnston was a student leader, a noted football player, and the business manager of Lehigh's yearbook, *The Epitome*. After graduation he took a job at the Bethlehem Iron Company, the predecessor of the Bethlehem Steel Corporation.

Archibald Johnston began his career at the Bethlehem Iron Company's physical testing department. Within a year his abilities and diligence had won for him the position of assistant to the company's general superintendent, John Fritz, and its chief metallurgist and assistant superintendent, Russell Wheeler Davenport. Working under Fritz and Davenport, Johnston helped to oversee the construction of Bethlehem's heavy steel forging plant. The first of its type to be built in North America, this plant was designed to produce large gun forgings and armor plate. After the forging plant's completion, he helped to supervise its operation and began to play an increasingly prominent role in the company's affairs. As early as 1894, Johnston had won the patronage of Bethlehem general manager and prominent stockholder Robert H. Sayre, whose son-in-law, Robert P. Linderman, also served as the company's president. In 1898 Johnston was appointed to the position of assistant general superintendent of the Bethlehem Iron Company.

In 1899 the Bethlehem Iron Company was reorganized as the Bethlehem Steel Company, and in 1901 Charles M. Schwab, the first president of the United States Steel Corporation, purchased control of it. A friendship soon developed between Schwab and Johnston. Schwab also admired Johnston's technical expertise and management skills. Within a year of Schwab's purchase Archibald Johnston was promoted to the position of general superintendent. He was also made a member of Bethlehem's board of directors.

In 1904 Schwab organized the Bethlehem Steel Corporation as a holding company to combine under sole management the operations of the Bethlehem Steel Company and the shipyards that he had salvaged from the financial collapse of the U.S. Shipbuilding Company. Under this new form of management, Archibald Johnston served briefly in 1906 as the president of Bethle-

hem Steel Company and then became first vice-president of the Bethlehem Steel Corporation in 1908. As first vice-president of the corporation, Johnston's primary responsibility was foreign sales. An expert in the production of cannon and armor, Johnston was able to greatly increase the sale of Bethlehem's military products abroad. In 1906 Johnston had won for Bethlehem a greatly increased share of the European armor market due to his negotiations with fellow members of the Harvey International Steel Corporation, a pool in which Bethlehem and the world's other armor producers were partners.

In 1913 Johnston traveled to China, where he successfully negotiated Bethlehem's large-scale participation in the building of a new naval base at Fuzhou. The outbreak of World War I opened other new opportunities for Bethlehem, and in October 1914 Schwab and Johnston traveled to London where they obtained massive new orders for war material, including submarines. This contract proved highly controversial, however, given American neutrality laws, and Bethlehem completed the contract in Canadian shipyards.

Beginning in 1916, civic affairs became the primary focus of Johnston's activities. In 1906 Johnston had served as chairman of the Joint Bridge and Highway Commission of Bethlehem. Although the stated purpose of this organization was the construction of a modern highway bridge to link the various boroughs of the Bethlehem area, it soon evolved into an advocacy vehicle for the political integration of these communities. By 1916 the bridge committee was reorganized as the Joint Bridge Commission, with Johnston remaining as chairman. Within a year the commission had raised the funds needed to construct the bridge, and its success brought about a formal campaign for the consolidation of the separate boroughs into the City of Bethlehem. Archibald Johnston played a prominent role in this campaign and, with the creation of the City of Bethlehem in 1917, was elected the first mayor of this consolidated municipality; he took office on January 7, 1918.

As mayor of Bethlehem, Johnston proved to be an active and dynamic leader. He established a city planning department and closely supervised the operations of all areas of city government. His primary concern was the welfare of Bethlehem's inhabitants. He had to overcome many problems caused by a housing boom and the expansion of Bethlehem's street network due to the massive influx of wartime workers at the Bethlehem plant. At the end of his term as mayor, Johnston retired from public life and constructed a large mansion on his wooded estate in the northern section of Bethlehem. He also became known locally as a philanthropist. In 1922 he donated a memorial bronze plaque to the largest Catholic church in Bethlehem to commemorate the military service of members of its congregation during World War I. He also served on the board of Moravian College, for which he funded the construction of an athletic center named in his honor.

Johnston was a member of the Iron and Steel Institute of Great Britain, the American Society of Mechanical Engineers, and the American Institute of Mining Engineers. He also took an active part in the affairs of the Bethlehem Club and the Northampton County Country Club. He married Estelle S. Borhek of Bethlehem in 1891, and their son, Archibald Johnston, Jr., became an executive with Bethlehem Steel Corporation. He died in Bethlehem in 1947.

Reference:

C. J. Stark, "Practical Ideas on Extending Foreign Trade," *Iron Trade Review,* 52 (January 2, 1913): 8–9.

Archives:

The Archibald Johnston biographical file, Bethlehem Steel Corporation, is at the Hugh Moore Historical Park and Museums, Easton, Pennsylvania.

John Johnston

(October 13, 1881 – September 12, 1950)

by Bruce E. Seely

Michigan Technological University

CAREER: Research associate, Massachusetts Institute of Technology (1907–1908); research staff, Geophysical Laboratory, Carnegie Institution (1908–1916); head of research department, American Zinc, Lead, and Smelting Company (1916–1917); U.S. Bureau of Mines (1917–1918); secretary, National Research Council (1918–1919); professor of chemistry and chairman of chemistry department, Yale University (1919–1927); director of research, United States Steel Corporation (1927–1946).

A distinguished chemist, John Johnston achieved distinction in the steel industry as the organizer and first director of the United States Steel Corporation's research and development laboratory. Although handicapped by limited resources and corporate support, Johnston assembled a distinguished staff that pursued a course of fundamental research into steel metallurgy and played a significant role in introducing academic science to the steel industry.

Johnston had a widely varied chemical career, with his attention often attracted to practical matters. Born October 13, 1891, in Perth, Scotland, he entered University College in Dundee, which was affiliated with the University of Saint

John Johnston

243

Andrews, and earned a bachelors degree in chemistry in 1903. After two additional years of postgraduate study as a Carnegie Scholar at Dundee, he traveled to Germany to study at the University of Breslau, before moving to the United States in 1907. He stayed for a year as research associate to Arthur A. Noyes, who headed a widely recognized laboratory in physical chemistry at Massachusetts Institute of Technology. In 1908 Johnston received his doctorate in science from Saint Andrews, but changed his mind about returning home and accepted a position with the Geophysical Laboratory of the Carnegie Institution in Washington, D.C. In 1915 he became a U.S. citizen.

Johnston's scientific work at the Geophysical Laboratory focused on the behavior of materials at high temperature and pressures; he published many articles on his work and began an association with the editorial board of the *American Chemical Society Journal*. His work had industrial ramifications, and in 1916 Johnston accepted the post of director of the laboratory for the American Zinc, Lead, and Smelting Company in Saint Louis. With America's entry into World War I in 1917, Johnston returned to Washington to take part in the U.S. Bureau of Mines research program on gas warfare. He became secretary of the newly organized National Research Council (NRC) in 1918 and served for one year, but maintained connections with the NRC's Division of Chemistry and Chemical Technology until 1943. With war's end, Johnston accepted the Sterling Professorship of Chemistry at Yale in 1919, where he directed graduate work and eventually became department head. While at Yale, he arranged the republication of the works of J. Willard Gibbs, earlier leader of chemistry at Yale, and he kept in touch with industry as a consultant for Bell Labs.

In 1927 he became director of the Department of Technology and Research at United States Steel Corporation, a momentous event in the corporation's history, for U.S. Steel previously had shown little interest in research. Although it had maintained small testing and quality-control labs, U.S. Steel, in keeping with Elbert H. Gary's emphasis on cooperation in the industry, regarded the research of advanced technology as potentially destabilizing. By the mid 1920s, 74 laboratories were scattered among the company's mills working on practical matters without central direction — an arrangement in keeping with the holding-company structure of the giant corporation. In 1926, however, Gary announced a change of corporate policy regarding research, a decision at least partially due to foreign competition and tighter profit margins. Moreover, Morgan partner and board member J. Percival Roberts had been promoting research for U.S. Steel because he had seen the value of research laboratories at General Motors, General Electric, and AT&T. An internal committee headed by George Crawford, president of the Tennessee Coal, Iron and Railroad Company, a U.S. Steel subsidiary, was formed to study the potential role of a laboratory and to seek names of potential directors. Crawford contacted Frank Jewett, research director at AT&T, and Robert Millikan, president of California Institute of Technology, for advice.

Crawford, significantly, was the only subsidiary president with a science or engineering background. Gary and most other top executives remained unsure of the value of research, and were daunted by the possible cost and fearful of the impact of a central research facility operating outside their control. But with Gary's tenuous backing, discussions continued into 1927. The committee agreed that a central laboratory would work on fundamental projects, while the small facilities at operating plants would continue to study practical problems related to production. After adopting this division of labor to protect the status quo, the research committee offered Millikan the post of research director. He declined; they turned to Jewett, but he, too, said no. Finally, the committee accepted the advice of Millikan and Jewett and approached Johnston.

Johnston's decision to join U.S. Steel was praised by the *New York Times,* although the article also noted the limited enthusiasm of many corporate officials. Johnston's first action was to tour the corporation's scattered labs, but even before he returned, he found himself in a highly precarious situation when Gary died in June 1927. Other corporate officials, but especially new president James Farrell, doubted that researchers could help produce steel, and Gary's plans for a research facility were soon scaled back. In 1928 a laboratory was temporarily located in a warehouse of the Federal Shipbuilding and Drydock Company, a U.S. Steel subsidiary in Kearney, New Jersey, and conversion was not completed until mid 1929. The location was symbolic of the isolation of the laboratory from the main steelmaking operations and signaled the start of a frustrating struggle for Johnston.

Interestingly, both Millikan and Jewett had favored Johnston because of his tenacity, a trait that certainly proved important as he worked to get the research program off the ground against internal opposition. First, he hired two key assistants: R. B. Sosman, a former colleague from the Geophysical Laboratory, and E. C. Bain, a rising young figure in the world of metallurgy from the Union Carbide & Carbon Company's laboratory. Second, Johnston mapped a research program that left routine matters to the oldest laboratory in the corporation — the American Sheet & Tin Plate facility in Pittsburgh, founded in 1915 — and to two other new labs. At South Works in Chicago, studies were conducted on fine steels, while National Tube in Pittsburgh studied pipe, especially corrosion.

In May 1930 Johnston published an article in *Iron Age* in which he stated that the main goals of the Kearney laboratory were to improve both the quality of steel and the uniformity of the quality. But he added that it was not easy to know where to begin: "There is no lack of problems — very much the contrary; but there is a lack of precise knowledge of the factors which, by their variability, constitute these problems. This goes back to the fact that the fundamental chemistry of the element iron and its compounds is in many respects less well established than is commonly thought." Johnston hoped to remedy this problem, a course that emphasized fundamental rather than applied research.

Despite the lukewarm support given by Farrell, the makeshift nature of the lab itself, and the onset of the Great Depression, Johnston moved ahead. He had a budget of about $100,000, and extra funds came for specific projects; by the end of 1929 he had a staff of 18. Two key areas dominated the lab's attention during the 1930s. First, Johnston launched investigations of the basic processes of iron and steel production. These included examinations of blast-furnace efficiency in the use of coke, and studies of the effect of moisture and blast temperature on furnace operation and the resulting iron. Johnston also focused on improving control over open-hearth furnaces, based on increased knowledge of the thermal efficiency of this process and its underlying chemistry. Researchers also reexamined the Bessemer process in searching for better control over this method of converting iron into steel.

The second general area of concern in Johnston's lab was the developement of a fundamental understanding of steel metallurgy. Perhaps the most important work involved studies of the precise temperatures at which changes occurred in both the chemical and physical properties of the metal. Johnston encouraged his metallurgists, especially E. C. Bain, in this painstaking work that placed the heat treatment of steel on a sound scientific basis. Other projects included studies of grain size, aging in steel, and ways of improving weldability.

Eventually, improvements in steel production emerged from this research. To promote such efforts, Johnston established a program to bring operating people into the lab for a year of special training. But he never doubted that fundamental research paid off, and Zay Jeffries, one of the leading metallurgists in the United States, stressed this point: "Perhaps his greatest contribution was in furthering an appreciation of the value of fundamental research in the steel industry and in imparting a sound scientific philosophy to those engaged in such research."

In his efforts to improve steelmaking through research, however, Johnston constantly faced obstacles within the corporation. He, for instance, never saw the new research laboratory promised him in 1927; his assistant, E. C. Bain, built the corporation's research center in the mid 1950s. Furthermore, lines of communication within the corporation were poor, and there was no coherent connection among the larger labs, much less between Johnston and the smaller testing centers in the mills. Johnston reported both active and passive resistance in some of those facilities to his research program, a continuing reflection of the organizational style and politics of the corporation. Indeed, the lab's near-total concern with fundamental metallurgical questions, while producing good science, demonstrated the relative isolation of researchers from steel production.

The underlying problem faced by Johnston was what historian Paul Tiffany described as U.S. Steel's management culture, which steadfastly resisted outside influences. Thus R. E. Zimmerman, an executive from one of the subsidiaries, was appointed vice-president for research and technology in 1929, a move that could be interpreted as insuring that a corporate insider controlled Johnston and his scientists. Johnston apparently worked well with his boss, but the newly established hierarchy to which Johnston had to answer and his lab's distance from the midwestern mills almost certainly diminished Johnston's influence and further insulated the laboratory from the corporation it was to serve.

By 1946, however, when Johnston retired to Bar Harbor, Maine, he had accomplished much; research was solidly entrenched at U.S. Steel with a group of renowned metallurgists. But through no fault of Johnston's, his efforts to firmly link research and production were largely thwarted.

Despite these frustrations, James B. Austin, a metallurgist whom Johnston brought to the Kearney laboratory from Yale, remembered his teacher and boss as devoted to science. By 1930 Johnston had published sixty technical papers, and he served as editor of the *International Critical Tables*. He was a member of several technical societies, including the American Chemical Society, the Franklin Institute, the British Iron and Steel Institute, the Institute of Metals, the Verein Deutscher Eisenhüttenleute, and the American Electrochemical Society, of which he was president in 1933–1934. He also served on the War Metallurgy Committee of the National Research Council during World War II. He died on September 12, 1950.

Selected Publications:

"On the Amphoteric Character of Cacodylic Acid," *Berichte der deutschen chemischen Gesellshaft* (1904);

and J. A. Noyes, "Conductivity and Ionization of Polyionic Salts," *American Chemical Society Journal,* 31 (September 1909): 987–1010;

"Change of Equivalent Conductance of Ions with Temperature," *American Chemical Society Journal,* 31 (September 9, 1909): 1010–1020;

and L. H. Adams, "Influence of Pressure on the Melting Point of Metals," *American Journal of Science,* 31 (June 1911): 501–517;

"A Correlation of the Elastic Behavior of Metals with Certain of their Physical Constants," *Washington Academy of Science Journal,* 1 (December 4, 1911): 260–267;

and Adams, "Density of Solid Substances with Especial Reference to Permanent Changes Caused by High Pressure," *American Chemical Society Journal,* 34 (May 1912): 563–584;

and Adams, "The Standard Scale of Temperatures Between 200° and 1100°," *Washington Academy of Sciences Journal,* 2 (June 19, 1912): 275–284;

and Adams, "Effect of High Pressures on the Physical and Chemical Behavior of Solids," *American Journal of Science,* 25 (March 1913): 205–253;

"Temperatures of Deep Wells at Findlay, Ohio," *American Journal of Science,* 36 (August 1913): 131–134;

and Adams "Capillarity in Rocks with Relation to Vulcanism," *Journal of Geology,* 22 (January–February, 1914): 1–15;

and Adams "Bore Hole Temperatures," *Economic Geology,* 11 (December 1916): 741–762;

and E. D. Williamson, "The Role of Inorganic Agencies in the Deposition of Calcium Carbonate," *Journal of Geology,* 24 (1916): 729–750;

"Some Aspects of Recent High-Pressure Investigation," *Journal of the Franklin Institute,* 183 (January 1917): 1–33;

"The Vapor Pressure and Volatility of Several High-Boiling Metals—A Review," *Journal of Industrial and Engineering Chemistry,* 9 (1917): 873–878;

"A Summary of the Proposals for the Utilization of Niter Cake," *Journal of Industrial and Engineering Chemistry,* 10 (1918): 468–471;

"The Volatility of the Constituents of Brass," *Journal of the American Institute of Metals,* 12 (1918): 15–26;

and Adams and Williamson, "Compressibility of Solids at High Pressure," *American Chemical Society Journal,* 41 (January 1919): 12–42;

"The History of Chemistry," *Scientific Monthly,* 13 (1921): 5–23, 130–143;

"The Mechanism of Corrosion," *Industrial and Engineering Chemistry,* 15 (1923): 904–905;

"The Partial Decomposition of Certain Solid Substances Effected by Grinding," *Revue travaux chemique,* 42 (1923): 850–854;

and D. H. Andrews, "Role of Absorption of Water by Rubber," *American Chemical Society Journal,* 46 (March 1924): 640–650;

and Andrews, "The Application of the Ideal Solubility Curve to the Interpretation of Equilibrium Diagrams in Metal Systems," *Institute of Metals Journal,* 32 (1924): 385–404;

"Solubility Relations of Isomeric Organic Compounds. I. Introduction," *Journal of Physical Chemistry,* 29 (July 1925): 882–888;

and A. C. Walker, "Preparation and Analysis of Constant Mixtures of Air and Carbon Dioxide," *American Chemical Society Journal,* 47 (July 1925): 1807–1817;

and Andrews and G. Lynn, "Heat Capacities and Heart of Crystallization of Some Isomeric Aromatic Compounds," *American Chemical Society Journal,* 48 (May 1926): 1274–1286;

and Norton, "Method of Determining the Equilibrium Pressure of Certain Hydrated Salts," *American Journal of Science,* 12 (December 1926): 467–476;

and Norton, "Transition Temperature and Solubility of Sodium Sulphate in the Presence of Sodium Chloride or Sodium Bromide," *American Journal of Science,* 12 (December 1926): 477–483;

and Walker and U. B. Bray, "Equilibrium in Solutions of Alkali Carbonates," *Journal of the American Chemical Society,* 49 (1927): 1235–1256;

and Harold Dietrich, "Equilibrium between Crystalline Zinc Hydroxide and Aqueous Solutions of Ammonium Hydroxide and of Sodium Hydroxide, *Journal of the American Chemical Society,* 49 (1927): 1419–1431;

and H. Geneva Leopold, "The Vapor Pressure of the Saturated Aqueous Solutions of Certain Salts,"

Journal of the American Chemical Society, 49 (1927): 1974–1988.

and L. E. Steiner, "Development of a Method of Radiation Calorimetry, and the Heat of Fusion or of Transition of Certain Substances," *Journal of the American Chemical Society,* 32 (June 1928): 912–939;

and E. P. Jones, "Solubility Relations of Isomeric Organic Compounds. VIII. The Mutual Solubility of the Dinitrobenzes with the Nitroanilines and of the Three Clorobenzoic Acids," *Journal of Physical Chemistry,* 32 (1928): 593–603;

and Leopold, "The Rate of Absorption of Water by Bakelite," *Journal of Physical Chemistry,* 32 (1928): 876–878;

and Charles L. Lazzell, "Solubility Relations of Isomeric Organic Compounds. VII. Solubility of the Aminobenzoic Acids in Various Liquids," *Journal of Physical Chemistry,* 32 (1928): 1331–1341;

"Josiah Willard Gibbs, An Appreciation," *Journal of Chemical Education,* 5 (1928): 507–514;

"Steel Corporation Research Program," *Iron Age,* 125 (January 2, 1930): 61–62;

"Chemistry of Steel Is Unduly Emphasized," *Iron Age,* 125 (May 22, 1930): 1536–1537;

"Some Comments on Steel Specifications," *American Iron and Steel Institute Yearbook* (1930): 63–81;

"Discussion to the Aging of Mild Steel Sheets," *American Iron and Steel Institute Yearbook* (1933): 160–161;

"Corrosion Problems," *Industrial and Engineering Chemistry,* 26 (1934): 1238–1244;

"Research in the Steel Industry," *Pennsylvania State College Bulletin,* 18 (1935): 11–23;

"Applications of Science to the Making and Finishing of Steel," *Mechanical Engineering,* 57 (1935): 79–86;

"Some Aspects of Steel Chemistry," *Industrial and Engineering Chemistry,* 28 (1936): 1417–1423;

and Florence Fenwick, "Steels Resistant to Scaling and Corrosion," *Industrial and Engineering Chemistry,* 28 (1936): 1374–1379;

"Steel," *Journal of the Franklin Institute,* 225 (1938): 373–398.

References:
Zay Jeffries, "John Johnston," *Yearbook of the American Philosophical Society* (1950): 309–312;
Edward C. McDowell, "John Johnston," *Metal Progress,* 43 (May 1943): 716–718;
Paul Tiffany, "The Origins of Industrial Research in the American Steel Industry: The Development of U.S. Steel's Research and Technology Department, 1928," paper presented at the Annual Meeting of Society for the History of Technology, Pittsburgh, Penn., October 25, 1986;
"To Direct Research Work," *Iron Age,* 119 (June 9, 1927): 1682.

Archives:
There is some correspondence regarding Johnston in the records of Yale University Secretary Anson Phelps and Yale President James Rowland Angell located in the Sterling Memorial Library, Yale University.

Logan T. Johnston

(September 1, 1900 – May 25, 1977)

by Carl Becker

Wright State University

CAREER: Salesman, Yale & Towne Company (1923–1925); salesman, Columbia Steel Company (1925–1927); manager, forged steel wheel division (1929–1930); manager, railroad sales division (1930–1945); administrative assistant, sales (1945–1947); general manager, sales (1947–1952); vice-president, sales (1952–1960); president (1960–1965), chairman, Armco Incorporated (1965–1971).

Logan T. Johnston, sales administrator and executive of Armco, was born September 1, 1900, in Pittsburgh, Pennsylvania. The son of

William C. Johnston, a Presbyterian cleric and missionary, and Emily Truex Johnston, he attended the public schools in Washington, Pennsylvania, and then served in the aviation section of the U.S. Army Signal Corps in France during World War I. After the war he entered the Carnegie Institute of Technology to study industrial engineering, breaking his family's traditional affiliation with Washington and Jefferson College, an institution that his great-grandfather had helped found.

While in college he spent his vacations working in the steel plants and oil fields around Pittsburgh. After graduating in 1923 he became a salesman of cranes and hoists for the Yale & Towne Company, his itinerary taking him through the iron range regions of Minnesota. In 1925 he took a sales position with the forged-steel wheel division of the Columbia Steel Company in Butler, Pennsylvania. At the time, Columbia and Armco were in a conflict over patents to the continuous hot-strip mill. Purchasing Columbia in 1927, Armco gained control of all the basic patents. Johnston remained with Armco, becoming manager of the wheel division in 1929 and moving to the company's general offices in Middletown in 1930.

For the next two decades, as Armco weathered the Great Depression and prospered through the wartime and postwar economy of the 1940s, Johnston gradually ascended the sales ladder. Certainly he had the personality for success in sales. Handsome and standing six feet, he was, as a prominent Armco executive recalled, "a great salesman" and was "very out-going, very charismatic." Early in the 1930s he organized the railroad sales division, then broke new ground in the sale of pig iron and coke for commercial use and became vice-president of all sales divisions in 1952.

In the 1950s, though Armco continued to record growth and profits, it was a company feeding on the momentum of the past without looking to the future: it was not open to innovate management and was subject to one-man rule. The board of directors, determined to develop a "more balanced structure," endorsed new policies late in the decade that, in part, led to the appointment of Johnston as president in 1960. Johnston called for the development of managerial talent within the company and resolved that, unlike recent "inside-minded" presidents, he would give Armco greater visibility nationally and internationally.

Johnston found himself serving as president through "the stormy years," a period of conflict with Ralph L. Gray, his predecessor and chairman of the board. Nonetheless, at his resignation in 1965, he could point to a record of substantial success. Early on, he effectively demonstrated his "outside bent" in a dramatic way. Addressing the General Meeting of the American Iron and Steel Institute in 1961, he called on the steel industry and the national government to develop and effect policies for "Going International." At the same time, he was presiding over an internal program known as "Project 600" for the expenditure of $600 million in new technologies of steelmaking and pollution control. He could also note that Armco had realized rapid growth during his tenure, with sales reaching $1 billion for the first time in its history. He also advanced a program for the training of a cadre in professional management who might give effective direction to Armco in the ensuing years.

Especially after 1950, as he entered the upper levels of management, Johnston gave vigorous support to professional and civic ventures. He played a leading role in the founding of the International Iron and Steel Institute in 1966 and became its first vice-president. At President Lyndon B. Johnson's request he accepted an appointment to the Public Advisory Committee on Trade Negotiations in 1964. In 1970 he received the Gary Medal, the prestigious award of the American steel industry, for his "industrial statesmanship . . . and leadership in the interest of a better understanding of the steel industry throughout the nation and the free world."

Johnston shared his talent and time with numerous educational institutions. He served on the boards of trustees of Carnegie Tech and Miami University and lent his support to the Air Force Museum in Dayton, Ohio, and the Cincinnati Museum of Natural History. He counted as a great success his part in locating a branch campus of Miami University at Middletown. He failed, though, in his efforts as chairman of the Southwestern Ohio Regional Airport Authority to establish a large international airport in southwestern Ohio.

At Johnston's death in 1977, one child, Logan T. Johnston, Jr., survived him. His wife, Janet Rutherford Johnston, had died the previous year. One of his associates remembered him as the "world's best steel salesman." Along with many other Armco men, he is buried at Woodside Cemetery in Middletown.

Publication:
"The Success of Our Steel Industry Depends on 'Going International,'" *The Commercial and Financial Chronicle*, 193 (June 1, 1961): 238+.

George H. Jones

(January 25, 1856 – July 6, 1941)

by Robert M. Aduddell

Loyola University of Chicago

CAREER: Clerk with Hall, Kimbark & Company (1871–1881); general sales manager, S. D. Kimbark and Company (1881–1893); vice-president and general sales agent (1893–1898), president (1898–1906), vice-president, Inland Steel Company (1906–1922); other presidencies: Hillside Fluorspar Mines, Pershing Quicksilver, and Midwest Forging Company.

George Herbert Jones, the second president of Inland Steel, was born in Brixton, England, on January 25, 1856. His early education was in private schools in England and on the Continent. He came to Chicago in 1871, where he married Myrtilla Colbert in 1876. They had two children, Ruth Caroline and Harold Colbert Jones, who served as a vice-president of Inland from 1919 until 1923; he died the following year.

Jones was an original member of the Inland firm, having learned the iron and steel business as a clerk for Hall, Kimbark & Company, Chicago iron merchants. At the time Inland was formed in 1893, Jones was general sales manager for S.D. Kimbark and Company, a wholesale hardware concern.

The Chicago Heights Land Association first offered 6 acres of land and $20,000 for buildings to Ross Buckingham, the brother of the bankrupt president of Chicago Steel Works. Buckingham bought the equipment of the defunct company with the intention of moving it to the proffered land, but he ran out of money. Ross then turned to his cousin, Clarence Buckingham, for financial assistance, but he refused. William Adams, secretary to Clarence, put Ross in contact with his friend and neighbor, George Jones. Jones and Adams then put together a group that purchased Ross Buckingham's interest in the equipment, erected it on the Chicago Heights site, and went into business as the Inland Steel Company. As one of the founders, Jones can be credited with imbuing Inland with the philosophy of stable management and growth through reinvestment of profits.

Jones first brought his father-in-law, Elias Colbert, into the group to provide Inland Steel with financial resources, although Colbert continued to work for S. D. Kimbark and his involvement with Inland proved limited. He, however, did serve on the board of directors until 1921.

Jones then called Joseph Porter, a farm implement manufacturer from Ottawa, Illinois, who did business with the Kimbark firm. Porter became Inland's first president and served in that capacity until 1898. Adams and Jones also persuaded John W. Thomas, the plant manager of Chicago Steel Works, to join the group. Thomas served as Inland's first superintendent, but he did not serve on the board of directors. He was the only original stockholder not to serve on the board.

At Chicago's Columbian Exposition in 1893, Jones approached Joseph Block, senior partner and general manager of the Block-Pollock Iron Company, Cincinnati iron merchants. The Block firm sold iron to the Chicago Steel Works. Although Block could not persuade his partners in Cincinnati to participate in the Chicago Heights project, he and his twenty-two-year-old son, Philip D. Block, joined the group, and Philip was sent to Chicago to serve as treasurer and plant manager.

George Jones acted as Inland's first vice-president and general sales agent. He received no salary as vice-president, but was paid $300 for acting as agent. In 1898 Jones replaced Porter as president, a position he occupied until 1906. Three years earlier, Adams had suggested Inland increase its dividends from 3 percent to 5 percent. Porter and Frank Wells, another stockholder with no managerial involvement, supported Adams's proposal. Jones, Colbert, and the two Blocks, however, disagreed. As a consequence, Adams left the company.

In 1906 Jones began to pursue other interests, such as the Hillside Fluorspar Mines, the Pershing Quicksilver, and the Midwest Forging Company, but he did not leave Inland. Jones ultimately became president of each of these enterprises, while continuing to serve Inland as vice-president until his retirement in 1922. Inland ultimately purchased the Hillside firm. After retirement, Jones continued to serve on Inland's board of directors until his death on July 6, 1941.

Publications:
"Under-Advertisement of the Steel Business," *Engineering,* 50 (December 1915): 357–363;

"Fluorspar and Its Uses," *Blast Furnace & Steel Plant,* 10 (November 1922): 563–567.

Reference:
William T. Hogan, S.J., *An Economic History of the Iron and Steel Industry in the United States,* 5 volumes (Lexington, Mass.: Lexington Books, 1971).

Jones & Laughlin Steel Company

by John A. Heitmann
University of Dayton

As one of America's major independent steel producers during the twentieth century, the Jones & Laughlin Steel Company (J & L) experienced a series of major organizational and structural changes that were reflective of the transitional nature of the industry. From a tightly held, family-dominated firm at the beginning of the century to its subsumption under a vast conglomerate by the late 1970s, J & L's story provides an excellent case study demonstrating the rise and subsequent decline of the "smokestack" industries in the United States.

While the origins of J & L can be traced back to the decade before the Civil War, the firm's modern period began in 1902 with its incorporation as the Jones and Laughlin Steel Company, an organization initially capitalized at $30 million. At the turn of the century the company's operations were centralized in Pittsburgh, and its steel-making plant consisted of two 10-ton Bessemer converters, six 40-ton basic open-hearth furnaces, and one 25-ton acid open-hearth unit. Beginning in 1903, J & L pursued a vigorous program of expansion that culminated with the erection of the Aliquippa Works in 1909, a facility located in Woodlawn, Pennsylvania, on the Ohio River, 26 miles from Pittsburgh. The decision to develop the 475-acre tract on the site of a former amusement park was in part the result of a managerial decision to diversify but was also the consequence of court decisions that had fined J & L for damage done to the neighborhood surrounding its Eliza furnaces on the Monongahela River in Pittsburgh. The construction of the Aliquippa plant included the erection of a blooming mill, Talbot furnaces,

Bessemer converters, a tin house, and a sintering operation — and the plant soon became the focal point of a company town. By the beginning of World War I, J & L was thus in a position to challenge the United States Steel Corporation in a few product lines and compete in a broad range of semifinished and finished goods with other major American steel manufacturers.

With increased wartime demand J & L prospered, and by 1920 the company had an annual capacity of 2.1 million tons of pig iron, 2.64 million tons of ingots, and 2.075 million tons of finished hot-rolled steel products. Furthermore, the company owned vast resources necessary for the integration of their process operations. By 1920 J & L owned the Vesta and Shannopian coal lands, the Martinsburg, West Virginia, limestone quarries, the Hill Annex Iron Ore Mine in Minnesota, and portions of the Breitung Iron Lands and Marquette Range in Michigan. In electing to expand gradually and internally in keeping with conservative financial and management strategies, J & L bucked the trend in the steel industry, which at that time called for expansion through acquisitions and mergers.

After the recession of 1921–1922, J & L was reorganized in 1923. The firm was now publicly owned, but in reality a small family group representing the Joneses and Laughlins possessed significant shares of stock and remained in control. J & L's chairman was B. F. Jones, Jr., his first cousin, William Larimer Jones was elected president, and G. M. Laughlin, Jr., was named vice-president. Indeed, the two families had seven representatives on the firm's board, and the power and conservative business

Jones & Laughlin executives and United Steelworkers representatives meeting at the first annual Labor-Management Participation Teams conference in 1983 (LTV Corporation)

stance of "The Family" during the 1920s were well portrayed in steelman Tom M. Girdler's autobiography, *Boot Straps* (1944).

The second half of the 1920s witnessed a marked expansion of J & L's capacity, and the accomplishment of a significant technological breakthrough that marked a better understanding of the relationship between nitrogen content and steel's physical properties. Previously it was thought that the presence of phosphorous in Bessemer steel was responsible for brittleness of the final product and the premature aging of cold-rolled materials. But J & L researchers not only demonstrated that the nitrogen introduced during the air blast was responsible for these undesirable properties but further proved that these conditions could be avoided by adding minute quantities of titanium, zirconium, and aluminum.

With the onset of the Depression, unbridled optimism that was in part the consequence of recent technological innovations gave way to shorter working hours and a more cautious modernization program. In 1931, however, began the operation of a massive blast furnace, measuring 28.5 feet in diameter with a 1,100-ton capacity. Furthermore, despite reduced revenues J & L gambled on the role that research and development could play during the second half of the decade. Until 1936 metallurgical research had been pursued in only a haphazard manner at J & L, but the company began to launch pioneering investigations with the opening of the Hazelwood Metallurgical Research Lab-

oratory in that year. The laboratory's first major payoff occurred in 1941 with the development of a Bessemer flame-control process that employed photoelectric cells and recording devices to determine with precision the completion of the Bessemer steelmaking process. J & L promptly licensed this method of making higher quality, more uniform material to Republic, Youngstown Sheet & Tube, and Wheeling Steel.

The Depression years also witnessed organized labor's successful challenge to J & L's managerial autonomy. J & L had taken a hostile stance to organized labor during the World War I era, and its company police and spies had been especially active in routing union activities at the Aliquippa facility in 1919. In an effort to maintain an open shop, J & L sponsored a company union and claimed that John L. Lewis's organizing efforts at Aliquippa that began in 1934 ultimately were controlled by communist activities. In 1937, however, two pivotal events shifted the balance of power between J & L management and organized labor. First, in a landmark Supreme Court case that affected labor relations for the entire American steel industry, J & L was found in violation of the National Labor Relations Act of 1935 by discriminating against ten employees who were members of the Amalgamated Association of Iron, Steel and Tin Workers of America. Second, in May 1937 J & L millworkers voted overwhelmingly in favor of the Steel Workers Organizing Committee (SWOC) to represent them in bargaining with management. The company recognized the union, but from this point forward, an increasingly adversarial relationship between management and labor developed that would in part contribute to the company's misfortunes during the 1960s and 1970s.

By the late 1930s defense-related manufacturing resulted in renewed prosperity for J & L, and in 1941 production reached 99 percent capacity. In 1944 the firm, with sales of more than $290 million, hit a peak of 4.3 million tons of pig iron and 5.1 million tons of ingots used in the fabrication of armor plate, shell forgings, shells and bombs, landing craft, mortar discs, powder boxes, bomb fins, and gun barrels. The war years also marked J & L's departure from its usual strategies of expansion, for in 1942 it acquired the Otis Steel Company of Cleveland. The merger was a natural; J & L now had a foothold in the Midwest market, access to Great Lakes shipping, and an entry into the lucrative sheet and strip business that figured so prominently in postwar projections of booming automobile and appliance industries. The Otis acquisition was arranged by J & L leader Horace Edgar Lewis, a Welsh-born steelman who had come up from the mills and who had ambitions of chasing J & L's competition, namely U.S. Steel, Republic, and Bethlehem.

Unfortunately Lewis's health gave way by the conclusion of World War II, and upon his retirement leadership fell upon former naval admiral Ben Moreell and C. L. (Lee) Austin. Moreell had built a strong reputation as a negotiator and reconciler with organized labor during 1945–1948, and his abilities in this area were soon tested at J & L. But of far greater concern to Moreell was J & L's outmoded equipment, and under his leadership he drove to improve productivity and efficiency by replacing antiquated furnaces, ovens, and mills. Moreell also directed J & L into new product areas, including semifinished, hot-rolled, and cold-finished steel as well as wire and tubular products. Some $683 million were spent during the 1950s on new plants and equipment, as obsolete Bessemer duplex and open-hearth furnaces were supplanted with new high- capacity designs. In 1957 J & L became one of the first American steelmakers to employ the basic oxygen furnace, and the company soon increased its capacity using this process.

Moreell continued to have significant influence within the corporation until the late 1950s, when Avery Adams became chairman of the board, president, and chief executive officer. Adams led J & L during the first few years of the 1960s, undoubtedly some of the most tumultuous in J & L's history. While the first half of the decade was most prosperous, J & L leadership began making decisions that ultimately worked against the fortunes of the company. In 1965 J & L embarked upon the largest single project in the company's history, the erection of a $600 million steel mill on 6,000 acres located in Hennepin, Illinois. While the facility was initially projected to finish rolled steel shipped from the East, it was hoped that eventually it would evolve into a completely integrated plant and would attract satellite industries that would use the product in fabricating finished goods. But this venture never fully realized its economic potential. By the late 1960s the Hennepin facility proved to be a giant millstone around the neck of J & L executives who were under enormous pressures after the takeover of the firm in 1968 by Ling-Temco-Vought Steel (LTV). As a result of the acquisition, J & L quickly went into an economic tailspin. One former director remarked that in 1969 J & L was run "like a company that isn't owned by anyone," and morale sank to an all-time low. With threats by LTV president James J. Ling to "shred J & L," uncertainty spread down the corporate ladder, perhaps contributing to the strikes of the period.

During 1969 and 1970, in part because of bad business decisions, the takeover, and the federal government's insistence that J & L plants comply with environmental legislation, the firm embarked upon a pivotal era in its history that culminated with its virtual demise by the mid 1980s. J & L had always been slow to diversify, and this long-standing corporate strategy was especially followed during the early 1960s, when, under the direction of Charles M. Beeghly rigid policies concerning price/earning rations, labor relations, and relative position within the industry precluded any realistic moves toward expansion. It was this inability to change with the times that was as much responsible for J & L's corporate dilemma as were the disruptions caused by the LTV takeover.

Despite the efforts of two of the steel industry's best managers during the 1970s and early 1980s, William R. Roesch and Thomas C. Graham, J & L's plants were shut down one by one, some to lay idle indefinitely, some to be demolished, a few continuing marginal operations with a skeletal workforce, the consequence of a global steel industry in transition. In 1984 LTV acquired Republic Steel, and the merger shifted the center of operations from Pittsburgh to Cleveland. Thus ineffectual management, high labor costs, and steady pressure from global competitors transformed a once significant American producer into a secondary player in the industry within the span of less than two decades.

References:

Richard O. Cortner, *The Jones and Laughlin Case* (New York: Knopf, 1970);

Tom M. Girdler, *Boot Straps* (New York: Scribners, 1944);

William T. Hogan, S.J., *An Economic History of the Iron and Steel Industry in the United States,* 5 volumes (Lexington, Mass.: Lexington Books, 1971);

John P. Hoerr, *And the Wolf Finally Came: The Decline of the American Steel Industry* (Pittsburgh: University of Pittsburgh Press, 1988);

New York Times, February 27, 1903, p. 1; January 31, 1907, p. 2; December 6, 1922, p. 2; January 23, 1941;

Gertrude G. Schroeder, *The Growth of Major Steel Companies, 1900–1950* (Baltimore: Johns Hopkins University Press, 1953).

Harvey Bryant Jordan

(July 21, 1895–April 8, 1965)

by Bruce E. Seely

Michigan Technological University

CAREER: Laborer, blast-furnace department, summer position (1914), chemical laboratory employee (1914), plant foreman and assistant superintendent (1915–1929), superintendent, Central Furnace and Docks division (1929–1933), vice-president's office (1933–1934), manager of metallurgical department (1934–1935), manager of operations, Cleveland division (1935–1937), assistant to the vice-president (1937–1938), assistant vice-president (1938–1939), vice-president for operations (1939–1950), president, American Steel & Wire Company (1950–1953); vice-president for production, executive vice-president, United States Steel Corporation (1953–1955).

Typical of many top executives at United States Steel Corporation and other major steel firms through the first half of the twentieth century, Harvey B. Jordan rose from humble origins to spend his entire career within one corporation. His career at the corporation began as a summer job and spanned forty years, most of them at American Steel & Wire Company, a subsidiary of U.S. Steel.

Jordan was born to Mary (Munger) and Harvey S. Jordan in Lansing, Michigan, on July 21, 1895; his father was a Presbyterian clergyman. Orphaned at fourteen, he lived with relatives and graduated from high school in Everett, Pennsylvania. He attended Pennsylvania State College (now

Harvey Bryant Jordan

Pennsylvania State University) from 1913 to 1915, earning a degree in industrial engineering. During his summer vacation in 1914 he worked on U.S. Steel's blast furnaces in Cleveland and in the chemical laboratory of the Central Furnaces and Docks division of American Steel & Wire. He returned as a clean-up foreman in 1915.

Jordan steadily climbed the management ladder in Cleveland, becoming general foreman, assistant superintendent, and, in April 1929, superintendent of the Central Furnaces plant. More impressively he found the time to attend the John Marshall Law College at night, receiving his law degree in 1926. The same year he was admitted to the Ohio bar, although he never practiced.

Jordan's rise through the company continued during the Depression. In May 1933 he was transferred to the vice-president's office of American Steel & Wire, where he performed a variety of activities that came under the heading of fieldwork. In 1934 he was made head of the metallurgical department, a post that came to include direction of manufacturing practice later that year. In December 1935 Jordan advanced to the post of manager of operations for the Cleveland division. Over the next three years he was assistant to the vice-president, assistant vice-president, and vice-president for operations of the American Steel & Wire Company. Don Watkins, editor of *Blast Furnace and Steel Plant,* commented in his "My Turn" column in the April 1939 issue of the journal, "check another mark up for a blast furnace superintendent who reached a high executive position, Harvey Jordan, now vice-president of American Steel & Wire Company came first into prominence as blast furnace superintendent at Central Furnaces at Cleveland, Ohio."

Jordan held this post through the war and was promoted to president of American Steel & Wire in 1950, replacing Clifford Hood, who moved over to head Carnegie-Illinois Steel. When all steelmaking subsidiaries were consolidated into a reformed U.S. Steel Corporation in 1953, Jordan was made vice-president for production, then executive vice-president of the new corporation. Jordan moved to Pittsburgh, where his main task was to chair the corporation's general administration committee. He brought to this task a reputation as "one of the industry's outstanding production specialists," according to his *New York Times* obituary.

In the early 1950s Jordan mostly concerned himself with the corporation's efforts to expand facilities to meet a rapidly growing market. But U.S. Steel executives, led by corporate chairman Benjamin Fairless, also kept up a drumbeat of criticism aimed at the federal government, criticisms sounded in an address delivered by Jordan at the 1951 regional meeting of the American Iron and Steel Institute (AISI) in Cleveland. Although admitting the industry's need to polish its own image by pursuing a more vigorous program of public relations, Jordan's remarks — titled "Silence is Not Always Golden" — denounced government intrusion in the steel industry and warned that the government had better cut waste so it could reduce taxes.

In 1955 Jordan received a gold watch from the corporation for his 40 years of service, and shortly thereafter he retired. From 1952 to 1959 he served on the board of trustees of Ohio University, chairing the board in 1957. While in Pittsburgh he was a trustee of Shadyside Hospital, a director of the Pittsburgh YMCA, and chairperson of the corporate gifts committee for the first successful campaign of the United Fund in Allegheny County. In 1957 he received Penn State University's distinguished alumnus award.

Like many steel men, Jordan's primary leisure activity was golf, and he was a member of the Fox Chapel Golf Club in Pittsburgh. He also belonged to the AISI, the Newcomen Society in North America, the Masonic Order, and Phi Delta Theta. In 1960 he retired to Naples, Florida, where he could pursue another interest — sailing — and joined the Naples Yacht Club. He also was a member of the Hole-in-the-Wall Club in Naples, the Union Club of Cleveland, and Pittsburgh's Duquesne Club, a favorite of U.S. Steel executives. Jordan married Ethel J. Mullin on May 6, 1919; they had one daughter. He remarried after his first wife died in 1940, wedding Marjorie Fuller in April 1943. Jordan died in Naples, Florida, on April 8, 1965.

Publications:
"Inventory Time," *Iron and Steel Engineer,* 27 (November 1950): 76–80;
"Silence Is Not Always Golden," *American Iron and Steel Institute Regional Technical Papers* (1951): 287–296.

References:
Edith Fisch, *Lawyers in Industry* (New York: Oceana, 1956), p. 23;
"Personals," *Iron Age,* 143 (March 16, 1939): 66;
Don Watkins, "My Turn," *Blast Furnace & Steel Plant,* 27 (April 1939): 371.

Henry John Kaiser

(May 9, 1882 – August 24, 1967)

by Mark S. Foster

University of Colorado at Denver

CAREER: Founder of many companies which later became publicly held corporations, president and chairman of the board of the following: Kaiser Paving (1914–1956); Six Companies, Inc. (1931–1945); Kaiser Steel (1942–1956); Kaiser-Frazer Corporation (1945–1955); Kaiser Aluminum & Chemicals (1946–1956); Permanente Cement (1939–1956); Kaiser Industries (1956–1967); and others.

One of the most prominent entrepreneurs in the United States, Henry John Kaiser played a major role in developing the economy of the American West. During World War II, when his West Coast shipyards employed more than 200,000 workers and built one-third of all American cargo ships, he also constructed the first large steel mill in the West. This mill was greatly expanded after the war. During the decade that followed World War II, Kaiser's aluminum operations challenged Alcoa's. Kaiser was also involved in home construction, cement and building materials, tourism and resort development, nuclear power, radio, television, and many other enterprises, including the Kaiser Health Plan, the largest health maintenance organization (HMO) in the United States.

The youngest of four children, Kaiser was born in Sprout Brook, a small village in upstate New York, on May 9, 1882. His parents, Frank and Mary, were German immigrants with a modest income. The son of a cobbler, Kaiser left school at thirteen and went to work full-time at a dry goods store in Utica, New York. During the three years he clerked and sold dry goods there, he became interested in cameras and worked part-time as a photographer.

When he was seventeen, Kaiser took up full-time photography just as Eastman-Kodak was pioneering advances in photographic equipment. In 1901 he became a partner in a photography studio in Lake Placid, New York. A year later the original owner sold his half to Kaiser. At that time

Henry John Kaiser

there was little business in Lake Placid during the winters. Kaiser decided to follow the tourists south; between 1903 and 1905 he opened a string of photography shops along the east coast of Florida.

While in Florida, he courted Bessie Hannah Fosburgh. Her father, however, disapproved of Kaiser's career and withheld his approval of a marriage until Kaiser found more suitable employment. Kaiser moved to Spokane, Washington, in 1906 and took a job as a clerk with McGowan Brothers, a wholesale hardware company. He was soon promoted to a lucrative sales position and married Bessie Fosburgh on April 8, 1907.

In 1913, while working as a salesman for the road-building and general construction company of J. B. Hill, Kaiser was sent to Canada to seek new contracts. After he was fired during managerial infighting, Kaiser found enough new business in Canada to start his own company. He formed Kaiser Paving in 1914.

Earning a reputation for fast, high-quality work, Kaiser was able to expand his operations into Washington, Oregon, and California. As his reputation grew, he won larger contracts, and in 1921 he moved his modest business headquarters to Oakland, California.

During the 1920s most of Kaiser's work was in road building, but he also built earthen dams and set up sand-and-gravel operations to provide materials for his jobs. In 1927 he was given a $19 million subcontract to build a 200-mile section of a trans-Cuba highway. The task was formidable, but Kaiser succeeded.

In 1931 Kaiser joined Six Companies, a consortium of contractors that won the Hoover Dam job with a bid of $48.9 million. Kaiser's main task was serving as liaison between the contractors and government bureaucrats, a public-relations role that was vital to the project. Between 1931 and the completion of the Hoover Dam superstructure in 1935, Kaiser spent many months in Washington, D.C., becoming expert at maneuvering himself and his partners into position to receive government contracts and loans. In the late 1930s Six Companies was twice reorganized under different combinations of partners and successfully bid major portions of the Bonneville and Grand Coulee dams on the Columbia River.

In 1939 failure to win the bid to build Shasta Dam in northern California launched the fifty-seven-year-old Kaiser into yet another career, production of construction materials. Having lost the job to build the Shasta Dam, he won the contract to supply cement for the dam. Kaiser did not have a cement plant yet, but within months Kaiser and his men built a huge plant south of San Francisco in Santa Clara County. Kaiser's Permanente Cement operation supplied much of the cement for military installations in the Pacific during World War II.

With the outbreak of war in Europe, Kaiser sensed much earlier than most American businessmen that U.S. involvement was nearly inevitable. Between 1939 and 1941 he became a public figure as he testified forcefully that the government should insist upon immediate plant expansion in magnesium, steel, aluminum, and other producer goods. If existing manufacturers failed to accept the challenge, Kaiser offered to help fill critical voids.

After 1939 Kaiser was heavily involved in the shipbuilding industry. He had met prominent shipbuilders during the late 1930s, when he and his construction partners had built millions of dollars' worth of harbor improvements on the West Coast. In late 1940 Kaiser, John A. McCone, and Stephen D. Bechtel were awarded a contract to build 30 cargo ships for Great Britain. By late spring 1941, Kaiser and his engineers had erected a huge shipyard in Richmond, California. The British ships were completed ahead of schedule, and the U.S. Maritime Commission began to give Kaiser's company a steady stream of contracts, mainly for cargo ships. By 1944 the Richmond shipyard employed nearly 100,000 men and women, and Kaiser's team developed similar operations in the Portland/Vancouver region on the Columbia River. During World War II the "Kaiser" yards on the West Coast built 1,490 vessels, including 50 small aircraft carriers and nearly one-third of the cargo ships produced in the United States. Journalists dubbed Kaiser the nation's "Miracle Man" and America's "Number 1 Industrial Hero." President Franklin D. Roosevelt seriously considered Kaiser as his running mate in 1944.

Before Pearl Harbor, when steel shortages had become a major bottleneck in meeting cargo-ship contracts, Kaiser had urged federal officials to grant him permits to build his own steel plant on the West Coast. He met opposition from eastern steel men, who had been incensed when Kaiser had publicly lectured them about their alleged dereliction of duty. In their eyes this upstart entrepreneur was a hypocritical publicity seeker whose success was due solely to large government contracts; he had never competed in "real" private enterprise. Viewing Kaiser as a potential threat, they vigorously opposed his repeated efforts to win government approval for his own steel plant. Throughout 1940 and 1941 Kaiser's "war" with Big Steel increased in intensity.

With America's entry in the war, however, Kaiser got the authorization he needed. He secured more than $100 million in loans from the Reconstruction Finance Corporation (RFC) to build a plant at Fontana, California, 50 miles east of Los Angeles. Setting aside their differences with Kaiser during the national emergency, Big Steel provided technical assistance. The Fontana plant was built in only nine months and began producing ingots.

Yet animosity continued between Kaiser and eastern steel men, who assumed that Kaiser had

"The Important Thing to Washington," a 1943 editorial cartoon praising Kaiser for refusing to let government rules and regulations slow his production of warships (Copyright © Chicago Tribune Company, all rights reserved, used with permission)

built the plant only as a wartime expedient and that his plant would be scrapped or sold to Big Steel interests after the war. Kaiser, however, showed no such intention. In fact, he talked publicly of creating his own West Coast steel empire, even discussing the possible purchase of the Geneva plant in Utah, which was operated by the United States Steel Corporation during the war. Between 1945 and 1950 hostilities intensified, with Big Steel complaining that Kaiser had received almost unlimited loans from the RFC and that he enjoyed "government favoritism" in dam and shipbuilding contracts. Kaiser, in turn, accused Big Steel of hypocrisy, claiming that they had resisted expansion until Pearl Harbor and then enjoyed the advantage of operating plants built at government expense.

The issue of whether Kaiser would have to repay the entire RFC loan, however, proved to be their most immediate battleground. Kaiser pleaded for a negotiated debt reduction, arguing that Big Steel had been permitted to buy surplus plants at one-third of government cost and accusing Big Steel of wanting to eliminate competition

and keep western steel prices artificially high. Big Steel countered that Kaiser had forced himself on the steel market and that he had borrowed the RFC money with his eyes wide open. Had the war lasted a year or two longer, they said, Kaiser would have easily recouped his investment. In their view he was asking for relief from a foolish commitment.

Kaiser's RFC loan ultimately became a political issue. Big Steel urged congressmen to pressure the RFC to demand repayment of the Fontana loan in full. Kaiser and his associates lobbied just as hard for relief. Meanwhile, Big Steel pressured western railroads to give the Geneva Steel plant rates for raw-material shipments that were better that those Kaiser paid.

After more than five years of fruitless negotiation, Kaiser finally relented and paid off the entire RFC loan, plus interest. In November 1950 he handed RFC officials a check for the entire balance. Kaiser's prediction that the West would need its own steel operations had proved spectacularly correct. The return of peace marked rapid economic and demographic growth on the West

Coast, and Fontana quickly expanded to meet demand. Furthermore, Kaiser had garnered another important, if intangible, asset; by arranging private financing to pay off the RFC loan, Kaiser earned the respect of many businessmen.

After 1950 relations between Kaiser and Big Steel gradually improved. The easterners realized that Kaiser would remain a major player; in the 1950s and 1960s Kaiser Steel became the pacesetter in the West. Reluctantly, eastern steel interests included Kaiser in conferences concerning national steel policy, particularly labor negotiations. For their part Kaiser and his men considered Big Steel old-fashioned in its production methods and attitudes toward labor. On several occasions Kaiser settled with the United Steel Workers of America (USWA) ahead of his peers.

Kaiser's postwar commitment to expanded steel production revealed the symbiotic relationships of many of his enterprises. In 1945 he and Joseph W. Frazer founded Kaiser-Frazer, which in 1948 became the fourth-largest automobile producer in the United States. Although based in Detroit and forced to rely on eastern producers for most of its steel, the automobile company occasionally turned to Kaiser Steel for critical shipments. Unfortunately, the automobile venture became Kaiser's most notable failure. Although marked by short-term growth and success, the company failed to raise sufficient capital when it had the opportunity, and it simply could not match the research and development programs, economies of scale, and other advantages enjoyed by the Big Three — General Motors, Ford, and Chrysler. In ten years of domestic automobile sales, Kaiser-Frazer lost $123 million.

While the automobile venture was unsuccessful, Kaiser Aluminum and Chemicals became an even more spectacular success than Kaiser Steel. The aluminum company evolved from Kaiser's experimentation in light metals, particularly magnesium, during World War II. In the postwar period the U.S. Justice Department was trying to encourage competition against Alcoa. Although federal officials were refusing concessions to Kaiser Steel, they granted Kaiser Aluminum some important favors. The aluminum company was Kaiser's most profitable enterprise.

The postwar years brought nearly geometrical expansion of Kaiser's corporate empire. He opened the Kaiser Permanente Health Care Program to the public and turned it into the largest HMO in the nation. He and his men diversified in many other fields, including housing and real-estate development, hotels and tourism, radio and television, and a variety of engineering ventures around the world. After his wife died in 1951, Kaiser promptly remarried. In 1954 he and his second wife, Alyce, moved to Hawaii, where he spent the remainder of his life. A workaholic, Kaiser never retired, and he enjoyed few hobbies; he had always been greatly overweight and had abused his body over the years. Nevertheless he enjoyed remarkably good health until his last months. He died in Hawaii on August 24, 1967, at the age of eighty-five.

Publications:

Building a Ship in 4 Days, 15 Hours, and 25 Minutes (N.p., 1943);

Management Looks at the Post-war World: An Address before the 47th Annual Meeting of the National Association of Manufacturers Held at New York on December Fourth, 1942 (Princeton: Princeton University Press, 1943);

Twenty-six Addresses Delivered During the War Years, September Thirtieth, Nineteen Hundred and Forty-two to July Nineteenth, Nineteen Hundred and Forty-five (San Francisco: Kaiser Family, 1945);

America, Look to Your Soul (New York: Newcomen Society of England, American Branch, 1945);

Imagine Your Future! A Philosophy of the Happy, Successful Life (N.p., 1948).

References:

Mark S. Foster, "Giant in the West: Henry J. Kaiser and Regional Industrialization, 1930–1950," *Business History Review,* 59 (Spring 1985): 1–23;

Foster, *Henry J. Kaiser: Builder in the Modern American West* (Austin: University of Texas Press, 1989);

Foster, "Prosperity's Prophet: Henry J. Kaiser and the Consumer/Suburban Culture, 1930–1950," *Western Historical Quarterly,* 17 (April 1986): 165–184;

"Kaiser-Frazer: The Roughest Thing We Ever Tackled," *Fortune,* 44 (July 1951): 74–77, 154–162.

Archives:

The Henry J. Kaiser Papers, at the Bancroft Library, University of California, Berkeley, include 700 linear feet of material, many boxes of which pertain to Kaiser Steel.

Kaiser Steel

by Mark S. Foster

University of Colorado at Denver

Renowned entrepreneur Henry J. Kaiser organized Kaiser Steel in 1942 for two reasons. First, with tens of millions of dollars in contracts to build cargo ships for the United States Maritime Commission, his West Coast shipyards required deliveries of large quantities of a specialized thickness of steel plate — requirements which Big Steel could not, or would not, meet. Kaiser often publicly chastised the eastern steel establishment for refusing to expand its capacity. Big Steel, in turn, relegated Kaiser to the end of the line when it came to meeting orders. Second, Kaiser had a long-term objective to create a postwar West Coast steel empire.

Without experience in steel and with only a handful of trained engineers, Kaiser needed technical assistance from Big Steel to get started. He also required large government loans from the Reconstruction Finance Corporation (RFC). The Japanese bombing of Pearl Harbor on December 7, 1941, provided the opening he needed. A few weeks later, Kaiser received government authorization and the first of a series of RFC loans which eventually totaled over $100 million. A state of national emergency brought grudging technical assistance from Big Steel executives, and Kaiser and his men built a steel plate mill in Fontana, California, 50 miles east of Los Angeles. Much to the surprise of industry experts, they learned the business quickly: the Fontana mill yielded its first ingots on December 30, 1942. Kaiser wisely kept his postwar plans to himself until early 1944; by then wartime production requirements for arms to defeat the Axis powers had created an industrial juggernaut.

Big Steel assumed that Kaiser would leave the steel business after the war. In 1945 Kaiser Steel produced only steel plate for ships; there was no postwar market for this specialized product. Eastern steel men, however, misread both their rival and postwar markets. They generally predicted shrinking markets for most steel products after the war, but Kaiser was convinced that consumer uses would greatly expand markets, particularly in the West. To capitalize on this opportunity, Kaiser needed to eliminate his debt to the RFC and modernize and expand Fontana. By 1945 his debt was $110 million, and interest was mounting. Between 1945 and 1950 Kaiser pleaded for a negotiated reduction, arguing that Big Steel had been allowed to purchase wartime plants built at government expense for one-third of their cost. The RFC, under intense pressure from eastern steel lobbyists, rejected Kaiser's request; he finally paid off the loan in full, plus interest, in 1950.

The two decades following World War II marked Kaiser Steel's heyday. Corporate officials quickly modernized the plant in the late 1940s. Ingot capacity and blooming and sheet-rolling output expanded quickly. In 1945 the Geneva mill in Utah, operated by United States Steel Corporation, had dominated production in the West; Kaiser Steel had only 60 percent of Geneva's ingot capacity, three-fourths of its blooming and sheet-rolling output. In 1954, however, Kaiser matched Geneva's ingot production; by 1960 Kaiser turned out 2.93 million tons to Geneva's 2.3 million. Kaiser grew dominant in blooming and sheet rolling; by 1965 it doubled Geneva's output. There were other important regional mills, including Colorado Fuel and Iron, and other scattered suppliers. But in the mid 1960s Kaiser Steel was the pacesetter in western steel. Just a few months before Henry Kaiser's death in August 1967 company president Jack L. Ashby observed that in the last ten years the company had sales of $2.5 billion, exceeding the value of all gold mined in California since 1848.

In retrospect, although nobody anticipated decline in 1967, factors leading to Kaiser Steel's decay had already been in place. Despite a booming western economy that meant healthy sales for many producers, a few shrewd industry analysts noted that foreign producers were competing effectively and that their percentage of total sales was mounting steadily. Ironically, Kaiser engi-

Kaiser Steel executives Eugene E. Trefethen, Jr., Jack Ashby, and Henry J. Kaiser discussing finances with Bank of America officials Fred Ferroggiaro, Atwood Austin, and Paul Marrin, 1951 (photograph by Dan Weiner)

neers helped a number of foreign companies become highly competitive in international steel markets. The engineers designed and helped build many foreign plants that used the oxygen thermal process, a more efficient and modern technique than the open-hearth method. They also helped the Japanese and other emerging steel producers develop dependable, economic sources of ore and energy.

By the mid 1970s the company faced a bleak future in its primary product, steel. Japanese producers were underselling the company even in Southern California markets. After World War II the company had become the only fully integrated steel operation in the West, having acquired considerable ore deposits, coal properties, and facilities turning out specialized forms of steel. To protect corporate assets, corporate officials quickened the pace of diversification, buying more coal resources, along with gas and oil reserves. They also invested in Southern California real estate. As two reporters noted, "Henry J's successors kept on doing more or less what they thought the old man did, but where he seized opportunities they played safe." Unfortunately, "safe" did not translate into profits. To keep the money-gobbling

steel facility operating, corporate officials gradually sold off their valuable resources. But a slump in oil prices, combined with the realization that they could not continue operating the steel plant in the face of rapidly mounting debt and interest, forced them to confront the inevitable. After forty-one years of continuous operation, the Fontana mill closed its doors in 1983.

The last six years have produced a complex series of corporate buyouts, as shifting groups of stockholders and outsiders have struggled over control of the once-proud company. In 1986 Kaiser Steel lost $96.8 million, and its valuable assets were reduced to Southern California real estate and a small hazardous-waste disposal operation. Kaiser Steel filed for bankruptcy. Stephen Girard, who labored effectively in the Kaiser organization for forty-six years, perhaps summed up recent events most succinctly: "Henry Kaiser would have seen this all coming. . . . He would have offloaded this years ago."

References:
Mark S. Foster, *Henry J. Kaiser: Builder in the Modern American West* (Austin: University of Texas Press, 1989);

"The 'Pugnacious Bulldog:' Fighting for Kaiser Steel,"
  Business Week (February 16, 1987): 72–73;
Allan Sloan and Peter Fuhrman, "An American Trag-
  edy," *Forbes* (October 20, 1986): 30–33;
Kenneth Warren, *The American Steel Industry, 1850–
  1970* (Oxford: Oxford University Press, 1973);
Peter Wiley and Robert Gottlieb, *Empires in the Sun:
  The Rise of the New American West* (New York:
  G. P. Putnam's Sons, 1982).

Archives:

The Henry J. Kaiser Papers, housed at Bancroft Library, Berkeley, California, contain 700 linear feet of material. Several dozen cartons document the rise of Kaiser Steel.

Kefauver Committee on Antitrust and Monopoly
by Paul Tiffany

Sonoma State University

The "Kefauver Committee" was the popular name given to the U.S. Senate Subcommittee on Antitrust and Monopoly — part of the Committee on the Judiciary — after 1956, when the subcommittee fell under the leadership of Democratic senator Estes Kefauver of Tennessee. Kefauver was a firm believer in the need to enforce the antitrust laws aggressively, as he felt that large, monopolistic firms were a detriment to the economic health of the nation. Accordingly, he used his influence on the antitrust and monopoly subcommittee to press his views.

From 1956 through 1961, while the Democrats controlled the U.S. Senate and Kefauver was chairman of the subcommittee, the unit conducted a series of hearings on "administrated pricing" as practiced by large firms. Administered prices reflected a theory first advanced by the economist Gardiner Means in the 1930s, who claimed that large, monopolistic firms did not set product prices according to the conventions of supply and demand, but rather set them arbitrarily according to their profit needs. The remedy for administered pricing, Means felt, was antitrust enforcement to break up the large firms. Kefauver led his subcommittee into an investigation of administered prices in several industries, including automobile manufacturing, pharmaceuticals, and steel.

Kefauver began his study by noting that "No better example of this behavior [of administered pricing] can be found than that exhibited by the steel industry itself. Because of this reason, and also because of its importance as the nation's basic materials industry, the subcommittee has chosen steel as its first specific field of examination." From August through November 1957 the Kefauver Committee questioned witnesses on the extent of industry concentration and price competition in steel. The clear intent of the subcommittee was to find evidence to justify an antitrust filing against the major integrated steel companies.

Testimony on a wide range of topics was solicited from company officials, including information on steel investments, costs, prices, profits, competition, inflation, technological change, and "corporate power." The press gave Kefauver a fair amount of publicity and brought the hearings to the attention of the nation's newspaper readers. Kefauver consequently gained status as a folk-hero of sorts for taking on the big "vested interests" and protecting the economic rights of the "little guy." Not only was the ongoing campaign of Senator Kefauver for high elective office enhanced, but the reputation of steelmakers as greedy plutocrats was also reinforced in the eyes of the nation through the congressional hearings. Obviously not conducive to the establishment of better business-government relations in the steel industry, the end result of the hearings was an unfortunate one for steelmakers. The industry was greatly in need of improved relations with the government as they prepared to compete with strengthened foreign steel suppliers who enjoyed increasing benefits from their home governments.

The Kefauver Committee issued its report on the steel industry in 1958, recommending that both the Department of Justice and the Federal Trade Commission undertake investigations into the pricing practices of the large integrated steelmakers. While no specific recommendations to break up the large domestic producers were forthcoming, the report worked to further divide the industry and the government at this critical time.

References:
Paul Tiffany, *The Decline of American Steel* (New York: Oxford University Press, 1988);

U.S. Congress, Senate, Committee on the Judiciary, Subcommittee on Antitrust and Monopoly, 85th Congress, 1st Session, Report, *Administered Prices, Steel* (Washington, D.C.: Government Printing Office, 1958).

Willis L. King

(February 14, 1851 – December 11, 1936)

by John A. Heitmann

University of Dayton

CAREER: Employee (1869–1902), Vice-president (1902–1934), Jones and Laughlin Steel Company.

Willis L. King was a unique figure in the early twentieth-century American steel industry, for rarely does a man's career in one manufacturing sector and with one firm span more than six decades. Not only was King a witness to the late nineteenth-century transition from iron to steel and from many small producers to a few corporate giants in the industry, he was also instrumental in establishing a new organization associated with the steel industry — the American Iron and Steel Institute (AISI).

King was born in Pittsburgh, Pennsylvania, on February 14, 1851, the son of Hugh Davidson and Eliza Ann (McMasters) King. He attended Washington and Jefferson College between 1867 and 1869. At age eighteen he joined what later became the Jones and Laughlin Steel Company (J&L) as a clerk and began an association with that company that would continue for more than sixty-five years.

During the last three decades of the nineteenth century, J&L expanded steadily under the primary direction of founder Benjamin Franklin Jones at its two Pittsburgh plants. The company's Eliza furnaces were situated on the north side of the Monongahela River, and directly opposite this facility the South Side Works housed some 75 puddling furnaces, 30 heating furnaces, 18 roller mills, and 73 nail machines that were in operation by 1875. In 1885 the company began steelmaking operations with the installation of a pair of 7-ton Bessemer converters at the South Side Works, and

this shift in product was so successful that a third Bessemer unit was erected in 1890. By 1896 the transition from iron to steelmaking was completed at Jones and Laughlin. In that year 42 basic open-hearth furnaces were put into production, and the puddling furnaces, so prominent in the iron-making operations of the past, were finally abandoned. Indeed, in the decade preceding World War I, J&L had grown impressively as the firm's rolled steel production more than doubled between 1901 and 1914, and a massive new plant and company town were built some 26 miles from Pittsburgh at Aliquippa, Pennsylvania. By the beginning of the war the South Side Works produced a broad range of products, including Bessemer and open-hearth steel, steel bars, sheet piling, power transmission machinery, cold-rolled steel, shafting, rope drives, concrete reinforcement bars, steel-wire nails, barbed wire nails, fence and special screw wire, tinplate, railroad spikes, light rails and connections, structural steel, and plates.

Although well into the twentieth century the highest levels of management at J&L remained reserved for family members, Willis King emerged by the late nineteenth century as one of the first professional managers within this tight-knit and conservatively run organization. One of King's responsibilities was to represent the firm in the always delicate government-industry negotiations associated with tariff levels, and on more than one occasion he testified before the House Ways and Means Committee concerning duty reductions. In addition, King played an active role in the establishment of the AISI in 1910 and served as the trade association's second vice-president, vice-president, and honorary vice-president between 1910 and his death in 1936. The AISI served to enhance stability within the American steel industry by normalizing business practices, and King emerged as a strong advocate of the sanctity of contracts between buyers and sellers, thereby attempting to prevent price undercutting during periods of market fluctuations and product gluts.

By the early 1920s King, then in his seventies, remained a powerful figure within the AISI and constantly spoke of strategies that would ultimately result in stability, whether it was of labor-wage rates, of freight rates, or in the creation of proper merchandising and distribution channels

for the manufacturers of steel. Ever an optimist, King remarked in 1928 that the "steel industry has no problem it cannot solve," but even during the 1920s boom he warned that foreign steel was a serious threat to the viability of the industry, an admonition that a future generation of leaders failed to heed.

References:

William T. Hogan, S.J., *An Economic History of the Iron and Steel Industry in the United States,* 5 volumes (Lexington, Mass.: Lexington Books, 1971);
Ben Moreell, *"J & L": The Growth of an American Business (1853–1953)* (New York: Newcomen Society, 1953).

Forrest H. Kirkpatrick

(September 4, 1905 –   )

by Elizabeth M. Nolin

Paeonian Springs, Virginia

CAREER: Professor and dean of students, Bethany College (1927–1940, 1946–1952); visiting professor / lecturer at New York University, Columbia University, University of Pittsburgh, University of Akron, University of Wisconsin, and Cornell University (1938–1954); general manager of personnel administration, Radio Corporation of America (1941–1946); assistant to the chairman, Wheeling Steel Corporation (1952–1964); vice-president of corporate and industrial relations, Wheeling Steel Corporation (later, Wheeling-Pittsburgh Steel Corporation; 1964–1970); visiting professor, West Virginia University (1970–1980); adjunct professor, Bethany College (1970–1987).

Forrest Hunter Kirkpatrick was born in Galion, Ohio, on September 4, 1905, to Arch M. and Mildred Hunter Kirkpatrick. In 1926 he attended the University of Dijon, France, and the following year he graduated from Bethany College in West Virginia with a bachelor of arts; as a university student, he had studied history and economics. Kirkpatrick's career was twofold, as he taught at several universities before interrupting his academic life to deal with industrial and corporate relations. During World War II Kirkpatrick estab-

lished himself in the arena of public service, and he continued this effort into the 1980s.

Kirkpatrick spent the early years of his career, from 1927 through 1952 (with the exception of the war years) at Bethany College, where he was dean of students and a professor in the business school. While at Bethany, Kirkpatrick received his master of arts degree in management from Columbia University in 1931 and two other professional diplomas in 1934 and 1936. In 1937 he studied industrial psychology at the University of London and pursued similar studies at the Universities of Pennsylvania and Pittsburgh and at Cambridge and Oxford.

Between 1938 and 1954, Kirkpatrick held visiting professor/lecturer status at New York University, the University of Pittsburgh, Columbia University, the University of Akron, the University of Wisconsin, and Cornell University, teaching courses in business and personnel management. During this period, Kirkpatrick held positions on the board of directors of Sharon Tube Company and Banner Fibreboard Company.

Between 1940 and 1946 Kirkpatrick worked for the Radio Corporation of America (RCA) in New York City, serving as general manager of personnel and labor relations. While at RCA, Kirkpatrick also maintained consulting positions in several government organizations, including the American Council on Education, 1938–1945; the War Manpower Commission, 1942–1944; the Department of State, 1944; and the U.S. Civil Service, 1945.

In 1952 Kirkpatrick joined Wheeling Steel Corporation — which in 1968 became the Wheeling-Pittsburgh Steel Corporation after merging with Pittsburgh Steel — as assistant to the chairman where he served in personnel and organization development until 1964, when he was promoted to the position of vice-president of corporate and industrial relations. He held this position until his retirement in 1970 at the age of sixty-five.

Kirkpatrick spent many hours of his career in public service. From 1948 to 1951 he was a member of the Educational Program Committee for the United States Air Force; consultant to the Post Office Department in 1953; member of the Mission to Sweden, operating under the Department of Labor in 1962; member of the Manpower Advisory Commission from 1963–1968; and consultant to the Department of Health, Education, and Welfare in 1970. Kirkpatrick also devoted many hours to educational, health, and state organizations within West Virginia.

He was also involved in many professional organizations and honorary fraternities. He had life memberships in the Industrial Relations Research Association, Academy of Political Science, American Personnel and Guidance Association,

and the National Education Association. He is an emeritus member of the American Association of University Professors and the American Management Association. He was director of the National Association of Manufacturers from 1965 to 1970 and metropolitan chairman of the National Alliance of Businessmen in Wheeling, West Virginia, in 1971–1972.

Honorary fraternities included Beta Gamma Sigma and Alpha Kappa Psi, both business fraternities, and Kappa Delta Pi and Phi Delta Kappa, educational fraternities.

After Kirkpatrick retired from Wheeling-Pittsburgh Steel in 1970, he served as a visiting professor at West Virginia University until 1980 and as an adjunct professor at Bethany College from 1970 through 1987.

References:

The International Year Book and Statesmen's Who's Who (West Sussex, England: Thomas Skinner, 1988);

Who's Who in Finance and Industry, 25th edition (Wilmette, Illinois: Macmillan, 1988);

Who's Who in America, 45th edition (Wilmette, Illinois: Macmillan, 1989), p. 1701.

Archives:

Wheeling Steel Corporation *Annual Reports* and Wheeling-Pittsburgh Steel Corporation *Annual Reports* are located in the West Virginia Collection, West Virginia University.

Bernard Kleiman

(January 26, 1928–)

by John P. Hoerr

Teaneck, New Jersey

CAREER: Attorney, law firm of Abraham W. Brussell (later, Kleiman, Cornfield & Feldman), Chicago (1957–1965); general counsel, United Steelworkers of America (1965–).

Bernard Kleiman, a general counsel of the United Steelworkers of America (USWA), was influential in the union's top-level decision-making apparatus for nearly three decades starting in the 1960s. Born January 26, 1928, in Kendallville, Indiana, he graduated from Purdue University in 1951 and earned a law degree at Northwestern

University in 1954. Kleiman began specializing in labor law in 1957 when he joined the Chicago law firm of Abraham W. Brussell, which represented USWA districts 31 (Illinois–Northern Indiana) and 32 (Minnesota–Northern Michigan).

When Brussell became a judge in 1960, Kleiman and two associates took over the practice and renamed it Kleiman, Cornfield & Feldman. Replacing Brussell as chief counsel for district 31, Kleiman came under the sponsorship of the district director, Joseph Germano, who had led the district since the USWA's founding. Because his was the union's largest district in terms of membership, Germano had a powerful voice on the union's international executive board. In 1965 Germano supported I. W. Abel in his successful bid to unseat USWA president David J. McDonald, and Abel later named Kleiman general counsel of the union.

As a presidential adviser, strategist, and negotiator during three successive USWA administrations, Kleiman could arguably be characterized as second only to the USWA president in his influence on union policy-setting. He participated on the union's top-level bargaining committee during each round of industrywide negotiations with major steelmakers from 1968 until the companies ended multiemployer bargaining in 1985. After that Kleiman retained a strong influence on union decisions in company-by-company negotiations. A tall, immaculately-dressed man, he had a deceptively soft-spoken manner. But he developed a reputation among company officials for relentlessly pushing the union's agenda in bargaining.

Kleiman played a major role in several milestone developments in steel. One was an incentive arbitration case of 1969 in which a panel of arbitrators sided largely with the union in extending incentive-pay coverage from an industry average of less than 70 percent of steelworkers to about 85 percent. In 1973 Kleiman assisted Abel in negotiating the Experimental Negotiating Agreement, which guaranteed automatic wage and cost-of-living raises for steelworkers in return for a no-strike pledge in years that steel labor agreements were renegotiated. Kleiman also headed a team of USWA lawyers who negotiated a 1974 consent decree with government agencies and major steel companies. This pact produced sweeping changes in seniority practices to redress past discrimination against black, female, and other minority steelworkers and to prevent future discrimination in hiring and promotion.

When the steel industry began to contract in the 1970s, Kleiman helped frame a so-called lifetime employment provision which, upon negotiation in 1977, provided early retirement and other income security benefits for high-seniority steelworkers affected by plant shutdowns. This provision aided many thousands of workers during the steel collapse of the 1980s — and also enormously increased labor costs.

Before Kleiman became general counsel, the steelworkers had no legal department but relied on outside lawyers to handle litigation and advise the officers on internal legal matters. Even the union's chief counsels were members of outside firms. Be-

Bernard Kleiman (photograph by Harry Coughanour, Pittsburgh Post-Gazette)

lieving that legal aid was too dispersed in this system, Kleiman established an in-house staff that became one of the strongest in organized labor. By the 1980s it consisted of ten lawyers based in Pittsburgh, Chicago, Detroit, and Toronto and made the USWA a formidable opponent in legal situations.

Kleiman's career epitomized the large influence of lawyers in steel labor relations from the days of the first formal meetings between union and industry in the 1930s. In the early years the industry's top-level negotiators tended to be lawyers. The union responded in kind. Kleiman was the fifth in the USWA's series of first-rate chief counsels, which also included Lee Pressman (1937–1946), Arthur J. Goldberg (1946–1960), and Kleiman's immediate predecessors, Elliot Bredhoff and David Feller.

Reference:
John P. Hoerr, *And the Wolf Finally Came: The Decline of the American Steel Industry* (Pittsburgh: University of Pittsburgh Press, 1988).

The La Follette Civil Liberties Committee
by Richard W. Nagle
Massasoit Community College

The La Follette committee was a New Deal–era Senate subcommittee that investigated corporate infringement of civil liberties. The committee had a broad directive to search for all violations of free speech and assembly, but it concentrated on business and interference with labor's right to organize and bargain collectively, as guaranteed by the Wagner Act of 1935. The committee's reports showed a pattern of intimidation of workers, some of it frighteningly brutal. The investigation coincided with the organizing drives conducted by the Congress of Industrial Organizations (CIO) in the 1930s and often assisted those campaigns. The La Follette Committee, coming on the heels of congressional investigations such as Ferdinand Pecora's exposure of the speculative activities of leading bankers and Senator Gerald Prentice Nye's arraignment of the "merchants of death" — American businessmen accused of profiting from World War I — put even more shine on the black eye of big business.

In March 1936 Robert M. La Follette, Jr., of Wisconsin, son of the legendary progressive senator, urged the Senate to probe the status of civil liberties. In June the Senate authorized its Education and Labor Committee "to make an investigation of violations of the rights of free speech and assembly and undue interference with the right of labor to organize and bargain collectively." La Follette chaired the investigating subcommittee through its four years of life; the only other member of the committee to play a sustained, active role in the investigations was Elbert D. Thomas of Utah, a courtly yet committed liberal. They recruited a staff of young, bright attorneys to pursue the investigations, often borrowing them from the Labor Department or the National Labor Relations Board (NLRB). The staff was passionately dedicated to unionism. (Over a decade later, at the time of the Alger Hiss case, some of the La Follette staff acknowledged that they were Communist party members or sympathizers in the 1930s.)

Robert La Follette, Jr.

The committee's first and most famous target was labor relations in the big auto and steel corporations. It amassed evidence, revealed in spectacular public hearings during 1937, that showed the underside of corporate America. Described by the committee as resolutely antiunion, the corporations — the evidence suggested — relied on four weapons to frustrate organizing drives among their workers: espionage, strikebreaking, stockpiling of munitions, and armed private police. The revelations about company arsenals and police forces were stunning. Each of the big steel companies regularly bought more tear gas guns and shells than most municipal police departments. Republic Steel, with 50,000 employees, possessed 552 revolvers, 64 rifles, 245 shotguns, and 143 gas guns. Corporate police, who were not responsible to elected officials or bound by the Constitution, regularly tapped telephones, read personal mail, confiscated literature, and attacked organizers.

The La Follette Committee's most dramatic moments came when it held a special investigation into the Memorial Day incident during the Little Steel Strike. Until La Follette, the prevailing story of the incident was of a rampaging mob of strikers dispersed by the beleaguered Chicago police, this despite the ten dead and hundred injured strikers. The committee researchers discovered eyewitnesses and a Paramount newsreel which told a different story: a peaceable parade of men, women, and children deliberately and violently attacked by reckless police.

The committee remained active for nearly three more years and investigated citizen groups that supported corporate attitudes toward the farm factories in California which John Steinbeck portrayed in fiction. However, with the big CIO strikes ending during the recession of 1938, the NLRB assuming the role of permanent investigator, and the committee itself coming under attack from conservatives, the committee's influence began to wane. In 1940, when La Follette proposed legislation to eliminate those corporate labor practices which the committee had denounced as injurious to civil liberties, the proposal failed and the committee went out of business without any legislative accomplishment. From labor's standpoint, its chief success came in 1937 when its revelations about corporate labor practices aided the CIO unions in their first successes.

References:

E. N. Doan, *The La Follettes and the Wisconsin Idea* (New York: Rinehart, 1947);

Stanley Vittoz, *New Deal Labor Policy and the American Industrial Economy* (Chapel Hill: University of North Carolina Press, 1987).

Lackawanna Steel Company

by Thomas E. Leary

Industrial Research Associates

and

Elizabeth C. Sholes

Industrial Research Associates

The Lackawanna Steel Company's brief history as an independent concern from 1902 to 1922 illustrates important trends in the economics, corporate organization, and technology of iron and steel production. Lackawanna Steel was incorporated in New York on February 14, 1902. Its Pennsylvania predecessors had tapped the anthracite coal fields around Scranton for smelting iron since the 1840s and had entered the Bessemer steel rail trade in 1875. By the 1890s, however, the westward shift of customers and raw material resources undermined the earlier locational advantages of northeastern Pennsylvania. Unsuccessful price competition with Carnegie Steel during 1897–1898 convinced Lackawanna's leadership

to erect a new plant on the Great Lakes. Buffalo was selected as the site in 1899 because of its strategic water and rail transportation routes.

The vacant 1500-acre expanse provided company engineers with an unprecedented opportunity to lay out rational product and energy flows while installing equipment of the most recent design. Steelmaking commenced on October 20, 1903, after three years of surveying and construction. The centerpiece of the new works was a rail mill boasting the largest capacity in the United States: 600,000 tons per year or approximately half of the plant's projected steel output. Noteworthy engineering innovations included substan-

Ingot production at the Lackawanna Steel Plant, 1908

tial provisions for recycling waste heat and energy.

The key figures who orchestrated Lackawanna's relocation included its president, Walter Scranton, whose family held extensive connections to the anthracite region. The descendants of banker Moses Taylor (1806–1882) were the leading financial players from New York City; their interests included National City Bank and the Delaware, Lackawanna & Western Railroad. Henry Wehrum, a German immigrant, furnished technical expertise; his designs for steelworks at both Scranton and Buffalo departed significantly from the standard American Bessemer rail plants of Alexander L. Holley.

Not all of the participants in the company's move flourished in their new environment. Lackawanna Steel's board of directors had be-

come increasingly influenced by New York banking circles, and as a consequence veteran steelmakers at the local management level were being phased out of the decision-making process. The board deposed Wehrum in 1903 for clinging to his centralized control over operations at the new plant. Scranton, whose métier was sales, was replaced as president in 1904 by E. A. S. Clarke, an experienced production manager.

Lackawanna diversified its original product line by adding mills for structural shapes, sheet piling, and merchant bars. Its pricing practices generally followed the oligopolistic leadership of U.S. Steel. The company's rates of return in the cyclical steel industry averaged 5 to 8 percent prior to World War I, with losses and low points in 1908, 1909, and 1911. Open-shop labor policies were occasionally ameliorated by corporate paternalism when reformers criticized long hours and hazardous conditions: examples of this trend included a company sociological department similar to the notorious bureau later associated with Henry Ford.

World War I brought increased profits and labor militance to Lackawanna. The company weathered the 1919 steel strike but posted a significant short-term loss in 1921. In May 1922 the board — which along with six other independent steel companies had been engaged in merger talks — responded to the loss by selling Lackawanna's assets to Bethlehem Steel.

References:
William T. Hogan, S.J., *An Economic History of the Iron and Steel Industry in the United States,* 5 volumes (Lexington, Mass.: Lexington Books, 1971);
Thomas E. Leary and Elizabeth C. Sholes, *From Fire To Rust: Business, Technology and Work At The Lackawanna Steel Plant, 1899–1983* (Buffalo: Buffalo and Erie County Historical Society, 1987).

Archives:
Historical files on the Lackawanna Steel Company are located in the Bethlehem Steel Collection, Hagley Museum and Library, Wilmington, Delaware, and in the Bethlehem Steel Historical Collections, Hugh Moore Historical Park and Museums, Inc., Easton, Pennsylvania.

Laclede Steel Company

by Thomas E. Leary
Industrial Research Associates
and
Elizabeth C. Sholes
Industrial Research Associates

The Laclede Steel Company is a semi-integrated producer based in St. Louis since its inception in 1911. As a site for steelmaking, St. Louis was disadvantaged by inferior local mineral supplies, but its position as a rail hub provided access to scrap for metal-melting operations. The large and relatively isolated St. Louis market also tempted new entrants because phantom freight rates under prevailing Pittsburgh-plus pricing policies gave local mills a cost advantage.

Laclede was originally organized by William E. Guy, a railroad promoter; the first president was Thomas R. Akin, whose family connections with the business continued for several decades. They started Laclede in Madison, Illinois, rerolling rails into hard steel bars. In 1915 the company acquired a plant from a bankrupt steelmaker in Alton, Illinois, including open-hearth furnaces and rolling mills.

The subsequent five decades were marked by efforts to expand capacity and diversify product lines for markets in the Mississippi valley and the Southwest. In 1927 Laclede applied the novel principle of four-high rolling to the production of narrow strip steel. Output of butt-welded pipe and conduit commenced in 1930. During the Depression, Laclede added facilities for rolling rods, drawing and coating wire, and fabricating mesh used in construction projects. By the 1950s the company operated auxiliary fabricating plants in five southern cities.

Addressing the American Iron and Steel Institute in 1952, Laclede president William M. Akin commented on the status of semi-integrated producers. After pointing out that over the long haul fixed charges on capital investments for modernization might outweigh potential savings, Laclede's chief executive speculated on the future of the steel industry and wondered " . . . what

the judgment of our successors will be 30 years from now when the Institute meets in 1982." Developments over those three decades verified some of Akin's analyses and premonitions.

By the mid 1960s Laclede faced competition in many product lines from imports and domestic minimills. In 1965 the company embarked on a five-year modernization program, adding electric furnaces to replace the open hearths, continuous bloom casting, and a bar mill complex with a swing forge. The caster plagued company engineers with design and operating problems for two years but eventually became an efficient unit; continuously cast steel at Alton alone increased from 20 to 35 percent by 1985. The expense and disruption of modernization plus a hefty rise in scrap prices conspired to tip Laclede into the red from 1968 through 1970.

Further competitive pressures during the 1970s and 1980s compelled Laclede to contemplate drastic remedies. Dividends were suspended between 1977 and 1985; the Madison mill was closed. After significant losses during 1982, management implemented a computerized financial system that became the basis of a consulting service. Laclede negotiated wage and work rule concessions with the United Steelworkers of America in return for repayment from future profits; the man-hours needed to produce a ton of steel were cut by 20 percent. In need of funds for reducing its long-term debt, Laclede sold a majority of its stock to Ivaco, a Canadian conglomerate, in 1983. These stringent measures allowed Laclede to weather the rampant downsizing of the domestic steel industry. At the cost of its independence, Laclede has managed to retain its niche in the North American market.

References:

William M. Akin, "A Semi-Integrated Steel Producer Views the Steel Industry," *American Iron and Steel Institute Yearbook* (1952): 25–30;

"Efficiency Pays," *Barron's,* 65 (June 24, 1985): 48–49;

Robert A. Garvey, "Bloom Casting at Laclede Steel Co.," *Iron and Steel Engineer,* 47 (December 1970): 61–77;

A. F. Kenyon, "Electric Drive for Ten-Inch Strip Mill at the Laclede Steel Company," *Electric Journal,* 24 (July 1927): 312–318;

Charles Longnecker, "Tube Mill of the Laclede Tube Company," *Blast Furnace and Steel Plant,* 18 (May 1930): 781–783, 824;

"A New Mill for Rolling Hard Steel Bars," *Iron Age,* 89 (March 28, 1912): 822–823;

"A New Steel Rolling Mill for St. Louis," *Iron Age,* 87 (April 6, 1911): 857;

"St. Louis — Its Place in the Steel Industry," *Iron Age,* 98 (October 19, 1916): 877–880;

"Scrappy Laclede," *Barron's,* 66 (September 1, 1986): 49–50;

H. S. Spitz, "Laclede Steel Co.," *Iron and Steel Engineer,* 52 (April 1975): 47–55.

Mack Clayton Lake

(April 29, 1890 – November 9, 1954)

by Terry S. Reynolds

Michigan Technological University

CAREER: Chief, field party for Wisconsin Geological Survey (1915); chief geologist, M. A. Hanna Company (1915–1949); consulting geologist (1915–1954); vice-president and director, Chapin Exploration Company (1929–1949); director, Consolidated Copper Mining Company (1932–1938); president and director, Calmich Mining Company (1940–1942); vice-president and director, Arisota Corporation (1941–1946); vice-president and director, Manganese Ore Company (1942–1945); vice-president and director, Hanna Development Company (1945–1949); consulting geologist, Oliver Iron Mining Company (1945–1949); president and director, Orinoco Mining Company (1949–1954).

An important geologist, mining engineer, and mining executive, Mack Lake's most notable accomplishment was the discovery and opening of the rich Venezuelan ore deposits near the Orinoco River in the late 1940s. Mack Lake was born in Broadhead, Wisconsin, on April 29, 1890, the son of William and Amanda (McNich) Lake. He studied mining engineering at the University of Wisconsin, graduating in 1914. After a year of postgraduate work in geology as an assistant to C. K. Leith, a well-known University of Wisconsin consulting geologist, he headed a field party for the Wisconsin Geological Survey. In 1915 he married Vera Alice Langdon. They had three sons. After his wife's death in 1946 he was to marry a second

time, in December 1947, to Edna Louise Moorehead.

In 1915 Lake took a position with M. A. Hanna Company, an important independent Cleveland ore company. He worked as Hanna's chief geologist and simultaneously as an independent consulting geologist until 1950. His early responsibilities with Hanna were primarily in the Lake Superior iron district, but in later years he worked with Hanna's interests in coal, manganese, copper, and other deposits in other locales. He often served as an officer in Hanna's subsidiary companies. Between 1929 and 1949, for example, he was vice-president and director of the Chapin Exploration Company. He was closely associated with Hanna's successful Labrador iron ore exploration work in the 1940s.

The high rate at which domestic iron ores were consumed in World War II led the larger American iron and steel companies to begin exploration for ore reserves abroad after 1945. In the summer of 1945 the Oliver Mining Company, a subsidiary of the United States Steel Corporation, engaged Lake to supervise its iron ore exploration plans in a region of Venezuela where scattered reports dating back to the 1920s suggested the existence of iron deposits. Lake, in turn, retained Earl K. Nixon, an engineer involved in Hanna company surveys in Venezuela in the 1920s, as his principal assistant and began work. The region where Lake's team began exploration was unmapped and uninhabited except along rivers and tributaries. Lake therefore used aerial surveys to locate possible deposits before sending ground parties through almost impenetrable jungles for surface exploration and test drillings. Covering an area roughly 80 by 200 miles, Lake's survey team located a mountain of iron ore 50 miles south of the Orinoco River and Ciudad Bolivar in early 1947. The deposit — some 4 miles long, 4000 feet wide, and 230 feet thick — contained at least a half billion tons of high-grade open-pit iron.

The deposits uncovered by Lake at Cerro Bolivar made Venezuela one of the United States' leading suppliers of imported iron ore in the 1960s and much of the 1970s. In 1949 Lake was appointed president and director of the Orinoco Mining Company, the U.S. Steel subsidiary formed to exploit these deposits. Shortly after production and shipping of Venezuelan ore began, Lake returned to the United States. He died in San Francisco, following a lengthy illness, on November 9, 1954.

Lake was active both professionally and socially. He was a member of the American Iron and Steel Institute, the American Institute of Mining and Metallurgical Engineers, the Society of Economic Geologists, the Lake Superior Mining Institute, and the Mining and Metallurgical Society of America. He was a Shriner as well.

Selected Publications:
"Cerro Bolivara — U.S. Steel's New Iron Ore Bonanza," *Engineering and Mining Quarterly,* 151 (August 1950): 72–83;
"The Future of Lake Ore Outlined," *Iron Trade Review,* 75 (December 25, 1924): 1701–1705.
"The Future of the Lake Superior District," *Proceedings of the Lake Superior Mining Institute,* 24 (1925): 48–67.

Reference:
"Mack C. Lake: Obituary," *Mining Engineering,* 7 (1955): 184;
William T. Hogan, S.J., *An Economic History of the Iron and Steel Industry in the United States,* 5 volumes (Lexington, Mass.: Lexington Books, 1971).

Thomas William Lamont

(September 30, 1870 – February 2, 1948)

by John N. Ingham

University of Toronto

CAREER: Reporter, *New York Tribune* (1893–1894); secretary, Cushman Brothers (1894–1898); director, Lamont, Corliss & Company (1898–1925); secretary, treasurer (1903–1905), vice-president, Bankers Trust Company (1905–1909); vice-president, First National Bank (1909–1911); partner, J. P. Morgan & Company (1911–1948); owner, *New York Post* (1918–1922); chairman of executive committee (1940–1943), chairman of the board, J. P. Morgan & Company, Incorporated (1943–1948); director and member of the finance committee, United States Steel Corporation (1911–1948).

Thomas Lamont was a major figure in the worlds of international finance and banking during the first half of the twentieth century. Although his involvement in the steel industry was indirect, Lamont managed to play a significant role in shaping the nature of the post–World War II industry.

Born in Claverack, New York, on September 30, 1870, Lamont was the son of a Methodist minister, Thomas Lamont, and grew up in a succession of parsonages. He attended Phillips Exeter Academy, graduating in 1888, and Harvard College on scholarships. He served as editor of the *Harvard Crimson,* which helped him get jobs with two Boston newspapers while still in college. Lamont graduated Harvard in 1892, and in 1893 he became a reporter on the *New York Tribune,* rapidly becoming a night editor. Lamont left the paper in 1894 and, with $5,000 he had borrowed, invested in Cushman Brothers, a New York import-export firm hobbled by the depression of the 1890s. As the firm's secretary, Lamont was greatly responsible for saving it through a clever advertising campaign. In 1898 he and his brother-in-law, Charles Corliss, reorganized the firm as

Lamont, Corliss & Company. Lamont's success brought him to the attention of Henry P. Davison of Bankers Trust Company and at Davison's invitation Lamont joined Bankers Trust in 1903 and advanced to vice-president in 1905. In 1909 J. P. Morgan asked Lamont to join J. P. Morgan & Company, replacing George W. Perkins, who had long functioned as Morgan's right-hand man.

As a Morgan partner, Lamont became an expert in international banking, helping to finance both industries and foreign governments and, in so doing, helped expand the role of the Morgan firm and the United States as major global creditors during the 1920s. After World War I, Lamont participated in developing the plans for German reparations. Having become one of the world's most influential bankers by the onset of the Depression, Lamont helped establish the Bank for International Settlements in 1931 and was a delegate to the World Economic Conference held in London in 1933. Early in his career at Morgan, he had also gained a measure of public notoriety as an outspoken defender of the firm against government accusations that it was part of the "money trust," the alleged collusion among the big banks.

When Lamont replaced George Perkins at the House of Morgan, he also assumed Perkins's responsibilities at United States Steel Corporation. When Perkins departed, however, the entire management structure of the steel firm was radically altered, as virtually all power was consolidated under Elbert H. Gary. Gary became not only chairman of the board of directors but also the clearly designated chief executive officer of the company. This began a period of one-man rule at the steel company that did not end until Gary's death in 1927. Upon Gary's death, the Morgan influence on the board was increased, with Myron C. Taylor becoming chairman of the finance committee and J. P. Morgan, Jr., assuming the mantle of chairman of the board. In 1932 Taylor replaced Morgan as chairman of the board and was also made chief executive officer, although most analysts thought that U.S. Steel was most likely run by committee during the 1930s. Lamont's role in all of this was quite muted. He served on the finance committee and the board of directors.

In 1938 Myron Taylor resigned from U.S. Steel, and the firm was radically reorganized. Since Taylor was replaced by three individuals, the role of the Morgan Bank on the board and the finance committee was enhanced, and Lamont emerged as one of the most influential Morgan participants during the 1940s, a critical period for

U.S. Steel due to wartime expansion and the changing face of labor relations. Whereas Taylor and his immediate successor, Edward R. Stettinius, Jr., demonstrated a fairly liberal attitude toward labor unions, government, and the role of steel in the national economy, Lamont gave strong support to the more conservative ideas of Benjamin F. Fairless, chief operating officer and president of the company.

The strike by the United Steel Workers — partly due to the pent-up demand for wage increases during the war years — was the critical issue for steel in 1946. Lamont and other U.S. Steel officials believed that, given its great deal of surplus capacity after the war, the company desperately needed labor stability if it was to reap the expected profits promised by the huge demand for consumer goods. This promise made steel company management somewhat more amenable to the union's demands. Nonetheless, the federal government became an important factor in the settlement. President Harry S Truman had set up a fact-finding board to deal with the strike in the auto industry. It recommended a generous 19.5 percent wage increase for the autoworkers, and a similar board was established for the steel strike. Steel management went along with government recommendations, and in return corporate officials got a major concession from union leaders. In the preamble to the U.S. Steel contract, company officials announced that they were not anti-union, but even more significant, union officials proclaimed they were "sincerely concerned with the best interests and well-being of the business." The unions had, to a certain extent, been coopted by management; they were no longer ideological outsiders. The problem came when U.S. Steel and other firms decided to raise their prices to meet the higher wages.

In July 1947 U.S. Steel announced that it was joining with the independent steel producers in raising the price of finished steel by 7 percent. The Truman administration, which was attempting to hold the line on prices during this inflationary period, attacked U.S. Steel for the increase. Lamont struck back at the government, proclaiming that "If there is one thing that [the government] ought to appreciate, it is that the U.S. Steel Corporation is the largest basic industry in the world . . . [and] that it should be kept sound, stable and liquid." His arguments, however, carried little weight outside U.S. Steel, and the government continued applying pressure.

Lamont died on February 2, 1948, in his winter home in Boca Grande, Florida, survived by his wife of 53 years, Florence Haskell (Corliss) Lamont. The conservative, antigovernment policies he advocated, however, continued to be espoused by Fairless and others for years to come.

Selected Publications:
Henry P. Davison: The Record of a Useful Life (New York & London: Harper, 1933);
My Boyhood in a Parsonage (New York & London: Harper, 1946);

Across World Frontiers (New York: Harcourt, Brace, 1951).

References:
John Brooks, *Once in Golconda: A True Drama of Wall Street, 1920–1938* (New York: Harper & Row, 1969);
Vincent P. Carosso, *The Morgans: Private International Bankers, 1854–1913* (Cambridge: Harvard University Press, 1987);
Paul Tiffany, *The Decline of American Steel* (New York: Oxford University Press, 1988).

R. Heath Larry

(February 24, 1914 –)

by John P. Hoerr

Teaneck, New Jersey

CAREER: Attorney, private practice (1937–1938); attorney (1938–1941), general counsel (1941–1944), secretary, director, National Tube Company (1944–1948); general attorney (1948–1952), assistant general solicitor (1952–1958), administrative vice-president of labor relations (1958–1966), executive vice-president, assistant to chairman (1966–1969), vice-chairman of the board, United States Steel Corporation (1969–1977); chairman and president, National Association of Manufacturers (1977–1980); counsel, Reed Smith Shaw & McClay (1980–1982).

R. Heath Larry, a labor-relations executive, played a major role in formulating steel-industry labor policies from the 1950s to the mid 1970s. Born on February 24, 1914, in Huntingdon, Pennsylvania, he was graduated from Grove City College and earned a law degree at the University of Pittsburgh in 1937. After a brief stint in private practice in Pittsburgh, Larry joined the legal department of National Tube Company, a subsidiary of U.S. Steel Corporation, in 1938 and rose to general counsel in 1941.

Larry was assigned labor-negotiating duties early in his career. In the 1940s he represented the company in a case before the National Labor Relations Board (NLRB) that helped establish the United Steelworkers of America (USWA) as the exclusive bargaining agent for all production and maintenance employees in the steel industry. Na-

R. Heath Larry

tional Tube joined with the USWA in successfully opposing efforts by the Bricklayers Union to represent workers who maintained brickwork on coke ovens and blast furnaces. The NLRB decision effectively ruled out further attempts by craft unions to represent steelworkers.

In 1948 Larry joined the U.S. Steel staff as a lawyer under General Counsel Roger Blough and in 1952 became assistant general solicitor, specializing in labor matters. In this capacity, he served as legal aide to George Stephens, the corporation's vice-president of industrial relations and chief labor bargainer. Starting in 1949 Larry assisted Stephens in negotiating steelworker contracts which, by virtue of U.S. Steel's dominant position in the basic steel industry, became the pattern for all other steelmakers. In 1956 U.S. Steel and other major companies formed, at first informally, a multicompany committee with responsibility for negotiating industrywide labor agreements with the USWA. This group was formally organized in 1959 as the Coordinating Committee Steel Companies. The actual bargaining was carried out by a four-man team consisting of two executives from U.S. Steel and one each from Bethlehem and Republic. Larry served on this team for seven bargaining rounds, from 1956 through 1974.

R. Conrad Cooper of U.S. Steel succeeded Stephens as chief negotiator in 1959. Larry, now an assistant vice-president, served under Cooper. Following the 116-day steel strike of 1959, the USWA and the industry formed the Human Relations Committee in an effort to avoid further strikes. Larry was cochairman of a subcommittee which attempted to develop guidelines for a peaceful, long-term bargaining relationship. This effort collapsed, however, because the two sides could not agree on a wage-productivity formula that encompassed both basic wage hikes and a periodic cost-of-living allowance (COLA) to protect the buying power of wages. The union had given up a previous COLA in settling the 1959 strike. Acting for the industry, Larry rejected the idea of incorporating a COLA in future wage agreements. The Human Relations Committee was disbanded in 1965 after I. W. Abel defeated David J. McDonald for the USWA presidency.

Despite continuing disagreements over the wage-productivity relationship, the two sides managed to negotiate peaceful settlements in the 1960s and 1970s. Larry assumed the chief negotiating role in 1971 after Cooper's retirement. His major regret, he later said, was granting a new COLA provision in 1971 bargaining under pressure from the Nixon administration to avoid a nationwide steel strike. In 1973 Larry and Abel negotiated the innovative Experimental Negotiating Agreement (ENA). Unprecedented in American labor relations, the ENA prohibited strikes and lockouts and instead provided for binding arbitration to settle unresolved issues in wage negotiations. To win the no-strike pledge, the companies agreed to pay 3 percent annual wage increases, plus COLA, and gave up the right to contest these terms in arbitration.

The ENA was an initial success. In 1974, negotiating for the first time under ENA rules, Larry and Abel consummated a new wage agreement without the need for arbitration. As intended, the ENA halted the costly practice of hedge-buying by steel users to build up stockpiles in the event of a strike. In two subsequent rounds of bargaining (1977 and 1980), however, the ENA committed the companies to high wage costs during a period of low productivity growth. Larry and other company officials, moreover, did not anticipate that oil prices would soar in the 1970s. Because of steeply rising inflation, steel's COLA alone raised hourly wages by $5.11, or 120 percent, between 1972 and 1982, pushing the industry's labor costs far above those of foreign competitors and contributing to the industry's spectacular decline in the 1980s.

Larry was named to the promotional job of chairman of the National Association of Manufacturers in 1976. Retiring from U.S. Steel in early 1977, he served as full-time president of the NAM until 1980. He retired permanently in 1982 after a brief period in private law practice.

Reference:
John P. Hoerr, *And the Wolf Finally Came: The Decline of the American Steel Industry* (Pittsburgh: University of Pittsburgh Press, 1988).

George McCully Laughlin, Jr.

(February 25, 1873 – March 9, 1946)   by John A. Heitmann
University of Dayton

CAREER: Various managerial positions, including plant superintendent (1893–1923), vice-president (1923–1928), chairman of the board, Jones & Laughlin Steel Corporation (1928–1936).

A third generation iron and steel manufacturer, George McCully Laughlin, Jr., was the grandson of Jones & Laughlin Steel Corporation founder James Laughlin and son of George McCully and Isabel Bowman (McKennan) Laughlin. Born in Pittsburgh, Pennsylvania, on February 25, 1873, Laughlin was educated at Saint Paul's School, in Concord, New Hampshire, and attended Yale University in 1892–1893, after which he began his career in the family business, Jones and Laughlin, Ltd. In 1902 this firm and Laughlin and Company, Ltd., were merged with an initial capitalization of $30 million to form the Jones & Laughlin Steel Company (J&L), America's foremost independent steel manufacturer during much of the twentieth century. Within this organization Laughlin played a significant leadership role until his retirement in the mid 1930s.

Although prior to World War II the firm remained a relatively conservative business — choosing to expand internally rather than through merger and acquisition — J&L experienced several phases of rapid growth during the first three decades of the twentieth century. In 1903 the company began a $10 million capital expenditure program in its Pittsburgh facilities that included the erection of several Talbot open-hearth furnaces, a continuous rod mill, and a combination rod and billet plant. It soon became clear, however, that despite the expansion J&L's product line remained limited and the existing facilities were not large enough to accommodate product diversification. Furthermore, residents of neighborhoods surrounding J&L plants in Pittsburgh had complained of air pollution, increasing the belief of management that a new plan was necessary. In 1905 J&L purchased 475 acres of land lo-

George McCully Laughlin, Jr.

cated in Woodlawn, Pennsylvania, 26 miles down the Ohio River from Pittsburgh; the site would become the Aliquippa Works, the company's operational hub until the late 1970s.

The first blast furnace was operational at Aliquippa in 1909, and by World War I the burgeoning plant possessed a blooming mill; several Talbot furnaces and open-hearth converters; a tin house for pickling, annealing, and rolling tinplate; and a sintering plant. With the addition of the Aliquippa Works, J&L began to challenge the United States Steel Corporation — the giant of the steel industry — in a host of expanding markets, including rod wire and tubular goods.

World War I dramatically increased demand for a variety of steel products. Production of semifinished billets, sheet bars, wire rods, hot-rolled bars, plates, and spikes rose more than 20 percent between 1914 and 1917, and the sales of cold-finished goods more than quadrupled. Having spent much time in the various operating units of the company, Laughlin gained in experience during these years of growth and expansion and was appointed superintendent of the Soho Works in Pittsburgh.

After the war J&L's output returned to 1915 levels. The firm, nevertheless, emerged from the war as a major steel company possessing an annual capacity of more than 6 million tons in the manufacture of pig iron, ingots, and hot-rolled steel products. It owned properties possessing a rich array of raw materials in West Virginia, Michigan, and Minnesota. After the recession of 1921–1922 the firm reorganized as a publicly held company, although the Jones and Laughlin families still held large amounts of stock and controlled upper management. The new corporation named B. F. Jones, Jr., chairman and his cousin William Larimer Jones president. Laughlin, now with the organization for thirty years and well acquainted with the intricacies of the steel business, was elected vice-president and a member of the executive committee in 1923, a position he held until his appointment as chairman of the board in 1928.

By the time Laughlin retired in 1936, J&L had become less a family-run enterprise, with increasingly more managers becoming involved in the company's decision-making process. The company also had begun to experience problems with labor relations; during the mid 1930s the firm was wracked with controversial disputes that ultimately resulted in the separation of management and labor into hostile, enemy camps.

Laughlin also served as director of the Pittsburgh Trust Company (later Pittsburgh National Bank). He was an enthusiastic sportsman, a member of the Allegany Country Club and the Racquet and Tennis Club of New York City. He was also president of the Pittsburgh Golf Club and a founder of the Fox Chapel Golf Club in Pittsburgh. He had married Henrietta Speer on January 10, 1895; the couple had four children. Laughlin died in Lake Wales, Florida, on March 9, 1946.

References:

Tom M. Girdler, *Boot Straps* (New York: Scribners, 1943);

William T. Hogan, S.J., *An Economic History of the Iron and Steel Industry in the United States*, 5 volumes (Lexington, Mass.: Lexington Books, 1971).

Albert Thomas Lawson

(February 22, 1898 – March 29, 1963)

by John A. Heitmann

University of Dayton

CAREER: Various positions (1912–1918), industrial engineer (1918–1926) assistant chief engineer, Aliquippa Works (1927–1929), chief industrial engineer, Aliquippa Works (1930–1946), chief industrial engineer (1946–1947), general superintendent, Pittsburgh Works (1947–1950), vice president, Jones & Laughlin Steel Corporation (1950–1963).

Albert Thomas Lawson, a key executive with the Jones & Laughlin Steel Company (J&L), played an important role in directing a program of technological modernization within the firm during the immediate post–World War II period. Lawson was born in Pittsburgh, Pennsylvania, on February 22, 1898, to Harry and Anna Harris Lawson. He attended the Alexander Hamilton Institute, where he received his training in business management, and in 1912 Lawson began a career that spanned more than 50 years with J&L when he joined the firm's Keystone Works, located in Pittsburgh. He rose through the ranks of the company's industrial engineering department and railroad subsidiaries, and in 1946 Lawson was appointed J&L's chief industrial engineer. A year later Lawson became general superintendent of the Pittsburgh Works, and in 1950 he was named J&L's vice-president of general services. Lawson was promoted to vice-president of production in 1953 and held this position until his retirement in 1963.

During the 1950s Lawson took the lead in transforming J&L's obsolete production facilities into one of the most efficiently run operations in the American steel industry. Lawson oversaw the continuous improvement of the materials used in J&L's blast-furnace operations and the modification of rule-of-thumb practices that had been handed down from one generation of furnace operators to another. Lawson's program, for instance, called for increased top pressures and blast temperatures, not only to speed up the process, but also to ensure the making of a higher quality steel. In addition to improving furnace practices, Lawson supervised the erection of new coke ovens, sinter plants, and beneficiating and pelletizing operations. As a result, blast-furnace efficiency increased 50 percent, while limestone and coke requirements were reduced significantly. Undoubtedly Lawson's major success directing J&L's efforts to become a technological leader in the steel industry was the introduction of the widespread use of the basic oxygen furnace (BOF) in 1957. This innovation proved so successful that it was subsequently imitated by other steel manufacturers.

While the idea of using oxygen in the steel-making process can be traced back to 1856 and the work of Henry Bessemer, the relatively high cost of oxygen prevented its use in commercial operations until the mid twentieth century. With the promising work during the 1920s of two German engineers, Linde and Frankl, and subsequent improvements in the process, the cost of oxygen eventually dropped to $12–15 per ton by the mid 1950s, and these price levels were a necessary precondition to the commercial viability of the basic oxygen process to make steel. Technological innovations related to the use of oxygen in steelmaking had been worked out and patented by the German engineers during the 1930s; therefore, by the 1950s the conditions were ripe for further development and application of the process on a large scale. Although the BOF was first used in Austria in 1949 and in the United States and Canada in 1954, J&L was the first major American steelmaker to gamble on the process. Although its development costs were high, the basic oxygen process had the potential of markedly reducing heating time, and it also offered the manufacturer a

Albert Thomas Lawson

wider range of raw materials that could be used in the furnace.

J&L's BOF was similar in shape to a conventional Bessemer converter, although tuyeres, wind boxes, and blast pipes were not a part of the design. Possessing a 20-inch-thick magnesite brick refractory lining, the furnace was initially tilted on its side to receive its charge of scrap, molten pig iron, roll scale, and lime. After being placed in an upright position, a water-cooled oxygen lance was lowered into position, and, with oxygen entering the vessel at 140 to 180 pounds per square inch, a vigorous reaction took place in which silicon, phosphorous, manganese, and carbon impurities were instantly oxidized. A drop in the flame at the mouth of the furnace indicated the completion of the oxidation reaction, and the operator responded by removing the lance and tilting the furnace to the horizontal position to skim the fluid slag. With the slag removed, the final product was poured into ladles.

The resounding success of the basic oxygen process culminated Lawson's long career at J&L. Despite his lack of formal technical education,

Lawson clearly recognized the value of new technology in steelmaking and used his authority to revolutionize production methods at J&L. Yet, the generation of technical experts that followed Lawson received their education in universities rather than on the shop floor and worked within an industry that had an increasingly mobile executive workforce. Corporate stability and gradual technological change, both characteristic of the

American steel industry of Lawson's era, gave way shortly after his retirement to complacency. Lawson died on March 29, 1963, in Tuscon, Arizona.

References:
Douglas Alan Fisher, *The Epic of Steel* (New York: Harper & Row, 1963);
William T. Hogan, SJ., *An Economic History of the Iron and Steel Industry in the United States,* 5 volumes (Lexington, Mass.: Lexington Books, 1971).

Ross Lillie Leffler

(August 7, 1886 – December 14, 1964)

by Richard W. Kalwa

University of Wisconsin — Parkside

CAREER: Salesman, Thayer's Jewelry Company (1905–1910); timekeeper, various positions (1910–1927), assistant superintendent, Duquesne Works (1927–1931), superintendent, Duquesne Works, the Carnegie Company (1931–1935); director of personnel and manager, industrial relations, Pittsburgh District (1935–1938), director, industrial relations (1938–1940), manager, safety division (1939–1944), assistant to manager of operations (1944–1947), assistant to president (1947–1951), assistant to executive vice-president, operations, Carnegie-Illinois Steel Corporation (1951–1956).

Ross Lillie Leffler supplemented a long career at the Carnegie-Illinois Steel Corporation with a lifetime of extensive involvement in the area of wildlife and natural conservation. His positions in governmental bodies as well as private and community organizations were testimony to his energy and public spirit.

Leffler was born August 7, 1886, to Emma Eliza Ouida Cole Leffler and John Robert Leffler, a physician, in Butte, Montana. He attended public schools in Bucksport, Maine, and Boston and studied at the University of Michigan from 1903 to 1905. He took a job as a salesman for Thayer's Jewelry Company in McKeesport, Pennsylvania, in 1905. He left Thayer's in 1910 to work as a timekeeper for the Carnegie Company, a subsidiary of United States Steel Corporation, in Duquesne, Pennsylvania. In that year he also married Erma M. Wernke of McKeesport.

Leffler worked at several positions in the

Ross Lillie Leffler

Duquesne Works, becoming assistant superintendent of the rolling mill in 1927 and receiving a promotion to superintendent four years later. In 1935 the company became the Carnegie-Illinois

Steel Corporation, and Leffler became director of personnel and manager of the newly renamed corporation. He was made director of industrial relations for Carnegie-Illinois in 1938 and held that position until 1940. From 1939 to 1944 he headed the safety division. He served as assistant to the manager of operations from 1944 to 1947 and as assistant to the president from 1947 to 1951. In 1951 he became the assistant to the executive vice-president of operations and served in that capacity until his retirement in 1956.

During and after his career in the steel industry Leffler actively participated in many organizations concerned with the conservation of wildlife and natural resources. He was a member of the Pennsylvania State Game Commission from 1927 to 1956 and was its president between 1928 and 1944. He served on the conservation advisory committee of the U.S. Department of the Interior from 1955 to 1961. As the department's first assistant secretary for fish and wildlife from 1957 to 1961, he established conservation areas in Alaska and Florida. He was commissioner of the U.S. Section of the International North Pacific Fisheries Commission from 1957 to 1961. He was an organizer of the Izaak Walton League and was its first president in 1922. He served as director of the National Wildlife Federation and was its president in 1963. He was an organizer of the Twentieth Century Minute Men, a lobbying group formed in 1962 to press for legislation to control strip-mining and to promote clean water. He was a member of the National Audubon Society, the Wilderness Society, the Camp Fire Club, and several other conservation organizations. For his activities in this area he received a total of 62 citations, awards, and medals. He also participated in many community groups, such as the Red Cross, Community Chest, Urban League, and Safety Council in Pittsburgh and Allegheny County. He remained active in the Boy Scouts of America for almost fifty years and served on its national executive board from 1944 until the end of his life. He died in Washington, D.C., on December 14, 1964.

The Little Steel Strike

by Richard W. Nagle

Massasoit Community College

The Little Steel Strike of 1937 was a failed attempt by the Steel Workers Organizing Committee (SWOC) to unionize the steel industry in one stroke. Prior to the strike, the SWOC's sponsor, the Congress of Industrial Organizations (CIO), with full government approval, had defeated all of its opponents in other industries and in the traditionalist American Federation of Labor, as well. But labor's string of victories ran out at the gates of Little Steel.

The major competitors of the United States Steel Corporation ("Big Steel"), the Little Steel companies — Bethlehem, Republic, Jones & Laughlin (J&L), Youngstown Sheet & Tube, National, and Inland — were in actuality large, economically integrated manufacturers with tens of thousands of employees working at mills located in many states. Annual dinners for industry executives — events which led to the founding of the American Iron and Steel Institute (AISI) — hosted by Judge Elbert Gary, U.S. Steel's first chairman, fostered cooperation between Big and Little Steel, sometimes in ways that mocked the antitrust laws. A pattern developed where U.S. Steel led the way in all matters — prices, working conditions, and labor policy — and the other companies followed. Such was the environment in the steel industry in 1936 when the CIO created the SWOC for a labor campaign against steel management.

In March 1937 U.S. Steel stunned the rest of the industry when it announced that it had come to terms with the SWOC. It at first appeared that Little Steel would naturally follow U.S. Steel's lead. SWOC membership had swelled to 300,000, and it had begun to sign contracts with dozens of firms. It even started to collect dues. A two-day strike in April forced J&L to negotiate; but Bethlehem, Republic, Youngstown, and Inland remained adamantly opposed to recognizing the SWOC, although they unilaterally put into effect all of the economic parts of the SWOC contracts. Tom Girdler, Republic's chairman — and once the head of J&L — proclaimed that he would shut Republic and "raise apples and potatoes" rather than recognize a union. In purchasing weapons for its company police, Republic spent $50,000 in May alone. It also distributed thousands of copies of a pamphlet titled "Join the CIO and Help Build a Soviet America." On May 20 Girdler closed Republic's mill in Massillon, Ohio, and locked out the workers.

The local union leaders beseeched SWOC head Philip Murray to call a strike against all of the recalcitrant firms. Murray knew this was risky because SWOC was not as strong as its gross numbers suggested. Its strength was in Pittsburgh and to a lesser extent in Chicago, but the Little Steel Mills were scattered throughout the Midwest. The need to support the Republic workers, however, overrode any doubts, and SWOC declared strikes against Republic, Youngstown, and Inland on May 26 and against Bethlehem two weeks later. The walkouts had a mixed effect: some mills experienced shutdowns while others were able to remain in full operation.

At Republic's mill in Chicago, operations continued with about half the crew working. The SWOC organized picketing at the factory gates, but the Chicago police dispersed the pickets on May 26 and 28. The SWOC called a mass meeting, and in the midafternoon on Sunday, May 30, about 2,000 strikers, their families, and supporters assembled a few blocks from the mill. They began to march toward the factory gates in order to picket, but were intercepted by a line of Chicago police. After a few minutes of talk between the strikers and the police, there was one shot fired; who fired it was never determined. The police then shot several volleys into the crowd, scattering the marchers, and the police gave chase, beating and kicking them. Ten marchers were shot dead; 30 others had bullet wounds and dozens more were hurt. Seven of the dead and 27 of the wounded were hit from behind. Thirty-five policemen were injured in the clash, although no policeman was shot and only three were hospitalized.

In initial coverage of what came to be known as the Memorial Day Incident, the press reported that the police repulsed an attack on the plant by armed strikers (the *New York Times* headline was "Steel Mob Halted"). A few weeks later, however, the La Follette committee — a Senate subcommittee examining infringements on civil liberties — conducted a special investigation into the incident and found many eyewitnesses who told a remarkably different story, claiming that the police had attacked a peaceful group of marchers. The committee located a previously suppressed Paramount newsreel of the entire encounter that substantiated these claims. The La Follette committee unequivocally blamed the Chicago police for provoking the confrontation, using unreasonable force, and treating the wounded abominably. The Memorial Day Incident was the most brutal event of the strike, but the La Follette committee also documented other examples of corporate-sponsored repression. In many mill towns, the companies secretly organized local officials, clergy, and businessmen into citizen committees which condemned the strike.

Violent clashes between plant guards and strikers continued in many areas. The companies, sometimes aided by martial law, kept open or reopened their mills, however. State governments and the federal government tried to mediate the strike, but all of these efforts came to nothing. On June 29 an exasperated President Franklin D. Roosevelt expressed public opinion of the strikers and the steel companies when he told reporters that "a majority of people are saying just one thing, a plague on both your houses." By early July, Little Steel crushed the strike, and the SWOC gave up.

The Little Steel strike deflated the heady spirit of SWOC and its sponsor, the CIO. It showed that America's corporations were not going to capitulate to industrial unionism. The Little Steel strike also marked a change in the country's attitude toward militant unionism; labor would not again see the absolute victories won by the CIO in its first year. Coinciding with Roosevelt's unsuccessful effort to "pack" the Supreme Court, organized labor's failure signaled the beginning of the end of the New Deal.

References:
Tom M. Girdler with Boyden Sparkes, *Boot Straps: The Autobiography of Tom M. Girdler* (New York: Scribners, 1944);
William T. Hogan, S.J., *An Economic History of the Iron and Steel Industry in the United States*, 5 volumes (Lexington, Mass.: Lexington Books, 1971).

Lone Star Steel Company

by H. Lee Scamehorn

Historic Learning and Research Systems

Faced with a shortage of pig iron in the early months of World War II, the Defense Plant Corporation (DPC) authorized the construction of a new production facility in Texas. At the urging of a group of local entrepreneurs headed by John W. Carpenter, president of the Texas Power and Light Company, the federal agency agreed to erect an integrated plant in northeast Texas, south of Daingerfield, where an abundance of iron ore could be obtained inexpensively by the open-pit method of mining.

The DPC advanced $26 million to build an ore-beneficiation plant, a 1,200-ton blast furnace, and Koppers-Becker by-product coking ovens. An additional $5 million was devoted to the development of coal mines at McAlester and in McCurtain County, Oklahoma, which also served the Sheffield Steel Company's plant in Houston. It was understood that an additional $35 million would be available at a later time for rolling mills and finishing facilities to turn out plates, pipe, and forging blooms.

In 1942 the Lone Star Steel Company was incorporated to operate the plant. Only the coke ovens were completed in time to contribute to the war effort; the blast furnace was not blown in until 1947. The property was acquired by the War Assets Administration and sold in 1948 to private interests headed by Eugene B. Germany of Dallas. The company began marketing cast iron pipe in 1951.

In 1953 Lone Star installed open-hearth furnaces and facilities for making steel tubular products, a product line that depended heavily, at times almost exclusively, on demand in the petroleum industry. By 1980 facilities included one blast furnace with an annual capacity of 800,000 tons, five open-hearth and two electric-arc furnaces, a continuous caster, and welded and seamless tube mills. When American oil production collapsed in 1982, Lone Star suspended most of its operations and eventually restructured in order to cut costs by streamlining operations.

The Texas steelmaker has a long history of corporate affiliations. The Lone Star Company was acquired by the Philadelphia Reading Corporation in 1966. Lone Star Steel was reincorporated that year. Philadelphia Reading was acquired by Northwest Industries, Incorporated, in 1968. Northwest was absorbed by Farley Metals, Incorporated, in 1985, at which time Lone Star was spun off by the holding company to its stockholders. A new holding company, Lone Star Technologies, Incorporated, redirected the efforts of the well-known manufacturer of oil field tubing to nonenergy products in 1986. These included environmental control devices and bearing-quality precision tubing for the automotive, construction-equipment, and machine-tool industries. In 1987 Lone Star moved away from steel, and in conjunction with other investors committed $48 million to the American Federal Bank, which assimilated 12 Texas savings and loan associations.

References:

William T. Hogan, S.J., *An Economic History of the Iron and Steel Industry in the United States,* 5 volumes (Lexington, Mass.: Lexington Books, 1971);

Hogan, *Steel in the United States: Restructuring to Compete* (Lexington, Mass.: Lexington Books, 1984);

Charles T. Post, "Texas Steel," *Iron Age,* 155 (June 28, 1945): 58–66.

Ludlow Massacre

by H. Lee Scamehorn

Historic Learning and Research Systems

The culminating event of the coal miners' strike in the southern fields of Colorado in 1913–1914 was called, by the union, a "massacre." The clash between strikers and state militia at Ludlow, 18 miles north of Trinidad, Colorado, claimed 19 lives — many of them women and children — and captured the nation's attention. Although it was unclear which side was responsible for the deaths, the mine operators and the militia were widely condemned, and Ludlow became a symbol of labor's struggle for recognition against corporate capitalism.

The strike, called by the United Mine Workers of America (UMWA) District 15, began on September 23, 1913. The UMWA hoped that the strike would force the mine operators to make a variety of concessions: recognition of the union, a 10-percent advance in wages, compensation for "dead" work (nonmining tasks for which workers were not paid), checkweighmen at all mines, the right of workers to trade at outlets other than company stores, enforcement of state mining laws, and the abolition of camp guards. Both sides understood that the central issue was acceptance of the UMWA as the exclusive bargaining agent for miners throughout southern Colorado.

Approximately 7,600 men, three-fourths of the work force in the mines of Fremont, Huerfano, and Las Animas counties, responded to the strike call. Those who walked out of the mines moved out of the company towns with their families and took refuge in tent colonies established by the UMWA at Ludlow, Primero, Segundo, and Sopris. The largest was Ludlow, with more than 100 tents. It was also strategically located at the head of the canyon leading to the principal mines of the region, giving the occupants an advantage in discouraging strikebreakers from taking jobs with the companies.

Each side, insisting that the other planned to use force, armed to defend its interests. Reports of several skirmishes, mostly in the vicinity of the tent colonies, convinced Colorado governor Elias

M. Ammons that local authorities needed help in maintaining order. National Guard units were dispatched to the strike zone on October 29, 1913. Order generally prevailed throughout the region until three-fourths of the troops were withdrawn on April 16, 1914, when the state no longer had the funds to keep them in the field.

The area north of Trinidad erupted on April 20. The so-called "Battle of Ludlow" commenced early in the morning and raged throughout most of the day. One militiaman and five miners were killed, and others were wounded. The tent colony was burned to the ground, after which it was discovered that two women and 11 children, who had taken shelter in shallow pits under the floors, had suffocated during the fire.

It is not known who started the fighting or who set fire to the tents. Each side blamed the other. The UMWA immediately called the incident a "massacre" and heaped blame on the operators and the militia. The deaths of the women and children focused national attention on the strike and on labor-management relations in southern Colorado.

News of the tragedy at Ludlow inflamed striking miners, who attacked the militia and company properties, imposing a reign of terror on the region. The governor responded to the growing violence by sending additional troops to the strike zone. When the fighting continued to intensify, Ammons asked President Woodrow Wilson to authorize federal troops to pacify the southern coalfield. When it became apparent that the two sides could not reach a settlement, Wilson authorized the movement of federal troops to Trinidad and Walsenburg on April 28. Within three days U.S. Army units began the task of disarming both sides and restoring peace to the strike zone.

The strike continued for another six months. While the mine operators steadfastly refused to negotiate, the union exhausted available funds. In December 1914 the UMWA quietly declared an end to the strike, but by that time many of the strik-

The striking coal miners' tent colony at Ludlow, Colorado, before and after it was burned to the ground during a violent confrontation between strikers and National Guard troops

ers had already returned to work without gaining any of the concessions they had demanded almost a year and a half earlier.

References:
Barron B. Beshoar, *Out of the Depths: The Story of John R. Lawson, a Labor Leader* (Denver: The Colorado Labor Historical Committee of the Denver Trades and Labor Assembly, 1942);

George S. McGovern and Leonard F. Guttridge, *The Great Coalfield War* (Boston: Houghton Mifflin, 1972);

Zeese Papanikolas, *Buried Unsung: Louis Tikas and the Ludlow Massacre* (Salt Lake City: University of Utah Press, 1982).

Lukens Steel Company

by Bruce E. Seely

Michigan Technological University

Lukens Steel Company is the longest continuously operating firm in the American iron and steel industry and a leader in the specialty steel business. The company's long success indicates that with sound management, a willingness to innovate, and attention to quality American steel companies could compete in the international steel market of the 1980s.

The venture that became Lukens Steel was started as the Brandywine Iron Works and Nail Factory by Issac Pennock in 1810. In 1817 Dr. Charles Lukens, who had married Pennock's daughter Rebecca, leased the mill, which was producing boiler plate. Until 1890 the works were operated under several partnerships, with Lukens's son-in-law, Dr. Charles Huston, playing the leading role after 1849. Throughout, the firm maintained a reputation as a technologically innovative producer of high-quality plate. In 1881 brothers Abram Francis and Charles Lukens Huston joined their father in the last partnership, Charles Huston & Sons. Continuing the innovative tradition, they first made steel in 1881 and introduced the production of metal bowl-like shapes called heads by spinning hot steel in 1885. In 1890 they incorporated as Lukens Iron & Steel Company; in 1897 Abram became president while vice-president Charles ran the mills. In 1899 Charles expanded the open-hearth furnaces first built in 1892 and added a universal rolling mill that could roll plates up to 48-inches wide. A massive three-high mill built in 1903 dwarfed all others in the country, producing plates up to 136 inches wide.

The firm entered the twentieth century with an enviable reputation for technical sophistication and quality and prospered despite fears that it could not compete with the emerging corporate giants. The keys to success remained developing or finding the best technology and producing large plates to demanding quality standards. True to that tradition, Charles Lukens Huston designed a 204-inch mill in 1915, which for 40 years stood as the largest plate mill in the world. Enlarged to 206 inches in 1919, the mill continued to operate through the late 1980s. The other feature of the Lukens tradition was good labor relations. Thanks to the paternalistic streak of Dr. Huston and his sons, the company had avoided labor friction during the nineteenth century and continued to do so after 1900.

In 1917 the brothers reorganized the firm as Lukens Steel Company to expand the company's capitalization. Abram retired as president in 1925 and was succeeded by engineer Robert F. Wolcott. He found troubles, for Lukens, like other small steel producers, struggled during the 1920s and 1930s. The company coped by adding two new subsidiaries. The By-Products Steel Company, founded in 1927, fabricated plates for such uses as diesel locomotive frames, gears, and marine items. In 1930 Lukenweld, the first commercial welding facility, began fabricating elevator-cage frames and locomotive underframes by welding plates. That same year Lukens introduced clad steels made by bonding two or more metals, opening additional specialty steel markets. But Lukens also coped with change in the industry by relying on its tradition of good labor relations. Thus, in 1937 Lukens recognized the Steel Workers Organizing Committee immediately after United States Steel Corporation announced its negotiations with the

A new flanging machine at the Coatesville, Pennsylvania, plant of Lukens Steel Company, 1950. At that time this machine, which makes flanged heads for industrial tanks, boilers, and pressure vessels, was the largest of its kind.

new union — a step that most of the Little Steel companies adamantly opposed.

World War II ended worries over profits and even brought an $18 million expansion program, including a 120-inch plate mill constructed by the navy. In 1942 the firm received the navy's "E" award for contributions to the war effort. But the conversion to peacetime was not easy, as profits disappeared in 1946. But to insure a steady supply of pig iron, Lukens purchased a share of E.&G. Brooke Iron Company in Birdsboro, Pennsylvania, a blast furnace operation. In 1949 Wolcott retired, and Charles Lukens Huston, Jr., replaced him as president.

Huston, Jr., presided over a return to prosperity in the 1950s that justified almost continuous expansion. New facilities were opened for the Cladding and Alloy Departments in 1953, while the navy completed a $10.5 million armor plate plant in 1956. Lukens steel went into the manufacturing of keel plates and hulls of many ships, including the nuclear submarine *Nautilus* and the N.S. *Savannah*. With sales exceeding $100 million in 1956, the company announced its largest ex-

pansion program ever in 1957. "Phase A" included a 100-ton electric furnace, opened in 1958, that boosted steel capacity to 930,000 tons. Also included was a versatile 140-inch slabbing and roughing mill, opened in 1959. The package cost $33 million and made Lukens the third largest plate supplier in the United States.

Lukens celebrated its one-hundred-fiftieth birthday in 1960, still adhering to the policies that had secured its niche in the industry. A second electric furnace became operational in 1962; a third two years later. Lukens installed the first computerized operator guidance system for electric furnaces the same year and added the first vacuum-degassing equipment in 1965. The 140-inch mill was modernized in 1966. Huston, Jr., became chairman at the end of the decade, but remained chief executive officer. He celebrated his new title by introducing a continuous strand caster in 1971, continuing the pattern of technological superiority at Lukens. Moreover, the company installed an electroslag remelt furnace the same year. In 1973 Lukens joined Fior de Venezuela, a company that constructed an iron-briquetting plant to reduce its dependence on scrap.

Since Huston's retirement in 1974 several men have headed the company. William E. Mullestein served as chairman until 1978, bringing on line the company's fourth electric furnace in 1975 and closing the last open-hearths. Charles A. Carlson, a veteran of Lukens's sales and purchasing operations, was chairman through 1981, overseeing the purchase of a plate mill from the bankrupt Alan Wood Steel Company for a bargain-basement price in 1978. W. R. Wilson left Inland Steel to become Lukens's chief operating officer in 1980, accepting the challenge of restoring eroding profit margins. He increased productivity, cut staffing requirements, and renegotiated union wage contracts. The company even purchased a 3.6-mile section of railroad to maintain connections to the Conrail system. Lukens diversified further in 1981, purchasing General Steel Industries of St. Louis, a company with a strong profit record in the production of heavy steel products.

In 1982 Lukens Steel became a subsidiary of a reorganized Lukens, Incorporated. It weathered the recession of the early 1980s that crippled the American steel trade and returned to profitability in 1984. Its reputation for producing specialty plate products has enabled the company to survive and maintain its technological position. Lukens's inclusion for the first time on the *Fortune* 500 list in 1985 was a fitting present for the company's one-hundred-seventy-fifth anniversary. Indeed, the continued success of Lukens stands as an exception to the problems experienced by most others in the contemporary American steel industry.

References:

Eugene DiOrio, *Lukens: Remarkable Past — Promising Future* (Coatesville, Pa.: Lukens Corporate Public Relations Division, 1985);

"Lukens Celebrates 125th Anniversary," *Iron Age,* 136 (July 4, 1935): 35–50;

"Lukens Completes Large Expansion Program," *Iron and Steel Engineer,* 36 (June 1959): 142, 144, 147, 148;

"Lukens Steel — The Heart and Lifeblood of a Town," *Iron Age,* 165 (March 30, 1950): 119, 225–228;

"One Hundred and Thirty Years of Iron and Steel Making," *Iron Trade Review,* 46 (June 30, 1910): 1263–1269;

Julian Skaggs, "Rebecca Lukens," in *Encyclopedia of American Business History and Biography: The Iron and Steel Industry in the Nineteenth Century,* edited by Paul Paskoff (New York: Facts on File, 1989) pp. 243–247.

Archives:

The Hagley Library in Wilmington, Delaware, is the repository for the records of Lukens Steel Company.

Reynold C. MacDonald

(October 7, 1918 –)

by Michael Santos

Lynchburg College

CAREER: Industrial engineer (1946–1948), shift foreman, general foreman, assistant plant superintendent, plant superintendent, assistant general superintendent, Kaiser Steel Corporation (1948–1963); vice-president of operations, Lone Star Steel Company (1963–1967); president, chief operating officer (1967–1969), director (1967–1969), president, chief executive officer, Interlake Steel Corporation (1969–1983); chairman of the board of directors, Acme Steel Company (1986–).

A self-made man with a single-minded drive to succeed, Reynold Coleman MacDonald wanted to enter the iron and steel business because the industry fascinated him. The son of a metallurgist, MacDonald grew up with men like Henry Clay Frick and Andrew Carnegie serving as his role models. Yet, by his retirement in 1983 as chief executive officer of Interlake, Incorporated, MacDonald would oversee his company's radical departure from the steel industry's accepted business methods, many of which dated back to Frick and Carnegie. Once the eleventh largest integrated steel producer, with 53 percent of its sales coming from iron and steel, Interlake was transformed into a multinational, diversified corporation that derived three-fourths of its revenue from nonsteel manufacturing operations. As such, MacDonald's strategies of diversification can be viewed as an anomaly in the twentieth-century steel industry, which by and large suffered widely for its adherence to the dated business philosophy of Carnegie: "Put all your good eggs in one basket, and watch the basket."

Born in Billings, Montana, on October 7, 1918, MacDonald and his family moved to California when he was very young. Growing up in Los Angeles, MacDonald attended Los Angeles City College in 1936–1937 and majored in engineering. He quit school to pursue a career as a race car driver and raced professionally until the outbreak of World War II. He moved on to a job in the defense industry, earning an A and E license, allowing him to work as an inspector of overhauled aircraft and engines. He then enlisted in the air corps with the hopes of training to be a pilot. MacDonald was instead placed in special teaching areas due to an injury he had sustained earlier as a race car driver. He was among the first to be trained in the air corps's jet propulsion training program and taught classes in the last few months of the war on jet engine theory and practice.

With the end of the war, MacDonald decided to pursue a career in steel. When later asked why, he remarked, "It was just a business that appealed to me." In 1946 he took a job as an industrial engineer with the Kaiser Steel Corporation in Fontana, California. After about a year and a half, MacDonald transferred into the operating end of the business, moving up the corporate ladder between 1948 and 1963 from shift foreman, to general foreman, to assistant plant superintendent, to plant superintendent, and finally to assistant general superintendent.

While happy with his progress at Kaiser, he had ambitions of running his own company and decided that the chances of his one day running Kaiser Steel were slim. He accepted a job with the Lone Star Steel Company in 1963 as that firm's vice-president of operations. Lone Star had been operating at a loss in the early 1960s and was facing a fiscal crisis. With MacDonald's help, the company turned into one of the more profitable small steel companies by 1967. The transformation was accomplished in large part because of MacDonald's management philosophy. Dedicated to doing what it took to get the job done, he used the talents of those at Lone Star. His strategy, whether at Lone Star or later

at Interlake, was to "put together a good management team. I made it a practice never to go in and fire a lot of people . . . I brought a few in . . . but normally I worked with the people that were there . . . and I found them to be receptive and helpful . . . particularly when they recognized that you knew exactly what you wanted to do, and you gained their respect for your capabilities and abilities."

Indeed, this philosophy was instrumental in winning Lone Star a major contract with a Spanish steel producer that eventually evolved into a mutually profitable relationship. When a bid for 100,000 tons of slab came on the market, MacDonald and Lone Star's sales vice-president went to Spain to make a bid. "All of the people were there," MacDonald remembered, "The Germans, the Japanese, you name it . . . and I was the only one who went up to the steel plant. The rest of them all did the fandango dances and so forth. I went up to the steel plant, looked at their facilities, what they needed, what the end product was, and the metallurgical department contacted Madrid and said that if we were competitive in price, that Lone Star should have the business because they knew what they needed."

Still wanting to run his own company, MacDonald moved to the newly created Interlake Steel Corporation in 1967 as that firm's president and chief operating officer. Within two years he became chief executive officer and by 1972 had been named chairman of Interlake's board of directors. When MacDonald became president, the company had a choice to make: it could invest in expanding its iron and steel facilities, but as MacDonald noted, that would be costly; or it could diversify. Under MacDonald's leadership, Interlake opted for the latter, becoming one of the first large steel corporations to successfully diversify its corporate holdings and interests. Throughout the 1970s and 1980s Interlake moved into a wide range of new areas, including material handling, metal powders, plastics, and aerospace components. It acquired firms and facilities in England, France, Germany, Australia, and Canada and increased its overall sales from $300 million in 1967 to $1 billion by the time MacDonald retired in 1983. Interlake became so diversified that it spun off its iron and steel division as a new firm, the Acme Steel Company. MacDonald was asked to come out of retirement, and in 1986 he became chairman of Acme's board of directors.

Reference:
The History of Interlake, Inc. (N.p., 1978).

Publication:
"Productivity, Profits, and the Engineer," *Iron and Steel Engineer* (March 1972): 63–66.

Elmer J. Maloy
(March 22, 1896 – 1970)

by Dennis C. Dickerson
Williams College

CAREER: Laborer, Carnegie Steel Company (1911–1936); member (1936–1937), president, Local 1256, Steel Workers Organizing Committee (1937).

Born on March 22, 1896, to Bridget Jane Tighe and William Patrick Maloy, in Pittsburgh, Pennsylvania, Elmer J. Maloy pursued a varied career as a steel unionist and politician. After attending public schools for nine years, Maloy took courses in accounting and engineering. In 1911 he took work at the Carnegie Steel Company plant in Duquesne, Pennsylvania. Except for a brief period of military service during World War I, he remained at the plant until 1936.

His father, a member of the United Mine Workers of America (UMWA), familiarized him with the labor movement and Maloy participated in the 1919 steel strike. In 1933, in response to Section 7A of the National Industrial Recovery Act, officials at the Duquesne steel works established an Employee Representative Plan (ERP), which many laborers pejoratively called a "company union." Though Maloy originally eschewed the ERP, fellow workers in 1935 elected him to represent them in the group. Immediately, he agitated for improvements in wages and working conditions, and with his counterparts at other plants, Maloy tried to organize a companywide central committee of ERP delegates. Maloy helped to spearhead a 1935 convention in New Castle, Pennsylvania, where various demands were made by the ERP representatives for higher wages, pensions, vacation time, arbitration, and the end of arbitrary dismissals of employees. Although by this time Maloy had come to believe that the ERPs could effectively serve the workers' interests, he was becoming increasingly allied with the Steelworkers Organizing Committee (SWOC).

Sponsored by the UMWA, the SWOC sought to introduce an independent union to the steel industry to serve as the workers' exclusive bargaining agent. Maloy, as chairman of the ERP central committee, entered negotiations with representatives of the United States Steel Corporation — Carnegie Steel's parent company — in an attempt to settle a disputed wage-increase formula. Behind the negotiating scenes, however, Maloy engaged in a series of public relations maneuvers in cooperation with the SWOC in order to discredit U.S. Steel's bargaining position. In 1937, in a move that stunned the rest of the industry, U.S. Steel agreed to recognize the SWOC. Having formally joined the SWOC in July 1936, Maloy became the president of Local 1256 at Duquesne.

Maloy rose rapidly in the SWOC and its successor, the United Steelworkers of America (USWA). He chaired the negotiating committee for the Elimination of Wage Inequities in the Steel Industry. Additionally, he met with the prime minister of Canada and his cabinet to settle a nationwide steel strike in that country. He served also as the chairman of the contract department of the United Steelworkers of America.

On the strength of his union activism, Maloy achieved a historic first in the Monongahela steel valley by winning the Duquesne mayoral election in 1937. In the past, Carnegie management in Duquesne, Mckeesport, Clairton, and other industrial communities in the valley controlled municipal politics. The 1936 and 1937 SWOC organizing drives, however, gave local union leaders a natural following, and Maloy successfully traded his union influence for political power. Mindful of the racial and ethnic diversity of his union constituency, Maloy chose Tony Salopek, a Slav, to be his campaign manager. Salopek was a financial secretary of Local

#1256 and an effective liaison with the Slavic community, a major voting bloc. During the campaign, Maloy continued to parlay the union's ethnic pluralism into an effective political strategy. He also marshaled significant Congress of Industrial Organizations support. Patrick Fagan, a district president of the UMWA, and Clinton Golden, the eastern director of SWOC spoke at a rally in his behalf.

Maloy married Ruth Anne Gilfoyle in 1922, and the couple had three children. Maloy died in 1970.

References:
Robert R. R. Brooks, *As Steel Goes, . . . Unionism in a Basic Industry* (New Haven: Yale University Press, 1940);
George Powers, *Cradle of Steel Unionism, Monongahela Valley, PA.* (East Chicago, Ind.: Figueroa Printers, 1972).

Edmund Fible Martin

(November 1, 1902 – January 11, 1993)

by Lance E. Metz

Canal Museum, Easton, Pennsylvania

CAREER: Management trainee (1922–1924), foreman, 42- and 48-inch structural mills, Saucon division (1924–1927), mill assistant superintendent (1927–1939), assistant superintendent, Saucon division (1934–1946), assistant general manager, Lackawanna Plant (1946–1950), vice-president, operations (1958–1960), president (1960–1964), director (1958–1974), chairman and chief executive officer, Bethlehem Steel Corporation (1970–1974).

During his term as chief executive officer of the Bethlehem Steel Corporation, Edmund Fible Martin helped to create the last fully integrated steel plant built in America. He was also a pioneer in adopting pollution abatement policies for his company.

Martin was born in Chicago, Illinois, on November 1, 1902, to Bettie G. (Fible) and Albert Martin. As a child, he exhibited a fascination for machines and an aptitude for mathematics. He attended elementary school at Hinsdale, Illinois, and received his secondary education at Chicago's prestigious University High School. In the middle of his high school years the family relocated to Orange, New Jersey. So outstanding were his abilities that he was able to win admission to the Stevens Institute of Technology at the age of 15. Despite his youth, Edmund took an active part in social and athletic activities at Stevens. He served as the manager of the yearbook and the baseball team. He also became a member of Tau Beta Pi National Engineering Hon-

orary fraternity. Martin compiled an outstanding academic record at Stevens and graduated in the upper fourth of his class, earning a degree in mechanical engineering in 1922.

A childhood visit to steel plants near Chicago had sparked Martin's interest in the steel industry. He chose to enter the employment of the Bethlehem Steel Corporation in 1922 because he believed that it was the most progressive and technologically innovative of America's steelmakers. He became one of the earliest members of Bethlehem's pioneering "loop" management training program, in which young college graduates spent several weeks in each of the Bethlehem Steel Corporation's operating departments to gain an overall understanding of how the corporation functioned. Much of Martin's time as a member of the loop course was devoted to manual labor and drafting. In 1924 he was promoted to foreman of the 42-inch and 48-inch rolling mills at the Saucon division of Bethlehem Steel's Bethlehem, Pennsylvania, plant. He soon gained a reputation as both a skilled engineer and an able manager. In 1927 he became assistant superintendent of his department, and in 1939 he was named the assistant superintendent of the entire Saucon division, which produced Bethlehem's most profitable and widely known products — the continuously rolled wide-flange structural beams and columns.

During World War II he was largely responsible for adapting Saucon's production facilities to the manufacture of wartime defense products, such as alloy shells, shell steel, airplane cylinder stock, and armor plate. Martin's success earned him a promotion to the position of assistant general manager of Bethlehem's steel mill at Lackawanna, New York, in 1946. Lackawanna was considered by Bethlehem's top executives to be its most troublesome plant due to a history of bad relations between management and employees. Martin, however, believed that this situation could be reversed through proper leadership. His philosophy of management was summarized when he stated, "You can spend all of the money for the finest plant in the world, but you can't make it work without the right people in the right jobs." His efforts to improve productivity at Lackawanna were successful. In 1950 he was promoted to general manager of the Lackawanna Plant, and within five years his management had helped to double annual production.

In 1958 Martin returned to Bethlehem's corporate headquarters to become vice-president of operations. He was also appointed a director of the Bethlehem Steel Corporation. In 1960 he became its president. During his tenure as president, Martin took an aggressive stand in favor of stricter federal import quotas on steel products.

In 1964 he became the chairman and chief executive officer of Bethlehem Steel. Under his leadership, the company completed a large new plant at Burns Harbor, greatly improving Bethlehem's competitiveness in the crucial midwestern steel market. It also proved to be the last fully integrated steel plant to be built in the United States. Martin also committed Bethlehem Steel to a massive effort to control the environmental damage that was caused by its operations. Under his leadership, Bethlehem spent over $200 million to install pollution-control equipment at its various plants. He also served as a spokesman for the steel industry and clashed with President Lyndon B. Johnson's attempts to lower steel prices, although he later supported President Richard Nixon's imposition of wage and price controls as a means of controlling inflation. At the time of his retirement in 1970, Bethlehem Steel had 125,000 employees and produced 25 million tons of steel annually.

Martin also took a leading part in the affairs of the American Iron and Steel Institute. He served as chairman of its Communications and National Affairs Coordinating committees and later as chairman of the entire organization. He received the institute's Gary Medal in 1969. The Bethlehem Steel Corporation named its new world headquarters building the Martin Tower.

During his retirement Martin focused much of his energies on Historic Bethlehem, Inc. This nonprofit organization was founded to preserve, interpret, and restore the industrial buildings of Bethlehem's Moravian community. Under his leadership during the early 1970s, Historic Bethlehem achieved many of its goals. Martin died in Bethlehem on January 11, 1993.

Publications:

and Roger Blough, "Steel Leaders See Gains Overshadowing Problems," *Iron Age,* 197 (January 6, 1966): 36;

Promise for the Future; Delivered to the National Newcomen Dinner, April 20, 1967 (South Bethlehem, Pa.: Bethlehem Steel, 1967);

with Daniel J. Morrison, *Bethlehem Steelmaker: My 90 Years in Life's Loop* (Bethlehem, Pa.: BMS Press Incorporated, 1992).

Archives:

The Martin biographical file is in the Bethlehem Steel Collection, Hugh Moore Historical Park and Museums, Inc., Easton, Pennsylvania.

Walter Emil Ludwig Mathesius

(August 20, 1886 – June 20, 1966)

by Robert Casey

Henry Ford Museum & Greenfield Village

CAREER: Metallurgist, American Steel & Wire Company (1911–1912); various positions, blast-furnace department (1912–1917), superintendent, blast-furnace department (1917–1925), assistant general superintendent (1925–1935), general superintendent, South Works, Illinois Steel Company (1935); manager of operations, Chicago district, Carnegie-Illinois Steel Corporation (1935–1937); vice-president of operations (1937–1942), director (1947–1951), United States Steel Corporation; president, Geneva Steel Company (1943–1951); consultant, Koppers Company (1952–1966).

Walter Emil Ludwig Mathesius was born in Hoerde, Germany, on August 20, 1886. After compulsory duty in the Imperial German Army in 1905, he attended Berlin's Institute of Technology, earning a degree in metallurgy in 1910. He served as an instructor at the institute while earning a doctorate of engineering in 1911.

Mathesius immigrated to the United States in 1911, becoming a naturalized citizen in 1919. His first job in America was as a metallurgist in the physical and research laboratory of American Steel & Wire Company, a U.S. Steel subsidiary in Worcester, Massachusetts. In 1912 he moved on to the South Works of Illinois Steel in South Chicago, where he rose steadily through the company ranks, becoming superintendent of blast furnaces in 1917 and assistant general superintendent of South Works in 1925. Mathesius was promoted to general superintendent in 1935 and manager of operations for the entire Chicago district in October of the same year. In 1937 Mathesius became vice-president of operations for U.S. Steel.

In 1942 U.S. Steel undertook the construction of a $180 million steel mill for the Defense Plant Corporation at Geneva, Utah. The following year Mathesius was selected to head the operation and remained president of Geneva Steel until his retirement in 1951. He also served as a director of U.S. Steel from 1947 to 1951.

Drawing on his academic training, Mathesius believed that blast-furnace operators could understand and apply theoretical explanations for metallurgical phenomena. He fought against what he called "the creed that blast furnace progress must be brought about through practice, leaving it to science to afterwards explain the 'whys and wherefores.'" He worked hard to ensure uniformity of raw materials, especially of coking coal, and thereby to improve the predictability of blast-furnace operations. In 1944 the Franklin Institute awarded him its Francis J. Clamer Medal for outstanding contributions in converting the art of blast-furnace operation to a science.

Mathesius was a member of the American Iron and Steel Institute, the American Institute of Mining and Metallurgical Engineers, and the American Institute for Metals. His many articles appeared in publications such as the *Journal of the American Iron and Steel Institute* and *Stahl und Eisen*. Mathesius died on June 20, 1966.

Walter Mathesius

Publications:

"Results with High Blast Heats," *Iron Trade Review*, 56 (February 18, 1915): 365–366, 368;

"Chemical Reactions of Ore Smelting," *American Iron and Steel Institute Yearbook* (1917): 169–182;

"The Principal Changes in Blast Furnace Lines during the Last Ten Years," *American Iron and Steel Institute Yearbook* (1918): 109–114;

"The Blast Furnace Hearth," *American Iron and Steel Institute Yearbook* (1920): 433–453;

"Uniform Coking Coal as a Factor in Blast Furnace Economy," *American Iron and Steel Institute Yearbook* (1924): 36–63;

"Stainless Steel Production-Equipment and Method of Manufacture," *American Iron and Steel Institute Yearbook* (1932): 230–238;

"Technological Advances in Steel Products," *American Iron and Steel Institute Yearbook* (1937): 125–128;

"Iron and Steel Production, 1851–1951, and the Coke Industry," *Blast Furnace and Steel Plant*, 40 (November 1952): 1305–1310;

"Steel Manufacturing," *Metals Progress*, 68 (September 1955): 77–81.

John A. Mathews

(May 20, 1872 – January 11, 1935)

by Geoffrey Tweedale

Manchester, England

CAREER: Metallurgist (1902–1904), assistant manager, Sanderson Brothers Steel Company (1904 –1908); operating manager (1908–1913), general manager (1913–1920), president, Halcomb Steel Company (1915–1920); president (1920 –1923), vice-president, director of research, Crucible Steel Company (1923–1935).

One of the foremost specialty steelmakers and metallurgists in the United States during the early twentieth century, John Alexander Mathews was born in the heart of America's steelmaking district in Washington, Pennsylvania, on May 20, 1872. He attended Washington and Jefferson College, receiving an undergraduate degree in 1893 and a master's degree in 1896. He took courses in chemistry and metallurgy at Columbia University, where he received a Ph.D. in 1898. He worked briefly as an instructor in chemistry at Columbia, during which time he was awarded the University Fellowship in Chemistry and the Barnard Fellowship for the Encouragement of Scientific Research. The fellowships allowed him in 1900 to travel to the Royal School of Mines, London, to study metallography under Sir William Roberts-Austen, who thought so highly of Mathews that he recommended him for one of the first Carnegie Scholarships offered by the Iron and Steel Institute (ISI).

England at that time practiced the most technologically advanced methods of steelmaking in the world. Centered in the district of Sheffield, the English steel industry provided America with specialty steel and innovations in production. At the turn of the century, however, metallurgy was still very much in its infancy, poised uneasily between the age of science and the rule of thumb. The industry stood on the verge of major advances in alloys — particularly tool steels — and production techniques were also being revolutionized with the advent of the electric furnace. Mathews was to make major contributions in all these areas.

In 1901 he returned to Columbia University and began researching alloy steels, which were

John A. Mathews

made for him at the Sanderson Brothers Steel Company at Syracuse, a firm founded by a Sheffield company of the same name in 1876. Mathews's report on his work received the first Carnegie Gold Medal awarded by the ISI in 1902. In the same year he was appointed metallurgist in charge of experimental work at Sanderson Brothers — which by then was a subsidiary of the Crucible Steel Company of America — and was assistant manager from 1904 to 1908. At Sanderson Brothers, Mathews devoted particular attention to the production of tool steel, especially the high-speed steels that had resulted from the research of Frederick W. Taylor and Maunsel White at the Bethlehem Steel Company. Their improved composition and heat treatment of alloy tool steels, which were made available after 1901, marked a major advance on the older carbon steel cutting tools. Intense research was soon underway in both England and America to improve further the Taylor-White steels. Mathews, no doubt spurred by British and French efforts along similar lines, had patented the use of vanadium in high-speed steel by 1905.

In 1908 Mathews became the operating manager of the Halcomb Steel Company in Syracuse, which had been founded in 1905 by a former Sheffield steelmaker. Once again Mathews found himself involved with alloy- and tool-steel manufacture, especially with their production in the electric furnace. This method of specialty steelmaking was still relatively untried, many firms preferring to melt their steels by the crucible method. In 1906 the Halcomb Steel Company became the first in the United States to install an electric furnace for steel manufacture, though initially the results were disappointing and it was considered a failure. It was Mathews's major achievement to bring this furnace back into regular production. By 1917 the firm had five electric furnaces in regular use, and the old crucible method was gradually phased out, as it was throughout the American steel industry.

Mathews became general manager of the Halcomb Steel Company in 1913 (which by then was part of Crucible Steel Company) and while still holding that position was elected president of the company in 1915. In 1920 he became president of Crucible Steel, a post he held from 1920 to 1923, a difficult period during which the firm's executive staff was thoroughly reorganized. After that task was completed, Mathews returned to his first love as director of research and vice-president of the corporation, the positions he held at his death.

Mathews's most important contributions were perhaps in the technical sphere. Besides his work on the electric furnace and vanadium additions to tool steels, his research spanned all the major aspects of specialty steels at that time: the development of various vanadium steels, especially spring steels, and of oil-hardening magnet steels of chrome vanadium, and of corrosion- and heat-resisting steels and the newly discovered stainless steels. A prolific technical writer, he contributed papers and discussions on these subjects to the various technical societies of which he was a member: notably the American Iron and Steel Institute, the American Society of Metals (of which he was an honorary member), the Ameri-

can Institute of Mining Engineers, and the British Iron and Steel Institute.

His contribution to the industry was recognized by an honorary Sc.D. from Washington and Jefferson College in 1903, the Hunt Gold Medal of the American Institute of Mining and Metallurgical Engineers in 1928, and in 1924 he was appointed the second Henry Marion Howe lecturer of the American Institute of Mining Engineers, succeeding Professor Albert Sauveur of Harvard.

Mathews took an active interest in local and municipal affairs in Syracuse and was, for example, a director of the First National Bank and first vice-president of the Syracuse Chamber of Commerce.

On January 29, 1903, he married Florence Hosmer, daughter of the Reverend I. F. King of Columbus. John A. Mathews died suddenly of a heart attack at his home in Scarsdale, New York, on January 11, 1935, and was survived by his wife and his son and daughter. He was recognized as one of a small number of metallurgists who made the U.S. specialty steel industry a force to be reckoned with internationally.

Publications:

and Howard J. Stagg, Jr., "Factors in Hardening of Tool Steel," *Iron Trade Review,* 57 (July 22, 1915): 185–187;

"Electric Furnaces in Steel Making," *Iron Trade Review,* 58 (June 8, 1916): 1264 –1267;

"Progress of the Electric Steel Industry," *Iron Age,* 99 (May 10, 1917): 1146–1148;

"The Present Status of the Electric Furnace in Refining Iron and Steel," *American Iron and Steel Institute Yearbook* (1922): 358–365;

"Comments on the Making and Use of Alloy Tool and Special Steels," *Transactions of the American Society for Steel Treating,* 7 (1925): 147–167;

"Silicon Steel," *American Iron and Steel Institute Yearbook* (1925): 259–261;

"Steel Used by the Automotive Industry," *American Iron and Steel Institute Yearbook* (1928): 486–490;

"Recent Developments in Nickel, Iron and Steel," *American Iron and Steel Institute Yearbook* (1930): 212–213;

"Tool Steel Progress in the 20th Century," *Iron Age,* 126 (December 4, 1930): 1672–1676.

References:

William Campbell, "John Alexander Mathews," *Science,* 81 (1935): 190–191;

Geoffrey Tweedale, *Sheffield Steel and America: A Century of Commercial and Technological Interdependence, 1830–1930* (Cambridge: Cambridge University Press, 1987).

James Lester Mauthe

(July 8, 1890 – January 2, 1967)

by Larry N. Sypolt

West Virginia University

CAREER: Furnace clerk (1914–1915); blast furnace superintendent, National Tube Company (1915–1917); superintendent of blast furnaces, Midvale Steel Company (1917–1918); assistant superintendent, coke ovens, Gary Works, Illinois Steel Company (1919–1920); superintendent of blast furnaces (1920–1930), assistant general superintendent, National Tube Company (1930–1935); assistant general superintendent (1935–1937); general superintendent, Youngstown District (1937–1943); vice-president for operations (1943–1950); director (1948–1963); president (1950–1956); chairman of the board, Youngstown Sheet & Tube Company (1956–1963).

The path James Lester Mauthe took in reaching the top of Youngstown Sheet & Tube Company as its president and chairman of the board was a traditional one, for Mauthe was a steel man with long experience in blast furnaces and pipe mills. His main contribution to Youngstown was his inauguration of an extensive improvement program after World War II.

Mauthe was born on July 8, 1890, in Turkey City, Pennsylvania, the son of Martin and Mary Margaret (Theiss) Mauthe. His father came from Württemberg, Germany, in about 1870 and settled in Erie, Pennsylvania. Mauthe received his early education in public schools at Dubois, Pennsylvania, and earned an undergraduate degree in metallurgy at Pennsylvania State College (later Pennsylvania State University) in 1913. He worked for two years as a blast-furnace laborer at the Duquesne Works of the Carnegie Steel Company — a United States Steel Corporation subsidiary — while attending college.

As a student at Penn State, he was captain and fullback of the school's first undefeated football team in 1912. He earned a reputation as one of college football's great placekickers and punters and at the time of his death still held the university's record for most points scored in a season (119) and in a career (171 between 1909 and

James Lester Mauthe

1912). In 1957 Mauthe became the first player from his alma mater to be named to the College Football Hall of Fame. After graduating from Penn State he coached football at Gettysburg College for one season.

In 1914 Mauthe entered the steel industry as a furnace clerk with the National Tube Company, a subsidiary of U.S. Steel. He advanced quickly and a year later was promoted to assistant superintendent of blast furnaces. He moved to the Midvale Steel Company in Coatsville, Pennsylvania, in 1917 as superintendent of blast furnaces. Two years later he was back with U.S. Steel as assistant superintendent of the coke plant and then assistant superintendent of blast furnaces at the Illinois Steel Company's plant in Gary, Indiana. He became superintendent of blast furnaces in 1920 for National Tube Company in Lorain, Ohio, and eventually advanced to assistant general superintendent for the entire operation in 1930.

In 1935 Mauthe joined Youngstown Steel and Tube Company as assistant general superintendent of the Youngstown District, which included the company's big Campbell Works and its Brier Hill plant. He was named general superintendent in 1937, and in 1943 Mauthe was made vice-president in charge of operations. He was chosen a director of the company in 1948, was elected president in 1950, and was made chairman of the board in 1956, a position he held until his retirement in 1963.

As president of Youngstown Sheet & Tube, Mauthe initiated an expansion program designed to modernize the Indiana Harbor plant. New ovens, furnaces, and mills were added at a cost of $663 million. He also inaugurated major construction and modernization projects to better accommodate customers' requirements, improve product mix, and reduce production costs. He initiated the rebuilding and modernization of the 79-inch hot-strip mill at the Campbell Works, the revamping of the blooming and cold-strip mills at Indiana Harbor (the world's first six-stand cold-reduction mill), and the construction of a second continuous annealing line and a new galvanizing line. During Mauthe's presidency Youngstown's annual ingot capacity rose from 4,000,000 tons in 1950 to 6,750,000 tons in 1956, while annual sales for that period increased from $410 million to more than $675 million.

In addition to these and other major renovation and construction projects undertaken during the Mauthe presidency, Youngstown also added to its holdings in raw materials, finished steel products, and transportation. Interests were acquired in the Mathies Coal Company, the Presque Isle Corporation (a limestone quarrying enterprise), four additional steamship companies, the Manganese Chemical Corporation, the Ozark-Mahoning Company, and the Youngstown Steel Door Company, among others. In 1953 Youngstown acquired an interest in the Perault Fibercast Corporation, which manufactured glass-fiber reinforced plastic pipe that resisted corrosion. During the 1950s the company also exercised its option on 1500 acres of coal land adjoining the Ontario Mine. In common with other American steel companies, Youngstown had launched efforts to enhance its ore holdings after the war, and its subsidiary, the Iron Ore Company of Canada, developed ore properties in Labrador and Quebec. Construction was completed on a 365-mile railroad from Seven Islands, Quebec, to the mine site in 1954. Ore-handling and other equipment was installed and the first shipment was made on July 31, 1954.

In directing the Youngstown expansion, Mauthe was met with difficulties beginning in 1954, when Youngstown and Bethlehem Steel made a second attempt at a merger. The Justice Department opposed the combination, which both Youngstown and Bethlehem believed would produce a strong company that served both eastern and midwestern markets. Furthermore, a merger, the two companies felt, would be more than geographically advantageous: Youngstown would bring to the combination certain product lines that Bethlehem did not carry — such as tubular steel — while Bethlehem had the financial resources Youngstown needed to fully modernize both its Indiana Harbor and Youngstown plants. Bethlehem, in fact, had pledged to spend $350 million for capital improvements at Youngstown's mills and raise ingot capacity by 3 million tons. But after several years of legal battles, the courts upheld the government, and the merger was abandoned in early 1959. Beyond the failure of the merger, the protracted litigation resulted in stalled growth at Youngstown, for programs of major improvements had been suspended during much of the court fight.

Mauthe's record as president was impressive enough from a technical standpoint, however. Ingot production averaged 4.5 million net tons per year; the works operated at a 94.95 percent capacity and had a net income of almost $187 million. Although too much capacity in the high-cost Youngstown plants remained an unresolved problem for the company, Youngstown was nevertheless profitable during much of the 1950s under Mauthe's leadership.

James Mauthe's technical background was evident in several articles he published on blast furnaces and open-hearth operations. A longtime member of the American Iron and Steel Institute (AISI), he and Youngstown's vice-president for research, Karl Fetters, won an AISI medal in 1948 for their article "The Mineralogy of Basic Open-Hearth Slags."

Like many corporate leaders in the steel industry, Mauthe also was interested in community affairs. He served as a trustee for the Youngstown

YMCA and Youngstown University. He was a member of the Mahoning Valley Management Association and Phi Sigma Kappa. But he reserved his greatest service for his alma mater, Pennsylvania State University. He was a member of the executive committee of Penn State from 1946 until his death and a member of the university's committees on research and on architecture, the Athletic Board of Control, and the Athletic Advisory Board. In 1964 he established the J. L. Mauthe Scholarship for a varsity athlete in the field of metallurgy, ceramic technology, or engineering.

Mauthe was married on May 17, 1918, in McKeesport, Pennsylvania, to Olive Kough, daughter of steel executive Howard Phillip Kough. The couple had two children. Mauthe died in Poland, Ohio, on January 2, 1967.

Publications:
"The New Technical and Economic Importance of Iron and Steel Scrap," *American Iron and Steel Institute Yearbook* (1936): 222–224;

and Karl L. Fetters, "The Mineralogy of Basic Open Hearth Slags," *American Iron and Steel Institute Yearbook* (1947): 264–297;
"Experience to Date on Iron Production by Methods Other Than Coke Blast Furnace," *American Iron and Steel Institute Yearbook* (1948): 343;
"Sinter Quality and Effect of Sinter in Blast Furnace Practice," *Blast Furnace and Steel Plant*, 36 (July 1948): 817–824.

References:
William T. Hogan, S.J., *An Economic History of the Iron and Steel Industry in the United States*, 5 volumes (Lexington, Mass.: Lexington Books, 1971);
Charles Longenecker, "Youngstown Sheet and Tube Increases Service to the Midwest," *Blast Furnace and Steel Plant*, 41 (August 1953): 898–925;
Kenneth Warren, *The American Steel Industry 1850–1970: A Geographical Interpretation* (London: Oxford University Press, 1973).

Lloyd McBride

(March 9, 1916 – November 6, 1983)

by John P. Hoerr

Teaneck, New Jersey

CAREER: Production worker, Foster Brothers Manufacturing Company (1930–1940); president, Lodge 1295 (1937–1940), staff representative, Steel Workers Organizing Committee (1940–1944); U.S. Navy (1944–1946); staff representative (1946–1965), director, District 34 (1965–1977), international president, United Steelworkers of America (1977–1983).

Lloyd McBride, the fourth president of the United Steelworkers of America (USWA), led the union during one of its most troubled periods. Born March 9, 1916 in Saint Louis, he went to work at the age of fourteen, replacing his ill father as a production employee at Foster Brothers Manufacturing Company, a steel fabricating plant in Saint Louis. In 1936 he helped the Steel Workers Organizing Committee (SWOC), forerunner of the USWA, organize the company's bedspring manufacturing plant. After leading a sit-down strike in 1938, McBride was elected by acclamation as president of SWOC Lodge 1295. He left the company in 1940, when SWOC chairman Philip Murray appointed him staff representative.

At the founding of the USWA in 1942, McBride was assigned as a field representative in District 34, which covered Missouri, southern Illinois, and neighboring states. He served two years in the U.S. Navy during World War II and returned to the USWA staff in 1946. For the next 19 years, McBride carried out the duties of a staff representative, organizing new members, negotiating contracts in a variety of industries, and servicing local unions. When the first chief of District 34 retired in 1965, McBride was elected to succeed him as director and member of the International Executive Board.

His many years in the field had prepared McBride well for the directorship. He ran the district in a businesslike way, working long hours, and expected much of his staff. His was not a fiery brand of unionism arising out of a deep belief in the necessity of class struggle;

United Steelworkers officials at union headquarters in Pittsburgh, September 1, 1981: Treasurer Frank McKee, Secretary Lynn Williams, President Lloyd McBride, and Vice-Presidents Leon Lynch and Joseph Odorcich (photograph: Pittsburgh Press)

only rarely did he engage in bristling oratory or anticompany polemics. For McBride, unions existed simply to protect workers from mistreatment, bargain a fair share of company profits, and influence the political process on behalf of members.

McBride's experiences in District 34, however, did not prepare him well for his eventual rise to the union presidency. It was a relatively placid USWA region with little steelmaking. The majority of its 28,000 to 34,000 members worked in small fabricating shops, foundries, and lead and zinc mines. Although McBride negotiated contracts at the small Granite City Steel Company in Illinois and served as secretary of the union negotiating committee at Armco Steel, he had little exposure to labor relations and politics in the basic steel industry. As director, he never faced an election challenge and did not have to deal with intense politicking of the kind that characterized USWA districts in basic steel regions such as the Pittsburgh and Chicago-Gary areas.

In 1976, as the tenure of two-term president I. W. Abel drew to a close, no clearly designated heir was named to succeed him. A deep political split developed. Many young militants and independent members backed Edward Sadlowski, the thirty-seven-year-old director of District 31 (South Chicago, Gary, and northern Indiana) and a vocal critic of Abel's cooperative policies vis à vis steel management, as exemplified by the union's adoption of a no-strike agreement in basic steel. McBride, a strong supporter of Abel's policies, eventually emerged as a compromise candidate of the pro-administration members of the International's executive board. Promising to continue the collective bargaining gains achieved under Abel, McBride obtained the backing of older, more satisfied steelworkers and, most important, of the union's administrative staff. After a long, rancorous campaign, McBride defeated Sadlowski by a vote of 328,861 to 249,281 and took office on June 1, 1977.

Almost immediately, the McBride administration ran into trouble. McBride's relations with steel industry officials were strained by a long strike of disgruntled iron ore miners. In late 1977 large steel plants in Youngstown and Johnstown were closed as overcapacity forced the industry to begin a severe contraction. The steel segment of the union remained divided between local union officers loyal to McBride and former pro-Sadlowski forces. When the recession of 1981–1982 devastated the steel industry and led to company demands for wage concessions, McBride had to contend with an emotionally charged opposition within the union. Although he believed that the union must accept concessions to keep the companies viable, McBride's prebargaining tactics aroused suspicion rather than support. He bluntly told local union leaders that they must conform to economic reality, but his leadership toward that goal was less than compelling. In addition, decades of alienation in the mills resulting from the industry's authoritarian management style, barred a sudden change in attitudes.

As a result, the USWA and the steel industry experienced two negotiating debacles in 1982. The first occurred in July when the major companies asked the union to replace an existing labor agreement with a less costly one. McBride, though he favored wage restraint, turned down an industry offer that he thought would fail in ratification, and the talks collapsed. Three months later, unemployment in the steel industry topped 40 percent, and the two sides resumed negotiations. This time, McBride himself proposed a large wage cut and eventually approved a $3.11 per hour reduction in labor costs. But the union's ratification body, consisting of local union presidents and named the Basic Steel Industry Conference, humil-

iated the president on November 19 by rejecting the agreement 231 to 141 votes. Only once before, in 1977, had a ratification committee turned down a steel pact approved by a USWA president, and in that instance President I. W. Abel had triumphed in a quick second vote.

Three months later, with the industry still in rapid decline, McBride mounted another effort to grant concessions to steelmakers. But he became gravely ill and was hospitalized. A USWA vice-president, Joseph Odorcich, took command of bargaining in February 1983. By then, job losses in the industry had dampened the anticoncession movement, and Odorcich managed to gain ratification of a new, three-year agreement with net savings to the industry of $2.15 per hour.

Recuperating partially from his illness, McBride returned to work for several months. But the strain of his job had taken a large toll on his health. In September 1983 he was hospitalized again and underwent multiple-bypass heart surgery. Released from the hospital, McBride died in his sleep on November 6 at the age of sixty-seven.

Reference:
John P. Hoerr, *And the Wolf Finally Came: The Decline of the American Steel Industry* (Pittsburgh: University of Pittsburgh Press, 1988).

Archives:
Historical records of the USWA are located in the Historical Collections & Labor Archives, Pennsylvania State University, University Park, Pennsylvania.

Elmer T. McCleary

(August 27, 1878 – April 22, 1930)

by Carol Poh Miller

Cleveland, Ohio

CAREER: Chemist, Diamond State Steel Company (1901); chemist, Carnegie Steel Company (190?–1906); chief chemist (1906–1909), assistant superintendent, blast-furnace and steel department (1909–1912), assistant general superintendent (1912–1923), manager, Youngstown district (1923–1924), assistant vice-president (1924–1925), vice-president, operations, Youngstown Sheet & Tube Company (1925–1928); president, Republic Iron & Steel Company (later Republic Steel Corporation, 1928–1930).

Elmer T. McCleary was born in Pine Grove Furnace, Pennsylvania, to John and Isabel Marks McCleary on August 27, 1878. His father, of Scottish descent, was a drummer boy in the Union Army during the Civil War; later, working as a locomotive engineer, he was killed in an accident on the Philadelphia and Reading Railroad. McCleary spent his early years in Carlisle and Harrisburg, Pennsylvania, where he attended public schools. He received an undergraduate degree in metallurgy from the Pennsylvania State College (later renamed Pennsylvania State University) in 1901.

After graduation, McCleary went to work as a chemist with the Diamond State Steel Company in Wilmington, Delaware. From Wilmington, he went to Youngstown, Ohio, where he worked in the laboratory of the Carnegie Steel Company. In 1906 he joined the Youngstown Sheet & Tube Company as chief chemist. He was promoted to assistant superintendent of the blast-furnace and steel department in 1909; assistant general superintendent of the plant in 1912; manager of the Youngstown district in 1923; assistant vice-president in 1924; and vice-president of operations in 1925. He had taken an active part in Youngstown Sheet & Tube's acquisitions of the Steel & Tube Company of America and the Brier Hill Steel Company in 1923. During World War I, as a member of the steel section of the advisory board of the U.S. Ordnance Department, McCleary had also helped to define the chemical and physical specifications for shell steel.

McCleary was an active inventor. He and Charles H. Elliot developed an open-hearth steel process, patented in 1918. Designed to overcome the difficulty of obtaining the desired carbon content in steel, this method called for the adding of silicon to the metal near the end of the heating process in order to keep carbon from escaping. He also patented a welding technique using super-heated metal with Henry W. Schoenfeldt and G. A. Reinhardt in 1923 and a double sheet drier and galvanizing apparatus — consisting of two sets of chain conveyors that increased galvanizing capacity and shortened the drying chamber — in 1928.

In April 1928 Cleveland industrialist Cyrus Eaton tapped McCleary as the new president of the Republic Iron & Steel Company, succeeding Thomas J. Bray. During his brief tenure with Republic, McCleary helped acquire companies — the Fretz-Moon Tube Company, Central Alloy Steel Corporation, and the Bourne-Fuller Company, among them — which were combined to form the new Republic Steel Corporation in 1930.

McCleary was a director of the American Iron and Steel Institute and a member of the

Elmer T. McCleary

American Institute of Mining and Metallurgical Engineers, the American Society for Testing Materials, and the Youngstown Chamber of Commerce. In addition, he was a director of the Aetna Standard Engineering Company, the First National Bank, and Dollar Savings and Trust Company, all of Youngstown. He had married Anna Jenkins of Youngstown on February 1, 1904, and the couple had two children. McCleary died following surgery at Youngstown Hospital on April 22, 1930.

Publication:
"The Pilger Tube Mill," *Year Book of the American Iron and Steel Institute* (1927): 132–134.

References:
Joseph G. Butler, Jr., *History of Youngstown and the Mahoning Valley*, 3 volumes (Chicago & New York: American Historical Society, 1921);
"Elmer T. McCleary New Republic Head," *Iron Age*, 121 (April 26, 1928): 1190–1191.

David J. McDonald

(November 22, 1902 – August 8, 1979)

by Dennis C. Dickerson

Williams College

CAREER: Secretary and assistant to the vice-president, United Mine Workers of America (1923–1936); secretary-treasurer, Steel Workers Organizing Committee (1936–1942); international secretary-treasurer (1942–1953), president, United Steelworkers of America (1953–1965).

Born on November 22, 1902, in Pittsburgh, Pennsylvania, David J. McDonald, unlike most union leaders, came to organized labor with superb academic credentials. He pursued theater studies at Duquesne University and graduated in 1932 from the Carnegie Institute of Technology. In later years McDonald earned a doctorate in industrial relations from St. Martin's College. Long before he earned his baccalaureate degree, how-

ever, McDonald commenced a career in the union movement. In 1923 Philip Murray, the vice-president of the United Mine Workers of America and future president of the United Steelworkers of America (USWA), tapped McDonald as his private secretary. Those ties would remain until Murray's death in 1952.

McDonald's closeness to Murray put him on the front lines of new developments in industrial unionism in the 1930s. When presidential backing and federal legislation emboldened John L. Lewis's Congress of Industrial Organizations (CIO) to recruit laborers in the mass production industries, McDonald went with Murray to serve as secretary-treasurer in the CIO-sponsored Steel Workers Organizing Committee (SWOC) and remained

United Steelworkers general counsel Arthur Goldberg, looking on as United Steelworkers president David J. McDonald and U.S. Steel vice-president John Stephens sign a wage agreement, 29 June 1954

in that position when the SWOC formed the USWA in 1942. A decade later Murray died, and the USWA, without a rank-and-file election, chose McDonald to succeed him as president on March 11, 1953.

McDonald faced the task of negotiating labor contracts which would ensure a rising standard of living for USWA members. While McDonald and United States Steel Corporation president Benjamin Fairless adopted cooperative postures to reduce confrontations, the USWA relied on its option to strike to win demands from steel employers. Despite early USWA successes in wage negotiations, a major strike in 1959 showed that McDonald, far from "cozying up" to industry leaders, was willing to fight for his 540,000 striking members. This 116-day strike resulted in the establishment of the Humans Relations Research Committee, which would arbitrate future labor-management conflicts.

Despite McDonald's national visibility as an important labor leader, he was vulnerable to grassroots unionists who viewed him as out of touch with the membership. It was well known that McDonald's union credentials had not been earned while laboring in a mill or foundry. His

rise in the SWOC and the USWA occurred because he was well placed within the bureaucracies of both organizations. Hence, Donald Rarick, a McKeesport, Pennsylvania, unionist, launched a serious challenge to his leadership in 1957. Although McDonald won the election with 404,172 votes, Rarick managed to garner 223,516 votes, an impressive showing. In 1964 USWA secretary-treasurer I. W. Abel challenged McDonald for the presidency. Losing 298,768 to 308,910 votes, McDonald relinquished his office in 1965 to the Canton, Ohio, steelworker. McDonald spent his retirement in Palms Springs, California, uninvolved with steel unionism. He died on August 8, 1979.

References:

Paul F. Clark, Peter Gottlieb, and Donald Kennedy, *Forging a Union of Steel: Philip Murray, SWOC, and the United Steelworkers* (Ithaca: New York State School of Industrial and Labor Relations, Cornell University, 1987);

John Herling, *Right to Challenge: People and Power in the Steelworkers Union* (New York: Harper & Row, 1972);

John P. Hoerr, *And the Wolf Finally Came: The Decline of the American Steel Industry* (Pittsburgh: University of Pittsburgh Press, 1988).

Archie J. McFarland

(July 2, 1883 – November 28, 1950)

by Elizabeth M. Nolin

Paeonian Springs, Virginia

CAREER: Bookkeeper, factory superintendent, Wheeling Corrugating Division (1905–1912), general superintendent, general manager, Portsmouth Works, Whitaker-Glessner Steel Company (1912–1923); general superintendent, La Belle Nailworks Division, Wheeling Steel Corporation (1923–1924); superintendent, Follansbee Steel Corporation (1924–1930); first vice-president (1930–1936), executive vice-president (1936–1940), president, Wheeling Steel Corporation (1940–1950).

Wheeling Steel Corporation remained the ninth-largest American steelmaker during Archie J. McFarland's tenure as its president in the 1940s. He guided Wheeling Steel through World War II as the company turned its finishing mills to mak-

ing items for the military. McFarland also made a start at preparing the company to grow and modernize in the postwar years, acquiring the Carnegie-Illinois Steel Company's plant in Mingo Junction, Ohio. Nonetheless, the company did not deviate from its traditionally conservative course during McFarland's presidency.

McFarland was born on July 2, 1883, on a farm in Pymatuning Township, Mercer County, Pennsylvania, the son of Walter R. and Ella Powers McFarland. In 1905, at the age of 22, he came to Wheeling, West Virginia, and worked as a bookkeeper and later a factory superintendent with the Wheeling Corrugating Division of Whitaker-Glessner Steel Company, a leading manufacturer in the production of corrugated steel for siding and roofing. During this period, he attended Wheeling Business School in the evenings.

In 1912 McFarland was transferred to the Portsmouth Division of Whitaker-Glessner as general superintendent; he later became general manager. At the Portsmouth Works, located 260 miles down the Ohio River from Wheeling, Whitaker-Glessner operated an open-hearth steelworks and various mills.

On July 2, 1920, Whitaker-Glessner, La Belle Nailworks, and Wheeling Steel & Iron Company merged to form the Wheeling Steel Corporation. For the first three years, Wheeling Steel acted as a holding company for the three original concerns. In 1923 McFarland was transferred to the La Belle Nailworks division of Wheeling Steel in Steubenville, Ohio, again as general superintendent. The Steubenville plant contained two blast furnaces, an open-hearth steelworks, sheet mills, plate mills, and a tube works; it became the company's main steelmaking facility. Between 1924 and

1930, McFarland worked as a superintendent with Follansbee Steel Corporation. He returned to Wheeling Steel in 1930 as first vice-president; in 1936 he was named executive vice-president, and in 1940 McFarland became the third president of Wheeling Steel. He retained this position until his death on November 28, 1950.

Publication:
"Imperfect Steel, or Heat Treating?" *Iron Trade Review,* 65 (July 31, 1919): 288–290.

Reference:
Earl Chapin May, *Principio to Wheeling, 1715–1945: A Pageant in Iron and Steel* (New York: Harper, 1945).

Archives:
Wheeling Steel Corporation *Annual Reports* are located in the West Virginia Collection, West Virginia University.

McLouth Steel Products Corporation
by Leonard H. Lynn
Case Western Reserve University

McLouth Steel Products, formerly McLouth Steel Corporation, is a medium-sized, integrated steelmaker that produces flat-rolled carbon steel. The company is noted for its pioneering role in the development of the basic-oxygen steelmaking process and continuous casting, two of the most important post–World War II advancements in steel production. McLouth was the first company in the United States (and fourth in the world) to introduce the basic-oxygen steelmaking process, and it was the first major U.S. steelmaker to adopt continuous casting for all of its steel production.

McLouth Steel was established in 1934 by Donald B. McLouth, a Detroit scrap dealer. Originally in the business of rerolling steel slabs purchased from other companies, McLouth Steel acquired government-surplus electric furnaces after World War II and started making steel from scrap. The company's location in Trenton, Michigan, near Detroit, tied its fate to that of the automobile industry. During the mid to late 1940s McLouth pros-

pered with the postwar boom in automobile sales. The company increased its assets from just over $6 million in 1945 to well over $50 million in 1951. The demand for steel from the automakers continued to be very high during the early 1950s, but a shortage of scrap both increased McLouth's costs and restricted its potential for further growth. Earnings dropped in 1951 and 1952. In late 1952 McLouth decided to reduce its dependence on scrap by becoming an integrated steelmaker. As originally planned, this was to entail building a blast furnace to smelt ore into molten iron and Bessemer converters to produce steel from the iron. The $105 million expansion program was budgeted at roughly three times McLouth's total capitalization at the time. General Motors purchased some $25 million worth of McLouth stock and agreed to buy at least 5 percent of its steel from McLouth, making it possible for McLouth to obtain the rest of the financing necessary for the expansion.

As planning for the expansion continued in 1953, McLouth's vice-president in charge of operations heard about the new basic-oxygen steel-making process which had recently come into operation in Austria. This new technology offered advantages over the Bessemer process with regard to steel quality and pollution control. The basic-oxygen steelmaking process, however, had only been used on a small scale, and considerable uncertainty about its viability remained.

McLouth Steel successfully introduced the new technology and prospered into the mid 1960s, when it again was faced with a major investment decision. Requiring new slabbing capacity, the company chose once again to invest a new, relatively untried technology. In 1967 McLouth began a program to invest more than $100 million in continuous slab casters. In becoming the first major integrated steelmaker to base all of its production on continuous casting, however, McLouth suffered some 40 straight months without profits; labor problems also contributed to the losses. In 1972 the company appeared to have re-bounded, but it soon faced a sharp reduction in demand for steel by the automakers and increased competition from imports. In December 1981, McLouth began operating under Chapter 11 of the Federal Bankruptcy Act. Its assets were acquired in November 1982 by a group headed by Cyrus Tang who kept the company in operation. Losses mounted again after a few years, however, resulting in necessary restructuring. In 1988 McLouth employees took ownership of 85 percent of the company under an employee stock ownership plan.

References:

William T. Hogan, S.J., *Steel in the United States: Restructuring to Compete* (Lexington, Mass.: Lexington Books, 1971);

Leonard H. Lynn, *How Japan Innovates: A Comparison with the U.S. in the Case of Oxygen Steelmaking* (Boulder: Westview Press, 1982);

Wickham Skinner and David C. D. Rogers, *Manufacturing Policy in the Steel Industry* (Homewood, Ill.: Richard D. Irwin, 1970).

Carl August Meissner

(September 29, 1859 – October 12, 1930)

by Robert Casey

Henry Ford Museum & Greenfield Village

Career: Chemist, assistant superintendent, Joliet Steel Company (1880–1881); chemist, assistant superintendent, Brier Hill Iron & Coal Company (1881–1884); chief chemist, Joliet Steel Company (1884–1886); general manager, Sterling Iron and Railway Company (1886–1889); vice-president, general manager, Vanderbilt Steel and Iron Company (1889–1895); general manager, Londonderry Iron Company (1895–1899); manager of the mineral department, acting assistant general superintendent, Dominion Iron and Steel Company (1899–1904); assistant to the executive, Canada Iron and Steel Company and Londonderry Iron and Mining Company (1904–1905); associate to the president's assistant, chairman of principal operating committees, United States Steel Corporation (1905–1930).

During a fifty-year career in the steel industry, Carl August Meissner became one of North America's leading experts on coke and pig-iron production. He was also an influential advocate of technological innovation and economies of scale.

Meissner was born September 29, 1859, on Staten Island, New York, and was educated in local public schools before entering the Columbia University School of Mines. Graduating with a degree in chemistry in 1880, Meissner joined the Joliet Steel Company as a chemist. For the next 25 years he held positions of increasing responsibility in mills all across North America. In 1881 he became chemist at the Brier Hill Iron & Coal Company in Youngstown, Ohio, but returned to Joliet as chief chemist in 1884. Stints as general manager of the Sterling Iron and Railway Company at Sterlington, New York, and vice-president and general manager of the Vanderbilt Steel and Iron Company in Birmingham, Alabama, followed. Between 1895 and 1905

303

Carl August Meissner

Meissner held high management positions at several Canadian iron and steel plants, including the Londonderry Iron Company, Ltd., at Nova Scotia; Dominion Iron and Steel Company, Cape Breton Island; and Canada Iron and Steel Company and Londonderry Iron and Mining Company at Montreal.

These years were a time of rapid progress in blast furnace and coke oven design. In 1880 the best American blast furnaces produced about 90 tons of iron per day. By 1905, blast-furnace production was five times as great. Larger furnaces, increased blowing rates, and mechanized raw-material handling made possible these improvements in productivity. The introduction of by-product coke ovens to the American steel industry further increased plant efficiency — the new ovens producing a higher quality coke than had the traditional beehive oven. Meissner became thoroughly versed in all these developments.

Having accumulated 25 years of expertise in the technology of iron and coke production, Meissner went to work for the United States Steel Corporation, becoming chairman of its committees on blast furnaces, open hearths, coke, limestone, and coal washing. He was instrumental in U.S. Steel's first installation of by-product coke ovens at Joliet in 1906, and in the construction of the plant at the Gary Works in 1911. His efforts at blast-furnace improvement culminated in the construction of the No. 2 furnace at Youngstown. Put into operation in 1929, No. 2's 25-foot hearth and 1000-ton-per-day capacity made it the largest furnace in the country. It exemplified Meissner's belief in economies of scale and became the model for other large furnaces.

Meissner married Clara Ayer on November 8, 1882, and they had seven children, two of whom died in infancy. Two of his sons, Charles Roebling and James August, also became steel mill engineers. Carl August Meissner died in Steubenville, Ohio, on October 12, 1930, while attending a meeting of the American Iron and Steel Institute. He was a leading member of the group of engineers who transformed ironmaking and coke production empirical activities to relatively controlled processes based upon a theoretical understanding of the phenomena involved.

Publications:
"The Modern By-product Coke Oven: A Practical Method of Conserving Our Coal Supply," paper read at the fourth general meeting of the American Iron and Steel Institute, May 23, 1913 (Washington, D.C.: United States Government Printing Office, 1913);
"Some Recent Developments in By-product Coke Ovens," *American Iron and Steel Institute Yearbook* (1914): 234–237;
"The Future of Oxygen Enrichment of Air in Metallurgical Operations," *American Iron and Steel Institute Yearbook* (1920): 135–138;
Robert Forsythe, *The Blast Furnace and the Manufacture of Pig Iron: An Elementary Treatise for the Use of the Metallurgical Student and the Furnaceman,* third edition, completely revised by C. A. Meissner and J. A. Mohr (New York: U.P.C. Book Company, 1922);
"Recent Observations on Some European Iron and Steel Conditions," *American Iron and Steel Institute Yearbook* (1928): 399–434; also in *Blast Furnace and Steel Plant,* 16 (November, December 1928): 1433–1435, 1574–1578; abstract as "Blast Furnace and Open Hearth Practice," *Blast Furnace and Steel Plant,* 17 (February 1929): 289–290;
"Use of Blast-Furnace and Coke Oven Gases," *Blast Furnace and Steel Plant,* 20 (June, July 1932): 506–509, 590–592.

References:
American Institute of Mining, Metallurgical, and Petroleum Engineers, *History of Iron and Steelmaking in the United States* (New York: American Institute

of Mining, Metallurgical, and Petroleum Engineers, 1961);

William T. Hogan, S.J., *An Economic History of the Iron and Steel Industry in the United States*, 5 volumes (Lexington, Mass.: Lexington Books, 1971).

Midvale Steel & Ordnance Company
by Gerald G. Eggert
Pennsylvania State University

The Midvale Steel & Ordnance Company was a highly profitable combine put together during World War I. For a brief period, its capacity and output was second only to that of the United States Steel Corporation. Although rumored to be merely a speculative war venture, the firm made serious efforts to carve out a substantial peacetime commercial market. That venture failed, and in 1923 Bethlehem Steel took over Midvale Steel & Ordnance's major properties except those at Nicetown, Pennsylvania.

The firm was formed by a syndicate of prominent manufacturers and bankers: William E. Corey, president of U.S. Steel from 1903 to 1911; Ambrose Monell, president of International Nickel; Samuel Vauclain, president of Baldwin Locomotive; New York investors Percy A. Rockefeller and M. Hartley Dodge; and Frank A. Vanderlip, president of National City Bank of New York. Midvale Steel was incorporated October 15, 1915, under Delaware law as a holding company with capital of $100 million. Syndicate members and other leading manufacturers and bankers constituted the board of directors chaired by Corey. Corey, in turn, recruited Alva C. Dinkey, then president of Carnegie Steel — a subsidiary of U.S. Steel — to be president the new firm and William B. Dickson, first vice-president of U.S. Steel from 1902 to 1911, to be vice-president and treasurer. Both were experienced steel producers who had risen with Corey through the ranks at Carnegie Steel.

Issuing $75 million worth of stock, the company acquired properties to produce a full range of war matériel: armor plate, big guns, shells, and rifles. It purchased the existing Midvale Steel Company of Nicetown, Pennsylvania, famous for specialized alloy steels, for $22 million. For 30 years the Nicetown plant had supplied armor plate to the U.S. government. Its 5500 employees also produced castings, finished guns of all calibers, ordnance material, marine engines, tool and automobile steel, axles, and pressed, steel-tired wheels. During World War I, although the firm sold their products to the American government, it declined to sell them to the warring European countries — the company's president having two daughters, one married to an Englishman and the other to a German.

The Remington Arms Company at Eddystone, Pennsylvania, a small arms producer, cost $20 million. It came with a contract for 2 million Enfield rifles, on which an expected profit of $10 each would by itself cover the purchase price. Worth Brothers Company and its copartner, the Coatesville Rolling Mill Company, required $18.5 million. Organized in 1896, its four-plate mills had successfully competed with Carnegie Steel for 20 years. The firm also had 15 open-hearth furnaces — six with 40-ton and nine with 50-ton capacities — two modern 500-ton blast furnaces, and facilities to produce 25,000 tons of boiler tube skelp annually. With a third of the 700-acre site occupied, there was plenty of room for expansion.

Offsetting rumors that it was incorporated only as a war venture, Midvale acquired 300 million tons of nickel-rich Cuban ore: a sixth of that amount would have allowed it to maintain current rates of operation, and all of it gave Midvale a reserve one-third as great as that of U.S. Steel. Taking Wall Street by surprise, in February 1916 Midvale purchased the Cambria Steel Company of Johnstown, Pennsylvania, for $71.9 million. This necessitated issuing the $25 million balance of its authorized stock. Cambria produced rails, structural shapes, plates, agricultural steel, spring steel, wire, and freight cars and had extensive holdings in iron and coal. Midvale, combined with Cambria, had a capacity of over 2 million tons per year, second only to U.S. Steel's 20 million tons. Clearly, Midvale sought a major role in the postwar commercial market.

In May 1916 Midvale purchased the Diamond State Steel Company of Wilmington, Delaware, which had been in receivership since 1904. Built originally in 1868 and rebuilt after a fire in 1891, the facility's puddling mills, heating furnaces, and bar mills produced steel that was shipped to Nicetown for use there. High operating costs led Midvale to close Diamond State in February 1918.

Midvale Steel & Ordnance made notable contributions to the war effort. Its Nicetown works produced 96 eight-inch howitzers and 110,300 12-inch shells for Great Britain; over 4,200 forged jackets for large guns for France; and 25,000 embrasured entrenching tools for Canada. Contracts with the U.S. government included 10,000 tons of steel for rifles, bayonets, and machine guns; 15,000 gross tons of heat-treated alloy steel for military cars and trucks; over 19,000 tons of heavy armor for the navy plus additional light armor for armored trucks; nearly 32,000 armor-piercing shells; completely finished or major forgings for 2,500 3-inch to 16-inch guns; and air flasks for torpedoes. The same plant also turned out 1,125 tons of high-speed tool steel for Midvale's own use and for other munitions makers; 1,200 gross tons of marine forgings; and 14,400 tons of steel wheels and tires for American and French railways. The Eddystone Works completed nearly 1.3 million Enfield rifles by the end of 1918. Coatesville produced ship and boiler plates, boiler tubes, and 380,000 shell forgings. Cambria, though chiefly a commercial plant, produced ship plates, shell steel, rails, gun mounts and trails, and barbed wire. Together, Coatesville and Cambria supplied 250,000 tons of plate and structural steel to private shipyards and 350,000 tons to the navy, the army, and locomotive builders holding priority government orders.

During the war Midvale enjoyed profits of nearly $95 million on net earnings of $153 million, while paying out $138.4 million in wages and $24 million (12 percent per annum) in dividends and accumulating a surplus of over $53.7 million. A sharp decline set in with the Armistice and postwar disarmament. Net earnings dropped to $16.7 million in 1919, rose to $22.2 million in 1920, then turned to substantial deficits in 1921 and 1922. Dividends fell from 8 percent in 1920 to 4 percent in January 1921; no dividends were paid that April. Midvale considered merging into a super combine with Lackawanna, Youngstown, Republic, Inland, Briar Hill, and the Sheet & Tube Company of America. A more modest plan of purchasing Inland Steel jointly with Republic Steel was shot down by the Federal Trade Commission. Finally, Bethlehem Steel purchased Midvale's Coatesville and Cambria plants for $95 million in 1923. The Nicetown plant continued operating independently under its original name, Midvale Steel Company.

Reference:

Gerald G. Eggert, *Steelmasters and Labor Reform, 1886–1923* (Pittsburgh, 1981).

Archives:

The Bethlehem Steel Corporation papers are at the Hagley Museum and Library, Wilmington, Delaware.

The diary and personal business papers of William Brown Dickson, held in Special Collections, Pattee Library, Pennsylvania State University, University Park, Pennsylvania, provide much on Midvale Steel and Ordnance.

Marvin Miller

(April 14, 1917 –)

by John P. Hoerr

Teaneck, New Jersey

CAREER: Clerk, U.S. Treasury Department (1939–1940); social investigator, New York City Department of Welfare (1940–1942); economist, War Production Board (1942–1943); economist, hearing officer, War Labor Board and National Wage Stabilization Board (1943–1946); commissioner, U.S. Conciliation Service (1946–1947); staff representative, International Association of Machinists (1947–1950); economist, associate director of research (1950–1960), chief economist and assistant to the president, United Steelworkers of America (1960–1966); executive director, Major League Baseball Players Association (1966–1983).

Marvin Miller, a top-level union negotiator who designed innovative wage and benefit plans in the steel industry, was born on April 14, 1917, in New York City. The son of a clothing salesman and a schoolteacher, he attended Miami University in Oxford, Ohio, and graduated from New York University in 1938 with a bachelor's degree in economics.

Miller entered federal service in the U.S. Treasury Department in 1939 and later specialized in industrial and labor-relations issues as an economist with various federal agencies during World War II. He served as a hearing officer for contractual disputes between labor and management with the War Labor Board and its successor, the National Wage Stabilization Board. After the war he helped rejuvenate the nearly defunct U.S. Conciliation Service (forerunner of the U.S. Mediation & Conciliation Service), serving as a training officer for mediators.

Having gained a broad knowledge of labor relations, Miller joined the labor movement as an economist, organizer, and publicist for the International Association of Machinists in 1948. Two years later he transferred to the United Steelworkers of America (USWA) as an economist in the research department. For much of the next decade he developed research data for steel bargaining and helped negotiate labor contracts in the can, aluminum, and copper industries. Miller became known as a "hard bargainer" with an ability to translate economic and social needs into contractual provisions.

In 1956 Miller was one of a team of union technicians who developed a new employee benefit known as supplemental unemployment benefits (SUB). Growing out of organized labor's decade-old demand for a "guaranteed annual wage," SUB provided a weekly layoff stipend for up to 52 weeks, in addition to government-paid unemployment compensation. Other key members of the team were Murray Lattimer, a consultant who originally designed the plan, and John Tomayko, director of the USWA Pension & Insurance Department. The union first negotiated SUB in the can-manufacturing industry in early 1956 and extended it later that year to the steel and aluminum industries.

Miller was promoted to assistant to USWA president David J. McDonald in 1960. Among the most important of his new duties was the coordination of union activities on the Human Relations Committee, a labor-management group. Formed after the 116-day strike of 1959 to solve troubling issues, the committee was largely responsible for negotiating peaceful steel-labor agreements in 1962 and 1963. In the latter year the committee developed a novel vacation plan under which senior employees in each company's workforce received a 13-week vacation once every five years.

Miller also played a key role in designing an innovative method of linking wages and productivity at Kaiser Steel Company. After breaking away from steel's industrywide bargaining group during the 1959 strike, Kaiser negotiated separately with the USWA. In 1963 the two sides established the Kaiser Long Range Sharing Plan, which tied group bonuses to output in an effort to gain worker commitment to technological change. Miller designed the plan in part with the hope of implementing it elsewhere in the steel industry. But other companies refused to consider the idea. Although the Kaiser plan experienced difficulties, it remained in effect until Kaiser sold its Fontana, California, plant in 1984.

By 1965 Miller was one of the best known of the USWA's widely heralded staff of skilled technicians. Like many other staff members, he supported McDonald in the split that developed when I. W. Abel challenged and defeated McDonald in the union's presidential election. Nevertheless, Abel recognized Miller's expertise by naming him to a top-level committee which negotiated a new steel agreement in 1965. This was Miller's last major role in the union. The fledgling Major League Baseball Players Association offered Miller the post of executive director. He resigned from the USWA in July 1966 and headed the players association until his retirement in 1983. Under his leadership the baseball union won a free agency provision, allowing certain players the right to move to other teams, and raised average salaries, making baseball players the highest paid professional athletes.

Thomas E. Millsop

(December 4, 1898 – September 12, 1967)

by Alec Kirby

George Washington University

CAREER: President, Weirton Steel Division, National Steel Corporation (1936–1954); president (1954–1957), chief executive officer (1957–1961), chairman of the board, National Steel Corporation (1961–1967).

Thomas E. Millsop, an executive with the National Steel Corporation, was born on December 4, 1898. As the son of poor Scottish immigrants, Millsop permanently abandoned his formal education after the eighth grade and took a job in the mill where his father was employed. Millsop earned ten cents an hour as an open-hearth laborer.

During World War I, Millsop joined the U.S. Marine Corps and became an aviator, piloting a Curtiss biplane from a base in the Azores. After the war he was discharged from military service as a captain and accepted a position as purchasing agent for the Standard Tank Company. His real interest, he quickly discovered, was selling, and he took every opportunity at Standard to learn sales techniques. Millsop was soon promoted to production manager for Standard, a position that did not satisfy his interest in sales. Resigning from Standard, Millsop sold metal scrap in Cleveland. This move enabled him to establish contact with E. T. Weir. In 1927 Millsop persuaded Weir to give him a "trial" as a salesman for the Weirton Steel Company, which was in the midst of a tremendous expansion. Weirton had just completed the installation of new open-hearth furnaces, which gave the company an ingot capacity of over 1 million tons — more than doubling Weirton's 1921 capacity. The company in 1927 had also acquired a major share in the Continental Can Company, to which it would provide tinplate. In the year after Millsop joined the firm, the company's net sales skyrocketed by over 31 percent.

Millsop proved successful as a salesman for the company: he booked an order worth $1 million within his first week on the job. From this auspicious beginning, Millsop blasted his way

Thomas E. Millsop

through the hierarchy of Weirton Steel, which in 1929 merged with the Great Lakes Steel Corporation and elements of the M. A. Hanna Company to create the National Steel Corporation. Weirton's product diversity, Great Lakes Steel's proximity to the Detroit steel market, and Hanna's ore resources combined to make a formidable new organization. The strength of National Steel was apparent as the corporation weathered the Great Depression without once running in the red — the only major steel producer to show a profit every year during the 1930s.

Within the Weirton Steel division, Millsop rose to the presidency in 1936. His management style reflected his sales background. He rejected arbitrary exercises of power, drawing on a collegial, team-oriented approach. He repeatedly asserted his conviction that no single individual "can make a success of an enterprise."

In 1954 Millsop replaced the retiring George Fink as president of National Steel during a critical period for the company. Anticipating a surging market for automotive steel, National's Great Lakes Steel division had accumulated a large inventory, which remained unsold. Overall, National's sales fell 24 percent while net income dropped 38 percent. Millsop, aided by a reviving U.S. economy, brought National Steel back to a more secure financial footing in 1955 as he reorganized the management of Great Lakes Steel. By 1956 sales stood at a record $664 million, with net income at a solid $52.5 million.

On April 24, 1957, E. T. Weir announced his retirement as chief executive officer of National Steel and recommended Millsop as a replacement. Assuming the post, Millsop operated under the shadow of George Humphrey, who became the chairman of the board upon his retirement as U.S. secretary of the treasury. Operating in the shadow of the more prominent Humphrey, however, suited Millsop's style of management, which was low-key. In February 1961 Millsop succeeded Humphrey as chairman.

Throughout his life, Millsop remained active in the Republican party, serving two terms as mayor of the town of Weirton. His domestic life was marred by the death of his first wife, Lauretta Brunswick, in a 1947 automobile accident. His second wife, Eleanor Marwitz, a stage and radio entertainer in Pittsburgh, divorced him in 1954. He died of a heart attack on September 12, 1967, survived by his third wife, Frances, and two daughters.

References:
"Strong Man in Shirtsleeves," *Fortune,* 57 (October 1957): 154–155;
"Weir Retiring, Proposing Millsop as Chief Officer of National Steel," *New York Times,* April 25, 1957, pp. 41, 49;
William T. Hogan, S. J., *An Economic History of the Iron and Steel Industry in the United States,* 5 volumes (Lexington, Mass.: Lexington Books, 1971).

Minimills

by Robert W. Crandall
The Brookings Institution

Traditionally, the production of carbon steel has involved the smelting of iron ore in blast furnaces and the refining of the resultant pig iron in a steel furnace using Bessemer, open-hearth, or basic-oxygen (LD) technology. Steelmakers using these technologies are generally referred to as *integrated* steel producers.

In recent decades, however, a different technology has been introduced that has allowed smaller firms to compete with the integrated steel producers. Rather than starting with iron ore, electric furnaces use electricity to melt iron and steel scrap and refine it into carbon steel. This technology has long been used to produce high-quality alloy and stainless steel and has even been used by integrated companies to augment their output of carbon steel. Indeed, the United States Steel Corporation installed one of the first electric furnaces in the United States at its Gary Works in 1909. But it has been the smaller steel firms — known as minimill producers — that have been responsible for the dramatic growth of electric steelmaking in recent years. Although these companies typically produce from 50,000 to 500,000 net tons of steel per year — far less than their integrated rivals who produce as much as 6 to 10 million tons in a single plant — they account for a large share of the steel market.

Minimills have assumed an important role in the carbon steel industry only in the past 30 years. Prior to 1960, electric-furnace steelmaking in the major industrialized countries was largely confined to the refining of higher-grade alloy and stainless steels. Italy, with its low-cost electric power, was an exception; many small companies in the Brescia region built electric furnaces in the early twentieth century, producing lower-grade bar products for the construction industry.

In the 1930s the Northwestern Barbed Wire Company, a U.S. wire producer in the Midwest, erected a mill to supply its wire-drawing machines with carbon steel. Further motivated by the desire to produce its own carbon steel rather than having to purchase it from the large steel companies, Northwestern built two small electric furnaces in Sterling, Illinois. In time Northwestern not only produced steel rods for its own consumption, but it began to roll bars and small-to-medium structural shapes for sale on the merchant market. Subsequently the company changed its name to Northwestern Steel and Wire, and became the first of the smaller carbon-steel electric furnace producers.

Northwestern's venture into steelmaking was not imitated by others until well after World War II. Even in the late 1950s, the large integrated companies accounted for about 98 percent of all carbon steel produced in the United States. At that time, however, the modern minimill began to develop. Stimulated by advances in electric furnace and continuous-casting technology, steel fabricators such as Florida Steel and the Nucor Corporation began to erect their own small steelmaking plants, which proved to be much more efficient than the integrated firms' facilities in producing reinforcing bars and merchant bar products.

By 1970 the number of minimills in the United States had grown substantially, but they still accounted for only about 6 percent of carbon-steel production. As steel demand stagnated and scrap prices fell in the late 1970s, however, the minimills continued to expand and prosper, spurred by much lower labor and capital costs.

At first minimills were confined to the production of the lower quality grades of steel and smaller-diameter steel shapes. Their lower costs forced the larger integrated companies virtually to abandon these product lines. During the 1970s the minimills began to expand into new steel product lines, such as higher-quality wire rod and medium-size structural shapes, and they controlled nearly 15 percent of the market by the end of the decade. By the early 1980s more than 60 minimills produced carbon steel in the United States and Canada. Most of these plants are owned by small private firms who operate only one facility. Yet there is an increasing number of multiplant firms, such as Nucor, North Star, Birmingham, and Co-Steel.

By the early 1980s the North American minimills dominated the domestic reinforcing bar, wire rod, and merchant bar steel markets, accounting for about 20 percent of all steel consumption. As electric furnace technology progressed, the typical minimill increased its production from an average of 200,000 to nearly 600,000 tons of raw steel per year. By the end of the 1980s one minimill — Chaparral — was producing at an annual rate of more than 1 million tons per year from a single plant in Texas. Flat-rolled products — sheets, strip, and plates — however, account for more than half of the North American market, a market which the minimills were unable to enter until the late 1980s due to two factors: steel quality requirements and the large scale required to support a single, efficient flat-rolling operation. Techniques involving the separation of the scrap, however, began to improve, as did steel-refining techniques. The minimills thus began to explore higher-quality markets. In 1988 Nucor announced construction of a minimill to produce hot- and cold-rolled sheets in Indiana, using an electric furnace and a new European thin-slab casting technology. This new plant was to produce between 800,000 and 1.2 million tons of steel per year in a continuous process that feeds approximately 1.5-inch slabs into a continuous hot-strip mill.

The Nucor flat-rolled minimill opened in 1989 and was operating near capacity by the end of 1990, although substantial questions remain as to the viability of the new thin-slab casting technology. At the end of 1990 Nucor was the only minimill producer using this technology, but other firms are poised to follow if it proves successful. Nucor has announced its intention to build at least one more such plant and perhaps a third.

The United States and Canada have become leaders in the development of steel minimills because of their relatively low scrap and electric energy prices and their willingness to allow competitive entry into steelmaking. By 1990 North American minimills were producing about 25 percent of all domestic carbon steel, a percentage that continues to increase. Minimills have also begun to invade the market for large structural shapes, previously dominated by large steel companies. Most other developed countries have lagged behind the United States and Canada in minimill development because of their higher scrap and/or energy prices — and because they face opposition from integrated steel producers who have suffered through a prolonged slump since the mid 1970s. Of these countries only Japan, Spain, and Italy have large minimill sectors.

Since 1980 few new integrated facilities have been built in developed countries, in part due to the prohibitive cost of new blast furnaces. Meanwhile, minimills continue to increase their production levels, with plants averaging between 500,000 and 1 million tons of steel per year. Continued growth, however, will ultimately depend on the development of new sources of iron other than scrap.

Reference:
Donald F. Barnett and Robert W. Crandall, *Up from the Ashes: The Rise of the Steel Minimill in the United States* (Washington, D.C.: The Brookings Institution, 1986).

Edwin Earl Moore

(January 20, 1894 – April 17, 1965)

by Robert Casey

Henry Ford Museum & Greenfield Village

CAREER: Assistant manager, Gary Works (1927–1932), assistant to operating vice-president, Illinois Steel Company (1932–1935); general superintendent, South Works, Carnegie-Illinois Steel Company (1935–1937); general superintendent, Gary Works (1937–1940), vice-president, industrial relations, Carnegie-Illinois Steel Company (1940–1950); vice-president, Industrial Relations Administration (1951–1953), vice-president and assistant to the president, United States Steel Corporation (1953–1959).

Edwin Earl Moore was one of the "home grown" group of executives who guided the United States Steel Corporation through the Depression, World War II, and into the era of postwar expansion.

Moore was born in Indianapolis on January 20, 1894. After attending Indiana University from 1913 to 1916, he served as quartermaster sergeant in the Indiana National Guard on the Mexican border. His unit was inducted into the U.S. Army in 1917, and Moore went to France with the Rainbow Division. Moore became a captain and served as a civil-court judge and a civil administrator in the Army of Occupation.

Mustered out of the service in 1919, Moore returned to Indiana and was hired as a construction machinist at the Gary Mills of the American Sheet and Tin Plate Company, a subsidiary of U.S. Steel. He spent the remainder of his career with the corporation.

A year after arriving at Gary he was made foreman of the machine shop, and in 1923 he was moved to the Pittsburgh office to do special work. Between 1922 and 1926 Moore served as acting manager of American Sheet and Tin Plate's Roll

and Machine Works in Canton, Ohio, and assistant manager of Sheet and Tin's Shenango Works in New Castle, Pennsylvania.

In 1927 Moore returned to Gary as assistant manager in charge of construction for American Sheet and Tin's $12.6 million, four-high continuous hot-strip mill and 42-inch, four-high cold-reduction mill. He was promoted in 1932 to assistant to the operating vice-president of the Illinois Steel Company. When Illinois Steel was merged into Carnegie-Illinois Steel Corporation in 1935, Moore became general superintendent of its South Works plant in Chicago. In 1937, 18 years after hiring on as a machinist, Moore returned to Gary as general superintendent of the entire Gary Steel Works.

Shortly before America's entry into World War II, Moore was elected vice-president of the industrial relations for the Carnegie-Illinois Steel Corporation, with offices in Pittsburgh. In this capacity he was especially concerned with rehabilitating and returning to productive work employees who had been injured on the job and soldiers who had been disabled in combat. In 1951 Moore became vice-president of the Industrial Relations Administration for U.S. Steel. In 1953 he was appointed vice-president and assistant to the president of U.S. Steel, a post he held until his retirement in 1959.

Moore's professional memberships included the American Iron and Steel Institute, the Association of Iron and Steel Engineers, and the Engineering Society of Western Pennsylvania. He was a director and regional vice-president of the National Association of Manufacturers and a trustee of the Historical Society of Western Pennsylvania.

Moore married Lillian Seaney in 1919, and they had two children. He died April 17, 1965.

Publications:
"Rehabilitation," *Yearbook of the American Iron and Steel Institute* (1943): 315–320;
"Open Hearth Steel-making Process," *Metals and Alloys,* 19 (May 1944): 1165;

"The Selection and Training of Personnel," *American Iron and Steel Institute Regional Technical Papers* (1952): 21–32.

Reference:
William T. Hogan, S.J., *An Economic History of the Iron and Steel Industry in the United States,* 5 volumes (Lexington, Mass.: Lexington Books, 1971).

William Henry Moore

(October 25, 1848 – January 11, 1923)

by Elizabeth C. Sholes

Industrial Research Associates

and

Thomas E. Leary

Industrial Research Associates

CAREER: Attorney (1872–1882); partner, W. H. & J. H. Moore (1882–1917); financier (1889–1917); director, United States Steel Corporation (1889–1917); founder, Moore Group (1899–1917).

William Henry Moore was cofounder of the Moore Group, one of the largest and most powerful early twentieth-century industrial conglomerates. Its most spectacular success involved effecting a merger of a group of steel companies with J. P. Morgan's new United States Steel Corporation, satisfying Moore's personal goal of creating a giant steel merger.

Moore was born in Utica, New York, to a family with a New England colonial background. He attended seminary in Oneida, New York, the Cortland Academy in Homer, then Amherst College, although ill health prevented his graduation. He moved to Wisconsin and studied law as an apprentice to W. P. Bartlett in Eau Claire. Moore was admitted to the bar in 1872.

Moore left Bartlett's practice to join leading Chicago corporate attorney Edward A. Small. The firm was soon renamed Small & Moore and continued until Small's death in 1882. Moore then recruited his brother, James Hobart, as partner, and together they began one of Chicago's leading corporate law firms, with clients such as American Express and Vanderbilt Fast Freight. Moore became the firm's primary trial lawyer, building an

William Henry Moore

enviable reputation as an expert in case and statute law. It was this superior knowledge that netted him the nickname "Judge."

The brothers began their forays into mergers and acquisitions with the 1889 consolidation of the Diamond Match Company. They reorganized and recapitalized the firm, transforming a small $3-million Connecticut operation into an Illinois venture worth $11 million. They promoted its stock on rumors of foreign contracts and nonexistent technical improvements. The stock soared from 120 in January to 248 in May 1896. Inside dumping, however, ended the boom and caused a major market collapse that closed the Chicago Stock Exchange for three months.

The Moores lost $4 million but remained undaunted and moved directly into yet another merger. In 1890 the brothers had created New York Biscuit Company and in 1896, with the support of Armour and Pullman capital, launched a price war against American Biscuit that resulted in the latter company's capitulation and merger with Moore to form the National Biscuit Company, or Nabisco. Nabisco began with $55 million in capital and garnered control over 90 percent of the market.

Moore's success at Nabisco renewed demand for the firm's services, but Moore turned his attention to the steel industry. In slightly more than one year he organized three steel operations: American Tin Plate (1898), National Steel (1899), and American Steel Hoop (1899). Prefiguring contemporary practice, Moore's stocks were valued on anticipation of future earnings rather than intrinsic worth. Nevertheless, the stock was rapidly absorbed and oversubscribed, thus watering the value further.

During these ventures, Moore and his brother acquired two new partners from within the steel industry, and together the four became the Moore Group, known in many circles as the "Moore Gang" or "The Big Four from the Prairies." In May 1899, Moore and his partners attempted a takeover of the Carnegie Steel holdings by acquiring a 3-month, $1-million option on the stock of Henry Frick and Andrew Carnegie. They failed to generate the necessary $320 million, and the option expired with Carnegie pocketing $1 million.

Moore, however, continued acquiring steel properties, buying American Sheet Steel and American Can Company. Their new five-company conglomerate made the Moores top contenders in the industry. The Moores' adversarial relationship with Morgan quickly turned cooperative; Mor

gan's own run at Carnegie Steel made an alliance with the Moores appealing, and the Moore Group sold all five steel operations to the Wall Street financier at a substantial profit. William Moore, leader of the negotiations, was retained as a director of the new company, U.S. Steel, but further relations between Moore and Morgan were strained. Morgan had no interest in using his steel company for speculation and curtailed Moore's opportunities in that regard. In short order, Moore turned his attention to more fertile grounds: railroads.

As in steel, certain American railroads were tightly controlled. Morgan directed eight railroads and had influence over a ninth. The remaining major railroads were divided among James J. Hill, the Rockefellers, the Harrimans, and Jay Gould. Only the Chicago, Rock Island & Pacific was independent. It was a well-managed, profitable road with dispersed stock holdings. Through buyouts the Moores managed to consolidate control, and Moore had himself and partner Daniel G. Reid added to the board of directors. They then placed four allies on the seven-member executive committee, and Moore's dominance was complete.

The Moore Group bought two more systems and leased the important Burlington, Cedar Rapids & Northern line, which grew from 3,600 to 15,000 miles, raising its value from $116 to $900 million. Moore became a director of the Delaware, Lackawanna and Western as well as the St. Louis & San Francisco and joined the board of the Harrimans' Chicago & Alton. By 1910 Moore had added the latter two lines plus the Lehigh Valley railroad to his own holdings and was vying for the Denver & Rio Grande, the Wabash, and the Missouri Pacific.

Moore's use of overcapitalization proved to be his undoing. The Rock Island's stock value was nearly tripled in 1902 in a massive stock watering scheme. In 1907 the Chicago & Alton was also made insolvent by overcapitalization and had to be sold at a total loss. In 1909 the St. Louis & San Francisco followed suit and cost the Moores $20 million in resale to the original owners. The Moore Group's pyramid structure proved too great a drain on operating assets, and by 1914 Moore was in receivership.

In June 1917, when receivership was complete, the Interstate Commerce Commission ousted Moore from control over any railroad activities. Throughout this period Moore had continued as a director of U.S. Steel, National Biscuit, the First National Bank of New York, and other companies. Following what has been described as "the most blatant and thorough piece of manipulation and looting in the history of the railroad industry," and with control over his railroad empire dissolved, Moore withdrew from all active business.

In retirement, Moore kept his memberships in several notable clubs, including the Metropolitan and Union, the New York Yacht Club, the Myopia Hunt Club, and the Racquet and Tennis Club. He was a renowned and avid breeder of harness horses, an interest that led to his investment in Madison Square Garden as an arena for equestrian shows.

He had married Ada W. Small, daughter of his first law partner, and they had three sons: Hobart, Edward, and Paul. William Moore died of heart disease at his home on January 11, 1923.

References:
Matthew Josephson, *The Robber Barons* (New York: Harcourt, Brace, 1934);
Gabriel Kolko, *Triumph of Conservatism* (Chicago: Quadrangle Books, 1967);
Naomi R. Lamoreaux, *The Great Merger Movement in American Business, 1895–1904* (New York: Cambridge University Press, 1985);
Mark S. Mizruchi, *The American Corporate Network, 1904–1974* (Beverly Hills, Cal.: Sage, 1982).

The Moore Group

by Thomas E. Leary
Industrial Research Associates
and
Elizabeth C. Sholes
Industrial Research Associates

One of the largest and most powerful industrial conglomerates at the turn of the century, the Moore Group, cofounded by William Henry Moore, worked between 1898 and 1900 to create a large steel merger.

The companies organized under the Moore umbrella included American Tin Plate, National Steel, American Steel Hoop, and American Sheet Steel. These mergers were part of a larger movement toward restructuring branches of the steel industry to insulate investors and managers from the rigors of competition. The alliances within the Moore combine also represented an attempt to emulate the vertically integrated structure of Carnegie and Federal Steel.

Small companies absorbed by the Moore Group shared common characteristics. Their manufacturing techniques in such flat-rolled lines as tinplate, sheets, and narrow strip lagged behind the technology used in the production of heavier products. Their skilled labor also remained indispensable, and proprietors usually had to cope with wage scales and work rules negotiated by the Amalgamated Association of Iron, Steel, and Tin

Workers. In the wake of the panic of 1893 such firms also faced virulent price competition. Conventional collusion practices of the time — such as companies cooperating in "pools" to decrease competition — proved ineffectual. By orchestrating the operations of different companies, the Moore Group kept pace with trends toward greater vertical integration. Larger mills in the Pittsburgh and Chicago areas were already discerning the value of linking raw materials production, steelmaking, and finishing operations.

William Moore's first plunge into steel mergers occurred in the tinplate sector. Measured in pounds, output of tin- and terneplate exploded from 42,119,192 in 1892 to 732,290,285 in 1898. As new plants flooded the markets, prices destabilized. Led by Daniel G. Reid and Warner Arms, the manufacturers sought to consolidate their operations and reduce overcapacity by closing redundant works. Harmonizing the interests of some forty competitors proved difficult, however. In September 1898 Reid and his colleagues turned to Moore, who undertook the delicate task of evaluating and purchasing individual plants. The result of his negotiations was the American Tin Plate Company, organized in December 1898 under New Jersey's hospitable general incorporation law for holding companies.

The new consolidation controlled more than 90 percent of domestic tinplate production; its 38 plants were concentrated in Indiana, Ohio, and western Pennsylvania. Moore, James B. Dill, and other lawyers responsible for the Tin Plate consolidation took unusual pains to insure that investors' interests were paramount. The corporation's charter severely restricted the authority of management to incur bonded debt or create other obligations that would infringe on the dividends of preferred stockholders.

Moore and his associates were propelled into the next phase of consolidation by the need to secure a dependable supply of semifinished steel for American Tin Plate. In February 1899 the formation of the National Steel Company brought together blast furnaces and steel works in the Mahoning, Shenango, and upper Ohio River valleys. Former proprietors of the principal firms — including William E. Reis of New Castle, Pennsylvania, and Henry Weir of Youngstown, Ohio — who survived the merger as National Steel officers, began to acquire raw material sources and transportation companies, thus linking the newly merged operations from mine to mill. National Steel's charter, like that of American Tin Plate, emphasized the prerogatives of stockholders over the powers of the company's officers and directors.

The formation and promotion of American Steel Hoop and American Sheet Steel in 1899–1900 followed a similar pattern. American Steel Hoop united nine makers of light strip products, merchant bars, cut nails, and iron skelp. The rationale underlying the merger included the centralized allocation of orders to previously-competing plants. These would specialize in particular sizes and shapes, reducing the need for expensive roll changes. The establishment of Na-tional Steel provided the impetus for a merger among unintegrated sheet producers. American Sheet Steel took in approximately 30 companies representing 70 percent of the capacity in this sector of the industry.

Operations of the Moore Group were closely coordinated but not legally affiliated. The firms possessed a common treasurer and joint office space in New York's Battery Park Building. A marked overlap among the various directors also ensured continuity: as the veteran industrialist John Stevenson, Jr., later observed, "We sold to the same old crowd that kept bobbing up." Moore, his brother James, William B. Leeds, William T. Graham, Frederick S. Wheeler, William E. Reis, and Daniel G. Reid served on the boards of all four holding companies.

Besides interlocking directorates, the Moore corporations shared a tendency toward overcapitalization. Moore's methods for enticing individual proprietors into the consolidation included liberal doses of common stock which, as even he noted, actually represented no tangible assets. The Stanley committee estimated the actual value of the four companies at about $67 million; the preferred stock alone amounted to $87 million, and total capitalization topped out at $194 million.

Moore's handiwork, with its noteworthy financial, technical, and organizational characteristics, was never tested over time. Beginning in 1901 the United States Steel Corporation absorbed the four corporations in yet another stage of the great merger movement.

References:

William T. Hogan, S.J., *An Economic History of the Iron and Steel Industry in the United States*, 5 volumes (Lexington, Mass.: Lexington Books, 1971);

"Iron and Steel Consolidations: Testimony Before the Federal Industrial Commission," *Iron Age*, 64 (October 19, 1899): 31–36;

Naomi R. Lamoreaux, *The Great Merger Movement in American Business, 1895–1904* (New York: Cambridge University Press, 1985).

Ben Moreell

(September 14, 1892 – July 30, 1978)

by John A. Heitmann

University of Dayton

CAREER: Resident and designing engineer, department of sewers, engineer, city of Saint Louis (1913–1917); lieutenant, vice admiral, United States Navy (1917–1946); president, Turner Construction Company (1946–1947); chairman, Jones & Laughlin Steel Company (1947–1958).

Adm. Ben Moreell made an important mark in twentieth-century American history not only for taking the lead in establishing the Naval Construction Battalions, known as the Seabees, during World War II but also for his aggressive leadership of the Jones & Laughlin Steel Company (J & L) between 1947 and 1958. As an advocate of production efficiency and plant modernization, Moreell directed a massive capital investment and expansion program at J & L during the immediate post–World War II period, transforming a tradition-bound firm into a dynamic organization that opened new markets and relied upon state-of-the-art production technology.

Ben Moreell was born in Salt Lake City, Utah, on September 14, 1892; when he was six years old his family settled in Saint Louis. He graduated from high school in 1908 and entered, as a scholarship student, Washington University, where he excelled in and out of the classroom, starring in both football and track. In 1913 he earned an undergraduate degree in civil engineering and subsequently found employment with the city of Saint Louis as a resident and design engineer, working on municipal sewer projects. Moreell joined the navy in 1917, was awarded an officer's commission, and held a series of assignments in the United States, the Azores, and Haiti. By the late 1920s Moreell had gained the reputation of a young engineering officer on the rise within the typically entrenched bureaucracy, and, while serving time at the Bureau of Yards and Docks in Washington, D.C., he found time to write *Standards of Design for Concrete* (1930).

Ben Moreell (photograph by H. Landshoff)

In the late 1930s Moreell was promoted to rear admiral at the age of forty-seven, one of the youngest naval officers to hold the rank. At the onset of World War II, Moreell was charged with forming the Seabees. Nicknamed by Moreell the "Can-Do Boys," the battalions were comprised of construction engineers who laid out roads, manned floating dry docks, and built advanced bases, airfields, and troop barracks. For his successes, Moreell was promoted to vice admiral in 1944 and was awarded the Distinguished Service Medal in 1945.

In the immediate postwar era Moreell won praise for his successful mediation of a dispute involving the Oil Workers International Union and the strikebound oil refineries and pipelines during autumn 1945. In May 1946 Moreell was named deputy coal mine administrator by Secretary of the Interior Julius A. Krug. Moreell's flexibility and sensitivity shown in overseeing government-seized bituminous coal mines and negotiating with the United Mine Workers gained him widespread public acclaim. In 1947, after having briefly served as president and a board member of the Turner Construction Company of New York, Moreell moved to J & L. Moreell's success in dealing with organized labor during the troubled years immediately following World War II made him the right man to lead J & L, a company with a long history of labor troubles originating with union organization efforts during the mid 1930s.

Moreell was a tireless worker who teamed with C. L. (Lee) Austin in revitalizing a company with an aging plant and declining profitability. Moreell was quick to point out that the low steel prices characteristic of the post–World War II era resulted in low dividends — which meant that fewer investment dollars entered the steel industry — and scant profits. Moreell thus believed that a modernization program must attract venture capital and introduce new process equipment. Between 1947 and the mid 1950s outdated coke ovens, furnaces, and mills were replaced, and dollar sales and net income quickly rose. Moreell pushed hard for increased product diversification as J & L entered new markets in semifinished, cold-finished, hot-rolled, and sheet steel, as well as wire, tubular, and tin mill products. During the 1950s more than $683 million was spent on new plant and equipment, and J & L made a foray into the stainless-steel-strip and electrical-sheet markets. Most significant, J & L installed one of the first basic oxygen furnaces in the United States in 1957, and its capacity was quickly increased from 56 to 80 tons. The basic oxygen process employed an egg-shaped vessel similar to that of a Bessemer converter, but, unlike the older method, oxygen was blown onto the surface of the molten-metal bath using a water-cooled lance thrust down through its top. Faster and more efficient than the best open-hearth practice, the oxygen converter marked the high point of the Moreell era at J & L.

It would be only a few years after Moreell's retirement in 1958 that J & L entered the most troubled phase in its long history. It is perhaps ironic that a Great Lakes ore freighter that bore his name sank in 1963, when troubling economic undercurrents began to undermine the position of the company he had worked so hard to modernize. He died July 30, 1978, in Pittsburgh, Pennsylvania.

Selected Publications:

Standards of Design for Concrete (Washington, D.C.: U.S. Government Printing Office, 1930);

"Suggests Ways to Expedite Steel Industry," *Commercial and Financial Chronicle*, 169 (February 3, 1949): 572;

"Plain Talk on Steel; Its Prices Are Too Low, Not Too High," *Barrons*, 30 (October 23, 1950): 5;

"The Challenge to American Business," *American Iron and Steel Institute Yearbook* (1952): 33–44;

"J & L": The Growth of an American Business, 1853–1953 (New York: Newcomen Society, 1953);

The Admiral's Log: God, Man, Rights, Government: A Compilation of Speeches (Philadelphia: Intercollegiate Society of Individualists, 1958);

The Admiral's Log II: In Search of Freedom: A Compilation of Speeches (Philadelphia: Intercollegiate Society of Individualists, 1960).

Reference:

William T. Hogan, S.J., *An Economic History of the Iron and Steel Industry in the United States*, 5 volumes (Lexington, Mass.: Lexington Books, 1971).

John Pierpont Morgan

(April 17, 1837 – March 31, 1913)

by John N. Ingham

University of Toronto

CAREER: Clerk, Duncan, Sherman & Company (1857–1860); manager, J. Pierpont Morgan & Company [New York agent for George Peabody & Company] (1860–1864); partner, Dabney, Morgan & Company (1864–1871); partner, Drexel, Morgan & Company (1871–1895); partner, J. P. Morgan & Company (1895–1913); director, United States Steel Corporation (1901–1913).

J. P. Morgan, as he is most often referred to, was the dominant financier of his generation and probably the most important Wall Street investment banker in history. In many respects, however, his most enduring achievement was the crucial role he played in the dramatic transformation of the American steel industry with the creation of the United States Steel Corporation.

Son of Junius Spencer Morgan, a prominent international banker who played a major role in opening the United States to European capital, Morgan graduated from the noted English High School in Boston in 1854 and studied for two years at the University of Göttingen in Germany. In 1856 J. P. Morgan entered his father's banking house, George Peabody Company, in London. A year later his father secured a position for him with the New York banking firm of Duncan, Sherman & Company, U.S. agents of the Peabody establishment.

In 1860 Morgan formed J. Pierpont Morgan & Company to serve as special agent in the United States for George Peabody & Company. During this period, Morgan concentrated primarily on foreign exchange. In 1864 Morgan became a member of the firm of Dabney, Morgan & Company and engaged Jay Gould and James Fisk in a struggle for control of the Albany and Susquehanna Railroad. Upon emerging victorious, Morgan's status in the banking and industrial communities was greatly enhanced, and he became an important voice for stabilization in an age noted for its volatile business sectors. In 1871 Morgan joined with Anthony J. Drexel to form

John Pierpont Morgan (Gale International Portrait Gallery)

Drexel, Morgan & Company. Two years after Drexel's death in 1893, the firm was reorganized as J. P. Morgan & Company.

Drexel, Morgan became a major factor in the financial world in 1873, when Morgan was able to break Jay Cooke's monopoly in government-bond sales. When Cooke's firm went bankrupt later that year, Drexel, Morgan's position was further enhanced. In 1879, in a deal that captured the public's attention and added to Morgan's growing prestige as a financial wizard, he sold a large block of William H. Vanderbilt's New York Central Railroad stock directly to English investors. The deal launched Morgan as a major player in the massive new American rail industry emerging at that time.

Despite his seemingly high-stakes deal making, Morgan's business philosophy was quite conservative. He abhorred the chaotic competition of the postbellum era and instead held a vision of the harmonious business community, one in which the forces of rationalization, centralization, and stabilization would bring about great efficiency and yield higher profits. He was convinced that even if prices did rise as the result of his efforts at consolidation, the consumer would still benefit as the result of more-stable and more-dependable operations. Morgan first brought his stabilizing influence — often called Morganization — to the railroad industry in the 1880s. He oversaw the reorganization of the New York Central, the Pennsylvania, the Philadelphia and Reading, and the Chesapeake and Ohio during this time. In the mid 1890s, when many railroads had been forced into bankruptcy by the depression, Morgan stepped in to consolidate and reorganize the industry into massive new railway systems. By the end of the decade, only the railway systems controlled by Edward H. Harriman and Gould operated outside his influence.

Morgan's success in reorganizing America's railways into a vast empire made him highly sought after by other industrial sectors. He helped in the organization of Edison General Electric in 1889, which was reorganized as General Electric in 1892, but it was in the steel industry that J. P. Morgan had his greatest impact and most enduring influence. Morgan's entry into the steel industry came in 1897, when Elbert H. Gary and John W. Gates approached him about providing the financing for a combination of wire and nail firms. A more unlikely pair of men could not have approached Morgan. While Gary was an austere, conservative lawyer, judge, and ex-mayor of Wheaton, Illinois, Gates was just the sort of gambler and plunger that Morgan abhorred. Fortunately, perhaps, it was not Morgan to whom they presented their case, but Morgan's partner, Charles H. Coster. The latter convinced Morgan of the worth of the enterprise, and the Morgan firm backed Gary and Gates in their consolidation, known as American Steel and Wire Company. To help provide capital for the $80-million

corporation, Morgan put together a syndicate to float $20 million. In February 1898, however, Morgan withdrew his support for the enterprise, leaving Gates and Gary to organize the firm on their own. The result was American Steel and Wire Company of Illinois, a $24-million corporation. Morgan, however, received his introduction to the steel industry and to Gary, whose business sense greatly impressed Morgan.

As a result, Morgan in the summer of 1898 quite willingly joined with Gary in another steel venture — Federal Steel Company. Gary had gotten involved with Illinois Steel, which traced its roots back to the North Chicago Rolling Mill Company; established by the legendary Eber B. Ward in 1857, it was one of the earliest Bessemer steel operations in America. In 1889 it had merged with two other large Chicago firms, Union Iron and Steel and Joliet Steel, to form Illinois Steel. The depression of the 1890s, however, was not kind to Illinois Steel, and by 1898 Gary, as general counsel of the firm, had become convinced that if Illinois Steel were to survive it had to acquire more ore lands and better manufacturing facilities and more effectively engage its competition, the aggressive Carnegie Steel Company.

Expansion of this magnitude, of course, could not come without a massive infusion of outside capital, and to this end Gary and his allies turned to J. P. Morgan, whose partner, Robert Bacon, was already a member of Illinois Steel's board. Gary convinced Morgan, Bacon, and Costner of the worth of the project, and in October 1898 J. P. Morgan & Company organized a syndicate to raise $14 million in new capital. Morgan's firm purchased the constituent companies and sold them to a newly created holding company called Federal Steel, to be capitalized at $100 million. Morgan's firm provided about $4.8 million in cash for the new venture, for which it received 14 percent of the stock, worth over $14 million. The new firm produced about 15 percent of the steel ingots in the country, second only to Carnegie's mammoth operations.

Morgan persuaded Gary to leave his lucrative Illinois law practice to become president of the new steel concern. Faced with a powerful new competitor, Andrew Carnegie in December 1898 — just two months after Federal Steel's organization — signed a truce in the ruinous price-cutting war in the steel-rail industry. Gary's faith in combination was shown to be well founded, at least for a time, and Morgan was greatly impressed. Partially as a result of that success, Morgan in 1899 sponsored another combination in the steel industry: the $80-million National Tube Company. This firm was a consolidation of 14 of the largest manufacturers in that branch of the industry and controlled over 85 percent of the output. The tube industry, like many finished-steel producers, had undergone enormous expansion in the late 1880s and early 1890s as a result of the falling price of crude steel and increases in tariff protection. The expansions resulted in severe competition characterized by falling prices.

But Morgan's efforts at stabilizing the steel industry were frustrated by several events occurring in 1900. Gates, still head of American Steel and Wire, informed Carnegie that his firm would no longer be purchasing crude steel from Carnegie Steel; it would, instead, begin producing its own semifinished steel. Within a short time, William H. Moore and James Moore sent Carnegie an identical notice on behalf of American Steel Hoop and American Sheet Steel. Soon after, Morgan's newly organized National Tube let it be known it would soon drop Carnegie Steel as a supplier and obtain its steel from Federal Steel. In retaliation, Carnegie immediately announced plans to construct a massive finished-steel operation to produce hoop, rod, wire, nails, and other items at Conneaut, Ohio. The purpose of his decision was clear: Carnegie sought to undersell all the newly organized finished-steel combinations and drive them out of business. Most of the steel fabricators turned to Morgan to help in dealing with the Carnegie "menace."

Although there had been some contact over the previous 18 months between Carnegie and Morgan's firm — Carnegie had made it known that he was willing to sell for "the right price" — nothing much came of these discussions until December 1900. The occasion was a dinner in New York City to honor Charles M. Schwab, Carnegie Steel's young president. Morgan was seated next to the guest of honor. Morgan was impressed with Schwab's vision of a fully integrated, centralized, and rationalized industry of finished- and crude-steel producers, and the two men agreed to meet again later to continue their talks concerning the fate of the steel industry. Three weeks later Schwab, along with Gates and Bacon, met in Morgan's library. At this meeting, which lasted until dawn, the outlines of a vast new steel consolidation, and the firms that were to comprise it, were drawn up. The key question, though, was whether Carnegie would actually sell to the new

group, and the task of pitching the idea to Carnegie fell to Schwab.

During a game of golf, Schwab received a commitment from Carnegie in terms of price and mode of payment. Carnegie gave Schwab a slip of paper on which he had written his demand of $480 million for his properties, and insisted that his personal share and those of his sister-in-law and of a close friend be paid entirely in the new corporation's first mortgage 5 percent gold bonds. Schwab delivered the news to Morgan, who replied simply, "I accept this price." The first step toward the creation of the United States Steel Corporation, the nation's first billion-dollar corporation, was complete. It was now up to Morgan to complete the difficult operation, a massive, unprecedented task.

To assist him, Morgan gathered around him experts from many fields. He worked most closely with Gary and Bacon, who had great experience in industrial ventures. In February 1901 J. P. Morgan & Company announced the capitalization of U.S. Steel, amounting to a whopping $1.154 billion. A few weeks later, additional properties were added, bringing total capitalization to $1.404 billion. The nation's press and political leaders were shocked at the amount, which was far greater than the worth of any past American business venture. Wall Street experts were also caught by surprise; Isaac Seligman commented that "I confess that it is enough to take one's breath away."

To ensure the success of the monstrous new enterprise, J. P. Morgan & Company organized a syndicate to underwrite $200 million of U.S. Steel's securities. Unlike most previous underwriters, the subscribers to this venture were not predominantly institutions but mostly individuals, many of whom were closely connected to the corporation's constituent companies. The underwriting was an unqualified success, and by March 1901 the new firm was operational. There was little question that the key to U.S. Steel's unprecedented market control was J. P. Morgan's backing. Industry experts generally approved of the formation of U.S. Steel, believing that the mammoth enterprise run in accordance with Morgan's conservative business philosophy would help stabilize the industry. Many in the federal government, however, were highly suspicious of U.S. Steel and charged that the corporation had been overcapitalized. The Bureau of Corporations, in a report filed in 1911, concluded that U.S. Steel's tangible assets in 1901 amounted to just $676 million; intangible assets brought the final total up to

$793 million. By the bureau's calculation, then, U.S. Steel had been watered to the tune of nearly $611 million. The Morgan firm itself received a fee of $12.5 million for arranging the merger. To some, this seemed a small reward for an enormous undertaking. To others, it appeared to be graft.

The resulting operation, watered or not, was enormous. The new company controlled 65.7 percent of the nation's production of steel ingots and castings, 60 percent of steel rails, nearly 65 percent of plates and sheets, 78 percent of wire rods, and 73 percent of tin and terne plates. The 1901 consolidation — the cobbling together of many semifinished- and finished-steel producers — however, did not guarantee a viable steel operation. Over the next several years Morgan had to deal with internal conflicts at the giant steel firm, as well as with a serious deterioration in the firm's profitability and market share. These concerns, though, were handled most directly by George W. Perkins, Morgan's representative as chairman of U.S. Steel's finance committee, and Gary. Morgan's own role was increasingly limited to providing general advice on policy and direction. One of Morgan's most direct involvements with U.S. Steel came in 1907 and resulted in one of the most controversial periods in his business career.

During the financial panic of that year, J. P. Morgan and his banking cohorts faced a serious crisis. In the fall the Trust Company of America and the Lincoln Trust Company were about to close their doors, and the Wall Street firm of Moore & Schley was also teetering on the brink of collapse. It had used its 157,000 shares of stock in the Tennessee Coal, Iron & Railroad Company (TCI & R) as collateral for loans it had granted to the firm's senior partner. If the banks were to sell stock to satisfy Moore & Schley's call notes, TCI & R's stock price would fall disastrously on the exchange and aggravate the nation's financial situation. A solution was worked out in Morgan's library. The proposal provided for U.S. Steel to buy TCI & R shares with 5 percent gold bonds. This would save Moore & Schley and help stabilize the entire market, which, in turn, would salvage the two trust companies. Morgan agreed to the plan, but Gary insisted that it be cleared with President Theodore Roosevelt, to avoid possible antitrust action. As a result, Gary and Henry Clay Frick called on the president, outlined the proposal, and Roosevelt agreed not to oppose the purchase. The transactions were completed, Wall Street was stabilized, and, perhaps for the only time in his career, Morgan was lionized as an economic savior.

The administration of William H. Taft, however, in 1911 brought suit against U.S. Steel for the TCI & R acquisition. This led to a two-year investigation of U.S. Steel's power in the industry and ultimately brought the firm before the courts in an antitrust case from which it finally emerged victorious in 1920. But charges continued to be leveled against Morgan, claiming that he was the center of a "money trust" in America, the symbol of concentrated economic power. In 1912 Congress formed the Pujo Committee to investigate the extent of Morgan's holdings and to determine the degree to which he was engaged in business manipulations. Morgan himself was called to testify, but he was generally handled with kid gloves by the committee's counsel, Samuel Untermeyer. Nonetheless, committee findings showed that 11 House of Morgan partners held 72 directorships in 47 major corporations worth billions.

A few months after his appearance in front of the Pujo Committee, Morgan died in Rome, Italy. Although vilified by many as a most dangerous influence on the American economy, he is generally credited for the transformation of the American steel industry whereby the personal management of the nineteenth century was replaced by the corporate bureaucracy of the twentieth-century. In that sense, J. P. Morgan was as revolutionary a force in the twentieth-century American steel industry as were Henry Bessemer and Andrew Carnegie in the nineteenth century.

Unpublished Documents:

U.S. Industrial Commission, "Report . . . ," 19 volumes, Washington, D.C., 1900–1901;

U.S. Bureau of Corporations, "Report . . . on the Steel Industry," 3 parts, Washington, D.C., 1911–1913;

U.S. House of Representatives, Subcommittee of the Committee on Banking and Currency, 62nd Congress, 3rd Session, Investigation of Financial and Monetary Conditions in the United States, "Hearings . . . ," 2 volumes, Washington, D.C., 1913.

References:

Frederick Lewis Allen, *The Great Pierpont Morgan* (New York: Harper & Row, 1949);

Vincent P. Carosso, *The Morgans: Private International Bankers, 1854–1913* (Cambridge: Harvard University Press, 1987);

Arundel Cotter, *The Authentic History of the United States Steel Corporation* (New York: Moody, 1916);

Robert Hessen, *Steel Titan: The Life of Charles M. Schwab* (New York, 1975);

John N. Ingham, *Biographical Dictionary of American Business Leaders* (Westport, Conn.: Greenwood Press, 1983);

Joseph M. McFadden, "Monopoly in Barbed Wire: The Formation of the American Steel and Wire Company," *Business History Review*, 52 (Winter 1978): 465–489;

Ida Tarbell, *The Life of Elbert H. Gary: The Story of Steel* (New York: Appleton, 1926);

Joseph F. Wall, *Andrew Carnegie* (New York: Oxford University Press, 1970).

Archives:

There are two large collections of Morgan family and business papers at the Pierpont Morgan Library in New York City.

Edmund W. Mudge

(January 12, 1870 – July 1, 1949)

by Glen V. Longacre III

Columbus, Ohio

CAREER: President, Edmund W. Mudge & Company (1905–mid 1920s); chairman of the board, Phillips Sheet & Tin Plate Company; chairman of the board and vice-president, Weirton Steel Corporation; president, Westmoreland-Connellsville Coal & Coke Company; president, Reliance Coke Company; president, Ella Furnace Company; president, Claire Furnace Company; president, Mudge Oil Company (1921–?).

As president of several steel companies in northern West Virginia and southwest Pennsylvania, Edmund Webster Mudge was known for his leadership of and dedication to the iron and steel industry. He was one of the founders of the Phillips Sheet & Tin Plate Company, which later became the Weirton Steel Corporation.

He was born to Mary Emma and Thomas Henry Mudge in Philadelphia, Pennsylvania, on January 12, 1870. Mudge came from an old New England family, whose first member, Thomas Mudge, had settled in Malden, Massachusetts, prior to 1654.

Mudge was educated at a private school in Philadelphia operated by the Society of Friends. Afterward he attended the Woodstown Academy located in New Jersey. In 1887, at the age of seventeen, he moved to Pittsburgh, Pennsylvania, and became involved in the steel industry. Twelve years later, on April 4, 1899, he married Pauline Seeley, the daughter of Pittsburgh business executive Leonard Perrin Seeley. The couple had three children: Edmund Webster, Leonard Seeley, and Mary Louise, the last of whom died in infancy.

Edmund Webster Mudge

Mudge founded his own iron business in Pittsburgh in 1905. The firm, Edward W. Mudge & Company, produced pig iron in its own blast furnaces. Under his direction and management the company became locally well known for its production of coke and pig iron. Mudge remained president of the company until its dissolution in the mid 1920s.

In addition to his own business ventures Mudge participated in several other steel-related businesses during his career. His most notable affiliation was with the Phillips Sheet & Tin Plate Company. Mudge's relationship with Phillips Sheet & Tin Plate began in 1905, when the thirty-five-year-old business executive along with other subscribers including James Phillips, Edward Kneeland, and Ernest T. Weir purchased approximately $200,000 in corporate stock to begin the company.

In 1912 Mudge joined with several stockholders of Phillips Sheet & Tin Plate, which was being managed by Ernest T. Weir following the untimely death of James Phillips, in incorporating the Weirton Steel Company. The company was located on the banks of the Ohio River in the northern panhandle of West Virginia. Two years later Mudge, as vice-president of Weirton Steel, participated in the merger of Phillips Sheet & Tin and Weirton Steel. By 1921 Mudge was chairman of the board, and, with more than $1 million invested, he was a leading stockholder in Weirton Steel. In September 1922 Weirton Steel's Board voted to purchase his remaining 50 percent interest in the Redstone Coal & Coke Company for

$635,000, which included $20,000 for an additional interest he maintained in the Oak Hill Supply Company. Mudge was paid in cash and in Weirton Steel stock, making him one of the five largest stockholders in the company.

In addition to his business interests in the steel industry, Mudge also served as senior vice-president of the Fidelity Trust Company, vice-president of the Hanna Furnace Company, and president of the Mudge Oil Company, which he founded in 1921. He was also director of the Columbia National Bank of Pittsburgh, the Edgewater Steel Company, Bell Telephone Company of Pennsylvania, the United States Radiator Company, the Mudge Oil Company of Texas, and the Union Fidelity Company.

Active in community affairs, he was vice-president of Allegheny General Hospital and a trustee of the Pittsburgh Associated Charities, YMCA, and Children's Service Bureaus. During World War I, he participated in the Pennsylvania Council of National Defense and was director of the Municipal Planning Association. Mudge died in Pittsburgh on July 1, 1949.

Reference:

William T. Hogan, S.J., *An Economic History of the Iron and Steel Industry in the United States*, 5 volumes (Lexington, Mass.: Lexington Books, 1971).

Walter F. Munford

(June 8, 1900 – September 28, 1959)

by Dean Herrin

National Park Service

CAREER: Entire career with United States Steel Corporation: open-hearth helper (1923–1924), superintendent, South Works, Worcester (1927–1930); open-hearth superintendent, Newburg Works, Cleveland (1930–1933); assistant superintendent, Steel Works, National Tube Company Division (1933–1934); assistant superintendent (1934–1937), superintendent, Cuyahoga Works, American Steel & Wire Division (1937–1939); assistant to vice-president (1939), assistant manager of operations, Pittsburgh district (1939–1942); manager of operations, Worcester district (1942); assistant vice-president of operations, vice-president of operations, president, American Steel & Wire Division (1942–1956); assistant executive vice-presi-

dent for operations (1956–1958); executive vice-president for engineering and research (1958–1959); president and chief administrative officer (1959).

Walter F. Munford was president and chief administrative officer of the United States Steel Corporation for only four months, from May to September of 1959, until his death. Despite his short tenure, Munford was forced to deal with a recession in the steel industry and the start of what would become the longest strike in the history of the industry.

Munford, a New Englander, was noted for his work habits and his sense of propriety. According to *Newsweek,* Munford once said, during an interview after he took office, "I understand a 40-hour week is in effect, but I can't say that I've ever experienced it." Colleagues praised Munford for his ability to delegate responsibility successfully and to instill confidence in his subordinates.

Born in Worcester, Massachusetts, in 1900, Munford attended Worcester Polytechnic Institute and M.I.T., paying his way through school by working at night in the Worcester steel mills of the American Steel & Wire Company, a subsidiary of U.S. Steel. He started work as an open-hearth helper in 1923 and advanced to superintendent of the open-hearth department only four years later. Munford was transferred to Cleveland's Newburg Works as superintendent of its open-hearth department in 1930, and he was next appointed assistant superintendent of National Tube Company's Lorain, Ohio, plant. Munford was promoted to superintendent of American Steel & Wire's Cuyahoga Works in Cleveland in 1937 and later to manager of the company's operations in the Pittsburgh district. Having gained a favorable reputation at these U.S. Steel subsidiaries, Munford was selected head of U.S. Steel's American Steel & Wire Division in 1953. He was brought to Pittsburgh in 1956 as an assistant executive vice-president for U.S. Steel and had advanced to executive vice-president for engineering and research when he was selected to replace the retiring Clifford Hood as president in May of 1959. Munford had earlier worked for Hood at American Steel and Wire.

The early 1950s were prosperous years for U.S. Steel. Between 1955 and 1958 profit levels were above $300 million — exceeding $272 million for the first time in the corporation's history — with the high mark of $419 million reached in 1957. But a recession hit the industry in 1958, and profits declined to $254 million in 1959; steel production at U.S. Steel was lower in 1958 and 1959 than at any other time of the decade.

Compounding the problems caused by the recession was a lengthy steel strike launched by the United Steelworkers of America in 1959. On Munford's second day in office, negotiations began between steel producers and the union on a new contract, with U.S. Steel's R. Conrad Cooper the chief industry spokesman. Demands were made on both sides covering most areas of employment, including wages, benefits, and work rules. Because of government pressures to reach an agreement that was noninflationary, the two sides were unable to come to terms, and the steelworkers on July 15 began the longest walkout in the industry's history. The strike lasted 116 days and ended only after an injunction ordering the steelworkers back to work was issued on October 19 and upheld by the U.S. Supreme Court on November 7. A new contract was finally hammered out in early 1960 after Secretary of Labor James Mitchell and Vice-President Richard Nixon intervened.

In the middle of the union walkout, while resting in his summer home on Cape Cod after treatment for nervous exhaustion and fatigue, Munford accidentally stabbed himself with a kitchen knife. After surgery for the knife wound, Munford suffered a stroke and died on September 28, 1959. He was succeeded as U.S. Steel's president by Leslie Worthington.

References:
John Hoerr, *And the Wolf Finally Came: The Decline of the American Steel Industry* (Pittsburgh: University of Pittsburgh Press, 1988);
William T. Hogan, S.J., *An Economic History of the Iron and Steel Industry in the United States,* 5 volumes (Lexington, Mass.: Lexington Books, 1971);
"How They Got There," *Newsweek,* 53 (May 18, 1959): 90.

Philip Murray

(May 25, 1886 – November 9, 1952)

by Elizabeth C. Sholes
Industrial Research Associates
and
Thomas E. Leary
Industrial Research Associates

CAREER: Coal miner, various mining companies in western Pennsylvania (1902–?); president, UMWA local, Horning, Pennsylvania (1905–?); member, UMWA international executive board (1912–1916); president, UMWA District 5 (1916–1920); vice-president, UMWA (1920–1940); vice-president, CIO (1935–1940); head of the Steelworkers Organizing Committee (1936–1942); president, CIO (1940–1952); president, United Steelworkers of America (1942–1952).

In June 1936 Philip Murray, the newly appointed head of the Steelworkers Organizing Committee (SWOC), began the steel organizing drive on the shores of the Youghiogheny River in McKeesport, Pennsylvania. The rally was held in a vacant field; the speaker's dais was the bed of a coal delivery truck. A cool reception was given the SWOC and Murray, not by the hundreds of steel laborers surrounding his makeshift platform, but by the town officials who were less than sympathetic to union causes. Despite this inauspicious beginning Philip Murray achieved high visibility during his tenure as union organizer and then as union president. He developed a reputation for combining a relentless negotiating style with a thorough knowledge of the steel industry and its financial resources. He maintained an unquestionably adversarial relationship with the heads of America's steel companies, yet Murray was respected and trusted throughout the industry. One company executive dubbed him "the most gifted triple-threat technician in the labor movement, as organizer, administrator, and negotiator."

Murray was born on May 25, 1886, in Blantyre, Lanarkshire, Scotland, to Irish Catholic immigrants William and Rose Ann Layden Murray. Blantyre was the scene of a great mine disas-

ter at Dixon's Colliery on October 22, 1877, in which 200 men had perished and from which ardent union sentiment had grown strong. Murray was schooled in strong union principles and organizational skills by his father, who was secretary for his local trade union.

Murray's mother died when he was two, and his father took him and his sister to live with their grandfather in Bothwell, where Murray first began work in the mines. At the age of seven he was already embroiled in union activism, collecting and distributing food for the families of miners engaged in the 17-week strike in 1893. After his father remarried the family returned to Blantyre where Murray quit school to work full-time in the mines.

In early 1902 Murray's father traveled to America to visit his brother in the coal region of Westmoreland County in western Pennsylvania. In December sixteen-year-old Murray accompanied his father on a return trip to America to find jobs in the Pennsylvania coal mines. Carrying international coal-union transfer cards, the two Murrays cleared Ellis Island on Christmas Day. They quickly found work in the coal pits, and the rest of the family joined the men in 1903. On the job Murray shoveled enough coal to fill three cars per day at $1 per car and saved enough money to invest $60 in economics and mathematics courses from the International Correspondence School, completing the courses in six months; his mental acuity in these two subjects would stand him in good stead over his years of union bargaining.

While working for the Keystone Coal and Coke Company in 1904, Murray became involved in a dispute with the company weighman. Murray was being paid by tonnage, and he accused the weighman, who was responsible for determining tonnage records, of cheating him out of nearly 40

U.S. Steel president Benjamin Fairless and CIO president Philip Murray signing the agreement that ended the steelworkers' strike of 1946

percent of the weight he had shoveled. After Murray demanded a union checkweighman on the tipple, the disagreement came to blows. Murray was fired, and he and his family were summarily evicted from their company housing. Nearly 600 miners went on a month-long strike in support of Murray and his call for a checkweighman, but Murray was forced out of town by deputy sheriffs and told never to return to Westmoreland County. Murray later declared that the confrontation proved to be significant in determining what he wanted to do with his life. The experience hardened his dedication to union principles, and his belief that organized labor was key to protecting the interests of the worker: "A coal miner has no money. He is alone. He has no organization to defend him. He has nowhere to go . . . it is not inadequacy of the State law. The law is there, but the individual cannot protect himself because he has no organization." State authorities and unions, Murray came to believe, needed to work together.

Despite his unceremonious ouster from Keystone, Murray went to work for pits in other southwestern Pennsylvania mining communities. In 1905 he was elected president of the United Mine Workers of America (UMWA) local in Horning, Pennsylvania. He became an American citizen in 1911 and advanced rapidly in the union, capturing the attention of UMWA president John White and winning a seat on the UMWA international executive board in 1912. In 1916 White and his chief organizer, John L. Lewis, arranged for Murray to occupy the presidency of District 5, the Pittsburgh area.

District 5 contained one of the nation's richest seams of bituminous coal and was an area where disputes between miners and coal company operators often ran bitter and violent. Murray suffered physical as well as mental challenges in organizing; once he was hit by a brick, another time pushed off a wagon by company men while addressing a union crowd. Nevertheless, Murray's intimate knowledge of the economics and politics of coal production made him a skilled organizer, gaining him recognition in the coalfields and at the highest levels of the UMWA.

Publicly he advocated labor-management cooperation as expressed in "The Labor Pope" Leo XIII's labor encyclicals and later the 1931 *Quadragesimo Anno* of Pope Pius XI. Privately, however, he was heard to lambaste company executives as "capitalist lackeys." His adviser in the steelworkers union, Harold Ruttenberg, maintained that Murray "could never bring himself personally to cooperate and collaborate with the capitalists and their 'factotums,' as he called them." Murray believed that the critical element in union advancement was intervention by the federal government to force companies into collective bargaining. Labor in turn would be compelled to honor contractual decisions. In this respect, Murray provided welcome relief to state and federal governmental officials who had seen their authority challenged by radical labor leaders such as Eugene Debs. Indeed, Murray soon became an accepted player in political circles. During World War I he served as a member of President Woodrow Wilson's War Labor Board, the Pennsylvania Regional War Labor Board, and in 1917–1918 was on the National Bituminous Coal Production Committee. He served as a mediator between government and union officials in reaching the 1918 Washington Agreement, which Murray presented to the UMWA convention. The agreement gave miners substantial wage hikes for their war work but also mandated strict discipline and fines for strikes. In 1920 the UMWA elected John L. Lewis its president. Murray had supported Lewis's candidacy against incumbent president Frank Hayes, whom Murray dubbed "a well meaning incompetent." Lewis in turn had seen in Murray a talent he wished to foster and placed Murray in the vice-president spot on his ticket. Murray at age thirty-three thus became second in command of the largest union in North America. The two worked together, complementing each other's strengths. Lewis dealt effectively with the rich and powerful; Murray "provided the economic facts to support Lewis' blustery rhetoric."

Murray was appointed by President Warren Harding in 1921 to quell the labor uprising in Logan County, West Virginia, where over 20,000 miners battled the company men of the Baldwin-Felt mines. Dissent was so widespread and the destruction of life and property so extensive that Harding placed all of West Virginia under martial law. As in 1919, Murray placed himself and the union on the side of the federal authorities. Murray hoped that the federal government would in turn intervene on behalf of the UMWA to halt deteriorating labor conditions in the mining industry. Company after company was breaking the hard-fought but tenuous agreements of the war years. Murray appealed to the Senate Committee on Education and Labor and to Secretary of Commerce Herbert Hoover, but to little avail.

Within the union, things were in disarray. As disparate as Lewis and Murray were in terms of

disposition and methods, they were united in a single cause with respect to the overall goals and composition of the union. Murray supported Lewis's centralization of power in the union and worked steadfastly with Lewis to purge the socialists and radicals who sought nationalization of the mines and democratization of the union. Lewis and Murray succeeded in eliminating the dissident forces and challenges to their own authority, but the union became bankrupt. The UMWA had been waging a costly struggle against mine companies that broke the Bituminous Coal Commission agreements, which outlined company-union relations. As cooperative agreements collapsed, the union's good relations with the Republican administrations ended, and districts began to succumb to the union-busting efforts of unregulated mining companies. By 1930 the UMWA was in tatters, and fundamental political and ideological differences between Lewis and Murray began to strain their relationship.

Unlike Democrat Murray, Lewis was strongly Republican. Philosophically, Murray drew from the work of Sidney and Beatrice Webb, British Fabian Socialist labor reformers. Their book, *Industrial Democracy* (1897), clearly influenced Murray's perspectives on relations between workers and companies, if not his views on the structures of the union itself. Murray argued that the "individual worker is helpless" and must have collective bargaining in which "employees actively participate, as equals, with employers in fixing the terms and conditions of their employment." Such bargaining would benefit both sides by allowing improvements in technology and organization and, therefore, expanding overall production. On these points Lewis and Murray were in agreement. Murray further argued, however, that collective bargaining had limits since it could not and did not cover all issues related to production: "State-wide and nation-wide standards are essential to put a floor under even the best contract terms" that unions and companies negotiated. It was Murray's stance on the strong role government should play in mediating labor agreements that ultimately divided the two union leaders.

Despite their differences, they worked closely together to rebuild the union and — partly due to the backing of the Democratic administration of President Franklin D. Roosevelt, with whom Murray was close — were successful by 1934.

In 1935 Lewis challenged the existing American Federation of Labor (AFL) practice of representing only the skilled trades and proposed organizing all mass-production workers in large-scale manufacturing. But the AFL executive council rejected the proposal. Consequently, Lewis and Murray founded the Committee for Industrial Organization (later Congress of Industrial Organizations, CIO) in November 1935, formally breaking with the AFL in 1936.

Political and economic circumstances in 1936 made conditions favorable for an organizing drive in the steel industry. By midyear several large U.S. industries, steel among them, were showing strong signs of recovery from the Depression. For steelworkers, the improving economy offered an opportunity to extract gains from steel companies experiencing an increase in production. In Washington, D.C., Congress had formed the LaFollette Committee to investigate antiunion practices by corporations, the results of which shocked the nation. Congress also began to lay the groundwork for the Wagner Act, which supported union efforts to organize and bargain collectively and established the National Labor Relations Board (NLRB) to oversee organizing disputes. By the end of 1936 the future of strong government support for unions was reinforced by the landslide reelection of Roosevelt, which was in part due to the labor vote.

Seizing on this pro-labor atmosphere, Lewis quickly appointed Murray to head the CIO's organizing campaign in the steel industry; the SWOC coalesced rapidly under Murray's guidance. Committee members were drawn from several CIO member unions, including the Amalgamated Association of Iron, Steel, and Tin Workers and the UMWA. Each member union pledged funds and personnel to support the organizing effort.

The committee met infrequently and left daily decisions to Murray alone. Murray, however, surrounded himself with an able staff drawn from universities, government, and law. Unfettered by AFL regulations, SWOC charted its own jurisdiction embracing as many iron and steel workers as possible. SWOC pursued its organizing goals in three distinct ways. First, the committee contacted workers through ethnic, fraternal, and religious organizations that comprised the hub of many workers' social lives. Second, SWOC propaganda exploited ties to the White House, much to the ire of the corporations. SWOC served as the eyes and ears for the La Follette investiga-

tion and reported on steel company malfeasance to the NLRB. Third, SWOC sought to capture the company unions, or employee representation plans (ERPs), which had been industry's way around Title I, Section 7(a) of the National Industrial Recovery Act of 1933 guaranteeing employees the right to organize. "Sweetheart" unions or employee plans that did deliver some improvements in working conditions — but provided no autonomy or independent bargaining base from which a worker could negotiate without fearing company retaliation — were created by the companies to keep "real" unions out of the shops. Worker dissatisfaction with company unions, however — particularly at U.S. Steel — led to an increasing willingness among workers to affiliate themselves with an independent union. Murray urged his supporters inside the company unions to "bite at the heels of management."

SWOC's first victory was also its largest. In early 1937 Lewis entered into formal negotiations with Myron C. Taylor, chairman of U.S. Steel. Lewis had been able to convince Taylor that the SWOC enjoyed widespread support among U.S. Steel workers, and Taylor had come to believe that any fight waged against the organizing movement would be a protracted, and ultimately futile, one. Many of U.S. Steel's ERPs had been successfully infiltrated by Lewis's and Murray's forces and were coming out in recognition of the SWOC. At U.S. Steel's Carnegie-Illinois subsidiary, workers were preparing to strike for recognition.

On March 2, 1937, Murray signed a preliminary agreement with Carnegie-Illinois providing for full union representation, an established hourly wage, and an 8-hour a day, 48-hour work week. A formal agreement was signed on March 17 and went into effect at U.S. Steel's five largest subsidiaries. Steelworkers, steel executives, and the general public as well were astounded that U.S. Steel, the nation's leader in the steel industry — an industry known for its virulent antiunionism — would concede to organized labor without a battle, particularly in light of hard-fought unionizing campaigns elsewhere. Despite the victory, the SWOC ran into stiff resistance in attempting to organize the "Little Steel" companies. Bethlehem, Inland, Youngstown Sheet & Tube, and Republic had repudiated the organizing guarantees of the Wagner Act, and the strike Murray called against these firms soon turned bloody. The conflict culminated in the notorious "Memorial Day Massacres," in which ten strikers were killed in Chicago and six others in Ohio, spelling the strike's defeat.

Murray refused to concede, however, and directed SWOC to continue organizing the small producers, fabricators, and specialty shops. Bolstered by successful court challenges and NLRB rulings, the smaller companies gradually fell in line. By October 1937 Murray announced that SWOC had organized 1047 locals and had secured 439 collective bargaining agreements in its 16-month history.

But the Little Steel companies remained defiant. Led by Republic's Tom Girdler, the companies refused compliance with the very NLRB ruling that had been so successful in turning the smaller companies into union shops; for the federal government had lacked the ability to enforce new labor regulations on a massive scale. It took another five years, a further succession of court battles and NLRB hearings, lawsuits from workers injured in organizing confrontations, and finally a world war to make the four fiercely antiunion companies capitulate to the law and their workers' demands.

Much of what SWOC accomplished owed a great deal to the ultimately successful intervention into the negotiating process by various government agencies. Yet, Lewis became increasingly distrustful of Roosevelt's wartime economic controls, fearing such legislation would severely interfere with labor's self-direction. Lewis toyed briefly with forming a labor party but ultimately reverted to his political roots and endorsed Republican Wendel Wilkie during the 1940 presidential election, convincing his UMWA cohorts — all save Murray — to do likewise. Lewis had promised to resign as president of the CIO if Roosevelt was reelected in 1940, and he kept that promise. His choice of experienced successors was limited to CIO's two other dominant figures, Murray and Sidney Hillman, head of the American Clothing Workers Association. Lewis picked Murray despite their differences — the philosophical gap between Lewis and Hillman being even wider.

Murray agonized over accepting the leadership of the CIO. He doubted that Lewis would genuinely relinquish control of decision making, a fear based in part on two decades of experience as Lewis's disciple. The deal called for Murray to continue to serve Lewis as vice-president of the UMWA, and Murray expected that he had been chosen simply to perpetuate Lewis's indirect control over the CIO; Murray believed Lewis was trying to interfere with CIO policy and practices. After Murray received the nomination for CIO president, he articulated his fear that his authority would be compromised: "I think I am a man, I think I have convictions, I think I have a soul and a heart and a mind. . . . With the exception of my soul, they all belong to me, every one of them." Despite his reservations, Murray accepted the nomination and was elected CIO president in 1940.

The personal contest between Murray and Lewis revolving around Roosevelt, federal interventionism, and America's entry into World War II was so great both men suffered heart attacks. Lewis thus chose to stay away from the 1941 CIO convention, where Murray endorsed the captive-mine strike at mines owned by steel companies, thereby undermining Lewis's authority in the UMWA. The endorsement caused conflict between opposing factions, particularly between miners and other industrial unionists, and led to fistfights in the halls of the convention hotel.

In January 1942 Lewis and Murray nearly came to blows after Lewis publicly upbraided Murray, claiming Murray had betrayed the union. When Murray attempted to rebut Lewis's charges, Lewis drew back his fist in a show of threat and force. Later that month Lewis proposed a "peace plan" to reunite the divided labor movement, and he and the UMWA executive board declared the UMWA vice-presidency vacant. The move was tantamount to ousting Murray from the ranks of the miners' union hierarchy. Murray was furious and declared in a letter to Lewis that in the future all arrangements on behalf of unity would "have to be initiated through the office of the President of the Congress of Industrial Organizations." The break between the two men was complete and never to be healed.

It was the advent of World War II and the creation of emergency rationing organizations that turned the tide in the industry and in the Little Steel conflict. In July 1942 the War Labor Board (WLB), which regulated all labor-management disputes under wartime production conditions, forced the four recalcitrant companies to comply with union demands for recognition. The WLB further compelled the four companies to confer a 15 percent pay hike to compensate the "Little Steel" employees for cost-of-living increases denied them between January 1941 and May 1942 and to grant a pay hike of $0.055 per hour to provide parity with other companies. Murray's greatest ambition, however, was to secure a WLB ruling creating a union shop for all government-ordered iron and steel production.

Murray increased his reliance on government to serve as the mediator and enforcer for union organizing and negotiations. Despite the fact that the WLB-recommended parity-wage increase was considerably less than what SWOC had desired, Murray abided by his no-strike wartime pledge. He did request, and receive, WLB permission to extend union elections to secure full union representation within U.S. Steel. The result was a great victory for SWOC that was soon followed by successful contract negotiations for employees at all the Little Steel plants.

Murray was not, however, a complete Roosevelt sycophant, as Lewis had charged. Murray's own program for the war economy proposed the formation of industrial councils with equal representation from unions and management and chaired by a federal official. The councils would determine market orders, set quotas, establish priorities, determine prices, and hammer out labor policies. This plan was greeted with horror by corporate executives and stockholders who repudiated outright the presence of union people on an equal footing with management. Roosevelt, needing business support for wartime rationing programs, also rejected Murray's proposals. Murray was outraged at the wholesale renunciation of labor's proposed contribution to national policy planning. "We want to be placed in positions of trust and responsibility where we can render a service to our nation," he thundered. "Why should the agencies of government in Washington today be virtually infested with wealthy men ... what are we running in Washington? A war production organization to win the war, or a war production organization to destroy labor?"

In May 1942, soon after his split with Lewis, Murray convened the SWOC constitutional convention. The moribund Amalgamated Association was annexed by the SWOC, which converted into the United Steelworkers of America (USWA). The first order of business was to elect Murray president. The union was in excellent shape, having paid off its organizing debts to the UMWA and created a healthy treasury of its own. Despite the appearance of solidarity, however, there was dissension among the members, some of whom were chafing over government controls and seeking to overturn the regulatory bonds.

In June 1944 Murray and CIO vice-president R. J. Thomas wrote a lengthy monograph published by the AFL and the CIO detailing the living costs in America between 1941 and 1944. Murray and Thomas asserted that the cost of living had in-creased 45.3 percent during the period, disputing a report by the Bureau of Labor Statistics, which claimed a 23.4 percent increase. In the report's summary, Murray and Thomas compared wartime living costs for farmers, steelworkers, and textile workers and concluded that living standards for those groups had seriously declined since the start of the war. In pressing the WLB for wage increases, Murray used the report as an important negotiating tool.

But the WLB withheld an increase in wages. Despite Murray's warning in his 1944 presidential address that breaking the no-strike pledge would be "regarded as an insult to the armed forces and to the balance of our union-minded population," the steelworkers voted 411,000 to 83,000 in favor of a work stoppage, and the union filed a strike petition with the WLB on October 29, 1944.

U.S. Steel, serving as representative for the industry, would continue to refuse further wage increases unless the Office of Price Administration, which oversaw industry pricing, allowed steel companies to raise their prices to offset a wage hike. In response, Murray declared that "the steel industry is clearly engaged in a brazen attempt to bludgeon the Government of the United States," and that the companies had huge "hidden profits" they refused to reveal. Disgusted by the industry's refusal to grant an increase and government's dithering over a solution, Murray finally acceded to strike demands from the rank-and-file and ordered the walkout to begin January 14, 1946. In an effort to head off the strike, President Harry S Truman suggested a compromise pay package, which Murray accepted but the steel companies did not. The strike, however, was relatively short-lived. On February 15 a settlement was reached, when U.S. Steel, again serving as industry envoy, accepted the Truman compromise increase of $0.185 per hour.

In 1947, with the cost of living escalating sharply after the cessation of wartime controls, Murray sought another wage increase. The USWA and the companies negotiated for three months on acceptable terms, reaching an accord in April on a general increase of $0.125 plus other adjustments that added another $0.025 to the pay packets. This negotiation was significant not just for the economic gains it secured to overcome the 1945–1946 earnings slump but also because it represented the first time steel labor and management bargained collectively without government intervention.

Continued steep increases in postwar prices, however, encouraged Murray to use a clause in the 1947 agreement that allowed wage negotiations to reopen after the first year. Discussions began again in April 1948, but the steel industry responded unfavorably to another wage increase. The union thus embarked on a nationwide advertising campaign telling the American public of the huge profits being generated by steel companies and detailing the rapid escalation in costs of living. In July the companies acceded to the union's demands, and workers gained a $0.13 hourly increase. From 1939 to 1948 steelworkers had moved from among the lowest paid industrial workers in the nation to being among the highest paid.

Unions had achieved unquestionable advances thanks to the Wagner Act, the NLRB, and the general climate of union acceptance during the war years. Postwar concerns about Soviet incursions into Eastern Europe, however, soon created an anticommunist fervor among many leaders in both the public and private sectors, and American unionism — with its connections to American Socialist movements — became a target.

American business leaders pressed Congress to find a means by which to curtail unions and limit their power. Nationally, the public resented the upsurge in strike activity during the immediate postwar years, believing that the walkouts contributed to domestic inflation. Partly to preserve domestic fiscal responsibility, and partly to protect against political subversion, Congress passed antiunion legislation, culminating in the Taft-Hartley Act of 1947. The act revived the federal injunction and imposed a "cooling off" period that could be implemented nationwide; it forbade access to the NLRB for any union whose officers had not signed an anticommunism affidavit; it restricted the scope of bargaining in which the unions could engage, outlawing union participation in issues of "investment," which could influence where workers might be employed. Finally, the act limited union support for political candidates.

Murray was outraged by the Taft-Hartley Act, calling it a "dastardly crime" perpetrated against labor. Despite his own fervent anticommunist sentiments, he initially refused to sign the affidavit, saying such a requirement was "very presumptuous." Only the specter of being locked out of the NLRB, the agency that had played a crucial part in so many CIO and USWA victories, led Murray finally to sign the hated document. He

blatantly violated the campaign restrictions, however, publishing support for a congressional candidate in the CIO newspaper. He did this, he told supporters, not as an act of bitter defiance but as a statement against congressional infringement on the basic constitutional right to free speech. His position was later validated by the Supreme Court. For the first time, Murray declared, the union was prepared to use its political machinery on a national level to remove antilabor candidates and encourage widespread labor registration and voting.

Despite Murray's antagonism to the antilabor forces within Congress, he exempted Truman from his full disdain. Like the Roosevelt administration, Truman called for steel price reductions, then in 1949 and again in 1951 he convened fact-finding boards to specify wage increases. The industry executives felt hounded by both the unions and the government since, as the keystone industry, it was being singled out for such intervention. Despite Murray's political moderation, a number of industry executives claimed that he was a socialist and that it was only a matter of time before the industry would be nationalized. In 1952 the industry launched a full-scale attack against Murray, the union, and Truman. They escalated their use of union-busting law firms to try to dismantle union support in their plants and attempt to expose the "collaboration" between Truman and the steelworkers.

Steel executives refused in late 1951 to grant any of the union's 22 demands. The nation was at war in Korea, and the executives believed the government would surely intervene to exert wartime controls over labor. The steel companies refused, however, even to consider wage and price recommendations made by the Wage Stabilization Board. After a five-month impasse, Truman acted. On April 8, 1952, he broadcast to the nation that he was seizing the nation's steel mills to avert a long and costly strike and blamed the corporations' refusal to bargain in good faith as the source of the crisis. Listening in a Manhattan hotel room, Murray cheered Truman's decision. The Supreme Court, however, later ruled Truman's steel seizure to be unconstitutional, and Murray in response led his men out on a protracted seven-month strike that ended in July 1952, the day Adlai E. Stevenson was nominated as Democratic candidate for president.

More than ever Murray was determined to consolidate pro-labor sympathies in government; he threw himself into Stevenson's election campaign with a vengeance and pinned his hopes for his union and the stability of the nation on Stevenson's victory. Murray tirelessly covered the nation by train campaigning for the Democratic ticket. Stevenson, however, lost to Republican Dwight D. Eisenhower, who had no obligation to labor, but who was politically indebted to Ohio Senator Robert Taft, author of the Taft-Hartley amendment.

The morning after the election Murray boarded a train for San Francisco where he would prepare for the annual CIO convention scheduled to begin November 17. Sometime in the early hours of November 9, alone in his hotel room at the Mark Hopkins, Murray succumbed to his second heart attack, this time fatal. He was buried in Shannon, Pennsylvania.

Publications:

Build the CIO: *Acceptance Speech of Phillip Murray upon his Election as President of the CIO, at the Third Convention of the Congress of Industrial Organizations, November, 1940* (N.p.: CIO Publication no. 50, 1940);

The CIO Defense Plan (Washington, D.C.: CIO Publication no. 51, 1940);

and Moris Llewellyn Cooke, *Organized Labor and Production: Next Steps in Industrial Democracy* (New York: Harper, 1940);

Your Wages and the War (Detroit: CIO, 1940);

and John Brophy, James Carey, I. F. Stone, *The CIO and National Defense* (Washington, D.C.: American Council on Public Affairs, 1941);

Wages and War Profits (Washington, D.C.: CIO Publication no. 53, 1941);

Charting the Victory: Excerpts from President Murray's Speeches at the CIO Executive Board Meeting, Cleveland, May, 1943 (Washington, D.C.: CIO, 1943);

CIO Re-employment Plan (Washington, D.C.: CIO Department of Research and Education, 1944);

Labor's Political Aims (Washington, DC: CIO Publication no. 102, 1944);

and R. J. Thomas, *Living Costs in World War II, 1941–1944* (Washington, DC: Congress of Industrial Organizations, 1944);

Our Pledge to the Nation (Washington, DC: CIO Publication no. 115, 1944);

The Harry Bridges Case: A Foreword to the Famous Dissenting Opinion of Mr. Justice William Healy and Mr. Justice Francis Garrecht of the United States Court of Appeals for the Ninth District (San Francisco: Harry Bridges Victory Committee, 1945);

"USWA Postwar Plans," *Iron Age,* 155 (March 29, 1945): 87;

"Steel Wage Controversy," *Commercial & Financial Chronicle,* 162 (September 9, 1945): 1341, 1367–1368;

The CIO Case for Substantial Pay Increases (Washington, D.C.: Congress of Industrial Organizations, 1945);

The Case for Labor: Summary of Testimony Presented to Senate Committee on Labor and Public Welfare on February 19, 1947 (Washington, D.C.: CIO Department of Education and Research, 1945);

"The Challenge to Labor," *Commercial & Financial Chronicle,* 170 (September 8, 1949): 6, 12;

The Steelworkers' Case for Wages, Pensions and Social Insurance, as Presented to President Truman's Steel Industry Board (Pittsburgh: United Steelworkers of America, 1949).

References:

"As Steel Goes . . . ," *Time* (January 21, 1946): 15–18;

Irving Bernstein, *The Turbulent Years: A History of the American Worker, 1933–1941* (Boston: Houghton Mifflin, 1970);

Paul F. Clark, Peter Gottlieb, and Donald Kennedy, eds., *Forging a Union of Steel: Phillip Murray, SWOC, and the United Steel Workers* (Ithaca, N.Y.: ILR Press, New York State School of Industrial and Labor Relations, Cornell University, 1987);

John P. Hoerr, *And The Wolf Finally Came: The Decline of the American Steel Industry* (Pittsburgh: University of Pittsburgh Press, 1988);

Charles A. Madison, *American Labor Leaders* (New York: Harper, 1950);

Clarence Randall, *Adventures in Friendship* (Boston: Little, Brown / Atlantic Monthly Press, 1965);

Ronald Schatz, "Philip Murray and the Subordination of the Industrial Unions to the United States Government," *Labor Leaders in America,* edited by Melvyn Dubofsky and Warren Van Tine (Urbana: University of Illinois Press, 1987);

Adlai Stevenson, "Philip Murray: The Nature of Leadership," *New Republic,* 127 (December 15, 1952): 10–12.

National Committee for Organizing Iron and Steel Workers

by James R. Barrett

University of Illinois at Urbana-Champaign

In June 1918, the national convention of the American Federation of Labor (AFL) unanimously endorsed William Z. Foster's resolution calling for the unionization of the steel industry, and the National Committee for Organizing Iron and Steel Workers was established late that summer. Circumstances were favorable for a successful organizing campaign. The war had created a severe labor shortage and tremendous steel demand. Both the federal government and steel employers were anxious to avoid a labor conflict. Samuel Gompers, president of the AFL, was elected chairman of the committee, though John Fitzpatrick, president of the Chicago Federation of Labor, soon assumed this role. William Z. Foster, a long-time labor radical, was appointed secretary-treasurer of the committee, and it was Foster who oversaw the actual organizing.

While the implications of the committee's plans for the industry were important, the enterprise also represented a major departure for the labor movement as a whole. While still firmly committed to the model of craft unionism on which the federation was founded, the AFL was searching for some way to break into the rising mass-production industries. Because of the composition of the labor force in steel, the committee's plans also confronted the labor movement with the problem of organizing across lines in the labor ranks defined by race and ethnicity, as well as skill.

Each of the 24 unions with jurisdiction in the steel industry had one representative on the committee. The constituent unions pooled organizing resources and agreed voluntarily to centralize direction of the organizing campaign in the committee. As workers were recruited, they were placed in a general pool and eventually assigned membership in the appropriate union. The failure of constituent unions to adequately fund the campaign, however, forced Foster to abandon his original concept of simultaneous campaigns in every steel center. He simply lacked adequate resources.

Still, the effort had some early successes in the midwestern steel centers and even scored breakthroughs in Pennsylvania's Monongahela Valley, where employers and local government officials bitterly resisted the committee's efforts. The committee had organized nearly 100,000 steelworkers by the spring of 1919. Immigrant workers were particularly responsive and formed the bulk of the committee's membership base. When negotiations with steel employers failed that summer, committee leaders called a national strike, which began on September 22, 1919. Organizers set up an elaborate welfare system to provide strikers and their families with food and other necessities. To maintain morale, they held frequent rallies and published a strike bulletin with vital information in several languages. The strike, which often involved violent confrontation with company guards and local and state law enforcement officers, was abandoned on January 8, 1920, and the committee disbanded that July.

References:
David Brody, *Labor in Crisis: The Steel Strike of 1919* (Philadelphia: Lippincott, 1965);
William Z. Foster, *The Great Steel Strike and Its Lessons* (New York: Huebsch, 1920).

National Steel Company

by Alec Kirby

George Washington University

The National Steel Company traces its origins to the Phillips Sheet and Tin Plate Company, founded in 1905 in West Virginia by James R. Phillips to manufacture tinplate from sheet bar purchased from other firms. By the 1920s, however, the firm had become a fully integrated steel company. By the 1990s National had diversified into a wide variety of industries and remains on firm financial footing.

Vigorous early expansion of Phillips came despite several setbacks. One month after the company's formal incorporation, James Phillips was killed in a train accident. Assuming leadership of Phillips at this critical juncture was E. T. Weir, the firm's secretary and plant manager. Weir quickly established himself as a dynamic and capable leader who engineered a remarkably swift expansion for the new firm. Recruiting large-scale corporate customers, Phillips's net sales surpassed $1 million in 1907 and doubled by 1909. The key to this growth was Phillips's versatility in the production of tinplate. Rather than specializing, Phillips committed the company to filling a variety of tinplate orders. It also expanded into new areas, such as the manufacturing of terneplate. To increase the versatility of its product line, Phillips had launched an intense program of capital improvement, making Clarksburg, West Virginia, the site of one of the foremost producers of tinplate in the world. Phillips's capitalization grew from $250,000 to $750,000 in 1908.

This capital expansion, however impressive, was not sufficient to allow Phillips to meet ever-increasing demand; by 1909 the firm was straining at capacity. The decision was made to construct a vast new facility, including 20 tin mills, in Hancock County, West Virginia, at a cost of $1.11 million. With the opening of the plant in Weirton, West Virginia, came attendant growing pains. Sales were initially sluggish, and the labor supply was tight in the rural surroundings. Indeed, in the first year of Weirton's operation Phillips's profitability was maintained only through the continued full-volume production of the Clarksburg facility. Yet the board of directors grimly stuck to its expansionist policies. Early in 1912 the firm purchased the Pope Tin Plate Company of Wheeling, West Virginia, a competitor based in Steubenville, Ohio, at a cost of $1.47 million. Several months later Weir launched an investigation into the feasibility of Phillips supplying its own steel. While the investigation proceeded, Weir and other Phillips shareholders formed their own steel firm, the Weirton Steel Company, on October 5, 1912. In December 1914 the Phillips company purchased Weirton Steel. Thus integrated, Phillips saw its profits rise sharply as World War I stimulated demand and drew the firm into the export trade.

Increasingly emphasizing steel production, Phillips was reorganized as the Weirton Steel Company in 1918. The firm's profitability continued apace until the jarring postwar recession and a series of labor strikes pushed Weirton into the red for both 1921 and 1922. Yet the directors continued their now-traditional policy of expansion, investing in by-product coke production as well as coal fields and railroad cars. This expansion was facilitated through development of Weirton's production capabilities as well as the outright purchase of independent firms. By 1924 Weirton declared a breather in expansion, paid off its corporate debt, and limited its efforts to the improvement of its existing plant. It was at this time that the board of directors took out a $1 million life insurance policy on Weir at a yearly cost of $35,000. The purpose of the insurance was to allow the firm to purchase Weir's stock after his death.

Demand, however, once again strained existing capacity, and Weirton once again launched a development program, coupling it with vertical integration by investing in ore production. By 1926, 55 percent of Weirton's ore was supplied from firms in which it had a financial interest. Still, the further acquisition of ore supplies was deemed

The Great Lakes Steel Corporation plant in Detroit, circa 1930, just after the company merged with the Weirton Steel Company to form the National Steel Company (photograph by Aiklee)

vital in Weirton's drive to become fully integrated. In 1929 Weirton officially launched negotiations to merge with the Great Lakes Steel Corporation of Detroit and the ore, freighter, and blast furnace components of the M. A. Hanna Company of Cleveland. Negotiations were completed in September 1929, and ratification of the articles of merger effected on December 4, 1929.

The new company, the National Steel Corporation, reflected the strengths of its parent companies. From Great Lakes Steel came a vast, new, steel-producing facility; from Weirton came product diversity, while Hanna brought large ore resources. E. T. Weir became chairman of the board of directors of National, with George Fink of Great Lakes Steel assuming the presidency. The net earnings of National Steel in its first year of operation stood at an impressive $12,573,000, making the fledgling company sixth largest among steel producers.

Compared to other steel companies, National weathered the Great Depression of the 1930s well. In 1932, as economic activity reached rock-bottom throughout the United States, National was the only major steel producer to operate in the black, although its profits were minimal. Profitability was maintained uninterrupted

through the 1930s. This continued strength through a turbulent decade reflected National's strategic geographical position and the highly efficient manner in which the merger had been effected. Great Lakes Steel received its ore from the M. A. Hanna Company, whose mines were concentrated in the Mesabi Range; ore was shipped in Hanna's freighters. The advantages of an integrated firm were apparent. National also benefited from constant improvements in facilities and equipment by the Weirton company, which had bequeathed a modern and technologically advanced plant to National. Finally, National's continued prosperity was due in part to its capability to produce light, flat-rolled steel, for the company was the first steelmaker to license ARMCO's new continuous hot-strip mill. This allowed National to produce tinplate, rolled strip, and auto body sheets, which together comprised almost 40 percent of the industry's shipments during the Depression.

The final jolt of the 1930s came with the recession of 1937–1938, which seriously affected National's financial standing, although a precarious profitability was maintained. Net earnings fell from $17.8 million in 1937 to $6.6 million in 1938, while earnings per share fell from $8.21 to

$3.03. The value of paid dividends plummeted from $7.5 million to $2.1 million. In the face of these reverses National stuck to its theme of continued investment and expansion and successfully positioned itself to share in the returning prosperity of the post-1938 period. At the heart of this prosperity was a further commitment to product diversification, which enabled National to profit from the pre–World War II growth in defense spending.

Although National benefited from government defense contracts, the firm's relations with the administration of President Franklin D. Roosevelt always remained strained. National vociferously resisted efforts to unionize its workers, leading to inevitable conflict with labor officials in the federal government. In July 1933 E. T. Weir, in the midst of a rash of strikes throughout the industry, established a company union in an apparent effort to placate Weirton steel employees. The effort failed. In October, Weirton employees went on strike in a bid for membership in an autonomous union. On October 16 Weir seemed to capitulate, signing an agreement with the National Labor Board (NLB) pledging to allow an election held under the auspicious of the NLB in return for an end to the strike. Yet once employees were back on the job, Weir repudiated the agreement, and a helpless NLB could only act to deny the National Recovery Administration's "Blue Eagle" logo from being used by National. The conflict, however, was far from over. In June 1936 the Congress of Industrial Organizations (CIO) launched a campaign to organize the steel industry into the Steel Workers Organizing Committee (SWOC). When the SWOC reached agreement with the mighty United Sates Steel Corporation in March 1937, Weirton was one of several "little steel" companies expected to follow suit. Yet Weirton resisted, and in May 1937 SWOC filed charges of unfair labor practices against Weirton with the National Labor Relations Board (NLRB). In June 1941 the NLRB ordered Weirton to cease discouraging membership in the SWOC. Weirton responded by signing an exclusive bargaining contract with the new Weirton Independent Union — a union with no CIO affiliation. Furious labor disputes followed, with the Third U.S. Circuit Court of Appeals ultimately ordering Weirton to cease practices in violation of the National Labor Relations Act. Thereafter the CIO intensified its recruitment drive, yet not until the postwar era would these efforts be successful.

Meanwhile, National prospered from defense contracts. By 1941, 90 percent of the firm's facilities were devoted to the war program, and expansion — notably for increasing pig-iron capacity — continued. Reflecting National's increased financial strength, this expansion — costing approximately $30 million — was financed entirely by the company in 1941 and 1942. By the end of World War II the company could boast 3.9 million tons of steelmaking capacity at its Great Lakes and Weirton plants. Its steel-producing units consisted of 28 open hearths, 16 of which were at Great Lakes with the remainder at Weirton. Each plant had three blast furnaces. Part of National's powerful postwar position was a result of its leadership in light flat-rolled products. Taking advantage of its strategic location near Detroit, Great Lakes concentrated on rolled sheets for the automobile industry, while the Weirton and Steubenville plants produced sheets and tinplate.

Postwar expansion centered on acquiring coal properties to ensure an adequate supply of coking coal for the future, while National also increased its ore supply. In 1946 a subsidiary corporation, National Mines Corporation, was established for the purpose of managing National's coal reserves in Pennsylvania, West Virginia, and Kentucky. National also worked with other firms to establish coal companies — in which National had financial interests — in order to be assured of supplies.

National kept pace with scientific advances in steelmaking and often was an innovative leader in industrial technology. At the Great Lakes plant, for example, open-hearth furnaces were constructed at unprecedented sizes. In 1967 what was called "The Steel Mill of the Future" was completed at Weirton, including two 300-ton basic oxygen converters, a vacuum degassing unit, and a continuous casting machine, which replaced eight of the twelve open-hearth furnaces in operation at that time.

National's innovations during the 1960s reflected the firm's traditional approach to vigorous expansion and development. By the 1970s, however, National had become increasingly willing to forgo new investment and technological improvement, with the result of declining productivity and profitability. In 1971 a merger with Granite City Steel proved disappointing as the newly acquired firm turned out to be in serious financial difficulty. Beginning in 1974 National began to scale down the size and scope of its operations. In 1981 the plant at Great Lakes was reduced to less than a four-million-ton capacity from its peak production of 6 million tons. On September 23, 1983, the Weirton Steel Division was sold to its employees as National sought to diversify out of its dependence on the steel industry. Reorganizing to create National Intergroup, the firm under Chairman Howard M. Love in 1984 attempted to sell National Steel to the USX Corporation, although the antitrust concerns of the federal government thwarted the plan. A later attempt was made to merge with Bergen Brunswig, a major distributor of health-care products. This effort, too, failed. Undaunted, National in 1986 borrowed heavily to acquire drugstores and drug distributors. By the end of the 1980s the vigorous commitment to expansion and diversification, dating back to the Phillips Company, could still be seen in National's policies.

References:
Irving Bernstein, *Turbulent Years: A History of the American Worker 1933–1941* (Boston: Houghton Mifflin, 1970);
William T. Hogan, S.J., *An Economic History of the Iron and Steel Industry in the United States,* 5 volumes (Lexington, Mass.: Lexington Books, 1971).
Hogan, *Minimills and Integrated Mills: A Comparison of Steelmaking in the United States* (Lexington, Mass.: Lexington Books, 1987).

National Tube Company

by Kevin M. Dwyer

George Washington University

The National Tube Company, a New Jersey firm, was a product of the turn-of-the-century merger trend in American business. Its history provides a good case study in the strategies of corporate consolidation that permanently reshaped the American steel industry. William N. Cromwell and Edward C. Converse acted in conjunction with the House of Morgan to engineer the merger of 14 steel tube, pipe, and wrought iron producers in June 1899. Capitalized at $80 million, National Tube held 15 plants capable of producing over 1 million tons of steel pipe and tube — 75 percent of the nation's capacity.

The largest component of the enterprise was the National Tube Works Company, founded in Boston in 1869 by John and Harvey Flagler. Originally retailers of iron pipe and tube, the Flagler brothers thrived after the company expanded into production, and in 1872 they established operations at a newly constructed plant in McKeesport, Pennsylvania. This mill put the enterprise closer to crude-iron producers, cutting down on transportation costs. The firm outfitted their plant with the latest technologies; the addition of Siemen's regenerative gas furnaces, which upped daily output by 350 percent over the standard coal furnaces, were the first of many such improvements made. An on-site foundry was also constructed, as were rolling mills — which by 1897 were converted to electricity — six new butt-weld mills, eleven lap-weld furnaces, two blast furnaces, and, with the conversion from iron to steel, a Bessemer plant in 1892. The McKeesport works became the most self-sufficient pipe plant in the country, a position further solidified by a merger with the Monongahela Furnace Company, also of McKeesport, in the early 1890s.

Although the leader in the production of tube and pipe, with an aggregate capacity of 180,000 tons in 1890, the National Tube Works faced increasing competition due to the booming demand for petroleum products throughout the 1880s. The Riverside Iron Works was the first in the nation to produce steel pipe, and its 90,000-gross-ton capacity ranked second to National Tube Works. Riverside's lucrative position was secured through its long-term sales agreement with Standard Oil, the single greatest consumer of steel pipe in the world. Both the Riverside and National works, with a combined capacity of 350,000 tons, came under the control of National Tube Company after its incorporation in 1899.

After its consolidation, the company adopted more sophisticated strategies to secure and retain larger shares of the tube market. National Tube became the leading producer of seamless tube in the country through the acquisition of the Standard Seamless Tube Company of Ellwood

City, Pennsylvania. The giant Shelby Tube Company remained its foremost competitor in this sector. In June 1901 National Tube was itself consolidated into the giant holding company, the United States Steel Corporation. National's two-year run as an independent, however, had been impressive, earning a total of $24.1 million in profits. A product of the great merger movement, National, too, lost its independence as investors in the steel industry sought to protect their capital from the effects of cutthroat competition.

References:

William T. Hogan, S.J., *An Economic History of the United States Steel Industry,* 5 volumes (Lexington, Mass.: Lexington Books, 1971);

Naomi R. Lamoreaux, *The Great Merger Movement in American Business* (Cambridge, U.K.: Cambridge University Press, 1985).

North Star Steel

by Robert W. Crandall
The Brookings Institution

North Star Steel is a private minimill steel company, wholly owned by Cargill Industries of Minneapolis, Minnesota. Established in 1966 by a consortium of investors that included Co-Steel International of Canada and Cargill, North Star began in Saint Paul, Minnesota, with a single minimill, which began producing steel bars in 1967. North Star currently operates seven minimill plants — each producing bars, small structural shapes, or tubular steel products — and one finishing plant.

In 1974 North Star negotiated to purchase Iowa Steel, a failing minimill in Wilton, Iowa. At the time Cargill bought out Co-Steel's interest in North Star because Iowa Steel was a direct competitor of one of Co-Steel's subsidiaries. As a result Cargill became the sole owner of the company.

In the 1980s North Star expanded by building a new minimill in Monroe, Minnesota, and acquiring four existing plants in Beaumont, Texas; Youngstown, Ohio; Milton, Pennsylvania; and Calvert City, Kentucky. These acquisitions and subsequent improvements gave North Star approximately 3 million tons of raw steel capacity annually, making it one of the two largest minimill companies in the United States. Like most minimills North Star produces only "long" products — bars, rods, small structurals, and tubular products. The Beaumont plant is capable of producing up to 700,000 tons of wire rods per year. The Youngstown plant is dedicated to producing seamless tubular products, and the Kentucky plant produces medium-size structural shapes. The remaining plants produce an array of merchant and special-quality carbon-steel bar products.

References:

William T. Hogan, S.J., *Minimills and Integrated Mills: A Comparison of Steelmaking in the United States* (Cambridge, Mass.: Lexington Books, 1987);

Richard Serjeantson, Raymond Cordero, and Henry Cooke, eds., *Iron and Steel Works of the World,* ninth edition (Surrey, England: Metal Bulletin Books, 1988).

Northwestern Steel and Wire

by Robert W. Crandall
The Brookings Institution

Northwestern Steel and Wire was an early pioneer in the use of electric furnaces to produce steel with a low carbon content. In 1936, as the Northwestern Barbed Wire Company, it opened an electric steelmaking facility in Sterling, Illinois, intending to produce wire rods to feed its wire-drawing facilities in direct contravention of National Recovery Administration rules. When the National Recovery Act was declared unconstitutional by the Supreme Court in 1936, Northwestern Barbed Wire was free to launch its new steelmaking operation. Subsequently, in 1938, it changed its name to Northwestern Steel and Wire.

Northwestern was an early forerunner to the modern steel minimill, producing small-diameter rods and bar products prior to World War II. Northwestern grew as a steel producer in the early post–World War II period, installing larger electric furnaces and rolling mills for relatively large structural shapes. The minimill revolution would follow a quarter of a century later, when firms such as Florida Steel, North Star Steel, and Nucor would move into the production of structural steel using electric furnaces.

The modern minimill generally employs one medium-sized electric furnace that feeds a continuous billet caster. By contrast, Northwestern Steel and Wire invested in increasingly larger furnaces that fed a conventional ingot-casting operation. These ingots then had to be reheated and rolled into blooms or billets which, in turn, were reheated and rolled into bars, rods, and structural steel. As a result, the company's production process was an inefficient one when compared to the post-1960 minimills, which eliminated the first reheating and rolling process by investing in continuous casters.

In the 1980s Northwestern Steel and Wire modernized its Sterling plant. Sterling's primary rolling facilities were replaced with a continuous billet caster and a "beam-blank" bloom caster, which could roll large structural shapes up to 24 inches in width.

In 1989 Northwestern purchased Armco's structural steel plant in Houston for $90 million and has invested another $60 million in modernizing the plant. Much of this investment has been directed toward building a modern beam-blank caster to feed the large structural mills. The plant is able to roll up to 600,000 tons per year of wide-flanged beams up to 24 inches in width.

Northwestern Steel and Wire continues to produce a variety of steel, wire and fabricated wire products. It still produces barbed wire, its original product, but it also produces a variety of other wire products, including fine steel wire, baling wire, electric fence wire, wire fencing, hardware clot, and other wire products.

Though a public company, Northwestern Steel and Wire has remained under the control of the Dillon family since its founding in 1879 by Washington M. Dillon. In 1920 Washington Dillon died and was succeeded by his son Paul W. Dillon, who served as chairman until 1979.

Reference:
William T. Hogan, S.J., *An Economic History of the Iron and Steel Industry in the United States* (Lexington, Mass.: Lexington Books, 1971).

Irving S. Olds

(January 22, 1887 – March 4, 1963)

by Bruce E. Seely

Michigan Technological University

CAREER: Member (1911–1917), partner, White & Case (1911–1963); director and member of executive committee (1936–1960), special counsel (1938–1940), chairman of the board, United States Steel Corporation (1940–1952).

Irving S. Olds was the last financial man connected with J. P. Morgan's investment banking house to head the United States Steel Corporation, replacing Edward R. Stettinius, Jr., just as World War II began. After a distinguished legal career, he guided the company through the war and into the prosperity of the 1950s.

Olds's family was well-to-do; his father, Clark Olds, was a successful lawyer in Erie, Pennsylvania. Olds graduated from Yale University in 1907 and Harvard Law School, where he finished near the top of his class, in 1910. Admitted to the Pennsylvania bar in 1910 and the New York bar in 1912, Olds clerked for Justice Oliver Wendell Holmes in 1910–1911. The New York law firm of White & Case hired Olds in 1911 and in 1915 named him counsel for the export department of J. P. Morgan & Company, which oversaw munitions purchases by the British and French governments. Two years later he was made a partner in the firm and, while still involved with munitions purchases as counsel to the British War Mission's purchasing department, served as special assistant to Stettinius, then assistant secretary of war.

After 1919 Olds practiced corporate law, gaining a reputation as an able reorganizer and administrator of corporations; he also opened an office for White & Case in Paris in 1928. In 1936 Olds was placed on the executive committee of U.S. Steel, an appointment in part explained by his long association with Morgan bankers, who dominated the U.S. Steel board. Also in Olds's favor were his relatively liberal social views, which he shared with board chairman Myron Taylor, fellow board member Stettinius, and others. Perhaps as important, Olds's career in law and business mirrored that of Taylor, who was at-

Irving S. Olds

tempting to introduce young blood to U.S. Steel management. In 1938 Olds was appointed the corporation's special counsel for the Temporary National Economic Committee's inquiry into monopoly practice. He won high remarks for his representation of U.S. Steel, often relying on his sense of humor. Olds provided what *Fortune* called "a model of how Big Business could defend its right to a profit and still be solicitous of the public interest. . . . There is no denying the favorable impression that the corporation made."

In June 1940 Olds was elected chairman of the board, succeeding Stettinius, and continuing the pattern of Morgan-connected chief executives. As chairman, Olds did not have a far-reaching agenda, and he continued Stettinius's plan of letting Benjamin Fairless control steelmaking operations. But a wartime building program and production effort soon had the corporation operating above 100 percent capacity, raising concerns over labor relations, government contracts, defense plant additions, and materials priorities.

The end of World War II brought further problems to U.S. Steel management. Labor contracts were among the first of Olds's postwar concerns. Some steel executives considered renewing their efforts to oust organized labor, but Olds negotiated contracts with the United Steelworkers of America in 1946, 1947, and 1948, establishing an industrywide trend. The only strike came in 1946, largely because government wage controls interfered with negotiations. The 1947 agreement provided a 15-cent-per-hour increase, quarterly meetings between labor and management, severance and vacation pay, and negotiations to discuss seniority. By 1948 Olds could talk of "a mutuality of interests between employer and employee."

Initially Olds's postwar relationship with the federal government was relatively harmonious, as U.S. Steel acquired the Geneva Steel Works in Utah, built by the corporation for the government during the war. Olds himself received a Treasury Department citation for his service on a committee that studied foreign trade. He also closed the long battle with the Federal Trade Commission over basing point pricing, adopting "f.o.b. mill" pricing. Olds's relationship with the administration of President Harry S Truman soon soured, however, as U.S. Steel sought to recoup labor costs through higher prices. Olds argued that the corporation needed higher profits to fund continued expansions of capacity. But the Truman administration was waging a war against inflation and was ready to blame high steel profits. Olds's commentary on government policy grew increasingly strident. He retired in 1952, on the eve of Truman's seizure of the steel mills, which he denounced as the first step toward socialism.

In retirement Olds remained on U.S. Steel's board until 1960, although after his wife, Evelyn Foster, died in 1957, he devoted more time to civic pursuits. In 1952, with Alfred Sloan of General Motors and Frank W. Abrams of Standard Oil of New Jersey, Olds organized the Council for Financial Aid to Education to encourage corporate support for private colleges. He also was a loyal Yale alumnus, chairing the Yale Council in 1959 and a fund drive in 1960. As a trustee of the New York Public Library, he chaired an industrial committee in a 1954–1955 fund drive. He also

was a trustee of the Pierpont Morgan Library; was a trustee of the Cooper Union; and after 1941 became a trustee of the New York Institute for the Education of the Blind.

Olds was also an accomplished amateur historian. He collected naval prints and authored two books on naval history. He wrote histories of Elbert Gary and Erie, Pennsylvania, as well as an account of U.S. Steel's first 50 years. His historical affiliations were many, including the Newcomen Society, the Naval History Association, the American Heritage Foundation, the American Academy of Political and Social Science, and the Bibliographic Association of America; he was also elected president of the New York Historical Society in 1962. Interest in the arts led him to serve as a trustee of the American Shakespeare Festival and the Metropolitan Museum of Art and as a director of the Metropolitan Opera.

Olds remained active in business late in life. In 1960 Howard Hughes's stock in TWA was placed in trust to satisfy the billionaire's creditors. As a TWA trustee, Olds completely reorganized the directorship of the company, against Hughes's will. Ill health forced Olds to resign in 1962, and he died on March 4, 1963.

Selected Publications:
"Cooperation of Steel Industry in Defense Program Pledged," *Commercial and Financial Chronicle,* 151 (November 21, 1940): 1644–1645;
"Position of Steel Industry to Meet Responsibilities in Defense Program," *Commercial and Financial Chronicle,* 152 (May 24, 1941): 3272–3273;
"American Steel Industry Has Capacity Adequate to Meet Demands of War," *Blast Furnace and Steel Plant,* 29 (June 1941): 603–605;
United States Navy, 1776–1815 (Np., Grolier Club, 1942);
"U.S. Steel Corporation Produces 30,000,000 Tons of Ingots in 1942," *Blast Furnace and Steel Plant,* 31 (January 1943): 132–133;
"Some Aspects of the Proposed Guaranteed Wage," *Vital Speeches,* 11 (August 1, 1945): 631–634;
Ten Momentous Years at Erie, 1753–1763 (New York: Newcomen Society, 1947);
Judge Elbert Gary: His Life and Influence upon American Industry (New York: Newcomen Society, 1947);
"Irving S. Olds," *Current Biography* (1948): 482–484;
"Government and the Steel Business," *Blast Furnace and Steel Plant,* 37 (February 1949): 224–225;
Erie: Historic City of the Keystone State (Erie, Pa., 1949);
Management's Responsibility to Industry (New York: U.S. Steel Corporation, 1949);

Pertinent Facts about Business (New York: U.S. Steel Corporation, 1949);

Half a Century at U.S. Steel (New York: Newcomen Society, 1951);

Our Privileged Millions (New York: U.S. Steel Corporation, 1951);

The Tax Education of Mr. Jones (New York: U.S. Steel Corporation, 1951);

Bits and Pieces of American History, as Told by a Collection of American Naval and Other Historical Prints and Paintings, Including Portraits of American Naval Commanders and Some Early Views of New York (New York, 1951);

What Price Controls? (New York: U.S. Steel Corporation, 1952);

The Price of Price Controls (Irvington-on-Hudson, New York: Foundation for Economic Education, 1952);

The Thousand Miles of Lao-Tse (New York: U.S. Steel Corporation, 1952);

"Two of Malbone's Miniatures," *Antiques,* 66 (October 1954): 288–289.

References:

Edith Fisch, *Lawyers in Industry* (New York: Oceana Publications, 1956);

William T. Hogan, S.J., *An Economic History of the Iron and Steel Industry in the United States,* 5 volumes (Lexington, Mass.: Lexington Books, 1971);

A. St. John, "U.S. Steel's New Chairman, Succeeding Stettinius," *Barron's,* 20 (July 29, 1940): 40.

Open-Hearth Furnace

by Robert Casey
Henry Ford Museum & Greenfield Village

The open hearth produces a relatively high-quality steel by removing carbon and other impurities from pig iron. The process originated in France, where Pierre and Emile Martin used William Siemens's regenerative furnace to make steel in 1864. Abram S. Hewitt brought the process to the United States in 1868. By 1900 one-third of all American steel was made in open hearths. Open-hearth production exceeded Bessemer production in 1908 and remained the dominant steelmaking technology for the next 60 years.

The open-hearth furnace had several advantages over the Bessemer converter: 1) the open-hearth process was slower, allowing better control over the final product; 2) it could use much more scrap, thereby recycling what was normally a waste product; 3) it could remove the high amounts of phosphorus typically present in American pig iron. Development of the process was characterized not by spectacular breakthroughs, but by incremental improvements. By 1900 Samuel Wellman had developed an electric charging machine and an electromagnet for handling pig and scrap iron. Water-cooled frames and doors — extending furnace life — and waste-heat boilers, which recaptured heat previously lost up the stack, were added by 1910. Straight-line reversing valves were added to the furnace in 1917 to improve its reliability; and the sloping backwall and the

An open-hearth furnace in a U.S. Steel plant (photograph by Aikins)

mechanical-bottom maker were incorporated by 1925 to simplify repairs. Dirty producer gas was replaced as a fuel by coke-oven gas, fuel oil, natural gas, and coal tar. In the 1930s improved measurement of furnace temperatures, pressures, and fuel flows gave operators more knowlege of and greater control over what was going on inside the furnace. In the 1950s the use of an explosive charge — the jet tapper — made tapping the furnace both simpler and safer. In the 1960s open-hearth productivity was virtually doubled by the use of oxygen lances extending through the furnace roof.

The size of the furnaces steadily increased during the twentieth century. A large furnace in 1900 had a capacity of 50 tons. By 1920, 100-ton furnaces were being built. In the 1930s, 150-ton furnaces were being used, and by the 1950s common furnace size increased to 350 tons.

The open hearth was eventually superseded by a descendant of the Bessemer converter, the basic oxygen furnace (BOF). First used in the United States in 1954, BOFs produced 58 percent of American steel by 1988, while open hearths produced only 5 percent.

References:

Bela Gold and others, *Technological Progress and Industrial Leadership: The Growth of the U.S. Steel Industry, 1900–1970* (Lexington, Mass.: Lexington Books, 1984);

William T. Hogan, S.J., *An Economic History of the Iron and Steel Industry in the United States*, 5 volumes (Lexington, Mass.: Lexington Books, 1971);

U.S. Steel, *The Making, Shaping, and Treating of Steel*, seventh edition (Pittsburgh: U.S. Steel, 1957).

Otis Steel Company

by John A. Heitmann

University of Dayton

The Otis Steel Company, one of America's most dynamic small producers of open-hearth steel during the late nineteenth and early twentieth centuries, had it origins in Charles A. Otis's pioneering efforts in the early 1870s to manufacture steel using new technology. Prior to the Civil War, Otis was the owner of a wrought-iron forge and rolling mill in Cleveland, Ohio, and in 1867 he sold his firm to the Cleveland Forge Company. Otis subsequently traveled to Europe, where he studied new steelmaking processes and received a license to use a technology developed by Dr. William Siemens. In 1856 Siemens and his brother had patented a steelmaking method that made a higher-quality steel with greater precision than the commonly used Bessemer process.

Otis hired engineer Samuel T. Wellman to help construct his Lakeside, Ohio, plant. Wellman supervised the erection of two 7-ton acid furnaces in 1874, the first open-hearth operation in the United States. In 1878 two more furnaces, each with a 15-ton capacity, were installed.

In 1889 Otis's firm was purchased by an English investment company, the Industrial and General Trust, Ltd.; for the next 30 years, however, management of the firm remained in the hands of Americans, under the leadership of George Bartol. Bartol had been hired as a chemist by Otis in 1879 and rose steadily through the ranks; he was appointed general manager in 1897 and elected president in 1898. Under Bartol the company expanded to a new 330-acre site in the Upper Cuyahoga River Valley in 1912. The new Riverside works was equipped with a modern 84-inch tandem plate mill and four jobbing mills. With the vast economic opportunities created by World War I the Otis Company prospered, and in 1919 Cleveland entrepreneur John Sherwin purchased the firm from its English owners. The late 1920s witnessed a reorganization at Otis and the rise of E. J. Kulas as president of the firm, followed by a steady expansion that brought capacity to 1 million tons on the eve of the Great Depression.

Otis's efficient operation, market niche, and geographical location on the Great Lakes made it an attractive target for a merger, and in 1942 Jones & Laughlin of Pittsburgh, eager to exploit midwestern markets, acquired the company. Despite the decline of the American steel industry in recent decades, the legacy of Otis Steel has persisted, since LTV Steel Company has made Cleveland its home office and the Riverside plant remains one of its integral manufacturing centers.

References:
Wilfred Henry Alburn and Miriam Russell Alburn, *This Cleveland of Ours* (Chicago: S. J. Clarke, 1933);

William T. Hogan, S.J., *An Economic History of the Iron and Steel Industry in the United States*, 5 volumes (Lexington, Mass.: Lexington Books, 1971).
Otis Steel Company, *The Otis Steel Company, Cleveland, Ohio* (Cleveland, 1929).

Thomas F. Patton

(December 6, 1903 –)

by Carol Poh Miller

Cleveland, Ohio

CAREER: Legal department, Union Trust Company, Cleveland (1926); associate, Andrews & Belden (1926–1932); partner, Belden, Young & Veach (1932–1936); legal department (1936), general counsel (1937–1944), director (1943–1976), vice-president and general counsel (1944–1953), assistant president and first vice-president (1953–1956), president (1956–1968), chief executive officer (1960–1971), chairman of the board (1963–1971), honorary chairman, Republic Steel Corporation (1971–1976).

Tom Patton "hitched his wagon to a steel ingot instead of a star," wrote the *Cleveland Press* in announcing his elevation to the presidency of the Republic Steel Corporation in August 1956. Patton's promotion followed 30 years in top executive positions, including service as general counsel during the formative years of the company. He was the first Republic president who did not come up through production.

Born in Cleveland, Thomas Francis Patton was the son of Irish immigrants John T. and Anna (Navin) Patton. He grew up on the city's West Side, attending Saint Colman's School and Saint Ignatius High School. He graduated from the College of Law of Ohio State University and was admitted to the Ohio bar in 1926. He worked briefly in the legal department of the Union Trust Company before joining the Cleveland law firm of Andrews & Belden.

In 1930, when the Republic Steel Corporation was organized after Central Alloy Steel Corporation merged with the Bourne-Fuller, Donner Steel, and Republic Iron & Steel Companies,

Thomas F. Patton (right), with William D. Martin, editor of the Republic Steel company magazine, March 1963 (Cleveland Public Library Picture Collection)

Patton's law firm served as counsel to the merger committee. Patton helped work out the complicated transaction, including financing, antitrust, and title problems. He gained considerable knowledge of the ore- and coal-mining industries through his legal work for such clients as Pickands, Mather & Company, the Cleveland-Cliffs Iron

Company, and the M. A. Hanna Company. As counsel for Cleveland's Corrigan, McKinney Steel Company, he handled extensive litigation, learning the economics and operations of the steel industry. In 1936, Republic president Tom M. Girdler invited Patton, then thirty-three, to join Republic to form a legal department; a year later, he was named general counsel for the corporation.

Patton's arrival at Republic coincided with John L. Lewis's drive to organize steelworkers, culminating in the "Little Steel" Strike of 1937. Patton represented Republic in the "Little Steel" hearings before the National Labor Relations Board in 1937 and before the War Labor Board in 1942 and 1944; he oversaw Republic's signing of its first labor contract with the United Steelworkers of America (USWA) on August 11, 1942. He was active in all of Republic's financing operations and in the negotiations that led to the firm's establishing iron ore operations in Liberia, Labrador, and Minnesota. In 1953 Patton was appointed to the specially created position of assistant president and first vice-president, Republic's third-ranking post; in that capacity he supervised the company's financial, legal, purchasing, traffic, and public and labor relations departments. "This is how a non-technical man who never poured steel became the president of a billion-dollar steelmaker," the *Cleveland Press* said in 1956.

A handsome six-footer with prematurely grey hair, Patton was a member of the four-man team that represented the steel industry in contract negotiations with the USWA in 1956. He succeeded the colorful Charles M. White as chief executive officer in 1960. In passing the baton, White also handed Patton a bottle of tranquilizers. "The steel business," he growled in his customary style, must "damn well get a hell of a lot better damn quick."

Profits had declined from the levels of 1956–1957, and foreign competition was stiffening. "We think it is neither flag-waving nor insular," Patton told an *Iron Age* reporter in 1964, "to say that our [nation's] first concern and our primary allegiance should be with our own industries [which] pay their own way and provide their own capital instead of receiving government grants and outlays." During the 1960s Patton presided over a major program of modernization and expansion of Republic's facilities and grappled with increasing steel imports, rising wages, and declining profitability. He retired from Republic in 1971 and the following year was appointed as a trustee to administer the bankruptcy of the Erie-

Lackawanna Railroad, a position he still held in 1990.

Between 1962 and 1965 Patton served as chairman of the board and chief executive officer of the American Iron and Steel Institute. He also served as a director of the American Telephone and Telegraph Company, Cleveland-Cliffs Iron Company, Union Commerce Bank (Cleveland), Standard Oil Company (Ohio), and the Ohio Bell Telephone Company.

Patton was a longtime member of the board of the Cleveland Chamber of Commerce and served as its chairman for two terms. Patton also served as chairman and trustee of the Cleveland Development Foundation, a group of government and business leaders that instigated slum-clearance and redevelopment efforts beginning in the 1950s.

Patton was married to Arline Everitt of Columbus, Ohio, in 1928. They had two daughters: Arline and Carol. Patton resides in Shaker Heights and still has an office in Cleveland.

Publications:
"Significant Gains to the Nation Resulting from Steel Strike," *Commercial and Financial Chronicle,* 176 (September 18, 1952): 1018;

"Steel Industry's Prospect and Blueprinting Tomorrow's Program," *Commercial and Financial Chronicle,* 185 (March 14, 1957): 1212;

"Value of Communication to Get Public Understanding," *Commercial and Financial Chronicle,* 191 (June 2, 1960): 2372–2373;

Business Survival in the Sixties [The Charles Moskowitz Lectures] (New York: Harper, 1961);

"Short-term Steel Outlook," *Commercial and Financial Chronicle,* 193 (May 25, 1961): 2282;

"Beyond Tomorrow: The Ultimate Force of Ideas," *Vital Speeches,* 28 (February 1, 1962): 226–229;

"Steel Industry's Deadly Roadblock; Broad Ignorance of its Problems," *Commercial and Financial Chronicle,* 195 (June 7, 1962): 2649;

"How Steel's Improving Fortunes Can Be Considerably Enhanced," *Commercial and Financial Chronicle,* 197 (May 30, 1963): 2205;

"Newly Emerging Steel Industry to Advance at a Lightning Pace," *Commercial and Financial Chronicle,* 199 (June 18, 1964): 2337.

References:
Edith L. Fisch, with Matthew Foner and Albert P. Blaustein, *Lawyers in Industry* (New York: Oceana Publications, 1956), pp. 25–26;

"How a Lawyer Became a Steel Corp. President," *Cleveland Press,* August 27, 1956;

"Steel Chief Dissects Industry's Future," *Iron Age,* 193 (February 27, 1964): 45.

Pennsylvania Steel Company

by Lance E. Metz

Canal Museum, Easton, Pennsylvania

The Pennsylvania Steel Company played a pioneering role in the development of the Bessemer steelmaking process in America. Through its subsidiary, the Maryland Steel Company, it created one of America's largest steel plants at Sparrows Point, Maryland.

Pennsylvania Steel was founded in Philadelphia on September 22, 1865. Its principal stockholders were J. Edgar Thomson and Tom Scott, president and vice-president of the Pennsylvania Railroad, Nathaniel Thayer of the Baldwin Locomotive Works, William Sellers, a Philadelphia tool manufacturer, and Samuel Felton, a former president of the Philadelphia, Wilmington, & Baltimore Railroad. Samuel Felton was elected Pennsylvania Steel's first president, and land was purchased along the Susquehanna River, east of Harrisburg, Pennsylvania. A steelworks and company town eventually known as Steelton were built.

Construction of the company's production facilities began in 1866, under the direction of noted engineer Alexander Holley. The initial plan was to import a Bessemer converter from England, but an American-built converter had to be substituted when the ship carrying the converter foundered off the coast of Ireland. On May 26, 1867, the converter was placed in operation, and three months later steel ingots were successfully rolled into rails at the Cambria Iron Company of Johnstown, Pennsylvania. These were among the earliest steel rails produced in America. On May 25, 1867, Pennsylvania Steel's own rail mill began production and, due to the quality of its products and the large orders that were placed by the Pennsylvania Railroad, Pennsylvania Steel soon became a leader of the nascent American steel industry. By 1872 the plant had added railroad switches and frogs to its product line. Prior to 1873 Pennsylvania Steel purchased pig iron for its converters from outside suppliers, but in that year it added a blast furnace to its operations. A blooming mill was installed in 1876

and a forge shop was also added. In 1876, the Bessemer converter was supplemented by an open-hearth furnace plant.

By 1880 Pennsylvania Steel was rolling 110,000 tons of rails annually and producing yearly profits of nearly $2 million. It dominated its eastern Pennsylvania rivals, the Bethlehem Iron Company and the Lackawanna Iron and Coal Company. In 1882 the Pennsylvania Steel Company in cooperation with the Bethlehem Iron Company began the development of large iron ore mines at Juraga, Cuba. To best use this new ore source, Pennsylvania Steel began to seek a site on the Atlantic Coast for a new steel plant. Sparrows Point near Baltimore, Maryland, was chosen, and work began on the plant in 1887. In 1891, the Maryland Steel Corporation was created to operate the Sparrows Point plant, which began operation in that year.

During the 1880s and 1890s Pennsylvania Steel continued to prosper under the direction of Luther Bent, the son-in-law of the company's first president, Samuel Felton. Due in large measure to the success of its Sparrows Point subsidiary, Pennsylvania Steel was able to maintain a substantial market share successfully despite increased competition from steel plants in western Pennsylvania and the Great Lakes area.

On February 16, 1916, Pennsylvania Steel and its subsidiary, the Maryland Steel Corporation, were purchased by the Bethlehem Steel Corporation; the plants at Steelton and Sparrows Point have since been operated by Bethlehem.

References:

Robert Hessen, *Steel Titan: The Life of Charles M. Schwab* (New York: Oxford University Press, 1975);

Jeanne McHugh, *Alexander Holley and the Makers of Steel* (Baltimore: Johns Hopkins University Press, 1980);

Mark Reutter, *Sparrows Point: Making Steel, the Rise and Ruin of American Industrial Might* (New York: Summit, 1988);

John B. Yetter, *Steelton, Pennsylvania* (Harrisburg, Pa.: Triangle Press, 1974).

Archives:
The Steelton history file is located in the Bethlehem Steel Corporation Collection, Hugh Moore Historical Park and Museums, Easton, Pennsylvania.

George Walbridge Perkins

(January 31, 1862 – June 18, 1920)

by John N. Ingham

University of Toronto

CAREER: Office boy (1877–1886), salesman, later district supervisor (1886–1892); first vice-president, New York Life Insurance Company (1892–1901); partner, J. P. Morgan & Company (1901–1910); chairman of finance committee and director, United States Steel Corporation (1901–1916).

Perkins came from an eminent American family, one which had come to New England in 1631. At the time of his birth on January 31, 1862, in Chicago, Illinois, his father was superintendent of a reform school; he later became warden of the state prison at Joliet, Illinois. His father believed the prison environment was not a good one in which to raise a son, and, as a result, he took over management of the business of the New York Life Insurance Company's Chicago offices. A poor student, Perkins left the Chicago public schools at the age of fifteen and became his father's assistant at New York Life. Perkins went to Cleveland when his father was made manager of the Ohio region.

Upon his father's death in 1886, Perkins made a bid to take over the agency, despite the fact that he was only twenty-four years old. The president of New York Life refused his entreaties and instead made him a salesman in Indiana. Within a short time, George Perkins had sold so many policies that he was made district supervisor. This was the beginning of Perkins's meteoric rise in the organization. In 1892 he was named third vice-president and given responsibility for all agency matters. In this post Perkins helped revolutionize New York Life and the insurance industry. It had long been an industry practice to farm out territory to middlemen or general agents, who then appointed other agents to do the actual

George Walbridge Perkins

soliciting. The result was a highly unprofessional operation, in which underpaid and untrained agents sold insurance, making spurious claims to secure the initial premiums. When Perkins took over, he quickly phased out the general agents and began hiring a permanent direct sales force. In 1896 Perkins introduced the "Nylic" system of benefits given to salesmen based on length of service and number of policies written. The system was later adopted by most other insurance firms. A few years later Perkins invaded the European market with the system.

In the later 1890s, Perkins began investing the massive sums controlled by New York Life in the profitable business of underwriting foreign securities. This naturally brought him into contact with J. P. Morgan, the eminent investment banker. Morgan, always on the watch for new talent, determined that Perkins would be a useful addition to the House of Morgan. He offered Perkins a partnership in the banking firm, but the young man at first refused. In 1901, however, when Morgan agreed to allow him to retain part-time vice-presidency at New York Life, Perkins joined the banking firm. Perkins soon became, in the words of his biographer, Morgan's "Right Hand Man" at the firm and was particularly important in the running of the United States Steel Corporation during the massive steel company's turbulent early years.

When Perkins joined J. P. Morgan & Company as a partner, the banking firm's most urgent task was to bring a sense of order to the newly organized U.S. Steel. A holding company presiding over still-independent finishing, semifinishing, iron ore, coal, and transportation firms, U.S. Steel was plagued by a power struggle between the old "Carnegie Men," allied with former Carnegie Steel head Charles M. Schwab, and the "bankers," allied with Elbert H. Gary. Head of the corporation's executive committee, Gary wanted to transform U.S. Steel into a highly centralized organization. Schwab, on the other hand, wished for more freedom and autonomy for the constituent firms comprising U.S. Steel.

Robert Bacon, as chairman of the finance committee, was J. P. Morgan's voice at U.S. Steel, and had to mediate between the two groups. The key player in the struggle prior to 1902, however, was Morgan himself. With both Schwab and Gary vying for Morgan's support, Bacon found himself pushed to the side by the three more powerful personalities. Ultimately, Gary brought Morgan his resignation, and when Morgan inquired why, Gary laid out his plans for a centralization. Morgan at that point gave his unqualified support to Gary; as a result, Schwab soon left U.S. Steel. When Schwab resigned, the executive committee was discontinued. Gary was named chairman of the board of directors, and power was centered in the finance committee. Bacon subsequently stepped down as chairman of that committee, and in replacing him Perkins was given the task of

working directly with Gary. Fortunately, the two men got on well, and together they ran U.S. Steel for the next eight years.

Although he was the bankers' representative at U.S. Steel, Perkins was not a banker by nature. He was an organizer, a person who was an expert in dealing with personnel; he represented the new world of professional management, which stressed the importance of working with individuals rather than with either materials or money. Perkins himself recognized his strengths: "In the ten years I was with Morgan's I never went behind the counter or examined into the bookkeeping end of the business. . . . My job was to assist in the physical organization of the great industrial combines which Mr. Morgan was then engaged in financing."

Perkins's first big task at U.S. Steel, however, involved money, not men. In 1902 the corporation needed to raise some $50 million in additional capital, and the first thought had been to issue new preferred stock. Others on the board advocated a new bond issue instead. Perkins came up with a novel and complicated alternative, referred to as the Bond Conversion Plan. Under this plan, the holders of the company's preferred stock would be offered an opportunity to convert 40 percent of their holdings into bonds. They also could purchase additional bonds up to 10 percent of their holdings for cash. If every stockholder exercised this option, the company would raise $50 million in cash for needed expansion and integration. The beauty of the plan, according to Perkins, was that the money would be raised at no cost to the steel firm, since the annual savings on dividends on the retired preferred stock would exceed the interest payments on the new 5 percent bonds.

Perkins's plan, however, ran into problems, and he was forced to modify it so that a syndicate headed by the House of Morgan would guarantee part of the transaction. In exchange, the banking firm would receive $10 million as a commission. This reduced the total amount of cash Perkins could raise to $40 million. Furthermore, a group of U.S. Steel shareholders objected strenuously to the plan and obtained an injunction enjoining the company from implementing it. Although the courts ultimately decided in favor of the plan, the long delay proved disastrous: when the time limit for converting the bonds expired, only $45 million worth had been exchanged; Perkins had figured on $250 million. Perkins, however, would prove much more successful in the field of labor relations.

The U.S. Steel finance committee in 1917: George F. Baker, Henry C. Frick, George W. Perkins, Percival Roberts, and Elbert H. Gary

He took Morgan's philosophy of the need for harmonious relations among large corporations and of the destructive nature of competition one step further, for Perkins also believed strongly in the need for harmonious labor relations. He thought this could best be achieved by giving the worker a stake in the corporate structure. To this end, he set up an innovative and important profit-sharing and stock ownership plan for U.S. Steel employees. Although not the first profit-sharing program in America, Perkins's plan attracted a good deal of attention, largely because of its size and comprehensiveness. There were two components: a bonus and a stock purchase plan. The bonus was a carryover from the Carnegie years and was available only to managerial personnel. The stock purchase plan, however, was open to all employees and was the portion of the plan that both caught the public fancy and invited a good deal of controversy.

Under the stock purchase plan, employees were invited to buy shares of stock in installments at prices slightly below market value. Although Perkins bragged that the plan was for everyone "from the President to the man with pick and shovel," critics pointed out that only a small portion of well-paid, privileged workers drew wages high enough to allow them to make these purchases even with installment payments. This criticism was leveled directly at Perkins when he testified before the Stanley Committee, which was investigating the steel industry in 1911. Perkins replied that 13,000 workers had subscribed to the plan, but critics noted that this total was only 15 percent of U.S. Steel's total workforce; the remaining 85 percent either chose not to participate or could not afford it.

Some members of the Stanley Committee wondered if Perkins had devised the plan to forestall unionization of U.S. Steel. Indeed, some recent analysts of labor reform in the steel industry have charged that the plan was nothing more than a public relations gimmick in lieu of more meaningful reforms. Samuel Gompers, head of the American Federal of Labor in the early twentieth century, was more charitable, asserting that Perkins "had a broad human understanding of the problems of industrial relations."

Perkins proved to be a major player in one of U.S. Steel's most controversial acquisitions — the purchase of the Tennessee Coal, Iron & Railroad

Company (TCI&R). U.S. Steel's intentions behind the purchase remain unclear. When the administration of President William Howard Taft later brought an antitrust suit stemming from the TCI&R incident, the corporation explained that the purchase was due to Gary's grudging acceptance of Morgan's plan to save a Wall Street brokerage firm. Moore & Schley held vast amounts of TCI&R stock but was on the verge of bankruptcy due to the Panic of 1907. Fearing that the unloading of TCI&R stock would further destabilize the market, Morgan suggested to U.S. Steel that it buy up the stock. Henry Clay Frick and Gary were hesitant, but were ultimately won over by Perkins' forceful presentation on the value of the property. President Theodore Roosevelt later testified that Gary had sought his approval prior to the purchase. Yet to many the purchase represented the dangerous ambitions behind overly aggressive trusts.

By 1909 Gary was once again feuding with the presidents of the subsidiary companies. William E. Corey, an old Carnegie employee who had succeeded Schwab as president of U.S. Steel, began to confront Gary over pricing policy. Gary and Perkins had long supported a policy of administered prices for their steel products — prices that were set by the "visible hand" of professional administrators rather than the "invisible hand" of the marketplace; they also believed that wages should also be held steady, even in bad times. Corey and the other subsidiary presidents had endured this policy for some time but with the financial Panic of 1907 began to argue strongly for changes.

In 1907 Gary and Perkins had worked to stabilize steel prices in the industry by holding informal meetings with the industry leaders. In what would come to be known as the "Gary dinners," they persuaded the heads of competing steel firms to follow U.S. Steel's leadership in price setting. As the market continued to deteriorate in 1909, however, and as the independent steelmakers began to slash both prices and wages, Gary and Perkins came under bitter attack from their own ranks. To preserve market share, Gary finally agreed to cut the prices of steel products but insisted on maintaining wages at the former levels. Corey led the charge against this plan and pleaded with the finance committee to allow him to reduce wages. By January 1910 a full-scale war had developed between Gary and Corey. Perkins sided firmly with Gary, and with the support of both J. P. Morgan and the powerful Frick, Perkins routed Corey's allies on the finance committee.

This power struggle, however, also marked the end of Perkins's association with U.S. Steel and J. P.

Morgan & Company. It is not clear just what problems developed between Morgan and Perkins, but there were rumors that the elderly banker had become irritated at some of Perkins's speculations with U.S. Steel stock; however, there has been no concrete evidence produced to substantiate the claims. Whatever the reason, Perkins resigned on December 31, 1910, stating that he wished to devote his time to "corporation work and work of a public nature."

During his years as a Morgan partner, Perkins had become increasingly interested in issues relating to America's emerging industrial structure. A pioneer in the movement to stabilize the growing industry, he worked to avoid strikes by establishing paternalistic labor policies and to eliminate what he viewed as wasteful competition by working out intraindustry agreements. He also attempted to strike a détente with the Justice Department to avoid antitrust suits. He firmly believed that the dominance of the big corporation in American industry was not only inevitable but desirable. This vast new corporate power, however, had to be controlled in the public interest, a belief that converted Perkins into a reformer and political leader during the last decade of his life.

The actions of the Stanley Committee in 1911 convinced Perkins to take a more active political stance. Perkins believed that the committee's antitrust investigations were a call "for the destruction of our great enterprises." As a result Perkins made speeches and wrote a series of magazine articles in which he put forward his idea of a new industrial commonwealth. When Theodore Roosevelt in 1912 announced his intention to seek the Republican nomination, Perkins became active in his campaign. William Howard Taft, however, became the Republican nominee, and Perkins and Roosevelt formed the Progressive or Bull Moose party to contest the election. Perkins became chairman of the party's executive committee and contributed large sums to the campaign. This action caused J. P. Morgan, Jr., to ask Perkins to relinquish his directorship at U.S. Steel on the grounds that the company should have no contact with "current politics." Perkins refused to resign.

In beating out Taft for second place in the election, Roosevelt's campaign seemed to promise a bright future for progressive politics. Perkins, therefore, remained active with Progressive party affairs. By 1920 Perkins's energy was spent, and although he was only in his late fifties, his health began to fail rapidly. Before he died, however, Perkins found vindication for his ideas and actions. On March 1, 1920, in a four-to-three decision, the Supreme Court ruled on the 1911 antitrust suit in favor of U.S. Steel. Perkins sent a joyous telegram to Gary: "OPINION ... PLACES MORAL CONDUCT ABOVE LEGAL TECHNICALITIES. AS THIS HAS BEEN OUR CONSTANT AIM NO GREATER ENDORSEMENT COULD BE MADE OF YOUR LEADERSHIP. HEARTIEST POSSIBLE CONGRATULATIONS." On June 18, 1920, he died in Stamford, Connecticut, of encephalitis complicated by a heart attack.

Publications:
The Currency Problem and the Present Financial Situation, a Series of Addresses Delivered at Columbia University, 1907–1908 (New York: Columbia University Press, 1908);
"Corporations in Modern Business," *North American Review,* 187 (March 1908): 388–398;
It Is the Function of Law to Define and Punish Wrong-Doing, and Not to Throttle Business (Detroit: Detroit Board of Commerce, 1911);
Wanted: A Constructive National Policy (Houghton: Michigan College of Mines, 1911);
New York's Responsibility (New York: Academy of Political Science, 1911);
"Business vs. Federal Control of Corporations," *Colliers,* 46 (March 11, 1911): 32;
"Business : The Moral Question," *World's Work,* 22 (June 1911): 4465–4471;
"Practical Profit-Sharing and its Moral," *World's Work,* 22 (July 1911): 4619–4625;
"Strangulation or Regulation," *World To-Day,* 21 (November 1911): 1298;
"Wanted — A National Business Court," *Independent,* 71 (November 20, 1911): 1173–1177;

"Business Man's View of the Progressive Movement," *Review of Reviews,* 45 (April 1912): 425–426;
"Our Big Business Blunder," *Independent,* 82 (April 12, 1915): 67;
The Sherman Law, Where it Failed, Why it Has Failed, and a Constructive Suggestion, Address before the Economic Club of Philadelphia, May 22, 1915 (Philadelphia, 1915);
We Are as Unprepared for Peace as for War, Address before the Bankers' Association at Indianapolis, October 12, 1915 (New York: Economic World, 1915);
"Germany's Example in Preparedness," *Scientific American,* 114 (February 5, 1916): 145;
"Our Present Economic and Social Conditions as Results of Applied Science and Invention," *Scientific Monthly,* 6 (March 1918): 223–230;
Profit-Sharing; or The Worker's Fair Share, Address before the National Civic Federation, September 11, 1919 (New York, 1919);
"Efficient Citizens," *Outlook,* 125 (June 30, 1920): 415–416.

References:
Vincent P. Carosso, *The Morgans: Private International Bankers, 1854–1913* (Cambridge, Mass.: Harvard University Press, 1987);
Arundel Cotter, *Authentic History of the United States Steel Corporation* (New York, 1916);
Gerald G. Eggert, *Steelmasters and Labor Reform, 1886–1923* (Pittsburgh: University of Pittsburgh Press, 1981);
John A. Garraty, *Right-Hand Man: The Life of George W. Perkins* (New York: Harper, 1960);
Ida Tarbell, *The Life of Elbert H. Gary: The Story of Steel* (New York: Appleton, 1926).

Phoenix Steel Company

by Bruce E. Seely

Michigan Technological University

When the Phoenix Iron Company celebrated its centennial in 1927, it was one of the oldest continuously operated firms in the American iron trade. Located in Phoenixville, Pennsylvania, the company was run by the Reeves family of Philadelphia for four generations. As early as 1855 the firm rolled structural iron beams; it patented the famous Phoenix column in 1862. The company also formed a successful bridge-building firm in 1871. The Phoenix Bridge Company became one of the

larger producers of structural steel in the East. By 1889 Phoenix Iron was running several open-hearth furnaces, making it one of the first fully integrated firms in the industry. It remained small and independent through World War I. By 1926 the bridge works was rebuilt and a program of electrification was near completion. But in 1944 Samuel Reeves died and the company passed out of the Reeves family control.

Stanley Kirk, chairman of Phoenix Steel, at the company's Claymont, Delaware, plant

The next 40 years were often a struggle. The reorganized company operated briefly as the Barium Steel Company, then was restructured again in 1949 as the Phoenix Steel Company. Performing well in the 1950s, Phoenix Steel developed a profitable product line with the installation of a seamless pipe and tube mill in 1956. In 1960 the company bought the Claymont Works of Worth Brothers in Wilmington, Delaware, to gain an entry into the specialty market of clad and stainless steels. The purchase backfired immediately, and losses in 1960, 1961, and 1962 totaled $12.1 million. Stanley Kirk, formerly with the Longines-Wittnauer Watch Company and Phoenix Steel's largest shareholder, took over as chairman in 1963. Kirk was an unwilling hatchetman as he tackled the problems at Claymont. The facility was plagued by production delays and inefficiencies because it was the only plate mill in the country without a separate slabbing mill. Moreover, Phoenix Steel's production was unbalanced — the open hearths at Phoenixville produced twice as much steel as could be rolled. Kirk brought in Ford, Bacon & Davis, a noted steel consulting firm, which developed plans to rebuild the rolling mill and install a separate slabbing mill and a continuous caster. A pair of electric furnaces were later added to the plan, bringing modernization costs to $55 million. The expansion was originally slated for completion in 1967, but the program dragged into 1970 due to a variety of construction delays.

Throughout the 1970s profits from the tube mill in Phoenixville failed to cover the losses at Claymont. In 1972 the pipe plant was damaged by flooding caused by Hurricane Agnes, and the firm struggled until April 1976, when the French specialty steelmaker Creusot-Loire purchased a controlling interest. This French connection provided Phoenix Steel with needed capital — $40 million was spent on improvements by 1982 — and a pool of technical talent. In 1978 both mills turned a profit, as Phoenix Steel found a niche filling small specialty-plate orders for nuclear plants and the military.

But the economic slump of the early 1980s crippled Phoenix Steel, which filed for chapter 11 bankruptcy protection in August 1983. Several reorganization efforts, one of which included a $1-million loan from the state of Delaware, were undertaken, and the company struggled on through 1985, when Creusot-Loire was liquidated. Land developer William Davidson and Detroit Pistons owner David B. Hermelin took over control under the name Guardian Industries, but this rebirth was short-lived. In 1987 the company again sought protection from its creditors. In March 1988 the tube mill was sold and reopened as the Phoenixville Pipe and Tube Company. But the most amazing turn of events in Phoenix's recent history occurred in July 1988, when the Chinese International Trust & Investment Corporation (Citic) and the Wai Hing Company, a Hong Kong–based trader in steel-mill equipment, bought the Claymont plate mill for $13.5 million. Now called CitiSteel Incorporated, the investors spent about $20 million on improvements to the plant and began producing plate in late 1989, operating as a mini-mill. The company continues to demonstrate an ability to rise from the ashes.

References:

Mary Beth Dougherty, "Rising from the Ashes," *Business Week* (September 26, 1964): 87, 89–90, 92;

E. C. Kreutzberg, "Phoenix: One Hundred Years Old, Pioneer in Industry," *Iron Age,* 81 (August 27, 1927): 433–436;

George J. McManus, "Has Phoenix Steel Finally Put It all Together?," *Iron Age,* 221 (July 31, 1978): 33, 36, 38;

"Phoenix Steel Shops for Acquisition," *Iron Age,* 225 (January 22, 1982): 16;

"Rising from the Ashes of Phoenix Steel," *Iron Age,* 4 (September 1988): 48–49, 51;

Norman L. Samways, "Citisteel U.S.A.: The Rebirth of a Plate Mill," *Iron and Steel Engineer,* 66 (November 1989): 17–25;

C. J. Suchocki, "Can Phoenix Steel Rise from the Ashes?," *Iron Age,* 205 (January 8, 1970): 54–55;

Frank Whelan, "Phoenix Iron Works," in *Encyclopedia of American Business History and Biography: Iron and Steel in the Nineteenth Century,* edited by Paul Paskoff (New York: Facts on File, 1989), pp. 277–278.

Pittsburgh Plus Pricing

by Donald F. Barnett

McLean, Virginia

Early in the twentieth century the large steel companies made many attempts to control steel prices through pools, agreements, and understandings. The term *Pittsburgh Plus Pricing* refers to an agreement reached in about 1903 as a mechanism to establish uniform and stable steel prices. Steel companies agreed to quote prices to buyers as the cost of production plus the cost of shipping the steel from Pittsburgh to its destination. All mills were encouraged to adhere to this formula, even if the mill were located closer to (or farther away from) a particular customer. For example, a Chicago mill would quote prices to Chicago customers as the Pittsburgh price plus transport to Chicago from Pittsburgh. The company would then pocket the phantom freight charge. Conversely, Chicago-area mills absorbed freight charges on any shipments east to Pittsburgh. After government investigation this pricing system was banned in 1924.

The Pittsburgh base price was usually established by "agreement" between the key suppliers and key buyers of steel products. Independent steel producers could not be forced to adhere to the Pittsburgh formula or to quote the Pittsburgh base. In times of severe market weakness or strength, prices would be cut or raised without reference to the formula. Occasionally, as in 1908, 1911–1912, and 1917–1918, Chicago plus pricing emerged for that area. But the industry leader, the United States Steel Corporation, led efforts — through various trade associations and meetings — to maintain price stability. Among these meetings were the notorious "Gary dinners" held at the Waldorf Hotel from 1907 to 1911; they were ended when the government began legal action in 1911.

From 1907 until 1919 the Pittsburgh Plus Pricing formula largely created price stability. As steel consuming and producing activity moved to the midwestern states, however, consumers in that area became increasingly unhappy. In 1919 they filed a complaint with the Federal Trade Commission (FTC), claiming (among other things) that Pittsburgh Plus Pricing resulted in discriminatory net prices for steel, since the steel shipped to a customer could come from a mill whose shipping cost could be higher or lower than the price of transporting it to the customer from Pittsburgh. In 1924, following a three-year investigation, the FTC directed U.S. Steel to abandon Pittsburgh Plus Pricing. U.S. Steel agreed to do so "only so far as it is practical." Pittsburgh Plus Pricing was immediately replaced by multiple basing-point pricing, a system that established freight charges from several different regional centers. This method did not entirely redress the inequity of the steel companies' pricing system, and, after World War II, all forms of basing-point pricing were abolished.

Reference:

"Tracing the History of Pittsburgh Basing," *Iron Age*, 110 (November 16, 1922): 1287–1289.

Pollution Control

by Terry S. Reynolds

Michigan Technological University

An early-twentieth-century blast furnace generated up to 200 tons of waste gas per day. Steel furnaces generated 10 to 40 pounds of dust per ton of steel. Water pollution has also been a major problem since nearly 40,000 gallons of water are used per ton of finished steel. At the turn of the century the wastes produced by iron and steel plants were all but completely ignored by the public, for steel mills brought work and community growth.

The iron and steel industry early in the twentieth century nonetheless began to install equipment to reduce wastes. Traditional beehive ovens converted coal to coke while releasing one-third of the weight of raw coal into the atmosphere. By-product coke ovens, which replaced them, trapped the gases for use as a fuel and reclaimed the pollutant chemicals. By 1920 more than half of American coke was produced in by-product ovens, and by 1950 the beehive oven was virtually extinct.

Similarly, American steel mills early in the century took steps to reclaim the emissions from blast furnaces. Prior to 1900 many American blast furnaces had dust catchers, but the finer texture of the Mesabi ores used in American furnaces after the 1890s significantly increased dust emissions. Tower washers were added to supplement dust catchers, so that particles too light to fall into dust catchers were captured in a downward spray of water that cleansed the air. The first tower washer was installed on an American blast furnace in 1906.

By the 1930s electrical precipitators were also being commonly used to cleanse the gases that fueled the furnace's engines. For blast gas to be used as gas-engine fuel it could contain no more than 0.03 grains of dust per cubic foot. The electric precipitator was based on the principle, forwarded in the nineteenth century, that gas particles when given an electrical charge would attract suspended dust and carry it to oppositely charged electrodes. Early in the twentieth century, F. G. Cottrell of the U.S. Bureau of Mines made the process commercially practical by using high-voltage alternating electrical current to charge the gas particles, and in turn large plates with an opposite charge collected the dust. The dust deposits would be dislodged from the plates by an automatic rapping device, and the dust would fall into bins placed at the bottom of the precipitator. By 1950 the industry was also collecting and reusing in some quantity a handful of other materials, including slag, soluble oils, mill scale, and scrap.

Before 1960 pollution-control equipment was installed for obvious economic reasons; the reclamation of some materials often produced profits. Iron and steel plants continued to release huge volumes of pollutants into the air and water, including spent pickling acids, plating solutions, slag, basic-oxygen-process dust, lubricating oils, coke dust, and nonprofitable coke-oven chemicals, such as sulfur. In the late 1960s the industry was responsible for about 10 percent of the air pollution in the United States and a third of all industrial waste-water discharge.

Growing public concerns in the 1960s were in the following decade transformed into legislation, such as the Clean Air Act of 1970, the Federal Water Pollution Control Act of 1972; both these acts were further amended in 1977, hitting the industry hard. For the first time, the steel industry had to consider, develop, and adopt technologies to reduce emissions without hope of any economic return. Although by-product coke ovens captured and used most of the gases produced by the coking process, the operation remained dirty; whenever coal was placed into the ovens, or coke was removed, there were large eruptions of smoke and dust. Pollution regulations forced the industry to begin to correct such problems by installing more-expensive larry cars that captured dust, by preheating coal, or by charging coal into the ovens via pipeline.

Economic factors combined with the changing shape of the steel market complicated the industry's response to pollution-control legislation. American iron and steel plants varied widely in age, in equipment, and in raw materials used; no standard pollution-control package could be applied. Making matters worse, the industry faced severe competition from foreign imports in the 1960s and urgently needed capital for modernization, but much of this capital had to be diverted into pollution-control equipment. In 1966, 2.9 percent of the industry's capital spending went into pollution-control machinery; by 1968 the figure was 4.4 percent. Total spending by steel companies for pollution-control equipment reached over $1.2 billion in 1972, more than any other industry. According to industry sources, between 20 and 25 percent of the industry's investments between 1975 and 1980 were environmentally related. Around 10 percent of the cost to build a new plant was wrapped up in environmental-control equipment. By 1984 the American Iron and Steel Institute estimated that U.S. steel mills had spent $5.8 billion on clean-up devices since 1951. Other estimates placed the cost at more than $10 billion by the late 1980s.

The installation of legislatively mandated pollution-control equipment sharply reduced the pollutants emitted by the industry. For example, the volume of particulates removed jumped from 77 percent in 1971 to 95 percent in 1979. The installation of water recycling and purification equipment — and the adoption of some of the water reclamation techniques used by towns and cities — similarly improved the industry's control of water pollutants. In 1963 the industry removed only 41.7 percent of the acid from effluent discharges; by 1977 the figure was 80.7 percent. In 1963 the industry removed only 3.8 percent of the soluble metals from its wastewater; by 1977 the figure had risen to 34 percent.

The installation of pollution-control equipment in the late 1960s and 1970s enabled the iron and steel industry by 1980 to meet most of the basic mandates of existing environmental legislation. During the 1980s the administration of President Ronald Reagan worked to ease regulations, enabling the industry to reduce sharply its spending on environmental equipment. By 1983 steel-industry expenditures for air- and water-pollution control had dropped to $140.2 million, approximately 25 percent of its peak 1979 expenditures.

In the end, the impact of environmental legislation of the 1960s and 1970s on the steel industry proved to be not entirely negative. Because it was often more costly to install pollution controls on older equipment than on new equipment or in new plants, the industry was provided with an incentive to modernize. Furthermore, the legislatively mandated pollution control equipment required in other industries provided a sizable new market for steel.

References:

(Note: There is no historical account of the industry's handling of its waste problems. However, the following items contain scattered information on the nature and magnitude of the problem.)

Donald F. Barnett and Louis Schorsch, *Steel: Upheaval in a Basic Industry* (Cambridge, Mass.: Ballinger, 1983);

Henry C. Bramer, "Pollution Control in the Steel Industry," *Environmental Science and Technology, 5 (1971): 1004–1008;*

W. H. Dennis, *A Hundred Years of Metallurgy* (Chicago: Aldine, 1964);

Emission Control Costs in the Iron and Steel Industry (Paris: Organization for Economic Co-operation and Development, 1977);

William T. Hogan, S.J., *The 1970s: Critical Years for Steel* (Lexington, Mass.: Lexington Books, 1972);

Hogan, *Productivity in the Blast-Furnace and Open-Hearth Segments of the Steel Industry: 1920–1946* (New York: Fordham University Press, 1950);

Shirley V. Margolin, "U.S. Environmental Laws and Their Impact on American Steel," in *Ailing Steel: The Transoceanic Quarrel,* edited by Walter H. Goldberg (New York: St. Martin's Press, 1986), pp. 351–360;

John J. Obrzut, "Can Industry Afford Pollution Control's Costs?," *Iron Age,* 221 (February 20, 1978): 39–50;

"Pollution Control: Is Steel Meeting the Challenge?," *Iron Age,* 202 (November 21, 1968): 95–102;

Clifford Russell and William Vaughan, *Steel Production: Processes, Products, and Residuals* (Baltimore: Johns Hopkins University Press, 1976);

Marshall Sittig, *Particulates and Fine Dust Removal: Processes and Equipment* (Park Ridge, N.J.: Noyes Data Corporation, 1977);

J. Szekely, ed., *The Steel Industry and the Environment* (New York: Marcel Dekker, 1973);

United States Steel and the Environment (Pittsburgh: United States Steel, 1980);

Roy E. Williams, *Waste Production and Disposal in Mining, Milling, and Metallurgical Industries* (San Francisco: Miller Freeman, 1975).

Frank Purnell

(October 11, 1886 – April 19, 1953)

by Larry N. Sypolt

West Virginia University

CAREER: Office boy, various positions, Youngstown Sheet and Tube Company (1902–1917); assistant to the director, Steel Division, War Industries Board (1917–1918); vice-president, Consolidated Steel Company (1919–1922); vice-president, Bethlehem Steel Export Company (1922–1923); vice-president (1923–1925), assistant president (1925–1930), president (1930–1950), chairman of the board, Youngstown Sheet and Tube Company (1950–1953).

Steel executive Frank Purnell was born in Youngstown, Ohio, on October 11, 1886, and would spend his entire career with the steel company headquartered there, the Youngstown Sheet and Tube Company. Beginning as an office boy, he rose, in Horatio Alger fashion, to become the chairman of the board in a career that spanned half a century.

Frank Purnell was the son of Edward and Ann Hanson Purnell; he was educated in the local public schools. The son of a steelworker in a family with a modest income, Purnell began as an office boy with Youngstown Sheet and Tube Company in 1902 in the company's city office. Purnell later told *Iron Trade Review* that he had wanted to work in the mill, but treasurer Richard Garlick kept him in the office. He advanced rapidly but may have gotten his big break when he became a member of the government's War Industry Board in Washington during World War I, working as assistant to J. Leonard Replogle at the steel division. The experience undoubtedly exposed Purnell to a broader view of the steel industry as he worked to coordinate production with other steelmen from across the nation to meet wartime needs.

Frank Purnell

In 1919 Purnell became vice-president in charge of sales of the Consolidated Steel Company, the export arm of Youngstown Sheet and Tube and other independent steel producers. In 1922, however, he pressed for the liquidation of Consolidated, arguing that it could not compete with European producers. When Consolidated ceased operation, Purnell became vice-president in charge of sales at the Bethlehem Steel Export Company.

In 1923 Purnell returned to Youngstown Sheet and Tube. He had caught Youngstown president James Campbell's eye with his advice to shut down Consolidated, and Campbell sought out Purnell for the vice-president's post. Purnell accepted the position even before discussing his salary with Campbell. He was elected assistant president in 1925 and for the next three years assisted Campbell in many of his responsibilities. His promotion at age thirty-nine was a mark of distinction as well as a sign of Campbell's personal interest in his career. Indeed, he was the heir-apparent to Campbell, who had run Youngstown Sheet and Tube since 1902. Purnell was quite modest about this promotion, explaining it with the comment, "I just worked hard." But a colleague told *Iron Trade Review* in 1925, "He is just a cork. You could not keep him down. He had so much ability, that he just floated always on top."

When Purnell became president of Youngstown Sheet and Tube on January 1, 1930, he inherited several problems, of which only one was the deepening depression. Youngstown was in the middle of fighting bruising court battles related to its attempt to merge with Bethlehem Steel. Moreover, while waiting for the mergers to be approved, the company had held back on plans for major improvements. With too much capacity in pipe and tubular products, the company lost money from 1931 through 1934.

Even without operating experience, Purnell saw a need to diversify Youngstown's product line. In 1934 he recommended the installation of a new 79-inch continuous hot-rolled strip and sheet mill at the Campbell Works. He was supported by board members, and the company borrowed $3.5 million to cover the investment. The bankers were hesitant to loan several million dollars when no one could see the end of the Depression. Purnell's wisdom, however, became evident when demand for sheet steel for automobiles and other commodities revived in 1935. The new mill quickly paid for itself, and in December 1935 the company paid back the first $500,000. The $3-million balance was paid in full in the next four months.

Purnell's 20-year presidency also spanned the World War II era. Demand for steel was great during the war years and the postwar recovery period. To keep up, Youngstown spent over $188 million in additions and improvements. Once again, as in the Depression,

the company was prepared to meet demands from the automobile and heavy-appliance industry during the postwar period.

Purnell also oversaw improvements made in employee-labor relations at Youngstown Sheet and Tube during the late 1930s. During that period, employee representation — a form of company union close to two decades old — was overhauled due to Purnell's interest in harmonious relations between employees and employers. Under Youngstown's plan, workers were represented by people they knew, chosen from their ranks. Having labor representatives who were familiar with existing conditions, Purnell believed, gave the employee-representation plan a decided advantage over organized labor unions.

The Steel Workers Organizing Committee (SWOC) began its organizing drive in the steel industry in 1936, and Purnell, serving as chair of the Labor Relations Committee of the American Iron and Steel Institute, pressed for some form of reconciliation: "It is my opinion that you cannot have our industry that is not prosperous and labor that is prosperous any more than you can have an industry that is prosperous and labor that is not; therefore, mutual understanding and harmony between the representative of industry and the representative of labor are absolutely necessary for satisfactory work relations and for the success and prosperity of both." Yet in the end Purnell's hopes for harmony failed, as Youngstown Sheet and Tube joined Bethlehem, Inland, and Republic in failing to recognize the SWOC after the United States Steel Corporation's surprising acceptance of the union. The resulting "Little Steel" strike in 1937 produced no violence at Youngstown, which staved off the union until the early 1940s. The government finally ordered the Youngstown employee-representation plan disbanded in 1941, and the company signed a contract with the union in August 1942.

Purnell also served as a director of the Youngstown Steel Door Company, Youngstown Foundry and Machine Company, Interlake Steamship Company, National City Bank of Cleveland, Emsco, and Derrick and Equipment Company of Los Angeles. On May 12, 1911, he married Anne Watkins. On April 25, 1950, Purnell became chairman of the board of Youngstown Sheet and Tube. He died on April 19, 1953.

Publications:

"Progress in Labor Relations," *American Iron and Steel Institute Yearbook* (1935): 56–62;

"Custom-Built Steel — More Value Per Ton," *Steel, 99* (October 12, 1936): 131–132;

Current Problems of Labor Relations, Address at 200th Meeting of the National Industrial Conference Board, Chicago, November 17, 1937 (N.p., 1937);

"Financial Problems in the Steel Industry," *American Iron and Steel Institute Yearbook* (1940): 42–49.

References:

50 Years in Steel (Youngstown, Ohio: Youngstown Sheet and Tube Company, 1950);

William T. Hogan, S.J., *An Economic History of the Iron and Steel Industry in the United States,* 5 volumes (Lexington, Mass.: Lexington Books, 1971);

"Industrial Men in the Day's News," *Iron Trade Review, 77* (December 31, 1925): 1685;

Kenneth Warren, *The American Steel Industry, 1850–1970: A Geographical Interpretation* (Oxford: Clarendon, 1973).

Clarence Belden Randall

(March 5, 1891 – August 4, 1967)

by Elizabeth C. Sholes

Industrial Research Associates

CAREER: Attorney, Belden & Randall (1915–1919); attorney, Berg, Clancey & Randall (1919–1925); assistant vice-president (1925–1930), vice-president (1930–1948), director (1935–1967), assistant to the president (1948–1949), president (1949–1953), chairman, Inland Steel (1953–1956).

Broadcast by radio and television to an estimated audience of 50 million Americans, Inland Steel President Clarence Belden Randall delivered a speech in April 1952 which challenged the right of President Harry S Truman to seize the nation's steel mills. His speech drew upon his formidable legal background as well as the conventional business ideology of the period to assert the unconstitutionality of the seizure. Randall came to this historic confrontation with Truman after a long and illustrious career with Inland. His career as a steel man, however, did not follow the traditional pattern; rather, he was part of a new breed of steel executives, whose backgrounds were in law, finance, or business rather than in the smoke and heat of the mills themselves.

Randall was born on March 5, 1891, in Newark Valley, New York, just outside of Binghamton, to a middle-class family. After graduating high school as valedictorian at age fifteen, Randall took his college preparatory courses at Wyoming Seminary in Kingston, Pennsylvania, and was admitted to Harvard University, graduating Phi Beta Kappa in 1912. Afterward he at-

Clarence Belden Randall

tended Harvard Law School where he was an editor for the *Harvard Law Review*. Earning his law degree in 1915, he graduated tenth in a class of

300. Randall began his legal practice with his cousin, William Belden, having worked during summer breaks at Belden's Ishpeming, Michigan, office. Randall had been offered a position with the prominent New York law firm of Cadwalader, Wickersham & Taft but shocked his Harvard colleagues by moving to Michigan's Upper Peninsula, arguing that Belden's firm offered a quicker avenue to direct practice and openings to a wider variety of cases. Proved correct, he had the opportunity early in his career to argue before the Michigan Supreme Court and to testify before committees in the Wisconsin legislature. In his autobiography, Randall noted that he was presenting cases to juries while his Harvard classmates "were still sharpening their pencils." He remained with Belden's firm until 1925, leaving Michigan only to serve in the infantry during World War I. He was briefly an aide-de-camp in the 169th Infantry Brigade, a unit in which Harry S Truman also served.

During the war Belden had moved to Cleveland to take care of family interests, and he left the Ishpeming office to Fred Berg and Thomas Clancey. Upon his return to Ishpeming from the war, Randall rejoined the firm, and Berg, Clancey & Randall quickly acquired the Cleveland-Cliffs Iron Company as its main client. Randall's success in negotiating the tricky issues affecting Cliffs Iron garnered him a favorable reputation, but he resisted offers to leave his small law practice. In June 1925, however, Randall received and accepted an intriguing invitation from Inland Steel president Phillip D. Block to join Inland as manager of their raw-materials acquisitions.

Randall's only qualification as assistant vice-president in charge of raw materials, and the only one required by Inland, was his legal experience with iron-ore operations in the Lake Superior district. As a Michigan lawyer he had only drawn contracts, having never seen the mines or the ore properties. He quickly set out to learn the various aspects of mining and transport. His job involved finding independent raw-materials sources for Inland that would free the company from reliance on purchases from larger competitors. Along with his wife, Emily, Randall personally scouted potential lakeside sites that provided easy loading for transport to Chicago. On the advice of his good friend at Inland, D. J. Thompson, Randall also ran a hot-roll sheet mill in Milwaukee for a year. Within a relatively short period he gained a thorough knowledge of most aspects of the steel industry.

In 1930 Randall was promoted to full vice-president at Inland, becoming a corporate director

in 1935. He formed a new subsidiary, Inland Lime & Stone Company, and became president of the new operation. During these years Randall developed limestone operations in northern Michigan and opened the Greenwood hard-iron mine. He also acquired the Morris, Sherwood & Iroquois Mines. Under his guidance, Inland supplemented its coal properties in Pennsylvania with large mines in Wheelwright, Kentucky, and expanded its Great Lakes fleet.

Randall's responsibilities as vice-president also included industrial and labor relations. In 1938 he began a recruiting program designed to attract to the firm eight to ten new college graduates as executive candidates. Randall stressed the importance of continual learning and broadening one's knowledge of the business. "Horizontally promoted" or "transplanted," executives were abruptly moved from one department to the next. Randall was quick to promote talent and equally quick to remove obstructions. He believed no one person should impede the progress of the company or other employees.

Randall's perspective on labor relations evolved over the years. He shared United States Steel Corporation president James A. Farrell's belief that workers have the right to unionize but went further than Farrell in asserting that the workers' right to strike is a necessary element in the free-enterprise system. Despite these beliefs Randall's experiences as an Inland top executive during the violent Little Steel strike of 1937 turned him decisively against what he termed the "monopoly" power of independent unions over the workers and led him to charge the Steel Workers Organizing Committee (SWOC) with using goon tactics against "innocent people." Believing that a worker's grievances were best handled man-to-man rather than through an independent bargaining agent, Randall argued that unions could not and would not represent issues fairly, for they were "organized not for worker betterment but for their own power." Their tactics interfered with human liberty — especially a worker's right not to join the union — by replacing reason with fear.

The antipathy Randall bore against union policies and practices was exceeded only by his animosity toward the role of government in industry. In his autobiography, *Over My Shoulder: A Reminiscence* (1956), Randall recalled the vigorous political discussions among customers in his father's general store. His father was a staunch Democrat given to deep distrust of the Republican administrations' support for business interests. It

A 1954 meeting of steelmen: (seated) Eugene Grace, Benjamin Fairless, and
Edward Ryerson; (standing) James Mauthe, Hiland Batcheller, Elton
Hoyt II, and Clarence Randall (Fortune Magazine *photograph*)

was his father's suspicion of government that
Randall, a Republican, apparently inherited from
those early civics lessons. Randall was a trenchant
critic of President Franklin D. Roosevelt's New
Deal and postwar governmental intervention in
planning and regulation. As a free-enterprise en-
thusiast, Randall distrusted the presence of government as either a mediator or regulator in the
private sector, whether disputes lay among business interests or between business and labor. In
1932 he wrote "Mining Taxation in the Lake Superior District," his first major critique of governmental incursions into the private sector. Awarded
the 1933 American Iron and Steel Institute medal,
the paper outlined the history of what Randall perceived to be governmental meddling in the lumber
and ore industries and documented the decline in
exploration and technological innovation that resulted from overtaxation. Randall continued to
denounce the corporate taxation policy that occurred at the federal and state levels in the postwar era.

During World War II, Randall had been
highly critical of the Roosevelt administration's
move to allow organized labor to participate on
wartime planning boards; he feared that govern-

ment would collude with unions to the detriment
of business and citizens. He also condemned the
wartime establishment of the Steel Fact Finding
Board as an example of government intrusion in
wage and price policy.

Randall's career at Inland progressed smoothly. He had the advantage of working for one of the
most competently run steel corporations in the nation, where his talents meshed well with those of
other officials. Inland had been founded in 1893 by
the Block family, which continued to have a presence within the corporate ranks, but important positions also were open to men such as Randall. Randall remained vice-president and corporate director
until 1948, when he became assistant to the president. He was elected president on August 27, 1949,
succeeding his friend and mentor Wilfred Sykes.

Under Randall's direction Inland had become the nation's seventh-largest steel company
by 1952. The company invested vigorously in
plant expansion and had added modern improvements such as continuous hot-strip rolling in
the 1930s. Furthermore, Randall had greatly expanded Inland's supply of raw materials. Because
Inland's main productive operations lay on Lake
Michigan, a short shipping distance from the

company's ore sources and from its main markets in Chicago, the company remained a low-cost producer. The mills ran at 85 percent of their maximum capacity, exceeding the industry average of 72 percent. Randall's reputation as an executive with uncommon business skills grew. A story told among Inland's executive circles held that Inland chairman Phillip D. Block was once introduced to a dinner companion as an employee of "Clarence Randall's company."

In June 1948 Randall wrote an article for the *Atlantic Monthly* surveying iron resources in America and disagreeing with propositions that the nation was dangerously short of raw materials. The article attracted government interest, and President Harry S Truman appointed Randall to serve with Averell Harriman as the first steel and coal consultant for the Economic Cooperation Administration (EAC), created by the Marshall Plan. Within days Randall was touring European steel-production sites and meeting with the heads of the industry, both former allies and opponents. He taught himself the intricacies of European economic and manufacturing operations and became the central public advocate for rebuilding Europe's steel industry.

On the heels of his work with EAC Randall became a member of the Business Advisory Council for the U.S. Department of Commerce. Appointed under Truman, Randall held that position into the Eisenhower years, despite the historic clash between Truman and Randall over the nationalization of steel. That confrontation brought together two of Randall's adversaries: organized labor and government. In the early 1950s the steel industry was once again being regulated by the government; America's entry into the Korean War revived the Wage Stabilization Board (WSB), with authority to set pay increases and recommend price increases. In 1952 the board accepted the union's request for wage increases, but the steel companies rejected the pay hikes on the grounds that an increase in wages would necessitate an increase in steel prices. The WSB argued that the wages could be absorbed by current profits plus the increase of $2–$3 per ton that the regulatory mechanisms allowed.

For weeks the debate raged among government, labor, and industry officials, with United Steelworkers' of America (USWA) leader Philip Murray standing firm on the WSB recommendations. During this time the steel mills were closed by a full-scale national strike. In the midst of the deadlock Truman feared that supplies for the war effort were imperiled, and in April 1952 he na-

tionalized the steel industry in an effort to force a resolution and restore steel production to serve the war effort.

Industry executives were appalled by Truman's action. They had considered, but dismissed, the possibility of a steel seizure, believing that such an executive decision would be beyond the limits of what a free-enterprise society would tolerate. On April 8 they had held a lengthy meeting over the strike issues and had been assured by sources close to the executive office that no such coup de main was forthcoming from Truman, so they adjourned for the night. At 9:00 P.M., however, Truman made his historic proclamation.

In his memoirs Randall stated that the declaration left him "physically ill." He equated Truman with Mussolini and Hitler. Randall and the other executives were immediately presented with a conundrum: nationalization might well mean workers would assume the strike was over and demand admittance to the mills. To open the gates, however, might erode the opportunity for any future assault on the constitutionality of Truman's action. Fortunately for the steel men, no labor confrontation occurred at Inland or other facilities.

The next morning Randall received a telephone call from U.S. Steel president Benjamin Fairless. The 92 companies that comprised the American steel industry needed an immediate rebuttal to Truman's declaration. The steel men had already arranged for every national radio and television network to carry their reply; Fairless and other members of the executive committee asked Randall to be their spokesman. The committee hammered out the issues that they wished to address, but the final speech was written by Randall alone and was accepted as representative of the industry position by consensus. In the speech, carried live over the national airwaves, Randall attacked Truman's action as an "evil deed" and asserted that Truman had sold himself to the labor community to repay their electoral support. He further alleged that the president had "transgressed his oath of office" and had abused his powers. He accused Truman of seizing private property "without the slightest shadow of legal right" and argued that the strike was a product of collusion between government and labor because some members of the WSB had been employed by labor unions.

The public response to Randall's remarks was overwhelming. In his autobiography Randall recalls that he became known as the man who "gave Harry hell" and received many letters and phone calls in support of his stance.

Despite having publicly lambasted the nation's president, Randall did not fall from grace with government in general. After former general Dwight D. Eisenhower became president in 1953, he appointed Randall to serve as chairman of the Commission on Foreign Economic Policy, which became popularly known as the Randall Commission. Randall shared the president's beliefs on matters of commerce and trade. To the steel industry in general Eisenhower's was a welcome alternative to past Democratic administrations. The steel men anticipated that there would be no more domestic interference in their business and that they would no longer be hindered by unnecessary regulation and price control.

On issues of trade, however, the industry was far less supportive. The steel industry was notoriously protectionist and defensive about its shaky relations with foreign competitors. In 1953 the debate over tariffs was heating up as the 1934 Reciprocal Trade Agreement Act (RTAA) was due to expire. Eisenhower, however, favored genuinely broadened free-trade terms and conditions. He nonetheless opted for a one-year extension of the protectionist measures that had existed under Truman. The extension would provide time to study how to liberalize trade without endangering industrial support for the Eisenhower administration. Eisenhower turned to Randall to direct the study.

In 1952, prior to the confrontation with Truman, Randall had debated trade and economic issues with Mortimer Adler at the Aspen Institute. Present in the audience was Ted Weeks, publisher of the *Atlantic Monthly,* for which Randall had written a few articles. Weeks was so impressed by Randall's knowledge of the issues that he immediately prevailed upon Randall to write a book on his economic beliefs that would be oriented to the general public. *A Creed for Free Enterprise* was published in that year.

Randall's position on free trade was partly attributable to Inland Steel's immunity from international competition. Located on Lake Michigan and isolated from the world market, Randall's company did not suffer from the effects of global rivalry. In addition, Randall's beliefs were also influenced by the devastation he witnessed in postwar Europe and by communist incursions into Eastern Europe. He argued that a strong free world could exist only within a broadened sphere of competition. This position coincided exactly with that of Eisenhower.

The Randall Commission was convened in September 1953 and was charged with completing its mission by January 1954. Commission members were recruited from business and labor sectors and from Congress. The commission drew upon the testimony of over 1,500 witnesses, including representatives from the steel industry's major trade organization, the American Iron and Steel Institute (AISI).

The AISI acknowledged that the age of pure protectionism was necessarily over. Counting on the Republicans' general respect for free enterprise, the AISI called for a new cooperative relationship between government and industry. But the steel industry stopped short of espousing free trade. Instead they proposed the creation of several industrial councils that would mediate each industry's particular trade requirements and grant protective relief where each industry deemed it essential.

Receiving increasing pressure from anti-free-trade groups, the Randall Commission began to place emphasis on creating overseas direct investment rather than domestic trade. Commission proposals included incentives to enhance overseas investment through tax credits and tax reductions. The one trade issue tackled by the commission was in keeping with Randall's own preoccupation with raw materials and natural resources. The commission devoted considerable attention to removing harmful tariffs that hindered the free flow of raw materials into America for industrial production. In the matter of general trade and tariff regulation, the commission did extend to the president the right to revise tariffs up or down by 5 percent and for Congress to abandon the 1930s "Buy American" act regulating governmental purchases and acquisitions.

The commission's report was not only challenged by the commision labor representative, USWA president David McDonald, but also by members of Congress, academics, and some industrialists. The report was considerably weaker than it had promised to be at the outset, as Randall sought compromises among opposing parties. The strong recommendations offered by the AISI were nowhere in evidence. The commission had also rejected subsidies for labor. The report did garner Eisenhower's wholehearted endorsement, but that alone could not save it. Both the long- and short-term results involved the extension of the RTAA and the continuation of a haphazard commercial balancing act oscillating between protectionism and free trade.

Randall once again appealed to the general public in his own exposition of the issues considered by his commission. In *A Foreign Economic Policy for the United States* (1954) he continued to argue for an expanded trade base that would benefit both the United States and its allies more effectively than the Marshall Plan. His economic theories anticipated those of a later decade. He proposed eliminating direct economic aid — which he believed contributed to centralized state control — and replacing it with private foreign investment. He called for the international convertibility of currencies to enhance the two-way flow of commerce and restated the commission's recommendation for the creation of tax benefits that would provide incentives for investment overseas. He was also an advocate of American investment in underdeveloped nations. He argued that such ventures would strengthen American interests in the third world. Randall believed that extending American influence through economic investment would slow the spread of communism to the underdeveloped nations and through Europe. To combat communism further by economic means, he advocated that agricultural surpluses generated in the United States be sold to Communist nations for gold. This would help unite America's agricultural policy with its foreign-trade policy and would also serve to break down communism's hold over its own economy.

In 1957 Randall was appointed to head the Council on Foreign Economic Policy. At the time he assumed this role, there was a national debate over Japanese purchases of American steel scrap. The purchases had resulted in higher scrap prices. Believing that U.S. trade with Japan and Germany would discourage Communist influence in Asia and in Europe, Randall disliked even voluntary restraints on Japan's purchases. He suggested to the Japanese that they revise their capital expenditures by adopting new steelmaking technology, such as the basic-oxygen process, and eliminate the use of scrap-hungry open-hearth furnaces. The Japanese took Randall's advice. By 1962 over 30 percent of Japanese steelmaking used the basic-oxygen process.

During these busy years in the public arena Randall had not neglected his responsibilities to Inland. In 1953 Randall had been promoted to serve as Inland's chairman of the board, at an annual salary of $100,000. Randall served as chairman of the steel corporation until 1956, when he became Eisenhower's special adviser on international economic affairs.

Randall continued to serve under President John F. Kennedy. In 1961 Kennedy sent Randall as special emissary to Ghana, where he met with and became a supporter of the new president,

Kwame Nkrumah. From 1962 to 1963 he served as chairman of a presidential panel to review federal pay schedules. The panel requested large pay increases for cabinet officers, congressmen, and military personnel. Randall also served on a State Department advisory committee studying international business problems. He never broke openly with Kennedy during the 1962 confrontation between the steel industry and the president over pricing, and in late 1963 President Lyndon B. Johnson acknowledged Randall's enduring loyalty and service to the nation by awarding the steel man and statesman the Presidential Medal of Freedom, the highest civilian award.

In addition to his role within the governmental advisory commissions Randall was also a member of key private-sector policy-planning groups, including the Business Advisory Council and the National Industrial Conference Board. Until his death he retained active interest in the private sector, serving on the board of directors of the Bell & Howell Company and the Chicago, Burlington & Quincy Railroad. An avid bird-watcher, he was also a member of the American Ornithological Association.

Between 1956 and his death in 1967 Randall wrote six books that were widely and well received. His subject matter ranged from a serious, if simplistic, rumination on the threat of communism to a lighthearted review of his own life and what the course of retirement means to active executives. In all, Randall was the author of 11 books and several articles for the *Atlantic Monthly, Saturday Review, Life, Saturday Evening Post,* and other popular periodicals.

Randall and his wife, Emily, whom he had married in 1917, traveled extensively during their nearly 50 years of marriage, visiting virtually every continent together. The Randalls had two daughters, Mary Fitch and Miranda Belden. The Randalls retained a family home in Ishpeming, where Randall died on August 4, 1967.

Selected Publications:
"Mining Taxation in the Lake Superior District," *American Iron and Steel Yearbook* (1932): 40–54;
and Roger N. Baldwin, *Civil Liberties and Industrial Conflict* (Cambridge, Mass.: Harvard University Press, 1938);
"Steel's New Challenge," *American Iron and Steel Institute Yearbook* (1949): 93–100;
"A Repeal of Collective Bargaining," *Commercial & Financial Chronicle,* 170 (August 18, 1949): 650;
A Creed for Free Enterprise (Boston: Little, Brown, 1952);
Freedom's Faith (Boston: Little, Brown, 1953);

and others, Commission on Foreign Economic Policy, *Report to the President and the Congress* (Washington, D.C.: U.S. Government Printing Office, 1954);

A Foreign Economic Policy for the United States (Chicago: University of Chicago Press, 1954);

Over My Shoulder: A Reminiscence (Boston: Little, Brown, 1956);

The Communist Challenge to American Business (Boston: Little, Brown, 1959);

The Folklore of Management (New York: New American Library, 1961);

Sixty-Five Plus: The Joy and Challenge of the Years of Retirement (Boston: Little, Brown, 1963);

Adventures in Friendship (Boston: Little, Brown, 1965);

The Executive in Transition (New York: McGraw-Hill, 1967).

References:

"Inland Steel," *Blast Furnace and Steel Plant,* 31 (August 1943): 926;

"Inland Steel," *Steel,* 123 (November 8, 1948): 84;

"Men of the Metal Industry for 1949," *Iron Age,* 165 (January 5, 1950): 127–130;

"Personals," *Iron Age,* 136 (October 17, 1935): 49;

Robert Sheehan, "Clarence Randall: Statesman from Steel," *Fortune,* 49 (January 1954): 120–122, 132, 134, 136, 138, 143;

"The Steel Industry," *Blast Furnace and Steel Plant,* 36 (December 1948): 1491;

Paul Tiffany, *The Decline of American Steel* (New York: Oxford University Press, 1988).

Archives:

Clarence Belden Randall's papers are at the Inland Steel Corporation in Chicago, Illinois.

Jacob Leonard Replogle

(May 6, 1876 – November 25, 1948)

by Elizabeth C. Sholes

Industrial Research Associates

CAREER: Office boy, shipper, timekeeper, assistant superintendent, superintendent, vice-president and general manager, Cambria Steel Company (1889–1915); vice-president and general manager, president, American Vanadium Company (1915–1917); director of steel supply, War Industries Board, Council on National Defense (1917–1919); president, Vanadium Corporation of America (1919–1923); chairman, Replogle Steel (1919–1924); president, Wharton & Northern Railroad Company (1919–1929); special partner, Harris, Upham & Company (1928–1932); chairman, Warren Foundry & Pipe Corporation (1932–1948).

During the course of his career Jacob Leonard Replogle directed several of America's steel and ferro-alloy companies. His most significant role in steel, however, was as head of the steel division of the War Industries Board as part of Bernard Baruch's "business cabinet" during World War I.

Born May 6, 1876, to the Reverend Rhinehart Z. and Mary Ann Furry Replogle in Bedford County, Pennsylvania, Replogle was a descendant of noted Huguenot immigrants who had settled in

Jacob Leonard Replogle

the county in 1760. In 1885 Replogle's family moved to Johnstown, where in May 1889 they lost their home and belongings in the great Johnstown flood. The disaster proved a turning point for Replogle, who left school to work as an office boy for the Cambria Steel Company at $5 per week. He ascended through the ranks, working successively as shipper, timekeeper, assistant superintendent, and superintendent of the forge, axle, and bolt department. While at Cambria he invented improved bolt-making machinery and championed the use of the Coffin heat-treating process that properly tempered experimental steel axles. His successes gained the attention of Cambria general manager Charles Price, ensuring his advancement.

In 1912 ill health forced Price's resignation, and Replogle was temporarily placed in charge of all operations. His outstanding performance led to permanent assignment as vice-president and general manager of sales in the Philadelphia headquarters, during which time Cambria enjoyed the greatest earnings in company history. In March 1915, despite his success at Cambria, Replogle resigned to accept a comparable position with American Vanadium Company in New York.

In October of that year he organized a syndicate with J. P. Morgan to purchase Pennsylvania Railroad's holdings in the Cambria Steel Company. Holding controlling interest, Replogle returned to Cambria and entered into merger discussions, selling Cambria to Midvale Steel & Ordnance Company in February 1916. By the time he entered government service during World War I, Replogle had served as an executive and a director of Cambria, Vanadium, Wharton Steel Company, the Wharton & Northern Railroad Company, and the Wabash Railway.

In September 1917 Replogle was summoned to Washington, D.C., by the financier Baruch, chairman of the advisory commission to the Council on National Defense under President Woodrow Wilson. The council was the precursor to the War Industries Board (WIB), which was established to regulate production for the war effort. Replogle resigned from the presidency of American Vanadium to become Baruch's director of steel supply at a dollar a year. His reputation in the steel industry was that of maverick — well respected by his peers but not part of the inner clique. This appealed to Baruch, as did Replogle's thorough knowledge of the industry and its supply requirements.

By fall 1917 the WIB had established a priorities division to assign ratings for steel orders under Replogle's direction. As his first official task Replogle called for a conference of steel leaders to devise an agreement on coordinated production. He met with more than 50 industry representatives in United States Steel Corporation chairman Elbert Gary's New York offices. Uncomfortable with his role as sole arbiter of production, Replogle acceded willingly to Gary's request for a delay in implementing regulation. Replogle and Gary solicited and obtained promises of voluntary compliance from the other steel producers, the first stages of what would be hailed as a model program of public and private cooperation.

The industry did not prove capable of total self-regulation, however, with domestic orders being indiscriminately filled. In response, Replogle's Steel Division sought cutbacks in military demands. But Wilson's executive order formalizing WIB operations made clear that WIB's function was to regulate the private sector, not to curtail military procurements. Replogle then tried restricting nonessential steel consumption but met with only mixed compliance, particularly from the automobile industry.

The WIB pursued its search for stable production guidelines in raw goods and finished commodities. By June 1918 a new requirements division was established to advise on shortages. Replogle issued priority certificates for all steel production with emphasis on governmental demands. During the height of the war he purchased 80,000 tons of steel per day for the military.

In addition Replogle negotiated the delicate issue of price regulation by establishing steel prices in conjunction with the American Iron and Steel Institute (AISI). As part of the Price Fixing Committee, Replogle could transform industry consensus on prices into governmental mandates, thereby preventing by law the exhausting price battles that had in the beginning hindered the war effort.

Industry leaders such as Gary were pleased to be relieved of competing demands. After the war the AISI applauded Replogle's leadership. Government was also pleased. At the end of World War I, Baruch and Replogle were each granted the Distinguished Service Award. His wartime service also brought him the Chevalier Legion of Honor from France, and he was made a commander in the Order of the Crown of Italy and a commander in the Order of the Crown in Belgium.

After the war Replogle returned briefly to American Vanadium; then in 1919 he formed his own company, Replogle Steel, which he directed until 1924. In 1928 he became a special partner in the Harris, Upham & Company brokerage firm and focused on Vanadium investments. In his later years Replogle retired to Palm Beach, Florida, where he served as Republican National Committeeman and pursued a love of tennis. Nevertheless, he remained chairman of the Warren Foundry and Pipe Corporation from 1932 until his death and was a director of John Wanamaker department stores and the Wabash Railway. He died on November 25, 1948, in New York City.

Publications:

"Conditions Causing the Adoption of Extra Prices for Special Sizes or Qualities of Steel," *Yearbook of the American Iron and Steel Institute* (New York: American Iron and Steel Institute, 1913), pp. 422–439;

"Extras — Reason and Necessity Therefore," *Iron Trade Review*, 53 (October 30, 1913): 785–786;

"Conditions Causing the Adoption of Extra Prices for Special Sizes or Qualities," *Iron Age*, 92 (October 30, 1913): 978–980;

"Government Steel Requirements," *Railway Review*, 62 (May 11, 1918): 682–683;

American Industry in the War: A Report of the War Industries Board (New York: Prentice-Hall, 1941);

The Public Years (New York: Holt, 1960).

References:

Robert D. Cuff, *The War Industries Board* (Baltimore: Johns Hopkins University Press, 1973);

E. C. Kreutzberg, "War Writes Steel's Brightest Page," *Iron Trade Review*, 64 (January 2, 1919): 42–51;

Melvin I. Urofsky, *Big Steel and the Wilson Administration* (Columbus: Ohio State University Press, 1969).

Archives:

The papers of Bernard Baruch, which are found in the Firestone Library, Princeton University, Princeton, New Jersey, contain information on Replogle's wartime government service.

Republic Steel Corporation

by Carol Poh Miller

Cleveland, Ohio

The Republic Steel Corporation was the brainchild of Cleveland financier Cyrus S. Eaton, who envisioned building a great midwestern steel company. Eaton embarked on his goal in 1927, when he emerged as the principal stockholder of the Republic Iron & Steel Company, which was organized in 1899. From 1927 to 1929 Republic acquired Trumbull Steel Company, Trumbull-Cliffs Furnace Company, Steel & Tubes, and Union Drawn Steel Company — acquisitions which significantly diversified Republic's outlets for its semifinished steel. On April 8, 1930, Republic merged with the Central Alloy Steel Corporation, the Donner Steel Company, and the Bourne-Fuller Company to become the Republic Steel Corporation.

Tracing Republic's "family tree back to the sub-soil," *Steel* magazine observed in 1935, "uncovers a luxuriant root growth which few other large producers of iron and steel can match." The old Republic Iron & Steel Company had repre-sented the consolidation of 34 independent iron and steel companies in the central and southern states. Republic immediately set about modernizing its scattered operations, either jettisoning the old bar-iron plants it had acquired or rebuilding them to produce steel. The other companies that would be joined together in the 1930 merger were equally impressive. The Central Alloy Steel Corporation, based in Canton, Ohio, was the largest producer of alloy and stainless steel in the United States and was well equipped with electric furnaces and a highly skilled workforce. The Donner Steel Company of Buffalo, New York, offered a fully integrated steel plant for the production of carbon and alloy steel, and the Bourne-Fuller Company in Cleveland had a large modern plant for the production of bolts and nuts. The merger, therefore, created the third largest steel producer in the country; the new Republic Steel Corporation was surpassed only by the United States Steel Corporation and the Bethlehem Steel Company.

The Republic Steel Corporation works in Youngstown, Ohio

Republic operated plants in 19 cities in eight states, with a plant in Canada. It counted 13 blast furnaces, 68 open-hearth furnaces, two Bessemer converters, seven electric furnaces, ten blooming mills, nine billet and bar mills, 26 merchant mills, 75 sheet mills, five hot-strip mills, 34 black-plate mills, and 536 by-product coke ovens. It also produced electric-, butt-, and lap-weld pipe.

A large share of its productive capacity was concentrated in northern Ohio at Youngstown, Canton, Massillon, Warren, Cleveland, Niles, and Elyria. Republic also had ore properties in the Lake Superior and Alabama districts, and coal mines in Pennsylvania, West Virginia, and Alabama. The corporation held 18.4 percent of the nation's capacity for making strip steel, 16.6 percent for pipe, and 13 percent for bars and allied products. With virtually no capacity for making rails and heavy structural shapes, Republic was a light-steel company, and it fabricated a large proportion of its own steel into products for thousands of manufacturers of consumer goods. As such, during the Depression years Republic was less restricted than other large steel companies, which saw their markets for heavy products become severely depleted.

Eaton had lured Tom Girdler from the Jones & Laughlin Steel Corporation to serve as Republic's chairman. Girdler assumed the additional title of company president — replacing E. T. McCleary, who had died soon after being elected to the office — and completed the task of

assembling a formidable group of Republic executives: Benjamin F. Fairless, former president of Central Alloy Steel, as executive vice-president; R. J. (Jack) Wysor, former assistant general manager and manager of properties at Jones & Laughlin, as vice-president of operations; Charles M. White, former superintendent at the Aliquippa Works at Jones & Laughlin, as assistant vice-president of operations; and, as vice-president of finance, Myron A. Wick, scion of an important Youngstown steel family and a former president of Steel & Tubes Company. Most of these men had followed traditional paths to the executive level, moving from the shop floor up through the ranks of management. All shared a devotion to Girdler, who fancied himself a no-nonsense steel man.

The executives' first task was to coordinate and consolidate the company's various operating units in the face of the severe economic depression that gripped the nation. Sales eroded and prices declined rapidly. Suffering losses in each of its first five years, Republic instituted stringent cost-cutting measures that included shutting down operations and laying off employees. The company survived the Depression and met payrolls by converting its inventories into cash.

The Depression underscored Republic's weaknesses. With most of its steelmaking facilities located in Ohio's Mahoning Valley, the company had to ship its Lake Superior ore an expensive 50 or 60 miles inland from the ports on Lake Erie. To solve the problem, Republic in 1935 acquired the Corrigan, McKinney Steel Company of Cleveland. The acquisition, which the Justice Department unsuccessfully challenged on antitrust grounds, gave Republic a modern steel plant with an ingot capacity of 1 million tons a year and dock facilities at Cleveland extensive enough to handle all of Republic's ore shipments; the purchase also doubled Republic's iron-ore reserves. During this period, Republic also took over the Truscon Steel Company, a leading fabricator of steel building materials with headquarters at Youngstown. Truscon would prove to be a large consumer of Republic's increased steel output.

Republic's organizational maneuvering placed Cleveland at the hub of a corporate wheel whose spokes radiated to Canton, Massillon, and Youngstown. In 1936 the company moved its general offices from Youngstown to Cleveland. Republic could now produce cheap, unfinished steel at its lakefront plants, allowing for its mills located further inland to make higher-priced specialty steels without incurring steep freight costs.

With business conditions showing a marked improvement in 1936, Republic embarked on an ambitious modernization program. Important projects included a major expansion of its steel plant at Cleveland, where it erected a 98-inch continuous strip mill, the world's widest. Specialty plants were also built, including a tinplate plant at Niles, Ohio; a wire mill at Chicago; and an electric-weld pipe works at Youngstown. Facilities for silicon strip steel at Warren, Ohio, were also expanded. In 1937 Republic acquired the Gulf States Steel Company and, in so doing, gained an integrated steel plant at Gadsden, Alabama, and strengthened its southern operations with the addition of the extensive iron-ore and coal properties of the Gulf States. By the end of its first decade Republic enjoyed a production capacity of 6 million tons and had emerged as one of the nation's major steel producers. Yet it continued to integrate and diversify. With the heart of its activity in northern Ohio, Chicago, and Buffalo, Republic was favorably situated to serve the Great Lakes region, where most of the nation's steel was consumed.

In 1937 Republic grappled with serious labor problems when, along with Youngstown Sheet & Tube and the Inland Steel Corporation, it refused to sign a contract with the Steel Workers Organizing Committee (SWOC). Despite the violence of the resulting strike and the intervention of a federal mediation board, Girdler steadfastly refused to bargain with the SWOC and continued to maintain that the majority of Republic workers did not want to join an independent union. Five years later, however, in August 1942, the "Little Steel" companies, including Republic, signed contracts with the United Steel Workers of America (USWA) following hearings by the War Labor Board.

During World War II Republic operated at capacity and set new production records. Ingot production rose from 6.1 million tons in 1940 to 8.6 million tons in 1943. In response to the demand for steel at home and abroad, the U.S. government established the Defense Plant Corporation (DPC) to boost the nation's steel production. Under the program, the government underwrote the installation of new facilities and leased them back to the steel company with an option to buy. Beginning in 1941 Republic entered into agreements with the DPC on iron-ore mines, coal mines, blast furnaces, steel plants, and finishing facilities — undertaking, in all, 28 projects in Alabama, Connecticut, Illinois, Kentucky, New York, Ohio, and Pennsylvania. At South Chicago, for example, Republic built an integrated open-hearth and electric-furnace steel plant for the production of steel for military aircraft and arms; in 1947 Republic purchased the plant for $35 million, greatly strengthening its position in the highly sought-after Chicago steel market. The Defense Plant Corporation invested a total of $188 million in steel and raw-material facilities at Republic, all of which the company eventually bought back. The war's impact on Republic personnel, meanwhile, was just as dramatic: by the close of 1943 approximately 17,000 Republic employees were serving in the armed forces. To meet expanded manufacturing operations, the company established extensive training programs for new employees and added some 8,000 women to the payroll.

In the immediate postwar years the company's transition to a peacetime economy was a relatively easy one. Thanks to pent-up consumer demand, production remained strong in the late 1940s. Price controls, imposed by the government in 1941, were lifted in 1946, and net earnings soared to an all-time high of $46.4 million in 1948. Meanwhile, Republic spent $171.6 million to expand and modernize further its facilities. At its Massillon plant, for example, the company installed new mills to roll and finish stainless-steel sheet and strip up to 60 inches wide. Capitalizing on the expansion of natural gas lines to the nation's population centers, Republic built a mill for the production of steel pipe at Gadsden, Alabama. The company also took steps to increase its holdings of raw materials. It acquired majority interest in the Liberia Mining Company, Ltd., which had a large iron-ore deposit in West Africa and joined other steel firms in developing iron-ore mines in Quebec and Labrador.

Republic prospered in the 1950s and 1960s and continued its expansion and modernization programs. Especially significant to Republic's success during this period was the major expansion of the Cleveland steel plant, which was well positioned to meet a large and growing demand for flat-rolled steel products. The company further augmented its ore reserves by purchasing half of the capital stock of the Reserve Mining Company, which controlled large taconite deposits in the Mesabi Range and had facilities in Minnesota to pelletize ore for use in the blast furnace. Republic also acquired large reserves of limestone and coking coal. Ingot production during the 1950s reached a high of 9.6 million tons in 1955; the following year net income reached a record high of $90.4 million.

Girdler retired in June 1956, at age seventy-eight. Charles M. White, Republic president since 1945, succeeded Girdler as chairman and controlled the company until his retirement five years later. Thomas F. Patton, longtime general counsel for Republic, and Willis Boothe Boyer, former company treasurer, were at Republic's helm in the 1960s and early 1970s, a period in which the company reached new highs in money earned and tons of steel produced. In 1965 Republic furthered its reputation as one of the most competitive firms in the steel industry by installing basic-oxygen furnaces — faster and more fuel efficient than open-hearth furnaces — at its Cleveland, Warren, and Gadsden plants.

During the 1970s, however, Republic began to feel the effects of inflation, record foreign steel imports, a costly labor settlement, and a protracted strike against its largest customer, General Motors. Even as the company completed installation of a new 84-inch hot-strip mill at its Cleveland plant, representing the largest capital project in company history, its workforce was reduced to its lowest level since 1938. Profitability returned, temporarily at least, and in 1974 earnings reached a new record high of $170.7 million.

The entire American steel industry faced hard times in the early 1980s, as an economic recession and the decline of the domestic automobile industry caused steel demand to plummet. Republic lost $239 million in 1982. Two years later, in June 1984, the Jones & Laughlin Steel Corporation, a wholly owned subsidiary of the LTV Corporation, merged with Republic, forming the LTV Steel Company, with headquarters at Cleveland. By 1986 LTV Steel had run up losses totaling nearly $1 billion. In July 1986 LTV Steel Company filed for reorganization under Chapter 11 of the Federal Bankruptcy Code.

References:

Tom M. Girdler and Boyden Sparkes, *Boot Straps: The Autobiography of Tom M. Girdler* (New York: Scribners, 1943);

William T. Hogan, S.J., *An Economic History of the Iron and Steel Industry in the United States*, 5 volumes (Lexington, Mass.: Lexington Books, 1971);

"Pruned Back, Republic Family Tree Is Still Luxurious," *Steel*, 96 (January 7, 1935): 122;

"Republic Steel," *Fortune*, 12 (December 1935): 76–83+.

David Milton Roderick

(May 24, 1924 –)

by Father William Hogan, S.J.

Fordham University

CAREER: Messenger (1941–1942), clerk, Gulf Oil Corporation (1946–1953); assistant comptroller, Bessemer & Lake Erie and Union Railroad companies (1953–1959); assistant to director of statistics, New York office (1959–1962), accounting consultant, international projects (1962–1964), vice-president, accounting (1964–1967), vice-president, international projects (1967–1973), chairman of finance committee (1973–1975), president (1975–1979), chairman, chief executive officer, United

States Steel Corporation (USX Corporation after 1986, 1979–).

David M. Roderick is chairman of the board and chief executive officer of the USX Corporation (formerly the U.S. Steel Corporation). USX is the parent company of four subsidiaries: the United States Steel Corporation, with headquarters in Pittsburgh; Marathon Oil Company, located in Findlay, Ohio; Texas Oil and Gas Corporation of Dallas; and the U.S. Diversified Group, located in Pittsburgh.

Roderick is a native of Pittsburgh and grew up on its north side, a predominantly industrial area. After graduation from high school in 1942, he joined the Marines and served as a platoon sergeant in the Pacific Islands until the end of the war. Upon return to civilian life he worked as a clerk at the Gulf Oil Corporation and attended night classes at Robert Morris Junior College and at the University of Pittsburgh, eventually earning a degree in economics and finance.

Roderick joined the Bessemer & Lake Erie Railroad and the Union Railroad subsidiaries of the U. S. Steel in 1953. Over the next 20 years he rose through various positions in U.S. Steel's financial departments. In 1973 Roderick was promoted to chairman of the finance committee, one of the three top jobs at U.S. Steel. In 1977 he was named president, and in 1979 he was made chairman of the board. He is the only person in the history of U.S. Steel to hold all three of the corporation's top positions.

The corporation — and the steel industry in general — was prosperous between 1979 and 1981. U.S. Steel's 1981 income ran close to $1.1 billion, a record figure for the corporation. Much of its income, however, was due to the sale of natural resources, the retention of which was no longer necessary with the capacity of the corporation beginning to decline. Indeed, by 1981 Roderick had become convinced that the steel industry was not growing as it had been in the previous two decades, and that the market was declining, especially the tonnage shipped to the automotive industry. With the production of smaller cars in the late 1970s and early 1980s, the automotive industry cut its steel purchases by some 10 million tons. As a consequence, Roderick reduced steel capacity and diversified into nonsteel activities.

For a corporation the size of U.S. Steel, diversification was not an easy task. Above all, it was necessary to acquire a company with substantial sales and profits to cushion the loss of steel-

David Milton Roderick (photograph courtesy of USX Corporation)

generated income. He chose the oil industry and, in a contest with Mobil Oil, was able to take control of the Marathon Oil Company, the seventeenth largest oil company in the United States, for approximately $6 billion. This proved to be a wise move, especially in the early 1980s, when the steel business dropped off precipitously.

In 1984 U.S. Steel made further deals in the oil business by acquiring the Texas Oil and Gas Corporation. The change in corporate structure by means of these acquisitions was dramatic; U.S. Steel had become an energy company, drawing 70 percent of its revenue from oil and gas. In 1986, as a result of this substantial change, the corporation changed its name to USX, and U.S. Steel became a wholly owned subsidiary.

In order to improve the efficiency and profitability of USX Roderick adopted a program to divest the corporation of facilities that were no longer considered essential to its operation. These included the American Bridge Company, which had been part of the original corporation in 1901; many chemical companies; and the United States

Steel Supply Company, a large service-center operation. In 1988 the transportation network consisting of railroads and barge lines was also sold. Throughout this period the company also closed or sold older steel-production facilities.

Roderick also pursued joint ventures, both domestic and international, often involving the United States Steel division's operations. The domestic joint ventures included an arrangement with the Rouge Steel Company to build a continuous electrolytic galvanizing line near Detroit to serve the automotive industry. Total capacity of the unit was 700,000 tons, the largest of several of these units installed throughout the industry. USX also joined with the Worthington Corporation to produce steel sheets for the automotive industry.

The international joint ventures involved the construction of a new cold-rolling facility at Pittsburg, California, to replace a plant that had been built in the early 1950s. Joining in on the contruction was the Pohang Iron and Steel Company of South Korea. Pohang, at the time of the joint venture's conclusion, was building a new integrated mill in South Korea that would produce an excellent hot-rolled band. The product would be shipped to the Pittsburg, California, facility for cold reduction. The result would be the highest quality product available. The total cost of the facility at Pittsburg was $400 million, shared jointly by USX and Pohang.

In 1989 a joint venture was undertaken with Kobe Steel of Japan, singling out the Lorain, Ohio, plant of U.S. Steel. Kobe, a producer of, among other things, high-quality bars for the automotive industry, was anxious to reach the growing market of Japanese automobile plants in the

United States. The Japanese steel producers were faced with deciding whether to build a new plant in the United States or form a joint venture. There were obvious advantages to the latter course of action, since the construction of a new plant would involve more capital and time.

Kobe worked out an arrangement with USX involving Lorain, which produced pipe as well as quality bars. With Kobe investing more than $300 million, USX detached Lorain as a separate company administered by both USX and Kobe Steel. Kobe also contributed its technology and expertise to the production and sales of high-quality bars for the Japanese automobile transplants. U.S. Steel was involved in the production of bars and pipe and accepted responsibility for selling the product.

Under the direction and guidance of Roderick, the company weathered the severe steel depression that took place during the 1980s. Moreover, his efforts at diversification brought USX out of this period an industrially and financially sounder institution. The American Iron and Steel Institute in 1984 awarded Roderick the Gary Medal, its highest award, for outstanding service to the American steel industry.

In addition to his positions at USX, Roderick serves as a director of Aetna Life and Casualty Company, Transtar, Inc., Procter & Gamble Company, and a general director of Texas Instruments. He is also chairman of the International Environmental Bureau, chairman of the U.S.-Korea Business Council, and a member of the policy committee of the Business Roundtable and the Business Council. He is also a member and past chairman of both the American Iron and Steel Institute and the International Iron and Steel Institute.

William Robert Roesch

(May 20, 1925 – December 2, 1983)

by John A. Heitmann

University of Dayton

CAREER: Various positions (1946–1970), president (1970–1972), chairman of the board and chief executive officer, Jones & Laughlin Steel Corporation (1972–1974); chief executive officer, Kaiser Industries (1974–1977); executive vice-president (1978–1979), president, United States Steel Corporation (1979–1983).

William Robert Roesch, a leading executive with the Jones & Laughlin Steel Company (J&L), Kaiser Industries, and the United States Steel Corporation (U.S. Steel), played an important role in managerial efforts to arrest the decline of the American steel industry during the 1970s and early 1980s. Born in Large, Pennsylvania, on May 20, 1925,

Roesch joined J&L as a mechanic in 1946. He attended college while working, studying engineering and management at the University of Pittsburgh. He graduated in 1960 and received graduate level training at Harvard's advanced management program. With Ling-Temco-Vought's acquisition of 80 percent of the J&L stock in 1968 and Charles M. Beeghly's decision to step down, Roesch gained the opportunity to participate in the highest levels of management. He became J&L's president in 1970 and its chief executive officer in 1972.

Nineteen-seventy was a difficult year for J&L. The company was plagued by labor strife, the breakdown of a major furnace, delays in the start-up of a continuous caster, and increased public criticism aimed at the firm's pollution of the air and water at its Pittsburgh and Cleveland facilities. Furthermore, the J&L of the early 1970s was one of the American steel industry's most lackluster performers, a firm possessing overlapping facilities, an inefficient managerial organization, and high overhead costs. The company's hopes for penetrating midwestern markets by setting up a fully integrated plant in the mid 1960s in Hennepin, Illinois, had never fully materialized. The plan had only managed to drain scarce capital resources while creating an awkward duplication of accounting, purchasing, scheduling, and personnel services.

In addition to the Hennepin plant, J&L had three significant production sites, two located in the Pittsburgh vicinity and one in Cleveland. The largest was its fully integrated mill in Aliquippa, Pennsylvania. J&L's subsidiary plants were geographically scattered, with sites in Michigan, Ohio, Indiana, and California. The company was unable to coordinate the activities of these subsidiaries effectively. With profitability on the decline, there was increasing pressure to turn the operation around quickly.

Roesch responded to these challenges by instituting a vigorous cost-cutting policy. While arguing for import protection, Roesch reduced capital outlays to $53 million in 1970. In addition, he initiated an aggressive campaign to eliminate overlapping services, to phase out outmoded equipment, to load up the more efficient operational facilities, and to reduce salaried employment by 20 percent. Roesch also consolidated the company's computer facilities and cut its research and development staff by almost 50 percent.

Roesch's efforts gained him the reputation of being a "white knight" within the ailing American steel industry; his leadership and cost-cutting abilities put him in the enviable position of being

William Robert Roesch (Fortune Magazine *photograph*)

wooed by several corporate giants. In 1974 he left J&L for a position at Kaiser Industries, which was followed by a stint at U.S. Steel beginning in early 1978. Roesch's first year at U.S. Steel was a tenuous one, not only because he was viewed by many as an outsider but also because the company lost $293 million in 1979. But within two years U.S. Steel earned more than $1 billion, the consequence of Roesch's cost-cutting policies combined with a national economic recovery.

But closing 13 U.S. Steel facilities and laying off more than 13,000 of its workers were only short-term answers to the problem of increasing steel imports. A recession plunged the company into a deeper crisis, and by 1982 U.S. Steel's losses totaled $361 million. Roesch died on December 2, 1983, at a time when the American steel industry was facing the worst crisis in its history.

References:
"For J&L, Cost-Cutting Is a Religion," *Businessweek* (June 3, 1972): 38-40;
William T. Hogan, *S.J., An Economic History of the Iron and Steel Industry in the United States,* 5 volumes (Lexington, Mass.: Lexington Books, 1971);
"J&L Pins Comeback Hopes on Versatile Production Man," *Industry Week,* 167 (October 5, 1970): 26;
"Roesch, William R," *New York Times Biographical Service* (1983): 1480;
"The Year that Ruines J&L's Profits," *Businessweek* (June 6, 1970): 110-112.

Rolling Mills

by Robert Casey
Henry Ford Museum & Greenfield Village

The most common method of shaping steel is by running it between rolls. By the end of the nineteenth century many specialized rolling mills had been developed: blooming, billet, and slab mills rolled ingots into square or rectangular shapes. These semifinished products were then processed by a host of smaller mills: rail mills, bar mills, plate mills, structural mills that rolled channels and angles, skelp mills that produced narrow strips from which pipe was made, and rod mills that made thin, round sections from which wire was drawn. During the twentieth century these mills were steadily improved to be made larger, more powerful, and more automated.

Throughout the nineteenth century, structural beams and columns were made by riveting plates and angles together. Attempts to roll beams in one piece met with little success until 1908, when a mill designed by Englishman Henry Grey went into operation at Bethlehem, Pennsylvania. Grey's ideas had found little favor among American steelmakers until Bethlehem Steel's Charles Schwab decided to stake his company's future on them. The gamble paid off, and Grey's beam was adopted by architects and engineers designing the new skyscrapers. In 1926 the United States Steel Corporation (U.S. Steel) erected a wide flange mill of its own design at South Works in Chicago. Progress in beam-mill design came slowly: the South Works mill was replaced in 1959; Bethlehem's pioneer Grey mills were not replaced until 1969.

Cold-rolling of flat steel was practiced in the nineteenth century, but the purpose was to give the steel a smooth, lustrous surface; there was little or no reduction in thickness. Cold reduction followed the introduction of the continuous hot-strip mill in the 1920s. Although faster than the old hand mills, hot-strip mills could not roll steel as thin as their predecessors could. The necessary reduction in the steel's thickness was best accomplished by cold-reduction mills. The cold-reduction process also improved the physical and mechanical properties of steel; therefore, after an-

nealing the steel could be stamped in large presses to form panels for auto bodies or home appliances. Like the structural mills, cold mills helped create new markets for steel.

Cold-rolling developed rapidly. In 1936 only 24 percent of tinplate stock was cold-reduced; the balance was rolled on hand-run hot mills. By 1939 75 percent of the tinplate was cold-reduced; by 1943, 100 percent. The mills themselves also developed rapidly from single-stand, nonreversing mills to single-stand, reversing mills to tandem mills, the current standard. A tandem mill is a group of two to six mill stands set in a row. The stands are four-high. They have two small-diameter work rolls that contact the steel, and two much larger backup rolls that keep the work rolls from bending under the tremendous pressures required for rolling the steel. The steel becomes progressively thinner as it passes from one stand to another.

The most important rolling-mill innovation in the twentieth century was electrification, for it affected every variety of mill. In the nineteenth century steam powered the mills and auxiliaries such as cranes, shears, and roll tables. As the century ended, Carnegie Steel pioneered the use of electricity for lighting the mills and driving auxiliaries. The first use of electric motors for main mill drives occurred in 1905, when U.S. Steel purchased two 1,500-horsepower motors for its rail mill at Braddock, Pennsylvania. In 1907 the successful installation of an electrically driven universal plate mill at U.S. Steel's South Works in Chicago disproved the notion that steam was superior for reversing mill drives. U.S. Steel's Gary Works was the first mill to be completely electrified.

Electric drives had several advantages over steam engines: lower maintenance costs, greater reliability, no standby losses (electric motors run only when in use; steam engines run all the time), and uniform torque over the entire rotation of the motor shaft. Furthermore, electric motors take up less floor space than steam engines and have bet-

A rolling mill

ter instrumentation to measure power and detect problems.

Although electric motors were being added to newly constructed plants, the industry was slow to replace existing steam engines with the newer technology. As late as 1950, nearly half of all blooming and slabbing mills were driven by steam.

References:
Arundel Cotter, *The Story of Bethlehem Steel* (New York: Moody Magazine and Book, 1916);

Bela Gold, William Peirce, Gerhard Resegger, and Mark Perlman, *Technological Progress and Industrial Leadership: The Growth of the U.S. Steel Industry, 1900–1970* (Lexington, Mass.: Lexington Books, 1984);

William T. Hogan, S.J., *An Economic History of the Iron and Steel Industry in the United States,* 5 volumes (Lexington, Mass.: Lexington Books, 1971);

William L. Roberts, *Hot Rolling of Steel* (New York: Dekker, 1983).

Edward L. Ryerson, Jr.

(December 3, 1886 – August 2, 1971)

by Robert M. Aduddell

Loyola University of Chicago

CAREER: Vice-president and general manager (1922–1929), president, Joseph T. Ryerson & Son, Incorporated (1929–1958); vice-chairman of the board (1935–1939), chairman of the board (1939–1952), chairman of the executive committee, Inland Steel Corporation (1952–1958).

Edward Larned Ryerson, Jr., was born in Chicago on December 3, 1886, the son of Edward L. and Mary (Mitchell) Ryerson. The family's association with the iron and steel industry in the United States dated back to the Revolutionary War; Ryersons forged the chain that spanned the Hudson River below West Point, closing the Hudson to British warships. Joseph T. Ryerson & Son, Incorporated, was founded in Chicago in 1842 and was the largest steel-warehousing operation in the country. Ryerson was associated with the family firm until it merged with Inland Steel Corporation in 1935; he served Inland until his retirement in 1958.

After graduating from Yale University in 1908, Ryerson spent a year taking a course in civil engineering at the Massachusetts Institute of Technology before joining the Ryerson firm in 1909. He started in operations and was promoted to works manager prior to World War I. In 1914 he married Nora Butler of Evanston, Illinois; the couple would have three children.

In 1917 Ryerson left the firm to work as a wartime production engineer for the Aircraft Production Board in Washington, D.C. He subsequently joined the U.S. Army Aviation Section and served as a first lieutenant in the Signal Corps. While serving, he received the rating of military aviator.

Ryerson returned to the family firm in 1919. In that year Joseph T. Ryerson & Son had turned seventy-seven years old. It was a unique, steel-warehousing firm serving nearly 40,000 customers from warehouses located in Chicago, Milwaukee, Saint Louis, Cincinnati, Detroit, Buffalo, Boston, Cleveland, and Jersey City. It purchased various shapes and sizes of iron and steel and held them for delivery. In addition to warehousing, the firm performed limited cutting and fabricating operations. Ryerson was elected vice-president and general manager in 1922. In 1929, his twentieth year with the firm, he was named president, replacing his grandfather, Joseph T. Ryerson, who remained on the board of directors and served as treasurer.

One of the firms from which Ryerson purchased steel was the Inland Steel Corporation. During the Depression, negotiations were conducted between Ryerson and Inland that resulted in a merger on September 30, 1935. The Ryerson firm received 240,000 shares of Inland common stock in exchange for the 400,000 outstanding shares of Ryerson stock. The Ryerson firm was to maintain its managerial autonomy and be operated as a subsidiary of Inland. Three Ryerson men were placed on Inland's board of directors. Edward Ryerson became vice-chairman of the board, while Joseph T. Ryerson and Everett D. Craft served as regular board members. Edward Ryerson also maintained his position as president of Joseph T. Ryerson & Son.

Edward L. Ryerson, Jr.

The two merging companies found the arrangement to be mutually beneficial. Inland ensured Joseph T. Ryerson & Son that it would have a source of supply for iron and steel; Ryerson & Son ensured Inland that it would have significantly more customers. At the time of the merger, Inland marketed 75 percent of its tonnage within 150 miles of its Indiana Harbor mill, its sole production site. Inland was also in the process of changing its product mix toward lighter products for the automobile and consumer appliance fields. With Ryerson & Son's warehouse chain covering a broad geographic area, Inland could introduce its products to new markets in new areas.

In 1939 Leopold Block resigned as Inland's chairman of the board, and Ryerson was elected to replace him. He served as chairman until 1952, when he relinquished the position to Joseph L. Block, Leopold's son. Ryerson assumed the position of chairman of the executive committee of the board, a position he retained until he retired from an active role on Inland's board in 1958. He continued as an honorary director until his death.

Ryerson assumed the top position at Inland during interesting times for the steel industry. As

World War II approached, labor's four-year effort to organize "Little Steel" was coming to a successful end. In accepting the independent steel union, Inland had to adjust its labor relations policies; at the same time wartime demand forced Inland to change its product mix back toward heavier products. Inland's mills worked beyond their capacity during the war; its operating rate between 1941 and 1946 was 104.8 percent. Inland's wartime income measured as a percent of long-term investment averaged 12.8 percent over the war years, the highest return earned by any of the top 12 steel firms.

In 1946 Inland purchased the Plancor 266 plant from the Reconstruction Finance Corporation for $13.5 million. In 1942 the government had begun construction on this facility on a site adjacent to the Indiana Harbor works as part of its program to help expand wartime industry. Plancor 266 consisted of 2 blast furnaces, 146 coke ovens, a power plant, and lakeside docks. The Ryerson subsidiary also continued to grow with Inland during the immediate postwar period. By the time Ryerson retired from Inland, the number of Ryerson & Son warehouses had doubled and the subsidiary had greatly expanded its fabricating operations.

Edward Ryerson was equally active outside of Inland. His activities in the industry included membership in the American Iron and Steel Institute (AISI), where he was an honorary vice-president, a member of the AISI board from 1939 to 1953, and chairman of the institute's committee on public relations between 1946 and 1953. In 1958 he headed the U.S. delegation on steel and iron ore mining at a meeting with its Soviet counterparts.

Ryerson had a deep and abiding involvement with education, becoming a Fulbright Lecturer to universities in Australia in 1958. Between 1953 and 1967 he was actively involved with the Chicago Educational Television Association, having served as honorary chairman in 1953, the association's inaugural year, and as president of the association from 1953 to 1967. He was a trustee of the University of Chicago from 1923 to 1956, becoming president of the university's board in 1953. Ryerson died in Chicago on August 2, 1971.

Publications:
"Commercial Problems of the Steel Industry," *American Iron and Steel Institute Yearbook* (1940): 51–57;

"Manufacture of War Materials in the United States," *Blast Furnace and Steel Plant,* 28 (October 1940): 1007–1008;

"Steel's Importance," *Blast Furnace and Steel Plant,* 33 (February 1945): 246;

"The Steel Industry's Public Relations Program," *American Iron and Steel Institute Yearbook* (1947): 655–663;

"Remarks on Public Relations Policies," *American Iron and Steel Institute Yearbook* (1950): 59–62;

"Old Challenges in Steel's Public Relations," *American Iron and Steel Institute Yearbook* (1954): 220–225;

A Businessman's Concept of Citizenship (Chicago: Privately printed, 1960);

The Ryerson Chronicle, 1886–1964 (Chicago: Privately printed by Johns-Byrne, 1964).

Reference:
William T. Hogan, S.J., *An Economic History of the Iron and Steel Industry in the U.S.,* 5 volumes (Lexington, Mass.: Lexington Books, 1971).

Edward Eugene Sadlowski

(September 10, 1938 –)

by Richard W. Kalwa

University of Wisconsin — Parkside

CAREER: President, Local 65 (1964–1968), staff representative (1968–1974, 1977–1984), director, District 31 (1974–1977), director, Subdistrict 3, District 31, United Steelworkers of America (1984–).

Edward Eugene Sadlowski represented a new generation of leadership within the United Steelworkers of America (USWA) and gives expression to many of the concerns of the union's rank and file. He challenged what he saw as a bureaucratic form of unionism that had diminished democracy and militancy among the membership.

Born September 10, 1938, in Chicago, Sadlowski is a third-generation steelworker. He began at the United States Steel Corporation's South Works in Chicago in 1956 and served as shop steward and grievance man in USWA Local 65, which represented workers in the plant. In 1964 Sadlowski assembled a multiethnic slate of candidates, made up of himself, two other Poles, two blacks, a Mexican, and a Serb, to defeat the long-standing incumbent leadership, and he became president of the Local 65. He was reelected three years later and in 1968 was appointed as a staff representative by Joseph Germano, director of the USWA Chicago-Gary district (District 31). At local union meetings and the national convention he introduced resolutions against the Vietnam War.

In February 1973 Sadlowski ran for director of District 31, the union's most populous district, with nearly 300 locals and 130,000 USWA members. His opponent, Sam Evett, was the chosen successor of Germano, who was retiring after more than 30 years as director. Although Sadlowski lost by almost 2,000 votes out of a total of about 45,000, he challenged the results of the election. The U.S. Department of Labor investigated, discovered evidence of interference and voting fraud, and ordered a new election in November 1974. Sadlowski won the rerun election by nearly 20,000 out of 60,000 votes cast.

During this campaign and afterward, Sadlowski criticized the leadership of the USWA for limiting membership participation in decisions regarding dues increases, contract ratification, and the elimination of strikes by the Experimental Negotiating Agreement. He also condemned the level of pay and benefits as well as working conditions in the steel mills and called for a more adversarial bargaining position toward the steel companies. In 1975 he and his associates formed Steelworkers Fight Back, an organization of dissenting USWA members.

In response to the impending retirement of USWA president I. W. Abel, the union leadership gave its support to Lloyd McBride, director of the Saint Louis district, as candidate for the top office. Sadlowski ran for president in opposition to McBride and what Abel called the "official family," the established hierarchy of the USWA. He resorted to lawsuits to stop the national office from withholding information regarding the location of local unions and to gain access to the union newspaper. His militant campaign attracted nationwide attention and drew contributions from supporters outside the USWA, including prominent liberals such as John Kenneth Galbraith. Denouncing Abel's overall strategy of cooperation with the steel companies, he stated, "I don't find anything compatible with the steelworkers union and United States Steel, nothing whatsoever."

In the February 1977 referendum, Sadlowski won only 43 percent of the total, losing 249,281 to 328,861. He achieved majorities in most of the heavily industrialized districts in the Northeast and Midwest, as well as in most large steel locals, but losses in Canada and in the South — and in industries outside of basic steel, which accounted for more than half of the union's membership by the late 1970s — swung the election in McBride's favor.

At the 1978 USWA convention the new McBride administration passed a constitutional amendment preventing contributions to union election campaigns from sources other than USWA members. Sadlowski questioned the legality of this rule, citing the limits placed on union democracy by the resources of incumbent officers. The Supreme Court, however, upheld that amendment in a 1982 decision. Following the death of McBride and the election of Lynn Williams to the presidency in 1984, Sadlowski was appointed director of the subdistrict representing South Chicago within District 31.

Publication:
"An Insurgent's View," *Nation,* 221 (September 6, 1975): 173–175.

References:
David Bensman and Roberta Lynch, *Rusted Dreams: Hard Times in a Steel Community* (Berkeley: University of California Press, 1987);
John P. Hoerr, *And the Wolf Finally Came: The Decline of the American Steel Industry* (Pittsburgh: University of Pittsburgh Press, 1988);
William Kornblum, *Blue Collar Community* (Chicago: University of Chicago Press, 1974);
Philip W. Nyden, *Steelworkers Rank-and-File: The Political Economy of a Union Reform Movement* (New York: Praeger, 1984);
Studs Terkel, *American Dreams: Lost and Found* (New York: Pantheon, 1980).

Charles M. Schwab

(February 18, 1862 – September 19, 1939)

by Robert Hessen

Hoover Institution, Stanford University

CAREER: President, Carnegie Steel Company (1897–1901); president, United States Steel Corporation (1901–1903); president (1901–1916), chairman, Bethlehem Steel Corporation (1916–1939); chairman, Emergency Fleet Corporation, U.S. Shipping Board (1918).

One of the most flamboyant and controversial of the American industrial giants during the early twentieth century, Charles M. Schwab to some of his contemporaries was the embodiment of the rags-to-riches story. To his critics he was a typical robber baron motivated by his desire for power.

He was born on February 18, 1862, in Williamsburg, Pennsylvania, to John and Pauline Schwab, a German Catholic couple. Suffering from bouts of dizziness and fainting, John was advised by a physician to move to a town with a higher elevation than Williamsburg's, and in 1874 he moved the family to Loretto, Pennsylvania — his and Pauline's birthplace — where John purchased a small livery stable. A deeply pious woman, Pauline was especially pleased by the move, for it allowed her sons, Charles and Joe, to receive intensive religious training in the Loretto grade school run by the Sisters of Mercury. After graduating from the grade school Charles entered Saint Francis College, founded by Irish Franciscans. During his years there, Schwab displayed flashes of a developing entrepreneurial talent, operating a wagon service from Cresson — the local railroad stop — to Loretto. He entertained his passengers with ballads, earning himself the nickname "the singing cabby."

Just before his seventeenth birthday he became infatuated with Mary Russell, a young actress from Pittsburgh who had come to Loretto to visit her older sister. Schwab's parents, fearing their son might elope with her and become a vaudevillian, arranged for him to go to Braddock to work as a clerk for McDevitt's, a grocery and dry-goods store located near the entrance of An-

Charles M. Schwab

drew Carnegie's Edgar Thomson Steel works. Schwab found his work as a store clerk and bookkeeper to be unchallenging, but he quickly became friends with a customer, William R. Jones — "Captain Bill" — the general superintendent of the steel works. Jones offered him a job, and Schwab began work at the steel works on September 12, 1879, in the engineering corps as a rodman, carrying the leveling rod used by the surveyors who were laying out plans for new furnaces. By parlaying a little knowledge and a lot of bluff, Schwab, who never graduated from Saint Francis, managed to create the impression that he was a highly qualified engineer and surveyor. He asked questions incessantly, without betraying how green and unseasoned he really was, and at night he read borrowed books to expand his knowledge.

His strategy worked: six months after having begun work Schwab was promoted over men older and far more experienced than he to serve as acting chief engineer in place of his boss, Pete Brendlinger, who had been temporarily transferred. Schwab subsequently began work as a draftsman at the Edgar Thomson works. Soon thereafter the chief draftsman held an unannounced competition for a promotion: the men were to work overtime on a special job, without additional pay, and the candidates for promotion would be those who worked unremittingly, without clock-watching or complaint. The chief draftsman sent Jones only one name: Schwab's.

Later, Jones rewarded Schwab's diligence by sending him as a messenger to Andrew Carnegie, who lived ten miles away in Pittsburgh. On one occasion Carnegie was delayed, and Schwab, who had become an accomplished musician as a student, sat down at Carnegie's piano and began to play. Carnegie, who loved music, heard him and asked him if he could play and sing at a party he was giving in a few days. Schwab agreed and returned to Braddock, where for the next three days he learned the music and lyrics of Carnegie's favorite Scottish ballads. He had caught the eye of one of America's greatest entrepreneurs, who would eventually regard him as a son.

Schwab never lost sight of his objective: to prove that he was indispensable to the Carnegie Steel Company and to seize every opportunity for advancement. When he recognized the increasing importance of chemistry in steelmaking, he rigged up a laboratory in his cottage and brought home specimens of steel to test for their carbon and phosphorus content. One of Carnegie's partners, Henry Phipps, heard about Schwab's studies and gave him a check for $1,000 to buy any equipment he needed. Meanwhile, Jones continued to offer Schwab challenging assignments. In 1885, when Schwab was twenty-three, Jones ordered him to design and build a bridge to carry hot metal over the Baltimore and Ohio Railroad tracks from the furnace area to the steelworks finishing departments. Schwab completed the job under budget and sooner than expected. As a reward, Jones gave him a diamond pin in the form of a spider, and Carnegie sent him a package containing ten $20 gold pieces.

The next year, at Jones's urging, Carnegie appointed Schwab general superintendent of the Homestead works, which Carnegie had bought in 1883 and was converting from rail to structural-steel production. Schwab's new salary was $10,000 a year. Three years later a furnace explosion killed Jones, and Carnegie had to choose a new general superintendent to run the works in Braddock. He was not inclined to choose Schwab because he did not want to disrupt the smooth operations Schwab had achieved at Homestead. But Schwab pressed for the appointment, and Carnegie relented. Only twenty-seven, he became the general superintendent of the Edgar Thomson works, the largest steelmaking facility in America.

As a boss, Schwab operated on the assumption that working men shared his desires and ambitions, and he offered them the same rewards for maximum effort that had appealed to him. At the Edgar Thomson works, for instance, Schwab had attempted to reduce the problem of "seconds" — substandard steel rails — by paying $20 cash bonuses to rail workers in return for first-class results. Competition for bonuses dramatically increased the number of quaity rails produced, resulting in company savings which greatly exceeded the bonuses paid. He used similar incentives throughout his career as a manager and business owner.

In June 1892 one of the most violent strikes in American industrial history erupted at Homestead, resulting in several deaths; the state militia was called out to restore order. In mid July, Carnegie suggested to Henry Clay Frick, president of Carnegie Steel Company, that Schwab be returned to Homestead to replace John A. Potter as superintendent. But Frick delayed action, fearing that any major change in personnel might be viewed by the strikers as a sign of weakening determination on the part of the company. Hence Schwab's return was delayed until October, when the strike was nearly over. Schwab had tried to decline the assignment, but Frick induced him by persuading Carnegie to give Schwab his first ownership interest in the company, one-third of 1 percent.

Schwab found Homestead in a general state of decay and did not set foot outside the plant during his first four months there. He personally greeted the returning strikers individually, calling many old-timers by their first names. He held a series of informal meetings with small groups of the leading workmen, hoping to win over their assistants and helpers. He went everywhere and spoke to everyone, on both the day and night shifts, compiling an inventory of physical and human assets on which he could base Homestead's reconstruction.

Schwab offered the returning strikers his vision of a new and greater Homestead, one which would surpass every other steelmaking facility. But even though he assured them that their grievances would be given a fair hearing, he stressed that the company was resolutely determined to be nonunion, claiming that otherwise Carnegie Steel would have to close down completely. This ploy — one he often used in later years — succeeded, enabling him to persuade the workers to give up their union and allowing him to proceed with the revitalization of Homestead. Schwab emerged from the Homestead conflict a hero, not only to Carnegie for having restored harmony but also to the workers for having saved them their jobs.

Less than a year later Schwab's reputation plummeted. He was accused of heading a conspiracy to defraud the United States Navy. Specifically, the charge was that Carnegie Steel had knowingly produced substandard armor for navy ships. In 1893 Secretary of the Navy Hilary A. Herbert ordered Carnegie Steel to pay a fine of $210,000. When Carnegie protested to President Grover Cleveland about the injustice of the secretary's action, the president reviewed the controversy and offered to reduce the fine to $140,484.94. But his decision pleased no one. Any fine, even a reduced one, was tantamount to a guilty verdict, so Carnegie and Frick rejected the offer. Neither Carnegie's congressional allies nor his enemies favored the president's compromise, so the House of Representatives passed a resolution authorizing the Committee on Naval Affairs to hold hearings on the charges. When Schwab began his testimony on July 6, 1894, he tried to absolve himself of any wrongdoing, denying that he had ever authorized any deviations from the contract specifications and claiming that he had no personal knowledge of any irregularities. He insisted that the navy's experts — the resident inspectors at the Homestead armor plant — were amateurs, filled with book knowledge but lacking the experience and expertise that only a professional steel man could possess. He argued that the navy should not set any procedural specifications for producing armor and should concern itself exclusively with the results as demonstrated in the ballistic test. His testimony made it clear that the root cause of the controversy was his unwillingness to deal with or obtain permission from the navy's inspectors.

Schwab's defense was wholly unsuccessful. The committee voted unanimously to reaffirm Herbert's original findings, and the reduced fine was not rescinded. In its report the committee hinted that Schwab's testimony was perjured, yet did not identify specific examples or suggest any indictments for perjury. The press jumped on the story and quoted the report's most lurid accusations and repeated its dramatic conclusions. It was Schwab, not Carnegie or Frick, whose reputation was most sullied by this incident. During the next 45 years of his career his role in the armor-plate scandal was repeatedly resurrected as evidence of his boundless capacity for criminal activity. Whenever he was accused of new acts of fraud or negligence his detractors cited the scandal of 1892–1893 as proof of his guilt.

But Schwab's position at Homestead was secure. Carnegie had remained loyal to Frick when Frick had been condemned by Congress and the press for his handling of the Homestead strike, and Carnegie was just as adamant in defending Schwab from criticism. To have replaced Schwab would have been tantamount to admitting that the charges of fraud were true. Besides, Carnegie regarded Schwab as irreplaceable. In 1897 Carnegie named the thirty-five-year-old Schwab president of Carnegie Steel, and Frick was elevated to the largely ceremonial role of chairman.

But Carnegie never would have promoted Schwab to the presidency if they had not been of one mind concerning cost cutting. Throughout the depression of 1893–1897 Carnegie Steel, having concentrated on diversification and cost cutting, was able to undersell its competitors and generate profits. Schwab began his presidency just as the depression was drawing to an end. By August 1897 there were signs of recovery.

In 1899 Carnegie and Frick clashed repeatedly over Frick's role as an intermediary for a syndicate seeking to buy out Carnegie's interest in the steel company; if successful, Frick and Phipps stood to reap a $5 million bonus. Carnegie was further outraged when it was disclosed that William H. Moore and John W. Gates, whom Carnegie despised as unscrupulous speculators and stock manipulators, were behind the syndicate. To oust Frick, who owned 6 percent of the Carnegie Steel Company and 23 percent of the Frick Coke Company, Carnegie invoked his ultimate weapon: the Iron Clad Agreement, which had been signed by every partner, including Frick. It required any partner to surrender his interest, at low book value, upon the request of three-quarters of the voting partners and three-quarters of the total voting shares. On Carnegie's behalf, Schwab agreed to collect the needed signatures from the other partners, but Frick filed suit in the Court of Equity to prevent his expulsion from the company. He was prepared to demand the dissolution of the company, if necessary.

Schwab, of course, would have viewed a court-ordered dissolution as a calamity, so acting as Carnegie's personal representative, he met Phipps and Francis T. F. Lovejoy, acting for Frick. They reached an agreement: Frick was to drop his legal action, and Carnegie was to agree to consolidate Carnegie Steel and Frick Coke into a single entity, the Carnegie Company, which would be capitalized at $320 million. Schwab was a major beneficiary of the settlement. Carnegie awarded him him a 2 percent increase in his partnership interest. Schwab, who had just passed his thirty-eighth birthday, now owned a 6 percent interest, a handsome return for 20 years of loyalty, ingenuity, and productive effort. He also gained new powers within the company and new prestige in the business community, both of which he found exhilarating. In later years he fondly recalled this period as the time when he was "number one."

A new crisis erupted early in 1900 when, in response to a declining demand for fabricated steel, two of Carnegie's biggest customers, Moore's National Steel and J. P. Morgan's Federal Steel, decided they had to cut prices to meet the market. In mid 1900 both companies advised Carnegie that they intended to expand their production facilities and would reduce future orders. In order to retain his dominant position in the industry Carnegie ordered Schwab to prepare designs and cost estimates for a full line of fabricating plants. The first plan Schwab completed was for a steel-tube plant at Conneaut Harbor, Ohio. The plant was to employ a new method for tube production that would save $10 a ton. Conneaut promised to be the most technologically advanced fabricating plant in America and would easily have eclipsed the older, less-efficient plants of the Morgan-controlled National Tube Company.

Once convinced that Carnegie was serious about expanding, Morgan knew that his steel holdings were endangered. In 1898 Judge Elbert H. Gary, president of Federal Steel, had urged Morgan to buy out Carnegie, but Morgan took no action. In 1900 Morgan was more amenable to the idea. It was widely known in the American business community that Carnegie might be willing to sell if a suitable buyer could be found.

At age sixty-five he was eager to devote his remaining years to distributing the vast fortune he had earned. A suitable buyer ultimately was found — J. P. Morgan — and Schwab played a central role in consummating the sale.

On the night of December 12, 1900, a dinner in Schwab's honor was held at the University Club in New York, attended by the city's leading bankers and businessmen. After the dinner Schwab spoke for a few minutes with no prepared text in hand. He sketched for the audience his hopes for a greater and more profitable steel industry, in which wasteful practices had been eliminated and each plant specialized in a single product and was located in the area in which its products were sold. If the steel industry were made as efficient as possible, if its plants were specialized, integrated, and centrally managed, and if its leaders were willing to cooperate for long-range mutual growth, then an ever-widening market for steel could be created. Everyone would benefit, he promised — consumers and producers alike. Afterward, Morgan called Schwab aside for a half-hour conversation.

Three weeks later Morgan told Schwab that his goal was to prevent Carnegie from expanding into fabricated products, not to buy him out. Schwab advised Morgan that Carnegie most likely would expand because he was determined to beat down all threats to his industrial supremacy. Morgan quickly became convinced that if an industrial war with Carnegie were to be avoided, a permanent merger among the key steel companies needed to be effected. Morgan and his partners began to explore with Schwab the possibility of forming a giant steel consolidation similar to that projected in Schwab's University Club speech.

Schwab spent the next few days preparing a memorandum for Morgan, listing the companies which should be included and the prices which ought to be paid for them. Then he faced an even more formidable problem: he had to convince Carnegie to sell. Schwab and Carnegie spent several pleasant hours playing golf, and by the time the game was over Carnegie had agreed to sell if Morgan could meet his price: $400 million. Morgan accepted Carnegie's terms, and the United States Steel Corporation was born.

On April 16, 1901, Schwab was named president of U.S. Steel. At thirty-nine he headed the first billion-dollar enterprise, which controlled nearly 50 percent of America's steelmaking capacity. Though Schwab had been little known outside of the business world, he now became a national celebrity. His next four years at U.S. Steel, however, were largely unproductive. He did not adjust well to his sudden rise to national prominence. He fought with his colleagues, made ill-considered speeches, allowed wild rumors about his salary and private life to go uncorrected, and generally comported himself in a way unlikely to inspire confidence in his maturity and sense of responsibility.

Schwab was not satisfied with his position in U.S. Steel's corporate hierarchy. He expected that he would have undivided authority — that he could be "an autocrat." He believed that his experience as a "practical steelman" should have given him preeminence over men whose training or background had been in law or finance. Yet U.S. Steel's organizational structure was deliberately arranged so that no one individual could have undisputed authority. From the very outset Schwab came into conflict with the board of directors and other committees. He was responsible to the executive committee, whose chairman was Judge Gary. The two men could scarcely have been more dissimilar. Schwab was dashing, ambitious and self-assertive, and fond of raucous horseplay and off-color stories; he had a passion for personal publicity that increasingly rankled the pious and staid chairman. In Gary's eyes Schwab also was tainted by his former association with Carnegie, whom Gary heartily disliked because of Carnegie's policy of aggressive price-cutting. Schwab, eager to be "number one" and believing that Gary was not knowledgeable about practical matters, tried to act without the committee's authorization, without realizing that he was putting his job in jeopardy.

What little Schwab accomplished as president of U.S. Steel occurred in his first months in office. He created an integrated foreign-sales department. He also formulated corporate policy on freight shipments by boat and rail, and he directed an extensive search for new sources of iron ore. Perhaps his most important innovation was to have each plant submit weekly cost sheets, which were compared in order to identify and reward the ablest superintendents. Outside of his office, however, Schwab was gaining notoriety, giving U.S. Steel the kind of publicity it could not afford. He provoked a storm of criticism when he told an audience of trade-school students that a college education was not essential for boys who planned a career in business and that they would profit more by the additional four years of work experience. He also was rumored to be paid a million-dollar annual salary, and he gave the story some credibility when, in 1901, he began building the mansion on New York's Riverside Drive that would be a monument to his success. The mansion, called Riverside, was just one of the early expressions of his growing passion for owning the biggest and the best homes, automobiles, and private railroad cars. His was the last, the largest, and the most lavish full-block mansion ever to be constructed in New York.

In late December 1901 Schwab sailed for Europe. Soon the cable wires to America were humming. Incredible tales of Schwab's breathtaking victories at roulette in Monte Carlo appeared on the front pages of daily newspapers. Carnegie, who felt personally betrayed because of his long-standing hostility to any form of gambling, bombarded Schwab with a barrage of stinging cables and censorious letters. Schwab, who had prized Carnegie's good opinion of him, became seriously ill because of Carnegie's unrelenting criticisms, so his doctors insisted that he should have a total change of scene and complete rest. But months of convalescence made Schwab restless, so in February 1903 he decided to return to America.

He did not immediately attempt to resume work as president of U.S. Steel. Instead he concentrated on various personal investments and served as a skilled negotiator for companies outside the steel industry which were interested in merging. Most of those mergers — International Nickel and American Steel Foundries, for example — proved successful and lasting. But one that did not was the United States Shipbuilding Company, a 1902 merger of seven shipyards on the Atlantic and Pacific coasts. By 1903, when the company was verging on bankruptcy, Schwab was accused of having deliberately engineered its collapse in order to obtain its assets for himself.

In May 1902 one of the promoters of the venture, Lewis Nixon, had contacted Schwab, hoping that he personally or as president of U.S. Steel might be willing to purchase some bonds. Schwab personally agreed to purchase $500,000 of bonds, provided U.S. Shipbuilding would purchase its steel from U.S. Steel. One month later Nixon asked Schwab if he would be willing to include Bethlehem Steel in the shipbuilding merger. Schwab had bought Bethlehem for himself, as a personal investment, in June 1901 — less than eight weeks after he became president of U.S. Steel. He now offered Bethlehem to the shipbuilding merger for $9 million in cash, but after negotiation, he accepted a counter-offer: $10 million in bonds, for which Bethlehem's plant and properties would be his collateral in the event of default. He also asked for and received a second mortgage on all the properties of the shipbuilding merger. Schwab, a shrewd and seasoned bargainer, obtained extremely favorable terms for himself. Whether the shipbuilding merger succeeded or failed, he would lose nothing. For that reason, when the company showed signs of imminent collapse in 1903, he was blamed.

U.S. Shipbuilding needed $2 million to provide working capital and to meet the interest payments due July 1, 1903, on the bonds. Schwab said that he would put up the money if, in return, his present second-mortgage bonds were replaced with new first-mortgage bonds, thereby giving him the primary lien on all the properties of U.S. Shipbuilding. But a group of existing first-mortgage bondholders, who had no intention of seeing Schwab awarded bonds with priority claims over their own, decided to give him a battle. On June 11, 1903, they filed suit in the U.S. Circuit Court of Trenton, New Jersey. Their objective was to prevent the execution of the planned reorganization of the company, which they said favored Schwab at their expense.

A few days later, on June 30, the Finance Committee of U.S. Steel elected William E. Corey to the post of assistant to the president. His promotion was deliberately intended to be the first step toward Schwab's resignation, but the initiative came from Schwab himself. He recognized that his notoriety had severely compromised his position as president. A month later, in August 1903, Schwab resigned, yet the shipbuilding scandal continued to be front-page news. A self-styled "practical man," Schwab was eager to end the controversy by making an out-of-court settlement with the bondholders. His losses were not merely monetary. In the yearlong press coverage of U.S. Shipbuilding's collapse, the publicity was almost unanimously unfavorable to him. His tarnished reputation, far more than his illness, had cost him the presidency of U.S. Steel. He had headed the world's two largest steel companies, so his new domain, Bethlehem Steel and its seven sister shipyards, was clearly a demotion. But to Schwab this defeat was not irreversible. Like his hero, Napoleon, he was now challenged to restore himself to his former stature and power. He did so by transforming Bethlehem Steel into the foremost rival of the giant corporation from which he had been ousted.

Before Schwab took control, Bethlehem Steel had been a small specialty producer; within a decade he had made it into the nation's second-largest and most diversified steel company. After his first official tour of the properties in August 1904, he initiated a series of sweeping changes. He announced plans for the plant to produce all types of guns, gun forgings, and tools. He strengthened the company by selling off its unprofitable properties. Then he turned to the shipyards worth saving, replacing the top managers and allocating funds for expanding and modernizing the facilities for shipbuilding and marine repair work. But his second annual report, despite its tone of optimism, read like a chronicle of disaster. In 1905 the corporation's net income had been $2,600,000; it fell to $762,000 in 1906. The problem lay with Bethlehem's largest single customer, the federal government, which in 1906 had sharply reduced its purchases of armor plate and ordnance. Schwab realized that Bethlehem's economic future would be precarious if it depended almost exclusively on government contracts. He also recognized that he could not effectively run a business in Bethlehem from an office on lower Broadway in New York City, so he moved to Bethlehem.

Schwab's plans for the growth of Bethlehem Steel centered on diversification. He decided to risk his fortune and the company's future by undertaking the production of Grey beams, a revolutionary wide-flanged structural beam which could be rolled as a single section instead of being riveted together. From his years at Carnegie Steel he knew that during a depression a company is in a better position to expand, because labor and raw materials cost less than in normal times. But he had no uncommitted cash reserves, so through a rather circuitous process, he arranged for his suppliers to underwrite partially his new Grey beam mill. In 1908, needing more money, Schwab asked Carnegie for help. Carnegie was impressed by the Grey beam, so he agreed to lend Schwab U.S. Steel bonds to use as collateral for bank loans. In October 1908 Schwab appointed Eugene G. Grace as general manager; then he turned his attention to finding clients for the Grey beam. He worked out a daring sales campaign, persuading architects to submit two cost estimates to their clients: one based on the use of conventional riveted beams, the other on Bethlehem beams — the name by which the Grey beam became more widely known. The new beam promised substantial savings in steel tonnage and in labor costs because it eliminated the overlapping sections of steel and the need for riveting. A promiment Chicago architect, Ernest R. Graham, was so convinced of the superiority of the beam that he used it on all of his major commissions. Chicago virtually became an advertisement for the Bethlehem beam. Once other architects and builders saw how good the Bethlehem beam really was, Schwab obtained many large orders, almost without effort.

Although the Bethlehem beam was Schwab's most successful innovation, it was not the only source of the company's rising sales and profits. Ironically, in 1909, after the new beam's commercial success was assured and the company no longer needed to depend so heavily on ordnance contracts, there was a sharp upturn in military orders. While rising sales contributed to Bethlehem's growing profits, so did increased efficiency. Schwab was an avid proponent of bonus plans; he believed that they encouraged extra effort and efficiency. Bethlehem's bonus system directly and promptly rewarded any employee who increased his productivity.

By 1910 Bethlehem Steel was flourishing. It once had been a sluggish company which depended heavily on military orders; now it was diversified and well managed. But once again — as had happened so often — Schwab's hard-won successes and achievements were suddenly overshadowed by a new crisis: a strike in 1910. Early in February a delegation of three workers, representing over 700 mechanics in two machine shops, presented their demands to C. A. Buck, the general superintendent of the Bethlehem plant. His response was direct and unequivocal: he fired the three men. A year before, he had fired another such delegation, which had made the same demands. This time the machinists rallied behind the men who had been fired. On February 3 they held a general work stoppage. Buck immediately alerted Schwab, who summoned the disgruntled workers to a mass meeting the next day. He promised that he and Grace would listen to their grievances and attempt to reach an amicable settlement. A spokesman presented the machinists' demands: the company should end Sunday work, pay extra for overtime, and rehire the delegates who had been fired. Schwab was conciliatory. He was willing to rehire the three men and examine the grievances of all the others, but on one condition: the strikers first must return to work.

Instead, the strikers concluded that the company could not dispense with their services even for a short period, and therefore it would be impelled to make concessions. They believed that

Schwab at Immergrun, his house in Loretto, Pennsylvania, circa 1920 (Hagley Museum and Library)

Schwab would not jeopardize his profits from armor and ordnance sales nor risk the penalties and forfeitures if Bethlehem failed to meet its contract delivery dates. The strike soon spread, as other groups of workmen, aware that Schwab wanted to avoid a general strike and shutdown, presented him with demands of their own. During the next few days the strikers waited, but Schwab gave no sign that he was willing to compromise or capitulate, and the arrival of mounted state police only exacerbated the tense scene. The constables, trying to keep crowds from forming, began to use their riot clubs. An innocent bystander was accidently killed by a stray bullet from a constable's gun. This shooting won new public sympathy for the strikers. They sent Schwab a new statement of their demands. He replied that he would not negotiate with them because they were not employees of his company.

Congress authorized an investigation of working conditions and wages at Bethlehem Steel, and President William Howard Taft gave the assignment to Charles Nagel, secretary of commerce and labor. He, in turn, delegated it to Ethelbert Stewart, an experienced investigator in the Bureau of Labor. Schwab issued a public statement assuring Inspector Stewart that he would be given full access to the company's premises and its payroll records. Throughout April 1910 both management and labor waited optimistically for the release of Stewart's report. On May 4 it was submitted to the Senate then released to the press. It was factual, neutral, and dispassionate; it neither praised nor criticized the company. But the strikers were pleased, even so, because it did document their claims about low wages and long hours. Schwab, by contrast, was incensed over one curious omission in the report: it had failed to state that these same hours, wages, and conditions existed throughout the American steel industry and were not unique to Bethlehem.

Following the release of the report, the strike continued. Through an intermediary Schwab indicated that he was willing to settle with the strikers, and on May 18 they agreed to return to work. But the only men who were permitted to return to their jobs were those who had neither damaged the company's property nor attacked its reputation. There were no wage increases and no recognition of any union. The only gains the strikers had made were that overtime and Sunday work were to be optional. But since there was no increase in wages, any man who did not work overtime and Sundays would lose a substantial amount of total income.

On the surface, Schwab had won. The strikers had obtained little, and what they did get had cost them 108 days' pay. Yet he could have offered them exactly the same terms at the outset, possibly avoiding a lengthy strike. He in fact *did* negotiate with the strikers — which he had said he would never do — before they returned to work. Bethlehem Steel obtained only one thing from the strike: adverse publicity. And Schwab's statements only served to undermine his own reputation and that of the company. In view of increased federal scrutiny of the steel industry, his success was at best a Pyrrhic victory. Thereafter steel men found it increasingly difficult to argue that high protective tariffs were a safeguard of high wages for American steelworkers.

In 1913 when Congress passed the Underwood Tariff, which reduced protection, Schwab predicted rising unemployment and falling wages and even threatened to retire from business. As he had foreseen, European rail imports did capture a larger share of the American market, but what he could not foresee — nor could anyone else — was that World War I would break out in August 1914. Once the war began, German and Belgian rail competition ended abruptly. Demand for

Bethlehem's products rose sharply; first came orders from Great Britain, then from the U.S. government. For the next four years Bethlehem made record profits and continually expanded its productive capacity to accommodate new orders. And for Schwab personally, the war brought two new opportunities to refurbish his reputation as a business leader.

First, in October 1914 the British Admiralty summoned Schwab to London. He had guessed correctly that the Admiralty needed shrapnel and shells; he received the largest order in Bethlehem's history: $135 million. But the last item on the British shopping list, by far the most important, was one he had not anticipated. Britain then had more than 50 submarines patrolling her home waters, but the Admiralty urgently wanted 20 more. Schwab offered them a daring proposal. Bethlehem usually spent 14 months constructing a submarine; he said he would guarantee to begin delivery in six months. He proposed that for every week Bethlehem beat its delivery dates, the Admiralty would pay a special premium: $10,000 per submarine. On November 7 the deal was closed.

Second, on December 1, Secretary of State William Jennings Bryan asked Schwab to come to Washington to discuss Bethlehem's contract with Britain. Bryan told him that President Woodrow Wilson would consider it to be a violation of American neutrality if Bethlehem shipped submarines to England. But, he added, Wilson had no objection to the submarines being built, so long as they were not shipped. Armed with this ambiguous advice, Schwab discovered that in Montreal a shipyard owned by the British firm Vickers contained the type of equipment required for rapid assembly of submarines and that most of its facilities were idle. By the summer of 1915 ten submarines built in Montreal were in active service. Schwab was so successful at supplying Great Britain with submarines and munitions that Germany tried to buy control of Bethlehem. In 1915 Count von Bernstorff, the German ambassador, sent a representative to Schwab, offering him $100 million for his controlling interest. When the British government learned of Germany's bid, it decided to make a counteroffer, but Schwab rejected both.

Not all of Schwab's wartime activities turned out as successfully. In 1916 Bethlehem became embroiled in a new controversy with the U.S. government, and again it involved armor plate. This time, however, it was not the quality of the armor which was called into question, but the price. To protect his investment and future profits, Schwab launched a costly campaign to persuade members of Congress and the public to take his side against the secretary of the navy, who said Bethlehem's prices were much too high, but in the end his campaign to save Bethlehem's $7 million facility was a failure. Congress authorized the navy to build a government-owned armor plant and stop buying from Bethlehem and Midvale Steel. In 1921, after the first plates were produced — at a cost nearly double the price per ton charged by private companies — the plant ceased operations. Hundreds of skilled workmen who had been hired to construct and operate the plant were dismissed, and the homes built for them remained empty for the next 20 years. This was a rare victory for Schwab in his lifelong battles with the navy.

When the United States entered the war in April 1917 it had few troop and cargo ships. The U.S. Shipping Board was legally empowered to seize all suitable ships, but these were insufficient to meet rising military needs, so its subsidiary, the Emergency Fleet Corporation, launched a massive shipbuilding program. Finding that they were unable to cope with the mounting demands made on them, the leaders of the Fleet Corporation decided to enlist the services of a major American industrialist to take charge of the program. They asked Henry Ford to do the job, but he refused. They summoned Schwab, their next choice, to Washington and told him he was "conscripted" to be director-general for the duration of the war. He declined, reminding them that he already was actively supporting the war effort by working full-time to complete crucial government contracts, and that he, personally, was risking millions of dollars if these contracts were not successfully fulfilled. But he did promise to think over their offer and return for another meeting.

Schwab reconsidered and accepted, but with three conditions, all of which were met. He wanted autonomy to run the shipbuilding program without being boxed in by elaborate rules and inflexible procedures; he wanted assurances that President Wilson would support him; and he wanted a written agreement with the government stating that he would not have to deal personally with any company of which he was an officer or owner, so that no charge of conflict of interest or financial favoritism could be made.

Schwab ran the Fleet Corporation just as he had run his steel mills, using the same methods he had employed so successfully throughout his career. Just as he sought only men of proven ability to work for the Fleet Corporation, so he also selected only companies of proven competence when he placed contracts. Believing that all men have a competitive spirit and that rivalry will drive them to do their best, he promised to publicize the work of the ablest managers and workmen. Yet he thought that money meant more to the men than flags or medals, so with his own funds he established a program of cash bonuses and awards. Within two months after he took charge of the Fleet Corporation there was already a marked increase in the number of keels laid, ships launched, and ships delivered for active service. His success as director-general, coupled with the fact that he had served at his own expense, without any salary except for the symbolic dollar a year, made him seem doubly deserving of praise. In the eyes of his erstwhile critics he had raised his moral stature through public service and financial self-sacrifice, or so it seemed for a while.

During the early 1920s Schwab developed diabetes, and thereafter he was required to stay on a sugar-free diet and receive daily injections of insulin. He had no adverse reaction to it, and generally speaking, his ability to work was not impaired by his condition. Both his doctor and his wife did urge him to take life easier, and for a time, he did. But he was not yet sixty, and he had no desire to retire; he was too active and vital ever to tolerate a sedentary existence. He readily agreed to rest for a few months at Immergrun, his newly built estate in Loretto. "I am nominally Chairman of the Board of the Bethlehem Steel Corporation; practically, I am a retired steel manufacturer," he said in November 1922. He should have said "semiretired." Although he was no longer actively involved in the day-to-day management of the company — that role was filled by President Eugene Grace — Schwab was its single largest owner. Although the two men were so different in personality and style, their fundamental business attitudes were very much alike. Grace's favorite slogan, "Always More Production," mirrored Schwab's own outlook.

Schwab was Bethlehem's roving goodwill ambassador. He was an immensely effective speaker, and business, civic, and academic groups were eager to hear him. Whenever possible, he accepted speaking engagements; he loved the challenge of capturing an audience's attention and winning its affection. But if any of his speeches contained a controversial idea, it was unintentional. He simply told his audiences what they wanted to hear. His frequent appearances as a speaker won him many new admirers — a factor of considerable importance to a man who drew emotional sustenance from other people's approval. Nonetheless, the final two decades of his life were not trouble-free. Behind him were his triumphs; ahead were new accusations which called his wartime service into question and new critics who challenged his management of Bethlehem.

The first criticism arose in January 1921 when a government auditor charged that of the $269,543 in expenses that Schwab incurred in a nine-month period, Bethlehem Steel had tried to recover $100,000 by billing the government for it as "cost of ship construction." Since Schwab had served as director-general without salary or an expense account, this charge, if shown to be true, would have been a serious blot on his recently enhanced reputation. But Grace was able to prove to the satisfaction of a congressional committee that the charge was unfounded, so Schwab received a public apology from the chairman, Congressman Joseph Walsh.

In March 1924 Congressman James F. Byrnes, a South Carolina Democrat, speaking on the floor of the House, claimed that Bethlehem had made $11 million in "excess profits" on its wartime shipbuilding contracts. According to Byrnes, grave misconduct by Schwab while he headed the Fleet Corporation had enabled Bethlehem to reap this wrongful gain. Eleven years later a special master appointed by the U.S. District Court completely exonerated Schwab, recalling that he was not responsible for the Fleet Corporation's contract arrangements with Bethlehem and that he could not have known precisely what profits Bethlehem was making on its shipbuilding contracts since final claims for payment, along with supporting cost sheets, were not filed until long after the Armistice in 1918. (The government appealed this decision to the U.S. Circuit Court of Appeals, and the case was not finally decided until 1941, more than two years after Schwab died, when the Supreme Court held against all of the government's allegations.)

Throughout the 1920s a tacit live-and-let-live agreement existed between Schwab and Gary of U.S. Steel, so the two largest steel companies were rarely in direct competition. But in 1926 Schwab made the startling discovery that U.S. Steel was in the process of building a mill to produce the patented Bethlehem beam, without his or Grace's knowledge or authorization. In the next few months no progress was made toward resolving the dispute. Early in 1927, however, Gary became ill. Perhaps sensing that he would not re-

cover, he decided to make his peace with Schwab. But after Gary died in August 1927, U.S. Steel broke the settlement agreement, so Bethlehem filed suit for patent infringement. In October 1929 the suit was withdrawn after U.S. Steel agreed to pay royalties for permission to produce the Bethlehem beam. But as a result of the patent fight, Schwab and Grace decided to expand Bethlehem's activities in the area west of Pittsburgh, which formerly was dominated by U.S. Steel.

After the war Schwab spoke regularly at meetings of the American Iron and Steel Institute, which Gary had organized in 1910. A gifted and genial speaker, Schwab found a solid niche for his anecdotes and reminiscences before this part-technical, part-social organization of steelmakers. At the banquets he delighted his listeners by appearing to speak impromptu, but actually he had memorized speeches that were ghostwritten for him. After he succeeded Gary as president in 1927, he thoroughly enjoyed the limelight twice yearly in his presidential addresses. When the group switched to a chairman in 1933, he served for two terms. In the eyes of the public and of his professional colleagues alike, Schwab was the voice of the American steel industry.

But at the same time he occupied this prestigious post, he came under new attack. In 1931 four Bethlehem shareholders filed suit in the New Jersey Court of Chancery, accusing Schwab of "favoritism" and "prejudiced judgment" in his administration of the bonus system and of failure to disclose even the aggregate amount of bonuses in Bethlehem's annual reports. He issued a vigorous defense of his system, arguing that Bethlehem owed its success to the large bonuses paid to its executives, but that secrecy was necessary to avoid invidious comparisons among the recipients. Settlement of the controversy involved four major changes: Schwab's exclusive power to administer the bonus plan was ended; the aggregate amount of bonuses paid would be published in the annual report; the bylaw providing that 8 percent of earnings was the maximum available for bonuses was maintained, but the basis of calculating earnings was changed so that depreciation, depletion, and obsolescence charges were deducted; and in anticipation of greatly reduced bonuses the executives were now to receive larger base salaries, and no bonuses at all were to be paid when dividends were passed on the common stock. Here again, Schwab, the practical man, believing that he knew the best way to get results, thought he had no need to consult or inform anyone about his policies, and once again his policy of secrecy backfired.

In 1934 Schwab was accused of being a "merchant of death," when a Senate committee headed by Gerald P. Nye of North Dakota began to probe the role played by munitions makers in America's entry into World War I. Another blow came in 1935, when a group of disgruntled Bethlehem stockholders complained that while they were receiving no dividends and workers were earning an average wage of only 67 cents per hour, Bethlehem's three top corporate officers were drawing salaries totaling $488,000. Schwab was told to his face that he had outlived his usefulness to Bethlehem and ought to step down. He was stung by the criticism, feeling that Bethlehem was *his* company, not theirs, and that they should not be second-guessing his business policies.

Throughout the Great Depression of the 1930s Schwab offered the ailing economy what he thought it needed most: frequent predictions of a bright future near at hand. He never changed his tune or his lifestyle. Even though the upkeep on his mansions, Riverside and Immergrun, were depleting his reserves, he felt there was no reason to retrench because, having no children, he was indifferent to the size of the estate he would leave at his death. So until 1936 he continued to spend freely, acting as a private welfare agency for friends and family members and making business loans which the recipients treated as if they were gifts. But since the economy did not recover, his financial situation continued to deteriorate. By 1936 the continuing drain was so great that he tried to sell Riverside, but without success. Although never willing to admit publicly that he was nearly bankrupt, he refused any new financial commitments.

In March 1936, a few months after her ninety-third birthday, his mother, Pauline, developed pneumonia and died before Schwab could reach her bedside. Three years later his wife Rana, whom he had married in 1883, died in her sleep. That summer he made his final trip to Europe. A bout of airsickness was more than a temporary discomfort; it had loosened a blood clot. On August 9, while at the Savoy Hotel in London, Schwab suffered a mild heart attack. Two weeks later he returned to New York aboard a transatlantic cruiser, a voyage he had made more than 100 times in his life. Schwab, a risk taker who so often had beaten the odds, did not recover; he suffered a second heart attack and died on September 19, 1939.

Selected Publications:

(Note: Almost all of the articles appearing in the trade press under Schwab's name were prepared by his publicist, Ivey Lee. Many of these articles are texts of speeches Schwab delivered.)

"The Huge Enterprises Built Up by Andrew Carnegie," *Engineering Magazine* (January 1901): 504 –517;

Succeeding With What You Have (New York: Century, 1918);

Andrew Carnegie: His Methods with His Men (Pittsburgh?, 1919);

"Steel Industry Will Disarm with World," *Iron Trade Review,* 69 (November 24, 1921): 1339, 1361;

Ten Commandments of Success: An Interview by B. C. Forbes with Charles M. Schwab (Chicago: LaSalle Extension University, 1924);

"Partnership Spirit Will Give Steel the Profit it Merits," *Iron Trade Review,* 81 (November 3, 1927): 1087–1135;

"Factors of Human Engineering," *Iron Age,* 120 (December 8, 1927): 1591–1592;

"Increased Steel Profits by Eliminating Waste in Distribution," *Iron Trade Review,* 82 (May 31, 1928): 1410–1411;

"Steel Distribution Must be Rationalized," *Iron Trade Review,* 84 (May 20, 1929): 1455, 1470–1471;

"Cost Sheet Should be Industry's Compass," *Iron Trade Review,* 85 (October 31, 1929): 1102–1103;

"Tendency to Big Units Aids Industry, Public," *Iron Trade Review,* 86 (May 15, 1930): 58–59, 98;

"Steel Industry Will Lead March to Prosperity," *Steel,* 87 (October 30, 1930): 24–25;

"Steel Wage Rates Should be Maintained," *Iron Age,* 127 (May 28, 1931): 1746–1748, 1780;

"Steel Institute to Broaden Scope as Leaders Assert Faith in Future," *Iron Age,* 129 (May 26, 1932): 1173–1176;

"Recovery's Biggest Hurdle — Legislative Restriction of the Flow of Capital," *Iron Age,* 133 (May 31, 1934): 15–17;

References:

Eugene G. Grace, *Charles M. Schwab* (Bethlehem, Pa.: Bethlehem Steel Corporation, 1947);

Robert Hessen, *Steel Titan: The Life of Charles M. Schwab* (New York: Oxford University Press, 1975);

Mark Reutter, *Sparrows Point: Making Steel — the Rise and Ruin of American Steel Might* (New York: Summit, 1988);

John Strohmeyer, *Crisis in Bethlehem: Big Steel's Struggle to Survive* (Bethesda, Md.: Adler & Adler, 1986).

Archives:

The major source for information on Schwab's early career is the Andrew Carnegie papers at the Library of Congress. Schwab's personal papers are in the Hugh Moore Historical Park and Museum in Easton, Pennsylvania, and in the Hagley Museum and Library in Wilmington, Delaware. A small collection of Schwab's papers can be found in the Pattee Library, Pennsylvania State University.

Sharon Steel Company

by Terry S. Reynolds

Michigan Technological University

The Sharon Steel Company was one of the smaller, independent, integrated steel companies in America in the twentieth century. In 1950, for example, it held only about 1.6 percent of industry capacity.

Sharon Steel was organized in October 1900 as the Sharon Steel Hoop Company. During the first half of the century it was managed by experienced steelmakers — Morris Bechman, Severn Kerr, and Henry Roehmer. From 1900 to 1917, however, Sharon's growth was slow. Gross fixed assets grew from slightly over $1 million in 1906 to only $2.25 million by 1916. But in 1917 and 1918 Sharon purchased companies which produced sheet steel and pig iron, raising its fixed assets to over $18 million by 1920 . For the next 25 years, however, the company resumed its traditionally slow rate of growth, largely due to very conservative management. For the pre-1950 period Sharon Steel nonetheless had a higher rate of return on fixed assets than any of the larger companies and was second among steel producers in rate of return on investment.

Sharon began to expand more rapidly after World War II. In 1946 it purchased the Farrell Works from the United States Steel Corporation. The newly acquired property contained blast furnaces, basic oxygen and electric furnaces, and rolling facilities, making Sharon Steel a fully integrated steel producer. Small integrated firms, like Sharon, could not afford much participating in the buying spree that characterized postwar growth among the larger iron and steel companies. But Sharon did acquire the Detroit Seamless Tube Company in 1945 and a steel-post firm in 1948. Sharon also could not afford to integrate backward into raw materials or transportation as extensively as its larger competitors. In 1970, for instance, Sharon was one of the few steel companies with no iron mines: it relied entirely on purchased ore. For years it also had to rely on outside purchases for its entire coke supply.

Sharon Steel began to encounter serious financial difficulties in the late 1950s. Its hand-me-down blast furnaces and older equipment were inefficient. The company's best product line — narrow, light-gauge "strip" steel — was being replaced in the market by sheet steel. Sharon thus fell from the tenth largest producer of steel in 1950 to sixteenth in 1968. The steps taken by management to correct the company's problems only worsened matters. In an attempt to replace the antiquated equipment, in 1962 Sharon installed two Kaldo oxygen furnaces, the first in the United States. Kaldo furnaces were supposedly more economical than conventional basic oxygen furnaces — and could also be fed with larger amounts of scrap. But the Kaldo steelmaking process reduced lining life of the furnace by 80 percent or more, greatly increasing production costs.

In the late 1960s steel firms generally faced high fixed assets with low market prices, making them attractive takeover targets. Sharon's precarious financial position made it more vulnerable than most. In 1968 Victor Posner, a corporate raider over the opposition of Sharon's management, acquired Sharon Steel for NVF, a producer of industrial plastics and one-seventh Sharon's size. Sharon Steel became the largest division of NVF.

The Sharon Steel division of NVF profited from the steel shortage of 1973–1974 and experienced some renewed growth. This resurgence, however, came not so much from increased production capacity or increased efficiency as from Posner's financial manipulations. In 1979 almost 40 percent of Sharon's total assets were minority investments in other companies. Sharon Steel had become almost as much an investment company as a steel company. Posner's use of borrowed money to finance these acquisitions created serious financial problems for Sharon Steel in the 1980s. In early 1985 Sharon failed to meet an interest payment, and in 1987 it filed for bankruptcy under Chapter 11. Posner was ousted. In 1991 Castle Harlan took over, selling off Sharon subsidiaries in 1992 and putting Sharon's future in doubt.

References:
Paul Blustein, "Victor Posner: Living on Borrowed Time," *Forbes,* 120 (September 1, 1977): 123–124;

William T. Hogan, S.J., *An Economic History of the Iron and Steel Industry in the United States,* 5 volumes (Lexington, Mass.: Lexington Books, 1971);

Hogan, *Steel in the United States: Restructuring to Compete* (Lexington, Mass.: Lexington Books, 1984);

Hogan, *The 1970s: Critical Years for Steel* (Lexington, Mass.: Lexington Books, 1972);

Gertrude G. Schroeder, *The Growth of Major Steel Companies, 1900–1950* (Baltimore: Johns Hopkins University Press, 1953);

"Too Little, Too Late?" *Forbes,* 87 (June 1, 1961): 31-32;

"The Woes of Sharon," *Institutional Investor,* 27 (April 1993): 58-59.

Shelby Steel Tube Company

by Thomas E. Leary

Industrial Research Associates

and

Elizabeth C. Sholes

Industrial Research Associates

The history of the Shelby Steel Tube Company is representative of the technological and economic forces that reshaped industry at the turn of the century. The company pioneered important innovations in the production of seamless tubing and was involved in the wave of mergers that culminated in the creation of the United States Steel Corporation.

In 1885 Europeans devised the Mannesmann machine, which allowed the efficient production of tubular steel without a welded seam. British capitalists using the Mannesmann process established a plant on American soil at Shelby, Ohio, in 1890. Founded as Lozier-Yost Seamless Tube Works, the firm subsequently changed its name to Shelby Tube Company and commenced production in 1891. The timing was perfect; the rapidly growing bicycle industry developed an insatiable appetite for seamless tubing used in frames of the popular recreational vehicle during the "Gay Nineties." A coalition of British investors and American bicycle manufacturers established control over seamless production by expanding Shelby and acquiring competitors' plants.

Despite the depression of 1893, annual capacity at Shelby more than tripled from 5.4 million feet in 1894 to 18 million feet by mid 1896.

In 1897 Shelby annexed five other makers of seamless tubing, one of which was the Ellwood Weldless Tube Company, where Ralph C. Stiefel had introduced an alternative to the Mannesmann method in 1894–1895. By 1899 additional consolidations brought all the major seamless tube plants in the United States under the management of Shelby's Cleveland headquarters. Aggregate capacity of Shelby-owned plants totaled 100 million feet.

Shelby became a New Jersey corporation in 1900. Despite its prominent position in the steel industry, Shelby felt threatened by National Tube's entry into the seamless market, handicapped by a lack of vertical integration, and afflicted with unspectacular profit margins. Shelby's directors sold out to U.S. Steel in 1901. Shelby, however, retained a measure of autonomy until 1917, when it was subsumed by National Tube. At that time, of Shelby's original empire only the Ellwood City works was still active as part of U.S. Steel.

References:
William T. Hogan, S.J., *An Economic History of the Iron and Steel Industry,* 5 volumes (Lexington, Mass.: Lexington Books, 1971);

"The Seamless Tube Consolidation," *Iron Age,* 64 (August 10, 1899): 28.

Sloss-Sheffield Steel and Iron Company
by Robert Casey
Henry Ford Museum & Greenfield Village

The Sloss-Sheffield Steel and Iron Company was one of the pioneer industries of Birmingham, Alabama, and was a major national producer of coal, coke, and merchant pig iron.

The Sloss Furnace Company was organized in 1881 by railroad promoter James Withers Sloss. Sloss had entered the iron industry in 1873, when he joined the effort to make coke iron at the Oxmoor Furnaces, southwest of Birmingham. Although the coke experiment succeeded in 1876, differences with the other investors lead Sloss to resign in 1881. Encouraged by Henry De-Bardeleben, who offered to supply him with coal at cost plus ten percent, Sloss formed the Sloss Furnace Company and erected two blast furnaces and 242 beehive coke ovens on the east side of Birmingham.

In 1886 Joseph Forney Johnston and John W. Johnston, backed by New York and Richmond investors, purchased Sloss Furnace. Reorganized in 1887 as the Sloss Iron and Steel Company, the firm bought the Coalburg Coal and Coke Company, built 300 new coke ovens, and added two additional blast furnaces in north Birmingham. The expansion, however, left the company strapped for cash, and by 1888 it was on the verge of bankruptcy. Johnston resigned, and Virginian Thomas Seddon assumed the presidency. By the time of Seddon's death in 1896, Sloss was on sound financial footing.

In 1899 Sloss expanded again, acquiring three furnaces in the Florence-Sheffield district, plus substantial coal and ore lands, and was reorganized as the Sloss-Sheffield Steel and Iron Company. At this time Sloss was the second-largest producer of pig iron in the district. This second expansion, however, produced a large debt. Virginia-born Wall Street financier John C. Maben became company president in 1902 and eliminated the debt by attracting eastern investors. Between 1902 and 1920 the Birmingham furnaces were substantially rebuilt, and 120 Semet-Solvay by-product coke ovens were installed in north Birmingham.

During 1923 and 1924 Sloss purchased the Sheffield Iron Corporation and the Alabama Corporation, acquiring furnaces in north Alabama and Gadsden as well as additional mineral lands. By 1927 all the outlying furnaces were shut down, and iron making was concentrated at the two Birmingham and two north Birmingham furnaces. All four had been thoroughly rebuilt by 1928. During this time, control of Sloss passed from the Virginia-based group that had guided it for decades to the Allied Chemical and Dye Corporation.

In 1942 United States Pipe and Foundry acquired majority control of Sloss, and in 1952 the two companies merged. Between 1952 and 1958 U.S. Pipe expanded the coke works and replaced the two old north Birmingham furnaces with a large modern one. Jim Walter Corporation, a major producer of building material and low cost homes, bought U.S Pipe in 1969.

In the ensuing years the former Sloss operations gradually fell victim to changing economic conditions. Low-grade Alabama ore could not compete with rich imported ores, and iron mining ceased in 1975. Imported pig iron and the increasing use of scrap iron by foundries gradually eroded the domestic pig-iron market. The two oldest Birmingham furnaces were shut down in 1970. The large north Birmingham furnace was shut down in 1980 and torn down in 1985. Only the coal mines and coke ovens remain in operation.

References:
Ethel Armes, *The Story of Coal and Iron in Alabama* (Birmingham, Alabama: Printed for the Chamber of Commerce, 1910);
W. David Lewis, "Sloss Furnaces and the Southern Foundry Trade: A Case Study of Industrialization in the New South," *Proceedings of the 150th Anniversary Symposium on Technology and Society* (Tuscaloosa: University of Alabama, 1988);
Marjorie L. White, *The Birmingham District* (Birmingham: Birmingham Historical Society, 1981).

William P. Snyder, Jr.

(August 8, 1888 – June 12, 1967)

by Geoffrey Tweedale
Manchester, England

CAREER: President, director (1918–1957), chairman, director, Shenango Furnace Company (1957); chairman and director, Shenango-Penn Mold Company (1927–1956); president and director, Snyder Mining Company (1930–); chairman, Crucible Steel Company of America (1944–1959).

William P. Snyder, Jr., iron and steel entrepreneur and chairman of the Crucible Steel Company of America, was born in Pittsburgh on August 8, 1888. His father, William P. Snyder, Sr., was one of the industry's major figures.

Snyder, Sr., had founded his own iron brokerage firm in Pittsburgh in 1888, and with close associate Henry W. Oliver he rapidly expanded it into mining, shipping and production. In 1906, with a capital of $5 million, Snyder incorporated the Shenango Furnace Company. With extensive blast furnace facilities, coal mines, coke ovens, Great Lakes ore properties, and shipping subsidiaries, the company was to emerge as one of the major integrated merchant iron producers in the Mahoning and Shenango Valley districts.

After graduating from Yale in 1910 with a degree in philosophy, Snyder, Jr., became a clerk in his father's business. He served a period of apprenticeship — which included working as his father's assistant — before becoming the vice-president of Shenango Furnace Company. Snyder, Sr., turned the duties of company president and director over to his son in May 1918 but stayed on as chairman of the board.

Snyder, Jr., continued to maintain the company's independence while expanding business. In 1927 the company purchased the Penn Mold and Manufacturing Company of Dover, Ohio, and incorporated it as the Shenango-Penn Mold Company. This firm also operated a brass works at Dover which produced large brass castings, paper mill rolls, and special alloys. In order to produce Lake Superior iron ore, Snyder, Jr., or-

William P. Snyder, Jr.

ganized the Snyder Mining Company of Pittsburgh and became its president.

Snyder's extensive interests led to his involvement with the specialty steel industry. In 1901 his father had begun a two-year term as president of the Clairton Steel Company, a wholly-owned subsidiary of the Crucible Steel Company of America, the country's leading specialty steel producer at that time. In 1930 Crucible acquired a 50 percent interest in the Snyder Mining Company to secure its ore supplies, and this brought Snyder, Jr., onto the Crucible board as a director.

In the 1930s Crucible was performing poorly. Although it had a diverse product line and many plants with high output, the company could claim only mediocre earnings. Top Crucible executives, such as Frederick B. Hufnagel, recognized that the company's problems were largely due to weaknesses in management. World War II, however, delayed reorganization. In 1944 Snyder became chairman of Crucible, and by the 1950s his demands for management reform had begun to effect change. In describing Crucible's record of improvement, *Fortune* in 1957 reported that Snyder, Jr., having recognized "that Crucible had become preoccupied with production at the expense of profits . . . roused other directors to the need for a change in management." Although Snyder was chief executive officer, he did not wish to be the sole decision maker; he chose William H. Colvin, Jr., to oversee company reforms. Calvin and Snyder created an entirely new executive group, cut the number of Crucible's plants from ten to six, and carried through a modernization program. In 1954 Colvin retired, handing over the presidency to Joel Hunter, who continued the reorganization under Snyder's chairmanship until 1959, when Snyder retired. By then Crucible's earnings had increased dramatically, and the company was once again paying dividends.

Snyder also served as a director of the Mellon National Bank & Trust Company, president and director of the Allegheny General Hospital, and trustee of the University of Pittsburgh. He was a member of the American Institute of Mining, Metallurgical and Petroleum Engineers, and the American Iron and Steel Institute. He married Marie Elsie Whitney on February 7, 1917, and they had two sons: William Penn III and George Whitney. Snyder died on June 12, 1967.

Publications:

W. P. Snyder's Shenango Companies: 75 Years in Iron Mining and Manufacturing (New York: Newcomen Society, 1956).

References:

William H. Colvin, *Crucible Steel of America: 50 Years of Specialty Steelmaking in U.S.A.* (New York: Newcomen Society, 1950);

Crucible Steel Company of America, *50 Years of Fine Steelmaking* (N.p., 1951);

Herrymon Maurer, "Crucible: Steel for Frontiers," *Fortune,* 55 (1957): 84.

Edgar B. Speer

(July 28, 1916 – October 13, 1979)

by Dean Herrin

National Park Service

CAREER: Metallurgical observer, superintendent of Bessemer department, superintendent of open-hearth department, Youngstown (1938–1951), assistant division superintendent of steel production, division superintendent of steel production, Gary (1951–1953), assistant general superintendent, Duquesne (1953–1955), assistant general superintendent, general superintendent, Fairless Works (1955–1958), general manager of operations for steel (1959), administrative vice-president of central operations (1959–1964), administrative vice-president of steel operations (1964–1967), executive vice-president of production (1967–1969), president (1969–1973), chairman of the board and chief executive officer (1973–1979), director, United States Steel Corporation (1968–1979).

As chairman and chief executive officer of the United States Steel Corporation from 1973 to 1979, Edgar Speer guided the giant steel company through a period in which its involvement in steel production declined. Under Speer's direction the groundwork was laid in the 1970s for the transformation of the U.S. Steel into USX, the diversified "energy" company of the 1980s.

Speer, like several of U.S. Steel's top executives, was born in Pittsburgh, spent his entire career with "The Corporation," and was groomed in the production side of the business. Born on July 28, 1916, he attended Pennsylvania Military College and the University of Pennsylvania before joining U.S. Steel in 1938. Speer spent the first 13 years of his career in the company's Ohio Works near Youngstown, first as a metallurgical ob-

Edgar B. Speer (Photograph by Ed Eckstein)

server, then as superintendent of the Bessemer department, and finally as superintendent of the open-hearth department. Promoted to successively higher management positions, he was made general superintendent of the company's Fairless plant in 1956. During the management reshuffling following U.S. Steel president Walter Mumford's sudden death in 1958, Speer was brought back to Pittsburgh as general manager of operations for steel. After serving in several vice-presidential posts for steel and central operations, he became the executive vice-president for production in 1967. With Edwin Gott's elevation to chairman and chief executive officer, Speer was selected as president of the corporation in 1969. Speer became chairman and chief executive officer in 1973 when Gott retired.

In 1972 the *New York Times* described him as a "steel man's steel man." Gott liked to say that no one in the world knew more about steel production than Edgar Speer. Some industry analysts speculated that one reason Speer was selected to fill the top position was that the organization, after several years of declining profits, expected more operation problems due to recently imposed pollution-control measures. Speer did make contributions in the area of steel production, espe-

cially in designs for new basic oxygen furnaces, but one of his more notable innovations at U.S. Steel came with the restructuring of the management hierarchy. Reversing a decades-old tradition of consolidating power at the top of U.S. Steel's chain of command, Speer decentralized the decision-making process in 1974 by creating four geographical divisions and a fifth division for raw materials. The organizational structure of U.S. Steel's international division and some of the corporation's nonsteel businesses served as models for Speer's changes, which he hoped would help cut through red tape. Unlike previous divisions in the U.S. Steel network — in which a division was assigned to oversee a single activity — the new divisions were responsible for all activity in their area: production, sales, marketing, and shipping. The division heads were thus given more responsibility for decision making and, ultimately, for controlling the profit margins. *Business Week* praised the reorganization in 1974 for "breathing new life into an enterprise that for years seemed on the verge of atrophy."

Never one to mince words, Speer was often very critical of the government regulatory policies of the 1970s, particularly pollution and price controls, and trade policies. He continued U.S. Steel's program of equipping plants with government-mandated pollution control devices but complained that "few regulatory agencies have been as arrogant in the use and abuse of power as the EPA."

Speer and other officials in the corporation also criticized government trade policies for allowing below-cost foreign steel to flood the domestic market. Critics of the company countered that U.S. Steel had been slow to modernize its plants and could never be competitive with the newer and more modern steel mill plants in Japan and Europe. Speer, however, continued to press for better enforcement of antidumping trade policies that would equalize the cost differential between domestic and imported steel. He derided the notion of quotas — which were practiced in the textile industry at the time — claiming that foreign steel producers had not earned the right to be given a guaranteed share of the American market automatically. He even threatened in 1978 to build plants in other countries and sell back into the U.S. markets if trade policies did not improve.

Speer helped to develop a "steel caucus" in Congress and tried to build public support for the domestic steel industry. Plant closings elicited a strong negative reaction from the public, and Speer and his colleagues were successful in portraying their companies as victims of foreign dumping. He reminded people that when a steelworker's job is eliminated, "the butcher, the banker, the baker, and the clergyman suffer, too."

Prompted by problems in the domestic steel industry and by gloomy forecasts for steel's future, U.S. Steel increasingly turned toward diversification in the 1970s. *Fortune* quoted Speer in 1973 as saying, "We could conceivably get to the point where steel would be a minor instead of a major segment of our business." U.S. Steel spent 20 percent of its capital expenditures in that year on nonsteel enterprises. Despite relatively prosperous years and record production levels in 1973 and 1974, profits from steel declined. U.S. Steel reported earnings in 1977 of $138 million, down 78 percent from profits earned in 1974. More significant, however, was that even though U.S. Steel still retained almost a quarter of the domestic steel market, none of its 1977 profit came from steel. The corporation relied on the profits generated by chemical and mineral production and other nonsteel businesses to meet its financial burdens. This situation was caused in part by decisions made earlier in the decade to stop spending money on plant modernization, eventually leading to the closing of the Ohio Works near Youngstown in November 1979 and the subsequent plant closings in Pittsburgh and elsewhere in the 1980s.

Even so, before he died in 1979, Speer did not think the domestic steel industry was completely dead, and at a time when most other steel executives were reluctant to expand, he urged construction of modern and efficient mills. He hoped to build such a mill at a site long owned by U.S. Steel in Conneaut, Ohio, that would be able to compete with foreign steel producers. Speer, a man who had spent a majority of his career solely concerned with the production of steel, and who had started his career in Youngstown, did not live to witness the shutdown of that plant. Treatment for cancer forced him to step down as chairman in April 1979, and he died that October.

Selected Publications:
"Responsiveness, Stress on Research Buoy Steel's Future," *Industry Week*, 171 (December 13, 1971): 14;
"New Profit Picture Brings out Steel Bulls," *Iron Age*, 214 (August 12, 1974): 31–32;
"After the Lean Years, The Good Years?," *Forbes*, 115 (January 15, 1975): 25–27;
A Very Simple Message (Pittsburgh: U.S. Steel Corporation, 1975);
Are the Regulatory Agencies Changing Our Government? (Pittsburgh: U.S. Steel Corporation, 1976);
"Hell With Expanding; We Won't Even Have the Money To Maintain," *Forbes*, 120 (November 15, 1977): 34–36.

References:
"Edgar Speer, Former Chairman of United States Steel, Dies at 63,"*New York Times*, November 30, 1972, p. 63; October 14, 1979, p. 44;
John Hoerr, *And the Wolf Finally Came: The Decline of the American Steel Industry* (Pittsburgh: University of Pittsburgh Press, 1988);
William T. Hogan, S.J., *An Economic History of the Iron and Steel Industry in the United States*, 5 volumes (Lexington, Mass: Lexington Books, 1971);
"An Old Hand Recasts Big Steel," *Fortune*, 87 (January 1973): 27;
"A Steelman Steps Up the Pace at U.S. Steel," *Business Week* (March 9, 1974): 154–164;
"A Steelman Strikes Back," *Fortune*, 97 (May 8, 1978): 47–48.

The Stanley Committee

by Paul Tiffany

Sonoma State University

In the 1910 congressional elections, the Democratic party captured a majority in the House of Representatives for the first time since 1892. Much of the momentum that brought about this change in politics was the growing influence of the Progressive movement and an emerging sense that the recent spate of large mergers and the growth of big business ran counter to American economic and political traditions. As a consequence, muckrakers and other reform-minded journalists increasingly attacked the trusts as being inherently un-American.

The House obliged the critics by passing a resolution calling for an inquiry into the "steel trust," the United States Steel Corporation, formed in April 1901. At its creation the firm became the world's first billion-dollar organization — capitalized at $1.4 billion — and held nearly two-thirds of the national crude steelmaking capacity. President William H. Taft responded to the House resolution by authorizing the establishment of a special Committee on Investigation, to be chaired by Representative Augustus Stanley of Kentucky. The Stanley committee began hearings in May 1911, and by the time they were completed in 1912, eight lengthy volumes of detailed testimony had been compiled.

The hearings produced an immense amount of data about the formation of U.S. Steel, as well as analyses of its financial and operating performance in the decade following its formation. Of particular interest was testimony concerning U.S. Steel's acquisition of the Tennessee Coal, Iron & Railroad Company in 1907. The testimony also produced some odd moments, such as U.S. Steel chairman Judge Elbert H. Gary's comment that he would welcome the U.S. Government as the official price-setter for the industry.

In the end, the Stanley committee's hearings helped create the atmosphere that led the Department of Justice to file suit in 1911 to break up U.S. Steel on antitrust grounds. The hearings thus marked one of the earliest confrontations between the steel industry and the government in what would prove to be an acrimonious relationship.

Reference:

U.S. House of Representatives, Hearings before the Committee on Investigation of the United States Steel Corporation, 62nd Congress, 1st Session, *United States Steel Corporation*, 8 volumes (Washington, D.C.: Government Printing Office, 1911–1912).

Steel Metallurgy in the Twentieth Century
by Thomas J. Misa
Illinois Institute of Technology

Metallurgy concerns the properties of metals related to structure. Steel metallurgy in the twentieth century can be seen as a product of three related developments: changes in the steel industry's production technology, the advent of new instruments and experimental techniques, and the sustained importance of metallurgy as a profession.

Changes in the steel industry's production technology profoundly shaped steel metallurgy. The introduction of Bessemer steelmaking in the late nineteenth century created a chemically oriented steel metallurgy and allowed for production control through analysis of raw materials and finished products. The early twentieth century once again saw a technological revolution in the industry with the introduction of the Siemens-Martin process — an open-hearth method of steelmaking that created a physically oriented metallurgy, or metallography. By 1910 open-hearth production surpassed Bessemer production. Steel production, however, continued to be plagued by problems: the hardening of foot-thick armor plates and failures in heavy railroad rails resisted chemical explanation, as did the behavior of Frederick Taylor's high-speed tool steels, which remained hard at elevated temperatures. In attempting to solve these problems, advocates of metallography — including the Americans Albert Sauveur (1863–1939) and Henry Howe (1848–1922) — offered the insight that a metal's properties depended not only upon its chemical composition but also upon its temperature history and heat treatment. Open-hearth steelmaking combined with metallography permitted steel firms to establish a diversified finished steel market.

That the industry was slow to adopt metallography is underscored by the virtual neglect, in America and elsewhere, of Henry C. Sorby (1826–1908), the British amateur petrographer credited with having first investigated and photographed the microstructure of polished iron samples. From 1863 until 1887, as Sorby remarked of his own work, "no one seemed to think it of any practical value." His studies were finally validated in 1887, when the British Iron and Steel Institute gave him recognition. French and German metallurgists adopted metallography in the 1890s; English and American metallurgists, in the 1900s.

Metallography permitted metallurgists to predict successful heat treatments by correlating a metal's temperature history, microstructure, and physical properties. Metallographers heated a sample to the temperature being studied, waited until thermal equilibrium was established, quenched the sample by cooling it rapidly, then examined a polished section of the quenched sample with an optical microscope. Quenching was believed to "freeze" the microstructure existing at high temperature. By 1901 the Englishman William C. Roberts-Austen (1843–1902) had drawn graphic curves that plotted the several microconstituents that resulted from varying a metal's quenching temperature (ordinate) and carbon percent (abscissa). These thermal-equilibrium diagrams were elaborated on in the next two decades, allowing metallographers to predict a metal's principal microconstituents, and hence its macrostructural properties. Metallographers were therefore able to suggest heating the metals at temperatures that were determined by scientific findings. This was a novel occurrence given that temperature settings had relied on a steelmaker's intuition. While "hot" rolling occurred at 1000°C or higher, Harvard's Sauveur showed that final rolling temperatures *below* 700°C yielded superior heavy railroad rails.

Steel metallurgists did not agree, however, on a theory of hardening. The "allotropic" theory, forwarded by Floris Osmond (1849–1912) in 1887, held that an element can assume different forms with changes in temperature and/or carbon content — much as carbon itself can exist as graphite or diamond. The advocates of this theory maintained that allotrope beta-iron determined steel's hardenability. The rival "carbon" theory of hardening held that the solidification of iron carbide during cooling determined steel's hardenability. This dispute was in part settled in the 1920s with the application of X-ray diffraction in the 1920s, allowing for a finer understanding of metallic structures and discrediting the beta-iron theory. It was found that iron could assume different elemental forms: *ferrite,* soft pure iron; *cementite,* a very hard and brittle iron carbide; *pearlite,* composed of strong layers of ferrite and cementite; *troostite,* an aggregate of ferrite and cementite, named for L. J. Troost (d. 1911); *austenite,* a nonmagnetic iron normally existing only above 720°C, named for Roberts-Austen; and *martensite,* a hard decomposition product of austenite, named for Adolf Martens (1850–1914).

Having had great successes using vanadium, tungsten, chromium, nickel, and molybdenum as alloying elements for automobile, airplane, and stainless steels, metallurgists after 1920 were challenged to explain the new steels and to exploit their remarkable properties. In 1928 the United States Steel Corporation founded its central research laboratory and placed Edgar Bain (1891–1971) in charge of physical metallurgy. Bain suggested the application of practical heat treatment processes and began to investigate hardenability quantitatively by studying the isothermal decomposition of austenite. His work, combined with similar studies in Germany, established the "S-curve," which correlated time, temperature, and transformation during the hardening process. Bain's colleague E. S. Davenport extended this analysis to alloy steels. The rise of electric steel furnaces, employed to make alloy steels after World War I, sharpened the need for a theory of hardening that would be applicable to alloy steels; by 1950 electric steel production had surpassed declining Bessemer production.

The advent of new instruments and experimental techniques furthered structure-oriented physical metallurgy. X-rays, discovered by W. C. Röntgen in 1895, could probe castings for macroscopic flaws. Physical metallurgy has more greatly benefited from X-ray diffraction, which permits analysts to infer a crystal's structure from its characteristic scattering pattern. As early as 1913 the technique's pioneers, W. L. Bragg (1890–1970) and W. H. Bragg (1862–1942), determined the lattice structure of metal crystals. In the early 1920s, metallurgists used X-ray diffraction to probe the structure of the allotropic forms of iron. Generally, the technique opened up new theoretical problems, including grain boundary properties, the effect of impurities, solid-state diffusion, and the effects of mechanical deformation on crystal structure.

The electron microscope has allowed metallurgists to examine many metallic structures and microstructural processes. Electron microscopy exploits the short wavelengths of electrons, either by transmission or reflection, to determine a structure's shape. In the early 1930s Max Knoll and Ernst Ruska, working at the Technische Hochschule Berlin, were able to magnify an electron image of an object. In 1933 Ruska demonstrated the first practical electron microscope, which exceeded the ability of an optical microscope in determining images; by the 1950s studies employing electron microscopy had matured.

Microscope technology complemented other investigative techniques, such as age hardening. First observed around 1906, certain alloys of aluminum, as well as of steel, became twice as strong and up to four times as hard simply by aging — or heating to 400° to 800°C for an hour or so. Metallographic studies revealed a solid precipitate phase toward the end of the age-hardening process. This finding suggested that the phenomenon relied on changes in the structure's solubility. X-ray diffraction studies detected the presence of submicroscopic aggregates in age-hardened materials, and in the 1950s electron microscopy revealed these aggregates. Subsequently, metallurgists have exploited the age-hardening properties of many ferrous and nonferrous alloys.

As a discipline, metallurgy has gained increasing acceptance as part of the professional scientific community. As such, metallurgy has benefited from the scientific investigations in other disciplines. In the 1910s, for instance, combined investigative efforts in physical metallurgy and quantum physics led to the electron theory of metals, which described the metallic state as a gas of electrons moving within a regular pattern created by positively charged atoms; electrical and thermal conductivity resulted from the drift of free electrons under an electric field from hot to cold regions. With the development of quantum mechanics came further metallurgical understandings of ways in which different alloys are formed. Metallurgy's influence has clearly extended beyond steelmaking and has contributed to science in general.

References:
Leslie Aitchison, *A History of Metals* (New York: Interscience Publishers, 1960);

Robert F. Mehl, *A Brief History of the Science of Metals* (New York: American Institute of Mining and Metallurgical Engineers, 1948);

Cyril S. Smith, ed., *The Sorby Centennial Symposium on the History of Metallurgy* (London: Gordon & Breach, 1965);

R. F. Tylecote, *A History of Metallurgy* (London: The Metals Society, 1976).

Steel Price Rollback Confrontation, 1962

by Amy E. Davis
University of California at Los Angeles

On April 10, 1962, President John F. Kennedy began a successful campaign to force the steel industry to roll back a price increase it had announced that day. Kennedy believed that the $6-per-ton hike — an increase of about 3.5 percent — threatened the domestic economy, and he wanted to prevent an inflationary wage-price spiral that would hurt U.S. exports and exacerbate the nation's balance-of-payments deficit. He also perceived steel's actions as an attack on his policies and the presidency and as an affront to him personally.

To help forestall inflation, the Kennedy administration had instituted a nonbinding wage-price policy. The administration hoped to avoid antagonizing both business and labor by providing direction without invoking direct controls. By setting "guideposts" enforceable only through persuasion, it wanted to keep the fluctuations of both prices and wages within bounds determined by the general trend of national productivity. The steel industry, however, had absorbed a series of pay increases scheduled by its 1959 labor contract, and industry officials suggested that a price hike might be needed to pay for the final wage increase, due to go into effect in October 1961. The Kennedy administration warned the producers not to raise prices, and they did not.

Soon after, the administration involved itself directly in the labor-management contract negotiations that began in February 1962. Despite Secretary of Labor Arthur Goldberg's union background, which included service as general counsel to the United Steelworkers of America, he emphasized the industry's need to maintain profit margins in order to modernize and urged labor to exercise restraint. When the talks collapsed in March, Goldberg pressed the union to ease its de-

Roger M. Blough (right), board chairman of U.S. Steel, speaking to reporters after meeting with President John F. Kennedy to discuss steel prices, 1962

mands, and on April 6 the two sides agreed to a 2.5 percent increase in wages and fringe benefits — an amount well within the wage-price guideposts and, in the administration's view, not enough to necessitate a price hike.

Just four days later, however, on April 10, United States Steel Corporation president Roger Blough stunned President Kennedy by handing him an announcement, already released to the press, that the firm would raise prices the next day. Kennedy was furious. He saw the hike as an abrogation of a tacit agreement and as a challenge to him, his office, and his entire economic agenda. He feared the hike stripped him of the ability to ask unions for restraint, and he worried about the effects of inflation on both the balance of payments and the newly recovered domestic economy. On the following day, after other major producers posted identical price hikes, Kennedy lambasted industry officials for price-fixing and anticompetitive behavior.

The White House mobilized members of Congress and Democratic party officials in a strategy combining persuasion and pressure. Even as

key figures privately requested that the industry respect the importance of price stability, the administration threatened steelmakers with investigations, punitive legislation, legal sanctions, and changes in the government's steel-purchasing policies. At Kennedy's urging, Tennessee senator Estes Kefauver pledged an inquiry along the lines of his 1957 hearings on administered steel pricing. Others, including Attorney General Robert Kennedy, promised separate investigations, and the Defense Department ordered its contractors to buy steel only from companies that had not raised prices.

In its attempts to divide and conquer, the administration tried to persuade the five companies that had not yet raised prices to hold the line and so force the others to remain competitive by rescinding their increases. It especially wanted to convince Inland Steel, the nation's eighth-largest steel manufacturer. Inland had closer links to the administration than the other producers and a history of public service. Most important, however, was that many economists believed Inland to be efficient enough to sell at a price lower than even some of the bigger firms.

The White House also leaned on Bethlehem Steel, which had already raised prices. Because Edmund F. Martin, a Bethlehem vice-president, had said on April 9 that there should be no increase, Bethlehem's decision to raise prices, the administration warned, suggested collusion. When Martin claimed he had been misquoted, the attorney general sent FBI agents out in the middle of the night to question reporters who had heard the original statement. In addition to making this implicit threat that Bethlehem executives might face antitrust charges if they did not rescind the hike, the Justice Department subpoenaed documents from several companies and announced a grand jury investigation. Since Bethlehem was a large shipbuilder, it also felt the loss of government contracts.

The battle ended abruptly. On Friday morning, April 13, Inland Steel and then others announced that they would not raise prices. That afternoon Bethlehem Steel backed down, and that evening U.S. Steel did too. The remaining firms soon followed.

Kennedy immediately tried to make peace with the steel producers, but many believed he had shown his true colors as an enemy of big business. An idle remark later printed in the *New York Times* seemed to cement the case, for Kennedy was overheard to have remembered that "My father always told me that all businessmen were sons-of-bitches." In late May business's anger with Kennedy soared when the stock market crashed. Critics argued that the plunge reflected perceptions of administration antagonism toward business and investors' fears that the White House might pursue policies to limit profits.

The steel crisis, however, strengthened Kennedy politically. He had shown that he was neither indecisive nor the servant of big business, and both domestic and foreign leaders reacted favorably to Kennedy's handling of the episode. Kennedy thought, however, that the greatest benefit came from what the nation had been spared. He told of union negotiators who had been ready to accept a moderate wage increase from the steel fabricators until they learned of Blough's announcement of a price hike. No longer "willing to settle for a dime," they confronted a mediator who had lectured them about inflation and said, "Don't give us that responsibility horseshit. We're going for a quarter, now." Had he not forced the rollback, Kennedy concluded, "we'd have the damnedest inflation ever, right now."

References:
William J. Barber, "The Kennedy Years: Purposeful Pedagogy," in *Exhortation and Controls: The Search for a Wage-Price Policy, 1945–1971*, edited by Craufurd D. Goodwin (Washington, D.C.: The Brookings Institution, 1975);
Wallace Carroll, "Steel: A 72-Hour Drama with an All-Star Cast," *New York Times*, April 23, 1962, p. 1;
Amy Elisabeth Davis, "Politics of Prosperity: The Kennedy Presidency and Economic Policy," Ph.D. dissertation, Columbia University, 1988;
Roy Hoopes, *The Steel Crisis* (New York: Day, 1963);
Grant McConnell, *Steel and the Presidency, 1962* (New York: Norton, 1963).

Steel Seizure Case of 1952

by Paul Tiffany

Sonoma State University

With the steel companies and the United Steel Workers of America (USWA) unable to renegotiate a labor contract, President Harry S Truman on April 8, 1952, seized the industry under the inherent powers doctrine of the Constitution. He had hoped that the "steel seizure" would circumvent the threat of a work stoppage and, in so doing, guarantee an uninterrupted supply of steel to the Korean War effort. Truman's act of executive intervention, however, resulted in controversy, creating one of the most serious tests of the power of the presidency in American constitutional history.

Relations between steelmakers and the USWA had been strained throughout the post–World War II years. In both 1946 and 1949, national strikes had been called by the union in order to improve the contract offered by producers. Moreover, most industry leaders perceived

Philip Murray, president of the United Steelworkers (left), and Benjamin Fairless, chairman of the board and chief executive officer of U.S. Steel (right) with President Harry S Truman (center) after the failure of Truman's attempt to nationalize the steel industry

the business and labor policies of the Truman administration as biased toward the demands of organized labor — and hostile to big business. Indeed, President Truman, in his 1949 State of the Union address, threatened to create a nationalized steel industry if the companies did not behave according to administration wishes. With the labor contract nearing its expiration date of December 31, 1951, tensions between the steelmakers and the USWA were running high. The USWA set a strike deadline of January 1, 1952.

By the winter of 1950 it had become apparent to many Americans that U.S. involvement in Korea would be protracted. With panic buying and the rationing of basic commodities for defense purposes having a potentially inflationary impact on the economy, President Truman had reorganized the federal stabilization agencies — which had been created during World War II — to control wages and prices. In late 1951, with the steel strike deadline fast approaching, Truman got both the USWA and the steelmakers to allow the Wage Stabilization Board and the Office of Price Stabilization to intervene in the bargaining. The USWA further agreed to extend its strike date. Steelmakers had countered the USWA call for higher wages

by arguing that more pay would mean a necessary hike in steel prices. Truman not only wanted to avoid an interruption in steel production but also wanted to head off a steel-labor contract that raised both wages and producer prices and, in so doing, contributed to already-mounting inflation.

The union immediately accepted the recommendations of the wage board made on March 20, 1952, which called for a 15-cents-per-hour wage increase with a corresponding $2- to $3-per-ton steel price hike. Yet, industry negotiators demurred, stating that they would have to have a price increase at least three times as large as the recommended figures in order to offset the higher wages. Negotiations were further hampered on March 28, when Charles E. Wilson, a Republican and president of the General Electric Corporation, resigned as head of Truman's stabilization programs. His resignation had been the result of poor communications between him and the White House regarding steel prices. Sensing that Wilson had been ready to grant them higher prices, the steel men saw Wilson's resignation as a victory for the union. Many perceived it as one more example of ineptness and partisanship in Truman's management of the economy.

With the new strike deadline of April 8, 1952, only days away, Truman and his advisers considered their options. The two parties were intransigent in their demands for higher wages and prices; imposition of the existing Taft-Hartley strike-injunction law was dismissed because of its perception as a Republican law geared to hurt organized labor, a major constituent of Truman's political coalition. Since the president believed that acquiescing to the companies' demands for higher prices would undermine his entire inflation-control program, he chose instead to invoke the constitutional right of the president to use the inherent powers of the office in a national emergency and seize the industry.

Truman's action immediately provoked a national debate. The steelmakers and the union each engaged in a high-stakes public relations campaign to convince the American public of the correctness of their respective positions. On June 2, 1952, the Supreme Court, quickly ruling on a suit filed by the industry to overturn the president's move, ruled the seizure to have been unconstitutional. The union immediately went on strike and stayed out for 52 days until an agreement — giving the industry the price hike it wanted — was reached.

The end result was a humiliating public defeat for Truman. Moreover, it also set back his wage-price control program. The court decision ultimately favored neither business nor government in their ongoing struggle to shape American economic policy. The decision, however, did provoke an outpouring of legal commentary, resulting in a more precise interpretation of the inherent powers doctrine of the Constitution.

References:
Maeva Marcus, *Truman and the Steel Seizure Case* (New York: Columbia University Press, 1977);
Grant McConnell, *The Steel Seizure of 1952* (Indianapolis: Bobbs-Merrill, 1960);
Paul Tiffany, *The Decline of American Steel* (New York: Oxford University Press, 1988).

Steel Strike of 1909

by Kevin M. Dwyer

George Washington University

Although the historical significance of the steel strike of 1909 is often overshadowed by the "great" strikes of 1892 and 1919, it nonetheless marks a pivotal time for unionism in the American steel industry. The primary cause of the strike was the mass production revolution; at the turn of the century, new technology was introduced in the manufacture of sheet steel and tinplate, decreasing the need for skilled labor. The Amalgamated Association of Iron, Steel, and Tin Workers of America, a collection of skilled laborers organized by craft, recognized its diminishing leverage — the workers it represented having become less important to steelmaking — and attempted to purchase its security with concessions to management. Bitterly antiunion since the bloody Homestead strike of 1892, however, steel management saw an opportunity to deal a crushing blow to Amalgamated. Sheet and tin workers responded with an unexpected show of determination that protracted the conflict over a period of 14 months.

On June 1, 1909, workers at 12 plants of the American Sheet and Tin Plate Company, a United States Steel Corporation subsidiary, found notices announcing that "all its plants after June 20, 1909, will be operated as 'open' plants." The notice also announced a decreased wage scale, with cuts ranging from 2 to 8 percent, a measure the Youngstown Sheet and Tube Company and several other steel firms were quick to adopt. As David Brody asserts in his classic study, *Steelworkers in America: The Nonunion Era* (1960), the large steel firms had adopted a strategy of directing work to nonunion mills to break Amalgamated. The declaration of open-shop policy was part of a wider antiunion campaign that hinged on company rhetoric proclaiming a worker's right to personal liberty: the shops would be open to union and nonunion men, alike. To undercut a union's collective bargaining power would be to destroy any influence a union might wield.

In response to the company announcement, the craftsmen immediately rallied around their union banner. Amalgamated, however, was in a weak position to sustain a strike; lax collection of dues combined with diminishing membership had left the union virtually without a strike fund. By the fall, however, the American Federation of Labor (AFL) and other sympathetic parties lent their moral and financial support to the strikers, especially in the pivotal Wheeling, West Virginia, and New Castle, Pennsylvania, districts. Moreover, the AFL public-relations apparatus helped paint a public image of Big Steel as an oppressor of its labor force, an image that took hold and thus became an important legacy of the strike. Amalgamated called off the strike in August 1910 after New Castle workers had returned to their jobs. The strike failed to alter company labor policy, and in capitulating, Amalgamated had proved an ineffective union. The company challenge to Amalgamated, however, convinced union members of the need to organize the entire corporation, opening up the union to all steel laborers. Yet, despite steps taken at its 1910 convention to

include unskilled workers among its membership, Amalgamated was unable to mobilize an effective organizing campaign. The AFL emerged as the primary organizer of steelworkers. Protests by iron workers at McKees Rocks and an unorganized strike at Bethlehem Steel's South Bethlehem mills in 1910 also presented the AFL with the opportunity to step up its activities in steel. The consequent demise of the craft union initiated the rise of the industrial union. By 1919, all steelworkers, immigrant and native, skilled and unskilled, would be targeted for organization under the direction of the AFL.

References:

David Brody, *Steelworkers in America: The Nonunion Era* (Cambridge, Mass.: Harvard University Press, 1960);

Gerald G. Eggert, *Steelmasters and Labor Reform, 1886–1923* (Philadelphia: University of Pennsylvania Press, 1981);

John A. Garrity, "The United States Steel Corporation vs. Labor: The Early Years," *Labor History,* 1 (Winter 1960): 3–38;

Charles A. Gulick, *Labor Policy in the United States Steel Corporation* (New York: Columbia University Press, 1924).

Steel Strike of 1919

by James R. Barrett

University of Illinois at Urbana-Champaign

On September 22, 1919, approximately 250,000 men, nearly half of the steelworkers in the United States, walked away from their jobs in the country's first nationwide steel strike. One of the largest labor conflicts in the nation's history, the strike signified a broad effort by the labor movement to organize not only in the steel industry, but in the newer mass-production industries as well. The large number of strikers surprised many steelmakers, who had convinced themselves that their workers were generally satisfied with company-sponsored labor-representation programs. Management felt that the strike thus not only threatened productivity and profits, but an entire labor-management relations philosophy — the open-shop doctrine.

Since the defeat of the Amalgamated Association of Iron, Steel and Tin Workers of America in the 1892 strike at Homestead, most steel companies had proclaimed their shops "open" to union and nonunion members alike. In so doing, management was sending a clear message that workers could not claim a union as a collective bargaining agent. Steel masters, like United States Steel Corporation head Judge Elbert H. Gary, hailed the open shop as necessary to protect a laborer's personal liberty. A laborer should have the right to bargain with his boss — so the management party line went — without having an outside agent presume to speak for him or force him into its ranks. In competing with the labor movement for worker loyalty, steel management had developed corporate welfare plans and company unions. By 1919,

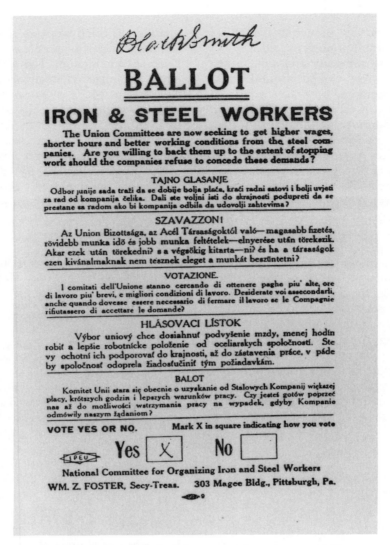

Strike ballot distributed nationwide in August 1919

however, many steelworkers had become increasingly dissatisfied with the 12-hour day and the seven-day week and felt that the steel companies had not let them share in the great profits gained during World War I.

Leading the unionization drive was the National Committee for Organizing Iron and Steel Workers, an umbrella organization representing the various unions with jurisdictional rights in the industry and sponsored by the American Federation of Labor (AFL). Launching its campaign in the summer of 1918, the committee made good headway, and by the spring of the following year it had organized almost 100,000 steelworkers. Ironically, its organizing success presented the committee with a serious strategic problem. Rank-and-file members demanded immediate action to improve working conditions in the industry. The

committee's leadership balked, fearing that the organization might not be ready for a confrontation with the "steel trust," and membership support began to slip away. A national ballot tabulated on August 20, 1919, showed that 98 percent of the voting members favored a strike should the companies not grant higher wages, shorter work hours, and improved working conditions.

The National Committee had adopted a series of twelve demands, including increased wages, the eight-hour day, a seniority system, one day off in seven, and the abolition of company unions. At the top of the list of demands, however, was the right to bargain collectively with the steel companies, removing any doubt that the critical issue in the confrontation was the industry's recognition of the unions. Gary, the key figure in the industry, steadfastly refused to meet with the labor leaders. He clearly felt that a majority of his employees would ignore a strike call and that in any case the principle of the open shop was worth the threat of a serious conflict. On September 17, National Committee leaders rejected President Woodrow Wilson's last-minute plea to postpone the strike — set for September 22. The committee had sought alternatives to the strike, including asking President Wilson to intervene. Gary, however, remained unmoved, and the committee feared that a strike postponement would break the momentum of the organizing drive. The conflict began as scheduled.

Workers' response to the strike was uneven — strong in the Gary–South Chicago region, weaker in and around Pittsburgh, negligible in the South. Still, the achievement was considerable; for the number of strikers was unprecedented in the American steel industry, and many steel mills stood silent.

William Z. Foster, a veteran AFL organizer, was clearly the mastermind of the operation. Working out of a small office in Pittsburgh, Foster maintained close contact with locals through a network of roving organizers. He worked closely with immigrant leaders and institutions to ensure the support of the various ethnic communities in the mill towns and issued strike bulletins in several languages. The bulletins and frequent mass meetings kept the strikers informed and helped to maintain morale and solidarity. Foster raised funds from other unions, voluntary organizations, and individuals; he also established relief stations to keep the strikers and their families fed. There is no doubt, however, that Foster's radical past, probed at length in the press and before a Senate investigating committee, damaged the cause of the strike. It was an era in which not only radical politics and strikes, but labor activity more generally, were seen to be a serious threat to the civil order. The steel companies played off of these public fears, suggesting that the strike was the fault of immigrants and Bolshevik agitators. The steel firms' repressive tactics in the mill towns soon turned the tide of the strikes. In Allegheny County, Pennsylvania, for example, the sheriff deputized some 5,000 men hired and armed by the steel companies. They were backed up by local policemen, mounted state troopers, and vigilante groups. In Clairton, Pennsylvania, a large outdoor meeting on the eve of the strike was attacked by a corps of mounted state policemen; men, women, and children were beaten. Organizers were routinely imprisoned, and strikers sometimes faced the choice of returning to work or going to jail. In some towns public meetings were simply outlawed.

During the fall of 1919 the companies recruited more than 30,000 black strikebreakers and brought them into the mills under armed escort. As winter set in and production levels rose, the ranks began to crack. By the end of the year it was clear that the strike had been defeated in all of the major districts. On January 8, 1920, the National Committee called an end to the struggle.

The strike did prompt investigations into industry conditions, causing public opinion to support a change to an eight-hour day. Under government pressure, the industry introduced the shorter day in 1923. From the perspective of the labor movement, however, the strike was a disaster that ended any hope of organization in the industry for almost two decades. Furthermore, as David Brody has argued, the strike experience set back the development of many of the more progressive labor-management concepts, and initiatives emerged only in the course of the twenties.

References:
David Brody, *Labor in Crisis: The Steel Strike of 1919* (Philadelphia: Lippincott, 1965);
Brody, *Steelworkers in America: The Nonunion Era* (Cambridge, Mass.: Harvard University Press, 1960);
William Z. Foster, *The Great Steel Strike and Its Lessons* (New York: B. W. Huebsch, 1920);
Interchurch World Movement, Commission of Inquiry, *Report on the Steel Strike of 1919* (New York: Harcourt, Brace & Howe, 1920).

Steel Strike of 1949

by Richard W. Kalwa

University of Wisconsin — Parkside

In May of 1949 the United Steelworkers of America informed the basic steel companies of its desire to bargain over wages, "social insurance" — life, accident, health, medical, and hospital insurance — and pensions. Although the agreement of the previous year had provided for a reopening of contract negotiations only on wages and insurance, the companies insisted that pensions were not a bargainable issue. A 1948 federal court ruling involving Inland Steel, however, had determined that the steel company had a duty to bargain over pensions under the National Labor Relations Act.

While contract talks in 1947 and 1948 had taken place during a period of low unemployment and rapid inflation, 1949 brought the first recession of the post–World War II era; with demand slackening, prices and profits fell in the steel industry. In 1947 Congress had passed the Taft-Hartley Act, which included provisions enabling the president to declare a national emergency in the case of a strike and, in accordance with the declaration, appoint a board of inquiry and obtain an injunction to halt the dispute. President Harry S Truman had vetoed the act — calling it anti-labor legislation — but his veto was overridden. In June 1949 the president attempted to revise Taft-Hartley by replacing its national emergency provisions with procedures that would call for the voluntary delay of labor disputes and create a fact-finding board to make public recommendations. Truman's proposed amendments, however, were defeated in Congress.

With negotiations between the union and the companies deadlocked and the threat of a strike looming, Truman nevertheless appointed a three-member Steel Industry Board, which would function according to his proposed amendments, shortly before the steel contract deadline in July. Truman requested that the union and companies postpone any work stoppage for 60 days and cooperate with the board. Philip Murray, head of the steelworkers and a strong supporter of Truman, readily agreed; the steel companies, in questioning the legitimacy of the board, cooperated only under protest.

After a month of hearings the board recommended (1) no wage increase due to business conditions, (2) social insurance at a cost of 4 cents an hour, and (3) a pension plan at a cost of 6 cents an hour. The board also determined that social insurance and pensions, which it termed "a part of normal business costs," be "noncontributory," meaning that the cost be paid for in full by the employer. The steel companies strongly opposed these recommendations. Negotiations having failed, nearly 500,000 workers in more than 37 steel companies went on strike on October 1. The Steel Industry Board's recommended "10 cent package," however, became a basis for settlement at Ford and several other companies, establishing a pattern in the pension and insurance areas.

After a month-long strike, Bethlehem Steel, which had long had a noncontributory pension plan, assumed the leadership role normally exercised by United States Steel Corporation and reached an agreement with the union. This settlement provided for a noncontributory pension with minimum benefits of $100 per month, a contributory insurance plan with half of the cost of 5 cents per hour to be paid by the employer, and a permanent pension and insurance committee. The master contract was extended to 1952 and the pension and insurance agreements made effective until 1954. The Bethlehem settlement was then adopted by the other steel companies, all of whom reached agreement with the steelworkers by mid November.

References:

Frederick H. Harbison and Robert C. Spencer, "The Politics of Collective Bargaining: The Postwar Record in Steel," *American Political Science Review,* 48 (September 1954): 705–720;

Harold M. Levinson, *Collective Bargaining in the Steel Industry: Pattern Setter or Pattern Follower?* (Ann Arbor and Detroit: Institute of Labor and Industrial Relations, University of Michigan – Wayne State University, 1962);

David J. McDonald, *Union Man* (New York: Dutton, 1969);

Paul A. Tiffany, *The Decline of American Steel: How Management, Labor, and Government Went*

Wrong (New York: Oxford University Press, 1988);

U.S. Department of Labor, *Collective Bargaining in the Steel Industry: A Study of the Public Interest and the Role of Government* (Washington, D.C.: U.S. Department of Labor, 1961).

Steel Strike of 1956

by Richard W. Kalwa

University of Wisconsin — Parkside

The demands presented to the steel industry by the United Steelworkers of America (USWA) in May 1956 included "a substantial wage increase," premium pay for shift and weekend work, and supplemental unemployment benefits (SUBs). A form of the union's long-standing demand for a guaranteed annual wage, SUBs had been introduced in the auto industry the previous year. Industry profits were favorable, but there were indications of increased resistance by the steel companies, linked perhaps to the replacement of Benjamin F. Fairless by Roger Blough as chairman of the board of the United States Steel Corporation (U.S. Steel), in the companies' desire for a long-term contract.

Negotiations in 1955 had for the first time brought representatives of the six largest companies to Pittsburgh to meet with their counterparts on the union bargaining committee, chaired by USWA president David McDonald. The steel companies presented even more of a united front during the 1956 talks. John Stephens, vice-president of U.S. Steel, served as chairman of a four-man management committee — which included representatives from Republic and Bethlehem — that was empowered by the "Big Twelve" steel companies to bargain on major issues.

The companies proposed a five-year contract with deferred increases in wages, benefits, premiums for shift and weekend work, and an SUB plan, adding up to an approximate 50-cents-per-hour increase over the term of the agreement. A cost-of-living adjustment was added to the proposal at the request of the steelworkers, who up to that time had relied on annual negotiations to maintain purchasing power of their wages. The union, however, characterized the companies'

offer as inadequate and demanded that the package be implemented over three years rather than five.

Further bargaining failed to produce compromise, with both the companies and the union resorting to the newspapers and television to air their respective views publicly. When the existing contracts expired on July 1, almost one-half million employees of the steel companies walked out.

President Dwight D. Eisenhower, in contrast to the previous Democratic administration, favored a policy of noninterference in labor disputes. The White House, however, became concerned with the strike's effect on defense programs and the presidential election in November. Eisenhower was hesitant to use an injunction provided by the Taft-Hartley Act and instead assigned Secretary of Labor James P. Mitchell and Secretary of the Treasury George M. Humphrey to mediate the dispute. Due to the administration's threat to appoint a public fact-finding board, U.S. Steel moved toward settlement.

The agreement, signed on 27 July, essentially included what the companies had offered over a five-year period in a three-year contract. Wages were increased by approximately 10 cents per hour in each year of the contract, a SUB plan and semiannual cost-of-living adjustments were introduced, and shift and weekend premiums were improved. The companies also granted the union shop, ending a long period of opposition.The wage and benefit increases of the 1956 contract were greater than those gained by other unions except for the United Mine Workers of America.

Picket line at the U.S. Steel works at Duquesne, near Pittsburgh, Pennsylvania, during the steel strike of 1956

References:

H. Scott Gordon, "The Eisenhower Administration: The Doctrine of Shared Responsibility," in Craufurd D. Goodwin, ed., *Exhortation and Controls: The Search for a Wage-Price Policy 1945–1971* (Washington: Brookings Institution, 1975): 95–134;

John Herling, *Right to Challenge: People and Power in the Steelworkers Union* (New York: Harper & Row, 1972);

Harold M. Levinson, *Collective Bargaining in the Steel Industry: Pattern Setter or Pattern Followers?* (Ann Arbor & Detroit: Institute of Labor and Industrial Relations, University of Michigan – Wayne State University, 1962);

David J. McDonald, *Union Man* (New York: Dutton, 1969);

Jack Stieber, "Company Cooperation in Collective Bargaining in the Basic Steel Industry," *Labor Law Journal,* 11 (July 1960): 614–621;

Jack Stieber, "Steel," in Gerald G. Somers, ed., *Collective Bargaining: Contemporary American Experience* (Madison, Wis.: Industrial Relations Research Association, 1980): 151–208;

Paul A. Tiffany, *The Decline of American Steel: How Management, Labor, and Government Went Wrong* (New York: Oxford University Press, 1988);

U.S. Department of Labor, *Collective Bargaining in the Steel Industry: A Study of the Public Interest and the Role of Government* (Washington: U.S. Department of Labor, 1961).

Steel Strike of 1959

by Richard W. Kalwa

University of Wisconsin — Parkside

Steel negotiations in 1959 were in part characterized by an extensive public-relations campaign launched by the steel companies in order to air publicly the corporate position in management-labor relations. The steel companies, like corporations in other industries, were concerned about wage inflation, lagging productivity, and increasing competition from imports. Given these economic conditions, the steel companies claimed that the 1956 steel agreement — which had raised labor costs by 25 cents an hour, or 8 percent, in each of its three years — could not be further improved upon without hurting the industry.

The industry granted authority to a four-man committee, chaired by R. Conrad Cooper of the United States Steel Corporation, to negotiate a complete contract for the 12 largest steel companies. The committee also included R. Heath Larry of U.S. Steel and representatives from Republic Steel Corporation and Bethlehem Steel Company. They resolved to remain united against union attempts to divide them and bargain with individual companies.

Among the rank and file of the United Steelworkers of America there was deep concern over unemployment in the steel industry due to the 1958 recession and technological advances displacing the workforce. Steelworkers president David McDonald, who had faced a strong challenge to his leadership in the 1957 union elections, thus felt obligated to demand "substantial" increases in wages and benefits.

In April 1959 the industry proposed that both the contract be extended for one year with no change in wages and benefits and that cost-of-living adjustments be eliminated. In responding to the industry's proposal the union also relied on a media campaign to present its views and demanded substantial improvements in wages and benefits.

In June the steel companies stated their willingness to grant modest increases in wages and benefits on the condition that contract language

be revised in several areas in the interest of efficiency. The companies were most concerned about Section 2B, which prevented management from modifying existing work practices at the plant level in the absence of union consent or technological change. The work rules issue became a matter of principle for both sides, with the companies stressing their need for flexibility in order to improve productivity, and the union defending its right to protect its members against managerial abuse and the loss of jobs. Following a two-week extension of bargaining at the request of President Dwight D. Eisenhower, a strike began on July 15, idling 519,000 workers and 87 percent of national steel capacity.

Eisenhower had earlier expressed hopes for a noninflationary settlement in steel, and industry prices had also been the object of hearings before Senator Estes Kefauver's antitrust subcommittee. At first, the government attempted to resolve the dispute through mediation, carried out in part by Vice-president Richard Nixon and Secretary of Labor James Mitchell, who made a fact-finding report. As the strike entered its third month, however, the president initiated the national emergency procedures of the Taft-Hartley Act by appointing a board of inquiry and seeking an injunction to end the stoppage. On November 7, after the union's appeal of the order was denied by the Supreme Court, workers returned to their jobs. The strike was the largest in U. S. history, lasting 116 days and accounting for over 42 million lost workdays.

The Taft-Hartley emergency procedures, however, provided for only an 80-day "cooling-off" period and could not prevent a resumption of the strike if agreement were not reached within that time. Nixon and Mitchell continued to meet with the parties. The companies rejected proposals for third-party recommendations or arbitration, but they also feared a hostile congressional reaction to their intransigence once the injunction expired. Meanwhile, the Steelworkers were estab-

Sign in the window of a Bethlehem, Pennsylvania, appliance store during the steel strike of 1959

lishing a pattern in negotiations elsewhere. Kaiser Steel had left the industry committee in October to sign a separate agreement, and the union reached settlements in the can and aluminum industries in December, each amounting to approximately 35 cents an hour over three years.

An agreement between the union and the major steel companies was finally reached on January 3, 1960. The 30-month contract was estimated to cost about 40 cents an hour over its duration. There would be no wage increase in the first year of the contract, but increases of nearly 10 cents an hour would occur in the second and third years. Cost-of-living adjustments would also be postponed for a year and thereafter made subject to limitations. Health and life insurance were improved and employee contributions eliminated; retirement benefits were also increased. The

agreement further provided for a joint Human Relations Committee to study problems concerning wages and benefits and a tripartite committee to report on the work rules issue.

In comparison to the 1956 contract, the new settlement provided labor with moderate gains, raising labor costs under 4 percent a year — less than half the increase of the earlier agreement. Public reaction to the settlement, however, was hostile. The 1959 conflict was widely perceived as an enormous waste of time and money, which had accomplished little or nothing. Perhaps in response to this mood, U.S. Steel departed from its usual practice of raising prices following labor agreements and announced no increase.

References:
H. Scott Gordon, "The Eisenhower Administration: The Doctrine of Shared Responsibility," in Craufurd D. W. Goodwin, ed., *Exhortation and Controls: The Search for a Wage-Price Policy, 1945–1971* (Washington: Brookings Institution, 1975), pp. 95–134;

John Herling, *Right to Challenge: People and Power in the Steelworkers Union* (New York: Harper & Row, 1972);

Irwin L. Herrnstadt and Benson Soffer, "Recent Labor Disputes over 'Restrictive' Practices and 'Inflationary' Wage Increases," *Journal of Business,* 34 (October 1961): 453–470;

Charles C. Killingsworth, "Cooperative Approaches to Problems of Technological Change," in Gerald G. Somers and others, eds., *Adjusting to Technological Change* (New York: Harper & Row, 1963), pp. 61–94;

Harold M. Levinson, *Collective Bargaining in the Steel Industry: Pattern Setter or Pattern Follower?* (Ann Arbor & Detroit: Institute of Labor and Industrial Relations, University of Michigan — Wayne State University, 1962);

David J. McDonald, *Union Man* (New York: Dutton, 1969);

George J. McManus, *The Inside Story of Steel Wages and Prices 1959–1967* (Philadelphia: Chilton, 1967);

Paul A. Tiffany, *The Decline of American Steel: How Management, Labor, and Government Went Wrong* (New York: Oxford University Press, 1988);

Lloyd Ulman, *The Government of the Steel Workers' Union* (New York: Wiley, 1962);

U.S. Department of Labor, *Collective Bargaining in the Steel Industry: A Study of the Public Interest and the Role of Government* (Washington: U.S. Department of Labor, 1961).

The Steel Workers Organizing Committee
by Richard W. Nagle

Massasoit Community College

The Steel Workers Organizing Committee (SWOC), which existed from 1936 to 1942, successfully unionized the nation's steel industry during the New Deal era. The creation of an industry-wide steel union was no small accomplishment; for the American steel industry had epitomized corporate antiunionism. The Carnegie Steel Company had broken the Amalgamated Association of Iron, Steel and Tin Workers of America at Homestead, Pennsylvania, in 1892. In 1919 the steel industry mounted a successful public-relations campaign to defeat a nationwide steel strike. Despite the passage of the Wagner Act by Congress in 1935 — a prounion bill pressed forward by the administration of Franklin D. Roosevelt — a broad labor movement in steel had seemed hopeless prior to 1936.

The American Federation of Labor (AFL) shied away from a campaign against steel, primarily because it did not want to threaten the autonomy of its craft unions. Furthermore, the old Amalgamated Association, although severely weakened by past defeats, guarded its industrial charter — one of the few in the AFL. The Amalgamated's membership, however, was less than 10,000, most of whom were skilled workers. Its treasury was minuscule, and its president, Michael F. Tighe, seventy-seven years old in 1935 and often referred to by the rank and file as a "Grandmother," did little to encourage a membership drive. Some labor leaders, however, were determined to take advantage of the Wagner Act and organize the mass production industries along industrial lines. John L. Lewis, president of the United Mine Workers (UMW), led this group. When the 1935 AFL Convention rejected their demands, they formed the Congress for Industrial Organizations (CIO).

Lewis's desire to organize steel stemmed not only from his belief in industrial unionism but also from his assertion that an unorganized steel industry presented a constant threat to the UMW. The miners' union had only a tenuous hold in the "captive mines," the mines owned by the steel companies. Lewis first set out to seize the Amalgamated with its valuable charter; he courted it, according to *Fortune,* "as a panther might stalk a moth-eaten alley cat." In June 1936 the Amalgamated joined the CIO and agreed to accept members recruited by the CIO-sponsored SWOC.

Lewis appointed UMW vice-president Philip Murray to serve as chairman of the SWOC and staffed the committee with other miners. UMW money was used to place more than 400 organizers into the field. The SWOC's first job was to infiltrate the steel companies' Employee Representative Plans (ERPs), which functioned as company-sponsored unions. In the first months of the drive the SWOC was successful in turning many of the ERPs against the companies. The biggest coup came when SWOC members captured U.S. Steel's ERP. On January 1, 1937, Murray claimed for the SWOC 125,000 members, many of them employees of U.S. Steel. A confrontation with U.S. Steel, the industry leader, seemed to be at hand.

It never happened. On March 1, 1937, Myron C. Taylor, U.S. Steel's chairman, shocked the industry by announcing his company's recognition of the SWOC. There were several reasons behind U.S. Steel's decision. The SWOC in six months had signed up an impressive number of U.S. Steel workers, and battling increasingly large numbers of organized workers would have been expensive and perhaps violent without a guarantee of victory. The sit-down strikes in the auto industry, which ended in February 1937 when General Motors signed a contract with the CIO's Auto Workers, were sobering and signalled a trend in American industry. Furthermore, a strike would retard the recovery of U.S. Steel, which by 1936 was finally showing profits. For Taylor, the company union approach had become outdated. He knew Lewis and Murray from mine negotiations and knew that the UMW honored all signed agreements. Finally, Taylor understood that the over-

411

Signing of the articles affiliating the Steel Workers Organizing Committee (SWOC) with the CIO, June 17, 1956: (seated) Joe Gaither and Tom Gillis of the Amalgamated Association of Iron, Steel, and Tin Workers; David J. McDonald and Philip Murray of the SWOC; Leo Krzycki of the Amalgamated Clothing Workers; and Julius Hochman of the International Ladies' Garment Workers; (standing) Clinton S. Golden and Van A. Bittner of the United Steelworkers of America; Pat T. Fagan of the United Mine Workers; attorney Lee Pressman; and John Brophy of the United Mine Workers (Pennsylvania State University Historical Collections and Labor Archives)

whelming reelection of Roosevelt in 1936 meant that the government would remain prounion.

On January 9, 1937, Taylor had unexpectedly met Lewis at the Mayflower Hotel in Washington where both were having lunch. A half-hour conversation led to a series of secret meetings wherein the basis of a labor contract was decided. U.S. Steel would recognize SWOC as the bargaining agent for its members, institute the eight-hour day with time-and-a-half paid for overtime hours, raise wages to a minimum of $5 a day, and establish grievance procedures. SWOC signed a formal agreement with the steelmaker on March 17. The country was aghast; Big Steel had surrendered without a fight.

In the wake of the U.S. Steel agreement Murray forecast that unionization of the rest of the industry would be a "mopping up" operation. At first his prediction appeared to be a correct one. Membership swelled to 280,000 by April, and more than 100 companies signed contracts. SWOC called its first strike in May against Jones & Laughlin Steel Company (J&L). After two days the company gave up and signed a contract even

better than the U.S. Steel pact; at J&L, SWOC was recognized as the exclusive representative for all workers, not just as the agent for its members.

The onslaught stopped at the gates of Little Steel, the steel firms that were little only when compared to U.S. Steel. Bethlehem, Republic, Youngstown Sheet and Tube, Inland, and a few others refused to bargain. SWOC called a strike against them in May, but the so-called Little Steel Strike was a violent failure. By July the companies reopened all their mills. SWOC could not mount another organizational drive with the country falling into recession and steel companies laying off workers. Instead, the committee went to the National Labor Relations Board (NLRB) and accused the companies of unfair labor practices. The board ruled against the companies, ordering reinstatement of dismissed workers, compensation for injuries and lost time, and dismantlement of the ERPs. The companies appealed, but the Supreme Court sustained the NLRB orders in 1941.

Murray started a new membership drive when the economic recovery of 1940 sent steel production climbing. In May 1941 SWOC petitioned for a representation election at Bethlehem. The results showed a large union majority. The other companies gave up and did not even demand an election but settled for a cross-check of SWOC membership lists with their payrolls. SWOC had majorities everywhere, and the NLRB certified SWOC as the exclusive bargaining agent for all the companies. With America's entry into World War II a War Labor Board was established, and in July 1942 the board announced the terms of a steel labor agreement, the Little Steel Formula. In August, Bethlehem, Republic, Youngstown, and Inland signed contracts, ending the open-shop era in the steel industry.

A few months prior to this achievement SWOC transformed itself into the United Steelworkers of America (USWA). For years it had operated as a de facto union, but Murray delayed a formal constitutional convention until he was certain of the organization's permanence. In May 1942, at Cleveland, a constitution was written, and Murray was elected president of 660,000 steelworkers.

John Allen Stephens

(May 1, 1895 – October 26, 1979)

by Richard W. Kalwa

University of Wisconsin — Parkside

CAREER: Manager, industrial relations, Chicago district, Carnegie-Illinois Steel Corporation (1935–1938); director, industrial relations, United States Steel Corporation of Delaware (1938–1943); vice-president, industrial relations, United States Steel Corporation (1943–1958).

John Allen Stephens played a key role in establishing the collective-bargaining policy of the United States Steel Corporation (U.S. Steel) during the 1940s. As the corporation's chief labor-relations executive he exercised an influence on collective bargaining not only within U.S. Steel but also across the steel industry.

He was born on May 1, 1895, in Albany, New York, to John Allen and Anna Mangan Stephens. His father was a law partner of David B. Hill, one-time governor of New York and U.S. senator. Stephens attended local schools, went to Wesleyan University and then to the School of Business and Finance at Columbia University, where he graduated with a bachelor's degree in 1917. In that year he joined the U.S. Army, and by the end of World War I he had reached the rank of major. In 1918 he married Mary Elizabeth Lathrop; the couple had three daughters: Marcia, Mary Elizabeth, and Nancy.

Stephens entered the business world and served as an officer of several companies between 1922 and 1933. In 1932 he was appointed director of the New York City Work Relief Program. In the following year the Bush Terminal Company, of which he was president, went bankrupt, leaving Stephens penniless and searching for what he called "something fundamental."

He took a sales job at the Gary works of the Illinois Steel Company in 1934 and rose rapidly through the ranks. In 1935 he was made manager of industrial relations for the Chicago district of the Carnegie-Illinois Steel Corporation, and in 1938 he became director of industrial relations for the U.S. Steel Corporation of Delaware. He began negotiating with the steelworkers union in 1941.

Stephens served as the U.S. Steel representative on a joint union-management commission set up by the 1942 collective agreement to study wage inequities. In 1943 Stephens became vice-president of industrial relations for U.S. Steel and was made chairman of the Cooperative Wage Study, a group established by 12 major steel companies to examine and reform the wage structure of the industry. In 1944 he served as chairman of the Steel Case Research Committee, which coordinated the presentation of 86 steel companies in hearings before the War Labor Board, and argued successfully against a wage increase.

In 1948 Stephens served as a management member of President Harry S Truman's Labor-Management Conference; he also served on the labor-policy committee, which was part of the business advisory council of the Department of Commerce. He was a member of the U.S. Chamber of Commerce and held labor-related positions in the American Iron and Steel Institute, the National Association of manufacturers, and the American Management Association.

In 1952, during the strike which followed Truman's short-lived seizure of the steel mills, Stephens and three representatives of the Bethlehem Steel Company and the Jones & Laughlin Steel Corporation took the unusual step of presenting the companies' proposals to the Wage Policy Committee of the Steelworkers. When the major steel companies coordinated their bargaining in 1956, Stephens chaired a four-man committee — with representatives from U.S. Steel, Bethlehem, and Republic Steel — which had authority to reach agreement on major issues for 12 companies. He publicly minimized the importance of this arrangement, however, claiming that the company spokesmen were negotiating simultaneously but independently. At the same time, the appointment of Roger Blough to succeed Benjamin F. Fairless as U.S. Steel chairman appeared to signal a new inflexible approach toward labor and constrained Stephens's freedom in negotiations until the threat of government intervention forced a settlement.

After Stephens retired from U.S. Steel in 1959 he was appointed by the Eisenhower administration to advise the Council on Foreign Economic Policy. His presence in Washington contributed to the feeling of confidence with which the steel companies entered labor negotiations in 1959. Stephens died on October 26, 1979, in Sewickley, Pennsylvania.

Publications:
"Jobs, Labor's First Postwar Aim; Manpower Stabilization in U.S. Steel Company's Plants," *Steel,* 113 (October 11, 1943): 158–159;
"The Evolution of Industrial Relations Between the Two Wars," *American Iron and Steel Institute Yearbook* (1943): 286–294;
and Howard M. Dirks, Forrest H. Kirkpatrick, and Samuel L. H. Burk, *Personnel Management and Professional Development* (New York: American Management Association, 1943);
Notes on My Russian Trip (Pittsburgh, 1958).

References:
David J. McDonald, *Union Man* (New York: Dutton, 1969);
Jack Stieber, *The Steel Industry Wage Structure* (Cambridge, Mass.: Harvard University Press, 1959);
Paul A. Tiffany, *The Decline of American Steel: How Management, Labor, and Government Went Wrong* (New York: Oxford University Press, 1988);
U.S. Department of Labor, *Collective Bargaining in the Steel Industry: A Study of the Public Interest and the Role of Government* (Washington, D.C.: U.S. Department of Labor, 1961).

Archives:
The speeches of John Allen Stephens are available in the Hagley Museum and Library, Wilmington, Delaware.

William J. Stephens

(December 16, 1906 – June 14, 1982)

by John A. Heitmann

University of Dayton

CAREER: Various positions, including director and assistant vice-president of sales, Bethlehem Steel Corporation (1921–1962); vice-president (1962–1963), president (1963–1969), chairman, Jones & Laughlin Steel Corporation (1969–1971).

A key figure in the American steel industry during the transitional decade of the 1960s, William J. Stephens was a keen observer of economic trends. He foresaw the shift in the industrial landscape that would ultimately undermine American steel's long-term dominance in the global marketplace, but could do little to arrest these changes.

Like many businessmen of his day, William J. Stephens gained his education not at the university but through practical experience. Beginning at age fifteen as a messenger boy in the Bethlehem Steel Corporation's New York City office, Stephens eventually became a salesman for the company and gradually rose through the ranks until he was appointed in the late 1950s as assistant vice-president of sales and a director of Bethlehem. Stephens was consumer-oriented in his business approach, always asking, "What does the customer want, what does he need, and what will that do to us in the marketplace." In 1961 Stephens moved to the Jones and Laughlin Steel Corporation (J&L) as executive vice-president, a position that in the past had frequently proved to be the crucial stepping stone to the executive suite.

During the early 1960s J&L had the reputation of being one of America's most dynamic steel companies, the consequence of a modernization program undertaken during the early 1950s under the leadership of Admiral Ben Moreell and C. L. (Lee) Austin. The firm was one of the first to develop the high-capacity basic-oxygen furnace and to employ computers to control them. With the most mechanized coal mine in the United States and a computer inventory system, the J&L of the early 1960s could afford to reduce its production of low-profit steel and concentrate on more-lucrative sheet and stainless steels.

The optimism of the early 1960s at J&L and elsewhere in the steel industry was soon tempered by the pressures of rising prices, increased foreign competition, and federal-government intervention. Stephens, who was promoted to president of the firm in 1963, replacing Charles M. Beeghly, was quickly faced with several formidable challenges.

Stephens's position at J&L was first tested in April 1964, when he, along with two other executives, were indicted for price rigging. Confronted with a strong government case against him, Stephens pleaded no contest and was forced to pay a fine of $25,000. Perhaps as a result of this episode, Stephens emerged as a strong critic of the Johnson administration and its policies toward

business. In a 1966 speech Stephens asserted that influential bureaucrats in Washington did not understand the establishment of prices in the marketplace and that prices and profits were key to the future of the Great Society. He declared that "profits are the bloodstream. Dry up the bloodstream and the patient dies. Dry up the profits and the corporation dies. Profit makes the difference between progress and stagnation."

Stephens survived the Ling-Temco-Vought (LTV) takeover of J&L in 1968 and became the chairman of the steelmaking firm in 1969. A super salesman, he was described by one former associate as possessing "a Machiavellian sense of commercial strategy." But until his retirement in 1971 he was faced with enormous internal and external pressures, the result of J&L's declining profitability in part brought about by the merger and in part due to the emergence of a new competitive structure in the industry. Stephens had inherited a situation that was to some extent brought on by the company's failure to diversify during the 1960s, and therefore the firm was especially vulnerable to the challenges posed by lower-priced steel made abroad and exported to the United States. Nearing retirement in 1971, Stephens, in one of his last public speeches, suggested that unless government policy changed with regard to international trade, taxation, and antitrust regulations, the American economy would "slowly and surely self-destruct." After stepping down, Stephens and his generation of leaders in the steel industry witnessed the dramatic and sudden decline during the 1970s of one of America's once-formidable sectors. Stephens died at his home in Hilton Head, South Carolina, on June 14, 1982.

Edward Reilly Stettinius, Jr.

(October 22, 1900 – October 31, 1949)

by Bruce E. Seely

Michigan Technological University

CAREER: Stockroom clerk, employment manager, Hyatt Roller Bearing Company (1924–1926); assistant to the vice-president, accessory division (1926–1930), assistant to the president (1930–1931), vice-president for industrial and public relations, General Motors (1931–1934); vice-chair, finance committee, (1934–1935), chair, finance committee (1935–1938), director (1936–1940), chairman of the board, United States Steel Corporation (1938–40); member, Council of National Defense (1940); Lend-Lease administrator (1941–1943); undersecretary of

Edward Reilly Stettinius, Jr.

state (1943–1944); secretary of state (1944–1945); U.S. representative to the United Nations (1945–1946); rector, University of Virginia (1946–1947).

Edward R. Stettinius, Jr., at the age of 37, became the youngest chairman in the history of the United States Steel Corporation. His youth and his lack of knowledge about steelmaking made him a surprise choice to succeed Myron Taylor, yet Stettinius's promotion symbolized a Taylor-inspired revolution at U.S. Steel in which the corporation's young blood with new business philosophies was rising to the top ranks. Like his father, Stettinius left the private sector during wartime to serve in government — a career in diplomacy that would include a stint as secretary of state under two presidents.

Stettinius was born in Chicago on October 22, 1900, with the proverbial silver spoon in his mouth. His father was president of Diamond Match Company, a partner in J. P. Morgan &

Company, and oversaw munitions purchases for the British and French governments during World War I. Stettinius attended the University of Virginia yet never received a degree. His business career began in the stockroom of the Hyatt Roller Bearing Company, a subsidiary of General Motors; his father was a Morgan representative on GM's executive committee. He rose quickly into the executive ranks, working in public and labor relations, areas in which his height, impressive good looks, and outgoing personality served him well. Yet underneath the suave exterior was a sharp mind that negotiated the largest group-insurance plan then written, covering 200,000 employees.

After Stettinius became a GM vice-president in 1931, he was involved with various relief programs — his first service for the government. In 1933, he became the liaison officer between the National Recovery Administration and its Industrial Advisory Board in 1933; by 1935, he was a special adviser to that board. Stettinius was sympathetic to the New Deal and was one of the very few top executives in an industry faced with unionization who was not virulently anti-Roosevelt.

In 1934 Stettinius left GM when Myron Taylor, chairman of U.S. Steel, made him vice-chair of the company's finance committee. To many, Stettinius seemed out of place in a steel company, given his political sympathies and youth. Two years later, however, Stettinius was chairing the finance committee and serving on the board of directors; he was involved in Taylor's stunning 1937 decision to recognize organized labor. Stettinius completed his rapid climb to the top in April 1938, when Taylor made him cochairman of the board with William Irvin.

Some observers saw Stettinius as only a continuation of the Morgan influence at U.S. Steel; financial writers dubbed him the "Crown Prince of Steel" in 1934. But the financial editor of the *New York Journal* noted that "the Morgan's don't go in for philanthropy when it comes to picking someone to run U.S. Steel." Taylor clearly wanted Stettinius because he was not hobbled by U.S. Steel traditions. Moreover, within the decentralized organization Taylor had built, Stettinius — whose policy it was to not interfere with the steelmen making the steel — could focus on public relations. Few industrialists were on better terms with the New Dealers in the Roosevelt administration, and Taylor wanted to erase the steel industry's terrible public image.

But the war in Europe drew Stettinius into government roles similar to his father's during World War I. In August 1939 he chaired the War Resources Board, surveying industry's crucial materials needs. A year later, he advised the Council for National Defense on materials. Although hampered by Roosevelt's penchant for ambiguous lines of authority, Stettinius decided to choose government work over running U.S. Steel and in March 1940 resigned as chairman. Stettinius denied rumors of a power struggle with U.S. Steel president Benjamin Fairless as the reason for his departure, noting that he had helped hire Fairless. Nonetheless, Stettinius completely severed his ties to steel.

Stettinius began his career in government managing industrial materials for the Council for National Defense, the U.S. Priorities Board, and, later, the Office of Production Management. In 1941 he became Lend-Lease administrator, winning high marks for managerial efficiency and congressional relations. Stettinius became undersecretary of state in 1943 despite questions raised about his lack of experience; the résumé of an undersecretary was traditionally that of a career diplomat. There were further complaints after he reorganized the department to eliminate redundancy and free upper-level officials from routine matters. Nonetheless, Stettinius played a central role in the creation of the United Nations, organizing the Dumbarton Oaks Conference and heading a fractious American delegation. He also fashioned the compromise on voting arrangements in the UN Security Council.

In late 1944 Stettinius became the second-youngest secretary of state in U.S. history. Roosevelt, however, handled foreign policy, while Stettinius administered the State Department. Historian Richard Walker labeled Stettinius "the Secretary who was not Secretary of State." He accompanied the president to Yalta and, after Roosevelt's death, managed the San Francisco Conference, which wrote the UN Charter, making fundamental contributions to that document. When President Harry S Truman wanted to name his own secretary of state, Stettinius resigned in 1945, accepting the assignment as first representative to the United Nations, where his organizational skills and backslapping style were highly successful in helping to shape the fledgling international body.

After leaving the United Nations post in July 1946, Stettinius became rector of the University of Virginia. A heart condition forced his resignation in 1947, but he remained active in world affairs and continued to involve himself in work for the civic and social organizations he had served since the 1930s. His work on the economic and social development of Liberia was consistent with his vision of a postwar world guided by free enterprise. Stettinius also prepared a book defending the Yalta Conference, which had come under increasing attack after the death of Roosevelt.

Stettinius married Virginia Gordon Wallace in 1926, and they had three sons. At his Virginia farm, the Horseshoe, he cared for his prize Belgian draft horses, white-faced Hereford cattle, and turkeys. He died in Greenwich, Connecticut, on October 31, 1949.

Publications:
"Lend-Lease Works Both Ways," *Saturday Evening Post,* 215 (September 12, 1942): 11;

"What Lend-Lease Taught Us," *New York Times Magazine* (January 9, 1944): 10;

"Lend-Lease and the Windsors," *American Mercury,* 59 (August 1944): 247;

"Working Together," *Vital Speeches,* 10 (September 15, 1944): 733–735;

Lend-Lease, Weapon for Victory (New York: Macmillan, 1944);

"Statement on U.S. Government's Position on Poland," *Current History,* 8 (February 1945): 165–166;

"What the Dumbarton Oaks Pew Plan Means," *Reader's Digest,* 46 (February 1945): 1–7;

"America's Role in World Peace," Address to Inter-American Conference on Problems of War and Peace, *Vital Speeches,* 11 (March 1, 1945): 292–295;

"Invitation to U.N. Conference," *Vital Speeches,* 11 (March 1, 1945): 336–337;

"San Francisco Conference Perspective," *Vital Speeches,* 11 (April 15, 1945): 388–389;

"Sound Foundation of World Organization," *Vital Speeches,* 11 (May 15, 1945): 455–457;

"Sovereignty of No Nation Is Absolute," *Vital Speeches,* 11 (June 15, 1945): 524 –528;

"Speech at Final Plenary Session of United Nation's Conference," *Vital Speeches,* 11 (July 15, 1945): 574;

"United Nation's Charter," *Vital Speeches,* 11 (July 15, 1945): 588–593;

"San Francisco: Gateway to Peace," *Rotarian,* 67 (August 1945): 8–10;

and others, *Partners for Peace: USA-USSR Nations United for Victory, Peace, and Prosperity* (New York: National Council of American-Soviet Friendship, 1945);

Human Factors in Economic Progress; Lecture to Accompany Modern Business Text on Economics (New York: Alexander Hamilton Institute, 1945);

Charter of the United Nations. Report to the President on the Results of the San Francisco Conference by

the Chairman of the United States Delegation, the
Secretary of State (Washington, D.C.: United
States Government Printing Office, 1945);

"Human Rights in the United Nations Charter," *Annals of the American Academy of Political and Social Sciences,* 243 (January 1946): 1–3;

"I Have Faith in World Peace," *Reader's Digest,* 49
(September 1946): 109–14;

"Leadership in World Peace," *Vital Speeches,* 13 (November 1, 1946): 41–44;

Report of the Delegation of the United States of America to the Inter-American Conference on Problems of War and Peace, Mexico, February 21–March 8, 1945 (Washington, D.C.: United States Government Printing Office, 1946);

Cooperation among Individuals and Nations (Chattanooga: University of Chattanooga, 1947);

Economics: The Science of Business (New York: Alexander Hamilton Institute, 1949);

Roosevelt and the Russians: The Yalta Conference (New York: Doubleday, 1949);

The Diaries of Edward R. Stettinius, Jr., 1943–1946, edited by Thomas M. Campbell and George C. Herring (New York: New Viewpoints, 1975).

References:
Rodney Carlisle, "The 'American Century' Implemented: Stettinius and the Liberian Flag of Convenience," *Business History Review,* 54 (Summer 1980): 175–191;

John Ingham, *Biographical Dictionary of American Business Leaders* (Westport, Conn.: Greenwood Press, 1983), III: 1358–1362;

"Managers of Steel," *Fortune,* 15 (March 1940): 64–67, 142, 145;

"More Life in the Steel Trade," *Literary Digest,* 120 (December 28, 1935): 36;

"The New Rector," *Time,* 48 (August 19, 1946): 57;

"The Optimist," *Time,* 54 (November 7, 1949): 21;

Dennis Plimmer, "Yeoman of Peace," *New Republic,* 114 (April 22, 1946): 572;

Graham H. Stuart, *The Department of State: A History of Its Organization, Procedure, and Personnel* (New York: Macmillan, 1949);

"U.S. Steel Corporation," *Steel,* 93 (December 18, 1933): 11;

Richard L. Walker, "E. R. Stettinius," in *The American Secretaries of State and Their Diplomacy,* volume 14, edited by Robert H. Ferrell (New York: Cooper Square Publishers, 1965), pp. 1–83, 397–404.

Archives:
The Edward R. Stettinius, Jr., Collection is located at the Alderman Library, University of Virginia, Charlottesville, Virginia.

Wilfred Sykes

(December 9, 1883 – May 3, 1964)

by Robert M. Aduddell

Loyola University of Chicago

CAREER: Engineer, Allegemeine Elektricitäts Gesellschaft, Australia division (1901–1906), Berlin division (1907–1909); engineer, Westinghouse Electric & Manufacturing Company (1909–1920); executive engineer, Steel and Tube Company of America (1920–1922); engineer (1923–1927), assistant general superintendent (1927–1930), assistant to the president, Indiana Harbor (1930–1941), president, Inland Steel Company (1941–1949).

The sixth president of Inland Steel Company, Wilfred Sykes, was born in Palmerston, North New Zealand, on December 9, 1883, to William J. and Emma Ricketts Sykes. Wilfred Sykes entered Melbourne Technical College in 1898, then transferred to Melbourne University

the following year. Upon graduation in 1902 with a degree in electrical engineering, he took a position as an electrical engineer with Allegemeine Elektricitäts Gesellschaft in Australia. In 1907 he was transferred to Berlin, where he worked until he came to the United States. Sykes was hired by Westinghouse Electric and Manufacturing Company in 1909. His specialty was the application of electrical power to manufacturing and industrial uses, with a particular focus on the steel industry.

Sykes left Westinghouse in 1920 and took an executive engineering position with the Steel and Tube Company of America in Chicago. Having gained a reputation as an efficient plant construction man, one possessing acute electrical engineering skills, Sykes joined Inland Steel in 1923.

During the early 1920s Inland was in the process of upgrading its facilities in the hope of expanding its markets over the next two decades. Upon joining Inland, Sykes immediately became involved in the several construction and conversion operations the company had under way. Two mills were modified to expand the production of railroad iron and structural shapes. A second phase of expansion began in 1923 and included the construction of a billet and slab mill, a merchant mill, and four new open-hearth furnaces of 100-ton capacity each. Plans were also under way to convert the entire Indiana Harbor works to electric power. Upon completion of the electrification program in 1926, Inland expanded its product line by adding high-quality rolled sheets to satisfy the growing demand from the automobile and appliance industries. The company's traditional heavy products were to remain an integral part of its product line, but the light products netted higher revenue per unit of sales.

To change the product mix in an integrated steel firm requires a substantial engineering input, and Wilfred Sykes deserved much of the credit for the technical aspects of this change at Inland. The speed and efficiency with which the electrification program was completed was a tribute to the technical and managerial virtuosity of Wilfred Sykes. In one instance, Sykes oversaw the conversion of three rolling mills in a one-week period when it had been assumed that the task would take at least three weeks. Due to his successful management of the construction and conversion programs, Sykes was promoted in 1927 to the position of assistant general superintendent of Inland's Indiana Harbor works. He held this position until 1930, when he was made assistant to the president. Five years later he was elected to the board of directors, and on April 30, 1941, Sykes was elected president of Inland, replacing Philip D. Block.

Having become familiar with virtually every aspect of Inland's operation, Sykes was well prepared as president to adopt the firm's philosophy of expansion to the exigencies of a wartime economy. The country's entry into World War II meant an increase of demand for heavy products and plates for shipbuilding, and Inland was forced to curtail its normal plans for expansion owing to shortages in materials and labor. The firm under Sykes's leadership, therefore, shifted its expansion efforts to a program of acquisition. In 1941 Inland acquired the Nevada Corporation, which gave them access to additional iron ore properties in Michigan. In 1943 the Hillside Fluorspar Mines, a Kentucky firm, was made a wholly

owned subsidiary. Milcor Steel Company, an Inland subsidiary, acquired the J. M. & L. A. Osborn Company of Cleveland, Ohio, in 1944. The Osborn firm was a fabricator of sheet metal products. Inland subsidiaries would continue a program of expansion and modernization after the war.

At war's end Inland sought to expand its capacity. Consequently, in 1946 Inland purchased Plancor 266 — a facility that sat on property adjacent to Inland's Indiana Harbor plant — from the Reconstruction Finance Corporation for $13.5 million. The plant consisted of two blast furnaces and 146 coke ovens, and had docking facilities and a power plant. During that same year, additional strip-mill capacity was added, and equipment was upgraded.

In the immediate postwar period Inland continued to integrate its operations through the acquisition of sources of raw material, acquiring coal reserves in eastern Kentucky and iron ore mined on the Cuyuna and Mesabi Ranges in Minnesota and the Marquette Range in Michigan.

The question of the adequacy of the American steel industry's capacity was an issue Sykes and others addressed before Congress in 1947 and 1948. As a result of these hearings Inland decided to increase its capacity in basic steel. This was, however, a task for the successor to Wilfred Sykes, Clarence Randall.

Inland was in a good position in 1949, when Sykes retired as president. Gross fixed assets had increased from $177.1 million in 1941 to $262.4 million in 1948. The firm's net income in 1948 was $63.5 million. Long-term debt decreased while improvements were made in the firm's light-products base.

Sykes continued to serve on Inland's board until, having reached the mandatory retirement age of seventy, he was given the title of honorary director in 1953. Inland had honored him in 1950 by naming its newest, largest, and most powerful ore carrier after him. Sykes had long been active in the Chicago Association of Commerce and Industry, serving as president of that organization in 1948. He was a longtime trustee of the Illinois Institute of Technology in Chicago and had been a trustee of the Glenwood School for Boys. His professional memberships included the American Institute of Electrical Engineers and the American Institute of Mining and Metallurgical Engineers. He was a longtime member of the American Iron and Steel Institute and served as a director of that organization. His British heritage was reflected in his having been made a fellow of the Royal Society of the Arts. He died in San Francisco on May 3, 1964.

Publications:

"An Electrically Driven Reversing Rolling Mill," *Iron Trade Review,* 50 (May 2, 1912): 960–965;

and G. E. Stolz, "Control of Rolling Mill Motors," *Iron Trade Review,* 55 (September 24, 1914): 590–591;

"Heavy Electric Reversing Mills," *Iron Trade Review,* 55 (December 24, 1914): 1181–1183;

and Brent Wiley, "Buying Power for the Rolling Mill," *Iron Trade Review,* 57 (September 16, 1915): 530–533;

"Electrically-Driven Reversing Rolling Mills," *American Iron and Steel Institute Yearbook* (1919): 393–405;

"The General Effect of Electrification on the Operation of Steel Mills," *American Iron and Steel Institute Yearbook* (1922): 121–149;

"A Modern Power and Blowing Plant," *American Iron and Steel Institute Yearbook* (1928): 212–230;

"Rolling Strip Steel at the Inland Steel Company's Plant," *Mining and Metallurgy,* 17 (October 1936): 471–476;

The Postwar Prospect: A Series of Seven Newspaper Articles (New York: National Association of Manufacturers, 1943);

"Steel: A Pioneer Chicago Industry," *Western Society of Engineers Journal,* special edition (December 1944): 5–14;

"Inland Steel Acquires Blast Furnaces and Coke Ovens," *Blast Furnace and Steel Plant,* 34 (December 1946): 1526;

"Steel Industry Outlook," *Commercial and Financial Chronicle,* 166 (December 11, 1947): 2365;

"Future of the Steel Industry," *American Iron and Steel Institute Yearbook* (1947): 68–83;

"Steel," *Midwest Engineer,* 2 (March 1950): 3–7.

Reference:

William T. Hogan, S.J., *An Economic History of the Iron and Steel Industry in the United States,* 5 volumes (Lexington, Mass.: Lexington Books, 1971).

Myron C. Taylor

(January 18, 1874 – May 6, 1958)

by Bruce E. Seely

Michigan Technological University

CAREER: Lawyer, DeForest Brothers & DeForest (1895–1920); textile executive (1910–1928); banker (1920–1931); chairman of finance committee (1927–1932), chairman of the board and chief executive officer (1932–1938), member of finance committee, United States Steel Corporation (1938–1956); personal representative of Franklin D. Roosevelt and Harry S Truman to the Vatican (1940–1949).

Myron Charles Taylor was an extremely successful lawyer, banker, and businessman whose greatest challenge came as head of the United States Steel Corporation during the Great Depression. During that period he guided U.S. Steel through a major reorganization and was the first steel executive to admit the rights of organized labor when he recognized the Steel Workers Organizing Committee (SWOC). He has been judged one of the best top executives in the history of U.S. Steel.

Born in Lyons, New York, on January 18, 1874, Taylor was raised in comfort; his father, William Taylor, had made a fortune in textiles and leather. Taylor attended Cornell University, where he was a student of Chief Justice of the Supreme Court Charles Evans Hughes. He earned a law degree from Cornell in 1894, was admitted to the New York bar in 1895, and joined the law firm of DeForest Brothers & DeForest. The Taylor family business interests led him to work with clients in the textile field.

His interest in mills near Boston drew Taylor into the problems of the Consolidated Cotton Duck Company in 1910, and he played the central role in saving the textile firm. Drawing on personal financial connections, Taylor managed a series of reorganizations that eventually created the International Cotton Mills, a combination of New England and southern mills, in 1912. Three principles guided Taylor's efforts: simplify financing, appoint progressive management, and install modern technology. Taylor soon sold his interests and formed a group that purchased several mills to produce cotton tire cord, betting on the growth of the automobile industry. The firm Taylor, Armitage, and Eagles prospered mightily, thanks to Henry Ford and World War I. But in 1920 Taylor wanted out of textiles, and he sold the firm in pieces to leading tire manufacturers. By 1923 Taylor had amassed a fortune of $20 million.

Taylor went into banking and soon became identified with J. P. Morgan & Company. Taylor served on the corporate boards of American Telephone & Telegraph; the Lehigh & Wilkes-Barre, Santa Fe, and New York Central Railroads; and several New York banks. Soon after Taylor arranged the merger of the Guaranty Trust Company with the National Bank of Commerce in 1928, George F. Baker, a Morgan banker and the largest stockholder in U.S. Steel, asked Taylor to become a director and chair the steel company's finance committee. The corporation faced serious challenges, and Taylor was reluctant to accept a commitment that would interfere with his travels to Europe, but Baker was persuasive, and Taylor launched into a career in steel in 1927.

U.S. Steel was J. P. Morgan's creation in 1901, and bankers remained a dominant presence on the board. Through Elbert Gary, Morgan's choice to run the steel holding company, the finance committee became the corporation's seat of power. Although chosen to replace Gary, who died in 1927, Taylor had no intention of solely running U.S. Steel. James Farrell ran steel operations; J. P. Morgan chaired board meetings; and Taylor minded the finances.

As head of finance, Taylor applied the formula for corporate rejuvenation that he had used in textiles. He first acted to lower the corporation's bonded debt. Worried about the market, he retired $340 million in bonds in March 1929. This action may have saved U.S. Steel during the Depression, as it lowered annual interest payments by $31 million. He also arranged the purchase of Columbia Steel in California to open West Coast markets. With the onset of the Depression, he instituted Herbert Hoover's share-the-work plan, and most workers went on part-time employment. By 1931 Taylor was running the corporation; almost naturally he became chairman of the board and chief executive officer in 1932.

Taylor spent six years tackling other inherited problems. *Fortune* argued in 1936 that Gary had tried to protect the status quo by stabilizing prices, shunning innovation, and undermining geographic advantage with the basing-point pric-

Myron C. Taylor

ing system. These strategies, prompted in part by fear of antitrust action, had created several problems. The company was saddled with old mills and obsolete equipment; it had ignored continuous-strip mills, which furnished steel for the market least affected by the Depression — automobile manufacture. Furthermore, management was weak, and the sales force was especially dismal. Worst of all, U.S. Steel was a very inefficient holding company, an assemblage of some 200 loosely linked and overlapping companies.

Taylor again applied his textile-mill strategy. With finances in order, he introduced a mandatory retirement age of seventy to bring younger blood into management. Taylor then promoted W. J. Filbert, fifty-one, to chair the finance committee; William A. Irvin, fifty-nine, replaced Farrell as U.S. Steel president; Arthur Young, fifty-three, became the first U.S. Steel vice-president for industrial relations; and Edward Stettinius, thirty-three, became vice-chair of the finance committee. Further indicating Taylor's willingness to look outside U.S. Steel, in 1935 he hired Ford, Bacon &

Davis, a management consulting firm that produced more than 200 reports on every phase of U.S. Steel by 1938. Taylor relied on this material as he reshaped the company, attempting to coordinate subsidiaries that had often worked at cross-purposes. In August 1935 the two steelmaking subsidiaries were merged into Carnegie-Illinois Steel, and, in an unheard-of move, Taylor tapped Benjamin Fairless of Republic Steel to run it. More important, the authority of the finance committee was reduced by installing a decentralized administration. For the first time, subsidiary presidents actually ran their companies.

Taylor slated sales for special attention. Robert Gregg, hired from Atlantic Steel to run the Tennessee Coal, Iron & Railroad Corporation in 1931, moved into the top sales spot. He inherited an archaic system whereby each subsidiary maintained a sales staff and jealously guarded its product line. Taylor began forming joint sales offices and retained an advertising company to coordinate a corporate program. Finally, Taylor attacked the corporation's technological obsolescence. He supported research efforts and had Ford, Bacon & Davis undertake an engineering survey of the corporation. This was followed by a capital-expenditures program that totaled more than $300 million from 1935 to 1938. In short, Taylor shook up U.S. Steel and left a very different company when he stepped down as chairman in 1938.

Taylor's most startling decision as head of U.S. Steel concerned labor. Organized labor made tremendous gains under the New Deal, and the steel industry — with its traditionally violent antipathy to unions — became a crucial battleground. Taylor at first resisted unionization. When National Recovery Act codes mandated that labor be allowed to bargain collectively, U.S. Steel followed other steel companies in developing an employee representation plan, in essence a company-sponsored union. Similarly, U.S. Steel battled the National Steel Labor Board through 1935, opposing prolabor legislation. But Taylor broke with his peers and refused to wage war against the SWOC, organized by John L. Lewis in 1936. Taylor doubted that a company union would forestall the SWOC. The union tapped a deep reservoir of resentment, including anger at Taylor's well-intentioned share-the-work plan of 1931 that had reduced hours even as pay was cut. The SWOC grew quickly and even turned the company union against the corporation. Pondering the situation, Taylor left for his annual trip to Europe in July 1936. He returned with a formula for recognizing the SWOC as a bargaining agent, but not the exclusive one.

After a chance meeting in January 1937, Taylor began a series of conversations with Lewis that produced an agreement. Taylor wrung concessions from Lewis by overstating the opposition of his board, but both sides compromised. U.S. Steel president Benjamin Fairless and Phillip Murray, head of the SWOC, hammered out a detailed agreement in early March 1937. By the end of the year the company had disbanded its employee representation plan.

Steel industry executives were stunned, including many within U.S. Steel's managerial ranks. Fairless, for example, had been notified of Taylor's tentative agreement with Lewis only on February 27. Other steel companies heard only as Murray and Fairless sat down to talk on March 1. According to one account, the portrait of Henry Clay Frick, the man who crushed the Homestead Strike in 1892, was removed from the room where the agreement was signed because "they did not think he could stand it." Yet in short order most smaller steel companies reached similar agreements with the SWOC. Those who held out — "little steel" — faced a major public-relations hurdle in justifying their resistance to unionization.

Taylor never fully explained his reasons for recognizing the SWOC, but several factors influenced his thinking. He did not want to pay the price of withstanding a strike called by the well-organized SWOC. The cost would be high, measured in lives, property damage, and ill will toward the corporation. Moreover, in 1937 U.S. Steel was enjoying its first heavy volume of orders — and profits — since 1929. There have also been rumors, repeatedly denied, that a large order from the British government was contingent on a no-strike guarantee. Another factor in the business equation was U.S. Steel's connections to an international steel cartel. He also knew that U.S. Steel might lose a strike. In the past workers had been divided by language and ethnic background, and tensions existed between skilled and unskilled laborers, making it difficult for workers to organize. Other factors, including the Red Scare in 1919, had crippled the appeal of unions. All this had changed by 1937. As Lewis and Taylor held their discussions, federal, state, and local officials sided with labor during the sit-down strike at General Motors. Public opinion was influenced by congressional revelations of steel-company espionage networks and other antiunion activities. Taylor had come to realize from his frequent contact with federal officials that the political climate had changed. He had served presidents Herbert Hoover and Franklin D. Roosevelt in various capacities during the Depression and met with Roosevelt to discuss several labor disputes; Taylor did not share the hatred many steelmen felt toward Roosevelt and the New Deal.

Taylor had also lost confidence in the company union; the turmoil over representation was impairing efficiency in the plants. He concluded that he could trust Murray and Lewis, who had honored an agreement reached in 1933 to end a strike by miners in steel-company mines. Finally, as historian Irving Bernstein asserts, Taylor had a sense of history and recognized that unions were inevitable. To a lawyer who prided himself on diplomatic skills, it made better sense to set the terms of engagement than to continue in a struggle that might stave off unionization for only a few years, and then at an unacceptable cost. As Taylor wrote in *Fortune* in 1938, "I felt it was my duty as a trustee for our stockholders and as a citizen to make any honorable settlement that would ensure a continuance of work, wages, and profit."

Taylor stepped down as chairman in April 1938, picking Edward Stettinius, Jr., as his replacement. He remained a guiding force on the board and finance committee for another 20 years. Soon after retirement from U.S. Steel, however, Taylor began a fourth career: diplomatic service. He already had a track record in public service, having assisted many relief and humanitarian efforts in New York during the Depression, and having aided in forming policy concerning refugees fleeing Nazi Germany. Leading the U.S. delegation at the Evian Conference in 1938, Taylor was credited by observers with the meeting's partial success. He continued to chair the Intergovernment Committee on Political Refugees until 1944.

Taylor became a full-fledged diplomat when Roosevelt asked him to become the president's personal representative to Pope Pius XII. Announced in late 1939, the decision angered many Protestant groups, already opposed to diplomatic ties to the Vatican in part because Roosevelt avoided congressional debate and ratification by not making Taylor an ambassador. Roosevelt had wanted access to the Vatican's information sources in Europe; Taylor knew Italy well, and his refugee work matched the church's interests. Moreover, the pope had stayed at Taylor's residence during a trip to the United States in 1936. Between 1940 and 1945 Taylor, an Episcopalian,

made five extended trips to the Vatican, conveying information between the American government and the church. He became a trusted confidant of the pontiff, so much so that other diplomats envied Taylor's access. The war prevented Taylor from reaching Vatican City from September 1942 to June 1944, so he served on several State Department advisory boards on postwar foreign and economic policy. Taylor returned to the Vatican and, despite renewed criticism from American Protestants, was Truman's personal representative until 1949. He also conducted several special missions for Truman through 1953.

Taylor was reserved, even aloof, but always involved in humanitarian and civic activities. In addition to performing relief work during the 1930s, he founded the American Relief Committee for Italy and the Italian National Committee for the Distribution of Relief in 1944. He reorganized the Italian Red Cross in 1946 and later served on the board of corporators of the American Red Cross. Taylor was a devoted alumnus of Cornell University, donating some $4.5 million, including the Myron Taylor Law School building in 1928. Taylor also was a trustee of Cornell, Wells College, Roberts College in Istanbul, the New York Public Library, and the Association for Improving the Condition of the Poor. He gave $1 million to the Episcopal church and turned his home in Lyons, New York, into a community center. He also served on the executive committee of Saint Luke's Hospital in New York. He died on May 6, 1958.

Publications:
The Iron and Steel Industry (New York: Columbia University Press, 1928);
"Machine Exists for Man — He Can and Does Control It," *Iron Age,* 127 (January 29, 1931);
"Steel Corporation's Employment Plan," *Review of Reviews,* 83 (March 1931): 61–62;
The Importance of Planning a Career (Geneva, N.Y.: Hobart College, 1931);
"Man and the Machine," *Review of Reviews,* 85 (January 1932): 37–38;
"Hold Fast to Established Principles," *Iron Age,* 129 (March 31, 1932): 788–790;
The Adventure of Living (New York, 1932);
"Key to Hidden Mysteries," *Golden Book Magazine,* 19 (January 1934): 21;
"Employment Record Is Bright Picture in Corporation's History," *Iron Age,* 133 (April 19, 1934): 33;
"Steel Gaining Impetus," *Iron Age,* 135 (April 4, 1935): 54;

"Carnegie and Illinois Steel Consolidation Plans Outlined," *Iron Age,* 136 (June 26, 1935): 150;

"Competition from Imports of Foreign Steel Products," *Iron Age,* 137 (June 4, 1936): 28–29;

"Year End Statement," *Blast Furnace and Steel Plant,* 26 (January 1938): 71, 117;

"Ten Years of Steel: How the U.S. Steel Corporation Earns Its Living and Distributes Its Income," *Iron Age,* 141 (April 7, 1938): 70; (April 14, 1938): 50–53, 92–94; (April 28, 1938): 36–38; (May 5, 1938): 47, 73–75; reprinted as *Ten Years of Steel* (Hoboken, N.J., 1938);

"Importance of the Refugee Problem," *Vital Speeches,* 5 (December 15, 1938): 157–158;

The Lawyer's Opportunity: Address Delivered to the Graduating Class of the Cornell Law School, May 29, 1941 (Ithaca: Cornell University, 1941);

Correspondence Between President Truman and Pope Pius XII, edited by Taylor (New York: Macmillan, 1947).

References:

Irving Bernstein, *The Turbulent Years: A History of the American Worker, 1933–1941* (Boston: Houghton Mifflin, 1970);

John S. Conway, "Myron C. Taylor's Mission to the Vatican," *Church History,* 44 (March 1975): 85–99;

"The Corporation," *Fortune,* 13 (March 1936): 59–67; (April 1936): 127–132, 134, 136; (May 1936): 93–97; (June 1936): 113–116;

John Tracy Ellis and John L. McMahon, "Our Envoy to the Vatican," *Catholic World,* 151 (August 1940): 573–581;

For Human Needs: The Story of Ford, Bacon and Davis (New York: Ford, Bacon and Davis, 1967);

"It Happened in Steel," Fortune, 15 (May 1937): 91+;

"Myron Charles Taylor," *Fortune,* 13 (June 1936): 117–120;

Paul Tiffany, *The Decline of American Steel: How Management, Labor, and Government Went Wrong* (New York: Oxford University Press, 1988);

"The Vatican Embassy Fraud," *Christian Century,* 63 (April 3, 1946): 422– 424.

Temporary National Economic Committee
by Paul Tiffany

Sonoma State University

In April 1938 President Franklin D. Roosevelt sent to Congress a message denouncing a "concealed cartel system" and the "disappearance of price competition" in the American economy. He appealed for the legislators to undertake a "thorough study of the concentration of economic power." Thus was launched the Temporary National Economic Committee (TNEC) and with it the triumph of those among Roosevelt's advisers who believed that "big business" was responsible for much of the economic agony of the 1930s through its usage of "administered prices" as a means to gouge profits from a helpless public.

Congressional approval of the TNEC in June 1938 called for six members of Congress and six representatives from different cabinet-level departments of the administration to make up the committee. Its chairman was Senator Joseph O'Mahoney, a populist Democrat from Wyoming who was highly suspicious of concentrated industry and big business in general. The committee slated as its first target of investigation the steel

industry, for it was not only large and oligopolistic in structure but also rigid in its pricing practices: the United States Steel Corporation usually established a price for a product, which the other producers quickly adopted. Yet, while pricing policy was the main focus of its investigations, the committee also looked into antitrust procedures, the merger movement, financial controls, trade association activities, patent laws, tax revisions, and the potential of establishing a permanent Bureau of Industrial Economics within government to better monitor private firm behavior.

Increasingly frustrated with the failure of any of Roosevelt's economic policies to cure the problems of the Great Depression, many New Dealers turned to antimonopoly strategies as a means to bring social improvement. In supporting the TNEC, they saw it as a vehicle to better enforce the antitrust laws and, with specific regard to the steel industry, initiate the breakup of the larger firms that dominated production. No recommendations of substance, however, came of the

TNEC studies. The committee itself was divided over just what the mission of the TNEC should be and, as a consequence, never developed a strategy of action. By early 1939 the domestic economy began to improve as contracts for war goods from Europe began to spill over into the United States, thus tempering the search for a solution to the problems of the Depression. As a result, the final report of the TNEC in 1941 made only a few minor recommendations for change in the anti-monopoly laws and in patent procedures, and most of these were never implemented.

The investigations of the TNEC had little effect on the steelmakers in the short term. They, however, furthered the suspicion and doubt that clouded nearly all of the interactions between industry and government. By World War II a culture of adversarialism between steel and the state had begun to establish itself, making it very difficult in later years for these two parties to find common ground when faced with the threat of encroaching foreign steel industries.

References:
Ellis W. Hawley, *The New Deal and the Problem of Monopoly* (Princeton: Princeton University Press, 1966);
Temporary National Economic Committee, 76th Congress, 2nd Session, *Investigation of Concentration of Economic Power*, Hearings (Washington, D.C.: Government Printing Office, 1940).

Michael Tenenbaum

(July 23, 1914 –　)

by Robert M. Aduddell

Loyola University of Chicago

CAREER: Metallurgist (1940–1956), superintendent, department of metallurgy (1956–1959), assistant general manager of technical services (1959–1966), vice-president of research (1966–1968), vice-president, steel manufacturing (1968–1971), president and chief operating officer, (1971–1978), director, Inland Steel Company (1971–1984).

Michael Tenenbaum was born to Harry and Ida Vivian (Kolohoski) Tenenbaum on July 23, 1914, in Saint Paul, Minnesota. He graduated from the University of Minnesota with a degree in metallurgical engineering in 1936 and returned for graduate study, earning a master's degree in 1937 and a Ph.D. in 1940.

After receiving his Ph.D. Tenenbaum joined the Inland Steel Company in Chicago as a metallurgist. In 1956 he was named superintendent of the department of metallurgy, and in 1959 he became the assistant general manager in charge of technical services. Tenenbaum became vice-president of research in 1966 and vice-president of steel manufacturing in 1968. In 1971 he became president and chief operating officer when Frederick Jaicks replaced Philip D. Block, Jr., as chairman of the board. With his promotion came his election as director of Inland Steel.

Tenenbaum's technical background made him well suited to oversee Inland's expanding research program. At the time Tenenbaum became president much of Inland's research focused on developing a corrosion-resistant steel. Progress was also made in improving sheets for the automobile industry and developing a quality machinable steel for cold extrusion. With Tenenbaum actively involved in the company's research, Inland increased its reliance on the basic oxygen furnace (BOF) and continuous casting — two new processes that, in improving the quality of product, made steelmaking more cost-effective.

Under Tenenbaum, Inland continued to upgrade its facilities. In 1973 the company began construction of its second BOF shop, which was expected to produce 2.2 million tons of steel per year. A second taconite ore mine and pelletizing facility was in the early phases of construction at the Monorca Ore property in Minnesota. Efforts to diversify Inland's product line that had begun under Jaicks continued under Tenenbaum. The Inland Steel Urban Development Corporation created the Inland Steel Finance Company in 1972 to help finance real estate construction. In addition, it opened a second plant for the manufacture of mobile homes. Inland, however, remained primarily a steel firm. Only 9 percent of its sales came from nonsteel production in 1973.

During Tenenbaum's tenure, raw steel production at Inland grew from 6.5 million to 8.6 million tons. In September 1974 Inland announced plans to increase production substantially by adding a new blast furnace, coke battery, and 160-inch plate mill and by upgrading the original BOF shop. These renovations would bring Inland's annual production to 10 million tons. Total assets rose from $1.38 billion in 1971 to $2.60 billion in 1978. Net additions were made in each year of Tenenbaum's presidency.

Even though net sales rose from $1.25 billion in 1971 to $3.25 billion in 1978, Inland's capital expansion created a demand for funds that could not be satisfied by internal sources. Thus, long-term debt increased from $327 million in 1971 to $652 million in 1978. The ratio of net income to sales rose above 6 percent only once during Tenenbaum's tenure; the average for the period was 4.3 percent, well below the company's historic average. Even so, Inland performed better than the average American steel firm during these difficult years, when U.S. consumption of steel per dollar of gross national product was declining.

Tenenbaum was a fellow of the American Institute of Mining, Metallurgical, and Petroleum Engineers, serving on its board of directors from 1964 to 1970 and as its vice-president in 1969. He was the U.S. representative to the International Iron and Steel Institute's Committee on Technology in 1970–1971. He served as an officer of both the Indiana Chamber of Commerce and the Chicago Association of Commerce and Industry, and he also served as a director of the Continental Illinois Corporation.

In 1978, at the age of sixty-four, he was succeeded as president by Frank Luerssen. He re-mained on the board of directors until he retired from Inland in 1984. He, however, continues to act as a consultant to Inland and other steel firms.

Publications:

"Anatomy of a Steel Shortage," *Iron and Steel Engineer,* 52 (January 1975): 45–47;

and T. L. Joseph, "The Iron Content and the Angle of Polarization of Chilled Surfaces of Open Hearth Slag," *Blast Furnace and Steel Plant,* 28 (December 1940): 1157–1159;

and Joseph, "Reduction of Iron Ores under Pressure by Carbon Monoxide," *American Institute of Mining and Metallurgical Engineers Transactions,* 140 (1940): 106–125;

and Joseph, "The Use of the Reflecting Microscope in the Examination of Open Hearth Slag," *Blast Furnace and Steel Plant,* 29 (April 1941): 403–407; (May 1941): 522–523, 545, 551;

and C. C. Brown, "Application of pH Slag-Basicity Measurements to Basic Open-Hearth Phosphorous Control," *Transactions of the American Institute of Mining Engineers,* 162 (1945): 60–72;

"Iron and Steel Process Metallurgy, 1944," *Mining and Metallurgy,* 26 (February 1945): 82–86;

"Structure, Segregation and Solidification of Semi-Killed Steel Ingots," *Iron Age,* 160 (October 30, 1947): 59–60;

"Effect of Sulphur on Quality and End Uses of Steel Products," *American Iron and Steel Institute Yearbook* (1949): 322–360;

and C. M. Squarcy, "Direct Reduction of Iron Ores Using Fluidized Solids Techniques," *American Iron and Steel Institute Yearbook* (1951): 208–240;

and C. B. Jacobs and J. F. Elliott, "Significance of Minor Elements in Iron Bearing Materials for Integrated Steel Plants," *American Iron and Steel Institute Yearbook* (1954): 123–150.

Reference:

William T. Hogan, S.J., *An Economic History of the Iron and Steel Industry in the United States,* 5 volumes (Lexington, Mass.: Lexington Books, 1971).

Tennessee Coal, Iron & Railroad Company

by Elizabeth C. Sholes

Industrial Research Associates

and

Thomas E. Leary

Industrial Research Associates

The history of the Tennessee Coal, Iron & Railroad Company (TCI & RR) dates back to 1852, when Nashville entrepreneurs established the Sewanee Mining Company. Facing bankruptcy in 1859, the Sewanee Mining Company was reorganized by New York investors as TCI & RR. By the end of the century northern financiers had gained a good deal of control over what was the South's largest producer of coal, iron, and steel, and in November 1907 TCI & RR was absorbed by the United States Steel Corporation. The local reaction to the takeover was generally favorable — many believing that the company's Ensley and Fairfield plants in and around Birmingham, Alabama, and in South Pittsburgh, Tennessee, might undergo expansion under U.S. Steel's leadership. The unclear reasons behind U.S. Steel's acquisition bid, however, made the takeover highly controversial, and it would later be a major subject in the antitrust action brought against the corporation in 1912.

The acquisition greatly enhanced U.S. Steel's resources in raw materials, for TCI & RR had 352,548 acres of coal- and ore-rich properties in Alabama and Tennessee. The takeover, however, was also seen as a way for the corporation to squelch a competitor which was selling rails to western railroads at prices that were lower than the ones dictated by industry leader U.S. Steel. Judge Elbert H. Gary, head of U.S. Steel, denied that the move had been made to reassert market control and argued that the corporation had acted to prevent the collapse of a leading investment banking house that had been caught up in the panic of 1907 and was looking to unload its TCI & RR stock. Initially reluctant to buy the company and risk drawing the attention of antitrust legislators, Gary agreed to the purchase only after he and Henry Clay Frick received President Theo-

dore Roosevelt's and Attorney General Charles J. Bonaparte's approval. Despite U.S. Steel's claims, for the next three decades the corporation directed business to its northern operations and away from TCI. U.S. Steel further undermined TCI's competitive stance by bringing its rail pricing in line with that of the northern-based industry.

Nevertheless, there was expansion during this time. In 1917 wartime demand led to the addition of structural and plate mills, and in 1924 TCI added a sheet mill. In 1936 U.S. Steel inaugurated expansion plans that included two coke batteries, a blooming mill, a continuous strip mill, two continuous cold-rolling mills, and tinning operations. TCI thus became one of the most fully integrated operations in the corporation's network.

World War II generated further expansion, including a third Fairfield blast furnace and a 140-inch plate mill. From 1946 through the early 1950s TCI built three open hearths and enlarged older furnaces. During the 1960s TCI added improvements at Fairfield, including a prepainted sheet mill and a 62-inch continuous hot dip twin galvanizing line. They upgraded the hot strip mill and, during the 1980s downturn, added a computerized seamless pipe mill and installed a continuous slab caster. Ensley received a twin process rail mill in 1970.

Changes were also made that directly affected labor. With works centered in largely rural areas and with the corporation still stinging from the criticisms raised by the antitrust investigations of the Stanley Commission, U.S. Steel quickly moved to extend housing, medical, and educational facilities designed for both black and white employees.

Economic difficulties and revised corporate priorities killed steelmaking at Ensley by 1976 and constricted secondary operations at Fairfield during 1980–1982. Labor-management relations were characterized by bitter antagonism, and along with Gary, Fairfield was the corporation's biggest loser at this time. Nevertheless, even as northern operations were closed outright, TCI successfully made the transition into the new era of steelmaking. Fairfield has continued operating within the restructured steel division of USX Corporation to the present day.

References:

William T. Hogan, S.J., *An Economic History of the Iron and Steel Industry in the United States*, 5 volumes (Lexington, Mass.: Lexington Books, 1971);

Charles Longenecker, "The Tennessee Coal, Iron and Railroad Company," *Blast Furnace and Steel Plant*, 28 (August 1939): 791–794;

Lloyd Noland, "Tennessee Coal, Iron & Railroad Company," *American Iron and Steel Institute Yearbook* (1915): 257–275;

"Tennessee Coal, Iron and Railroad Company, Birmingham, Ala.," *Monthly Bulletin of the American Iron and Steel Institute*, 5 (March–April 1917): 63–69;

Kenneth Warren, *The American Steel Industry, 1850–1970* (Oxford: Clarendon Press, 1973).

Alexis W. Thompson

(November 30, 1850 – February 8, 1923)

by Carol Poh Miller

Cleveland, Ohio

CAREER: Secretary and treasurer, Etna Iron Works, Ltd. (1875–1893); vice-president, Atlantic Iron & Steel Company (1893–1899); vice-president (1899–1901), president, Republic Iron & Steel Company (1901–1905); chairman of the board (1906–1919), president, Inland Steel Company (1908–1919).

Alexis Wellington Thompson was born on November 30, 1850, to Marion J. (Webster) and David Thompson in Mercer County, Pennsylvania. After receiving his education in the public schools, he found employment with the Etna Iron Works, Ltd., of New Castle, Pennsylvania. He served as the company's secretary and treasurer from 1875 to 1893. He then served as vice-president of the Atlantic Iron & Steel Company until 1899, when he became vice-president of the newly organized Republic Iron & Steel Company. Thompson became president of Republic in 1901 but was ousted four years later when an investors' group led by John Warne Gates — who regarded Thompson as "unimaginative" and a "patient plodder" with a good if dull record of accomplishments — seized control of the company.

In 1906 Thompson became the first chairman of the board of the Inland Steel Company of Chicago. In 1908 he became president. During the next decade Thompson oversaw an ambitious pro-

Alexis W. Thompson

gram of acquisitions and expansion. Acquisitions included new ore mines in Minnesota — one of which was named after Thompson — and a vast coal reserve in the Freeport vein near Pittsburgh, while new construction focused on Inland's Indiana Harbor plant. By 1910 Inland had increased its tonnage sales 100 percent over 1905 levels. Thompson resigned as president in 1919 but remained a director of the company until his death in 1923. Thompson was vice-president and a director of the Wilmington & Springfield Coal Company and also served as a director of the Sharpsville Furnace Company, United Iron & Steel Company, Benson Mines Company, Annie Laurie Mining Company, and the Continental Bank of Chicago.

In 1877 Thompson was married to Eva Pearson of New Castle, Pennsylvania. They had two sons: David — who joined his father in the steel industry, first at Republic Iron & Steel and later at Inland Steel — and George. Alexis Thompson died at Battle Creek, Michigan, on February 8, 1923.

References:

William T. Hogan, S.J., *An Economic History of the Iron and Steel Industry in the United States,* 5 volumes (Lexington, Mass.: Lexington Books, 1971);

Lloyd Wendt and Herman Kogan, *Bet a Million! The Story of John W. Gates* (Indianapolis & New York: Bobbs-Merrill, 1948), pp. 258–259, 269–270.

John A. Topping

(June 10, 1860 – August 24, 1934)

by Carol Poh Miller

Cleveland, Ohio

CAREER: Payroll clerk (1878), president, Aetna Iron & Nail Company (1898–1900); first vice-president, American Sheet Steel Company (1900–1903); chief executive, La Belle Iron Works (1903–1904); president, American Sheet & Tin Plate Company (1904–1906); chairman (1906–1907), president, Tennessee Coal, Iron & Railroad Company (1906–1907); chairman, Republic Iron & Steel Company (1908–1930).

John Alexander Topping spent more than half a century in the iron and steel industry, capping his career as chairman of the Republic Iron & Steel Company, a post he held from 1908 until his retirement in 1930. He was born in Saint Clairsville, Ohio, the son of Henry and Mary Tallman Topping, and was educated in the public schools of Kansas City, Missouri. After working briefly as a bank clerk, he entered the iron and steel business in 1878 as a payroll clerk with the Aetna Iron & Nail Company; he became the company's president in 1898. The company later merged with another to become the Aetna-Standard Iron & Steel Company. Following still another merger, it became part of the American Sheet Steel Company, a holding company with 27 plants in Pennsylvania, Ohio, Indi-

ana, and Kansas. Topping became first vice-president of the company, but resigned in 1903 to reorganize the La Belle Iron Works in Wheeling, West Virginia. A year later he became president of the American Sheet & Tin Plate Company, a House of Morgan subsidiary.

Steel financier John W. Gates had become interested in Republic Iron & Steel Company, an amalgamation of 32 small iron and steel firms scattered across the Midwest and South, following its organization in 1899. He began buying shares in Republic, as well as in a southern steelmaking firm, the Tennessee Coal, Iron & Railroad Company (TCI). Gates planned to unite the two firms to form a southern steel empire. He succeeded in taking control of Republic and, early in 1906, tapped John Topping — a "glum, sad-eyed man," according to Gates's biographers — as president of Republic and chairman of TCI. Gates's dream of an empire, however, was thwarted by the panic of 1907, when ownership of TCI passed to the United States Steel Corporation. In 1908 Topping was named chairman of Republic. He served in that capacity from the firm's executive offices in New York City until he retired from active business in 1930, only months before the formation of the new Republic Steel Corporation.

Topping was a founding director and vice-president of the American Iron and Steel Institute. He was married in 1883 to Minnie C. Junkins of Bridgeport, Ohio. Following her death, he was married a second time, to Louise Johnston Manning, in 1914. He died of pneumonia at Belle Haven, his home in Greenwich, Connecticut, on August 24, 1934.

Publications:
"Competition: Its Uses and Abuses," *American Iron and Steel Institute Yearbook* (1912): 44–46;

"Four Prerequisites for Home Production," *Iron Trade Review,* 56 (January 7, 1915): 21–22;

"Not Dependent upon War," *Iron Trade Review,* 59 (December 28, 1916): 1284–1285;

"How Industrial Leaders Face the War: Iron and Steel," *World's Work,* 34 (May 1917): 25–26;

"Cooperation and Mobilization of Public Sentiment," *American Iron and Steel Institute Yearbook* (1917): 197–203;

"World's Steel Output Below Need," *Iron Trade Review,* 67 (May 6, 1920): 1321–1322;

"Topping Urges Cut in Upper Lake Rates," *Iron Trade Review,* 69 (December 15, 1921): 1542–1544;

"Is American Steel Capacity Over-rated?," *Iron Trade Review,* 77 (August 6, 1925): 301–302;

"Prosperous Year in Steel: The Outlook for 1926," *Annalist,* 27 (January 8, 1926): 63.

References:
William T. Hogan, S.J., *An Economic History of the Iron and Steel Industry in the United States,* 5 volumes (Lexington, Mass.: Lexington Books, 1971);

"Steel Head Discusses Outlook for 1927: Interview with J. A. Topping," *Iron Age,* 119 (January 13, 1927): 139;

Lloyd Wendt and Herman Kogan, *Bet a Million! The Story of John W. Gates* (Indianapolis & New York: Bobbs-Merrill, 1948).

Walter Sheldon Tower

(July 26, 1881 – February 5, 1969)

by Bruce E. Seely

Michigan Technological University

CAREER: Professor, Wharton School, University of Pennsylvania (1906–1911); professor, University of Chicago (1911–1917); trade expert, U.S. Shipping Board (1918–1919); special adviser, Paris Peace Conference (1919); trade adviser, Consolidated Steel Corporation (1919–1921); commercial attaché, U.S. Embassy, London (1921–1924); Bethlehem Steel Corporation (1924–1933); executive secretary (1933–1940), president, American Iron and Steel Institute (1940–1952).

Walter S. Tower was not an orthodox steel man, for few leaders of this industry had experience as a college teacher, a government adviser, diplomat, and an expert on foreign trade. Not until 1919, at the age of thirty-eight, did he first join a steel company, and not until 1924 did he settle more permanently at Bethlehem Steel Corporation. Tower's major contribution to the industry, however, came during his 20 years as executive secretary and president of the American Iron and Steel Institute (AISI). Through the New Deal and World War II and into the early 1950s —

tumultuous years for the steel industry — he often was the steel industry's spokesperson, while building the AISI into an important trade association.

Born in West Bridgewater, Massachusetts, in 1881, Tower earned his undergraduate degree from Harvard in 1903, a master's degree from the same school a year later, and his Ph.D. from the University of Pennsylvania in 1906. His field was economic and commercial geography, and he taught these subjects at the University of Pennsylvania's Wharton School from 1906 to 1911. In 1911 Tower moved to the University of Chicago to serve as professor of economic geography. With the American entry into World War I, Tower joined the U.S. Shipping Board as a trade expert from 1918 to 1919, when he was made a special adviser to the Paris Peace Conference. At war's end Tower had his first contact with the steel industry, as trade adviser for Consolidated Steel Corporation in 1919. The firm was organized by several steel companies to expand foreign markets. Until this point only the United States

Walter Sheldon Tower (right) with U.S. Steel vice-president Max D. Howell (left) and American Iron and Steel Institute secretary George Rose (center) at the 1952 AISI banquet, where Tower received the Gary Medal

Steel Corporation had shown serious interest in overseas sales, but Bethlehem Steel, Inland, Republic, and Youngstown Sheet and Tube wanted to share the costs of establishing sales offices outside North America. This cooperative approach reflected both the style of doing business within the industry and the relatively amicable relationship between the industry and government that was characteristic of the 1920s, and it hinged on President Woodrow Wilson's acceptance of steel's collaborative efforts in price-setting and in opening overseas markets. But the steel firms' overseas interests waned, especially after investment bankers turned their attentions to rebuilding European steel mills. Plans for U.S.-owned mills in Europe were shelved, and efforts such as Consolidated Steel were soon abandoned.

In 1921 Tower left Consolidated Steel and spent the next three years in London as commercial attaché at the U.S. Embassy. Tower returned to the steel industry in 1924 and began to help Bethlehem Steel pursue markets outside the United States.

Then, in 1933, Tower joined the AISI. The *New York Times* reported later that the move seemed unexpected: "Never an engineer, metallurgist, financier, or salesman, Dr. Tower, a former college professor, stepped in 1933 into a $100,000-a-year job as executive secretary of the Institute." But there were reasons for Tower's appointment. Judge Elbert H. Gary of U.S. Steel had organized the AISI in 1909, and through the 1920s the institute remained Gary's personal instrument for promoting stability — both as a forum for communication and as a mechanism for disseminating technical information — in the steel industry. When Gary died in 1927, Charles Schwab of Bethlehem Steel assumed leadership of the AISI, and this connection helped account for Tower's appointment. The institute was not a strong trade association, but as the administration of Franklin D. Roosevelt unveiled the New Deal, steelmen such as Schwab began to see a larger role for the AISI. Eugene Grace, second in command behind Schwab at Bethlehem Steel, apparently was a central figure in convincing Tower to take the job at AISI and represent the industry in government circles.

Tower found that job a very interesting one. Steel executives, like most business leaders, were wary of President Roosevelt and his New Deal, but their cautious acceptance of the administration's policies was a mark of desperation. The AISI and the New Deal forces, on the other hand, joined other trade associations in tackling many issues, but especially the development of the National Recovery Administration (NRA) code for steel, which established prices and wages in an effort to restore economic stability and prosperity. Indeed, Grace noted that this chore was the main reason he asked Tower to move to the AISI. But Tower also used the opportunity to expand the AISI's statistical collection efforts, first launched in 1921, and generally to strengthen the institute. One of Tower's first moves was to develop a public relations program, a step that was especially important as the industry and the Roosevelt administration began to lock horns on a variety of issues. In 1934 Tower launched the institute's magazine, *Steel Facts,* and this effort at disseminating the industry's viewpoints was soon supplemented by press releases on an array of subjects and controversies.

As a spokesperson for the steel industry, Tower lectured and spoke widely, and he published articles describing the state of the industry and its position on many fronts. Increasingly, Tower's pronouncements reflected the growing hostility of steel leaders intent on challenging New Deal labor policies, although Tower observed in 1935 that steel firms and the industry as a whole had benefited from the administration's policies. But relations between the steelmakers and Roosevelt continued to deteriorate, and Tower emerged as a leading opponent of government regulation. In appearance before Congress, Tower adamantly, but without the rancor and open hostility that characterized so many pronouncements by steel executives, repeated the industry's views on labor and regulatory policy.

With the onset of the war in Europe, the issue of preparedness moved to the forefront. Tower's concerns changed as well. In 1940 he became the president of the AISI, a promotion that reflected the respect he had won for his efforts made in defending steel's position. But now his main concern was responding to charges that steel firms were not moving quickly enough to increase production. By 1944 he began to address industry concerns about the postwar world, even as the institute oversaw efforts to develop special wartime steels that conserved scarce alloying materials. The attention to the postwar situation was evident in the inauguration of a general-information public-relations magazine, *Steelways,* in 1944. The magazine regularly called for restraining government interference in industry and limiting the gains of organized labor.

By the end of the war the AISI was one of the leading trade associations in the country. The strengthened statistical, public-relations, and technical programs were firmly in place. Moreover, the industry was more firmly united than at any time since U.S. Steel had splintered the industry by signing a labor contract with the United Steelworkers in 1936. Tower continued in his role as spokesperson for the views of steelmakers on such issues as production capabilities and expansion, and government regulation. He opposed the government's desire to increase capacity rapidly, but in the end his cautious position was given little attention.

Upon his retirement in 1952 Tower became the eleventh recipient of the AISI's Gary Medal, its most prestigious honor. Tower remained an occasional adviser and consultant to the institute during the early 1950s but moved to a home in Carmel, California. He married twice — first in 1906 to Lurena Wilson, who died in 1917, and again in 1919, to Edith Jones. He had two sons with his first wife and another with his second. Tower died on February 5, 1969.

Selected Publications:
"Labor Chief Beneficiary of Steel Code," *Iron Age,* 133 (January 25, 1934): 36;

"Steel Code Had Given Strong Impetus to National Recovery," *Iron Age,* 133 (May 31, 1934): 23–26;

"The Operation of the Steel Code," *American Iron and Steel Institute Yearbook* (1934): 44–57;

"Report on Institute Activities," *American Iron and Steel Yearbook* (1935): 67–78;

"Review of Code and Institute Activities in the Past Year," *Iron Age,* 135 (May 30, 1935): 29–32;

"Report on Institute Activities," *American Iron and Steel Yearbook* (1936): 83–91;

"What Is the Institute and Why?," *American Iron and Steel Institute Yearbook* (1937): 64–72;

"Steel Defended against Its Critics," *Iron Age,* 141 (January 27, 1938); 64–66;

"Activities of the American Iron and Steel Institute," *American Iron and Steel Institute Yearbook* (1939): 62–76;

"Steel Industry's Preparedness," *Steel,* 107 (October 14, 1940): 75–76;

"Address of the President," *American Iron and Steel Institute Yearbook* (1941): 28–42;

"American Steel Industry Has Capacity Adequate to Meet Demands of War," *Blast Furnace and Steel Plant,* 29 (June 1941): 603–605;

"Speech to American Iron and Steel Institute," *Blast Furnace and Steel Plant,* 30 (June 1942): 636–638;

"Address of the President," *American Iron and Steel Institute Yearbook* (1943): 29–41;

"Address of the President," *American Iron and Steel Institute Yearbook* (1944): 29–36;

"Steel Faces the Post War Years," *American Iron and Steel Institute Yearbook* (1945): 11–19;

"Address of the President," *American Iron and Steel Institute Yearbook* (1946): 31–39;

"Address of the President," *American Iron and Steel Institute Yearbook* (1947): 628–637;

"Steel's Problems and Prospects," *American Iron and Steel Institute Yearbook* (1948): 67–76;

"Steel in an Abnormal Year," *American Iron and Steel Institute Yearbook* (1949): 67–75;

"Fifty Years of Steel," *American Iron and Steel Institute Yearbook* (1950): 25–33;

"Steel and National Defense," *American Iron and Steel Institute Yearbook* (1951): 13–21;

"Subjugation or Liberation," *American Iron and Steel Institute Yearbook* (1952): 15–23;

"Steel Industry; Where Is It Headed?" *Commercial and Financial Chronicle*, 175 (May 29, 1952): 2248–2249.

References:

"Award of Medals," *Yearbook of the American Iron and Steel Institute* (1952): 59–63;

Max D. Howell, "The Institute's First Fifty Years — and a Look Ahead," *Yearbook of the American Iron and Steel Institute* (1958);

"Tower Receives Gary Memorial Medal," *Iron Age*, 130 (June 2, 1952): 65.

Archives:

Copies of many of Tower's speeches and published papers can be found in the W. S. Tower File, Speeches by Steel Men, American Iron and Steel Institute Collection, Hagley Library, Greenville, Delaware.

Robert Carroll Tyson

(August 13, 1905 – January 2, 1974)

by Dean Herrin

National Park Service

CAREER: Accountant, Remington Rand, Incorporated (1928); junior accountant, later senior supervising accountant, Price, Waterhouse and Company (1929–1939); assistant audit supervisor (1939–1941), general accountant (1941–1947), assistant comptroller (1947–1950), comptroller (1950–1955), vice-president (1951–1952), vice-chairman, finance committee (1952–1956), director (1952–1970), chairman, finance committee, United States Steel Corporation (1956–1970).

Robert Tyson served as chairman of the United States Steel Corporation's finance committee from 1956 to 1970. As chief financial officer for the corporation, he was one of the triumvirate — which also included the president and the chairman of the board — that established policies for running U.S. Steel.

Tyson was born on August 13, 1905, in Thurmont, Maryland. After graduating from Princeton University in 1927, he began his career as an accountant with Remington Rand, Incorporated. The following year he joined Price, Waterhouse and Company as a junior accountant. In 1937 he was placed in charge of the U.S. Steel audit, and two years later he joined the steel company as an assistant audit supervisor. His impact on the corporation was immediately felt: he or-

Robert Carroll Tyson

ganized an auditing department and helped create a new central accounting department. He was promoted to general accountant in 1941, assistant comptroller in 1947, and comptroller in 1950. He was made a company vice-president in 1951. The following year he was elected to the board of directors and was promoted to vice-president of the finance committee. In 1956 he was given leadership of the finance committee, in effect placing him in a position of shared responsibility with the president and chairman of the board.

As part of U.S. Steel's leadership, Tyson was one of the architects of various programs designed to restore the company's declining profit margin during the 1960s. Profits had fallen due to foreign steel imports, outdated mills and equipment, increased domestic competition, and rising costs of production. The restoration programs, however, were successful in raising U.S. Steel's annual capital budget from $200 million in 1962 to $350 million in 1965 — and to more than $600 million a year later. A $1.8-billion capital expenditure campaign was also started in 1966 to modernize aging plants.

In 1964–1965 Tyson participated in the reorganization of U.S. Steel's hierarchy, placing wholly owned subsidiaries under the direct managerial control of the parent company. He was also a major player in the reincorporation of U.S. Steel in 1966, in which the parent company, a corporation chartered in New Jersey, merged into its Delaware subsidiary, allowing U.S. Steel to benefit financially from an exchange of preferred stock.

While most steel industry officials in the 1960s were beginning to focus on foreign steel imports as a serious problem, Tyson maintained that the industrywide decline in profit margins was mostly due to labor's demand for wage increases without a corresponding increase in productivity. Even without the threat of foreign steel, reasoned Tyson, wages and other rising costs would still plague steel producers.

When Tyson retired in 1970, a reporter for the *New York Times* described him as "completely and cheerfully aware that he has been regarded as a corporate tightwad, even by fellow officers of the company." Tyson kept a close watch on the company's liquid assets, such as cash and marketable securities, and was more reluctant than financial officers of other companies to spend a large portion of these assets on capital expansion.

In addition to his duties with U.S. Steel, Tyson was a director of Uniroyal and the Chemical New York Corporation, a member of the board of trustees of the Tax Foundation, a member of the advisory board of the Hoover Institution in Stanford, California, and a national director of Boys' Clubs of America. He was appointed in 1971 as head of the American delegation to the world administrative radio conference for space telecommunications. Tyson died in Norfolk, Virginia, on January 2, 1974.

Publications:
"The Cost Facts of U.S. Steel," in *Steel and Inflation: Fact vs. Fiction* (New York: Public Relations Department, United States Steel Corporation, 1958), pp. 27–41;

The Private Impact of Public Spending (New York: Public Relations Department, United States Steel Corporation, 1964);

Freedom and Enterprise (New York: Public Relations Department, United States Steel Corporation, 1965);

"Views Competition," *Iron Age*, 198 (September 22, 1966): 26–27;

Federalism or Federalization (New York: Public Relations Department, United States Steel Corporation, 1967).

References:
John P. Hoerr, *And the Wolf Finally Came: The Decline of the American Steel Industry* (Pittsburgh: University of Pittsburgh Press, 1988);
William T. Hogan, S.J., *An Economic History of the Iron and Steel Industry in the United States,* 5 volumes (Lexington, Mass.: Lexington Books, 1971);
New York Times, July 30, 1970, p. 43.

John Butler Tytus

(December 6, 1875 – June 2, 1944)

by Carl Becker

Wright State University

CAREER: Superintendent, Zanesville Mill (1906–1910), superintendent, Sheet Mill Department, Middletown Division (1910–1914), assistant general superintendent, Middletown Division (1914–1927), vice-president, process developments (1927–1938), vice-president, operations (1938–1943), vice-president, technical development, American Rolling Mill Company (1943–1944).

John Butler Tytus, the leading figure in the development of the continuous hot-strip rolling mill, came to prominence in steelmaking through paper manufacturing.

His grandfather Francis Jefferson Tytus in 1827 moved from Hampshire County, Virginia, to Middletown, Ohio, then a village of a few hundred people. He took employment as a clerk in a dry-goods store and soon became a partner in the business. Late in the 1830s he also became the proprietor of a pork-packing firm. In 1854 he became a partner in the operation of two paper mills, and in 1873 he had his own paper company, incorporated as the Tytus Paper Company; his son, John Butler Tytus, was its secretary. The secretary's son, John junior, often visited the mill to see paper made and began to assist in the repair of machinery.

Young Tytus attended the local public school and at the age of fourteen entered the Westminister preparatory school at Dobbs Ferry, New York. He attended Yale, taking a bachelor's degree in English literature in 1897. While in college he had worked in the family's mill during his summer vacations, and at graduation accepted a rather menial position in the shipping department. In the late 1890s, after his father's death, the family sold the company, and Tytus went to work for a contractor who was constructing a bridge in Dayton, Ohio. Placed in charge of a labor gang, Tytus displayed his ability to deal with engineering problems. At the completion of construction he found employment briefly in a photographic advertising firm in Dayton.

His interest in steelmaking evidently whetted, Tytus returned to Middletown in 1904 seeking employment at American Rolling Mill Company (Armco), which in 1901 had begun production of black and galvanized sheet steel. The founder, George Verity, knowing that Tytus was a college man, sent him to the superintendent, Charles Hook, assuming that Hook would convince the young man that a steel mill was not a proper place for a man with a liberal arts education. But Hook heard out the persuasive Yale man and obliged him, hiring him as a "spare hand" to drag steel sheets to the shears and performing other toilsome tasks. While observing workers passing and repassing bars through stands of manual mills, Tytus remembered the Fourdrinier machine that he had seen in the Tytus mill rolling out a continuous sheet of paper and speculated on the prospect of "making sheets in long strips like they make paper." He knew, of course, of steel men who had attempted to effect such production — from John Hazelden late in the eighteenth century to engineers at United States Steel Corporation early in the twentieth century — and had failed.

As he contemplated a continuous mill, Tytus was rising to new positions at Armco. In 1906 he became superintendent at a mill in Zanesville, Ohio, recently purchased by the company. He returned to Middletown in 1910 as superintendent of the Sheet Mill Division and as the engineer building a new works — the East Works — in the city. Meanwhile, he dreamed of the continuous mill, looking especially at sheets coming off manual mills to determine the precise shape of the rolls required for continuous rolling. At the same time, electrical engineers were developing large-horsepower motors that might power a continuous mill. Verity and Hook engaged Tytus in long conversations about such a mill and encouraged him in developing his idea.

During World War I Armco became a leading producer of shell forgings — output that did not use sheet mills for the Allies and the United States — Tytus, now assistant general superintendent of the Middletown Division, used the idle mills for experimentation in continuous rolling,

John Butler Tytus

concluding that "sudden changes in the shape of the rolls" in stands of mills constituted the greatest barrier to success. By 1919 he believed that he had surmounted that problem and had blueprints in hand for the installation of a continuous mill at Middletown. Unfortunately, domestic producers of consumer goods using sheet steel — automobile and refrigerator manufacturers, for example — and manufacturers abroad using speciality steels were demanding sheets; thus Tytus, though still resolving to work with a continuous mill, had to attend to production on conventional manual mills.

In 1921, almost serendipitously, Verity set the stage for Tytus to act out his drama. Primarily because of an attractive price, Verity pushed Armco to purchase the Ashland Iron and Mining Company in Ashland, Kentucky, where the company intended to turn out steel from six existing open-hearth furnaces. The purchase raised the question of whether to continue conventional methods of steel production or construct a continuous mill. From the outset Verity and Hook saw a fair prospect for the latter alternative. At Hook's recommendation Armco directors authorized an expenditure of $10 million for Tytus's version of a continuous mill.

Tytus, receiving the faith and support of his company, now either faced glory or bankruptcy. No longer, then, could he "sit patiently, and idly ruminate." He organized an office in Middletown, gathered engineers around him to churn out blueprints, ordered building materials, and, after two years of preparation, led his men to Ashland. At Ashland he had 14 stands of rolls constructed in a straight line

ready to reduce 5-inch-thick slabs of steel into 36-inch-wide sheets of steel. He incorporated into his design knowledge derived from failures of men in the past and from his own research and experience. If he adopted a singular principle, it was that "at each successive pass the sheet would have progressively less convexity." Early in 1924 he directed the first run of slabs through the mill, which, despite breakage of machinery and buckling of sheets, proved practicable. Production then began in earnest.

Armco gained productivity, profits, and prestige out of the continuous mill. With Tytus making improvements the mill was soon turning out 40,000 tons of sheets a month, far exceeding the 18,000 tons that engineers estimated was necessary to meet its cost. At Tytus's suggestion, Armco permitted competitors to install continuous mills under licensing fees. From 1927 to 1940 U.S. steel companies constructed at least 26 mills, costing over $500 million. Costs of production in the industry declined, and output rose as a result. Trade magazines celebrated the triumph — *Iron Age* called it "epoch-making" — and rival steel men masked their envy in the rhetoric of praise.

Tytus had his rewards, too. In 1927 he was promoted to the position of vice-president of Process Developments. His name became a household word in the steel industry, and for his contribution to steel production he received the Gary Medal of the American Iron and Steel Institute in 1935. A shy, even introverted man, he probably took his greatest pleasure out of simply seeing his dream of steel running through a continuous mill realized. Though surely no great genius, John Butler Tytus demonstrated that imagination and relentless determination joined to an advancing technology and material support could yield a triumph in technology.

Ever an Armco man, Tytus returned to Middletown from Ashland. Late in the 1930s he displayed no little ingenuity and exertion in rehabilitating the hot-strip mill at Middletown. In 1938 he became vice-president in charge of operations and in 1943 vice-president over technical development.

Tytus did not long continue in his third vice-presidency. On June 2, 1944, he died, having suffered a coronary thrombosis. Surviving him were his widow, Marjorie Denny Tytus, and three children, Elizabeth Tytus Jones, and twin sons John Butler Tytus and Francis Jefferson Tytus. Like many other pioneers in steelmaking at Armco, he was buried in Woodside Cemetery in Middletown.

Publications:
"As It Was in 1915 and As It Is in 1940," *Metals Progress,* 37 (March 1940): 316.

References:
Christy Borth, *True Steel: The Story of George Matthew Verity and His Associates* (New York: Bobbs-Merrill, 1941);

George C. Crout and Wilfred D. Vorhis, "John Butler Tytus: Inventor of the Continuous Steel Mill," *Ohio History,* 76 (Summer 1967): 132–145;

Douglas Alan Fisher, *The Epic of Steel* (New York: Harper & Row, 1963);

R. C. Todd, Bennett Chapple, and Vorhis, eds., *Fifty Years at ARMCO: A Chronological History of the Armco Steel Corporation* (Middletown, Ohio, 1950).

U.S. Steel Antitrust Suit, 1911

by Paul Tiffany

Sonoma State University

The formation of the United States Steel Corporation (U.S. Steel) by J. P. Morgan & Company in 1901 raised many questions about the changing structure of American industry. Since 1894 several large mergers had been completed, often resulting in a single firm dominating the market. The creation of the "Steel Trust" was the largest merger of the time and resulted in the world's first firm capitalized at more than $1 billion. The merger brought together 213 different manufacturing plants and transportation companies; 41 mines; 1,000 miles of railroad; 112 ore vessels; and 78 blast furnaces. The corporation's holdings in coal, coke, and iron ore were vast. It commanded a total of more than 43 percent of the national pig iron capacity and nearly 60 percent of the production of basic and finished steel.

From the outset Morgan and his associates worried that U.S. Steel's dominance of the industry would attract an antitrust suit, forcing a dissolution of the firm. But Judge Elbert H. Gary, the representative chosen by Morgan to oversee the new combine, stated that if the corporation "is allowed to continue in business until it has been proven that the intentions of its managers are good, that there is no disposition to exercise a monopoly or restrain legitimate trade," then U.S. Steel was to be judged to have been legal and in good standing. Gary managed U.S. Steel accordingly over the first decade of its life; the corporation's market share diminished as other firms entered the market and expanded operations, and Gary consistently pursued a policy of "conciliation and cooperation" between his firm and his industry rivals. What Gary sought was stability in the steel industry, and he used a variety of means to achieve it. He hosted series of dinners at which all the important leaders of the industry met to discuss issues such as prices and capacity expansion. He later formalized these gatherings by creating the American Iron and Steel Institute.

The first decade of the century also saw the growth and expansion of the "Progressive Movement" of reform, however, and it was perhaps inevitable that the huge new corporations would be attacked for undermining the nation's fundamental economic and political traditions — competition being chief among them. Both Presidents Theodore Roosevelt and William Howard Taft initiated antitrust suits under the 1890 Sherman Act, moribund for the first decade of its life. In 1905 the U.S. House of Representatives — also worried about the problems of size in the economy — endorsed a resolution calling upon the newly created Bureau of Corporations specifically to investigate U.S. Steel. Finally, in 1910, the House passed another resolution calling for yet another investigation; the result was the formation of the Stanley committee in 1911 to probe the activities and dealings of the firm from the time of its inception to the present.

With the Progressive movement in full swing, there was increasing pressure on the federal government to do more than investigate U.S. Steel. In the spring of 1911 President Taft ordered the Justice Department to prepare a case against the firm, based upon the antitrust statutes. The suit was formally brought on October 26, and the government — charging U.S. Steel as an unlawful monopoly — asked the court to break up the firm into its original operating units.

Gary was shaken by this turn of events. Believing that he had not aggressively pursued larger market shares for U.S. Steel, he was convinced that his management of the firm had been proper and that the suit was unwarranted. Indeed, he strongly felt that U.S. Steel was a positive force in the economic health of the nation. Accordingly, he resolved to fight the suit rather than seek a compromise with the government — as had Standard Oil Trust when it was sued by the Department of Justice on similar antitrust grounds.

After voluminous testimony and much analysis, the Federal District Court of New Jersey, represented by Justice Joseph Buffington, announced in June 1915 that it found U.S. Steel not guilty of any of the accusations brought by the Department of Justice. While the original framers of the Steel Trust may have intended to monopolize the steel business through creation of the firm, U.S. Steel, the court found, was not guilty of carrying out those intentions. The Justice Department, disappointed in the outcome, decided to appeal to the U.S. Supreme Court. After a delay brought on by World War I, the Court finally ruled on March 1, 1920; in a close four-to-three decision (two justices had to excuse themselves due to earlier activities related to the firm), the Court upheld the lower court's decision. Reasserting the "rule of reason," the Court said that size per se was not illegal, only abuses which resulted from size. Finding no such abuses in the behavior of the firm, it denied the complaint.

Judge Gary thus won his long battle in the courts to prove that his firm was not an evil trust but rather an economic force that was beneficial to the nation as a whole. Yet the victory would remain complete. Congress and other units of the federal government remained suspicious of U.S. Steel and continued to file a variety of legal charges against it. A climate of distrust evolved that in the end benefited neither party.

References:
United States v. *United States Steel Corporation et al.,* 223 Fed. 35 (1915);
United States v. *United States Steel Corporation et al.,* 251 U.S. 417 (1920);
Melvin I. Urofsky, *Big Steel and the Wilson Administration* (Columbus: Ohio State University Press, 1969).

United States Steel Corporation

by Bruce E. Seely

Michigan Technological University

In 1901 several iron and steel companies merged to create the United States Steel Corporation, the largest business enterprise in the world. Capitalized at more than $1.4 billion, this consolidation became a focal point for public concern about the increasing number of big businesses and monopolies, and the company has never escaped its controversial origins. Moreover, U.S. Steel's history has been marked by intriguing contradictions. For much of its history the company was the industry's indisputable leader in setting prices and voicing the concerns of steelmakers; yet, the company's performance in the steel market has often been lackluster. Heir to Andrew Carnegie's

U.S. Steel plant and company town at Gary, Indiana, in 1936 (photograph by Sarra)

legacy of technological efficiency that characterized the Carnegie Steel Company, U.S. Steel nevertheless avoided innovation. Although fiercely antiunion, U.S. Steel became the first major steel company to recognize the steelworkers' right to unionize in the 1930s, yet labor relations have remained contentious. Finally, despite a history mostly defined by its reluctance to change, the corporation in 1986 adopted the name USX, symbolizing its transition from a steelmaker to a highly diversified company.

The formation of U.S. Steel involved the most spectacular merger in a wave of combinations that occurred at the turn of the century. Other steel trusts, including the Moore Group and John W. Gate's American Steel & Wire, had their origins in the speculative concerns of primary investors. U.S. Steel, however, was a product of J. P. Morgan's fear that competition was running rampant in the steel industry — a fear partly fueled by Andrew Carnegie's love of competition and willingness to undersell all competitors.

The specific catalyst to U.S. Steel's creation was the decision in 1899 of several makers of finished steel products — wire, sheet, hoops, and pipe — to erect steel plants and bypass their main supplier, Carnegie Steel. Carnegie retaliated by

threatening to build finishing mills that would run them out of business. In late 1900 he took aim at Morgan's National Tube Company, which was building a steelworks at Benwood, West Virginia; Carnegie announced plans for a tube plant at Conneaut Harbor, Ohio. To forestall a return to ruthless competition, Morgan bought Carnegie Steel, abandoning his earlier doubts about creating a steel trust.

Accounts of Morgan's purchase of Carnegie Steel differ slightly, with both John Gates and Elbert H. Gary claiming they gave Morgan the idea. Also instrumental in the purchase was Charles Schwab, Carnegie's main assistant, who supplied Morgan with a vision of a stable, profitable steel industry dominated by a single firm. Moreover, Carnegie wanted to sell, so when Schwab approached Carnegie at Morgan's request, an agreement on price of approximately $480 million was easily reached. On February 25, 1901, a New Jersey charter was granted to the United States Steel Corporation. The banquet celebrating the deal featured a table shaped like an I-beam.

The first billion-dollar company in the world, U.S. Steel could operate on a scale which had never before been seen. Morgan's steel corporation brought together Carnegie Steel, Federal

Steel, National Steel, National Tube, American Steel & Wire, American Sheet Steel, American Hoop Steel, American Tin Plate, American Bridge, and the Lake Superior Consolidated Iron Mines. By April 1 the Shelby Steel Tube Company and the Bessemer Steamship Company were added, creating an integrated giant. The company controlled massive iron deposits on the Mesabi Range in Minnesota, coal reserves in western Pennsylvania, the largest shipping line on the Great Lakes, 80 blast furnaces, and 149 steel plants and mills. Its market share of finished steel products stood at slightly more than 50 percent. The steel corporation's mills had the capacity to produce 60 percent of the nation's Bessemer steel, 75 percent of the pipe, 70 percent of the sheet steel, 68 percent of the structural steel, 60 percent of the rails, 68 percent of the nails, and 77 percent of the wire rod.

The organization's sheer size, however, presented many headaches. First, there was the problem of how this huge enterprise should be managed. U.S. Steel remained a holding company, with existing local management continuing to run the plants. Second, the company's size drew intense public scrutiny due to the public fear of trusts. Yet the corporation was slow to respond to these fears. Indeed, Charles Schwab, U.S. Steel's first president, endorsed Carnegie's aggressive business approach: cut production costs and run the mills full, slashing prices when necessary. Carnegie's "boys" at the new company soon began to round out U.S. Steel's holdings, purchasing Union Steel, Sharon Steel, and Saint Clair Steel in Clairton, Pennsylvania, by 1904. Older, less-efficient plants were closed, while improvements were made at others.

Yet U.S Steel never became a hard-driving, price-cutting competitor. In fact, the bankers who dominated the board — the composition of which led *Fortune* to remark in March 1936 that U.S. Steel was "perhaps *the* Morgan Company" — abandoned Carnegie's methods of doing business. Instead, Elbert H. Gary, former president of Federal Steel, chair of U.S. Steel's finance committee, and U.S. Steel's board chairman from 1903 to 1927, crafted a new approach. Gary believed that U.S. Steel could avoid antitrust action only if it was seen as a "good trust." Carnegie's men resisted; they even defied the board. But with Morgan's support, Gary won every skirmish, and Schwab resigned as president in 1903. His successor, William E. Corey, remained until 1911; Gary then passed over another Carnegie lieutenant,

William Dickson, and chose James A. Farrell, a Federal Steel man, to be president.

From the outset, the bottom line of Gary's corporate policies was clear — give the government no cause for a lawsuit. In 1905, for example, he cooperated with an investigation by the commissioner of corporations, despite the objections of Corey and other directors. Similarly, Gary horrified the board by cultivating a friendship with the trustbuster himself, President Theodore Roosevelt. He also angered some directors by demanding moral rectitude in an era of shady business dealings; one board member complained that U.S. Steel's board room resembled a Sunday school. To stop stock speculation, Gary issued quarterly and annual reports to shareholders that directors did not see in advance. Nor was he content with modifying the standards of behavior at U.S. Steel. He also wanted to bring stability to the industry. At the informal "Gary dinners" and later in the American Iron and Steel Institute (AISI), which he founded in 1908, Gary pressed other companies to replace cutthroat tactics with cooperation, or as he called it, "friendly competition." He announced U.S. Steel's prices in January and stuck by them in spite of cuts by competitors.

U.S. Steel also tried to treat labor more fairly, although Gary never recognized labor's right to organize; strikes were broken in 1901 and 1909. In response to public criticism aimed at U.S. Steel labor relations, however, Gary launched a publicized welfare program that improved safety and sanitary conditions, built company houses, offered pension and stock-purchase programs to workers, and paid death benefits. But he refused to end the 12-hour workday and seven-day workweek. Of this limited sense of social responsibility, *Fortune* remarked in March 1936, "If his Directors were Neanderthal in their social ethics, Judge Gary may be assigned at least to the Cro-Magnon period."

Other changes that Gary chose not to make also reflected his conservative management strategy. U.S. Steel remained a holding company with some 200 subsidiaries and no rational managerial structure. Only the all-powerful chairman coordinated the corporation's efforts, basing his authority on Morgan's and the finance committee's support. Unlike Carnegie Steel, which had exploited every market advantage, U.S. Steel under Gary moved ponderously. It lost market share deliberately; U.S. Steel's share of most steel products fell to between 35 and 45 percent of the national total.

Benjamin Fairless, chairman of the U.S. Steel executive committee, appearing before the Joint Committee on the Economic Report in January 1950 (photograph by Fons Iannelli)

Although incomprehensible to Carnegie's men, this approach led the press and other steel executives to praise U.S. Steel for fairness to workers, stockholders, and competitors. Some observers, both old-time steel men in the company and later business analysts, indicted Gary for weakening the corporation. But Gary's caution almost certainly insured U.S. Steel's survival. His friendship with Roosevelt, for example, may have prevented release of the 1905 probe by the commissioner of corporations, who concluded that U.S. Steel's initial capitalization was heavily watered. The report did not appear until 1911; earlier disclosure might have damaged the new enterprise.

In the end the antitrust action Gary feared was brought. The Justice Department, under Roosevelt's successor, William Howard Taft, filed suit against U.S. Steel in 1911. But the outcome largely justified Gary's policy. A central complaint involved U.S. Steel's purchase of the Tennessee Coal, Iron, and Railroad Company (TCI) during the panic of 1907, which the government

argued was proof of the corporation's monopolistic intent. Gary's defense was that U.S. Steel acted only to prevent an economic collapse, the bankers on the board fearing an investment house holding TCI stock would fail if TCI had gone under. The key to this defense was Gary's decision to buy TCI only after clearing the move with Roosevelt, who also testified on behalf of the steel corporation. In 1920 the Supreme Court upheld the legality of the TCI purchase. The Gary dinners, however, were judged illegal, but they had ceased before the suit was brought, and no other illegal acts could be found. U.S. Steel, ruled a Court majority, could not be dissolved simply because of its size. Underlying this judgment was the testimony of many executives from competing steel companies who praised the corporation's behavior. Of crucial importance, however, was the decline in U.S. Steel's share of the market, hardly evidence of predatory behavior.

Court evidence, however, hardly painted Gary a corporate altruist. It was shown that Gary had failed to give full information to Roosevelt about the TCI deal. In March 1936 *Fortune* observed that it was dangerous to accept "'friendly competition' or any other Gary policy at face value," arguing that Gary avoided only the *appearance* of monopoly. Even without Carnegie-style aggressiveness, U.S. Steel dominated the industry. In 1906 the corporation flexed its muscle with a massive expansion program at Gary, Indiana — a new steel city with the world's largest, most efficient steel mill — improving its position in the rapidly growing midwestern market. By 1910 U.S. Steel had invested $78 million in the Gary project. Gary also championed basing-point pricing for the industry, a system that favored U.S. Steel's Pittsburgh and Youngstown mills by skewing freight charges to eliminate the pricing advantages of midwestern mills.

The steel corporation's dominance showed again during World War I. From April 1917 through 1918 U.S. Steel spent $202 million on capital projects, emphasizing by-product coke ovens, and produced 12.4 million tons of steel and 27 million gallons of benzol and toluol. Moreover, U.S. Steel officials dominated the AISI, which coordinated business-government relations in steel during World War I. At war's end Gary served as the industry's spokesman and leader during the 1919 steel strike. Refusing to meet with striking steelworkers, Gary orchestrated a brutal program of resistance. The strike was broken, although public sentiment often favored the workers.

By the 1920s, however, Gary's strategy aimed at forestalling antitrust action was also responsible for increasing corporate sluggishness. Although U.S. Steel's internal expansion program — $719 million during the 1920s — doubled capacity at Gary to 4.4 million tons and added a wide-flange beam mill at Homestead in 1926–1927, the corporation lagged well behind other companies in terms of growth. *Fortune* observed in March 1936 that "a phrase [was] once widely quoted as the Steel Corporation's policy: 'No inventions, no innovations.'" The corporation was slow to adopt continuous hot-strip mills, pioneered in the 1920s. Moreover, Gary's smothering control left no room for ambitious managers. Nor could subsidiaries communicate or cooperate: Illinois Steel once refused to roll plates at the price suggested by Carnegie Steel, so the order went to Lukens Steel. *Fortune* concluded that the corporation was driven by a banker's vision of protecting investment, not an industrialist's vision of increasing production; Gary had attempted to freeze the steel industry. Historians Thomas McCraw and For-

est Reinhardt agreed that "Gary's pursuit of 'stability' had drifted into an obsession with maintaining the status quo — an ominous development for any firm, no matter how powerful."

Gary died in August 1927. A triumvirate succeeded him: President James Farrell ran steel operations, J. P. Morgan, Jr., was board chairman, and Myron Taylor chaired the finance committee. But with no one person in charge, U.S. Steel remained sluggish, and Myron Taylor eventually took control. Another Morgan banker, he had a reputation for reviving troubled corporations but had no experience in steel. He set out to undo some of Gary's legacy. He acquired Columbia Steel in 1929, the corporation's first production facility on the West Coast, and added the Oil Well Supply Division in 1930. He retired the long-term debt in early 1929, a move that saved millions of dollars in the early 1930s. By 1932 Taylor was chairman of the board.

The Depression hit U.S. Steel hard. Profits disappeared in 1933 and 1934; Wall Street's faith in the corporation, however, had disappeared even earlier. U.S. Steel's share of the product least affected by the economic crisis — light-rolled sheet — fell dramatically because the corporation had hesitated to adopt the continuous hot-strip mills. Employees suffered from wage cuts and Taylor's well-intentioned share-the-work program that cut hours, not jobs. Taylor, however, continued in his efforts to breathe new life into U.S. Steel and generally hired young outsiders, men like Edward Stettinius from General Motors, Arthur H. Young from Industrial Relations Counselors, and Enders Voorhees from Johns-Manville.

A key step was hiring an engineering consulting firm — Ford, Davis & Bacon — in 1935. After examining the hot-strip mill at Gary and Universal Atlas Cement, the firm began a complete study of the corporation, producing more than 200 reports by 1938. The consultants condemned the absence of a rational management structure, the cause of such redundancies as two separate raw-steel operations, Carnegie Steel and Illinois Steel. In 1935 Taylor followed their advice and created Carnegie-Illinois Steel, the largest steel producer in the world; American Sheet and Tin Plate joined the new subsidiary in June 1936. Taylor further shook up the corporate management by hiring Benjamin Fairless from Republic Steel as president of the new company. Ford, Davis & Bacon also recommended forming a management company to stand between the holding corporation and the operating subsidiaries. The United States Steel Corporation of Delaware, formed in 1938, ended the rigid control of the finance committee and gave operating officials autonomy while coordinating their efforts. Finally the consultants proposed an expansion program to build continuous hot-strip mills, and beginning in 1935 modern plants were erected at Gary, Youngstown, and the new Irvin Works in Pittsburgh.

Taylor was less successful in attacking problems of location. Pittsburgh and Youngstown were high-cost production areas far removed from growing markets. Reluctant to close mills, Taylor did not touch the company's geographic structure.

He showed much more imagination in labor relations. With the support of New Deal labor policy, industrial unions targeted the antiunion steel industry; John L. Lewis underwrote an organizing campaign by Philip Murray and the Steel Workers Organizing Committee (SWOC) in 1936. Initially U.S. Steel hoped to avoid unionization with an employee representation plan formed in 1934. But unlike most steel executives Taylor was unwilling to wage war against unions, recognizing that a confrontation would further damage U.S. Steel's already tarnished image and hurt the slow recovery from the Depression. In early 1937 he met informally with Lewis and agreed to recognize the SWOC, a move that stunned the industry.

Not sharing his peers' virulently anti-Roosevelt sentiments, Taylor tried to improve the industry's image in Washington. His agreement with the SWOC enhanced his reputation among New Dealers, and he and finance committee chairman Edward Stettinius had frequent contacts with Roosevelt administration officials. Taylor retired in 1938, leaving a reinvigorated company. Stettinius, a friend of Secretary of Commerce Harry Hopkins, became U.S. Steel's chairman.

In managing U.S. Steel Stettinius worked with two other Taylor appointees — company president Benjamin Fairless and finance committee chairman Enders Voorhees. All were young and new to U.S. Steel, with only Fairless having a working knowledge of the steel industry. But the triumvirate worked well together: Fairless, the steelman and dynamic salesman, continued to upgrade facilities; Voorhees reshaped accounting procedures and became an industry spokesman on financial matters; Stettinius was the corporation's voice in Washington, although other steel executives did not share his mildly liberal views.

Stettinius, however, resigned in 1940 to join Roosevelt's defense mobilization program. Irving

S. Olds, a member of the board and finance committee since 1936, took his place. Like Taylor, Olds was an attorney with a financial background and Morgan connections; he had ably represented U.S. Steel during the Temporary National Economic Committee investigation of monopoly in the American economy in 1938. Olds's first concern as chairman was preparing U.S. Steel for wartime production. Between January 1, 1940, and August 15, 1945, U.S. Steel spent $928 million on improvements; half of that sum came from the Defense Plant Corporation for projects such as the Geneva, Utah, works. During World War II, U.S. Steel output included 124.5 million tons of steel ingots and 109 million tons of finished steel, including 1.2 million tons of armor plate, 911 ships, 16.4 million shells, 2.9 million bombs, 1,600 miles of pipe for the Big and Little Inch pipelines, 92 million square feet of airplane landing mat, and 290 Bailey bridges.

Yet, even though U.S. Steel operated near full capacity throughout the war, profits fell from $100 million in 1937 on sales of about $1 billion to $60 million on sales of $2 billion in 1944. Higher taxes, inflation, steady increases in wages, and rigid government price controls all accounted for the decline. In the immediate postwar period, Olds found it difficult to reverse this trend, mostly because the federal government seemed reluctant to reduce its regulatory involvement in the industry. Although hardly a reactionary, Olds grew increasingly hostile toward government interference. Yet he opposed the plans of some steelmen to break the union and maintained good relations with Philip Murray and the steelworkers union; he also ended a lengthy dispute with the Federal Trade Commission by abolishing basing-point pricing.

During this period U.S. Steel worked to shed the legacy of being a "trust," the nation biggest business. Gary and Taylor had viewed the company as more than a business, believing that the company's size coupled with steel's importance to the economy meant that profit considerations alone could not guide their actions. Partly public-relations rhetoric, partly a response to antitrust fears, this sentiment of social responsibility disappeared after 1940, coinciding with the formation of the steel union. The shift was most apparent in U.S. Steel's strategy of paying for wage settlements by increasing steel prices. Federal officials, however, did not automatically accept U.S. Steel as just another firm, and several congressional committees probed accusations of steel price-fixing and monopoly after 1946. In defense, steel executives pointed to low profits, and Fairless, Voorhees, and Olds denounced government tax policies, interference in labor settlements, and price controls. The Harry S Truman administration, in turn, blamed steel-price hikes for inflation. The two camps had sharp exchanges about capacity in the steel industry, and Olds's anger peaked after Truman seized the steel industry in 1952 in order to ensure steel for the Korean War effort, a move that Olds labeled creeping socialism.

Yet, as relations soured with the government, U.S. Steel entered what otherwise appeared to be a golden era. Company officials again assumed leadership roles in steel, especially as industry spokesmen in labor relations. Initially U.S. Steel set the pattern in postwar labor contracts; after 1956 the company's executives chaired the industrywide contract negotiating team. Furthermore, U.S. Steel's modern research center was one of the best industrial laboratories in the world. Best of all, profits approached 10 percent of sales in the mid 1950s, setting a record of $419.4 million in 1957. U.S. Steel finally seemed to be performing as analysts thought it should.

The dominant theme at U.S. Steel in the 1950s was expansion. In 1946, $875 million was allocated for new continuous hot-strip mills, pipe mills, open hearths at Homestead, and blast furnaces at Edgar Thomson works in Braddock, Pennsylvania. The company also purchased the Geneva, Utah, mill from the government for $47.5 million. During 1950–1969, U.S. Steel spent the incredible sum of $7.696 billion on capital projects, setting records of $515 million in 1957 and $1.7 billion in 1967–1969. The corporation developed expensive iron-ore projects in Canada, Brazil, and Venezuela, and constructed a huge greenfield plant — the Fairless Works. Then the largest steel plant built at a single time, the Fairless plant was the company's first fully integrated plant in the eastern United States. Ground was broken on March 1, 1951, at Morrisville, Pennsylvania, 30 miles north of Philadelphia, for coke ovens, two blast furnaces, nine open-hearth furnaces, a universal slabbing mill, two billet mills, a 10-inch bar mill, a pipe mill, a continuous hot-strip mill, and two cold-reduction facilities. With an annual capacity of 1.8 million tons of rolled products, the plant cost $500 million.

The corporation also initiated reorganization in 1951: U.S. Steel of Delaware, Carnegie-Illinois Steel, H. C. Frick Coal and Coke Com-

pany, and the United States Coal and Coke Company were united into a new United States Steel Corporation. Operations were further streamlined on December 31, 1951, by folding the other steel subsidiaries into a single company, the United States Steel Company. This was a transitional step to the merger of all steelmaking operations into the U.S. Steel Corporation in 1952.

With reorganization completed, Benjamin Fairless replaced Olds as chairman; C. F. Hood became president, and Enders Voorhees continued as chairman of the finance committee. Fairless dominated the circle of leadership, but other key figures included Roger Blough, David Austin, Malcolm Reed, and George Rooney — policy makers working out of the New York headquarters. They pushed diversification programs and expanded existing ventures in prefabricated houses, cement, and chemicals, while adding titanium production and financial services. With a Republican, Dwight D. Eisenhower, in the White House, governmental relations improved.

Almost every top executive during this period had spent his career within the steel corporation, including Roger Blough, chairman from 1956 through 1969; Leslie Worthington, president from 1959 though 1967; Ed Gott, president in 1967 and chairman after 1969; Edgar Speer, president after 1969; and Wilbert Walker, chair of the finance committee in 1970 and later president. This seemed to indicate an air of self-reliance at U.S. Steel that was further evidenced by the popular company attitude: only ideas originating at U.S. Steel are worthwhile. The strategy of recovering the cost of wage settlements through price hikes largely eliminated pressure to lower production costs with new techniques; therefore the company was slow to incorporate technology developed by others. Basic-oxygen furnaces, developed in Europe in the late 1940s, were not considered until the late 1950s; instead, U.S. Steel built larger open-hearth furnaces. The company also lagged behind the rest of the industry in developing continuous casting.

Although experts at the time disagreed about the technical processes the industry should pursue, U.S. Steel's failure to innovate was mostly driven by arrogant self-confidence, an attitude that influenced much of the corporation's decision making. Justifiably proud of the corporation's postwar status, the men running U.S. Steel assumed their position in the industry was fixed. They viewed the postwar suspension of international competition, the pattern of administered pricing in the domestic market, and technological stasis as permanent. But by the late 1950s problems appeared that soon grew into major headaches.

In 1958 the corporation's profits fell. Although the 116-day strike in 1959, inspired by efforts to amend union work rules, accomplished little for either side, it allowed foreign steel to gain a toehold in American markets, and prestrike hedge buying to build inventories dampened demand after the settlement. Finally, for the first time, the company was unable to raise prices to compensate for the cost of the new contract. These circumstances resulted in a pattern of lower profits, and the company encountered difficulty financing expansion with retained earnings. A clash in 1962 between President John F. Kennedy and Roger Blough over steel prices compounded the corporation's difficulties. Blough attempted a price hike after wage negotiations, violating what government officials had thought was a commitment not to raise prices. Kennedy attacked the move as inflationary and moved vigorously against the steel companies, which soon rolled back the increase. It was a humiliating loss for Blough and marked a return to confrontational government-business relations.

The Blough / Kennedy confrontation ended the company's postwar strategy of meeting higher costs with higher prices, and after 1960 U.S. Steel failed to generate what it considered adequate profits. Some analysts blamed the high start-up costs associated with the company's new facilities, as well as the inability to pass on higher labor costs. Another factor was the piecemeal nature of the improvement program. Only the Fairless Works was completely new; elsewhere facilities were only updated, thereby limiting the gains in productivity and efficiency made by the new equipment. This set up a dangerous spiral, however, as the company's financial situation constrained expansion, further hurting the company's cost picture, which cut further into profits, and so on. In 1969 the company borrowed $1.4 billion dollars to finance expansion. Analysts, however, blamed management for poor investment choices in which the money was poured into brandnew open hearths and old plants instead of the construction of greenfield mills.

In 1964 the corporation had attempted to remedy the situation through restructuring. TCI, National Tube, Columbia / Geneva Steel, and other operating companies were brought within the corporation, and divisions were established along product lines — heavy products, sheet and tin products, tubular products, and wire products. A year later the U.S. Steel Corporation of Delaware replaced the

holding company established in New Jersey in 1901, in part to replace preferred stock with long-term debentures. In 1968 the company moved its headquarters to a 64-story building in Pittsburgh. But performance problems continued, and in 1970 the product-oriented structure was abandoned in favor of two geographic divisions, western and eastern steel operations.

The years since 1970 have been very difficult for U.S. Steel. Although strikes were avoided until 1986, labor relations steadily disintegrated at the plant level. Imported steel steadily claimed a larger share of the market, while the increasing number of domestic minimills threatened the large integrated producers. Even if U.S. Steel had wanted, it could not raise prices to recover rising labor costs. As profits contracted, the company introduced basic oxygen furnaces and continuous casters into existing plants, tabling plans for a major new mill at Conneaut, Ohio.

Government policies offered little assistance against imports, thanks in part to the legacy of bitter controversies over wages and prices. Import restrictions and tariff protection were eventually negotiated in the early 1970s, but in other areas federal officials remained at loggerheads with steel executives. Government efforts to modify hiring practices and reduce pollution, for example, hit U.S. Steel especially hard. In Pittsburgh the Clairton Coke Works attracted special attention from the Environmental Protection Agency, which ignored polluting coke plants at less-healthy steel companies. U.S. Steel's response, however, struck many observers as arrogant. The corporation litigated every pollution complaint, while its lawyers fought charges of employment discrimination. Yet, after denouncing governmental interference, U.S. Steel responded to rising steel imports with demands for Congress to provide tariff protection.

Demand for steel declined after 1975, and in 1979 U.S. Steel wrote off obsolete plants in Youngstown, signaling a contraction in steel capacity that continued into the 1980s. The situation worsened during the recession of the early 1980s. David Roderick, president of U.S. Steel since 1975 and chairman since 1979, became convinced that the company could not survive by simply making steel. U.S. Steel had always been diversified, but Roderick placed new emphasis on nonsteel activities. In 1982 U.S. Steel acquired Marathon Oil and, two years later, Texas Oil & Gas. These moves angered many steelworkers, who believed the funds could have modernized steelmaking facilities. But steelmaking capacity continued to decrease as plants with capacities totaling 5.2 million tons were closed in December 1983. Other plants with a total of 7.2 million tons of capacity were placed on indefinite layoff following a 184-day strike in 1986–1987.

These changes were applauded by Wall Street and bitterly denounced by workers and communities whose mills were closed. Union officials felt betrayed because wage and benefit concessions had not saved many jobs. By the late 1980s the company had cut raw-steel capacity to half that of 1975 — about 19 million tons — but at a high social cost. Historic plants at Duquesne, Pennsylvania, and Homestead were closed in 1984 and 1986; the Texas Works was shuttered; the South Works was left with only an electric furnace; the Geneva and Johnstown Works and the American Bridge Company were sold. Only five fully integrated steel mills remained — Gary, Lorain, Fairfield, Fairless, and Edgar Thomson in Pittsburgh. The bitter strike in 1986, fought over the issue of wage and benefit concessions, further strained already tense labor relations and led journalist John Hoerr and others to criticize sharply the corporation's apparently callous treatment of workers and communities.

Corporate officials argued they were making hard choices to remain competitive. They developed joint ventures with American and foreign producers to build continuous electrolytic galvanizing lines, and pursued international joint ventures with Kobe Steel of Japan and Pohang Iron and Steel of South Korea. One highly controversial plan that failed to materialize was an attempt to import slabs from British Steel for rolling at the Fairless plant. Moreover, in 1984 U.S. Steel attempted to acquire National Intergroup, which had a plant in Detroit, but the deal was abandoned because of restrictions placed by the Justice Department's antitrust division.

Steelmaking recovered somewhat in the late 1980s; profits returned as productivity jumped sharply, due mostly to modern equipment and pared-down union and management crews. But steel's position within the corporation has changed permanently. During the 1986 strike the company was renamed USX, with David Roderick arguing that the corporation was now a diversified energy company; steel operations accounted for only 25 percent of sales. In 1987 all steelmaking operations were placed under the U.S. Steel Company, an independent subsidiary expected to survive without assistance from other subsidiaries. U.S. Steel retains a presence in the world steel market, but it bears little resemblance to the company that stunned the world when it was formed in 1901.

References:

Donald Barnett and Louis Schorsch, *Steel: Upheaval in a Basic Industry* (Cambridge, Mass.: Ballinger, 1983);

Irving Bernstein, *The Turbulent Years: A History of the American Worker, 1933–1941* (Boston: Houghton Mifflin, 1970);

"The Corporation," *Forbes,* 87 (February 1, 1961): 11–15;

" 'The Corporation,' " *Fortune,* 13 (March 1936): 59–67; "The U.S. Steel Corporation: II," (April 1936): 127–132, 134, 136; "The U.S. Steel Corporation: III," (May 1936): 93–97, 134, 136, 138, 141–142, 144, 147, 148; "The U.S. Steel Corporation: IV," (June 1936): 113–116, 164, 166, 170, 172;

John A. Garraty, "The United States Steel Corporation versus Labor: The Early Years," *Labor History,* 1 (Winter 1960): 3–38;

John P. Hoerr, *And the Wolf Finally Came: The Decline of the American Steel Industry* (Pittsburgh: University of Pittsburgh Press, 1988);

William T. Hogan, S.J., *An Economic History of the Iron and Steel Industry in the United States,* 5 volumes (Lexington, Mass.: Lexington Books, 1971);

Hogan, *Minimills and Integrated Mills: A Comparison of Steelmaking in the United States* (Lexington, Mass.: Lexington Books, 1987);

Hogan, *The 1970s: Critical Years for Steel* (Lexington, Mass.: Lexington Books, 1972);

Hogan, *Steel in the United States: Restructuring to Compete* (Lexington, Mass.: Lexington Books, 1984);

"It Happened in Steel," *Fortune,* 15 (May 1937): 91–94, 176;

Thomas McCraw and Forest Reinhardt, "Losing to Win: U.S. Steel's Pricing, Investment Decisions, and Market Share, 1901–1913," *The Journal of Economic History,* 49 (September 1989): 593–619;

"Myron Charles Taylor," *Fortune,* 13 (June 1936): 117–120, 172ff;

Myron Taylor, "Ten Years of Steel," *Iron Age,* 141 (April 7, 1938): 70–71; (April 14, 1938): 50–53, 92–94; (April 28, 1938): 36–38; (May 5, 1938): 47, 73–75;

Paul Tiffany, *The Decline of American Steel: How Management, Labor, and Government Went Wrong* (New York: Oxford University Press, 1988);

"Transformation of U.S. Steel," *Fortune,* 53 (January 1956): 88–95, 198–200, 202–204;

Melvin Urofsky, *Big Steel and the Wilson Administration* (Columbus: The Ohio State University Press, 1969);

"U.S. Steel: Break It Up?," *Fortune,* 41 (April 1950): 90–93, 157–158, 160, 162, 164.

United Steelworkers of America
by Dennis C. Dickerson
Williams College

The United Steel Workers of America (USWA) grew out of the Steel Workers Organizing Committee (SWOC), a constituent part of the Congress of Industrial Organizations (CIO). Established in 1936, the SWOC took primary responsibility for recruiting steelworkers into the CIO. John L. Lewis of the United Mine Workers of America was the SWOC's principal sponsor, and he assigned Philip Murray, his vice-president, to head the group.

In 1937 the SWOC won collective bargaining agreements with the United States Steel Corporation and Jones & Laughlin Steel. In May 1942 at a constitutional convention in Cleveland, the SWOC dissolved and the union became the United Steelworkers of America. The USWA succeeded in reaching bargaining agreements with Bethlehem Steel and Republic Steel. Steel unionism had thus firmly established itself in the industry, with the USWA representing 660,052 members in 903 plants. Although such firms as Weirton Steel and Armco Steel remained unorganized, the USWA by 1945 represented three quarters of all steel-industry employees.

The need for industrial peace and labor-management cooperation during World War II restricted the bargaining flexibility of the USWA. The union's agreement with the "little steel" companies, which the National War Labor Board approved, held wage increases to 5 cents and recognized the USWA as the exclusive bargaining agent for steelworkers. Continued inflation, however, weakened the union's commitment to this wartime pact. Hence, a December 1943 walkout by 150,000 workers met with little opposition from USWA president Philip Murray. With negotiations at a standstill, President Franklin D. Roosevelt intervened and a retroactive raise for steelworkers was agreed upon. But, the issue of wage scales was ultimately unresolved. A major strike in 1946 occurred when steel-company executives refused to accept President Harry S Truman's proposed 18.5-cent wage increase for union

members. Truman already had convinced the USWA to abandon its original demand for a 25-cent increase. The walkout by laborers ended when the industry officials gained a price increase of $5 per ton of steel. Negotiated contracts in 1947 with Carnegie-Illinois Steel and other companies produced a consensus on wage scales and various benefits. These talks further established the USWA as the steel industry's singular negotiating partner. The fragile truce between the union and the steel companies only held until 1952, however.

Since its founding, the USWA, through its strong identification with and financial support of the Democratic party, gained favorable treatment from the Roosevelt and Truman administrations, and industry executives became increasingly critical of this alliance. New contract demands in 1951 further angered industry officials, leading to a seven-month-long strike which lasted until July 1952. Truman assumed control over the nation's steel plants lest a work stoppage impair the nation's war effort in Korea. The Supreme Court, however, nullified this action. Nonetheless, Murray won a contract favorable to his members, but tensions between the companies and the union continued unallayed.

Murray's death on November 9, 1952, brought the union's secretary-treasurer, David J. McDonald, to the presidency of the USWA on March 11, 1953. He negotiated a contract with U.S. Steel that ended regional wage differentials, and in 1955 he successfully pressed for supplementary unemployment benefits in labor contracts. Despite his victories, McDonald faced increasing dissatisfaction from the rank and file. At the 1956 USWA convention, for instance, McDonald's men proposed increases in union officers' salaries and in union dues. A challenge in 1957 from Donald Rarick, a U.S. Steel employee from McKeesport, Pennsylvania, while unsuccessful, emboldened others to contemplate opposition to McDonald's presidency. The rank and file were

Joseph Odorcich, United Steelworkers vice-president, J. Bruce Johnston, senior vice-president of U.S. Steel, and George A. Moore, vice-president of industrial relations at Bethlehem Steel, announcing agreement on a new contract, April 1980

further embittered when a 1960 settlement — following a 116-day strike — fell well below union expectations. In 1965, Iorwith Wilbur Abel of U.S. Steel's Timken Bearing Division in Canton, Ohio, defeated the longtime USWA official.

Abel held office from 1965 until 1977. During his tenure the USWA expanded its jurisdictions through mergers with other unions. Consolidation with the International Union of Mine, Mill, and Smelter Workers; the United Stone and Allied Products Workers; and with a part of the Allied Technical Workers of America expanded the membership of the USWA. Moreover, under Abel's leadership the USWA in 1973 signed a contract with steel producers which granted to workers an annual 3-percent wage increase and cost of living adjustments. In return for these generous concessions, the USWA consented to a no-strike pledge.

Abel's successors include Lloyd McBride, Lynn Williams, and Frank McKee. These leaders had to handle the fallout from escalating decline in the steel industry. Intense competition from foreign producers and lagging technological innovation resulted in massive layoffs and the eventual closing of many plants starting in the 1970s and

culminating in the 1980s. In 1946, for example, America manufactured over 54 percent of the world's steel. In 1970 it dropped to about 20 percent and that drop continued. Between 1975 and 1981, steel-union membership declined from 1,300,000 to 1,042,730. The decrease continued through the remainder of the 1980s. As major plants in Pittsburgh, Gary, and other crucial centers for steel production commenced closing, it was clear that the heyday for the USWA had passed.

References:

Irving Bernstein, *Turbulent Years: A History of the American Worker, 1933–1941* (Boston: Houghton Mifflin, 1970);

Paul Clark, Peter Gottlieb, and Donald Kennedy, *Forging a Union of Steel* (Ithaca, N.Y.: ILR Press, Cornell University, 1987);

John Herling, *Right To Challenge: People and Power in the Steelworkers Union* (New York: Harper & Row, 1972);

John P. Hoerr, *And The Wolf Finally Came* (Pittsburgh: University of Pittsburgh Press, 1988);

Colston E. Warne, ed., *Labor in Postwar America* (Brooklyn, N.Y.: Remsen Press, 1949).

C. William Verity

(January 26, 1917 –)

by Carl Becker

Wright State University

CAREER: Director, organization and planning, director, public relations (1957–1964); vice-president and general manager, Armco Division (1963–1964); vice-president, steel division (1964–1965); executive vice-president (1965); president (1965–1971); chairman, Armco Incorporated (1971–1982); U.S. secretary of commerce (1987–1988).

C. William Verity, a grandson of Armco's founder, entered his family's company rather hesitantly, made his mark as a manager, and then pursued a career in public service.

As a teenager in the 1930s, William Verity spent many hours with his grandfather George M. Verity, cutting trails through a wooded canyon in a five-hundred acre timber and farm tract. The grandson shared, too, in the social and material status that the grandfather had created as founder of the American Rolling Mill Company. When he graduated Yale University in 1939 with a degree in history, however, he chose not to go to work in the family business. "I wasn't sure," he recalled many years later, that "I wanted to be the third generation of my family to enter the steel business." He joined an advertising agency in New York City and churned out copy for the firm's clientele. Though he enjoyed the writing, he decided to depart Madison Avenue a year later and continue the Verity line at Armco. As he explained it, the decision did not spring out of an artificial loyalty to his family but rather derived from his deep ties to the steel industry and Armco.

At Armco, young Verity received, so he asserted, a "thorough grounding in all phases of steel working there." For a while he worked as a laborer at the Armco mill in Hamilton, Ohio; then, turning to a less strenuous assignment, one more in consonance with his formal education, he began to work in public relations. With the entry of the United States into World War II, he joined the navy and served on the USS *Botetourt* in the Seventh Fleet. After his discharge in 1945, he re-

C. William Verity

turned to Armco at Middletown, Ohio, and became assistant safety adviser. Next he went to Ashland, assigned to the personnel office. In 1957 he returned to Middletown and served as the director of organization and planning in what appeared to some observers as a "make-work" assignment. In 1961 he became director of public relations.

Verity had as yet to make a distinctive imprint on Armco. He had, as it were, a birthright to prominence, but not to achievement. But as the director of planning he attained a reputation as a strong leader. At the time strategic and organizational planning was drifting. Executive management was aging and almost willing to believe that profitable growth would continue routinely. Thus, as one internal report had it, a vacuum was developing that might not accommodate onrushing ideas and events. Paradoxically, under Ralph Gray, a powerful, strong-minded man of unlimited energy, the presidency was dominant. No counterbalancing force was available to contest the style and power exerted by his office.

Verity emerged, either by force of his name or personality, as the countervailing power. Ironically, he began his ascent working with Gray as an adviser on organizational problems. Championing the idea of "management by objectives," Verity advanced a comprehensive internal training program designed to correct deficiencies in managerial perspectives. First developed by the managerial specialist Louis Allen and popularly known as POM (Profession of Management), it was offered as a crash course in management to over 5,000 Armco personnel within a two-year period. It was more than a perfunctory or ritualistic program, for it gave managers a greater sense of responsibility and provided Verity with additional leverage for becoming a vital figure in the life of Armco.

Verity, identified with a plan for the reduction of concentrated power, now had new status at Armco. His view of a "balanced structure" in management prevailed at the selection in 1960 of Logan Johnston as president; like Verity, he was prepared to develop managerial talent and to become more "outside-minded." Together, Johnston and Verity assumed primary responsibility for "Project 600," the program for modernizing Armco facilities for production and pollution control. Rather curiously, in view of the purported retreat from a one-man style of management, the Armco board in 1965 appointed Verity to the position of executive vice-president in charge of all steel-operating activities and soon assigned five officers to his domain. In so doing, the board had virtually declared Verity in line for president — and shortly thereafter appointed him to that post, eliminating his previous position as vice-president of steel operations.

From the outset Verity had to deal with diverse and vexing problems — some inherited, some of his own making. The board seemed to give its primary allegiance to Johnston, who remained as chairman. Their evaluations of managers often at variance, Gray and Johnston had allowed executives to take positions unsuited to their abilities, with the consequence that Verity spent an inordinate share of his time shifting their assignments. Though Project 600 was nearing completion late in the 1960s and promised substantial returns in the future, it had created a financial burden for the company. Verity also felt

that he had to make an unusual effort to rehabilitate dormant programs, such as strategic planning and international competition, that had deteriorated for want of attention. A man of exceptional skills in human relations, he found many groups in the community — in banking and education, for example — seeking his support. Altogether, he faced too many problems spread over too little time.

Verity made time, nonetheless, to define and develop his long-range objectives for Armco. The company, he insisted, had to take two steps concurrently for growth: increase sales and production of improved products in existing lines and enter into new markets with new products and services. Though several directors had serious reservations about Verity's goals, Verity doggedly pursued opportunities for diversification and ultimately added to Armco's traditional production ventures in the manufacturing of nonmetallic wares, the leasing of airplanes and railroad cars, the sale of industrial insurance — and more — until he could count 150 profit centers.

As Verity looked to the future, he concluded, too, that all aspects of strategy and development had to be fully integrated into all levels of the managerial processes. To that end he initiated a series of organizational changes. His most important proposal, approved by the board in 1970, had five significant goals: the reduction of the president's responsibilities, the need to find a strong executive to direct all steel activities, the assignment of "new business frontiers" to a senior vice-president for his full attention, the definition of corporate planning as a distinct function, and the movement of Armco into financial enterprises under the direction of a strong executive.

Following approval of the plan, Verity vacated his positions of president and chief executive officer but stayed on as chairman of the board. Donald Reichelderfer took on the two open offices. Commenting on the change, Verity declared that "the many responsibilities of managing a large, diversified, multinational company requires the attention of at least two principal executives." But the board did not act on his suggestion.

Neither satisfied nor satiated by the changes that he had wrought, Verity began a "continuous" conversation with the board in the mid 1970s on an array of problems. He worked to change policies in order to accommodate the chairman's growing need to deal with governmental relations and public affairs. The chairman, he believed, also had to be better apprised on all important corporate matters — but not at the price of inhibiting executive managers in the performance of their duties. Verity also felt that too many top executives did not have responsibilities commensurate with their abilities. The board looked at these issues and others but took little action, owing in part to their disquiet over problems in the industry.

As Verity proposed and the board disposed, Armco was riding a roller coaster. Profits had fallen early in the 1970s, but by the end of 1977 sales and profits were rising rapidly. At Verity's assumption of the presidency in 1965, sales had stood at about $1 billion; now they had risen to $3.5 billion. Profits had increased from about $80 million to $120 million. Verity took cold comfort in such statistics when he looked at the return in shareholders' equity, which had declined from 10.7 percent in 1960 to 8.4 percent in 1977. He was certain that through continued diversification and exploitation of world markets Armco could measurably improve equity. Late in the 1980s, recalling the record of the previous decade, Verity praised the management team that had effected high profits. But he faulted managers of the insurance division, saying that they had not acted with sufficient wisdom and vigor in meeting a downturn in business. He lamented, too, what he saw as his successor's failure to direct Armco in an energetic fashion.

In 1982, at age sixty-five, Verity resigned from the board. His official life at Armco ended, he began a second career, moving into ventures in the public arena. Having long been interested in business-government relations, he became chairman of the United States Chamber of Commerce in 1980–1981 and embarked on a nationwide speaking tour to initiate the chamber's program, "Let's Rebuild America." Repeatedly he called for businesspeople to participate more vigorously in governmental life. In 1981 President Ronald Reagan appointed him to a one-year term as chairman of the Task Force on Private Sector Initiatives. The task force, composed of representatives from business corporations, foundations, and voluntary and religious organizations, had as its purpose the promotion of private-sector leadership in the solution of public needs.

In 1984, as cochairman of the U.S.–Soviet Union Trade and Economic Council, a private organization of businessmen seeking greater trade between the two superpowers, Verity found himself in a rather controversial position. Some observers suggested that Verity was "mindlessly in favor of promoting Soviet trade." He denied that

characterization, insisting that he intended to "prevent strategic materials from leaking to the Soviets."

Verity reached a pinnacle in his public career as the secretary of commerce in 1987 and 1988. Appointed by Reagan to that post at the death of Malcolm Baldridge, he proved to be an effective administrator, especially in promoting reforms generally favorable to the nation's commerce. Though he rehearsed some traditional platitudes about the partnership between private enterprise and government — and though his department kept a low profile in the media — he addressed himself to some narrow and concrete goals. He pushed forward the automation of the patent system, streamlined the bureaucratic processes for issuing export licenses, urged his former comrades in manufacturing to improve their products, and — ever the organizer — reorganized his department to give greater strength to offices related to technology.

Less specific but perhaps more important was the role Verity played in articulating economic policies of the Reagan administration. He gave strong support to the nonprotectionist agreements with Canada and generally championed free trade. During his management of the nation's commerce, American exports rose about 30 percent, and the trade deficit fell by $35 billion. He could not claim credit for such improvements in trade, but he had been a leader in creating the ambience for success.

All the while, Verity was active in civic affairs. Among his many posts, he has been chairman of the Board of Trustees at Ford's Theater and a trustee of the Colgate Darden Graduate School of Business at the University of Virginia and Phillips Exeter Academy. He has also served as a director of Eli Lilly and Company, the Mead Corporation, and other companies. In 1979 he was awarded the Benjamin F. Fairless Award of the American Institute of Mining Engineering.

Verity is married to Peggy Wymond Verity of Louisville, Kentucky. They have three children: Jonathan Verity, Happy Verity Power, and William Wymond Verity. Verity is now retired and lives in Beaufort, South Carolina.

References:

"Good Will and Progress Mark a Brief Tenure," *New York Times,* November 29, 1988;

"Reagan Choice for Commerce is Called a 'Team Player,'" *Christian Science Monitor,* August 12, 1987;

"Reluctant Hero," *Forbes,* 96 (October 15, 1965): 50.

George Matthew Verity

(April 22, 1865 – November 6, 1942)

by Carl Becker

Wright State University

CAREER: Bookkeeper and manager, W. C. Standish Grocery Company (1884–1887); manager, L. L. Sagendorph Iron Roofing and Corrugating Company (1888–1889); vice-president and general manager, American Steel Roofing Company (1891–1899); president (1899–1930), chairman of the board, American Rolling Mill Company (1930–1942).

George Matthew Verity, founder of the American Rolling Mill Company, was born April 22, 1865, in East Liberty, Ohio, the son of Jonathan Verity, a Methodist minister, and Mary Ann Deaton Verity. Jonathan Verity was the descendant of a Huguenot family that had immigrated to England and then to the United States in the 1830s.

Following his mother's death in 1867, Verity lived for two years with his maternal grandparents, George and Hannah Deaton, at their farm about 10 miles north of Springfield, Illinois. In 1869 his father married Louise King, a former schoolteacher who was prim, exacting, and willing to use the rod on students and stepsons. Ministering to a round of congregations, Jonathan Verity usually left his wife to her own devices in regard to discipline, which could be severe.

Verity's father accepted one new assignment after another in Ohio. The Veritys lived no more than three consecutive years in any one community — four different times residing in Cincinnati — in his first 20 years. In 1883 Verity graduated from the high school in Georgetown, Ohio, and his father again moved the family to Cincinnati, this time accepting a call to McKendree Chapel in the city's east end. With hundreds of machine shops, foundries, clothing factories, and other manufacturing establishments the city was also a production center. Seeking to enter a profession of some kind, Verity decided to enroll in a business course at Richard Nelson's Business College in order to pursue a career in banking. On vacation from Nelson's in the summer of 1884 he found

George Matthew Verity, circa 1900

employment in Epworth Heights — a Methodist summer colony — as a room clerk in a hotel and as an assistant in a store operated by McKendree Chapel. There he met his future wife, Jennie Standish, the fifteen-year-old daughter of W. C. Standish, a Cincinnati grocer. He began working for the W. C. Standish Grocery full-time as a bookkeeper and manager.

At age nineteen he was supervising older employees and purchasing goods for the store. Despite running a successful enterprise, by 1887 Verity feared that the store might go under in face of competition offered by Bernard "Barney" Kroger, whose rapidly growing chain of stores known as the Great Western Tea Company could buy groceries in very large quantities and drive prices down. He urged the widowed Mrs. Standish to sell the store; she reluctantly agreed, and Verity found a buyer.

Verity married Jennie Standish in October 1887, and in 1888 he went to work for the L. L. Sagendorph Iron Roofing and Corrugating Company, which turned iron sheets into roofing and siding. The company's owner, Harlan Page Lloyd, had been looking for a manager and had learned from Nelson's College that Verity was able and available. Verity's new salary of $1,200 a year was almost twice what he had been earning at the grocery.

Verity knew little about the sheet-iron business. He managed a workforce of about 20 men who knew more about iron than he did. In his first days on the job he spent long hours taking care of neglected correspondence, reorganizing records — which had been in disarray — and learning how to purchase basic sheet metal and sell the company's finished product. He also successfully sold off the company's overly large inventory of sheet metal. What distinguished him as a manager — and what was to become typical of his managerial style through the years — was his ability to gather a coterie of able and loyal employees: a bookkeeper, an office boy, a stenographer, and a salesman.

After a fire nearly destroyed the company, it reorganized as the American Steel Roofing Company in 1891. After three years as manager Verity had turned the company into a profitable enterprise and was putting a portion of his salary into savings. His eyes open to new opportunities, he borrowed $5,000 for the purchase of a patent for a spiral pipe known as Polygon, a stronger and more attractive product than conventional corrugated downspouts.

During the 1890s the competition in virtually all of America's business sectors grew fierce. Though he had once resolved to meet competitors on their own terms, Verity became a member of the Sheet Metal Roofing Association, which sought to avoid ruinous competition among the fabricators by resorting to "cooperative" agreements on prices and production. He served as secretary and treasurer of the association.

Late in the decade he played a leading role in a plan to combine the nation's fabricating factories into four strategically located plants. For prospective participants he painted a portrait of business life made less rigorous. Cooperatively pur-

chasing their materials in volume from sheet-rolling mills and reducing the mileage of transportation, participants would realize large savings and greater profits. He and Lloyd went to New York City to meet financiers and nearly had an underwriting agreement in hand, but the agreement fell apart when the stock market fell.

No longer wanting to rely on a volatile market to supply his company with sheet steel, Verity decided to erect his own steel-producing mill. He found a site for the mill in Middletown, Ohio, and sought the help of friend William T. Simpson in attracting investors. Simpson owned a galvanizing shop in Cincinnati and in 1890 had bought an old rolling-mill to produce his own steel. Simpson also introduced Verity to veteran steelmakers, and Verity began visiting steel mills at Pittsburgh. After having observed the changing of ore into pig iron, pig iron into ingots of steel, and ingots into bullets and bars, he decided that his proposed plant should do more than roll sheets made at another steel mill: it should make the steel and the bars for milling. But that required the installation of a Bessemer converter or an open-hearth furnace in his mill. To add even the least expensive steel furnace would raise the cost of construction of the plant from about $300,000 to $500,000.

With Lloyd's and Simpson's help he was able to find subscribers to more stock and to an issue of bonds. Capitalized at $200,000, the American Rolling Mill Company (Armco) was incorporated in December 1899. By March 1900 capital had been increased to $500,000 — $350,000 in common stock and $150,000 in bonds. Verity became president, and Simpson was named vice-president. The new company had 35 stockholders.

Verity's plans for the construction of the mill called for the installation of a 30-ton open-hearth furnace, a bar mill for conversion of ingots into sheet bars, four sheet mills for turning bars into sheets, and galvanizing facilities. He hoped to become one of the first men in steel to create an integrated plant encompassing production from ingots to finished steel products. (Manufacturing processes were usually divided among separate companies.) Construction of the plant began in the spring of 1900. At the laying of the cornerstone of the mill building on July 12, 1900, Middletown stores and schools were closed, and most town members gathered to celebrate what they hoped would mean a new era of growth and prosperity for Middletown.

By February 1901 the mill was ready for steel operations. His workforce comprised melt-ers, pitmen, rollers, and heaters recruited from various steelmaking centers; many had worked on the construction of the plant. Verity also brought Robert Carnahan to Middletown from the Homestead Works of the Carnegie Steel Company. A graduate of the University of Pittsburgh and an outstanding metallurgist, Carnahan presided over the installation and operation of the furnace. James Strawbridge, who had helped in the construction of the plant, was the first general superintendent of the works, and Tom Rogers, a burly steel veteran from Pittsburgh, ruled over the rolling mill. Along with Verity these three men constituted the company's first "board of strategy." Altogether about 200 men were ready to work in the plant.

On the morning of February 7, 1901, Carnahan issued the order to "Tap the heat!" According to Orley Moles, a ladle man who witnessed the event, it was hardly an auspicious beginning. The man in charge of the open hearth had set up the ladle to catch the molten steel. The heat was tapped, but the released molten steel passed through the ladle. The stopper in the ladle had not been closed, and steel flew in every direction, scattering the visitors who had come to see the tapping of the furnace.

Though its low capacity required Armco to carve out a niche in speciality steels, Verity and his men pushed the plant to the limit to produce steel sheets, overburdening facilities and damaging equipment. Men from other steel towns joined the workforce, which soon numbered about 400. By the end of 1901 the company had produced over 10,000 tons of ingots and recorded sales of over $600,000 and a net profit of $56,000.

Verity, however, paid a price for Armco's success. He had been working long hours at the mill and suffered an emotional and physical collapse early in 1902. At his physician's order he went to convalesce at a resort in French Lick, Indiana. He began following a strict dietary regimen and took up horseback riding; he rode every morning, a practice that he continued for years.

After regaining his health, Verity gave increasing attention to two problems: management-labor relations and Armco's product line. Rogers, a stern taskmaster, employed a simple creed supervising the shop: a man did what he was told to do, without explanation; if he failed, he was fired. Out of his experience in the Cincinnati plant Verity had developed a different view on management-labor relations: a man worked more productively

if managers shared knowledge of their problems with him, building a spirit of cooperation. Rogers and Strawbridge, whom Verity found to be "faint-hearted" and pessimistic, soon departed Armco, and Verity replaced them with men — notably Charles Hook — attuned to his concepts of leadership. Concerning the matter of matching a product line to a market, Verity knew that Armco could not compete with the giants in large-tonnage production but instead might seek out relatively small consumers who were looking for improved quality in the steel that they used. Verity approached the electrical-equipment industry, which, although not a large consumer of sheet steel, was growing rapidly. Electrical engineers required "magnetic permeability" in the steel used in armatures. At first they had resorted to use of solid-iron forgings. Finding that magnetic friction stole power from the solid iron, they turned to the use of laminated sheets. But laminated sheets — usually rolled from Bessemer steel — often did not possess adequate uniformity; and the large-tonnage producers, seeing a limited market for the sheets, had little inclination to engage in the vigorous research that might yield an improved product.

Verity sent a representative to the Westinghouse Electric and Manufacturing Company in 1902 with an offer to initiate experimental production at Middletown as a cooperative project. The Westinghouse people, though surprised at the offer, readily explained the problem: the need for a wafer-thin, low-sulphur and low-phosphorus steel with a carbon content of .08 percent and manganese content of .35 to .50 percent. Westinghouse sent an engineer, Wesley Beck, to work with Carnahan in Middletown. Carnahan and Beck experimented with raw materials, temperatures, and fuel. The tests often damaged equipment. In 1903, however, they were able to produce a nonsilicon steel meeting Westinghouse's requirements. It was a dramatic breakthrough that would enable the electrical industry to develop smaller and more-efficient motors capable of many industrial and domestic uses.

In the meantime Verity was learning that an integrated plant presented problems of balancing production. As sheet rolling improved and more melting facilities were required, Verity had the original furnace enlarged to a 50-ton unit and had two furnaces added. But these additions meant that sheet-rolling capacity needed to be increased. Rather than adding a mill at Middletown, Verity and his associates, through various means of financing, acquired the Muskingum Valley Steel Company in Zanesville, Ohio in 1905. By that year capitalization had increased from $500,000 to $1,400,000; employment from 350 to 1,000; capacity from 1,000 tons a month to 3,300. Verity next turned his attention to producing a corrosive-resistant steel. Verity had become interested in the work of Dr. Allerton Seward Cushman, head chemist of the United States Office of Public Roads. Cushman had studied the problem of corrosion of woven-wire fence and called for the elimination of impurities from steel as a solution. Verity invited Cushman to serve as an Armco consultant, and in 1905 Carnahan and his colleagues soon began to conduct experiments on the reduction of impurities from steel. His men carefully fed selected ore and scrap into furnaces and oxidized the metal bath in order to burn out the manganese and carbon — a procedure steelmakers had always shunned, for temperatures going beyond 3,000 degrees did terrible damage to the furnaces and ladle linings and caused shutdowns. During one of the trials Carnahan's chief assistant, who was working in one of the furnaces, was severely injured, and Carnahan slowed down the pace of experimentation.

In 1907 Verity moved east to assist George Westinghouse, whose firm had gone into receivership. Armco had a claim of $160,000 against the electrical company and hence had a substantial interest in its fate. Verity worked out of a New York office for a year developing a plan of reorganization to satisfy the demands of Westinghouse creditors. The Panic of 1907, however, had also damaged the steel market, and his own company faced an enveloping financial crisis. Twice in 1908 he returned to Middletown to persuade workers to permit the company to withhold a portion of their pay as long as necessary. But on both occasions the payroll was saved by the receipt of a large order, and workers lost neither hours nor pay.

Verity returned to oversee Carnahan's experiments. The mill was finally able to produce a "perfect" ingot and a pack of perfect sheets. According to the patent issued to the company in 1909, the new steel was 99.94 percent pure. At the suggestion of a customer in Chicago, Verity decided to call it Armco Ingot Iron; it would become the signature product of the company.

The new iron found a market throughout the nation and world. Wire fencing made from it enclosed a large sheep ranch in Australia; acted as a barrier against sharks at a beach near Durban, South Africa, and protected a sultan's palatial

Verity when he was chairman of the board at Armco (photograph by Blank & Stoller, New York)

grounds in East Africa against beasts of the jungle. Irrigation flumes manufactured from the iron carried water from mountain reservoirs to deserts in the western United States. Underground tanks, garbage cans, and railroad cars were made from the new iron. It also became a base metal for refrigerators and other household appliances. Most important, the iron was used to construct culverts for the expanding roadway system.

Verity was soon directing an expansion necessitated by increased production and earnings. He decided to build a new plant on a site east of the existing mill. The plant would enable the company to employ so-called big-mill practices, roughing down ingots in a large blooming mill. Completed in 1911 at a cost of nearly $5 million, the new mill became known as the East Works, the old mill being referred to as the Central Works. The East Works employed more than 1,200 men and more than doubled Armco's aggregate capacity.

But as Armco built a reputation for specialty steels, and as the frequency of steel experiments began interrupting normal production, Verity became convinced of the need for a more systematic approach to research. By 1910, after consultations with a special committee of managers, the company had begun erecting a research building where chemical and electrical experiments could be conducted. Beck was appointed director of research and development. Special facilities were provided for: among other things, heat-treating research and chemical testing. Part of the plant grounds were used to test the effect of weather on materials.

After 1911 Verity turned Armco into a major supplier to the automobile industry, which was on the verge of revolutionary change. Before 1911, automobiles were constructed of cold-rolled steel sheets fixed on wooden frames. The sheets, pitted and rough, required grinding and buffing before they could be painted. Armco began producing steel sheets with a silvery finish. Although the new steel was expensive, it was adopted by automobile manufacturers because its smoothness reduced production time and labor. It has often been asserted that Armco furnished half of the steel used in Ford bodies — enough, remarked one observer, to put a roof over Detroit. Stove manufacturers also began to place orders for the new steel because they needed sheets of the same quality.

Verity was also pushing Armco into new directions in marketing. At the news of the development of Armco Ingot Iron, a Brazilian engineer had ordered culverts made of the iron. His order prompted Verity to send sales representatives to several South American nations in 1912. He later appointed a sales agent in Buenos Aires and established a branch office in Rio de Janeiro. The next few years saw the organization of an export department in Middletown and the employment of agents in Great Britain, France, Russia, and Australia. By license, mills in Great Britain, Australia, and elsewhere were producing Armco sheets. Eventually, in 1924, the company created a subsidiary, Armco International, to direct all company activities abroad.

Verity pioneered a more innovative piece of marketing in 1914 with the use of "collateral advertising." At the suggestion of associates he had placed in the *Saturday Evening Post* a double-page advertisement extolling the virtues of Armco Ingot Iron used in the manufacturing of household appliances and directing consumers to purchase products made from Armco metals. Advertising men often ridiculed such copy, but Armco continued running it in a wide range of popular magazines and soon saw competitors following suit. A semi-integrated company prior to World War I, Armco had to purchase pig iron in the open market, and this requirement always presented the threat of shortened supplies and the rapid escalation of prices. Verity, determined to reduce dependence on the market, charged a special committee with the task of finding and acquiring a company that operated blast furnaces. By 1917 he had purchased the Columbus Iron and Steel Company and owned two blast furnaces, three coal mines, interests in ore mines, about 10,000 acres of timberland, 19 ore carriers on the Great Lakes, a sinter plant, coke ovens, and contractual access to limestone flux. The acquisition transformed Armco into a fully integrated steel company.

Prior to World War I, Hook and another Armco supervisor, John Butler Tytus, had begun discussing the possibility of developing a continuous hot-strip mill that might turn out steel sheets in much the same manner as a Fourdrinier machine rolled paper. Since early in the eighteenth century, steel men had turned out sheets on manual or hand-sheet mills usually consisting of two rollers between which sheet bars about 18 inces wide and 30 inches long were passed and repassed by workers. They had repeated the operation until they had attained the required gauge. It was a lengthy and laborious process. Men had long entertained the idea of rolling sheets on tandem or consecutive mills in a continuous operation; and indeed they had developed a semicontinuous mill for narrow strip steel, but the method was not applicable to wide sheets because of various engineering problems: proper heating furnaces and rollers were unavailable, and rope drives and steam engines did not provide the necessary power and variability of speed for a continuous mill. Continuous-mill engineers had to wait for improved electric motors.

A flood of orders for specialty steels, however, meant postponing production at the experimental mill. Verity chose instead to install new manual mills at Zanesville and Middletown. Still, Verity, Hook, and Tytus clung to their vision of an operational continuous mill. In 1921 Armco purchased the Ashland Iron and Mining Company in Ashland, Kentucky, which had two blast furnaces, six open-hearth furnaces, a blooming mill, and coal mines — but no finishing facilities. Verity saw Ashland as a possible site for the construction of a continuous mill, and he authorized the expenditure of $10 million for the experiment.

Construction of the mill began in 1922, and in late 1923 Verity sent 100 men to Ashland to begin the installation of equipment, admonishing them to keep their work secret. Fourteen stands of rolls were built in a straight line to reduce red-hot, 5-inch slabs into thin steel sheets 36 inches wide.

Tytus was ready to run the first bars through the mills early in 1924. A coupling on the mill broke as a bar slid through, and the run was halted. In subsequent runs, more machinery broke, and sheets buckled; but Verity persisted in repair and refinement of the mill, and by the end of February — after the first full month of production — 9,000 tons of sheet steel had been rolled. Engineers had estimated that the mill would have to produce 18,000 tons a month to pay for itself. With Tytus almost obsessively effecting improvements in the machinery and production practices, the mill was yielding 40,000 tons a month within three years. Later, Verity assessed the importance of the mill to Armco: "So at the end of all these years, we knew where we stood. We risked the whole future of our company on this experiment. Now Armco knew it would be repaid for pioneering." Seeing the mill in a more dramatic context, the June 1927 issue of *Iron Age* portrayed it as "epoch-making" and as "a monumental example of the scientific approach to a manufacturing problem."

In 1926 the Forged Steel Wheel Company in Butler, Pennsylvania, put into operation a semicontinuous mill that rolled 48 inches wide; the four-high stands employed in the mill gave it some mechanical superiority over the Ashland mill. But Verity and his men believed that the mill constituted an infringement of Armco patents. The Butler people, short of capital and fearing costly litigation, agreed to sell their plant and patents to Armco.

Verity accepted Tytus's view that Armco should license its patents to competing firms. The principal steel companies in the nation constructed 26 mills costing over $500 million from 1927 to 1940. Armco, of course, installed mills at Middletown and at plants that it acquired in later years. Undoubtedly the continuous mill contributed significantly to reduction of costs in the industry and the expansion of output in the following decades.

Approaching retirement late in the 1920s, Verity continued to press for innovative solutions to Armco's production problems. The company needed more pig iron for the open-hearth furnaces in Middletown, and the blast furnaces in Columbus were not turning out a wholly satisfactory iron. Verity had long lamented the practice of shipping pig iron from blast furnace to open hearth, which robbed the heat from the pig iron.

The blast furnace, he thought, ought to be joined to the open-hearth furnace. At New Miami, a small community between Middletown and Hamilton, about 12 miles from the East Works, there were inoperative blast and coke ovens. Armco bought the facility and moved the Columbus works to the site, and the Baltimore and Ohio railroad constructed a line between New Miami and the East Works. At New Miami the blast furnace poured molten iron into "thermos-bottle" railroad cars capable of carrying 300 tons each. At the East plant the cars were rolled on their sides and emptied into ladles at an open-hearth furnace, with the heat only minimally diminished.

As the 1920s drew to a close Verity sought a foothold in the West for his company. At his direction Armco acquired a subsidiary, the Sheffield Steel Company, which had plants in Kansas City, Missouri; Oklahoma City, Oklahoma; and Saint Louis, Missouri. The company specialized in the production of bars, wire, and rods, which were in demand in the Southwest.

Verity let Sheffield stand as his final mark as president. He turned his post over to Hook in 1930 but remained with the company as chairman of the board. Thirty years before, Armco had but 200 men at work at one site producing about 10,000 tons of steel that sold for $600,000 in the United States. In 1928 and 1929 over 11,000 men and women at plants scattered throughout the Midwest were turning out over 1.3 million tons of ingots that sold for about $70 million in markets throughout the nation and the world.

Verity often took an innovative approach to labor relations. Early in his career at Armco, Verity had begun to shape a welfare program for his employees. In 1903 the workers organized the American Rolling Mill Mutual Benefit Association. Originally intended to provide medical and funerary benefits to its members, it soon became a social and recreational club, in part because Verity encouraged members to broaden their programs and voluntarily committed Armco to match members' dues dollar for dollar. The association sponsored activities such as sports programs, study clubs, and music festivals. With Armco providing the facilities, workers formed baseball, football, and basketball teams to compete against other industrial teams.

Verity carried his concern for workers into his plants, keeping a step ahead of the workplace-reform legislation of the day. After about 1900 the national and state governments had begun to establish agencies and commissions to examine the problem of safety in the nation's mills and factories. Working with Hook, Verity also pursued programs to make Armco mills safer. Safety committees organized by the workers awarded cash prizes to departments with good safety records. Working conditions in the mills improved, and by 1917 the company became one of the first in the industry to purchase accident insurance for employees under a group policy. Verity insisted, too, on the establishment of medical facilities at his mills. If the company demonstrated its concern for the workers' health and safety, then, in Verity's view, they would become more satisfied, cooperative, and productive.

Verity was also at the forefront in reducing hours worked by his employees. For years men in the steel industry had worked the 12-hour day, sometimes seven days a week. Occasionally steel entrepreneurs had experimented with the eight-hour day, but not until the 1920s, with the Federal Council of Churches and the Federated American Engineering Societies leading the way, did a widespread call for the shorter workday develop. Verity, however, by 1913 initiated a program for reduction of hours, believing that a shorter workday would promote a healthier and more productive workforce. The galvanizing department first went on a three-shift, eight-hour schedule. By 1916 the blooming mill was on three shifts, and by 1920 all operations requiring continuous work were on the eight-hour day.

Verity also tried to improve conditions for immigrant laborers, who had come to work in Armco mills in increasing numbers before World War I and were often the victims of discrimination. He insisted that they be accorded humane treatment in the mills, and he took measures to provide decent housing for them. Verity ordered bunkhouses built when he learned that landlords in the community often refused to rent property to them. Ultimately he had company houses constructed in units of four for immigrant families, with a community toilet, garden plot, and children's playground in each unit. He organized a Foreign Club as a social center for immigrant workers and employed teachers to instruct them.

Verity later formalized his beliefs on management-labor policy in a statement known as the "Armco Policies." Approved by the board of directors in 1919, it constituted a bill of rights and code of ethics for management and labor. Some employers damned Verity as a "socialist" and "visionary." He argued, in response, that his fairness and support of workers represented a sound investment in the resources of production; altruism, he asserted, had nothing to do with Armco policies — a desire for higher net profits dictated policy. His beliefs were in keeping with many business progressives of the period, who saw in welfare programs the basis for a stable workforce. It was also believed that company-sponsored welfare programs would discourage unionization.

In the 1920s, with the installation of a continuous mill at Middletown, local and national officers of the Amalgamated Association of Iron, Steel and Tin Workers called for an alteration of wage scales to reflect the increased productivity of the new operation. But Verity and Hook, arguing that an amended agreement would compromise the secrecy of mill operations, refused to negotiate the question. The national officers called for a strike, but only a handful of local workers walked out; the local chapter collapsed.

Almost at the moment of his arrival in Middletown, Verity had become concerned about the public schools. He saw poorly equipped school buildings where underpaid teachers labored long hours and where children learned little or nothing. Accepting an appointment to the board of education, he led a campaign for a bond issue for the construction of a new high school. After the issue passed, Verity pressed for a new city hospital. He feared, though, that people in many sections of Middletown might resist the idea. To reach his goal, he helped create a public nursing bureau, pledging, as did Hook and others at Armco, ten dollars a month to support it.

He constantly encouraged Armco employees to take up civic causes. Residents, however, complained that he and his company were attempting to dominate life in the city — that they were paternalistic monarchs. Opponents of a proposed Carnegie library used this argument in defeating the proposal. Verity assumed blame for the defeat, saying that he and his friends had failed to convince residents that they had no ulterior motives in supporting the construction of the library. Local lore, however, still has it that Verity did not want a library built that bore the name of another steel man.

For all his interest in the community, Verity did not fully integrate his family into Middletown life. Dissatisfied with the state of public education, he sent his children to private schools in Cincinnati and New York after they completed eighth grade.

Verity and his wife toured Europe in 1935. After his return he persuaded Armco's board to turn 500 acres of Armco property into a recreational park for employees and the entire community. The community, paying tribute to his life and work, in return organized a "Verity Day" in 1936.

Verity suffered a stroke and died on November 6, 1942. He is buried in Woodside Cemetery in Middletown, not far from the steel plants that he had built years before.

Publications:

"Remarks," *American Iron and Steel Institute Yearbook* (1914): 275–277;

"Reciprocity Necessary to Prosperity," *Iron Trade Review,* 56 (January 7, 1915): 30;

"Remarks," *American Iron and Steel Institute Yearbook* (1917): 380–385;

Papers by the President of the American Rolling Mill Company, George M. Verity (Middletown, Ohio, 1922);

"Steel Industry Must Needs," *Iron Trade Review,* 73 (August 16, 1923): 469– 471;

"The Human Element," *Blast Furnace and Steel Plant,* 12 (January 1924): 2– 4;

"Remarks," *American Iron and Steel Institute Yearbook* (1928): 40– 42, 284–286;

"Antiquated Antitrust Laws Must Be Modified," *Steel,* 87 (October 30, 1930): 44;

"Nation's Latent Buying Power Is Ready to Come into Action," *Steel,* 88 (February 5, 1931): 37;

"Human Forces Are Dynamic," *Iron Age,* 127 (February 12, 1931): 560;

"A Rift in the Clouds," *Iron Age,* 134 (November 1, 1934): 34;

"Must Charity Begin — and End — at Home?," *Iron Age,* 136 (August 1, 1935): 49.

References:

Christy Borth, *True Steel: The Story of George Matthew Verity and His Associates* (Indianapolis: Bobbs-Merrill, 1941);

"Exceeds Ton of Sheets Per Minute," *Iron Age,* 119 (June 16, 1927): 1731+;

William T. Hogan, S.J., *An Economic History of the Iron and Steel Industry in the United States,* 5 volumes (Lexington, Mass.: Lexington Books, 1971);

John Tebbel, *The Human Touch in Business: The Story of Charles R. Hook, Who Rose from Office Boy to Internationally-Known Business Leader* (Dayton, Ohio: Otterbein Press, 1963);

R. C. Todd, Bennett Chapple, and W. D. Vorhis, *Fifty Years at ARMCO: A Chronological History of the Armco Steel Corporation* (Middletown, Ohio, 1950).

Voluntary Restraint Agreements, 1968 and 1972
by Donald F. Barnett

McLean, Virginia

From the late 1800s until the 1950s, the American steel industry was unchallenged as the world's leading steel producer. Though the U.S. remained a net exporter until the late 1950s, international demand for its steel began to slow in the 1950s. Following World War II, the U.S. steel industry also became entangled in disputes with government and labor over prices, wages, and capacity — leading to higher costs and lower profitability. Steel producers in Europe and Japan were making rapid technological strides, benefiting from government assistance and adding substantial new capacity. Foreign steel was becoming larger and more competitive.

In 1959 a dispute over wages in the American steel industry resulted in a long strike. Imports increased dramatically; domestic steel consumers realized that they could depend on foreign steel. The increase in imports continued through the 1960s,

passing 10 million tons — or over 10 percent of steel consumption — in 1965, rising to 12 percent by 1967 and to almost 18 million tons in 1968.

The U.S. steel industry demanded protection against foreign producers. In 1966 the American Iron and Steel Institute (AISI) called for temporary import tariffs, and congressional hearings were held. By late 1967 a temporary quota was favored over tariffs, and a bill to limit imports to less than 10 percent of the market was introduced. At hearings in 1968 both the industry and its unions supported quotas. During these hearings both the Japanese and German steel producers offered an alternative to the quota bill, proposing export restrictions on their shipments to the United States beginning in 1969. Following discussions, a three-year period of restrictions was decided upon in December 1968, and 14 million tons was made the voluntary limit for 1969, rising to 14.7 million tons in 1970 and 15.4 million tons in 1971. Japan and the European Coal and Steel Community (ECSC) were each to get 41 percent of this total, the balance (18 percent) was left for the rest of the world. Canada, the United Kingdom, and the rest of the world were not parties to the agreement.

Under the voluntary restraint agreements (VRAs), steel imports fell in 1969 and 1970 to less than the VRA limits. This likely had more to do with the exchange rate and other economic circumstances than with strict compliance. Both Canadian and United Kingdom tonnages also dropped, but imports from other noncompliance countries rose. Higher-valued products, such as stainless steel, made up most of the imports, and import prices rose.

In 1971 negotiations began on a three-year extension of the VRAs, which were due to expire at the end of the year. In mid 1971, however, due to turmoil in the international exchange market, the United States imposed a temporary 10 percent surcharge on all imports. This was perceived by foreign suppliers as a violation of the VRA agreement, and imports of steel into the United States, from both VRA and non-VRA countries, surged

in 1971 to over 18 million tons, well over the agreed totals. This brought additional pressure for renewed and stronger VRAs. In May 1972 the renewed three-year VRAs were agreed to retroactively for the year, with Japan restricted to 6.5 million tons in 1972 (with no more than one-third to enter West Coast ports) and the European Economic Community (now to include the United Kingdom) to 8.0 millions tons, each to rise no more than 2.5 percent a year for 1973 and 1974. Stainless and specialty steels were given separate VRA limits.

In 1973–1974 the U.S. and world steel industries had strong markets, and import pressure weakened. The VRAs begun in 1968 were not renewed in 1975. However, with weaker steel markets worldwide in the mid 1970s, new forms of protection emerged for the American steel industry. VRAs were reimposed in the mid 1980s but suspended in 1992.

Imports increased and U.S. producers filed "anti-dumping" and unfair trade charges against the foriegn suppliers. In 1993 the Federal Trade Commission found that dumping, but not injury, had occurred. The decision left the direction of U.S. trade policy unsettled.

Enders McClumpha Voorhees

(April 28, 1891 – April 12, 1980)

by Bruce E. Seely

Michigan Technological University

CAREER: Junior accountant, Hollis H. Sawyer & Company (1914–1915); auditor, Pacific Commercial Company (1915–1916); U.S. Army (1917–1918); assistant comptroller, U.S. Rubber Company (1918–1920); general auditor and assistant to the president, Ajax Rubber Company (1921–1925); auditor, W. A. Harriman & Company and Sanderson & Porter (1926–1927); general auditor, secretary-treasurer and vice-president, Johns-Manville Corporation (1927–1937); vice-chairman, finance committee (1937–1938), finance-committee chairman and member, executive committee (1938–1956), director and member of finance committee, United States Steel Corporation (1956–1965).

Enders McClumpha Voorhees was one of the least visible and most powerful top executives at

the United States Steel Corporation from the mid 1930s through the mid 1950s. As chair of the finance committee, the traditional seat of power at the corporation, he guided U.S. Steel's financial affairs. Voorhees also played a key role in the reorganization of U.S. Steel under Myron Taylor in the 1930s and Benjamin Fairless in the 1950s. He was passionately committed to accounting and to giving meaning to the numbers he generated.

Voorhees was born April 28, 1891, in Amsterdam, New York, into a family with ties to the early Dutch settlers. His grandfather, father, and uncle all were presidents of Amsterdam's largest bank, and it was assumed that he would follow in that tradition. But he disappointed family expectations by going to Boston at age seventeen after having graduated from high school. After two

Enders McClumpha Voorhees appearing before the Joint Committee on the Economic Report, 1950 (photograph by Fons Ianelli)

years selling real estate and working in a brokerage firm, he entered Dartmouth College in 1911 to study chemical engineering. During his junior year, however, a conversation with his father about an investment that failed because of an accounting error led Voorhees to shift from chemistry to business. He graduated from Dartmouth in 1914 but had already begun studies at the Amos Tuck School of Finance. He also attended the Bentley School of Accounting in Boston from 1914 to 1916.

After Dartmouth Voorhees immediately landed a job as junior accountant with the Boston accounting firm Hollis H. Sawyer & Company. Once he had passed the CPA exam, Voorhees began a colorful career marked by frequent job changes. In 1916 he joined a trading company in Manila, then enlisted in the army a year later. After he was demobilized Voorhees became assistant comptroller at the U.S. Rubber Company. In 1921 he moved up to general auditor and assistant to the president at the Ajax Rubber Company but

left in 1926 to join W. A. Harriman & Company. While with that banking firm he liquidated the Cramp shipyards, a difficult task because of the glut of shipping. His success attracted the attention of Sanderson & Porter, an industrial engineering firm, and Voorhees changed jobs again. One assignment was to study and revamp the financial methods of Johns-Manville Corporation. Only after examining company operations did he look at the books; his report was only five pages long. Impressed by Voorhees's novel approach to solving a problem, the executives at Johns-Manville hired Voorhees in 1927 as general auditor.

Voorhees settled down at Johns-Manville, advancing to secretary treasurer and vice-president. But in April 1937 Myron Taylor, chairman of U.S. Steel, lured Voorhees away to become vice chairman of the steel corporation's finance committee. Taylor was bringing in younger managers to help transform the steel holding company into a more efficient operation. Taylor wanted to coordinate the efforts of the subsidiaries while at the same time letting them enjoy a degree of autonomy. Business historian Alfred D. Chandler, Jr., asserts that Voorhees played a central role in making this management plan work.

A year after Voorhees arrived Taylor retired and chose Edward Stettinius as U.S. Steel's new chairman. Voorhees succeeded Stettinius as the chairman of the finance committee, while Benjamin Fairless, himself a U.S. Steel man only since 1935, became the new president. Taylor thus had his young management team intact — although of the three newcomers only Fairless was a trained steel man. *Fortune* reported in 1940 that Voorhees was the "most mystifying of the triumvirs"; he gave no interviews and disliked being photographed. But *Fortune* added, "Accountants are something like priests; they are dwarfed by the rites they perform, and in a corporation as big as U.S. Steel only the logic of figures can supply a Stettinius or a Fairless with beacons by which to navigate." Voorhees supplied figures for almost twenty years and outlasted both Fairless and Stettinius.

Voorhees first revamped the corporation's accounting procedures, determined to give meaning to the mass of numbers U.S. Steel generated. He tested his new cost-accounting system at the Tennessee Coal, Iron & Railroad subsidiary and implemented it across the corporation. Among the changes was handling inventory on a last-in, first-out basis, saving both excessive inventory profits and losses. In keeping with his expertise Voorhees later advised the Defense Department on fiscal and organizational procedures in the late 1940s and early 1950s.

But Voorhees soon was more than an accountant, for he became U.S. Steel's defender on such issues as wage and steel-price increases. A pair of articles that he prepared for *Nation's Business* in 1943 and 1944 sketched what became long-lasting positions. The first, "What Business Does with Money," called for clear accounting procedures so the public could understand how businesses operated. Many business executives, but especially those at U.S. Steel, were opposed to government intervention in industry affairs, and in the second article, "Management Holds the Scales," Voorhees argued that government could not maintain jobs that were not there. Only management could do that — and only if profits allowed for it.

Voorhees often had to serve as the defender of the corporation after World War II, as

the financial affairs of steel were frequently questioned. The administration of Harry S Truman blamed rising steel prices for inflation, while labor clamored for higher wages; many demanded that the industry expand to prevent steel shortages. Voorhees, in turn, criticized federal tax policies as examples of government interference. Voorhees testified frequently before Congress, repeating what became his litany: wage increases could be absorbed only if accompanied by productivity increases; only profits allowed the replacement of the tools that gave workers jobs. An urbane witness, Voorhees appeared and sounded more reasonable than many steel executives. He never completely opposed union demands for better wages, pensions, or vacations; and what objections he did raise were often financially oriented. In 1944, for example, Voorhees argued that a union request for a 17-cent-per-hour increase would wipe out 1945 profits if the economy slowed down. In 1949 he claimed that funding a pension system would cost $1 billion, a sum equal to the value of all U.S. Steel stock. These objections led the union to accuse him of hiding profits.

When not dealing with the government, Voorhees oversaw the enormous capital expenditures of U.S. Steel after the war. He also helped Fairless create a multidivisional corporate structure that finally replaced the holding company in 1951–1952. Voorhees chaired the finance committee until 1956, a year after Roger Blough replaced Fairless as chairman. Then Robert Tyson took over the finance committee, although Voorhees remained an executive committee member and director until 1965.

Unlike many U.S. Steel executives, Voorhees kept a low civic profile. A loyal Dartmouth supporter, he funded a $26,000 scholarship in 1953. His only service, however, came as a member of the board of Roosevelt Hospital in 1954. Voorhees also had limited outside business interests. In 1940 he was elected to the executive committee of Johns-Manville. In 1943 Voorhees was the leading investor in a Long Island real-estate project. He also was elected to the board of the Electric Bond and Share Company in 1957. He had married Pauline Andrews of Boston in 1923, and they had a boy and a girl. Voorhees died in New York City on April 12, 1980.

Publications:
"What Business Does with Money," *Nation's Business,*
  31 (November 1943): 23, 56, 58, 60, 62;

"Management Holds the Scales," *Nation's Business,* 32 (October 1944): 23–24, 94.

"The State vs. Customers," *American Iron and Steel Institute Yearbook* (1946): 56–62.

References:

Alfred D. Chandler, Jr., *Strategy and Structure: Chapters in the History of the Industrial Enterprise* (Cambridge, Mass.: M.I.T. Press, 1962);

Arundel Cotter, "Steel's Voorhees," *Barron's,* 30 (September 18, 1950): 7–8;

"Managers of Steel," *Fortune,* 21 (March 1940): 64–67, 142, 146, 148, 150.

Wagner Act

by Richard W. Nagle

Massasoit Community College

The Wagner Act, named for its sponsor, Democratic senator Robert F. Wagner of New York, is the basic law governing labor organization and collective bargaining in the United States. President Franklin D. Roosevelt signed it into law on July 5, 1935, at the high tide of New Deal legislation. Organized labor enthusiastically proclaimed the Wagner Act to be the "Magna Carta of labor"; John L. Lewis and other militant labor leaders saw it as a boon to recruiting new members. Business leaders, on the other hand, denounced it as "un-American" and quickly challenged its constitutionality in the courts. When the Supreme Court decided *National Labor Relations Board v. Jones & Laughlin Steel* on April 12, 1937, it upheld Wagner in a five-to-four decision. In doing so, the Court reversed generations of its own probusiness rulings.

The act declared it public policy to encourage unionism and collective bargaining: "Employees shall have the right to self-organization, to form, join, or assist labor organizations, to bargain collectively through representatives of their own choosing, and to engage in concerted activities, for the purpose of collective bargaining or other mutual aid or protection." Furthermore, for employers "to interfere with, restrain, or coerce employees in the exercise of the rights guaranteed" was an unfair practice. Specific unfair practices were yellow-dog contracts, blacklisting, and discrimination based on union activity. Company unions were virtually outlawed. The act also established a National Labor Relations Board to order and conduct elections for union representation, to issue cease-and-desist orders, and, in general, to administer labor relations.

The Wagner Act was significantly amended by the Taft-Hartley Act in 1947. Most notable, Taft-Hartley added a list of unfair practices by labor, outlawed closed shops and allowed states to outlaw union shops, and outlined a procedure for the courts to delay a national emergency strike.

Alexander E. Walker

(November 12, 1887–April 19, 1960?)

by Carol Poh Miller

Cleveland, Ohio

CAREER: Timekeeper, Riter-Conley Company (1910–1911); salesman, La Belle Iron Works (1911–1916); salesman (1916–1919), assistant general manager of sales (1919–1928), general manager of sales, Republic Iron & Steel Company (1928–1930); general manager of sales, Republic Steel Corporation (1930–1936); president, Trucson Steel, Niles Steel Products (1936); executive vice-president, director, Pittsburgh Steel Company (1937–1939); vice-president and director (1939–1940), president (1940–1954), chairman, National Supply Company (1943–1958); director, Armco (1958–1960).

Alexander Edward Walker was born in Detroit, the son of John and Isabella Paton Walker. His father emigrated from Scotland in 1857, settling in Detroit, where he became a business executive. The young Walker was educated in the Detroit public schools and at the Detroit University School. He received an undergraduate degree from the University of Michigan in 1910 and joined the Riter-Conley Company of Pittsburgh as a timekeeper.

In 1911 Walker became a salesman for the La Belle Iron Works in Steubenville, Ohio, but was soon transferred to the Chicago district sales office of that firm. Five years later, in February 1916, he joined the Republic Iron & Steel Company of Youngstown as assistant manager of pipe sales. In 1919 he was made assistant general manager of sales in charge of pipe and sheets. When the Republic sales department was reorganized following the absorption of the Trumbull Steel Company in 1928, he was made general manager of sales, a position he retained following the formation of the Republic Steel Corporation in 1930. He served briefly in 1936 as president of two Republic subsidiaries, Trucson Steel and Niles Steel Products, before leaving Republic in 1937 to become executive vice-president and director of sales for the Pittsburgh Steel Company.

Alexander E. Walker

In 1939 Walker joined Pittsburgh's National Supply Company — a large manufacturer and distributor of pipe and oil-field machinery and equipment — as vice-president and director. A year later he was elected president of the firm, taking on the additional title of chairman of the board in 1943. During Walker's tenure as president, National Supply's sales increased from $60.6 million to $231.7 million. Walker retired as president in 1954 and as chairman of the board in 1958, when National Supply became a subsidiary of the Armco Steel Corporation. Walker continued as a director of Armco until his death two years later.

Walker was a director of the Blaw-Knox Company of Pittsburgh; the Firth-Sterling Steel & Carbide Corporation of McKeesport, Pennsylvania; the Fretz-Moon Tube Company of Butler, Pennsylvania; the American Welding & Manufacturing Company of Warren, Ohio; the First National Bank of Pittsburgh; and the Oil Well Engineering Company, Ltd., in England. He was also a director of the American Iron and Steel Institute.

Walker was married to Marie Gaston of Meadville, Pennsylvania, in 1912. They had two children: Frances Marie and Alexander Edward. Walker died in Pittsburgh. The year of his death is given variously as 1960 and 1961.

Publication:
"The Problem of the Small Order," *Yearbook of the American Iron and Steel Institute* (1938): 246–248.

Reference:
Iron Age, 122 (October 18, 1928): 995.

Wilbert A. Walker

(July 26, 1910–)

by Dean Herrin

National Park Service

CAREER: Accountant, Oswald and Hess (1931–1934); accountant, Ernst and Ernst (1934–1940); sales and system staff, International Business Machines Corporation (1940–1941); vice-president of accounting, director, Carnegie-Illinois Steel division (1949–1956), vice-president, comptroller (1956–1959), administrative vice-president, comptroller (1959–1967), executive vice-president, comptroller (1967–1969), vice-chairman of finance committee (1969–1970); chairman of finance committee (1970–1973); president (1973–1975); director, U.S. Steel (1968–1983).

Wilbert Walker, known to his friends as Wib, was president of the United States Steel Corporation from 1973 to 1975. Unlike most previous U.S. Steel presidents, who had extensive backgrounds in steelmaking, Walker's career at the corporation was spent solely on financial operations.

Walker was born in Pittsburgh on July 26, 1910, graduated from the University of Pittsburgh in 1931, and started his career as an accountant, first with the Pittsburgh firm Oswald and Hess and moving to Ernst and Ernst in 1934. In 1940 Walker joined the International Business Machines Corporation's sales and system staff but switched to U.S. Steel the following year. Walker

rose steadily within the hierarchy of the corporation's financial offices, and in 1969 he was appointed vice-chairman of the finance committee. The committee's chairman, Robert Tyson, retired in 1970, and Walker assumed the position. In accordance with U.S. Steel's organizational structure at the time, the chairman of the finance committee was one of the top three officers of the corporation; along with the president and the chairman of the board, he was responsible for the overall management of the firm. In 1973, when Edwin Gott retired as chairman, Edgar Speer replaced Gott, and Walker replaced Gott as president of U.S. Steel. Walker retired in 1975, when he reached U.S. Steel's mandatory retirement age of 65, but remained a director of the corporation until 1983.

Walker was a dedicated financial officer, spending most of his company time and even his time outside the office on financial matters. He once told the *New York Times* that being a "financial man" had always been his "one big interest" and joked that his wife complained that he had no special interests outside of financial management. Like Robert Tyson, his predecessor on the finance committee, Walker believed that rising foreign steel imports damaged the domestic steel industry but felt that the most serious problem in the industry was the continually increasing cost of producing steel in America. During Walker's tenure on the finance committee and as president, U.S. Steel made more of a commitment to diversify its interests. By 1973, 20 percent of the company's capital expenditures were in nonsteel areas.

Walker, under the leadership of Edgar Speer, helped reorganize U.S. Steel's operational structure in 1974. Whereas previous leaders of the company had believed in consolidating power at the top, Speer and Walker, along with finance committee head David Roderick, redesigned the hierarchy to give more responsibility to their lieutenants. Five "super" divisions were organized, four broadly covering the geographical regions of the United States and a fifth responsible for raw materials. Unlike the old operating divisions of the company — which were strictly confined to one area, such as production or sales — each of the new divisions combined all aspects of production and sales. Each division thus had a greater degree of autonomy, and it was up to local management to make the important decisions and create

healthy profit margins. The reorganization was in part a response to criticisms that U.S. Steel was too slow to respond to market concerns and too caught up in red tape. In the old structure issues were bottled up in committees, and it often took months to reach decisions. Walker summed up the rationale for change to *Business Week* in 1974: "Our objective is to get the right guy in the right job; and if he has the incentive he does not need minute-to-minute supervision."

Publication:

Industrial Internal Auditing (New York: McGraw-Hill, 1951).

References:

John Hoerr, *And the Wolf Finally Came: The Decline of the American Steel Industry* (Pittsburgh: University of Pittsburgh Press, 1988);

William T. Hogan, S.J., *An Economic History of the Iron and Steel Industry in the United States*, 5 volumes (Lexington, Mass.: Lexington Books, 1971);

New York Times, October 4, 1970, III, p. 5;

"A Steelman Steps Up the Pace at U.S. Steel," *Business Week* (March 9, 1974): 154.

Ernest Tener Weir

(August 1, 1875 – June 27, 1957)

by Alec Kirby

George Washington University

CAREER: Office boy, Braddock Wire Company (1890–1891); clerk, Oliver Wire Company (1891–1899); chief clerk (1899–1903), plant manager, superintendent, Monessen Tin Plate Mills, American Tin Plate Company (1903–1905); secretary and plant manager (1905–1908), president, Phillips Sheet and Tinplate Company (1908–1918); president, Weirton Steel Company (1918–1929); chairman of the board, National Steel Company (1929–1957).

Ernest Tener Weir, industrialist and conservative commentator on public affairs, was born on August 1, 1875, to James and Margaret Manson Weir. James Weir was a day laborer, and his son grew up under harsh economic conditions. At age fifteen he began full-time employment as an office boy at the Braddock Wire Company. The

metal trade appealed to him, which helped to compensate somewhat for his menial pay of $3 a week. The next year he was hired as a clerk at the Oliver Wire Company. His managers were soon impressed with his drive and ability and envisioned him as a future manager of the firm. Yet Weir had become convinced that the wire industry was teetering on the brink of overproduction. In 1899 he moved to the American Tin Plate Company, where he became chief clerk. His economic and social status rising, he married Mary Kline of Pittsburgh in 1903. By that year he had reached the position of superintendent of the Monessen mills — subsidiaries of the American Tin Plate Company — and his future seemed secure.

Still Weir was restless. In 1905 he and his brother David joined James R. Phillips, who was organizing a new tinplate mill in Clarksburg, West Virginia. The firm, the Phillips Sheet and

Tin Plate Company, had had a rocky start; shortly after its founding James Phillips was killed in a train crash. As secretary and plant manager, however, Weir quickly brought the company to profitability by attracting a clientele of large corporations. Phillips's net sales surpassed $1 million in 1907 and reached more than $2 million in 1909. In that year, with his company straining at capacity, Weir helped arrange for the purchase of 400 acres of land in Hancock County, West Virginia, where a vast new facility boasting 20 tin mills was constructed — along with a company town, Weirton — at a cost of $1.11 million. Phillips suffered growing pains from this expansion, however; sales were initially sluggish and there was a shortage of labor in the rural surroundings of Weirton. Yet Weir was committed to the expansion and stressed to his associates the desirability of Phillips supplying its own steel. Weir formed the Weirton Steel Company on October 5, 1912. In December 1914 the Phillips Company purchased Weirton Steel, and the newly integrated firm saw its profits soar as European countries preparing for war increased their demands for steel, thus drawing the company into the export trade. In 1918 the Phillips Company reorganized into the Weirton Steel Company, with a capitalization of $30 million.

Although his corporate interests continued to grow, Weir's management style changed little. Rejecting detailed organization charts, he ran his firm like a family proprietorship, delegating generous amounts of authority yet maintaining the flexibility to intervene. Weir took a deep, personal interest in the decision-making process — an interest that often resulted in sudden inspections or hastily called meetings.

At the end of World War I, Weir recognized that the automobile industry based in Detroit held vast potential for profits. By selling Weirton Steel stock directly to the public, Weir raised funds to expand the firm's facilities to meet the automotive industry's needs. Weirton became the first company to license Armco's revolutionary continuous hot-strip mill, opening its facility in 1926. Yet Weir was growing worried that Weirton lacked sufficient ore reserves. As a result, in December 1929 he sought and successfully concluded negotiations to merge his firm with the Great Lakes Steel Corporation of Detroit and the ore, freighter, and furnace components of the M. A. Hanna Company of Cleveland. The merger created the National Steel Corporation, boasting large ore reserves, a new steel-producing facility, and access to Detroit markets. Weir became the new corpora-

Ernest Tener Weir (portrait by Ernest Hamlin Baker)

tion's chairman of the board, and George Fink of Great Lakes Steel assumed the presidency.

Reflecting the exceptional manner in which the merger fit together resources, National weathered the Great Depression without a single unprofitable year; in 1932 it was the only major steel producer to avoid a loss. Its profitability coincided with major — and expensive — programs to expand capacity, as Weir ambitiously planned new facilities in Detroit and Gary, Indiana. As a result of these initiatives, National became the fifth-largest steel producer in the United States. Part of Weir's corporate strategy involved increasing wages, and he pleaded with other firms to follow suit. After 1929 Weir insisted that his support for higher wages was offered as an antidepression tonic, not as an effort to preempt demands for unionization. In an address in early 1931 Weir stated his conviction that the worst of the Depression was over and warned that wage cutting could stifle the nascent recovery. He added that "modern thought" dictated that high wages produced efficiency, leading to lower unit costs and thus to higher profits. As vice-chairman of the Maurice Falk Foundation, established for "the encouragement, improvement and betterment of mankind," Weir financed and endorsed a Brookings Institution study which asserted that better distribution of purchasing power in a free-competition environment represented the best hope for capitalism. He further called for unemployment insurance and old-age pensions.

Yet these seemingly liberal positions were, to Weir, advisable only in the context of strictly free competition — including competition among workers. Labor unions, he believed, impinged on the freedom of the individual and hampered economic progress. In a 1937 address to the Economic Club of Chicago, Weir made clear his conviction that, should the "Little Steel" companies yield to the Steelworkers Organizing Committee (SWOC), unionization throughout the industry would follow. Widespread unionization, he further asserted, would empower a small group of "labor politicians," who "display an utter disregard for life, property, law or anything that stands between them and their ambition." Only free competition, unfettered by union or government, could determine the economic character of an industry, which in turn would determine wages and working conditions.

Given these views, the 1930s were troubling for Weir. He was deeply disturbed by the policies of President Franklin D. Roosevelt, which he felt undermined free enterprise and interfered with economic growth. He was particularly agitated by the prolabor posture of the federal government. Weir's antipathy toward unions had first become apparent shortly after World War I, when he closed the profitable Weirton plant in Steubenville, Ohio, as a strike-breaking measure. The Roosevelt administration thus proved to be an obstacle to Weir's antiunion policies. During the brief life of the National Recovery Administration, Weir became embroiled with the National Labor Board (NLB) over the implementation of section 7(a) of the National Industrial Recovery Act, which guaranteed labor the right to bargain collectively. In the first major proceeding relating to 7(a), a federal court rejected government charges that the company had discriminated among its employees and interfered in union elections. The court upheld company-sponsored elections, asserting that the NLB had no authority to conduct an election of its own. The board, the court stated, was only "acting as a group of individuals and not as a lawfully constituted body."

The passage of the National Labor Relations Act signaled further complications. In June 1936 the Congress of Industrial Organizations (CIO) sponsored the SWOC in an attempt to unionize the steel industry. After the SWOC reached an agreement with U.S. Steel in March 1937, Weir worked stridently to avoid becoming the next union conquest. Part of his strategy involved preempting employee demands; hours before U.S. Steel and SWOC signed a union contract, Weir announced a major wage increase at National. Nevertheless, SWOC efforts to unionize National continued. In May 1937, SWOC filed charges of unfair labor practices against the company with the National Labor Relations Board, which in June 1941 ordered National to cease discouraging membership in SWOC. Fuming that the National Labor Relations Act was unconstitutional, Weir signed an exclusive bargaining contract with the newly established Weirton Independent Union — a union with no CIO affiliation. Furious labor disputes followed, with the U.S. Third Circuit Court of Appeals ordering the company to comply with the Labor Relations Act. A reinvigorated CIO drive was not successful until the post–World War II era.

Weir believed that the Roosevelt administration's policies created nothing but conflict and enmity between employers and employees. Speaking to the West Virginia Chamber of Commerce in May 1938, Weir asserted that if the administration were to change its policies business recovery would follow within six months. Yet Weir did more than bash White House policies; he also submitted his own depression-fighting proposal. A five-point plan, it began by suggesting that the federal government take a pledge to "retain unaltered the principles and structure of the American systems of business and government." He pleaded for consistency in federal economic controls, a revision of the tax code to promote growth, and a "balanced" labor law. Point five insisted that the president send his advisers "and their kind back where they came from. Cease attacks on business."

Weir's outrage against the New Deal led him increasingly toward right-wing political philosophies and activities. He blasted the Roosevelt administration for aligning itself against productive industry. He became an active member of the Liberty League, founded in 1934 by a bipartisan group of conservatives to thwart the New Deal. In 1940 Weir served as chairman of the Republican National Finance Committee.

As European war orders began to stimulate the American economy, and as the United States itself girded for World War II, National benefited from its ability to meet defense orders. By 1941, 90 percent of the company's facilities were devoted to the war program, and continued expansion — especially in pig-iron capacity — took place. Not all of National's impressive profits of this period could be attributed to generous government contracts, for the firm was remarkably successful in controlling costs during both stringent and profitable times. When the recession of 1937–1938 resulted in a decline of $50 million in sales, National was able to cut costs by a roughly equivalent amount. After the war National reduced costs in 1946, 1949, and 1954, when sales also declined. With controlled costs, National also helped the war effort with increased production, prompting Gen. Douglas MacArthur to send a congratulatory message, which noted "the magnificent record you are making on behalf of our beloved country."

Weir's firm belief in vigorous expansion guided National as it became a leader in light, flat-rolled products. This effort allowed the firm to take advantage of its strategic location near the Motor City to supply the postwar automobile industry. Weir, however, remained concerned that his burgeoning company be guaranteed enough coal supplies to meet future needs. In 1946, under Weir's guidance, National established a subsidiary firm, the National Mines Corporation, in order to manage National's coal reserves in Pennsylvania, West Virginia, and Kentucky.

Despite his uninterrupted business success, Weir's personal life was marred by the death of his wife, with whom he had twin sons and a daughter. In 1941 his second marriage, to Aeola Dickson Siebert, ended in divorce. Later that year he married Mary Hayward, with whom he had a son.

In the last decade of his life Weir gave increasing attention to public affairs. He expressed delight with the Taft-Hartley Act and took great satisfaction from the Supreme Court decision that declared President Harry S Truman's 1952 seizure of steel mills unconstitutional. It was also in 1952 that he published his book *Progress Through Productivity,* a testimonial to free enterprise. He served as director of the American Iron and Steel Institute (he was president in 1939–1940) and on the advisory board of the Transportation Association of America. Maintaining homes in West Virginia, Florida, and Pittsburgh, Weir served as vice-president of the Pittsburgh Symphony Society and as a trustee of the East End Christian Church of Pittsburgh. In his spare time he enjoyed golf and horseback riding.

Early in 1952 Weir suffered a heart attack, and in April he announced his decision to retire from National Steel. The decision was particularly painful for him, as he emphasized that National was his primary interest in life. Two months later Weir suffered a cerebral hemorrhage and died in Philadelphia.

Selected Publications:

"Steel Industry Needs," *Iron Trade Review,* 73 (September 20, 1923): 805–806, 811;

"Selling Practices in the Steel Industry," *American Iron and Steel Institute Yearbook* (1924): 75–85;

"Who's to Blame, Machines or Men?" *American Magazine,* 116 (August 1933): 60–61;

"New Responsibilities of Industry and Labor," *Annals of the American Academy of Political and Social Science,* 172 (March 1934): 76–87;

"Hold Fast to Tried and True Principles . . . Fight for Them If Necessary," *Iron Age,* 133 (May 31, 1934): 22;

"Present Relations of Business to Government," *Vital Speeches,* 1 (April 22, 1935): 476–480;

"I Am What the President Calls an Economic Royalist," *Fortune,* 14 (October 1936): 118–23+;

"Labor Relations in This Administration Founded on Basis of Conflict, Says E. T. Weir," *Iron Age,* 149 (December 2, 1937): 78, 80–83;

"E. T. Weir Urges Country to Try Collective Cooperative," *Iron Age,* 140 (December 16, 1937): 74–75;

"Roosevelt Caused Depression but Won't Change, Weir Finds," *Iron Age,* 141 (May 26, 1938): 65–66;

"Profits and Patriotism," *Iron Age,* 144 (October 19, 1939): 95–98;

"New Deal 'Theories' and Big Spending Curb Sound Recovery, Says E. T. Weir," *Iron Age,* 144 (November 30, 1939): 73–74;

"F. D. R.: A Laggard in Defense, While Producing Bad Foreign Relations," *Steel,* 106 (May 27, 1940): 23–24;

"Steel Industry and World Economic Conditions," *Commercial and Financial Chronicle,* 166 (October 23, 1947): 1641;

"After the Turn of the Year," *Commercial and Financial Chronicle,* 167 (January 22, 1948): 388–389;

"Steel Capacity Surpasses Need," *Iron Age,* 165 (April 6, 1950): 114–115;

Facts about Competition in the Steel Industry: Statement before the Special Subcommittee of the Judiciary Committee, House of Representatives, Washington, D.C., May 4, 1950 (N.p.: National Steel Corporation, 1950);

Statement on Our Foreign Situation (N.p., 1951);

Progress through Productivity (New York: Newcomen Society, 1952);

Notes on the Foreign Situation Based on a Trip Abroad (Pittsburgh, 1953);

"Importance of Steel: Metallurgical and Human Problems Ahead," *Commercial and Financial Chronicle,* 179 (June 3, 1954): 2428–2429+;

"Importance of the Steel Industry," *Iron and Steel Engineer,* 31 (August 1954): 144–147;
Importance of the Steel Industry (American Iron and Steel Institute, 1954).

References:
Irving Bernstein, *Turbulent Years: A History of the American Worker, 1933–1941* (Boston: Houghton Mifflin, 1970);

William T. Hogan, S.J., *An Economic History of the Iron and Steel Industry in the United States,* 5 volumes (Lexington, Mass.: Lexington Books, 1971);
Hogan, *Minimills and Integrated Mills: A Comparison of Steelmaking in the United States* (Lexington, Mass.: Lexington Books, 1987).

Archives:
Weir's speeches are found in the American Iron and Steel Institute Collection in the Hagley Museum and Library, Wilmington, Delaware.

Weirton Steel Company

by Glen V. Longacre III

West Virginia University

Having been in operation for over 75 years, the Weirton Steel Company's roots date back to 1905, when James R. Phillips purchased the Jackson Iron and Tin Plate Company for $127,000. Located in Clarksburg, West Virginia, the mill had been constructed in 1901–1902 and consisted of five tin mills, three sheet mills, and a bar mill. Several other businessmen — Ernest T. Weir, Edmund W. Mudge, Edward Kneeland, and W. H. Baldridge among them — joined Phillips in the purchase.

The new plant had hardly begun operation when Phillips was killed in a train crash near Harrisburg, Pennsylvania. The mill's daily operation and management was taken over by Ernest T. Weir, the plant manager and company secretary. Weir changed the mill's name to Phillips Sheet and Tin Plate Company in memory of the founder.

From 1907 to 1909 Phillips experienced astounding growth. Weir had instituted an aggressive program of expansion, and the mills produced tinplate for a general rather than a specific market — reasons which explain a net sales increase of 150 percent during that period. By 1909 expansion had reached its limit at the Clarksburg plant, and water resources had become scarce. Weirton, having convinced major stockholders of the necessity of building a new plant, began a search for a suitable location. An ideal site was found on the banks of the Ohio River in Hancock County, West Virginia; the land was close to road, rail, and river-transportation systems. Originally eight tinplate mills were planned for construction,

but after construction commenced in June 1909, it was decided to build 20 tin mills and company houses to accommodate 75 families. The new plant and the surrounding area would be called Weirton.

The new tin mills were completed by January 1911, but in the following years the Phillips mill encountered problems, plagued by natural disasters, labor disputes, and increasing competition. In 1911 Phillips purchased a competitor firm, the Pope Tin Plate Company in Steubenville, Ohio, for $1.47 million. By the end of 1911 Phillips had three tin-mill plants in operation at Clarksburg, Steubenville, and Weirton, and in 1913 the Weirton plant expanded to include the production of its own hot- and cold-rolled sheet steel. By 1918 the Phillips Company had more than 50 tinplate hot mills in operation and was making plans for further expansion of pig-iron and open-hearth capacity. At a stockholders' meeting in July 1918 the company's name was officially changed to the Weirton Steel Company.

The early 1920s saw Weirton Steel continue to expand at a rapid rate, even though the steel industry was in a depression. By 1928 Weirton Steel had purchased or acquired holdings in, among other firms, the Wheeling Steel Corporation, the Easton Can Company of San Francisco, the Bates Iron Company, and the Holman Cliffs Mining Company.

In 1929 Weirton Steel, looking to become an integrated steel producer, merged with the Great Lakes Steel Corporation and the M. A. Hanna

Weirton Steel Company plant on the Ohio River in Hancock County, West Virginia, in 1932 (photograph by Aiklee)

Company of Cleveland, Ohio, to form the National Steel Corporation, and the facilities at Weirton became National Steel's flagship plant. Throughout the Depression of the 1930s the National Steel Corporation operated in the black, the only major steel company to do so. Factors contributing to National Steel's success were its determination to install modern and technically advanced equipment, its proximity to raw materials and transportation systems, and its concentration on light, flat-rolled steels — the only product line little affected by the economic downturn.

The Weirton plant continued to expand through the Depression, with more coke ovens added in 1930, one Bessemer converter in 1936, and a second in 1941. During World War II National Steel's production was devoted entirely to the war effort, as the plants ran at full capacity. Similarly, the late 1940s and 1950s brought continued expansion and improvements to plant facilities for National Steel's Weirton Division.

With previous decades having been devoted to increasing National Steel's steelmaking capacity, the 1960s were a period in which the company advanced technologically. In 1967 Weirton Steel introduced several significant new facilities to house a vacuum degasser, basic-oxygen furnaces,

and a continuous slab caster, all of which improved product quality while greatly reducing both the time and expense that steelmaking had once required.

Unfortunately, these additions proved inadequate to keep the company — and the Weirton plant — viable in the changing steel world of the 1970s and 1980s, and competitive problems set the stage for perhaps the most significant occurrence in the history of Weirton Steel in the early 1980s. Labor costs, a shrinking market, and stiff competition from imported steel forced National to reexamine its operations. In 1982, after having operated in the red for several years, National Steel's Weirton Division recorded approximately $104 million in losses. The Weirton Division had clearly become cost-prohibitive, and it was decided that it would be slowly phased out. In March 1982, however, National Steel chairman Howard Love offered to sell the Weirton Division to the employees, and an Employee Stock Ownership Plan (ESOP) was proposed. On September 23, 1983, the employees at the Weirton Steel Division voted overwhelmingly to accept the plan, which included a 20 percent cut in pay to help finance the ESOP. Four months later the ceremonial signing was held, making Weirton Steel Corporation the largest employee-owned company in the United States, and the nation's ninth-largest integrated steelmaker.

Since that time, Weirton Steel has recovered under the guidance of president and chairman Robert Loughhead. The company borrowed money for modernization; shifted its product mix from strictly tinplate; and equally as important, installed a genuinely participatory style of labor-management relations. A return to profitability followed immediately — fed partly by labor costs that were $3 per hour below the industry average — and Weirton became an efficient producer of quality steel with outstanding labor relations. Symbolic of the new life at Weirton was a $52 million grant from the U.S. Energy Department in 1987 to build facilities for the KR process, a direct-reduction procedure for making iron.

The renaissance of Weirton Steel is one of the few success stories in the basic American steel industry since the mid 1970s. It is also seen by industrial analysts, such as John Hoerr, as demonstrating the crucial importance of cooperative labor relations to the survival of American steelmakers.

References:
John T. Hoerr, *And the Wolf Finally Came: The Decline of the American Steel Industry* (Pittsburgh: University of Pittsburgh Press, 1988);
William T. Hogan, S.J., *An Economic History of the Iron and Steel Industry in the United States,* 5 volumes (Lexington, Mass.: Lexington Books, 1971);
Hogan, *Minimills and Integrated Mills: A Comparison of Steelmaking in the United States* (Lexington, Mass.: Lexington Books, 1987);
Charles Longenecker, "Weirton Steel Company," *Blast Furnace and Steel Plant,* 28 (August 1940): 773–777.

Jesse Floyd Welborn

(March 9, 1870 – September 30, 1945)

by H. Lee Scamehorn

Historic Learning and Research Systems

CAREER: Bookkeeper, Colorado Fuel Company (1890–1892); sales agent and general sales agent (1893–1903), vice-president for sales and traffic (1903–1907), president (1907–1929), chairman (1929–1933), member of board, Colorado Fuel & Iron Company (1934–1945).

Jesse Floyd Welborn was associated with the Colorado Fuel & Iron Company (CF&I) for more than half a century. The son of John Wesley and Jennie Roberts Welborn, he was born at Ashland, Nebraska, on March 9, 1870, and raised on the family farm. He left at the age of 17 to work as a cashier for the state bank at Indianola. He remained there for three years before relocating to Denver, where he was hired by the Colorado Fuel Company as a clerk in the sales department. In 1892 that firm merged with the Colorado Coal & Iron Company to form the Colorado Fuel & Iron Company, the region's premier producer and distributor of iron, steel, coal, and coke.

Under the tutelage of Alfred C. Cass, the corporation's vice-president for sales, Welborn

developed what proved to be acute marketing skills. In 1899 he was named general sales agent in charge of a newly established office in New York City. Four years later Cass died, and Welborn was promoted to vice-president for sales and traffic, serving under Frank J. Hearne, president and chief executive officer. Following Hearne's death in 1907, Welborn was elected president of the enterprise.

As the head of Colorado's leading heavy industry, Welborn was widely admired and respected for his leadership qualities. A man of imposing build and vigorous personality, he was modest almost to a fault. He readily responded to requests for interviews but gained a reputation for giving out little to no information about himself. Reporters called him "Welborn the Silent."

He served as vice-president during a time when George Jay Gould and John D. Rockefeller jointly controlled CF&I. In 1907 financial problems forced Gould to surrender his interest in the company, and Rockefeller quickly appointed his own man, Lamont Montgomery Bowers, to take charge of the firm as vice-president and chairman of the board. Although Welborn assumed the company presidency, he played a subordinate role to Bowers. The violent coal miners' strike of 1913–1914 forced John D. Rockefeller, Jr. — who by then held control of his family's business interests — to change leadership. Welborn, however, remained president and became the chief executive officer, a position he filled until 1929.

Now under Welborn's direction, CF&I experienced growth and relative prosperity — conditions which had eluded the company in previous years. Capital improvements included an expanded open-hearth department, by-product coke ovens, and an enlarged rail mill. Welborn also sought, with limited success, to broaden the company's product line.

Welborn was named chairman of the board in 1929 and served in that capacity until the Depression economy forced the firm to reorganize in 1933. Afterward, Welborn remained at the firm as a member of the company's directorate, largely due to his close relationship with Rockefeller, Jr. The Rockefellers sold their interest in January 1945 to a syndicate headed by New York investment banker Charles Allen.

During the miners' strike of 1913–1914, Welborn led the coal operators in their steadfast refusal to recognize the United Mine Workers of America. After the miners' walkout ended, however, he played a key role in implementing the Employee Representation Plan (ERP). Better known as the Rockefeller Plan, the ERP was an attempt to form a company union, ostensibly to provide for harmonious relations between labor and management. The scheme was devised by William L. Mackenzie King, former labor minister of Canada (later prime minister) at the request of Rockefeller, Jr. The plan was adopted in the coal and iron mines, at the Pueblo, Colorado, steelworks, and in the administrative offices at Denver and Pueblo. Although resented by coal miners, the plan was eventually accepted by a majority of employees.

The plan was abandoned in the coal camps in 1933 but remained in effect at the Pueblo works until 1942.

The Rockefeller Plan produced major changes in the company's labor policies. Workers were given a voice in determining wages and conditions of labor. Mechanisms were created for the airing of grievances. The plan also provided welfare programs to improve the quality of life in the mining camps and the mill town. Housing was improved, and social, recreational, and educational programs were expanded for workers and their dependents. The instrument for many of these reforms was the industrial department of the Young Men's Christian Association, which maintained "Ys," or clubhouses, at the steelworks and in the mining camps.

After leaving CF&I management, Welborn devoted much of his time to community services. he was a trustee for a local theater group and for many years served as president of the Denver Community Chest. During the Depression he directed the Denver Unemployment Council's efforts to counter the impact of economic stagnation.

Welborn married Ada Elizabeth Baker in Milwaukee on June 2, 1903, and the couple had three children. Welborn's wife died in 1936. In spite of a heart attack in 1936, Jesse Welborn remained active in Denver until shortly before his death on September 30, 1945.

Publication:
"Problems of Capital and Labor," *Iron Trade Review,* 55 (October 8, 1914): 690.

References:
"Jesse Floyd Welborn," in *History of Colorado,* edited by James B. Baker and LeRoy R. Hafen (Denver, Colo.: Linderman, 1927);
"Jesse Floyd Welborn," *Sketches of Colorado,* edited by William C. Ferrill (Denver, Colo.: Western Press Bureau, 1911).

Welfare Capitalism

by Kevin M. Dwyer

George Washington University

A series of reforms from within carried out by a company can be defined as welfare capitalism. Various systems by which management provided employee benefits gained wide currency during the Progressive Era in America. In the steel industry the United States Steel Corporation set the standard for this type of reform.

The great merger movement in the American steel industry brought about a significant shift in labor-management relations. The billion-dollar merger that created the giant U.S. Steel Corporation was the consequence of banker J. P. Morgan's desire to stabilize the industry at large. As part of its program of stabilization the new corporation enacted a series of labor reforms to improve labor-management relations. U.S. Steel executives also realized that the corporation could reap public-relations benefits from labor reform. Public discontent over the trusts and the willingness of the federal government to break them up were increasing, and it was hoped that labor reform would serve as evidence that U.S. Steel was a "good" trust.

Elbert H. Gary, chairman of U.S. Steel, gradually instituted policies to improve the financial security and environmental safety of steelworkers after 1902, calling the reforms "a simple duty that management owes labor." On other labor matters, however — specifically working hours and company acceptance of unions — Gary and his colleagues held the line. In 1922 the administration of President Warren G. Harding began to press the steel industry to cut short its working day to eight hours. The industry reluctantly adopted the short day in 1923. Muckraking journalists, privately funded public-protection associations, and "radical" industrialists further pressed for the seven-day-a-week labor practices in the industry. In the hopes of preempting independent organized labor, some steel companies — notably Bethlehem and Midvale — formed company unions, called Employee Representation Programs (ERPs), during World War I. The ERPs were ostensibly im-plemented to create a more harmonious working relationship between labor and management. Although the ERPs were often successful in instituting social improvements in steel towns, they generally were not effective instruments for genuine employee representation.

The steel industry's welfare capitalism began in the steel towns that sprang up around the mills. By 1900 three-fifths of America's steelworkers lived in steel towns, often in housing rented or mortgaged from the company. The system of social organization in these cities and towns — many of which were named after the local steel master — were dependent upon the steel plant, and, as such, a company's relationship to the steel town was often a patriarchal one. "The ruling presence of the mill" dominated not only the economy but "the mentality of the steel centers," as industry historian David Brody notes; and steel executives established close ties with community leaders because "local support for management was vital" to normal plant operations. The steel executives often held public office in the town.

Steel companies built many of the steel towns' schools, parks, and playgrounds and generously funded local churches and chapters of the YMCA. "Community health" initiatives — an integral part of welfare capitalism — provided health care through company-sponsored hospitals and clinics. Such measures insured generally favorable relations between the firms and the town leaders, providing the companies with a valuable means of leverage against dissident workers.

In 1902 U.S. Steel announced a stock-purchase plan for employees "from the president down to the man with the pick and shovel," offering corporate stock at a discount, with low-interest installment plans available to production workers. The program was designed, Gary said, to bind "the interests of capital and labor . . . more closely together." U.S. Steel board member George W. Perkins, an architect of the plan, stated that he believed stock ownership made workers "more

An on-the-job safety class at U.S. Steel, part of the company program to reduce occupational injuries

competent" to analyze labor-management issues, since it gave them "the standpoint of a partner rather than . . . the standpoint of a mere hireling." Participation in the program among production workers was for the most part limited to skilled laborers, however, as they were more likely to have some portion of disposable income; common laborers living payday to payday could not afford the benefits. As Gerald Eggert has observed, only a sixth of the labor force took part in the stock plan.

Wage rates continued to be a divisive issue for steel management. Gary's larger program for price stability in the industry hinged on standardized production costs — which, in Gary's view, meant having to find a way to standardize the wage rate. But Gary encountered deep philosophical resistance from ex–Carnegie Steel executives — such as William Corey, Alva Dinkey, and William Dickson — who had moved to U.S. Steel. They had been trained to value aggressive competition, not price stability. Disagreement between the Gary and Carnegie camps came to a head during the 1904 and 1907 recessions. In complete violation of the "cut costs and run full" competitive ideology of Carnegie, Gary initiated a program to stabilize the wage structure and cut back on ca-

pacity utilization, which meant laying off workers. To the dismay of the Carnegie men, Gary had frozen costs and run at half speed. After 1908 steel prices, costs, and wages developed along the lines that Gary sought.

Skilled workers, generally the first hired and the last fired, approved of Gary's program, feeling that wage stabilization in return for running the mills at half speed provided a measure of security during the slow times. Unskilled laborers, whose jobs would be sacrificed in return for a promise of no wage cuts, were not as thrilled, and approval of wage stabilization was confined mostly to those plants that could run at half speed without stifling production for long periods of time.

By 1908 U.S. Steel was seeking to make its plants safer, an effort mostly driven by a desire to upgrade its image, which had been tarnished by muckrakers, industrial watchdog groups, disaffected workers, and the federal government. William Hard's November 1907 article, "Making Steel and Killing Men," graphically illustrates the hazards common to steelworkers for readers of *Everybody's Magazine*. Hard reported 46 deaths and almost 2,000 serious injuries at U.S. Steel's South Works at Chicago in 1906. Although officials had already resolved to redress occupational

...anded quick action, ...y" became the motto ... new $500,000 safety ...ed in August 1908. In four ... were cut by 43 percent.

...ary 1910 U.S. Steel established a ... in which workers were entitled to ... medical treatment for work-related inju-...s; at least one-third wages during convales-cence; between six and 18 months' full wages for permanent disability; and at least 18 months' wages paid to a worker's dependents in case of his death. The corporation added a $12 million pension fund in January 1911. As with many of U.S. Steel's welfare programs, however, the pension plan primarily benefited skilled workers, who enjoyed longer, steadier employment with the company.

Another change came in plant facilities — toilets, clean and separate eating areas, and cold, in-shop drinking water — which were upgraded or established in cases where none existed. "The enforcement of cleanliness and order, fencing, painting, cutting weeds and collecting garbage" were, by Gary's orders, matters requiring the "consideration even of the presidents of the subsidiary companies," as one U.S. Steel assistant put it before a meeting of the American Iron and Steel Institute in 1912.

Worker unrest and complaints about hours and working conditions continued, and the era of welfare capitalism ended after the great strike of 1919. Substantial changes in working conditions, such as shorter hours, would be effected by the government.

References:
Stuart Brandes, *American Welfare Capitalism, 1880–1940* (Chicago: University of Chicago Press, 1976);
David Brody, *Steelworkers in America: The Non-union Era* (Cambridge, Mass.: Harvard University Press, 1960);
Gerald Eggert, *Steelmasters and Labor Reform, 1886–1923* (Pittsburgh: University of Pittsburgh Press, 1981);
Gabriel Kolko, *The Triumph of Conservatism* (New York: Free Press of Glencoe, 1963).

Wheeling-Pittsburgh Steel Company

by Elizabeth M. Nolin

West Virginia University

and

Bruce E. Seely

Michigan Technological University

In December 1968 the Wheeling Steel Corporation and the Pittsburgh Steel Company — each dating back to early-twentieth-century consolidations of small regional iron and steel firms — merged to form the Wheeling-Pittsburgh Steel Company. The merging companies had had reputations for conservative corporate policy, leading both companies into difficulties after World War II. The 1968 merger was largely inspired by problems at Wheeling Steel, but it was hoped the combination would produce a strong corporation. Unfortunately, the merger was effected just as the American steel industry was beginning to falter in a changing competitive environment.

Wheeling Steel Corporation

The Wheeling Steel Corporation grew out of three venerable iron companies — the Wheeling Steel & Iron Company, the La Belle Iron Works, and the Whitaker-Glessner Company — in the Wheeling, West Virginia, area. By the turn of the century these three companies had greatly expanded their operations, each having added steelmaking facilities to supply their sheet, plate, and pipe mills. Business relations among the three companies were strong and — given the trend toward increasing company size in the industry — their merger in July 1920 seemed to be a natural one.

Allison Maxwell, Jr., president of the Pittsburgh Steel Company, and Donald Duvall, executive vice-president of Pittsburgh Steel, after they became chairman of the board and president, respectively, of Wheeling Steel in 1967 (Fortune Magazine photograph)

The new company began with ten coal mines, two iron mines on the Mesabi Range, several large steamers, and three main plants. The largest was in Steubenville, Ohio, where La Belle had constructed blast furnaces; open hearths; and sheet, plate, and tube mills. In Portsmouth, Ohio, the company had an open-hearth plant and finishing mills that had been built by Whitaker-Glessner. In Benwood, just south of Wheeling, a Bessemer converter and pipe mill were later expanded by acquisition of an adjacent National Tube plant. Several smaller works were scattered for 25 miles along the Ohio River between Steubenville and Wheeling.

By 1926 Wheeling Steel had an ingot capacity of 1.25 million tons, making it the sixth-largest steelmaker in the country. The company emphasized production of semifinished items — sheet, tinplate, pipe, and rods — and the fabrication of finished steel products, including nails, steel drums, and fencing. Its strength in light-rolled steel production brought Wheeling Steel through the Depression in better shape than some of the giant firms that depended on the heavy-steel market, which had shrunk in the economic downturn.

Light-steel production also gave Wheeling an edge in the specialty markets created by World War II, as the company made bomb fins, landing mats, "blitz" cans, and other wartime items.

Wheeling Steel, however, faced a difficult postwar transition; the company had resisted technological change during its 25-year history, and its plants and equipment had become obsolete. Wheeling management's suspicions of innovation were reflective of general attitudes in the steel industry; however, company president John Neudoerfer recognized that to allow Wheeling to remain technologically stagnant threatened its future. In the late 1940s he launched an extensive modernization program — a program that by 1958 had cost the company $75 million. Yet the return on its investment seemed to be a negligible one, as Wheeling continued to be outpaced by other large steel companies. By the early 1960s Wheeling had slipped to twelfth-largest steel company, and Neudoerfer had been succeeded by William Steele — a former Wheeling blast-furnace superintendent with little preparation for top management. By the mid 1960s the company was in trouble.

Pittsburgh Steel Company

The Pittsburgh Steel Company was organized in 1901 by combining two small companies — the Pittsburgh Steel Company, Ltd., and the Pittsburgh Steel and Hoop Company in Glassport, Pennsylvania. Pittsburgh Steel's initial plant was in Glassport and contained two hoop mills producing 30,000 tons of hoops and cotton ties annually. A wire mill was opened in Monessen, Pennsylvania, in 1903, and eight open-hearth furnaces and rolling operations were added to it in 1907. A pair of blast furnaces and partnerships in mining companies were added in 1913; additional open hearths were built in 1917 — bringing the company's ingot capacity to 600,000 tons per year; and in 1919 coal mines were acquired, completing the integration of the company.

Pittsburgh Steel made two important purchases during the 1920s to provide outlets for its finished steel products. In 1921 the company acquired the National Steel Fabric Company — and in 1925 it bought the Pittsburgh Steel Products Company, which provided the automotive industry with seamless tubing. During World War I the company built a second plant on the Monongahela River, and in 1924 it licensed a mill from the Demag Company in Germany to produce large-diameter drilling pipe and casing.

Afterwards, Pittsburgh Steel seemed content with its position, making no other acquisitions or additions to its capacity. Pittsburgh Steel's management adhered to basic conservative policies through 1945; the company's share of American steel production held at about 1 percent.

As had Wheeling Steel, this small, sleepy steel company found that it could not continue as it had after World War II. It made its first expansion in productive capacity in 1946, buying an uncompleted blast furnace from the government and two finished-steel-products companies. But this move left the company with one-third more ingot capacity than finishing capacity, so that profits depended on the sale of semifinished steel, which in turn depended on the state of the economy. In 1949 the company embarked on a three-year expansion program under a new president, Avery C. Adams. Finishing capacity was expanded by 82 percent in order to bring it into balance with steel production. Essential to this expansion was the purchase of the Thomas Steel Company of Warren, Ohio, a producer of cold-rolled strip steel. The purchase not only placed Pittsburgh Steel in the sheet steel market but also provided a guaranteed outlet for its larger steel capacity, which at the time was approximately 1.6 million tons annually.

Adams's successor, Allison R. Maxwell, continued the capital improvements drive, promising "everything new but the name." His program included the addition of hot and cold sheet mills at the Allenport plant, a new billet mill, and rebuilt blast furnaces and open hearths in Monessen. Pittsburgh Steel ranked about fourteenth in size among American steel companies, and industry observers praised Maxwell for his vigorous management and attention to cost-cutting. Yet the result of the improvements — totaling about $120 million during the 1950s — was limited. In 1959 *Forbes* reported that Wall Street experts regarded Pittsburgh Steel as " 'a company that prosperity forgot.' " The company's relatively inefficient facilities made money only in the best years, and it paid common stock dividends only three times between 1932 and 1961. In 1961 Maxwell hoped to bring the company in step with the industry leaders by adding two new oxygen converters, rolling mills, furnaces, and a sintering plant. The plan's financing, however, required the sale of stock at a time when the stock was worth 25 percent of its book value.

The Merger

In 1963 Wheeling was shaken from its complacency by the arrival of a new director, Norton Simon, whose Hunt Foods had bought a large block of Wheeling stock. According to a 1965 *Fortune* report, upon his arrival at Wheeling, Simon found "a corporate world unto itself, where the doors were shut, the clocks had run down, and life followed the remembered patterns of the past." Simon had the reputation of a corporate raider and immediately launched an aggressive challenge to the company's management. By 1964 Simon had declared war on Steele, who eventually resigned with most of the board. An outsider — Robert Morris from Monsanto — was appointed to replace Steele as president, and Simon took over the chairmanship. Simon's coup followed a sharp drop in company profits, and one of Simon's first acts as chairman was to tell stockholders that the paying of dividends might be suspended for four years.

Simon and Morris worked hard to restore the company to profitability, although it lost money in both 1965 and 1966. They launched an award-winning advertising campaign built around the slogan "Hustle!" that painted Wheeling as eager and aggressive compared to the other stodgy firms in the industry. Simon's brusque style, however, combined with his lawsuits brought against several former Wheeling directors, antagonized other steel companies, as well as the bankers. Failing to secure needed loans in 1966 for further modernization, Simon finally gave up and accepted a plan developed by Wheeling's lenders and sold his interest in the company. Simon was replaced; Pittsburgh Steel's Allison Maxwell, Jr., and his assistant, Donald Duvall, became the new chairman and president, respectively, of Wheeling Steel in April 1967.

A merger was not necessarily envisioned at the outset of these developments, although Maxwell noted that he had contemplated a merger with Wheeling in 1958. Yet it was clear that Pittsburgh Steel knew it faced serious difficulties — Duvall even quipped at a meeting in May 1967, "Allie, why did we get ourselves into this, anyway?" — given the problems facing both companies. The board of directors from both companies set up merger committees in April 1968, and the merger was approved in September.

Despite initial sounds of approval from most industry observers, who felt that the Wheeling operations would flourish under Pittsburgh Steel's competent management team, the merger was soon judged a disappointment. The merger made Wheeling-Pitt (as it came to be known) the ninth-largest steelmaker; however, it did little more than survive in the early 1970s. Conditions improved somewhat under Robert Lauterbach, who became president in 1970 and chairman in 1973. His strategy was to place a renewed emphasis on the company's essential product — flat-rolled carbon steel. He sharply pruned back marginal product lines, top-heavy management, and debt, and the company profitably responded when worldwide steel demand jumped in 1973.

But the improvements had only a temporary effect, and the firm joined the ranks of many other troubled American steel producers by 1980. In 1978 former company president Dennis Carney had succeeded Lauterbach as chairman of the struggling company. With a production and research background at the United States Steel Corporation, Carney was a brilliant manager determined to save the company with a thorough modernization program. By 1982 the plan included the installation of continuous casters at Steubenville and Monessen and the construction of a rail mill with federal loan guarantees. He moved ahead with his $500 million program thanks in part to wage concessions from the company's workers; he also negotiated a joint venture with Nisshin Steel of Japan. But by 1985 Carney's relations with the steelworkers' union had soured — in part because Carney, who had a reputation as an autocrat, refused to listen to union ideas on how the company could escape debt. The union refused more wage concessions; the company sought Chapter 11 bankruptcy protection and shed its pension obligations. The union retaliated with a strike that it successfully portrayed as a lockout. In the end, Carney was forced out, and the company closed the Monessen mill.

The company survived, however, because even after the strike its wage contracts were below the national level. With an increase in steel demand in the late 1980s, the company emerged from bankruptcy in the second quarter of 1989 as the eighth-largest American steelmaker. With its headquarters still in Wheeling, it employed 6,500 workers.

References:

Duncan Burn, *The Steel Industry, 1939–1959: A Study in Competition and Planning* (Cambridge: Cambridge University Press, 1961);

Dan Cordtz, "Antidisestablishmentarianism at Wheeling Steel," *Fortune,* 76 (July 1967): 105–109, 133–136;

Stanley Devlin, "Investment Audit of Wheeling Steel," *Magazine of Wall Street,* 76 (April 28, 1945): 86–87, 112–13;

"Geometry Lesson," *Forbes,* 83 (June 15, 1959): 23–24;

John A. Heitmann, "La Belle Iron Works," in Paul Paskoff, ed., *Iron and Steel in the Nineteenth Century, Encyclopedia of American Business History and Biography* (New York: Facts on File, 1989), pp. 228–229;

"Hobson's Choice," *Forbes,* 88 (September 15, 1961): 53–55;

John P. Hoerr, *And the Wolf Finally Came: The Decline of the American Steel Industry* (Pittsburgh: University of Pittsburgh Press, 1988);

William T. Hogan, S. J., *The 1970s: Critical Years for Steel* (Lexington, Mass.: Lexington Books, 1972);

George J. McManus, "Has Hustle Helped?" *Iron Age,* 198 (August 25, 1966): 28–29;

Earl Chapin May, *From Principio to Wheeling, 1713–1945: A Pageant of Iron and Steel* (New York: Harper, 1945).

Archives:

The West Virginia Collection at West Virginia University holds the Wheeling Steel Corporation Annual Reports and *Management Memo* and the Wheeling-Pittsburgh Steel Company Annual Reports.

Charles M. White

(June 13, 1891 – January 10, 1977)

by Carol Poh Miller

Cleveland, Ohio

CAREER: Millwright helper, American Bridge Company (1913– circa 1915); various supervisory positions, Jones & Laughlin Steel Corporation (circa 1915–1930); assistant vice-president, operations (1930–1935), vice-president, operations (1935–1945), president (1945–1960), chairman of the board, Republic Steel Corporation (1956–1960).

Spanning nearly half a century, Charles McElroy White's career in the steel industry was mostly spent in top executive positions with the Republic Steel Corporation. By his own admission, he patterned his life and business philosophy after that of Republic Steel chairman Tom Girdler, who had recruited him. Like Girdler, White was a feisty individualist whose language in and out of the boardroom the *Cleveland Plain Dealer* once described as "variously literate and as salty as that of a longshoreman." His motto was "never duck anything that comes your way."

Born on June 13, 1891, in Oakland, Maryland, to Charles F. and Estella Jarboe White, "Charlie" White spent most of his youth in nearby Hutton, where his father worked in the lumber business. By age twelve White was working in a lumber camp, driving a mule team. He later worked on a railroad-construction gang and in a tannery. Though not a high-school graduate, White was admitted to the University of Maryland at Baltimore, from which he received a bachelor's degree in mechanical engineering in 1913. Upon graduation White worked as a millwright helper with the American Bridge Company in Ambridge, Pennsylvania, then joined the Jones & Laughlin Steel Corporation (J&L) in Pittsburgh, where he advanced quickly, being promoted to successively higher supervisory positions. He became general superintendent of J&L's Aliquippa works in 1929. While at J&L, White attended the Carnegie Institute of Technology between 1914 and 1917.

In 1930 Tom Girdler, who had left J&L the previous year to become chairman of the newly

Charles M. White (Cleveland Public Library)

organized Republic Steel Corporation, recruited White as assistant vice-president in charge of operations. By the time he became vice-president in charge of operations in 1935, White, according to *Fortune,* had built a reputation as "one of the smartest production men in the industry." That year Republic was facing mounting pressure from the Steel Workers Organizing Committee (SWOC) to recognize labor's right to form a union. White adopted Girdler's adamant stand against unionization, and he soon had the reputation as steel's most militant and articulate spokesman.

On the eve of the violent and ultimately unsuccessful Little Steel Strike of 1937 White spoke before the American Iron and Steel Institute (AISI).

He claimed that history had shown that unions tend to "straitjacket an industry and seriously retard the development of new and better methods." In 1938 White defended Republic's employee-representation plan — a kind of company-sponsored union — before the Senate Civil Liberties Committee, which was investigating Republic's record on collective bargaining. But despite White's and Girdler's militant stand on the issue of independent unions, the National Labor Relations Board finally compelled Republic to recognize the union in 1942.

When not berating union movements and the government for intervening in such matters, White worked to upgrade Republic's technology. An authority on air and hydraulic pressure, White contributed articles to many technical journals. As vice-president of operations he armed Republic with more-efficient top-pressure blast furnaces, electric furnaces, and continuous steel-casting equipment and stressed further study of alloy metallurgy. The lesson White had learned from the 1930s, when Republic lost some $30 million, was that the steel industry had entered the Depression "a technological laggard."

On May 9, 1945, White succeeded R. J. Wysor as president of Republic. He was named chairman of the board in 1956 following Girdler's retirement. During White's tenure the company acquired stock interest in the Liberia Mining Company, Ltd. (which held iron-ore properties in Liberia, West Africa), and a variety of wholly owned mining and metal-fabricating subsidiaries. With Armco Steel, Republic organized the Reserve Mining Company in 1950 to begin commercial exploitation of low-grade iron ore in the Mesabi Range in Minnesota. Under White's leadership Republic became the country's third-largest steel producer, with a total annual capacity of 12.5 million tons of steel ingots in 1960, compared to 8.6 million tons in 1945, the year White assumed the presidency. Annual sales, meanwhile, more than doubled, from $498 million in 1945 to more than $1.05 billion in 1960. Following his retirement in 1960, White continued to serve as a director of Republic. In honor of his contributions to the firm's success, he was designated honorary chairman of the board.

White's community activism earned him special recognition on April 16, 1961. At "Charles M. White Recognition Day," held at Cleveland's Case Institute of Technology, White was honored for his contributions to, among other institutions, the YMCA, the YWCA, the Boys Club of Cleveland, the Boy Scouts of America, and the Maternal Health Association. White served as trustee of the National Industrial Conference Board, serving as chairman from 1959 to 1961. He also was a director of the Cleveland Trust and the Sherwin-Williams companies and was a vice-president and director of the AISI.

White was the recipient of many awards, including the AISI Gary Memorial Medal in 1960 and a half dozen honorary degrees. A staunch Republican (he once apologized to friends for having been photographed at a business conference with President John F. Kennedy, whose election he had vigorously opposed), White also served as vice-chairman of the party's National Finance Committee.

On September 4, 1918, he was married to Helen Gordon Bradley. They had one daughter, Jean Bradley. White died in his home in West Palm Beach, Florida, on January 10, 1977.

Selected Publications:
"Relations Between Foremen and Workers," *Blast Furnace and Steel Plant,* 24 (October 1936): 906–907, 913;

"Technological Advances in Steel Products," *American Iron and Steel Institute Yearbook* (1937): 105–125;

"SWOC Will Ask Checkoff, Closed Shop, White Warns," *Iron Age,* 140 (December 9, 1937): 93–95;

Blast Furnace Blowing Engines; Past, Present, and Future (New York: Newcomen Society, American Branch, 1947);

A Realistic Appraisal of Steel Capacity, Present and Future; Address before the Economic Club of Detroit (Detroit, 1950);

"Why Assail the Steel Industry?" *Commercial and Financial Chronicle,* 172 (December 7, 1950): 2185;

"Capital Requirements and Productivity," *American Iron and Steel Institute Yearbook* (1951): 23–34;

Of, By and For the Government: A Speech (Public Relations Department, Republic Steel Corporation, 1952);

"Address," *American Iron and Steel Institute Regional Technical Papers* (1952): 377–383;

"Steel Industry Has its Neck Out," *Commercial and Financial Chronicle,* 177 (February 5, 1953): 590;

Republic Steel Today (Cleveland: Public Relations Department, Republic Steel Corporation, 1954);

"Long-Range Outlook for Steel," *Commercial and Financial Chronicle,* 188 (November 27, 1958): 2198;

"Prospects for an Industry That Has Found Itself," *Commercial and Financial Chronicle*, 190 (October 29, 1959): 1797.

References:
"The 'If' in Steel Expansion: An Interview with Charles M. White," *U.S. News and World Report* (December 1, 1950): 34–38;

"Republic Steel," *Fortune*, 12 (December 1935): 76–83+;
"Sees Better Steel Year in 1955;" *Iron Age*, 175 (January 13, 1955): 37;
"White of Republic Steel," *Fortune*, 40 (September 1949): 10–11.

Witherbee, Sherman & Company

by Bruce E. Seely

Michigan Technological University

Founded in 1849 at Port Henry, New York, Witherbee, Sherman & Company was one of several small companies working high-quality iron deposits (up to 65 percent iron) near Lake Champlain in northern New York State. After 1860 the company became a supplier of high-quality charcoal iron for crucible-steel makers in Pittsburgh. The company built one of the first laboratories connected to a blast furnace, for its Fletcherville plant in 1867. Its Cedar Point plant, which became operational in 1875, was the first in the country with a Whitwell hot-blast furnace.

After Frank S. Witherbee succeeded his father, founder J. A. Witherbee, in 1875, Witherbee, Sherman & Co. became one of the largest eastern iron producers. Frank S. Witherbee expanded operations at Mineville, west of Port Henry, and experimented with magnetic-ore concentrators. Two years after the company was incorporated in 1900, the Lackawanna Steel Company acquired an interest in Witherbee, Sherman. Facing a secure future, Frank Witherbee further expanded and upgraded the facilities. He began electrifying mining operations in 1903 and continued pioneering work in the production of magnetic concentrates. By 1912 the company's separation facilities were perhaps the most modern in the country. Annual production of crude iron reached 1 million tons in 1912, even as the company acquired ore property in Cuba. In 1913 Witherbee announced plans to construct a blast furnace and steel plant near New York City. He had served on a state commission studying the barge-canal system, which could provide essential cheap transportation. Although the canal was deepened to 12 feet, and modern ore docks and

traveling cranes were built in Port Henry, the grand scheme never came to fruition.

Frank S. Witherbee died in 1917, and Lewis Witherbee Francis, grandson of J. A. Witherbee, became company president. Francis faced difficult times. In 1923 he built a new blast furnace that doubled the company's output to 500 tons a day. The company was reorganized in 1926, and a second reorganization followed in 1933, but the Depression had brought financial failure. One furnace ceased operation in 1934. Republic Steel purchased the property in 1937. Production continued, but only the mines and concentrating facilities were left intact. Prospects brightened with World War II, as Republic added a 500,000-ton concentrator, and the government replaced old mining machinery and built a new sintering plant. Highly efficient magnetic separators developed by Jack and Robert Finney permitted the deep mines to continue to earn a profit. In 1965 the mines produced 650,000 tons of processed concentrates, but mining operations finally ceased in 1971, when Republic's lease expired.

References:
H. Comstock, "Concentrating at Barton Hill," *Iron Trade Review*, 55 (August 6, 1914): 253–256;
"Frank S. Witherbee," *Iron Age*, 99 (April 19, 1917): 969;
"The Historic Port Henry Furnace Dismantled," *Iron Trade Review*, 51 (September 26, 1912): 573–574;
Floy S. Hyde, *Adirondack Forests, Fields, and Mines* (Lakemont, N.Y.: North Country Books, 1974), pp. 154–155;
"A Large Modern Iron Ore Concentrating Plant," *Iron Trade Review*, 49 (November 9, 1911): 825–829;

John Liston, "Electric Power in Eastern Iron Ore Mines," *Iron Trade Review,* 50 (March 21, 1912): 649–653;

Frank L. Nason, "The Importance of the Iron Ores of the Adirondack Region," *Yearbook of the American Iron and Steel Institute* (1922): 169–207;

Frank S. Witherbee, "The Iron Ores of the Adirondacks," *Iron Trade Review,* 59 (November 2, 1916): 891–894;

"Witherbee, Sherman & Co. Readjustment," *Iron Age,* 117 (March 18, 1926).

Harleston Read Wood

(October 18, 1913–　)

by Bruce E. Seely

Michigan Technological University

CAREER: Steel apprentice, Alan Wood Steel Company (1938–1942); lieutenant U.S. Navy (1942–1946); sales-development engineer (1946–1950), manager for planning and development (1950–1952), assistant vice-president (1952–1954), vice-president, planning and development (1954–1955), president, chief executive officer (1955–1972), chairman of the board, Alan Wood Steel Company (1962–1982).

Harleston Read Wood was a fifth-generation steelmaker and head of the Alan Wood Steel Company between 1962 and 1982. Under Wood's direction Alan Wood Steel gained a reputation as an innovator in the industry, but the company nevertheless fell victim to poor market conditions during the 1970s. Wood's struggle in 1977–1978 to save the company failed, and it closed in February 1978.

Born October 18, 1913, Wood grew up in comfortable surroundings in Philadelphia and attended the prestigious Haverford School. He graduated from Princeton University with a bachelor's degree in economics in 1936. While in college, Wood won a spot on the *Daily Princetonian* editorial board when his interview with Huey Long was carried by several major papers.

Two years after graduating, Wood took a job with Alan Wood Steel. His family exercised little control over Alan Wood Steel at this time, having sold the firm to the Koppers Company in 1929. Nonetheless, Harleston was targeted for a management position. He worked briefly in every part of the company, starting as a helper at an open-hearth furnace. His beginning at Alan Wood Steel was not an auspicious one. The open hearths were experiencing problems with the molten metal eating through the furnace bottoms, and

Harleston Read Wood

word was passed that the next man with bottom trouble would be laid off. Sure enough, Harleston Wood spent several weeks out of work.

Wood entered the navy in 1942, emerging as a lieutenant in 1946 to find exciting developments at Alan Wood Steel. The company had languished through the 1920s and 1930s and entered the war with all the problems of an aging plant. Alan

Wood Steel was facing a doubtful future when Wood's father, Alan Wood III, a board member of the company, proposed regaining control from Koppers. He brought the firm back into the family's control in late 1945.

John T. Whiting, president since 1939, stayed on; Wood became sales-development manager. To lessen the burden posed by preferred stock, a recapitalization plan was completed in 1948. With greater financial freedom a modernization program was launched that year, and over the next decade $56 million in capital improvements — including the installation of a 30-inch hot-strip mill — was made. In 1950 Wood was named manager for planning and development and became involved in efforts to renovate the blooming mill, reconstruct the plate mill, and increase annual steel-ingot capacity. Wood also diversified the firm's product line, beginning with the construction of a modern cold-rolling department in 1952. Completed in 1954 — the year Wood became vice-president in charge of planning and development — the cold-rolling facility permitted production of light sheet products. Moreover, in 1955 the company acquired a metal-fabricating division — Penco Metal Products Division, a maker of steel lockers, cabinets, and shelving.

In 1955 Wood became president. He continued to modernize facilities, although the company's small size limited his options. In 1960, for example, Alan Wood Steel opened an H-iron fluid-bed direct-reduction facility to produce iron powder. The firm used some of this material as a protective coating on its products but sold most of the iron powder to producers of welding rods. But this type of innovation did not address a fundamental problem at Alan Wood Steel: blast furnaces produced more pig iron than the open hearths could refine, and the rolling mill could not roll all of the steel the company produced. In 1960 Wood borrowed $30 million with the help of First Boston Corporation to build a new slabbing mill and a new plate mill.

Wood had mortgaged the firm's future, making it one of the most leveraged companies in the industry. The move paid off in the short run, allowing Alan Wood Steel to build modern, efficient facilities. The profits during the 1960s paid for the construction of two 150-ton basic-oxygen furnaces in 1968; these replaced the open-hearth furnaces where Wood had worked in 1938. Through the early 1970s Alan Wood Steel prospered, posting record profits in 1974.

The oil crisis in 1973–1974, however, resulted in Alan Wood Steel's decline , which ended in bankruptcy. A costly labor settlement, rising imports of steel plates, and the escalation of oil prices turned profits to red ink overnight; losses in 1975 and 1976 totaled $25.1 million. The oil increase alone erased the company's $40-per-ton cost advantage over electric furnaces. In an attempt to cut losses Wood closed the hot-strip mill in late 1976 and sought protection from creditors under Chapter 11 in February 1977. The court made any effort at reorganization contingent on raising $15 million by June. Abandoning attempts to operate an integrated facility, Wood attempted to sell the plate-mill shop. When no buyer was found, the mill closed in August. In early 1978 the assets of Alan Wood Steel were liquidated. Lukens Steel acquired the 110-inch plate mill for $3.7 million; Sharon Steel purchased the basic-oxygen furnaces for $2.3 million; and Alabama By-Products Company bought the coke plant. The buildings and equipment which remained were sold at a six-day public auction. Wood watched the disappearance of a venture that had endured 152 years. Only the metal-fabricating division survived the bankruptcy, adopting the name Vesper Corporation in 1979.

Wood remained chairman of Vesper until 1982, when he retired and moved to Florida. During his career he served on the boards of the Budd Company and Fidelity Mutual Life; he was a long-time member of the National Industrial Conference Board and the Council on Foreign Relations and a director of the American Iron and Steel Institute.

Publications:

"Year-end Review Statement," *Blast Furnace and Steel Plant,* 45 (January 1957): 93–94;

Alan Wood: A Century and a Half of Steelmaking (New York: Newcomen Society in North America, 1957);

"The Measurement of Employment Costs and Prices in the Steel Industry," *Review of Economics & Statistics* (1959);

and others, *Study Tour of USSR Steel Industry, October 3–11, 1966,* United Nations, Economic Commission for Europe, Steel Committee (N.p., 1966);

"Leadership Difficult for Labor," *New York Times* September 6, 1970, III: 6.

References:

Tom Dulaney, "Alan Wood Begins a 'Life or Death' Sale," *Iron Age,* 221 (March 20, 1978): 34–36;

Dulaney, "Ask Not for Whom the Gavel Tolls," *Iron Age,* 221 (May 24, 1978): 33–35;

T. J. Ess, "Alan Wood Steel Company," *Iron and Steel Engineer,* 28 (May 1951): 1–18;

Ess, "Alan Wood Steel Company," *Iron and Steel Engineer,* 39 (March 1962): 1–25;

William T. Hogan, S.J., *Minimills and Integrated Mills: A Comparison of Steelmaking in the United States* (Lexington, Mass.: Lexington Books, 1987).

Alan Wood Steel Company

by Bruce E. Seely

Michigan Technology University

In 1826 James Wood leased a mill near Wilmington, Delaware. Naming the operation the Delaware Iron Works, he and his son, Alan, began making nail plates. In 1832 Wood moved the company to Conshohocken, Pennsylvania, where a rolling mill turned out high-quality sheet iron. For the next 20 years Wood and his sons operated both mills under the company name J. Wood & Bros. In 1857 Wood and Lewis Luken incorporated the Schuylkill Iron Works — which they built adjacent to the Conshohocken mill — as Alan Wood & Company. A third Wood company emerged when Alan's son, W. Dewees Wood, began making Russian sheet iron at McKeesport, Pennsylvania, in 1851. Incorporated as W. Dewees Wood Company in 1888, this largest of the Wood ventures was bought by J. P. Morgan in 1900.

Two of W. Dewees Wood's sons, Richard G. and Thomas D. Wood, had run W. Dewees Wood and returned to Conshohocken in 1901, determined to make Alan Wood & Company an integrated steel producer. They built five open hearths and a blooming mill at Ivy Rock, 2 miles from Conshohocken, tapping their first steel on June 3, 1903. Renamed the Alan Wood Iron and Steel Company, the firm grew steadily under Richard Wood, acquiring Richard Hecksher and Sons in 1911. Hecksher's merchant blast furnaces across the Schuylkill River from Ivy Rock had supplied Alan Wood with pig iron; now a bridge was built to move molten iron directly to the open hearths. New blast furnaces were added, as well as a state-of-the-art plate mill, and a new, partially owned subsidiary, Rainey-Wood Coke Company. Ingot capacity in 1920 was 500,000 tons, about 1 percent of the nation's total. The firm's specialty was plate, including floor plates.

After 1920, however, Alan Wood & Company stagnated, and in 1929 Koppers Company bought the plant and renamed it Alan Wood Steel

Open-hearth furnaces in the Alan Wood Steel plant at Ivy Rock, Pennsylvania, circa 1903 (Hagley Museum and Library)

Company. Koppers added a third battery of coke ovens and leased iron mines at Dover, New Jersey, that were purchased outright in 1941. Performance was sluggish at best in the 1930s, and high production costs limited profits during World War II. Prospects were bleak when Alan Wood III bought a controlling interest from Koppers in 1945. He had worked as an engineer at Alan Wood and after 1937 served as a company director. With other family members, including his son

Harleston, in the company's management, he began a modernization program highlighted by the installation of a 30-inch hot-strip mill that began operation in 1950. The blooming mill was upgraded in 1953, and in 1955 a fabricating division — Penco Metal Products Division — was acquired. Harleston R. Wood became president in 1955 and continued to modernize facilities. In 1960 the firm opened a 50-ton-per-day direct-reduction facility to make iron powder. In 1968 two 150-ton basic-oxygen furnaces raised ingot capacity to 1 million tons. In 1974 Alan Wood Steel recorded record profits of $8.3 million.

Then, almost overnight, Alan Wood Steel fell apart. A costly labor settlement in 1974 combined with the growing flood of plate imports and rising oil prices brought losses in 1975 and 1976 totaling $25.1 million. After seeking Chapter 11 bankruptcy protection in February 1977, Wood tried to raise $15 million by selling the plate mill. He found no buyers, and in August 1978 Alan Wood Steel closed. The 3,000 people who lost their jobs received federal assistance for workers hurt by imports. For the first time in American history, in-dustrial assets were liquidated at a public auction. Only the fabricating operations remained, becoming the Vesper Corporation in 1979.

References:
Tom Dulaney, "Alan Wood Begins a 'Life or Death' Sale," *Iron Age,* 221 (March 20, 1978): 34–36;
Tom Dulaney, "Ask Not for Whom the Gavel Tolls," *Iron Age,* 221 (May 24, 1978): 33–35;
T. Joseph Ess, "Alan Wood Steel Company," *Iron and Steel Engineer,* 28 (May 1951): 1–18;
Ess, "Alan Wood Steel Company," *Iron and Steel Engineer,* 39 (March 1962): 1–25;
William T. Hogan, S.J., *Minimills and Integrated Mills: A Comparison of Steelmaking in the United States* (Lexington, Mass.: Lexington Books, 1987);
Charles Longenecker, "Alan Wood Steel Company," *Blast Furnace and Steel Plant,* 21 (January 1933): 53–60;
Frank H. Taylor, *History of the Alan Wood Iron and Steel Company, 1792–1920* (Philadelphia: privately printed, 1920);
Harleston R. Wood, *Alan Wood: A Century and a Half of Steelmaking* (New York: Newcomen Society in North America, 1957).

Woodward Iron Company

by Robert Casey

Henry Ford Museum & Greenfield Village

From 1881 until the collapse of the merchant iron industry in the 1970s, Woodward Iron Company was one of the major independent iron producers in the Birmingham, Alabama, area.

The company traces its history to 1869, when Wheeling, West Virginia, ironmaster Stimson H. Woodward purchased 550 acres of ore deposits on Red Mountain. Woodward kept tabs on the development of the Birmingham iron industry but did not decide to move until his son William visited Henry DeBardeleben's successful Alice Furnace operation in 1880.

Stimson Woodward died in 1881, but sons William and Joseph purchased several hundred acres southwest of Birmingham, erected a blast furnace and 100 beehive coke ovens, and opened coal mines, ore mines, and a limestone quarry. All the raw materials were within 5 miles of the furnace itself and were carried on the company's own railroad. Woodward was the most self-contained of all the Birmingham operations, a condition that helped the company remain independent until the 1960s.

Woodward added a second blast furnace in 1887 and a third in 1905. By 1909 the company employed 2,000 people, many of whom lived in the growing company town. In 1912 Woodward underwent a major expansion. The company purchased the Birmingham Coal and Coke Company, whose assets included two blast furnaces (known as the Vanderbilt furnaces) in north Birmingham and major coal and ore reserves. A new blast furnace was also built, along with by-product coke ovens to replace the beehive batteries.

During World War I, Woodward opened the Pyne mine, one of the few shaft ore mines in the Birmingham District. By the 1960s this was one of the largest iron ore mines in the country. In 1929

Woodward closed both of the Vanderbilt furnaces.

Throughout much of its history Woodward had concentrated on one product: merchant pig iron, sold to foundries. In the 1950s the company began to diversify. Since most of Woodward's iron output went into cast-iron pipe, the company sought to gain captive markets by purchasing pipe plants in Texas, Virginia, and Alabama. Woodward also acquired the National Cement Company in Ragland, Alabama, and began manufacture of cement, lime, cement-asbestos pipe, chemicals, and ferro-alloys. In order to reflect this broadened scope more accurately, the name Woodward Iron Company was changed to Woodward Corporation in 1968.

Woodward's high earnings during this period made it attractive to larger firms. Eighty-seven years of independence ended when the Mead Corporation, a paper and pulp producer, acquired Woodward in 1968. Mead's timing was poor, however, for the merchant pig-iron business was dying. Low-grade Birmingham ore could not compete with rich imported ores, and foundries were replacing domestic pig iron with imported pig and domestic scrap iron. In 1971 Mead closed the Pyne mine, the last operating red-ore mine in the district. The last Woodward blast furnace was blown out in 1973, and all furnaces were demolished. Koppers Company bought the coke plant in 1974 and continues to operate it. The United States Steel Corporation bought the Woodward railroad. The Drummond Coal Company now operates the coal mine at Short Creek, and Mead continues to operate the Mulga coal mine. The company town was dismantled in 1968.

References:

Ethel Armes, *The Story of Coal and Iron in Alabama* (Birmingham: Birmingham Chamber of Commerce, 1910);

Earl Chapin May, *From Principio to Wheeling, 1715 – 1945* (New York: Harper, 1945);

Marjorie L. White, *The Birmingham District* (Birmingham: Birmingham Historical Society, 1981).

World War I: Government and the Steel Industry
by Kevin M. Dwyer

George Washington University

After twenty years of unprecedented corporate expansion and consolidation and highly discretionary but at times virulent federal antitrust prosecution, relations between the U.S. government and the steel industry returned to what could be called a state of peaceful coexistence during World War I. The U.S. government began projecting global influence sufficient to alter the world balance of power; however, U.S. strength was dependent upon Big Steel's production and cooperation. The industry, meanwhile, proved in court during the war that its internal, oligopolistic balance of power complied with the federal restrictions defined in the Sherman Anti-Trust Act of 1890. Massive wartime industrial development in accordance with new economic thinking insured cooperation among the steelmakers themselves as well as between the industry and the federal government. Events following President Woodrow Wilson's war message to Congress on April 2, 1917, presaged the trickle-down economies of the 1920s New Era, the heyday of American steel manufacturing.

Political and business interests soon struck upon the theme of cooperation in their public relations following America's entry into the Great War. *Iron Age* celebrated the "smoothness and efficiency of the cooperation between the Government and the iron and steel industry" while Wilson praised the "spirit of cooperation . . . manifested by the steelmen." But these declarations of harmony disguised an ambivalence that grew as the war forged more interdependent industry-government relations. Melvin I. Urofsky saw this process as an interaction between "blocs" — one public and one private — "carving out specific spheres of authority" in an increasingly centralized urban and industrial society. This process, part collusion and part tug-of-war, was not without friction.

Before the war ambiguous federal antitrust legislation and selective government enforcement had made Judge Elbert Gary's plans to stabilize the industry under the influence of his United States Steel Corporation uncertain. Public opinion had seemed to favor the busting of trusts, and President Theodore Roosevelt's vigorous attack on the National Securities Trust in 1902 evoked popular calls for an encore. But Roosevelt appreciated the order of the Gary system, and it was left to his successors, who were more inclined to question the legality of the formation of U.S. Steel, to challenge the corporation. On October 26, 1911, the Justice Department under the administration of President William Howard Taft filed for a federal district court injunction to dissolve U.S. Steel for conspiring to restrain trade via monopoly.

In the ten years prior to the suit, Gary had run the corporation expressly to prevent a monopoly in the steel industry. There was, as he said, enough "but only so much business to go around." The old-style "destructive competition" of the Gilded Age was in his view "not reasonable, not desirable, and never beneficial in the long run to anyone." The cooperation he sought — best coordinated through his establishment of the American Iron and Steel Institute (AISI) in 1908 — certainly implied collusion among the producers; but Gary believed that to be altogether different from monopoly. U.S. Steel's competitors testified under oath that they had grown and profited under Gary's system, and in June 1915, weeks after the sinking of the *Lusitania,* the judiciary found for the defense and denied the federal government the right to dissolve U.S. Steel.

Woodrow Wilson had taken the oath of office in March 1913, and Wilson's New Freedom — his campaign promise to end monopolies and institute labor's right to bargain collectively — proved easier to preach than to practice but committed him to accept reform-minded Attorney General Thomas Gregory's appeal of the antitrust decision to the Supreme Court. By that point, however, the war in Europe had begun to reorder national priorities. While Wilson called for neutrality, U.S. steelmakers sold the Allies war matériel to the tune of $10 million a day during 1917. In his insistence on maintaining neutrality, Wilson had made no contingencies for steel production in the event of war, which suited Big Steel's more or less laissez-faire purposes. With Judge Gary presiding, the industry undertook the bulk of its colossal wartime expansion before the United States actually entered the war. Ingot capacity nearly doubled from 1914 to 1916 — 23, 513,000 to 42,774,000 gross tons — and pig

iron capacity rose almost 60 percent between 1914 and 1917, from 23,332,000 tons to 39,435,000 tons. Bethlehem Steel, clearly the big winner and a challenger welcomed by U.S. Steel, earned over $60 million in 1916, becoming the world's largest munitions maker with arms sales to the Allies of $246 million before April 1917. All told, neutrality brought over a billion dollars worth of steel orders to U.S. industry.

Defiant collusion on bids from the "armor trust" — Bethlehem, Midvale, and U.S. Steel — convinced Wilson to support efforts by Senator Benjamin Tillman and Secretary of the Navy Josephus Daniels to build a government-operated armor plant to keep armor bidding competitive. The ground for the plant was broken by 1917, but after war was declared, "adequate production supplanted adequate price controls as the main concern of government," as B. F. Cooling noted. A "world safe for democracy" demanded cooperation with the steel industry. By and large, it was a cooperation defined on the industry's terms to industry's benefit. Industry considered cooperation to be a lack of government intrusion; government policy by and large fit that bill in the name of the war effort, although the case of the *U.S. v. U.S. Steel* was not dropped but only suspended in 1917.

This wartime cooperation required a mending of the fences between government and the industry, which meant the reopening of diplomatic channels severed since 1911. A month before his reelection in 1916, Wilson named a seven-man advisory committee of businessmen to join Secretary of War Newton D. Baker, Daniels, and several other cabinet members on the newly created Council for National Defense (CND), a World War I prototype for the National Security Council. If not up to the task of full-scale mobilization, the CND nonetheless established a coordinated national effort to win "the war of production."

Bernard Baruch was the CND's man in charge of raw materials, minerals, and metals. A Wall Street financier and veteran player in New York politics, Baruch knew business as well as government and was fiercely loyal to Wilson. He tapped Jacob Leonard Replogle, an AISI outsider who had been a highly placed executive in the Cambria and Vanadium Steel companies and had a commanding knowledge of the steel business, to coordinate government steel supply for a dollar a year in August 1917. Baruch and Replogle established close relations with the AISI and Gary, who firmly controlled production. Judge Gary headed a powerful executive committee that controlled a string of AISI products committees. The executive committee comprised James A. Farrell, Republic Steel's John Topping, and E. A. S. Clarke of Lackawanna Steel. The AISI, naturally, recommended a run at full capacity for as long as government promised to pay for it.

The scale of U.S. government-mandated wartime operations, unprecedented in history, entailed more than unleashing Big Steel for production. The logistics of putting sustained, record-breaking outputs to efficient uses demanded a more centralized government apparatus. The hastily conceived CND and other committees of limited authority gave way to the sprawling War Industries Board (WIB) in June 1917. The strain of running the WIB, however, broke Frank Scott — and frustrated the Baltimore & Ohio Railroad's Daniel Willard, his replacement — because Secretary of War Baker, wary of the WIB acquiring unwarranted influence, resisted the board's centralized command structure. The effective chain of command for steel supply remained intact; Gary extended the reach of the AISI's general committee in May, and he, Replogle, and Baruch came to a price-fixing agreement on September 24, 1917.

Business, particularly steel, ran full as the WIB, finally centralized under Bernard Baruch after March 1918, awarded cost-plus contracts, bought millions of tons of steel at high WIB-set prices, underwrote expansion, assessed priorities, and coordinated distribution. Antitrust policies were also suspended on the home front, and those that applied to foreign trade under the 1918 Webb-Pomerene Bill were made less stringent. Acting in conjunction with the WIB, the War Labor Board determined labor policy.

By the end of the war, with federal authority harnessed to a business experiment described by Arthur Jerome Eddy as the "New Competition" of oligopoly, the basic structures of America's twentieth-century political economy were determined. The primary impact of World War I on the steel industry was found in the completion of a fundamental socio-economic change: progressive reform movements within the industry died out, and Wall Street bulls cheered the strangulation of steel unions during the Red-Scare Steel Strike of 1919.

References:

William H. Becker, *The Dynamics of Business-Government Relations: Industry and Exports, 1893–1921* (Chicago: University of Chicago Press, 1981);

B. F. Cooling, *Gray Steel and the Blue Water Navy: The Formative Years of America's Military-Industrial Complex, 1881–1917* (Hamden, Conn.: Archon, 1979);

William H. Hogan, S.J., *An Economic History of the Iron and Steel Industry in the United States*, 5 volumes (Lexington, Mass.: Lexington Books, 1971);

Gabriel Kolko, *The Triumph of Conservatism: A Reinterpretation of American History, 1900–1916* (New York: Free Press of Glencoe, 1961);

E. C. Kreutzberg, "War Writes Steel's Brightest Page," *Iron Trade Review*, 64 (January 2, 1919): 42–51;

Mark Reutter, *Sparrows Point: Making Steel* (New York: Summit, 1989);

Paul Tiffany, *The Decline of American Steel: How Management, Labor and Government Went Wrong* (New York: Oxford University Press, 1988);

Melvin I. Urofsky, *Big Steel and the Wilson Administration* (Columbus: Ohio State University Press, 1969);

Robert H. Weibe, *The Search for Order, 1877–1920* (New York: Hill & Wang, 1967).

World War II: Government and the Steel Industry
by Paul Tiffany

Sonoma State University

The American steel industry, at first glance, would have appeared to have responded with enthusiastic vigor to the calls of the U.S. government to provide the vital metal to the national war effort. Crude steelmaking capacity, after all, stood at 81.6 million tons in 1940 but had expanded to 95.5 million tons by 1945; production, meanwhile, grew from 67 million tons in 1940 to a wartime peak of 89.6 million tons in 1944.

A look beneath the surface, however, reveals a business-government relationship fraught with tension and distrust, and one in which the industry agreed to expand capacity only after a strong push from the government buttressed by a liberal amount of public financial assistance provided to the steelmakers. Yet, rather than being due to any lack of patriotism, the cautious approach taken by the industry was rooted in historical events that would have given any industry pause before plunging ahead with a massive investment program for expansion.

Modern steelmaking is a capital-intensive business in which the assets for the creation of the product — plants and mills — have an unusually long life. Moreover, during much of the twentieth century steel had been a globally traded commodity, which meant that capacity expansion in one geographic area had an economic impact on steelmakers and steel buyers in other areas. As such, steel-industry decision makers were generally quite cautious when dealing with the issue of capacity expansion, some even viewing it as the most critical choice they would make in terms of the long-run financial health of their firm.

Thus when the administration of President Franklin D. Roosevelt began to call for more domestic production of steel in late 1940, industry officials hesitated to respond immediately. After all, they noted, most experts believed the war in Europe would be a short-lived affair, and they were concerned over what would happen at war's end with Europe's new steelmaking capacity, which had been steadily growing since the mid 1930s. American experts also fretted over the expansion in steel that had occurred in other parts of the world, such as in Japan, during the same period. Industry officials concluded that once the war was over the steel that had to be produced from the new capacity would find its way into world markets, where it would ultimately challenge domestic markets. Accordingly, they doubted the wisdom of investing in added capacity when world competition would be increasing in but a few years.

The federal government thus found that it had to offer financial inducements in order to provoke the steelmakers into expansion. By September 1941 the government had pledged some $1.1 billion in public funds that eventually added 15.2 million tons of new capacity. While the government funded the construction and owned the new mills, the contracts called for the private steelmakers to actually operate them. At war's end this new capacity was sold to its operators at a consid-

A navy gun being forged at a Bethlehem Steel plant, 1941 (engraving by Rudolf von Ripper)

erable reduction in cost. The necessary additions to steel capacity were thus made, and of course it contributed significantly to the success of the national war effort.

This public-private cooperation in steel, however, was often characterized by distrust and a lack of understanding, and nearly all of the interaction was conducted from an adversarial rather than a cooperative stance. The resumption of the debate over steel-capacity expansion in the immediate post-World War II years, when industry leaders resisted public calls for more steel for the same reasons they did in the early 1940s, would eventually result in a competitive situation that was favorable neither to the producers nor to the national interest.

References:

Richard A. Lauderbaugh, *American Steel Makers and the Coming of the Second World War* (Ann Arbor, Mich.: UMI Research Press, 1980);

Paul Tiffany, *The Decline of American Steel* (New York: Oxford University Press, 1988).

Leslie Berry Worthington

(June 22, 1902 –)

by Dean Herrin
Historical Society of Western Pennsylvania

CAREER: Sales apprentice, sales clerk, salesman, South Chicago works (1923–1933), assistant manager of sales, Chicago district (1933–1935), manager of sales, Saint Paul district (1936), sales executive, Detroit office (1936–1941), manager of sales, bar and strip steel and semifinished materials, Carnegie-Illinois Steel Division (1941), vice-president, president, United States Steel Supply Company (1942–1957), president, Columbia-Geneva Steel Division (1957–1959), president, chief administrative officer (1959–1967), director, United States Steel Corporation (1959–1975).

Leslie Berry Worthington spent his entire business career with the United States Steel Corporation (U.S. Steel), rising from a sales apprentice to become the eighth president of one of the largest corporations in America.

Worthington's family emigrated from England to the small coal-mining town of Witt, Illinois, in 1907. He worked in the local general store after school and during summer vacations to supplement the family income. He worked his way through the University of Illinois, where he majored in business.

After graduation in 1923 Worthington began his career with U.S. Steel as a sales apprentice in the company's South Chicago works.

As Worthington was transferred to such cities as Chicago, Saint Paul, and Detroit, he rose through the sales-office ranks until he was brought to Pittsburgh in 1941 to oversee the sales of bar and strip steel and semifinished materials for the Carnegie-Illinois subsidiary of U.S. Steel. In 1946 he was appointed president of the United States Steel Supply Company (another U.S. Steel subsidiary) and 11 years later moved to San Francisco to head U.S. Steel's Columbia-Geneva Steel Division.

Worthington was selected to serve as president and chief administrative officer of U.S. Steel in November 1959. Walter Munford, who had taken over the top post in May of that year, died

Leslie Berry Worthington

suddenly in September. Worthington, with his sales background, was considered a dark-horse candidate for the presidency, since most U.S. Steel presidents had had operating backgrounds. His selection at a special board meeting was thus a surprise. Most business and industry analysts explained Worthington's selection as a signal that U.S. Steel was placing a renewed emphasis on sales and marketing to combat intense domestic and foreign competition.

Worthington's tenure as president began under very unfavorable circumstances, the least of which was the sudden administrative transition caused by Munford's death. A recession hit the steel industry in 1958, affecting both the production and the profit margin of U.S. Steel. Profits had fallen from $419 million in 1955 to $254 million by the time Worthington took office. Of more long-term significance, Worthington had become president during a 116-day industrywide strike. The strike ended in January 1960 only after Vice-president Richard Nixon and Secretary of Labor James Mitchell intervened. By that time, however, many steel customers had switched to foreign steel, and steel imports continued to rise after 1959, especially during periods of labor negotiation when customers sought insurance against steel shortages and price hikes. By 1968, 16 percent of the American steel market was supplied by foreign steel.

In response to these problems Worthington and his colleagues — Chairman Roger Blough, Chief Executive Officer Robert Tyson, and Chairman of the Finance Committee Robert Tyson — tried a variety of tactics. Worthington proposed voluntary restrictions on those steel imports he called "unfair and unrestricted." But the *New York Times* reported in 1966 that opponents of such a program complained that the domestic industry sought to interfere with "normal" trade rather than compete. In 1962, when profits reached a ten-year low of $164 million, U.S. Steel announced a price increase on most products averaging $5 a ton. The response by the administration of President John F. Kennedy was overwhelmingly negative and unusually strong, with Kennedy holding a news conference to call the proposed increase "a wholly unjustifiable and irresponsible defiance of the public interest." With public opinion against it, U.S. Steel backed down from raising steel prices.

Worthington and his colleagues also tried cost-cutting methods, eliminating company programs and projects and retiring employees before they reached the age of sixty-five. Several of the company's subsidiary divisions were reorganized in 1963–1964 as true operating divisions of the parent corporation, a move which consolidated more of the decision-making responsibility at the top levels of the corporation. To produce steel more efficiently and cheaply, Worthington sought to modernize aging plants. New mills were installed in the company's Gary and Fairfield operations, two basic-oxygen furnaces were planned for the Duquesne works, improvements were made in

raw-material facilities, and a $1.8 billion capital spending program was launched in 1966.

Rising steel imports, strong governmental opposition to steel-price increases, aging plants, and increased domestic competition, however, combined to undermine the American steel industry, and U.S. Steel continued its decline. Even though the corporation was the sixth largest in total assets among all U.S. companies in 1967, U.S. Steel's share of the domestic steel market dropped from 29.3 percent in 1957 to 24 percent in 1966. The company's 1967 profit earnings of $172 million was the second-lowest figure recorded since 1952.

Despite the many problems that went unsolved during his presidency, Worthington was considered by colleagues and employees a warm and "people-oriented" leader. His broad knowledge of the steel trade impressed colleagues, who remarked that Worthington was as comfortable discussing issues with warehousemen, machinists, automobile manufacturers, and other customers as with his own managers. After his retirement in 1967 Worthington continued to serve on the company's board of directors until 1975. He also served as chairman of the American Iron and Steel Institute from 1965 to 1967.

Publications:
"Challenge to Engineers," *Iron and Steel Engineer,* 36 (July 1959): 136–138;
"Can't Keep on Selling the Same Steel Products," *Commercial and Financial Chronicle,* 191 (June 2, 1960): 9;
"Wise Man's Eyes," *Iron and Steel Engineer,* 37 (August 1960): 183;
" 'An Agreement We Need,' " *Vital Speeches,* 28 (May 15, 1962): 469–472;
"Come and Get It!," *Vital Speeches,* 33 (November 15, 1966): 78–81;
"Steel in a Competitive World," *Iron and Steel Engineer,* 44 (June 1967): 143–145.

References:
"Big Steel Picks a New Boss," *Business Week* (November 14, 1959): 156–158;
"Debate Over Price Rise — Kennedy and the Steel Industry," *U.S. News and World Report,* 52 (April 23, 1962): 86–92;
John Hoerr, *And the Wolf Finally Came: The Decline of the American Steel Industry* (Pittsburgh: University of Pittsburgh Press, 1988);
William T. Hogan, S.J., *An Economic History of the Iron and Steel Industry in the United States,* 5 volumes (Lexington, Mass.: Lexington Books, 1971).

Rufus J. Wysor

(December 8, 1885 – June 11, 1967)

by Carol Poh Miller

Cleveland, Ohio

CAREER: Employee, Virginia Iron, Coal & Coke Company (1906); assistant chemist, Carnegie Steel Company (1906–1910); partner, Kahn & Wysor (1910–1911); metallurgical inspector, Isthmian Canal Commission (1911); chief chemist, Bethlehem plant (1912–1914), engineer (1914–1916), superintendent, blast-furnace department (1916–1918), superintendent, service department (1918–1919), assistant general manager, Sparrows Point plant (1919–1923), assistant general manager, Cambria plant, Bethlehem Steel Corporation (1923–1925); assistant general manager general manager of properties, Jones & Laughlin Steel Corporation (1925–1929); vice-president, operations (1929–1935), executive vice-president, general manager, and director (1935–1937), president, Republic Steel Corporation (1937–1945); chief, metals branch, Office of Military Government, Germany (1945–1946); private consultant (1947–1949); special consultant to the United States Steel Corporation (1949–1955).

Rufus Johnston Wysor worked for four of the biggest steel companies in the United States, becoming president of the Republic Steel Corporation. He then embarked on a career as an international consultant, helping to restore German and Japanese steel production following World War II. "He knows as much about steel as any steel man alive," *Fortune* said of him in December 1935.

Called "Jack" by friends and family, Rufus Johnston Wysor was born in Dublin, Virginia, the son of Henry Charlton and Mary Elizabeth Shipp Wysor. His paternal ancestors (who spelled their name "Weiser") had immigrated to the United States from Württemberg, Germany, in the early 1700s. Henry Wysor was a farmer and an insurance agent. After receiving his early education in a private school on his father's farm, Rufus Wysor earned a bachelor's degree in chemistry and metallurgy from the Virginia Polytechic Institute in 1906. He worked briefly for the Virginia Iron,

Rufus J. Wysor

Coal & Coke Company in Foster Falls, Virginia. Later in 1906 he became assistant chemist at the Duquesne, Pennsylvania, steelworks of the Carnegie Steel Company. In 1910 he became a partner in Kahn & Wysor, a small consulting laboratory in New York City. The following year he served as a metallurgical inspector with the Isthmian Canal Commission, inspecting material for the locks and gates of the Panama Canal.

Wysor joined the Bethlehem Steel Corporation in 1912. Until 1919 he worked at their Bethlehem, Pennsylvania, plant, advancing from chief chemist, to engineer of tests (1914), to superintendent of the blast-furnace department (1916–1918), to superintendent of the service department (1918–1919). From 1919 to 1923 Wysor was assistant general manager of Bethlehem's Sparrows Point, Maryland, plant. In 1923–1925 he was assistant general manager of the Cambria plant in Johnstown, Pennsylvania, then the second-largest steel plant in the country. He joined Jones & Laughlin Steel Corporation, Pittsburgh, in 1925. There he advanced from assistant general manager to general manager of properties.

In 1929 Tom M. Girdler, who recently had resigned as president of Jones & Laughlin, persuaded Wysor to help him organize the Republic Steel Corporation, in Youngstown, Ohio. "Nothing was too big for him," Girdler later wrote of Wysor, who was named vice-president of operations at Republic. In 1935 Wysor was appointed executive vice-president, general manager, and a director. In 1937, at age fifty-one, he was named president of Republic Steel, succeeding Girdler (who remained chairman of the board). Wysor served as president until 1945. During this period Republic became the third-largest steel company in the United States, with 65,000 employees in 76 plants. It also coped with economic depression, serious labor unrest, and often-violent strikes as John L. Lewis and the CIO worked to organize workers at Republic and other "Little Steel" companies in the late 1930s. Hoping to stave off unionization, Wysor convinced Republic to adopt an employee representation plan; nevertheless, the National Labor Relations Board compelled Republic to recognize the Steelworkers Union in 1942.

In 1945 Wysor resigned from Republic to supervise metallurgical operations in Germany for the Allied Control Commission. Working with representatives of other steel and metals industries, he helped rebuild German industrial production, and he received the Medal of Freedom for his efforts in 1946. Following a short stint as a consultant for Republic Steel in Panama, Chile, Argentina, and Brazil, Wysor went to Japan in 1947 and conducted a ten-month survey of heavy industry in cooperation with Gen. Douglas MacArthur's staff. As a consultant to the U.S. State Department in 1949, he made a similar study of economic conditions in Greece and Turkey. In that same year Wysor joined the United States Steel Corporation as a special consultant on raw materials and a member of its long-range planning committee. During the next six years he visited all domestic operations of the corporation and traveled to Venezuela to oversee development of the Cerro Bolivar ore deposit. Wysor wrote numerous articles on the steel industry. His memoir, *Boyhood Days in Southwest Virginia,* was privately published in 1961.

Wysor served on regional executive committees of the Boy Scouts of America, the steering committee of the Virginia Polytechnic Institute, and the board of governors of Youngstown College. He was married three times. In 1914 he married Mary Elizabeth MacFate, with whom he had two daughters, Jeanne MacFate and Carolyn Elizabeth, and one son, Rufus Johnston. After a divorce in 1943 he married Nina Douglas, who died in 1947. The next year he married Helen Johnston Marsh. Wysor died at his home on Sea Island, Georgia, in 1967.

Selected Publications:

"Loss of Heat in Hot-Blast Mains," *Iron Trade Review,* 58 (February 24, 1916): 435–437;

and others, "Progress in Hot Blast Stove Design," *Year Book of the American Iron and Steel Institute* (1916): 390–395;

"Progress in Hot Blast Stove Design," *Iron Trade Review,* 60 (January 25, 1917): 259–260;

and T. M. Girdler, "Republic Steel Corporation Operates at Record Breaking Capacity," *Blast Furnace and Steel Plant,* 31 (January 1943): 131;

Boyhood Days in Southwest Virginia (New York: Vantage, 1961).

References:

Boys Grow Tall: A Story of American Initiative (Cleveland: Cleveland Plain Dealer, 1944), pp. 120–131;

Tom M. Girdler and Boyden Sparkes, *Boot Straps: The Autobiography of Tom M. Girdler* (New York: Scribners, 1943), pp. 185–187, 194–195;

"Republic Steel," *Fortune,* 12 (December 1935): 76–83+;

"R. J. Wysor Elected President of Republic Steel Corp.," *Iron Age,* 139 (April 22, 1937): 79.

Frank J. Yaklich, Jr.

(February 20, 1936 –)

by H. Lee Scamehorn

Historic Learning and Research Systems

CAREER: Mechanical engineer, Inland Steel Company (1960–1962); industrial engineer, Colorado Fuel & Iron Corporation (1962–1973); general superintendent, operations (1974–1976), works manager (1976–1979), vice-president of operations (1979–1980), president, CF&I Steel Corporation (1980 –).

Frank J. Yaklich, Jr., was elected president of the CF&I Steel Corporation of Pueblo, Colorado, on September 1, 1980. Born in Pueblo on February 20, 1936, he earned two associate's degrees from Pueblo Junior College (1956, 1958), a bachelor's degree in engineering from the University of Colorado in Boulder (1960), and a degree in business administration from Southern Colorado State College (1969; now the University of Southern Colorado). Upon graduation he went to work in 1960 as a mechanical engineer for the Inland Steel Company in Chicago, Illinois.

In 1962 he returned to Pueblo to work as an industrial engineer for the Colorado Fuel & Iron Corporation (CF&I). In 1968 he was promoted to assistant superintendent of the seamless-tube mill at CF&I, and two years later he was named a superintendent of the rolling and finishing mills. He was made general superintendent of rolling and finishing in 1974, works manager in 1976, and vice-president of operations, a newly created position, in 1979. He continued to be responsible for all rolling, finishing, and services departments, coke and steel production, quality control, industrial engineering, safety and security, planning, and shipping activities at the Pueblo plant. He was also assigned the task of making internal improvements in productivity and efficiency to match recent modernization and expansion of operating facilities.

Throughout the 1960s and much of the 1970s the company enjoyed a large measure of prosperity. The market for tubular goods in oil-producing countries, rails and accessories, and a variety of products for construction and agriculture remained active. Profits from these markets

Frank J. Yaklich, Jr.

funded ongoing improvement programs. By applying state-of-the-art technology, the corporation maintained a firm grip on its traditional share of the regional market.

Yaklich took charge of CF&I in 1980, when Robert J. Slater, president since 1976, was elected to a similar position in the parent enterprise, the Crane Company of New York. The Pueblo firm recorded all-time-high earnings in its first year under Yaklich's direction. In the second year, however, CF&I losses also set new company records, as the American integrated steel industry experienced a depression. American steel companies suffered from intense competition from low-cost minimills at home and imports from Europe and Japan.

495

Yaklich had to adapt CF&I to new conditions of competition. He curtailed production and reduced costs as deficits rose to alarming levels. Output was slashed. Some mills were closed. But a stagnant market required still more reductions in 1982. The Pueblo plant shut down the last of four blast furnaces in June, leaving only two electric-arc units to make steel from scrap metal.

In December 1983 the Pueblo steelmaker began drastic restructuring in an effort to make it competitive in a global market. Operations were cut approximately in half. Some 2,400 employees, many of whom had been without work since 1981, were terminated because the company no longer made pig iron as the intermediary step in the production of steel. In the future the plant would make steel only from scrap. This meant that blast and basic-oxygen furnaces, by-product coke ovens, and related facilities were to be permanently shut down and some of them dismantled.

The decision to abandon integrated operations was prompted by a deficit that approached $50 million in 1983. That loss was compounded by the write-off of another $80 million in discarded production facilities. Some of these units were subsequently transferred to the parent enterprise to retire outstanding debts. Crane's officials organized a subsidiary, Evergreen Land and Resource Company, to dispose of approximately 500 square miles of agricultural, mountain-recreational, timber, and mineral land located mostly in Colorado, with small amounts in New Mexico, Montana, and South Dakota.

After trying unsuccessfully to sell CF&I, the Crane Company spun it off to shareholders. Owners of each share of the holding company's stock received four-tenths of a share of the Pueblo firm's equity capital. With the completion of that transaction in 1985 Yaklich became president of an independent steelmaker. The board of directors remained under the influence of the men who had large interests in the parent company. Slater resigned as CF&I chairman and was succeeded by Crocker Neven, a senior adviser at Drexel Burnham Lambert, in New York. He had been a director of the company since 1968.

In order to strengthen the company's market position, management asked employees to accept concessions on wages and benefits. Workers agreed to reductions in compensation in 1982 and again in 1983. In October 1986 the workers ratified a new contract that precluded wage increases over a period of three years in return for a 38-percent-ownership stake in the form of an Employee Stock Ownership Plan (ESOP).

Yaklich has pointed out that CF&I is saddled with a heavy pension and group-insurance plan, and the company has faced an unstable market for its tubular goods. Given its streamlined restructuring as a minimill, however, CF&I has regained the ability to compete successfully with domestic and foreign producers. CF&I in 1988 recorded a profit of $2.3 million, its first in seven years.

Yaklich is active in the Pueblo Chamber of Commerce, the Rotary Club, the American Iron and Steel Institute, and various other community and professional organizations. He received a distinguished-alumnus award from the University of Southern Colorado in 1978 and from the University of Colorado in 1982.

Publication:
and D. Farries and V. S. C. Porter, "Rolling and In-Line Finishing of Bars, and Shapes at CF&I Steel's Pueblo Plant," *Iron and Steel Engineer*, 53 (August 1976): 33–38.

Arthur Howland Young

(December 19, 1882 – March 4, 1964)

by Bruce E. Seely

Michigan Technological University

CAREER: Messenger boy and laborer, Illinois Steel Company (1894–1902); laborer, Colorado Fuel & Iron Company (1902–1905); timekeeper, chief safety inspector (1905–1912), supervisor of labor and safety, Illinois Steel Company (1912– 1917); director, American Museum of Safety (1917–1918); chief safety expert, U.S. Employees Compensation Commission and U.S. Arsenals and Navy Yards (1917–1918); manager of industrial relations, International Harvester Corporation

(1918–1924); industrial-relations counselor, Curtis, Fosdick & Belknap (1924–26); secretary, Industrial Relations Counselors, Inc. (1926–1934); vice-president of industrial relations, United States Steel Corporation (1934–1937); lecturer on industrial relations, California Institute of Technology (1938–1952).

Arthur H. Young began and ended his business career in the steel industry. One of the leading industrial-relations experts in the country, he built his reputation as an advocate of company unions.

Young was born in Joliet, Illinois, on December 19, 1882, and by age twelve was working as a messenger and leverman at the Joliet and South Chicago plants of the Illinois Steel Company. He completed high school in 1901 and moved to the Minnequa plant of the Colorado Fuel & Iron Company (CF&I) as a rigger and roll hand. He returned to Joliet in 1905, securing a timekeeper's slot, and married Edith Kellogg in 1906. He worked as a chief payroll clerk, statistician, assistant supervisor of labor, and chief safety inspector before moving to the South Chicago Works as supervisor of labor and safety in 1912. Young's new position quickly grew in importance, for the United States Steel Corporation — the industry leader and trendsetter — had launched its corporate welfare program. A combination of housing, sanitation, and safety reforms — U.S. Steel had coined the slogan "Safety First" — the program reflected U.S. Steel head Elbert H. Gary's interest in improving both the working conditions and the corporation's poor public image: the program was initiated soon after an antitrust suit was brought against U.S. Steel in 1911. Young's position thus not only covered plant-safety issues but virtually all other aspects of industrial relations.

He left U.S. Steel in 1917 to direct the American Museum of Safety in New York and serve as expert adviser to the U.S. Arsenal and Navy Yard safety program and the U.S. Employees Compensation Commission. He also supported the public employment program of the U.S. Employment Service. In 1918 Young became industrial-relations manager at International Harvester Corporation. His rapid rise in the ranks of industrial management owed much to the sudden popularity of welfare capitalism, as changing public attitudes, government regulation, and immigration restrictions led to more "enlightened" corporate behavior toward wage earners. Young became identified with the development of employee representation plans

Arthur Howland Young

(ERPs) or company unions, similar to a plan developed in 1916 at CF&I following a bloody workers' strike later labeled the Ludlow Massacre. CF&I owner John D. Rockefeller had ordered the plan, which did not contain a mechanism for negotiating wages. The CF&I plan, however, did offer means for resolving worker complaints about safety and working conditions, and Young devised a similar program for Harvester.

The big step for Young came in 1924, when Rockefeller's law firm — Curtis, Fosdick & Belknap — hired him as industrial-relations counselor. Rockefeller had bankrolled a group to institute the employee representation in his industrial empire, but Young worked to spread the plan to outside companies as well. In May 1926 the group was split off from the law firm as a nonprofit organization, Industrial Relations Counselors, Inc. Young was secretary of this pioneer labor relations consulting firm, although consulting contracts only slowly replaced Rockefeller support.

Through the 1920s Young's stature rose steadily in the young field of industrial relations and personnel administration. He headed the National Safety Council in 1921–1922 and helped form the American Management Association (AMA) in early 1923. He presented papers at the AMA, chaired a committee studying army-style organization in 1926, served as vice-president for programs and publications from 1932 to 1934, and chaired the executive committee in 1935 and 1936. He lectured at the Wharton School and the Harvard Graduate School of Business Administration. After Industrial Relations Counselors was formed Young also became the American delegate to the International Labor Office (ILO) in Geneva, serving from 1925 to 1934; Industrial Relations Counselors opened a Geneva office to investigate European corporate welfare programs.

Young believed that company unions opened opportunities for labor and management to engage in a discussion that would allow industry to be self-governing. The ERPs failed in practice, largely because of corporate attitudes. After one of Young's papers on employee representation, an industrial executive announced that the audience had just heard a prescription for socialism. Yet Young persisted and argued there was much to learn from European old-age pensions and other government-backed welfare schemes, which in 1929 he correctly labeled as an inevitability in corporate practice. There were limits to his liberalism: he denounced the five-day work week in 1927, for instance. During the Depression, however, Young promoted public employment services. As chair of Governor Franklin D. Roosevelt's New York State Advisory Committee on Employment in 1930, he proposed one of the first statewide systems of employment offices; Frances Perkins called it a laboratory study of the labor market. A staff member of Industrial Relations Counselors directed the program.

Ironically, the Governor Roosevelt who implemented this plan became the president who signed legislation abolishing Young's favorite program — company unions. Young found himself in the middle of the rancorous debate over the status of company unions when he became vice-president for industrial relations at U.S. Steel in 1934. He devised a plan to set up company unions in response to National Industrial Recovery Act requirements for collective-bargaining mechanisms and to the steel companies' desire to preserve the open shop. The Amalgamated Association of Iron, Steel and Tin Workers condemned company unions as shams. A protracted legal battle erupted in 1934, with the steel companies resisting every move, legislative or otherwise, to form independent unions. Before Roosevelt's Steel Labor Board, which was set up to resolve the situation, Young defended U.S. Steel's plan, noting that William Green of the American Federation of Labor (AFL) had called it one of the fairest devised.

Young was stung by the union attacks and responded with what *Fortune* called "die-hard opposition to organized labor" in 1935. Speaking at a banquet at which he received the Henry Gantt medal from the AMA "for outstanding and creative work in the field of industrial relations," Young told his audience that he "would rather go to jail or be convicted as a felon" than accept "any formula for the conduct of human relationships imposed on us by demagogues."

To Young's credit, U.S. Steel's plan approached the ideal Young had envisioned in the 1920s — that company unions might bring together the representatives of all plants in the company. In 1935, however, the Steel Workers Organizing Committee (SWOC) was successful in turning the ERP against the company. In the end, company control of the plan's finances and corporate unwillingness to take into consideration the recommendations of union officers doomed the plan. By 1936 Young was supporting the employees' call for a pay raise in a last-ditch effort to prove the plan's integrity. But the raise came too late, and in 1937 U.S. Steel recognized the SWOC. Young left U.S. Steel shortly thereafter.

Young retired to his southern California fruit ranch and lectured on industrial relations at California Institute of Technology until 1952; he was also a consulting professor of industrial relations at Stanford. He remained on the advisory committee of the President's Council on Personnel Administration and served on the industrial-relations committee of the business advisory and planning council for the U.S. Department of Commerce. During World War II Young was an adviser to the Office of Production Management in Los Angeles and a consultant to the secretary of war in 1944. Later he was a civilian member of the Navy's Manpower Survey Board in San Diego. After 1945 he served as a member of the board of the First National Trust & Savings Bank (later United California Bank) and advised the First Western Bank and Trust Company. Young died in Santa Barbara, California, on March 4, 1964.

Publications:
"Harmonizing the Man and His Job," *Iron Trade Review,* 60 (February 15, 1917): 425–427;

"Goals of Employment Management," *Industrial Management,* 57 (June 1919): 491;

"How Industrial Accidents Are Being Prevented," *The Nation's Health* (February 1923);

"Is the Works Council a Success?," *100% Management* (March 1922): 52–56;

"What Management Problems Loom up for '23?," *Factory* (January 1923): 17–19, 54–72;

"The Occupational Rating Plan of the International Harvester Company," *Management Engineering* (May 1923): 301–306;

Evaluating Personnel Work in Industry (New York: American Management Association, 1924);

"Tribute to Intelligence of the Working Man," *Iron Age,* 113 (January 3, 1924): 44;

Relations of the Supervisor to His Men (New York: American Management Association, 1926);

and others, *Business and the Church: A Symposium,* edited by Jerome Davis (New York: Century, 1926), pp. 131–152;

Army Organization in Industry; Report of a Committee (New York: American Management Association, 1926);

"Developments in Industrial Pension Plans," *Iron Age,* 127 (February 26, 1931): 686–689;

"Should America Adopt the Five-Day Week?," *Congressional Digest,* 11 (October 1932): 235.

References:

"Americans Extend Labor Office Work," *New York Times,* October 31, 1929, p. 10;

"Arthur Young, 81; Ex-U.S. Steel Aide," *New York Times,* March 7, 1964, p. 23;

Irving Bernstein, *The Lean Years: A History of the American Worker, 1920–1933* (Boston: Houghton Mifflin, 1960);

Robert R. R. Brooks, *As Steel Goes . . .* (New Haven: Yale University Press, 1940);

"Faces of the Month," *Fortune,* 12 (July 1935): 140;

John A. Fitch, "A Man Can Talk in Homestead," *Survey Graphics,* 25 (February 1936): 75–76;

"Personals," *Iron Age,* 129 (February 8, 1934): 129;

"Steel Challenges the Labor Board," *New York Times,* October 3, 1934, pp. 1, 2;

Bryce M. Stewart, *Unemployment Benefits in the United States: The Plans and Their Setting* (New York: Industrial Relations Counselors, 1930).

Youngstown Sheet and Tube Company

by Larry N. Sypolt

West Virginia University

and

Bruce E. Seely

Michigan Technological University

The Youngstown Sheet and Tube Company was, for most of its history, a success story in the steel industry. In 1900 James A. Campbell and George D. Wick resigned from the Republic Iron and Steel Company to run their own iron company and founded Youngstown Sheet and Tube, which by 1926 was the third-largest steel producer in the country. More than did most steel companies of the time, Youngstown relied on mergers to expand its operations. Unsuccessful mergers in the late 1920s and late 1950s, however, weakened the company. By the mid 1960s Youngstown was encountering severe difficulties, eventually leading to the firm's disappearance as a separate company in late 1978.

On November 23, 1900, Wick, Campbell, scrap dealer William Wilkoff, Youngstown merchant George L. Fordyee, and iron maker Edward L. Ford met in the office of attorney C. D. Hines and signed the articles of incorporation forming the Youngstown Iron Sheet and Tube Company. The capital stock was $600,000, divided into 6,000 shares of $100 each. Wilkoff and Wick each took 500 shares and Campbell another 150 shares. Within five days all but 96 shares had been sold among friends and merchants in the Mahoning Valley. Wick and Campbell split these remaining shares among themselves. The ease with which the stock was subscribed was itself an achievement.

The Indiana Harbor Works of the Youngstown Sheet and Tube Comapny, circa 1950

Republic Iron and Steel had been organized in the same geographic region only a year earlier, and in April 1900 the United States Steel Corporation had purchased the Ohio Works in Youngstown. There seemed to be no urgent need for more iron production, but Youngstown-area merchants backed anything that might benefit the region. They also had faith in both Wick and Campbell, the guiding lights in the new company. The stockholders met on November 28, 1900, and elected a board of directors. Wick became the company's first president, Campbell was elected vice-president, and Robert Bentley was made secretary. A committee composed of Wick, Campbell, and Wilkoff was created to select a site for the proposed mill. Soon after this meeting, however, the directors realized that their original plans had been too conservative. In December they increased the capital to $1 million and created a $100,000 reserve.

The most pressing concern of the new company was construction of a facility to make iron and steel. In February 1901 the directors authorized the acquisition of a 117-acre tract on the east end of Youngstown between Haselton and Struthers. This site was bordered by the Mahoning River and the tracks of the Baltimore & Ohio and Pittsburgh & Lake Erie railroads; additional land was available for expansion. Construction began in

April 1901, and by February 1902 the puddling furnaces and sheet mills were in operation; the tube mills were working six months later.

Youngstown Iron Sheet and Tube, however, encountered difficulties in its first years of operation. New units did not come on-line simultaneously, and it was therefore not an easy task to synchronize the operations of the company's different facilities. Capitalization had to be increased again to meet the cost of construction. In addition Wick's health began to suffer, and in May 1902 he stepped down as president. Campbell served as acting president for the next two years before the board formally appointed him president.

Prior to the appointment Campbell had already taken a leadership role in his determination to make the company self-sufficient in materials and in pig-iron capacity. In December 1901 Campbell arranged the acquisition of the Little Alice furnace in Sharpsburg, Pennsylvania, to supply pig iron. At about the same time, Youngstown purchased the Ohio Galvanizing Company in order to move into the manufacture of finished products, and another tube mill was added to Youngstown's facilities to produce lap-weld pipe for the oil industry. By 1904 it had become clear that steel rather than iron would be Youngstown's main product, and in May 1905 the word *Iron* was deleted from the company title. In 1905–1906 two

12-ton Bessemer converters were built to supply steel, and in late 1906 the company began construction of two blast furnaces.

Key to the company's success during its early years was its close tie to Pickands, Mather and Company, a Cleveland-based iron-ore concern. In 1902 Youngstown Iron Sheet and Tube had purchased the Little Alice Furnace in Sharpsville, Pennsylvania, by borrowing the needed money from Pickands, Mather in return for Youngstown stock. In that year Henry G. Dalton and James A. Parmalee of Pickands, Mather joined the Youngstown board. Thereafter, the two companies were involved in the acquisition of several iron and coal mining operations. By the 1920s Samuel Mather was Youngstown's largest stockholder, and Dalton served a short term as chair of Youngstown Sheet and Tube in the 1930s.

After 1906 Campbell continued his aggressive policy of expansion, and by 1911 the main facility in Youngstown, called the Campbell Works, was able to finish its own steel into marketable products. The plant included three blast furnaces, two Bessemer converters, 50 double-puddling furnaces, a blooming and slabbing mill, a sheet-bar mill, three skelp mills, a billet mill, ten tube mills, 14 sheet mills, six galvanizing mills, and a plant to make roofing sheet. Subsidiary operations produced spring steel, rod, and wire. Youngstown also purchased several coal and iron properties and added an open-hearth unit in 1912–1913 with six 100-ton furnaces. In 1919 Youngstown joined 11 other steel companies in the formation of Consolidated Steel Company. Consolidated handled the export sales of Youngstown and the other member companies.

Two acquisitions made in 1923 capped the growth of Youngstown Sheet and Tube under Campbell. In January Campbell announced the purchase of Brier Hill Steel Company at Youngstown for $30 million and the Sheet and Tube Company of America in Chicago for $55 million. Brier Hill traced its origins to the mid nineteenth century and owned coal and ore lands as well as modern steelmaking facilities, giving the company much-needed sheet-steel capacity. The Steel and Tube Company, the other acquisition, owned seven ore and fuel companies and two zinc producers as well as additional plants in Mayville, Wisconsin; Zanesville, Ohio; Evanston, Illinois; and Kalamazoo, Michigan. Most important to the purchase were two Sheet and Tube–owned plants on the Lake Michigan waterfront. The Iroquois Iron Company operated a blast furnace in South

Chicago, and a modern blast furnace and steel plant were located in nearby Indiana Harbor. These low-cost plants, serviced directly by Lake Michigan freighters, afforded Youngstown an entrance to the rapidly growing midwestern steel market.

These purchases vaulted the company from a regional to a national leader in the American steel industry. Ingot capacity jumped from 1.68 million tons in 1920 to more than 3 million in 1925. By 1926, however, further growth at Youngstown was stalled by failed merger negotiations. In the early 1920s Youngstown had been involved in an attempt to merge with Republic Iron and Steel, Inland Steel, Midvale Steel & Ordnance, Lackawanna Steel, and Brier Hill. But the merger fizzled after Bethlehem Steel Corporation bought first Lackawanna Steel and then Midvale Steel. Although these moves led Youngstown to buy Brier Hill and Steel & Tube, Campbell continued to worry that Youngstown could not compete alone against U.S. Steel.

In 1928 Inland and Youngstown began new merger talks. Their combination would have rivaled Bethlehem, but negotiations stalled. In 1930 yet another merger was proposed to link Youngstown and Bethlehem. The move would have permitted Bethlehem to expand into the Midwest, and Campbell was enthusiastic about the merger as a solution to Youngstown's competitive problems. Unfortunately, the plan met with significant opposition, especially from Cyrus Eaton, who was at that time organizing Republic Steel and had designs on Youngstown himself. The majority of the stockholders supported Campbell, but lawsuits were brought by Eaton's interests. Courtroom disclosures of the enormous financial expenditures by both sides made spectacular headlines at a time of falling public confidence in big business. Youngstown and Bethlehem lost the suit, then won the appeal in August 1931. But the possibility of a merger had been ruined by the Depression. It was Campbell's only setback as the chief executive of Youngstown Sheet and Tube. He had resigned as president in 1930, and Frank Purnell had assumed the presidency. Campbell remained chairman until 1933.

For almost three years Youngstown management's attention had been diverted from operations. The company had hesitated to install a new continuous hot-strip mill, and the absence of this new technology left Youngstown poorly prepared for a depressed steel market. Steel-sheet products were least affected by the economic disaster, but Youngstown had 41 percent of its capacity in pipe. Indeed, in 1930 Youngstown had installed a new electric-welding process for making pipe at Brier Hill. When the pipe market collapsed, the company operated at 18 percent of capacity in 1932; it lost money from 1931 through 1934. Youngstown finally took an enormous risk in 1934 and borrowed $3.5 million to install a continuous sheet mill. The recovery of steel markets in 1935 permitted repayment of this investment, and efforts were made to shift production toward sheet products and away from pipe. But by this time Youngstown had fallen to sixth place among American steelmakers in terms of ingot capacity. Dalton replaced Campbell as chairman in 1933.

As was typical throughout most of the steel industry during World War II, Youngstown ran at full capacity and made record sales, but profits declined. Rising labor rates and the inability to boost prices due to government restrictions accounted for the poor profit performance. But in 1946 the company launched an impressive improvement program, an investment of more than $72 million. The program included construction of a new blast furnace at Indiana Harbor.

During the 1950s the company continued to upgrade facilities. In 1956 J. L. Mauthe became chairman, and Alfred S. Glossbrenner became the company's fourth president. These men placed emphasis on the Indiana Harbor plant, adding new coke ovens, blast furnaces, open hearths, a 45-inch blooming mill, and a new conduit department. Improvements also were made at Brier Hill and the Campbell Works. The company spent $535 million on improvements between 1950 and 1959, and ingot capacity was raised from about 4 million tons to 6.75 million.

But Youngstown again faced problems in the 1950s. The company had fallen to eighth place in terms of ingot capacity, continuing the erosion that had started in the 1930s. The company's attention was diverted by labor strife and by confrontation with the government occasioned by President Harry S Truman's seizure of the steel industry in 1952. The company's name appeared on the related Supreme Court decision, *Youngstown Sheet & Tube Co. v. Sawyer*. Youngstown again found itself involved in merger discussions. Again, Bethlehem was the proposed partner; and, again, the argument was that the two firms complemented each other in terms of geography, product mix, and resources — neither competed directly with the other. But the attorney general did not re-

lent, and the merger was abandoned in 1959 after the courts upheld the government's position.

Successful completion of the merger might have helped both companies. Bethlehem would not have needed to spend millions on a plant at Indiana Harbor during the 1960s. The worst damage to Youngstown was the disruption of planning during the five years of waiting; the company also fell behind its rivals in the installation of new equipment. Finally, Youngstown lost a much-needed source of capital for improvements. As it was, Youngstown spent $784.4 million for additional capital improvements during the 1960s. Most of the money went for new equipment at Indiana Harbor; the company had too much steel capacity at Youngstown. But modernization efforts to eliminate this imbalance and the heavy shipping costs it brought were piecemeal and failed to provide maximum cost reductions or improvements in efficiency. Moreover, the company accumulated a staggering debt, even as the strategy of paying for improvements by raising prices failed due to rising volumes of imported steel.

By the mid 1960s Youngstown Sheet and Tube looked to be in better shape, thanks to Glossbrenner's efforts and changes in the executive ranks, a departure from the company's traditional pattern for Youngstown executives. Glossbrenner, who had replaced Mauthe as chairman and chief executive officer in 1963, passed the CEO's duties in 1965 to Robert E. Williams. Williams was not only an outsider — he joined the company from U.S. Steel in 1961 — but also a salesman, the first chief executive not out of the production ranks. The company reported record profits in 1964, and the business press praised the turnaround, which included several new finishing and cold-reduction mills that opened new product lines. However, the company remained the eighth-largest steel producer, its position since the late 1940s. The worst problem continued to be the piecemeal modernization of old plants. That Youngstown was the last major steel company to adopt basic-oxygen furnaces was a symptom of the company's difficulties.

By 1969 *Business Week* reported that Youngstown Sheet and Tube was "in a situation increasingly familiar for steel companies — a low growth rate, a high cash flow, and a sizable gap between per-share market price and book value." This situation led to another round of merger discussions, this time with conglomerates. Beginning in 1968 Youngstown fought off Ling-Temco-Vought (LTV), Signal Companies, Avnet, and

Lykes Industries; in January 1969 the board accepted Lykes's third offer, becoming a subsidiary of a company one-sixth its size, with Frank Nemec as president and Joseph Lykes as chairperson. Williams remained as chief administrative officer.

Unfortunately for the conglomerate, Lykes-Youngstown encountered enormous difficulties during the 1970s. Intended to generate cash for the conglomerate, Youngstown instead produced red ink. Despite immense expenditures during the 1960s, the company's facilities were not competitive. Only the Indiana Harbor plant had modern facilities — a new blast furnace, hot-strip mill, and basic-oxygen furnace installed in 1968. The recession of 1975 almost overwhelmed Youngstown, which lost $7 million in 1976 and $58 million in 1977. Lykes stock fell from $24 to $6 a share in the 18 months before September 1977. The company began closing parts of the Brier Hill plant and the Campbell Works in Youngstown as Youngstown Sheet and Tube edged toward bankruptcy.

Once again, merger talks were opened, this time with Jones & Laughlin Steel Corporation, itself owned by another conglomerate, LTV. The Justice Department's antitrust division was opposed, but the prospect of Youngstown's failure inclined Attorney General Griffin Bell to accept the $7 billion merger in December 1978, after a year of negotiations. Only the company's Indiana Harbor Works made the transition. The Brier Hill plant was completely closed, and by late 1979 only a pipe mill at the Campbell Works remained open; it closed in the 1980s. In the end, Youngstown Sheet and Tube ceased to exist, a dramatic change for the city that had supported the steel company for 80 years.

References:

"Bethlehem — Youngstown Merger Trial," *Barron's,* 10 (November 10, 1930): 24;

"Better than a Savings Bank," *Forbes,* 103 (April 1, 1969): 30–31;

"Facing a Fight," *Forbes,* 95 (February 1, 1965): 28–29;

F. A. Fall, "Battle of Youngstown," *Outlook,* 157 (January 21, 1931): 102;

"The Hard Lesson Youngstown Taught Lykes," *Business Week* (October 3, 1977): 83, 86;

John Lee Higgs, *The Iron and Steel Industry in the Mahoning Valley,* M.A. Thesis, West Virginia University, 1934;

William T. Hogan, S.J., *An Economic History of the Iron and Steel Industry in the United States,* 5 volumes (Lexington, Mass.: Lexington Books, 1971);

Hogan, *Minimills and Integrated Mills: A Comparison of Steelmaking in the United States* (Lexington, Mass.: Lexington Books, 1987), pp. 31–34;

Charles Longenecker, "Youngstown Sheet and Tube Increases Service to the Midwest," *Blast Furnace and Steel Plant,* 41 (August 1953): 898–925;

"Lykes-Youngstown Merger Nears," *Business Week* (February 15, 1969): 106;

Staughton Lynd, *The Fight Against Shutdowns: Youngstown's Steel Mill Closings* (San Pedro, Cal.: Singlejack Books, 1982);

Maeva Marcus, *Truman and the Steel Seizure Case: The Limits of Presidential Power* (New York: Columbia University Press, 1977);

"Restructuring Figures Heavily in J&L's Plans after Its Merger with Youngstown Steel," *33 Metal Producing,* 17 (January 1979): 57;

Gertrude G. Schroeder, *The Growth of Major Steel Companies, 1900–1950* (Baltimore: Johns Hopkins University Press, 1953);

Kenneth Warren, *The American Steel Industry, 1850–1970: A Geographical Interpretation* (London: Clarendon, 1973);

Youngstown Sheet and Tube Company, *Fifty Years in Steel* (Youngstown, Ohio: Youngstown Sheet and Tube Company, 1950).

Contributors

Robert M. Aduddell — *Loyola University of Chicago*
Donald F. Barnett — *Economic Associates, Inc.*
James R. Barrett — *University of Illinois at Urbana-Champaign*
Carl Becker — *Wright State University*
Louis P. Cain — *Loyola University of Chicago*
Robert Casey — *Henry Ford Museum & Greenfield Village*
Robert W. Crandall — *The Brookings Institution*
Stephen H. Cutcliffe — *Lehigh University*
Amy E. Davis — *University of California at Los Angeles*
Dennis C. Dickerson — *Williams College*
Kevin M. Dwyer — *George Washington University*
Gerald G. Eggert — *Pennsylvania State University*
Mark S. Foster — *University of Colorado at Denver*
John A. Heitmann — *University of Dayton*
Dean Herrin — *National Park Service*
Robert Hessen — *Hoover Institution, Stanford University*
John P. Hoerr — *Teaneck, New Jersey*
Father William Hogan, S.J. — *Fordham University*
John N. Ingham — *University of Toronto*
Richard W. Kalwa — *University of Wisconsin — Parkside*
Alec Kirby — *George Washington University*
Thomas E. Leary — *Industrial Research Associates*
Glen V. Longacre III — *Columbus, Ohio*
Leonard H. Lynn — *Case Western Reserve University*
George McManus — *Pittsburgh, Pennsylvania*
Lance E. Metz — *Canal Museum, Easton, Pennsylvania*
Carol Poh Miller — *Cleveland, Ohio*
Thomas J. Misa — *Illinois Institute of Technology*
Richard W. Nagle — *Massasoit Community College*
Elizabeth M. Nolin — *Paeonian Springs, Virginia*
Terry S. Reynolds — *Michigan Technological University*
Michael Santos — *Lynchburg College*
H. Lee Scamehorn — *Historic Learning and Research Systems*
Bruce E. Seely — *Michigan Technological University*
Elizabeth C. Sholes — *Industrial Research Associates*
Larry N. Sypolt — *West Virginia University*
Paul Tiffany — *Sonoma State University*
Geoffrey Tweedale — *Manchester, England*

Index

The following index includes names of people, corporations, organizations, laws, and technologies. It also includes key terms such as *basing point pricing, labor relations,* etc.

A page number in *italic* indicates the first page of an entry devoted to the subject. *Illus.* indicates a picture of the subject.

Index